Beginning
Algebra

Stefan Baratto

Stefan began teaching math and science in New York City middle schools. He also taught math at the University of Oregon, Southeast Missouri State University, and York County Technical College. Currently, Stefan is a member of the mathematics faculty at Clackamas Community College where he has found a niche, delighting in the CCC faculty, staff, and students. Stefan's own education includes the University of Michigan (BGS, 1988), Brooklyn College (CUNY), and the University of Oregon (MS, 1996).

Stefan is currently serving on the AMATYC Executive Board as the organization's Northwest Vice President. He has also been involved with ORMATYC, NEMATYC, NCTM, and the State of Oregon Math Chairs group, as well as other local organizations. He has applied his knowledge of math to various fields, using statistics, technology, and web design. More personally, Stefan and his wife, Peggy, try to spend time enjoying wonders of Oregon and the Pacific Northwest. Their activities include scuba diving and hiking.

Barry Bergman

Barry has enjoyed teaching mathematics to a wide variety of students over the years. He began in the field of adult basic education and moved into the teaching of high school mathematics in 1977. He taught high school math for 11 years, at which point he served as a K-12 mathematics specialist for his county. This work allowed him the opportunity to help promote the emerging NCTM standards in his region.

In 1990, Barry began the next portion of his career, having been hired to teach at Clackamas Community College. He maintains a strong interest in the appropriate use of technology and visual models in the learning of mathematics.

Throughout the past 32 years, Barry has played an active role in professional organizations. As a member of OCTM, he contributed several articles and activities to the group's journal. He has presented at AMATYC, OCTM, NCTM, ORMATYC, and ICTCM conferences. Barry also served 4 years as an officer of ORMATYC and participated on an AMATYC committee to provide feedback to revisions of NCTM's standards.

Don Hutchison

Don began teaching in a preschool while he was an undergraduate. He subsequently taught children with disabilities, adults with disabilities, high school mathematics, and college mathematics. Although each position offered different challenges, it was always breaking a challenging lesson into teachable components that he most enjoyed.

It was at Clackamas Community College that he found his professional niche. The community college allowed him to focus on teaching within a department that constantly challenged faculty and students to expect more. Under the guidance of Jim Streeter, Don learned to present his approach to teaching in the form of a textbook. Don has also been an active member of many professional organizations. He has been president of ORMATYC, AMATYC committee chair, and ACM curriculum committee member. He has presented at AMATYC, ORMATYC, AACC, MAA, ICTCM, and a variety of other conferences.

Above all, he encourages you to be involved, whether as a teacher or as a learner. Whether discussing curricula at a professional meeting or homework in a cafeteria, it is the process of communicating an idea that helps one to clarify it.

Dedication

We dedicate this text to the thousands of students who have helped us become better teachers, better communicators, better writers, and even better people. We read and respond to every suggestion we get—every one is invaluable. If you have any thoughts or suggestions, please contact us at

Stefan Baratto: sbaratto@clackamas.edu
Barry Bergman: bfbergman@gmail.com
Don Hutchison: donh@collegemathtext.com

Thank you all.

Letter from the Authors

Dear Colleagues,

We believe the key to learning mathematics, at any level, is active participation! We have revised our textbook series to specifically emphasize *GROWING MATH SKILLS* through active learning. Students who are active participants in the learning process have a greater opportunity to construct their own mathematical ideas and make stronger connections to concepts covered in their course. This participation leads to better understanding, retention, success, and confidence.

In order to grow student math skills, we have integrated features throughout our textbook series that reflect our philosophy. Specifically, our *chapter-opening vignettes* and an array of section exercises relate to a singular topic or theme to engage students while identifying the relevance of mathematics.

The *Check Yourself* exercises, which include optional calculator references, are designed to keep students actively engaged in the learning process. Our exercise sets include application problems as well as challenging and collaborative writing exercises to give students more opportunity to sharpen their skills.

Originally formatted as a work-text, this textbook allows students to make use of the margins where exercise answer space is available to further facilitate active learning. This makes the textbook more than just a reference. Many of these exercises are designed for insight to generate mathematical thought while reinforcing continual practice and mastery of topics being learned. Our hope is that students who use our textbook will grow their mathematical skills and become better mathematical thinkers as a result.

As we developed our series, we recognized that the use of technology should not be simply a supplement, but should be an essential element in learning mathematics. We understand that these "millennial students" are learning in different modes than just a few short years ago. Attending course lectures is not the only demand these students face—their daily schedules are pulling them in more directions than ever before. To meet the needs of these students, we have developed videos to better explain key mathematical concepts throughout the textbook. The goal of these videos is to provide students with a better framework—showing them how to solve a specific mathematical topic, regardless of their classroom environment (online or traditional lecture). The videos serve as refreshers or preparatory tools for classroom lecture and are available in several formats, including iPOD/MP3 format, to accommodate the different ways students access information.

Finally, with our series focus on **growing math skills,** we strongly believe that ALEKS® software can truly help students to remediate and grow their math skills given its adaptiveness. ALEKS is available to accompany our textbooks to help build proficiency. ALEKS has helped our own students to identify mathematical skills they have mastered and skills where remediation is required.

Thank you for using our textbook! We look forward to learning of your success!

Stefan Baratto
Barry Bergman
Donald Hutchison

About the Cover

A flower symbolizes transformation and growth—a change from the ordinary to the spectacular! Similarly, students in a beginning algebra math course have the potential to grow their math skills with resources to become stronger math students. Authors Stefan Baratto, Barry Bergman, and Don Hutchison help students to *grow their mathematical skills*—guiding them through the different stages to mathematical success!

"This is a good book. The best feature, in my opinion is the readability of this text. It teaches through example and has students immediately check their own skills. This breaks up long text into small bits easier for students to digest."

— Robin Anderson, Southwestern Illinois College

Grow Your Mathematical Skills With Baratto/Bergman/Hutchison!

Helping students develop study skills is critical for student success. With over 80 years in the classroom, Stefan Baratto, Barry Bergman, and Don Hutchison have helped students sharpen their mathematical skills and learn how to use their mathematical knowledge in everyday life! The Hutchison Series helps *grow mathematical skills* to motivate students to learn!

Grow Your Mathematical Skills Through Better Conceptual Tools!

Stefan Baratto, Barry Bergman, and Don Hutchison know that students succeed once they have built a strong conceptual understanding of mathematics. *"Make the Connection"* chapter-opening vignettes help students to better understand mathematical concepts through everyday examples. Further reinforcing real-world mathematics, each vignette is accompanied by activities and exercises in the chapter to help students focus on the mathematical skills required for mastery.

Make the Connection	*Learning Objectives*
Chapter-Opening Vignettes	*Self-Tests*
Activities	*Cumulative Reviews*
Reading Your Text	*Group Activities*

Grow Your Mathematical Skills Through Better Exercises, Examples, and Applications!

A wealth of exercise sets is available for students at every level to actively involve them through the learning process in an effort to grow mathematical skills, including:

Prerequisite Tests	*End-of-Section Exercises*
Check Yourself Exercises	*Summary Exercises*
Career Application Exercises	

Grow Your Mathematical Study Skills Through Better Active Learning Tools!

In an effort to meet the needs of the "millennial student," we have made active-learning tools available to sharpen mathematical skills and build proficiency.

ALEKS	*Conceptual Videos*
MathZone	*Lecture Videos*

"The Baratto/Bergman/Hutchison textbook gives the student a well-rounded foundation into many concepts of algebra, taking the student from prior knowledge, to guided practice, to independent practice, and then to assessment. Each chapter builds upon concepts learned in other chapters. Items such as ... Check Yourself exercises and Activities at the end of most chapters help the student to be more successful in many of the concepts taught."

— *Karen Day, Elizabethtown Technical & Community College*

Grow Your Mathematical Skills

"Make the Connection"—**Chapter-Opening Vignettes** provide interesting, relevant scenarios that will capture students' attention and engage them in the upcoming material. Exercises and *Activities* related to the *Opening Vignette* are available to utilize the theme most effectively for better mathematical comprehension (marked with an icon).

Activities are incorporated to promote *active learning* by requiring students to find, interpret, and manipulate real-world data. The activity seen in the chapter-opening vignette ties the chapter together by way of questions to sharpen student mathematical and conceptual understanding, highlighting the cohesiveness of the chapter. Students can complete the activities on their own, but they are best worked in small groups.

NEW! Reading Your Text offers a brief set of exercises at the end of each section to assess students' knowledge of key vocabulary terms. These exercises are designed to encourage careful reading for greater conceptual understanding. *Reading Your Text* exercises address vocabulary issues, which students often struggle with in learning core mathematical concepts. Answers to these exercises are provided at the end of the book.

98. How long ago was the year 1250 B.C.E.? What year was 3,300 years ago? Make a number line and locate the following events, cultures, and objects on it. How long ago was each item in the list? Which two events are the closest to each other? You may want to learn more about some of the cultures in the list and the mathematics and science developed by that culture.

> chapter 1 > Make the Connection

    Inca culture in Peru—A.D. 1400

    The *Ahmes Papyrus,* a mathematical text from Egypt—1650 B.C.E.

    Babylonian arithmetic develops the use of a zero symbol—300 B.C.E.

    First Olympic Games—776 B.C.E.

Activity 1 ::
An Introduction to Searching

Each activity in this text is designed to either enhance your understanding of the topics of the preceding chapter or provide you with a mathematical extension of those topics, or both. The activities can be undertaken by one student, but they are better suited for a small group project. Occasionally it is only through discussion that different facets of the activity become apparent.

There are many resources available to help you when you have difficulty with your math work. Your instructor can answer many of your questions, but there are other resources to help you learn, as well.

Studying with friends and classmates is a great way to learn math. Your school may have a "math lab" where instructors or peers provide tutoring services. This text provides examples and exercises to help you learn and understand new concepts.

Another place to go for help is the **Internet.** There are many math tutorials on the Web. This activity is designed to introduce you to searching the Web and evaluating what you find there.

Reading Your Text

The following fill-in-the-blank exercises are designed to ensure that you understand some of the key vocabulary used in this section.

SECTION 1.7

(a) When multiplying expressions with the same base, _____ the exponents.

(b) When multiplying expressions with the same base, the _____ does not change.

(c) When multiplying expressions with the same base, _____ the coefficients.

(d) To divide expressions with the same base, keep the base and _____ the exponents.

Through Better Conceptual Tools!

Self-Tests appear in each chapter to provide students with an opportunity to check their progress and to review important concepts, as well as to provide confidence and guidance in preparing for exams. The answers to the *Self-Test* exercises are given at the end of the book. Section references are given with the answers to help the student.

Cumulative Reviews are included, starting with Chapter 2, and follow the *self-tests*. These reviews help students build on previously covered material and give them an opportunity to reinforce the skills necessary to prepare for midterm and final exams. These reviews assist students with the retention of knowledge throughout the course. The answers to these exercises are also given at the end of the book, along with section references.

Group Activities offer practical exercises designed to grow student comprehension through group work. Group activities are great for instructors and adjuncts—bringing a more interactive approach to teaching mathematics!

CHAPTER 1 **self-test** 1

The purpose of this self-test is to help you assess your progress so that you can find concepts that you need to review before the next exam. Allow yourself about an hour to take this test. At the end of that hour, check your answers against those given in the back of this text. If you miss any, go back to the appropriate section to reread the examples until you have mastered that particular concept.

Name _____

Section _____ Date _____

Answers

Evaluate each expression.

1. $-8 + (-5)$
2. $6 + (-9)$
3. $(-9) + (-12)$
4. $-\dfrac{5}{3} + \dfrac{8}{3}$
5. $9 - 15$
6. $-10 - 11$
7. $5 - (-4)$
8. $-7 - (-7)$
9. $(8)(-5)$
10. $(-9)(-7)$

1. _____ 2. _____
3. _____ 4. _____
5. _____ 6. _____
7. _____ 8. _____
9. _____ 10. _____

---

cumulative review chapters 1-2

The following exercises are presented to help you review concepts from earlier chapters. This is meant as review material and not as a comprehensive exam. The answers are presented in the back of the text. Beside each answer is a section reference for the concept. If you have difficulty with any of these exercises, be certain to at least read through the summary related to that section.

Name _____

Section _____ Date _____

Answers

Perform the indicated operations.

1. $8 + (-4)$
2. $-7 + (-5)$
3. $6 - (-2)$
4. $-4 - (-7)$
5. $(-6)(3)$
6. $(-11)(-4)$

1. _____ 2. _____
3. _____ 4. _____
5. _____ 6. _____
7. _____ 8. _____

Activity 5 ::
Determining State Apportionment

The introduction to this chapter referred to the ratio of the people in a particular state to their total number of representatives in the U.S. House based on the 2000 census. It was noted that the ratio of the total population of the country to the 435 representatives in Congress should equal the state apportionment if it is fair. That is, $\dfrac{A}{a} = \dfrac{P}{r}$, where A is the population of the state, a is the number of representatives for that state, P is the total population of the U.S., and r is the total number of representatives in Congress (435).

Pick 5 states (your own included) and search the Internet to find the following.

1. Determine the year 2000 population of each state.
2. Note the number of representatives for each state and any increase or decrease.
3. Find the number of people per representative for each state.
4. Compare that with the national average of the number of people per representative.

Grow Your Mathematical Skills with Better Worked Examples, Exercises, and Applications!

"Check Yourself" Exercises are a hallmark of the Hutchison series; they are designed to actively involve students in the learning process. Every example is followed by an exercise that encourages students to solve a problem similar to the one just presented and check, through practice, what they have just learned. Answers are provided at the end of the section for immediate feedback.

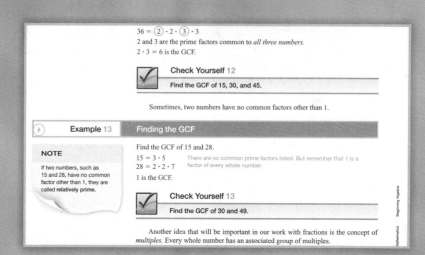

$36 = ②·2·③·3$

2 and 3 are the prime factors common to *all three numbers*.

$2 · 3 = 6$ is the GCF.

✓ Check Yourself 12

Find the GCF of 15, 30, and 45.

Sometimes, two numbers have no common factors other than 1.

▶ **Example 13** Finding the GCF

NOTE

If two numbers, such as 15 and 28, have no common factor other than 1, they are called relatively prime.

Find the GCF of 15 and 28.

$15 = 3 · 5$ There are no common prime factors listed. But remember that 1 is a
$28 = 2 · 2 · 7$ factor of every whole number.

1 is the GCF.

✓ Check Yourself 13

Find the GCF of 30 and 49.

Another idea that will be important in our work with fractions is the concept of *multiples*. Every whole number has an associated group of multiples.

"I like the placement of the 'check yourself's.' Students are confronted with thought-provoking questions to answer without mindlessly proceeding through the text."

— *Byron D. Hunter, College of Lake County*

End-of-Section Exercises enable students to evaluate their conceptual mastery through practice as they conclude each section. These comprehensive exercise sets are structured to highlight the progression in level, not only providing clarity for the student, but also making it easier for instructors to determine exercises for assignments. The *application exercises* that are now integrated into every section are a crucial component of this organization.

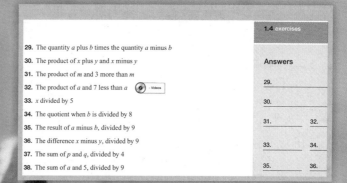

1.4 exercises

29. The quantity *a* plus *b* times the quantity *a* minus *b*

30. The product of *x* plus *y* and *x* minus *y*

31. The product of *m* and 3 more than *m*

32. The product of *a* and 7 less than *a*   ▶ · Videos

33. *x* divided by 5

34. The quotient when *b* is divided by 8

35. The result of *a* minus *b*, divided by 9

36. The difference *x* minus *y*, divided by 9

37. The sum of *p* and *q*, divided by 4

38. The sum of *a* and 5, divided by 9

Answers

29. _____
30. _____
31. _____   32. _____
33. _____   34. _____
35. _____   36. _____

Summary and Summary Exercises at the end of each chapter allow students to review important concepts. The *Summary Exercises* provide an opportunity for the student to practice these important concepts. The answers to odd-numbered exercises are provided in the answers appendix.

summary :: chapter 1

Definition/Procedure	Example	Reference
Properties of Real Numbers		Section 1.1
The Commutative Properties		
If *a* and *b* are any numbers,		p. 3
1. $a + b = b + a$		
2. $a · b = b · a$		
The Associative		
If *a*, *b*, and *c* are a		
1. $a + (b + c) =$		
2. $a · (b · c) = (a$		
The Distributive		
If *a*, *b*, and *c* are a		

summary exercises :: chapter 1

Use a calculator to perform the indicated operations.

31. $489 + (-332)$ 32. $1,024 - (-3,206)$ 33. $-234 + (-321) - (-459)$

34. $981 - 1,854 - (-321)$ 35. $4.56 + (-0.32)$ 36. $-32.14 - 2.56$

37. $-3.112 - (-0.1) + 5.06$ 38. $10.01 - 12.566 + 2$ 39. $13 - (-12.5) + 4\frac{1}{4}$

40. $3\frac{1}{8} - 6.19 + (-8)$

1.3 *Multiply.*

41. $(10)(-7)$ 42. $(-8)(-5)$ 43. $(-3)(-15)$

44. $(1)(-15)$ 45. $(0)(-8)$ 46. $\left(\frac{2}{3}\right)\left(-\frac{3}{2}\right)$

Tips for Student Success offers a resource to help students learn how to study, which is a problem many new students face—especially when taking their first exam in college mathematics. For this reason, Baratto/Bergman/Hutchison has incorporated *Tips for Student Success* boxes in the beginning of this textbook. The same suggestions made by great teachers in the classroom are now available to students outside of the classroom, offering extra direction to help improve understanding and further insight.

Notes and Recalls accompany the step-by-step worked examples helping students **focus on information critical to their success.** *Recall Notes* give students a *just-in-time* reminder, reinforcing previously learned material through references.

Cautions are integrated throughout the textbook to alert students to common mistakes and how to avoid them.

2 > Recognize applications of the associative properties

3 > Recognize applications of the distributive property

Tips for Student Success

RECALL

The first Tips for Student Success hint is on the previous page.

Over the first few chapters, we present you with a series of class-tested techniques designed to improve your performance in your math class.

Become familiar with your syllabus.

In your first class meeting, your instructor probably gave you a class syllabus. If you have not already done so, incorporate important information into a calendar and address book.

1. Write all important dates in your calendar. This includes the date and time of the final exam, test dates, quiz dates, and homework due dates. Never allow yourself to be surprised by a deadline!

2. Write your instructor's name, contact information, and office number in your address book. Also include your instructor's office hours. Make it a point to see your instructor early in the term. Although not the only person who can help you, your instructor is an important resource to help clear up any confusion you may have.

NOTE

We only work with real numbers in this text.

3. Make note of other resources that are available to you. This includes tutoring, CDs and DVDs, and Web pages.

Given all of these resources, it is important that you never let confusion or frustration mount. If you "can't get it" from the text, try another resource. All of these resources are there specifically for you, so take advantage of them!

Everything that we do in *algebra* is based on the **properties of real numbers.** Before being introduced to algebra, you should understand these properties.

Example 11 | **Writing a Percent as a Decimal**

Write each percent as a decimal.

RECALL

Multiplying by $\frac{1}{100}$ is the same as dividing by 100.

(a) $25\% = 25\left(\frac{1}{100}\right) = 0.25$ — The decimal [...] the 5.

(b) $4.5\% = 4.5\left(\frac{1}{100}\right) = 0.045$ — We must add [...]

(c) $130\% = 130\left(\frac{1}{100}\right) = 1.30$

NOTE

A percent greater than 100 gives a decimal greater than 1.

Check Yourself 11

Write as decimals.

(a) 5%   (b) 3.9%

Writing a decimal as a percent is the oppo[...] We simply reverse the process. Here is the rule:

Example 6 | **Evaluating Expressions**

Evaluate each expression if $a = -4$, b [...]

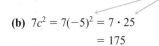

> **CAUTION**

When a squared variable is replaced by a negative number, square the negative.

$(-5)^2 = (-5)(-5) = 25$

↑
The exponent applies to −5!

$-5^2 = -(5 \cdot 5) = -25$

↑
The exponent applies only to 5!

(a) $7a - 4c = 7(-4) - 4(-5)$

$= -28 + 20$

$= -8$

(b) $7c^2 = 7(-5)^2 = 7 \cdot 25$

$= 175$

grow your math skills with ALEKS®

Experience Student Success!

ALEKS® ALEKS is a unique online math tool that uses adaptive questioning and artificial intelligence to correctly place, prepare, and remediate students . . . all in one product! Institutional case studies have shown that **ALEKS has improved pass rates by over 20% versus traditional online homework, and by over 30% compared to using a text alone.**

By offering each student an individualized learning path, ALEKS directs students to work on the math topics that they are ready to learn. Also, to help students keep pace in their course, instructors can correlate ALEKS to their textbook or syllabus in seconds.

To learn more about how ALEKS can be used to boost student performance, please visit www.aleks.com/highered/math or contact your McGraw-Hill representative.

ALEKS Pie
Each student is given their own individualized learning path.

Easy Graphing Utility!
Students can answer graphing problems with ease!

Course Calendar
Instructors can schedule assignments and reminders for students.

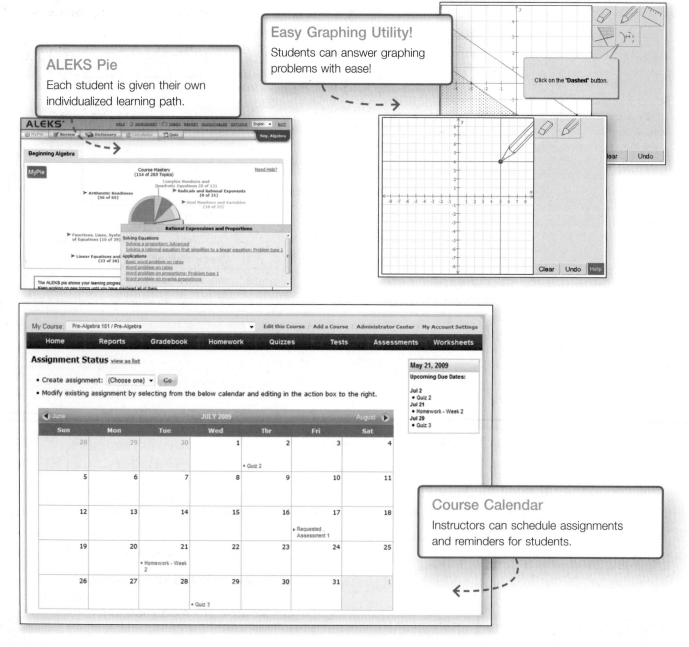

New ALEKS Instructor Module

Enhanced Functionality and Streamlined Interface Help to Save Instructor Time

ALEKS® The new ALEKS Instructor Module features enhanced functionality and a streamlined interface based on research with ALEKS instructors and homework management instructors. Paired with powerful assignment-driven features, textbook integration, and extensive content flexibility, the new ALEKS Instructor Module simplifies administrative tasks and makes ALEKS more powerful than ever.

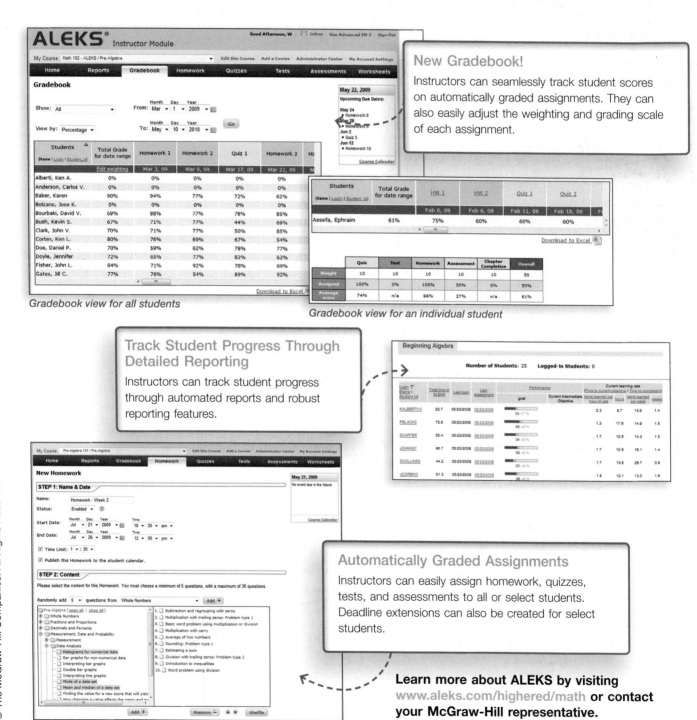

Gradebook view for all students

Gradebook view for an individual student

New Gradebook!

Instructors can seamlessly track student scores on automatically graded assignments. They can also easily adjust the weighting and grading scale of each assignment.

Track Student Progress Through Detailed Reporting

Instructors can track student progress through automated reports and robust reporting features.

Automatically Graded Assignments

Instructors can easily assign homework, quizzes, tests, and assessments to all or select students. Deadline extensions can also be created for select students.

Select topics for each assignment

Learn more about ALEKS by visiting www.aleks.com/highered/math or contact your McGraw-Hill representative.

360° Development Process

McGraw-Hill's 360° Development Process is an ongoing, never-ending, market-oriented approach to building accurate and innovative print and digital products. It is dedicated to continual large-scale and incremental improvement driven by multiple customer feedback loops and checkpoints. This is initiated during the early planning stages of our new products, and intensifies during the development and production stages, then begins again upon publication, in anticipation of the next edition.

A key principle in the development of any mathematics text is its ability to adapt to teaching specifications in a universal way. The only way to do so is by contacting those universal voices—and learning from their suggestions. We are confident that our book has the most current content the industry has to offer, thus pushing our desire for accuracy to the highest standard possible. In order to accomplish this, we have moved through an arduous road to production. Extensive and open-minded advice is critical in the production of a superior text.

Listening to you...

This textbook has been reviewed by over 300 teachers across the country. Our textbook is a commitment to your students, providing clear explanations, concise writing style, step-by-step learning tools, and the best exercises and applications in developmental mathematics. How do we know? You told us so!

Teachers *just like you* are saying great things about the Baratto/Bergman/Hutchison developmental mathematics series:

"Well written and organized. It tends to get directly into what the students actually are required to do without lengthy preamble. Nice division of exercises into basic/advanced suitable for our students and personal finance exercises they can relate to."

– Jonathan Cornick, Queensborough Community College

"This text is clearly written with developmental students in mind, including numerous examples and exercises to reinforce the concepts presented. The examples are thorough, including step-by-step guidance to students."

– Shelly Hansen, Mesa State College

"A very readable and comprehensive textbook that does a great job of presenting and describing the basic and more advanced concepts. "Practice" and "learning by doing" are themes that penetrate throughout the text, for there is a large volume of (homework) problems from which both teachers and students can choose. Each problem section builds upon the concepts learned previously: a very sound pedagogical approach."

– Bob Rhea, J. Sargeant Reynolds Community College

Acknowledgments and Reviewers

The development of this textbook series would never have been possible without the creative ideas and feedback offered by many reviewers. We are especially thankful to the following instructors for their careful review of the manuscript.

Symposia

Every year McGraw-Hill conducts general mathematics symposia, which are attended by instructors from across the country. These events are an opportunity for editors from McGraw-Hill to gather information about the needs and challenges of instructors teaching these courses. This information helped to create the book plan for *Beginning Algebra*. They also offer a forum for the attendees to exchange ideas and experiences with colleagues they might have not otherwise met.

Napa Valley Symposium

Antonio Alfonso, *Miami Dade College*

Lynn Beckett-Lemus, *El Camino College*

Kristin Chatas, *Washtenaw Community College*

Maria DeLucia, *Middlesex College*

Nancy Forrest, *Grand Rapids Community College*

Michael Gibson, *John Tyler Community College*

Linda Horner, *Columbia State College*

Matthew Hudock, *St. Phillips College*

Judith Langer, *Westchester Community College*

Kathryn Lavelle, *Westchester Community College*

Scott McDaniel, *Middle Tennessee State University*

Adelaida Quesada, *Miami Dade College*

Susan Schulman, *Middlesex College*

Stephen Toner, *Victor Valley College*

Chariklia Vassiliadis, *Middlesex County College*

Melanie Walker, *Bergen Community College*

Myrtle Beach Symposium

Patty Bonesteel, *Wayne State University*

Zhixiong Chen, *New Jersey City University*

Latonya Ellis, *Bishop State Community College*

Bonnie Filer-Tubaugh, *University of Akron*

Catherine Gong, *Citrus College*

Marcia Lambert, *Pitt Community College*

Katrina Nichols, *Delta College*

Karen Stein, *University of Akron*

Walter Wang, *Baruch College*

La Jolla Symposium

Darryl Allen, *Solano Community College*

Yvonne Aucoin, *Tidewater Community College*

Sylvia Carr, *Missouri State University*

Elizabeth Chu, *Suffolk County Community College*

Susanna Crawford, *Solano Community College*

Carolyn Facer, *Fullerton College*

Terran Felter, *Cal State Long Bakersfield*

Elaine Fitt, *Bucks County Community College*

John Jerome, *Suffolk County Community College*

Sandra Jovicic, *Akron University*

Carolyn Robinson, *Mt. San Antonio College*

Carolyn Shand-Hawkins, *Missouri State*

Manuscript Review Panels

Over 150 teachers and academics from across the country reviewed the various drafts of the manuscript to give feedback on content, design, pedagogy, and organization. This feedback was summarized by the book team and used to guide the direction of the text.

Reviewers of the Hutchison/Baratto/Bergman Developmental Mathematics Series

Board of Advisors

Robin Anderson, *Southwestern Illinois College*

Elena Bogardus, *Camden County College*

Dorothy Brown, *Camden County College*

Kelly Kohlmetz, *University of Wisconsin–Milwaukee*

Kathryn Lavelle, *Westchester Community College*

Karen Stein, *University of Akron*

Reviewers

Nieves Angulo, *Hostos Community College*

Arlene Atchison, *South Seattle Community College*

Haimd Attarzadeh, *Kentucky Jefferson Community & Technical College*

Jody Balzer, *Milwaukee Area Technical College*

Rebecca Baranowski, *Estrella Mountain Community College*

Wayne Barber, *Chemeketa Community College*

Bob Barmack, *Baruch College*

Chris Bendixen, *Lake Michigan College*

Karen Blount, *Hood College*

Donna Boccio, *Queensborough Community College*

Steve Boettcher, *Estrella Mountain Community College*

Karen Bond, *Pearl River Community College-Poplarville*

Laurie Braga Jordan, *Loyola University-Chicago*

Kelly Brooks, *Pierce College*

Michael Brozinsky, *Queensborough Community College*

Amy Canavan, *Century Community & Technical College*

Faye Childress, *Central Piedmont Community College*

Kathleen Ciszewski, *University of Akron*

Bill Clarke, *Pikes Peak Community College*

Lois Colpo, *Harrisburg Area Community College*

Christine Copple, *Northwest State Community College*

Jonathan Cornick, *Queensborough Community College*

Julane Crabtree, *Johnson County Community College*

Carol Curtis, *Fresno City College*

Sima Dabir, *Western Iowa Tech Community College*

Reza Dai, *Oakton Community College*

Karen Day, *Elizabethtown Technical & Community College*

Mary Deas, *Johnson County Community College*

Anthony DePass, *St. Petersburg College-Ns*

Shreyas Desai, *Atlanta Metropolitan College*

Robert Diaz, *Fullerton College*

Michaelle Downey, *Ivy Tech Community College*

Ginger Eaves, *Bossier Parish Community College*

Azzam El Shihabi, *Long Beach City College*

Kristy Erickson, *Cecil College*

Steven Fairgrieve, *Allegany College of Maryland*

Jacqui Fields, *Wake Technical Community College*

Bonnie Filer-Tubaugh, *University of Akron*

Rhoderick Fleming, *Wake Tech Community College*

Matt Foss, *North Hennepin Community College*

Catherine Frank, *Polk Community College*

Matt Gardner, *North Hennepin Community College*

Judy Godwin, *Collin County Community College-Plano*

Lori Grady, *University of Wisconsin-Whitewater*

Brad Griffith, *Colby Community College*

Robert Grondahl, *Johnson County Community College*

Shelly Hansen, *Mesa State College*

Kristen Hathcock, *Barton County Community College*

Mary Beth Headlee, *Manatee Community College*

Kristy Hill, *Hinds Community College*

Mark Hills, *Johnson County Community College*

Sherrie Holland, *Piedmont Technical College*

Diane Hollister, *Reading Area Community College*

Denise Hum, *Canada College*

Byron D. Hunter, *College of Lake County*

Nancy Johnson, *Manatee Community College-Bradenton*

Joe Jordan, *John Tyler Community College-Chester*

Eliane Keane, *Miami Dade College–North*

Sandra Ketcham, *Berkshire Community College*

Lynette King, *Gadsden State Community College*

Jeff Koleno, *Lorain County Community College*

Donna Krichiver, *Johnson County Community College*

Indra B. Kshattry, *Colorado Northwestern Community College*

Patricia Labonne, *Cumberland County College*

Ted Lai, *Hudson County Community College*

Pat Lazzarino, *Northern Virginia Community College*

Richard Leedy, *Polk Community College*

Jeanine Lewis, *Aims Community College-Main Campus*

Michelle Christina Mages, *Johnson County Community College*

Igor Marder, *Antelope Valley College*

Donna Martin, *Florida Community College-North Campus*

Amina Mathias, *Cecil College*

Jean McArthur, *Joliet Junior College*

Carlea (Carol) McAvoy, *South Puget Sound Community College*

Tim McBride, *Spartanburg Community College*

Sonya McQueen, *Hinds Community College*

MariaLuisa Mendez, *Laredo Community College*

Madhu Motha, *Butler County Community College*

Shauna Mullins, *Murray State University*

Julie Muniz, *Southwestern Illinois College*

Kathy Nabours, *Riverside Community College*

Michael Neill, *Carl Sandburg College*

Nicole Newman, *Kalamazoo Valley Community College*

Said Ngobi, *Victor Valley College*

Denise Nunley, *Glendale Community College*

Deanna Oles, *Stark State College of Technology*

Jean Olsen, *Pikes Peak Community College*

Staci Osborn, *Cuyahoga Community College-Eastern Campus*

Linda Padilla, *Joliet Junior College*

Karen D. Pain, *Palm Beach Community College*

George Pate, *Robeson Community College*

Margaret Payerle, *Cleveland State University-Ohio*

Jim Pierce, *Lincoln Land Community College*

Tian Ren, *Queensborough Community College*

Nancy Ressler, *Oakton Community College*

Bob Rhea, *J. Sargeant Reynolds Community College*

Minnie M. Riley, *Hinds Community College*

Melissa Rossi, *Southwestern Illinois College*

Anna Roth, *Gloucester County College*

Alan Saleski, *Loyola University-Chicago*

Lisa Sheppard, *Lorain County Community College*

Mark A. Shore, *Allegany College of Maryland*

Mark Sigfrids, *Kalamazoo Valley Community College*

Amber Smith, *Johnson County Community College*

Leonora Smook, *Suffolk County Community College-Brentwood*

Renee Starr, *Arcadia University*

Jennifer Strehler, *Oakton Community College*

Renee Sundrud, *Harrisburg Area Community College*

Sandra Tannen, *Camden County College*

Harriet Thompson, *Albany State University*

John Thoo, *Yuba College*

Fred Toxopeus, *Kalamazoo Valley Community College*

Sara Van Asten, *North Hennepin Community College*

Felix Van Leeuwen, *Johnson County Community College*

Josefino Villanueva, *Florida Memorial University*

Howard Wachtel, *Community College of Philadelphia*

Dottie Walton, *Cuyahoga Community College Eastern Campus*

Walter Wang, *Baruch College*

Brock Wenciker, *Johnson County Community College*

Kevin Wheeler, *Three Rivers Community College*

Latrica Williams, *St. Petersburg College*

Paul Wozniak, *El Camino College*

Christopher Yarrish, *Harrisburg Area Community College*

Steve Zuro, *Joliet Junior College*

Finally, we are forever grateful to the many people behind the scenes at McGraw-Hill without whom we would still be on page 1. Most important, we give special thanks to all the students and instructors who will *grow* their *Beginning Algebra Skills*!

Supplements for the Student

 www.mathzone.com

McGraw-Hill's **MathZone** is a powerful Web-based tutorial for homework, quizzing, testing, and multimedia instruction. Also available in CD-ROM format, MathZone offers:

- **Practice exercises** based on the text and generated in an unlimited quantity for as much practice as needed to master any objective
- **Video** clips of classroom instructors showing how to solve exercises from the text, step by step
- **e-Professor** animations that take the student through step-by-step instructions, delivered on-screen and narrated by a teacher on audio, for solving exercises from the textbook; the user controls the pace of the explanations and can review as needed
- **NetTutor** offers personalized instruction by live tutors familiar with the textbook's objectives and problem-solving methods

Every assignment, exercise, video lecture, and e-Professor is derived from the textbook.

ALEKS Prep for Developmental Mathematics

ALEKS Prep for Beginning Algebra and Prep for Intermediate Algebra focus on prerequisite and introductory material for Beginning Algebra and Intermediate Algebra. These prep products can be used during the first 3 weeks of a course to prepare students for future success in the course and to increase retention and pass rates. Backed by two decades of National Science Foundation funded research, ALEKS interacts with students much like a human tutor, with the ability to precisely assess a student's preparedness and provide instruction on the topics the student is most likely to learn.

ALEKS Prep Course Products Feature:

- Artificial Intelligence Targets Gaps in Individual Students Knowledge
- Assessment and Learning Directed Toward Individual Students Needs
- Open Response Environment with Realistic Input Tools
- Unlimited Online Access-PC & Mac Compatible

Free trial at www.aleks.com/free_trial/instructor

Student's Solutions Manual

The *Student's Solutions Manual* provides comprehensive, worked-out solutions to the odd-numbered exercises in the Pre-Test, Section Exercises, Summary Exercises, Self-Test and the Cumulative Review. The steps shown in the solutions match the style of solved examples in the textbook.

New Connect2Developmental Mathematics Video Series!

Available on DVD and the MathZone website, these innovative videos bring essential Developmental Mathematics concepts to life! The videos take the concepts and place them in a real-world setting so that students make the connection from what they learn in the classroom to their experiences outside the classroom. Making use of 3-D animations and lectures, Connect2Developmental Mathematics video series answers the age-old questions "Why is this important?" and "When will I ever use it?" The videos cover topics from Arithmetic and Basic Mathematics through the Algebra sequence, mixing student-oriented themes and settings with basic theory.

Video Lectures on Digital Video Disk

The video series is based on exercises from the textbook. Each presenter works through selected problems, following the solution methodology employed in the text. The video series is available on DVD or online as part of MathZone. The DVDs are closed-captioned for the hearing impaired, are subtitled in Spanish, and meet the Americans with Disabilities Act Standards for Accessible Design.

NetTutor

Available through MathZone, NetTutor is a revolutionary system that enables students to interact with a live tutor over the web. NetTutor's Web-based, graphical chat capabilities enable students and tutors to use mathematical notation and even to draw graphs as they work through a problem together. Students can also submit questions and receive answers, browse previously answered questions, and view previous sessions. Tutors are familiar with the textbook's objectives and problem-solving styles.

Supplements for the Instructor

 **www.mathzone.com**

McGraw-Hill's **MathZone** is a complete online tutorial and course management system for mathematics and statistics, designed for greater ease of use than any other management system. Available with selected McGraw-Hill textbooks, the system enables instructors to **create and share courses and assignments** with colleagues and adjuncts with only a few clicks of the mouse. All assignments, questions, e-Professors, online tutoring, and video lectures are directly tied to **text-specific** materials.

MathZone courses are customized to your textbook, but you can edit questions and algorithms, import your own content, and **create** announcements and due dates for assignments.

MathZone has **automatic grading** and reporting of easy-to-assign, algorithmically generated homework, quizzing, and testing. All student activity within **MathZone** is automatically recorded and available to you through a **fully integrated gradebook** that can be downloaded to Excel.

MathZone offers:

- **Practice exercises** based on the textbook and generated in an unlimited number for as much practice as needed to master any topic you study.

- **Videos** of classroom instructors giving lectures and showing you how to solve exercises from the textbook.

- **e-Professors** to take you through animated, step-by-step instructions (delivered via on-screen text and synchronized audio) for solving problems in the book, allowing you to digest each step at your own pace.

- **NetTutor,** which offers live, personalized tutoring via the Internet.

Instructor's Testing and Resource Online

Provides a wealth of resources for the instructor. Among the supplements is a **computerized test bank** utilizing Brownstone Diploma® algorithm-based testing software to create customized exams quickly. This user-friendly program enables instructors to search for questions by topic, format, or difficulty level; to edit existing questions or to add new ones; and to scramble questions and answer keys for multiple versions of a single test. Hundreds of text-specific, open-ended, and multiple-choice questions are included in the question bank. Sample chapter tests are also provided. CD available upon request.

Grow Your Knowledge with MathZone Reporting

Visual Reporting

The new dashboard-like reports will provide the progress snapshot instructors are looking for to help them make informed decisions about their students.

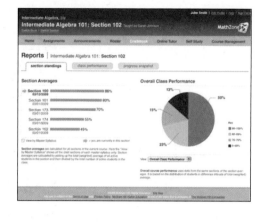

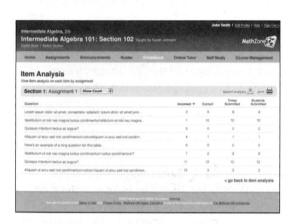

Item Analysis

Instructors can view detailed statistics on student performance at a learning objective level to understand what students have mastered and where they need additional help.

Managing Assignments for Individual Students

Instructors have greater control over creating individualized assignment parameters for individual students, special populations and groups of students, and for managing specific or ad hoc course events.

New User Interface

Designed by You! Instructors and students will experience a modern, more intuitive layout. Items used most commonly are easily accessible through the menu bar such as assignments, visual reports, and course management options.

The Streeter/Hutchison Series in Mathematics   Beginning Algebra

New ALEKS Instructor Module

The new ALEKS Instructor Module features enhanced functionality and a streamlined interface based on research with ALEKS instructors and homework management instructors. Paired with powerful assignment-driven features, textbook integration, and extensive content flexibility, the new ALEKS Instructor Module simplifies administrative tasks and makes ALEKS more powerful than ever. Features include:

Gradebook   Instructors can seamlessly track student scores on automatically graded assignments. They can also easily adjust the weighting and grading scale of each assignment.

Course Calendar   Instructors can schedule assignments and reminders for students.

Automatically Graded Assignments   Instructors can easily assign homework, quizzes, tests, and assessments to all or select students. Deadline extensions can also be created for select students.

Set-Up Wizards   Instructors can use wizards to easily set up assignments, course content, textbook integration, etc.

Message Center   Instructors can use the redesigned Message Center to send, receive, and archive messages; input tools are available to convey mathematical expressions via email.

Baratto/Bergman/Hutchison Video Lectures on Digital Video Disk (DVD)

In the videos, qualified instructors work through selected problems from the textbook, following the solution methodology employed in the text. The video series is available on DVD or online as an assignable element of MathZone. The DVDs are closed-captioned for the hearing-impaired, are subtitled in Spanish, and meet the Americans with Disabilities Act Standards for Accessible Design. Instructors may use them as resources in a learning center, for online courses, and to provide extra help for students who require extra practice.

Annotated Instructor's Edition

In the *Annotated Instructor's Edition (AIE),* **answers to exercises and tests appear adjacent to each exercise set,** in a color used *only* for annotations. Also found in the *AIE* are icons within the Practice Exercises that serve to guide instructors in their preparation of homework assignments and lessons.

Instructor's Solutions Manual

The *Instructor's Solutions Manual* provides comprehensive, worked-out solutions to all exercises in the Pre-Test, Section Exercises, Summary Exercises, Self-Test, and the Cumulative Review. The methods used to solve the problems in the manual are the same as those used to solve the examples in the textbook.

A commitment to accuracy

You have a right to expect an accurate textbook, and McGraw-Hill invests considerable time and effort to make sure that we deliver one. Listed below are the many steps we take to make sure this happens.

Our accuracy verification process

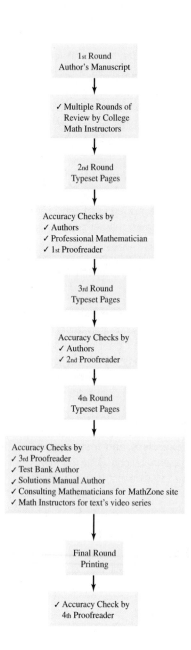

First Round

Step 1: Numerous **college math instructors** review the manuscript and report on any errors that they may find. Then the authors make these corrections in their final manuscript.

Second Round

Step 2: Once the manuscript has been typeset, the **authors** check their manuscript against the first page proofs to ensure that all illustrations, graphs, examples, exercises, solutions, and answers have been correctly laid out on the pages, and that all notation is correctly used.

Step 3: An outside, **professional, mathematician** works through every example and exercise in the page proofs to verify the accuracy of the answers.

Step 4: A **proofreader** adds a triple layer of accuracy assurance in the first pages by hunting for errors, then a second, corrected round of page proofs is produced.

Third Round

Step 5: The **author team** reviews the second round of page proofs for two reasons: (1) to make certain that any previous corrections were properly made, and (2) to look for any errors they might have missed on the first round.

Step 6: A **second proofreader** is added to the project to examine the new round of page proofs to double check the author team's work and to lend a fresh, critical eye to the book before the third round of paging.

Fourth Round

Step 7: A **third proofreader** inspects the third round of page proofs to verify that all previous corrections have been properly made and that there are no new or remaining errors.

Step 8: Meanwhile, in partnership with **independent mathematicians,** the text accuracy is verified from a variety of fresh perspectives:

- The **test bank author** checks for consistency and accuracy as he/she prepares the computerized test item file.
- The **solutions manual author** works every exercise and verifies his/her answers, reporting any errors to the publisher.
- A **consulting group of mathematicians,** who write material for the text's MathZone site, notifies the publisher of any errors they encounter in the page proofs.
- A video production company employing **expert math instructors** for the text's videos will alert the publisher of any errors they might find in the page proofs.

Final Round

Step 9: The **project manager,** who has overseen the book from the beginning, performs a **fourth proofread** of the textbook during the printing process, providing a final accuracy review.

⇒ What results is a mathematics textbook that is as accurate and error-free as is humanly possible. Our authors and publishing staff are confident that our many layers of quality assurance have produced textbooks that are the leaders in the industry for their integrity and correctness.

The Streeter/Hutchison Series in Mathematics   Beginning Algebra

brief contents

contents

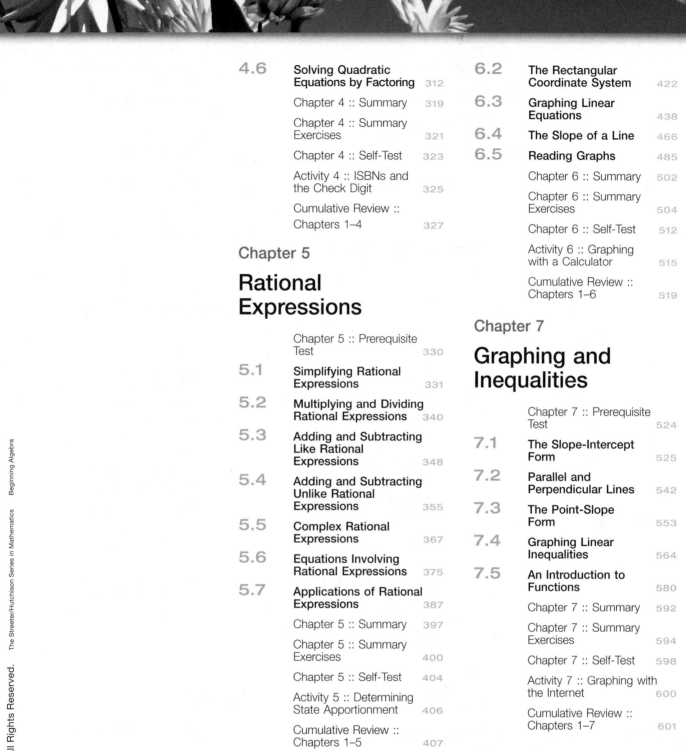

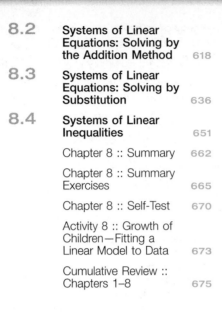

Chapter 9

Exponents and Radicals

Chapter 10

Quadratic Equations

applications index

CHAPTER

1

chapter 1 > Make the Connection

INTRODUCTION

Anthropologists and archeologists investigate modern human cultures and societies as well as cultures that existed so long ago that their characteristics must be inferred from buried objects. With methods such as carbon dating, it has been established that large, organized cultures existed around 3000 B.C.E. in Egypt, 2800 B.C.E. in India, no later than 1500 B.C.E. in China, and around 1000 B.C.E. in the Americas.

Which is older, an object from 3000 B.C.E. or an object from A.D. 500? An object from A.D. 500 is about $2{,}000 - 500$ years old, or about 1,500 years old. But an object from 3000 B.C.E. is about $2{,}000 + 3{,}000$ years old, or about 5,000 years old. Why subtract in the first case but add in the other? Because the B.C.E. dates must be considered as *negative* numbers.

Very early on, the Chinese accepted the idea that a number could be negative; they used red calculating rods for positive numbers and black rods for negative numbers. Hindu mathematicians in India worked out the arithmetic of negative numbers as long ago as A.D. 400, but western mathematicians did not recognize this idea until the sixteenth century. It would be difficult today to think of measuring things such as temperature, altitude, and money without negative numbers.

The Language of Algebra

CHAPTER 1 OUTLINE

1000 B.C.E. = −1000   A.D. 1000 = +1000

◄──── Count   Count ────►

Name _____

Section _____ **Date** _____

Answers

This prerequisite test provides some exercises requiring skills that you will need to be successful in the coming chapter. The answers for these exercises can be found in the back of this text. This prerequisite test can help you identify topics that you will need to review before beginning the chapter.

Write each phrase as an arithmetic expression and solve.

1. 8 less than 10

2. The sum of 3 and the product of 5 and 6

Find the reciprocal of each number.

3. 12

4. $4\dfrac{5}{8}$

Evaluate, as indicated.

5. $\left(\dfrac{3}{2}\right) \times \left(\dfrac{2}{3}\right)$

6. $(4)\left(\dfrac{1}{4}\right)$

7. $\dfrac{2}{2}$

8. $5 + 2 \times 3^2$

9. 8^2

10. $3 + 2 \cdot (2 + 3)^2 - (4 - 1)^3$

11. **BUSINESS AND FINANCE**   An $8\dfrac{1}{2}$-acre plot of land is on sale for \$120,000. What is the price per acre?

12. **BUSINESS AND FINANCE**   A grocery store adds a 30% markup to the wholesale price of goods to determine their retail price. What is the retail price of a box of cookies if its wholesale price is \$1.19?

▶ **Tips for Student Success**

Over the first few chapters, we present a series of class-tested techniques designed to improve your performance in this math class.

Become familiar with your textbook.

Perform each of the following tasks.

1. Use the Table of Contents to find the title of Section 5.1.

2. Use the Index to find the earliest reference to the term *mean*. (By the way, this term has nothing to do with the personality of either your instructor or the textbook author!)

3. Find the answer to the first Check Yourself exercise in Section 1.1.

4. Find the answers to the Self-Test for Chapter 2.

5. Find the answers to the odd-numbered exercises in Section 1.1.

6. In the margin notes for Section 1.1, find the formula used to compute the area of a rectangle.

7. Find the Prerequisite Test for Chapter 3.

Now you know where some of the most important features of the text are. When you have a moment of confusion, think about using one of these features to help you clear up that confusion.

1.1

Properties of Real Numbers

< 1.1 Objectives >

1 > Recognize applications of the commutative properties

2 > Recognize applications of the associative properties

3 > Recognize applications of the distributive property

▶ **Tips for Student Success**

RECALL

The first Tips for Student Success hint is on the previous page.

Over the first few chapters, we present you with a series of class-tested techniques designed to improve your performance in your math class.

Become familiar with your syllabus.

In your first class meeting, your instructor probably gave you a class syllabus. If you have not already done so, incorporate important information into a calendar and address book.

1. Write all important dates in your calendar. This includes the date and time of the final exam, test dates, quiz dates, and homework due dates. Never allow yourself to be surprised by a deadline!

2. Write your instructor's name, contact information, and office number in your address book. Also include your instructor's office hours. Make it a point to see your instructor early in the term. Although not the only person who can help you, your instructor is an important resource to help clear up any confusion you may have.

NOTE

We only work with real numbers in this text.

3. Make note of other resources that are available to you. This includes tutoring, CDs and DVDs, and Web pages.

Given all of these resources, it is important that you never let confusion or frustration mount. If you "can't get it" from the text, try another resource. All of these resources are there specifically for you, so take advantage of them!

Everything that we do in *algebra* is based on the **properties of real numbers.** Before being introduced to algebra, you should understand these properties.

The **commutative properties** tell us that we can add or multiply in any order.

Property	
The Commutative Properties	If *a* and *b* are any numbers,
	1. $a + b = b + a$ Commutative property of addition
	2. $a \cdot b = b \cdot a$ Commutative property of multiplication

You may notice that we used the letters *a* and *b* rather than numbers in the **Property** box. We use these letters to indicate that these properties are true for any choice of real numbers.

▶ **Example 1**   Identifying the Commutative Properties

< Objective 1 >

(a) $5 + 9 = 9 + 5$

This is an application of the commutative property of addition.

(b) $5 \cdot 9 = 9 \cdot 5$

This is an application of the commutative property of multiplication.

Check Yourself 1

Identify the property being applied.

(a) $7 + 3 = 3 + 7$ **(b)** $7 \cdot 3 = 3 \cdot 7$

We also want to be able to change the grouping when simplifying expressions. Regrouping is possible because of the **associative properties.** Numbers can be grouped in any manner to find a sum or a product.

Property

The Associative Properties

If a, b, and c are any numbers,

1. $a + (b + c) = (a + b) + c$ Associative property of addition

2. $a \cdot (b \cdot c) = (a \cdot b) \cdot c$ Associative property of multiplication

---

    Example 2   Demonstrating the Associative Properties

< Objective 2 >

RECALL

Always do the operation in the parentheses first.

(a) Show that $2 + (3 + 8) = (2 + 3) + 8$.

$2 + \underbrace{(3 + 8)}_{\text{Add first.}}$ $\underbrace{(2 + 3)}_{\text{Add first.}} + 8$

$= 2 + 11$ $= 5 + 8$

$= 13$ $= 13$

So

$2 + (3 + 8) = (2 + 3) + 8$

(b) Show that $\dfrac{1}{3} \cdot (6 \cdot 5) = \left(\dfrac{1}{3} \cdot 6\right) \cdot 5$.

$\dfrac{1}{3} \cdot \underbrace{(6 \cdot 5)}_{\text{Multiply first.}}$ $\underbrace{\left(\dfrac{1}{3} \cdot 6\right)}_{\text{Multiply first.}} \cdot 5$

$= \dfrac{1}{3} \cdot (30)$ $= (2) \cdot 5$

$= 10$ $= 10$

So

$\dfrac{1}{3} \cdot (6 \cdot 5) = \left(\dfrac{1}{3} \cdot 6\right) \cdot 5$

Check Yourself 2

Show that the following statements are true.

(a) $3 + (4 + 7) = (3 + 4) + 7$ **(b)** $3 \cdot (4 \cdot 7) = (3 \cdot 4) \cdot 7$

(c) $\left(\dfrac{1}{5} \cdot 10\right) \cdot 4 = \dfrac{1}{5} \cdot (10 \cdot 4)$

NOTE

The area of a rectangle is the product of its length and width:

$A = L \cdot W$

The **distributive property** involves addition and multiplication together. We can illustrate this property with an application.

Suppose that we want to find the total of the two areas shown in the figure.

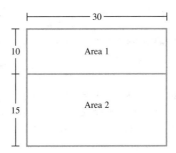

We can find the total area by multiplying the length by the overall width, which is found by adding the two widths.   [or]

We can find the total area as a sum of the two areas.

		(Area 1)		(Area 2)
Length	Overall width	Length · Width		Length · Width
$30 \cdot$	$(10 + 15)$	$30 \cdot 10$	$+$	$30 \cdot 15$
$= 30 \cdot 25$		$= 300 + 450$		
$= 750$		$= 750$		

So

$$30 \cdot (10 + 15) = 30 \cdot 10 + 30 \cdot 15$$

This leads us to the following property.

Property

The Distributive Property

If a, b, and c are any numbers,

$$a \cdot (b + c) = a \cdot b + a \cdot c \quad \text{and} \quad (b + c) \cdot a = b \cdot a + c \cdot a$$

Example 3

Using the Distributive Property

< Objective 3 >

NOTES

You should see the pattern that emerges.

$a \cdot (b + c) = a \cdot b + a \cdot c$

We "distributed" the multiplication "over" the addition.

It is also true that

$\frac{1}{3} \cdot (9 + 12) = \frac{1}{3} \cdot (21) = 7$

Use the distributive property to remove the parentheses in the following.

(a) $5 \cdot (3 + 4)$

$5 \cdot (3 + 4) = 5 \cdot 3 + 5 \cdot 4$ We could also say $5 \cdot (3 + 4) = 5 \cdot 7 = 35$

$= 15 + 20 = 35$

(b) $\frac{1}{3} \cdot (9 + 12) = \frac{1}{3} \cdot 9 + \frac{1}{3} \cdot 12$

$= 3 + 4 = 7$

Check Yourself 3

Use the distributive property to remove the parentheses.

(a) $4 \cdot (6 + 7)$ **(b)** $\frac{1}{5} \cdot (10 + 15)$

Example 4 requires that you identify which property is being demonstrated. Look for patterns that help you to remember each of the properties.

| **Example** 4 | **Identifying Properties** |

Name the property demonstrated.

(a) $3 \cdot (8 + 2) = 3 \cdot 8 + 3 \cdot 2$ demonstrates the distributive property.

(b) $2 + (3 + 5) = (2 + 3) + 5$ demonstrates the associative property of addition.

(c) $3 \cdot 5 = 5 \cdot 3$ demonstrates the commutative property of multiplication.

Check Yourself 4

Name the property demonstrated.

(a) $2 \cdot (3 \cdot 5) = (2 \cdot 3) \cdot 5$

(b) $4 \cdot (-2 + 4) = 4 \cdot (-2) + 4 \cdot 4$

(c) $\frac{1}{2} + 8 = 8 + \frac{1}{2}$

Check Yourself ANSWERS

1. (a) Commutative property of addition;
(b) commutative property of multiplication

2. (a) $3 + (4 + 7) = 3 + 11 = 14$ **(b)** $3 \cdot (4 \cdot 7) = 3 \cdot 28 = 84$
$(3 + 4) + 7 = 7 + 7 = 14$ $(3 \cdot 4) \cdot 7 = 12 \cdot 7 = 84$

(c) $\left(\frac{1}{5} \cdot 10 \right) \cdot 4 = 2 \cdot 4 = 8$

$\frac{1}{5} \cdot (10 \cdot 4) = \frac{1}{5} \cdot 40 = 8$

3. (a) $4 \cdot 6 + 4 \cdot 7 = 24 + 28 = 52;$ **(b)** $\frac{1}{5} \cdot 10 + \frac{1}{5} \cdot 15 = 2 + 3 = 5$

4. (a) Associative property of multiplication; **(b)** distributive property;
(c) commutative property of addition

Reading Your Text

The following fill-in-the-blank exercises are designed to ensure that you understand some of the key vocabulary used in this section.

SECTION 1.1

(a) The _____ properties tell us that we can add or multiply in any order.

(b) The order of operations requires that we do any operations inside _____ first.

(c) The _____ property of multiplication states that $a \cdot (b \cdot c) = (a \cdot b) \cdot c$.

(d) The _____ of a rectangle is the product of its length and width.

Basic Skills | Challenge Yourself | Calculator/Computer | Career Applications | Above and Beyond

< Objectives 1–3 >

Identify the property illustrated by each statement.

1. $5 + 9 = 9 + 5$

2. $6 + 3 = 3 + 6$

3. $2 \cdot (3 \cdot 5) = (2 \cdot 3) \cdot 5$

4. $3 \cdot (5 \cdot 6) = (3 \cdot 5) \cdot 6$

5. $\dfrac{1}{4} \cdot \dfrac{1}{5} = \dfrac{1}{5} \cdot \dfrac{1}{4}$

6. $7 \cdot 9 = 9 \cdot 7$

7. $8 + 12 = 12 + 8$

8. $6 + 2 = 2 + 6$

9. $(5 \cdot 7) \cdot 2 = 5 \cdot (7 \cdot 2)$

10. $(8 \cdot 9) \cdot 2 = 8 \cdot (9 \cdot 2)$

11. $7 \cdot (2 \cdot 5) = (7 \cdot 2) \cdot 5$

12. $\dfrac{1}{2} \cdot 6 = 6 \cdot \dfrac{1}{2}$

13. $2 \cdot (3 + 5) = 2 \cdot 3 + 2 \cdot 5$

> Videos

14. $5 \cdot (4 + 6) = 5 \cdot 4 + 5 \cdot 6$

15. $5 + (7 + 8) = (5 + 7) + 8$

16. $8 + (2 + 9) = (8 + 2) + 9$

17. $\left(\dfrac{1}{3} + 4\right) + \dfrac{1}{5} = \dfrac{1}{3} + \left(4 + \dfrac{1}{5}\right)$

18. $(5 + 5) + 3 = 5 + (5 + 3)$

19. $7 \cdot (3 + 8) = 7 \cdot 3 + 7 \cdot 8$

20. $5 \cdot (6 + 8) = 5 \cdot 6 + 5 \cdot 8$

Boost *your* GRADE at
ALEKS.com!

ALEKS

- Practice Problems   • e-Professors
- Self-Tests   • Videos
- NetTutor

Name _____

Section _____ Date _____

Answers

1. _____
2. _____
3. _____
4. _____
5. _____
6. _____
7. _____
8. _____
9. _____
10. _____
11. _____
12. _____
13. _____
14. _____
15. _____
16. _____
17. _____
18. _____
19. _____
20. _____

Answers

21. _____

22. _____

23. _____

24. _____

25. _____

26. _____

27. _____

28. _____

29. _____

30. _____

31. _____

32. _____

33. _____

34. _____

35. _____

36. _____

37. _____

38. _____

39. _____

40. _____

41. _____

42. _____

43. _____

44. _____

45. _____

46. _____

47. _____

48. _____

49. _____

50. _____

51. _____

52. _____

53. _____

54. _____

Verify that each statement is true by evaluating each side of the equation separately and comparing the results.

21. $7 \cdot (3 + 4) = 7 \cdot 3 + 7 \cdot 4$

22. $4 \cdot (5 + 1) = 4 \cdot 5 + 4 \cdot 1$

23. $2 + (9 + 8) = (2 + 9) + 8$

24. $6 + (15 + 3) = (6 + 15) + 3$

25. $\dfrac{1}{3} \cdot (6 \cdot 3) = \left(\dfrac{1}{3} \cdot 6 \right) \cdot 3$

26. $2 \cdot (9 \cdot 10) = (2 \cdot 9) \cdot 10$

27. $5 \cdot (2 + 8) = 5 \cdot 2 + 5 \cdot 8$

28. $\dfrac{1}{4} \cdot (10 + 2) = \dfrac{1}{4} \cdot 10 + \dfrac{1}{4} \cdot 2$

29. $(3 + 12) + 8 = 3 + (12 + 8)$

30. $(8 + 12) + 7 = 8 + (12 + 7)$

31. $(4 \cdot 7) \cdot 2 = 4 \cdot (7 \cdot 2)$

32. $(6 \cdot 5) \cdot 3 = 6 \cdot (5 \cdot 3)$

33. $\dfrac{1}{2} \cdot (2 + 6) = \dfrac{1}{2} \cdot 2 + \dfrac{1}{2} \cdot 6$

34. $\dfrac{1}{3} \cdot (6 + 9) = \dfrac{1}{3} \cdot 6 + \dfrac{1}{3} \cdot 9$

35. $\left(\dfrac{2}{3} + \dfrac{1}{6} \right) + \dfrac{1}{3} = \dfrac{2}{3} + \left(\dfrac{1}{6} + \dfrac{1}{3} \right)$

36. $\dfrac{3}{4} + \left(\dfrac{5}{8} + \dfrac{1}{2} \right) = \left(\dfrac{3}{4} + \dfrac{5}{8} \right) + \dfrac{1}{2}$

37. $(2.3 + 3.9) + 4.1 = 2.3 + (3.9 + 4.1)$

38. $(1.7 + 4.1) + 7.6 = 1.7 + (4.1 + 7.6)$

39. $\dfrac{1}{2} \cdot (2 \cdot 8) = \left(\dfrac{1}{2} \cdot 2 \right) \cdot 8$

40. $\dfrac{1}{5} \cdot (5 \cdot 3) = \left(\dfrac{1}{5} \cdot 5 \right) \cdot 3$

41. $\left(\dfrac{3}{5} \cdot \dfrac{5}{6} \right) \cdot \dfrac{4}{3} = \dfrac{3}{5} \cdot \left(\dfrac{5}{6} \cdot \dfrac{4}{3} \right)$

42. $\dfrac{4}{7} \cdot \left(\dfrac{21}{16} \cdot \dfrac{8}{3} \right) = \left(\dfrac{4}{7} \cdot \dfrac{21}{16} \right) \cdot \dfrac{8}{3}$

43. $2.5 \cdot (4 \cdot 5) = (2.5 \cdot 4) \cdot 5$

44. $4.2 \cdot (5 \cdot 2) = (4.2 \cdot 5) \cdot 2$

Use the distributive property to remove the parentheses in each expression. Then simplify your result where possible.

45. $3 \cdot (2 + 6)$

46. $5 \cdot (4 + 6)$

47. $2 \cdot (12 + 10)$

48. $9 \cdot (1 + 8)$

49. $0.1 \cdot (2 + 10)$

50. $1.2 \cdot (3 + 8)$

51. $\dfrac{2}{3} \cdot (6 + 9)$ > Videos

52. $\dfrac{1}{2} \cdot \left(4 + \dfrac{1}{3} \right)$

53. $\dfrac{1}{3} \cdot (15 + 9)$

54. $\dfrac{1}{6} \cdot (36 + 24)$

Basic Skills | **Challenge Yourself** | Calculator/Computer | Career Applications | Above and Beyond
▲

Use the properties of addition and multiplication to complete each statement.

55. $5 + 7 = \quad + 5$

56. $(5 + 3) + 4 = 5 + (\quad + 4)$

57. $(8) \cdot (3) = (3) \cdot (\quad)$

58. $8 \cdot (3 + 4) = 8 \cdot 3 + \quad \cdot 4$

59. $7 \cdot (2 + 5) = 7 \cdot \quad + 7 \cdot 5$

60. $4 \cdot (2 \cdot 4) = (\quad \cdot 2) \cdot 4$

Use the indicated property to write an expression that is equivalent to each expression.

61. $3 + 7$ (commutative property of addition)

62. $2 \cdot (3 + 4)$ (distributive property)

63. $5 \cdot (3 \cdot 2)$ (associative property of multiplication)

64. $(3 + 5) + 2$ (associative property of addition)

65. $2 \cdot 4 + 2 \cdot 5$ (distributive property) > Videos

66. $7 \cdot 9$ (commutative property of multiplication)

Basic Skills | Challenge Yourself | Calculator/Computer | Career Applications | **Above and Beyond**
▲

Evaluate each pair of expressions. Then answer the given question.

67. $8 - 5$ and $5 - 8$
Do you think subtraction is commutative?

68. $12 \div 3$ and $3 \div 12$
Do you think division is commutative?

69. $(12 \overset{4}{-} 8) - 4^{\,0}$ and $12^{\,12} - (8 \overset{4}{-} 4)\ 8$
Do you think subtraction is associative?

70. $(48 \div 16) \div 4$ and $48 \div (16 \div 4)$
Do you think division is associative?

71. $3 \cdot (6 - 2)$ and $3 \cdot 6 - 3 \cdot 2$
Do you think multiplication is distributive over subtraction?

72. $\dfrac{1}{2} \cdot (16 - 10)$ and $\dfrac{1}{2} \cdot 16 - \dfrac{1}{2} \cdot 10$
Do you think multiplication is distributive over subtraction?

55. _____

56. _____

57. _____

58. _____

59. _____

60. _____

61. _____

62. _____

63. _____

64. _____

65. _____

66. _____

67. _____

68. _____

69. _____

70. _____

71. _____

72. _____

Answers

73. _____

74. _____

75. _____

76. _____

77. _____

78. _____

Complete the statement using the

   (a) Distributive property
   (b) Commutative property of addition
   (c) Commutative property of multiplication

73. $5 \cdot (3 + 4) =$

74. $6 \cdot (5 + 4) =$

Identify the property that is used.

75. $5 + (6 + 7) = (5 + 6) + 7$

76. $5 + (6 + 7) = 5 + (7 + 6)$

                                               > Videos

77. $4 \cdot (3 + 2) = 4 \cdot (2 + 3)$

78. $4 \cdot (3 + 2) = (3 + 2) \cdot 4$

Answers

1. Commutative property of addition   **3.** Associative property of multiplication   **5.** Commutative property of multiplication
7. Commutative property of addition   **9.** Associative property of multiplication   **11.** Associative property of multiplication
13. Distributive property   **15.** Associative property of addition
17. Associative property of addition   **19.** Distributive property
21. $49 = 49$ **23.** $19 = 19$ **25.** $6 = 6$ **27.** $50 = 50$
29. $23 = 23$ **31.** $56 = 56$ **33.** $4 = 4$ **35.** $\dfrac{7}{6} = \dfrac{7}{6}$

37. $10.3 = 10.3$ **39.** $8 = 8$ **41.** $\dfrac{2}{3} = \dfrac{2}{3}$ **43.** $50 = 50$ **45.** 24
47. 44 **49.** 1.2 **51.** 10 **53.** 8 **55.** 7 **57.** 8 **59.** 2
61. $7 + 3$ **63.** $(5 \cdot 3) \cdot 2$ **65.** $2 \cdot (4 + 5)$ **67.** No   **69.** No
71. Yes   **73. (a)** $5 \cdot 3 + 5 \cdot 4$; **(b)** $5 \cdot (4 + 3)$; **(c)** $(3 + 4) \cdot 5$
75. Associative property of addition   **77.** Commutative property of addition

1.2

Adding and Subtracting Real Numbers

< 1.2 Objectives >

1 > Find the sum of two real numbers

2 > Find the difference of two real numbers

We should always be careful when performing arithmetic with negative numbers. To see how those operations are performed when negative numbers are involved, we start with addition.

An application may help, so we represent a gain of money as a positive number and a loss as a negative number.

If you gain $3 and then gain $4, the result is a gain of $7:

$$3 + 4 = 7$$

If you lose $3 and then lose $4, the result is a loss of $7:

$$-3 + (-4) = -7$$

If you gain $3 and then lose $4, the result is a loss of $1:

$$3 + (-4) = -1$$

If you lose $3 and then gain $4, the result is a gain of $1:

$$-3 + 4 = 1$$

A number line can be used to illustrate adding with these numbers. Starting at the origin, we move to the *right* when adding positive numbers and to the *left* when adding negative numbers.

	Example 1	Adding Negative Numbers

< Objective 1 >

(a) Add $-3 + (-4)$.

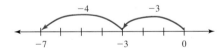

Start at the origin and move 3 units to the left. Then move 4 more units to the left to find the sum. From the number line we see that the sum is

$$-3 + (-4) = -7$$

(b) Add $-\dfrac{3}{2} + \left(-\dfrac{1}{2}\right)$.

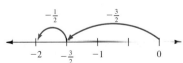

As before, we start at the origin. From that point move $\frac{3}{2}$ units left. Then move another $\frac{1}{2}$ unit left to find the sum. In this case

$$-\frac{3}{2} + \left(-\frac{1}{2}\right) = -2$$

Check Yourself 1

Add.

(a) $-4 + (-5)$ \qquad (b) $-3 + (-7)$

(c) $-5 + (-15)$ \qquad (d) $-\frac{5}{2} + \left(-\frac{3}{2}\right)$

You have probably noticed some helpful patterns in the previous examples. These patterns will allow you to do the work mentally rather than with a number line.

We use absolute values to describe the pattern so that we can create the following rule.

Property

Adding Real Numbers with the Same Sign

If two numbers have the same sign, add their absolute values. Give the sum the sign of the original numbers.

In other words, the sum of two positive numbers is positive and the sum of two negative numbers is negative.

We can also use a number line to add two numbers that have *different* signs.

Example 2   Adding Numbers with Different Signs

(a) Add $3 + (-6)$.

First move 3 units to the right of the origin. Then move 6 units to the left.

$3 + (-6) = -3$

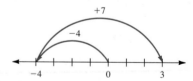

(b) Add $-4 + 7$.

This time move 4 units to the left of the origin as the first step. Then move 7 units to the right.

$-4 + 7 = 3$

Check Yourself 2

Add.

(a) $7 + (-5)$ \quad (b) $4 + (-8)$ \quad (c) $-\frac{1}{3} + \frac{16}{3}$ \quad (d) $-7 + 3$

You have no doubt noticed that, in adding a positive number and a negative number, sometimes the sum is positive and sometimes it is negative. This depends on which of the numbers has the larger absolute value. This leads us to the second part of our addition rule.

Property

Adding Real Numbers with Different Signs

If two numbers have different signs, subtract their absolute values, the smaller from the larger. Give the result the sign of the number with the larger absolute value.

Example 3   **Adding Positive and Negative Numbers**

(a) $7 + (-19) = -12$

Because the two numbers have different signs, subtract the absolute values $(19 - 7 = 12)$. The sum has the sign $(-)$ of the number with the larger absolute value.

(b) $-\dfrac{13}{2} + \dfrac{7}{2} = -3$

Subtract the absolute values $\left(\dfrac{13}{2} - \dfrac{7}{2} = \dfrac{6}{2} = 3\right)$. The sum has the sign $(-)$ of the number with the larger absolute value: $\left|-\dfrac{13}{2}\right| > \left|\dfrac{7}{2}\right|$.

(c) $-8.2 + 4.5 = -3.7$

Subtract the absolute values $(8.2 - 4.5 = 3.7)$. The sum has the sign $(-)$ of the number with the larger absolute value: $|-8.2| > |4.5|$.

Check Yourself 3

Add mentally.

(a) $5 + (-14)$ **(b)** $-7 + (-8)$ **(c)** $-8 + 15$

(d) $7 + (-8)$ **(e)** $-\dfrac{2}{3} + \left(-\dfrac{7}{3}\right)$ **(f)** $5.3 + (-2.3)$

In Section 1.1 we discussed the commutative, associative, and distributive properties. There are two other properties of addition that we should mention. First, the sum of any number and 0 is always that number. In symbols,

Property

Additive Identity Property

For any number a,

$a + 0 = 0 + a = a$

In words, adding zero does not change a number. Zero is called the **additive identity**.

Example 4   **Adding the Identity**

Add.

(a) $9 + 0 = 9$

(b) $0 + \left(-\dfrac{5}{4}\right) = -\dfrac{5}{4}$

(c) $(-25) + 0 = -25$

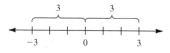

Check Yourself 4

Add.

(a) $8 + 0$ **(b)** $0 + \left(-\dfrac{8}{3}\right)$ **(c)** $(-36) + 0$

NOTES

The opposite of a number is also called the **additive inverse** of that number.

3 and -3 are opposites.

Recall that every number has an *opposite*. It corresponds to a point the same distance from the origin as the given number, but in the opposite direction.

The opposite of 9 is -9.

The opposite of -15 is 15.

Our second property states that the sum of any number and its opposite is 0.

Property

Additive Inverse Property

For any number a, there exists a number $-a$ such that

$$a + (-a) = (-a) + a = 0$$

We could also say that $-a$ represents the opposite of the number a. The sum of any number and its opposite, or additive inverse, is 0.

Example 5   **Adding Inverses**

(a) $9 + (-9) = 0$

(b) $-15 + 15 = 0$

(c) $(-2.3) + 2.3 = 0$

(d) $\dfrac{4}{5} + \left(-\dfrac{4}{5}\right) = 0$

Check Yourself 5

Add.

(a) $(-17) + 17$ **(b)** $12 + (-12)$

(c) $\dfrac{1}{3} + \left(-\dfrac{1}{3}\right)$ **(d)** $-1.6 + 1.6$

To begin our discussion of subtraction when negative numbers are involved, we can look back at a problem using natural numbers. Of course, we know that

$$8 - 5 = 3$$

From our work in adding real numbers, we know that it is also true that

$$8 + (-5) = 3$$

NOTE

This is the *definition* of subtraction.

Comparing these equations, we see that the results are the same. This leads us to an important pattern. Any subtraction problem can be written as a problem in addition. Subtracting 5 is the same as adding the opposite of 5, or -5. We can write this fact as follows:

$$8 - 5 = 8 + (-5) = 3$$

This leads us to the following rule for subtracting real numbers.

Property

Subtracting Real Numbers

1. Rewrite the subtraction problem as an addition problem.
   a. Change the operation from subtraction to addition.
   b. Replace the number being subtracted with its opposite.
2. Add the resulting numbers as before.
   In symbols,

$$a - b = a + (-b)$$

Example 6 illustrates this property.

Example 6   **Subtracting Real Numbers**

< Objective 2 >

Simplify each expression.

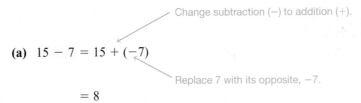

(a) $15 - 7 = 15 + (-7)$ Change subtraction (−) to addition (+).

Replace 7 with its opposite, −7.

$= 8$

(b) $9 - 12 = 9 + (-12) = -3$

(c) $-6 - 7 = -6 + (-7) = -13$

(d) $-\dfrac{3}{5} - \dfrac{7}{5} = -\dfrac{3}{5} + \left(-\dfrac{7}{5}\right) = -\dfrac{10}{5} = -2$

(e) $2.1 - 3.4 = 2.1 + (-3.4) = -1.3$

(f) Subtract 5 from -2. We write the statement as $-2 - 5$ and proceed as before:

$-2 - 5 = -2 + (-5) = -7$

> **CAUTION**

The statement "subtract b from a" means $a - b$.

 Check Yourself 6

Subtract.

(a) $18 - 7$ (b) $5 - 13$ (c) $-7 - 9$

(d) $-\dfrac{5}{6} - \dfrac{7}{6}$ (e) $-2 - 7$ (f) $5.6 - 7.8$

The subtraction rule is used in the same way when the number being subtracted is negative. Change the subtraction to addition. Replace the negative number being subtracted with its opposite, which is positive. Example 7 illustrates this principle.

Example 7   **Subtracting Real Numbers**

Simplify each expression.

Change subtraction to addition.

(a) $5 - (-2) = 5 + (+2) = 5 + 2 = 7$

Replace −2 with its opposite, +2 or 2.

(b) $7 - (-8) = 7 + (+8) = 7 + 8 = 15$

(c) $-9 - (-5) = -9 + 5 = -4$

(d) $-12.7 - (-3.7) = -12.7 + 3.7 = -9$

(e) $-\dfrac{3}{4} - \left(-\dfrac{7}{4}\right) = -\dfrac{3}{4} + \left(+\dfrac{7}{4}\right) = \dfrac{4}{4} = 1$

(f) Subtract -4 from -5. We write

$-5 - (-4) = -5 + 4 = -1$

Check Yourself 7

Subtract.

(a) $8 - (-2)$ **(b)** $3 - (-10)$ **(c)** $-7 - (-2)$
(d) $-9.8 - (-5.8)$ **(e)** $7 - (-7)$

In order to use a calculator to do arithmetic with real numbers, there are some keys you should become familiar with.

The first key is the subtraction key, $\boxed{-}$. This key is usually found in the right column of calculator keys along with the other "operation" keys such as addition, multiplication, and division.

The second key to find is the one for negative numbers. On graphing calculators, it usually looks like $\boxed{(-)}$, whereas on scientific calculators, the key usually looks like $\boxed{+/-}$. In either case, the negative number key is usually found in the bottom row.

One very important difference between the two types of calculators is that when using a graphing calculator, you input the negative sign before keying in the number (as it is written). When using a scientific calculator, you input the negative number button after keying in the number.

In Example 8, we illustrate this difference, while showing that subtraction remains the same.

▶ **Example 8**   **Subtracting with a Calculator**

Use a calculator to find each difference.

(a) $-12.43 - 3.516$

Graphing Calculator

$\boxed{(-)}$ 12.43 $\boxed{-}$ 3.516 $\boxed{\text{ENTER}}$ The negative number sign comes before the number.

The display should read -15.946.

Scientific Calculator

12.43 $\boxed{+/-}$ $\boxed{-}$ 3.516 $\boxed{=}$ The negative number sign comes after the number.

The display should read -15.946.

(b) $23.56 - (-4.7)$

Graphing Calculator

23.56 $\boxed{-}$ $\boxed{(-)}$ 4.7 $\boxed{\text{ENTER}}$ The negative number sign comes before the number.

The display should read 28.26.

Scientific Calculator

23.56 $\boxed{-}$ 4.7 $\boxed{+/-}$ $\boxed{=}$ The negative number sign comes after the number.

The display should read 28.26.

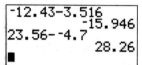

Check Yourself 8

Use your calculator to find the difference.

(a) $-13.46 - 5.71$ (b) $-3.575 - (-6.825)$

Example 9   An Application Involving Real Numbers

Oscar owned four stocks. This year his holdings in Cisco went up $2,250, in AT&T they went down $1,345, in Texaco they went down $5,215, and in IBM they went down $1,525. How much less are his holdings worth at the end of the year compared to the beginning of the year?

To find the change in Oscar's holdings, we add the amounts that went up and subtract the amounts that went down.

$$\$2,250 - \$1,345 - \$5,215 - \$1,525 = -\$5,835$$

Oscar's holdings are worth $5,835 less at the end of the year.

Check Yourself 9

A bus with fifteen people stopped at Avenue A. Nine people got off and five people got on. At Avenue B six people got off and eight people got on. At Avenue C four people got off the bus and six people got on. How many people were now on the bus?

Check Yourself ANSWERS

1. (a) -9; (b) -10; (c) -20; (d) -4 2. (a) 2; (b) -4; (c) 5; (d) -4
3. (a) -9; (b) -15; (c) 7; (d) -1; (e) -3; (f) 3
4. (a) 8; (b) $-\dfrac{8}{3}$; (c) -36 5. (a) 0; (b) 0; (c) 0; (d) 0
6. (a) 11; (b) -8; (c) -16; (d) -2; (e) -9; (f) -2.2 7. (a) 10; (b) 13;
(c) -5; (d) -4; (e) 14   8. (a) -19.17; (b) 3.25   9. 15 people

---

Reading Your Text

The following fill-in-the-blank exercises are designed to ensure that you understand some of the key vocabulary used in this section.

SECTION 1.2

(a) When two negative numbers are added, the sign of the sum is

_____.

(b) The sum of two numbers with different signs is given the sign of the number with the larger _____ value.

(c) _____ is called the additive identity.

(d) When subtracting negative numbers, change the operation from subtraction to addition and replace the second number with its _____.

1.2 exercises

Name _____

Section _____ Date _____

Answers

1. _____	2. _____
3. _____	4. _____
5. _____	6. _____
7. _____	8. _____
9. _____	10. _____
11. _____	12. _____
13. _____	14. _____
15. _____	16. _____
17. _____	18. _____
19. _____	20. _____
21. _____	22. _____
23. _____	24. _____
25. _____	26. _____
27. _____	28. _____
29. _____	30. _____

< Objective 1 >

Add.

1. $3 + 6$

2. $8 + 7$

3. $\dfrac{4}{5} + \dfrac{6}{5}$

4. $\dfrac{7}{3} + \dfrac{8}{3}$

5. $\dfrac{1}{2} + \dfrac{4}{5}$

6. $\dfrac{2}{3} + \dfrac{5}{9}$

7. $-4 + (-1)$

8. $-1 + (-9)$

9. $-\dfrac{1}{2} + \left(-\dfrac{3}{8}\right)$ > Videos

10. $-\dfrac{4}{7} + \left(-\dfrac{3}{14}\right)$

11. $-1.6 + (-2.3)$

12. $-3.5 + (-2.6)$

13. $3 + (-9)$

14. $11 + (-7)$

15. $\dfrac{3}{4} + \left(-\dfrac{1}{2}\right)$

16. $\dfrac{2}{3} + \left(-\dfrac{1}{6}\right)$

17. $13.4 + (-11.4)$

18. $5.2 + (-9.2)$

19. $-5 + 3$

20. $-12 + 17$

21. $-\dfrac{4}{5} + \dfrac{9}{20}$

22. $-\dfrac{11}{6} + \dfrac{5}{12}$ ⊙ > Videos

23. $-8.6 + 4.9$

24. $-3.6 + 7.6$

25. $0 + (-8)$

26. $-15 + 0$

27. $7 + (-7)$

28. $-12 + 12$

29. $-4.5 + 4.5$

30. $\dfrac{2}{3} + \left(-\dfrac{2}{3}\right)$

< Objective 2 >

Subtract.

31. $82 - 45$

32. $45 - 82$

33. $18 - 20$

34. $136 - 352$

35. $\dfrac{8}{7} - \dfrac{15}{7}$

36. $\dfrac{17}{8} - \dfrac{9}{8}$

37. $5.4 - 7.9$

38. $11.7 - 4.5$

39. $-3 - 1$

40. $-15 - 8$

41. $-14 - 9$

42. $-8 - 12$

43. $-\dfrac{2}{5} - \dfrac{7}{10}$

44. $-\dfrac{7}{18} - \dfrac{5}{9}$

45. $-3.4 - 4.7$

46. $-8.1 - 7.6$

47. $5 - (-11)$

48. $8 - (-4)$

49. $12 - (-7)$

50. $3 - (-10)$

51. $\dfrac{3}{4} - \left(-\dfrac{3}{2}\right)$ > Videos

52. $\dfrac{11}{16} - \left(-\dfrac{5}{8}\right)$

53. $8.3 - (-5.7)$

54. $14.5 - (-54.6)$

55. $-28 - (-11)$

56. $-11 - (-16)$

57. $-19 - (-27)$

58. $-13 - (-4)$

59. $-\dfrac{3}{4} - \left(-\dfrac{11}{4}\right)$

60. $-\dfrac{5}{8} - \left(-\dfrac{1}{2}\right)$

Answers

31. _____

32. _____

33. _____

34. _____

35. _____

36. _____

37. _____

38. _____

39. _____

40. _____

41. _____	42. _____
43. _____	44. _____
45. _____	46. _____
47. _____	48. _____
49. _____	50. _____
51. _____	52. _____
53. _____	54. _____
55. _____	56. _____
57. _____	58. _____
59. _____	60. _____

Answers

61. _____

62. _____

63. _____

64. _____

65. _____

66. _____

67. _____

68. _____

69. _____

70. _____

71. _____

72. _____

73. _____

74. _____

75. _____

76. _____

| Basic Skills | **Challenge Yourself** | Calculator/Computer | Career Applications | Above and Beyond |

Solve each application.

61. BUSINESS AND FINANCE Amir has $100 in his checking account. He writes a check for $23 and makes a deposit of $51. What is his new balance?

62. BUSINESS AND FINANCE Olga has $250 in her checking account. She deposits $52 and then writes a check for $77. What is her new balance?

Bal: _____ 250
Dep: _____ 52
CK # 1111: _____ 77

63. STATISTICS On four consecutive running plays, Duce Staley of the Philadelphia Eagles gained 23 yards, lost 5 yards, gained 15 yards, and lost 10 yards. What was his net yardage change for the series of plays?

64. BUSINESS AND FINANCE Ramon owes $780 on his VISA account. He returns three items costing $43.10, $36.80, and $125.00 and receives credit on his account. Next, he makes a payment of $400. He then makes a purchase of $82.75. How much does Ramon still owe?

65. SCIENCE AND MEDICINE The temperature at noon on a June day was 82°. It fell by 12° over the next 4 h. What was the temperature at 4:00 P.M.?

66. STATISTICS Chia is standing at a point 6,000 ft above sea level. She descends to a point 725 ft lower. What is her distance above sea level?

67. BUSINESS AND FINANCE Omar's checking account was overdrawn by $72. He wrote another check for $23.50. How much was his checking account overdrawn after writing the check?

68. BUSINESS AND FINANCE Angelo owed his sister $15. He later borrowed another $10. What integer represents his current financial condition?

69. STATISTICS A local community college had a decrease in enrollment of 750 students in the fall of 2005. In the spring of 2006, there was another decrease of 425 students. What was the total decrease in enrollment for both semesters?

70. SCIENCE AND MEDICINE At 7 A.M., the temperature was $-15°$F. By 1 P.M., the temperature had increased by 18°F. What was the temperature at 1 P.M.?

Evaluate each expression.

71. $9 + (-7) + 6 + (-5)$

72. $(-4) + 6 + (-3) + 0$

73. $-8 - 4 - 1 - (-2) - (-5)$

74. $6 - (-9) - 7 - (-5)$

75. $3 - 7 + (-12) - (-2) - 9$

76. $-12 + (-5) - 7 - (-13) + 4$

77. $-\dfrac{3}{2} + \left(-\dfrac{7}{4}\right) + \dfrac{1}{4}$

78. $-\dfrac{1}{2} + \dfrac{1}{3} + \left(-\dfrac{5}{6}\right)$ ⊙ > Videos

79. $2.3 + (-5.4) - (-2.9)$

80. $-5.4 - (-2.1) + (-3.5)$

81. $-\dfrac{1}{2} - \left(-\dfrac{3}{4}\right) + (-2) - 3\dfrac{1}{2} + \dfrac{3}{2}$

82. $0.25 + 0.7 - 1.5 - (-2.95) + (-3.1)$ ⊙ > Videos

Basic Skills | Challenge Yourself | **Calculator/Computer** | Career Applications | Above and Beyond

Use your calculator to evaluate each expression.

83. $-4.1967 - 5.2943$

84. $5.3297 - (-4.1897)$

85. $-4.1623 - (-3.1468)$

86. $-3.6829 - 4.5687$

87. $-6.3267 + 8.6789 - (-6.6712) + (-5.3245)$

88. $32.456 + (-67.004) - (-21.6059) - 13.4569$

Basic Skills | Challenge Yourself | Calculator/Computer | **Career Applications** | Above and Beyond

89. MECHANICAL ENGINEERING A pneumatic actuator is operated by a pressurized air reservoir. At the beginning of the operator's shift, the pressure in the reservoir was 126 pounds per square inch (psi). At the end of each hour, the operator recorded the change in pressure of the reservoir. The values recorded for this shift were a drop of 12 psi, a drop of 7 psi, a rise of 32 psi, a drop of 17 psi, a drop of 15 psi, a rise of 31 psi, a drop of 4 psi, and a drop of 14 psi. What was the pressure in the tank at the end of the shift?

90. MECHANICAL ENGINEERING A diesel engine for an industrial shredder has an 18-quart oil capacity. When the maintenance technician checked the oil, it was 7 quarts low. Later that day, she added 4 quarts to the engine. What was the oil level after the 4 quarts were added?

ELECTRICAL ENGINEERING Dry cells or batteries have a positive terminal and a negative terminal. When the cells are correctly connected in series (positive to negative), the voltages of the cells can be added together. If a cell is connected and its terminals are reversed, the current will flow in the opposite direction.

For example, if three 3-volt cells are supposedly connected in series but one cell is inserted backwards, the resulting voltage is 3 volts.

3 volts + 3 volts + (−3) volts = 3 volts

The voltages are added together because the cells are in series, but you must pay attention to the current flow.

Now complete exercises 91 and 92.

Answers

77. _____

78. _____

79. _____

80. _____

81. _____

82. _____

83. _____

84. _____

85. _____

86. _____

87. _____

88. _____

89. _____

90. _____

Answers

91. _____

92. _____

93. _____

94. _____

95. _____

96. _____

91. Assume you have a 24-volt cell and a 12-volt cell with their negative terminals connected. What would the resulting voltage be if measured from the positive terminals?

92. If a 24-volt cell, an 18-volt cell, and 12-volt cell are supposed to be connected in series and the 18-volt cell is accidentally reversed, what would the total voltage be?

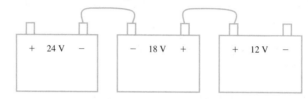

MANUFACTURING TECHNOLOGY At the beginning of the week, there were 2,489 lb of steel in inventory. Report the change in steel inventory for the week if the end-of-week inventory is:

93. 2,581 lb   **94.** 2,111 lb

Basic Skills | Challenge Yourself | Calculator/Computer | Career Applications | **Above and Beyond**
                                                                                         ▲

95. En route to their 2006 Super Bowl victory, the game-by-game rushing leaders for the Pittsburgh Steelers playoff run are shown below, along with yardage gained.

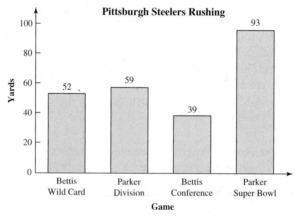

Source: ESPN. com

Use a real number to represent the change in the rushing yardage given from one game to the next.

(a) From the wild card game to the division game

(b) From the division game to the conference championship

(c) From the conference championship to the Super Bowl

96. In this chapter, it is stated that "Every number has an opposite." The opposite of 9 is −9. This corresponds to the idea of an opposite in English. In English, an opposite is often expressed by a prefix, for example, *un-* or *ir-*.

(a) Write the opposite of these words: unmentionable, uninteresting, irredeemable, irregular, uncomfortable.

(b) What is the meaning of these expressions: not uninteresting, not irredeemable, not irregular, not unmentionable?

(c) Think of other prefixes that *negate* or change the meaning of a word to its *opposite*. Make a list of words formed with these prefixes, and write a sentence with three of the words you found. Make a sentence with two words and phrases from each of the lists. Look up the meaning of the word *irregardless*.

What is the value of $-[-(-5)]$? What is the value of $-(-6)$? How does this relate to the previous examples? Write a short description about this relationship.

Answers

97. _____

98. _____

97. The temperature on the plains of North Dakota can change rapidly, falling or rising many degrees in the course of an hour. Here are some temperature changes during each day over a week.

Day	Mon.	Tues.	Wed.	Thurs.	Fri.	Sat.	Sun.
Temp. change from 10 A.M. to 3 P.M.	$+13°$	$+20°$	$-18°$	$+10°$	$-25°$	$-5°$	$+15°$

Write a short speech for the TV weather reporter that summarizes the daily temperature change.

98. How long ago was the year 1250 B.C.E.? What year was 3,300 years ago? Make a number line and locate the following events, cultures, and objects on it. How long ago was each item in the list? Which two events are the closest to each other? You may want to learn more about some of the cultures in the list and the mathematics and science developed by that culture.

chapter 1 > Make the Connection

Inca culture in Peru—A.D. 1400

The *Ahmes Papyrus,* a mathematical text from Egypt—1650 B.C.E.

Babylonian arithmetic develops the use of a zero symbol—300 B.C.E.

First Olympic Games—776 B.C.E.

Pythagoras of Greece is born—580 B.C.E.

Mayans in Central America independently develop use of zero—A.D. 500

The *Chou Pei,* a mathematics classic from China—1000 B.C.E.

The *Aryabhatiya,* a mathematics work from India—A.D. 499

Trigonometry arrives in Europe via the Arabs and India—A.D. 1464

Arabs receive algebra from Greek, Hindu, and Babylonian sources and develop it into a new systematic form—A.D. 850

Development of calculus in Europe—A.D. 1670

Rise of abstract algebra—A.D. 1860

Growing importance of probability and development of statistics—A.D. 1902

Answers

99. _____

100. _____

99. Complete the following statement: "$3 - (-7)$ is the same as ____ because . . . ." Write a problem that might be answered by doing this subtraction.

100. Explain the difference between the two phrases: "a number subtracted from 5" and "a number less than 5." Use algebra and English to explain the meaning of these phrases. Write other ways to express subtraction in English. Which ones are confusing?

Answers

1. 9   **3.** 2   **5.** $\dfrac{13}{10}$ **7.** -5 **9.** $-\dfrac{7}{8}$ **11.** -3.9 **13.** -6

15. $\dfrac{1}{4}$ **17.** 2   **19.** -2 **21.** $-\dfrac{7}{20}$ **23.** $-3.\overline{7}$ **25.** -8

27. 0   **29.** 0   **31.** 37   **33.** -2 **35.** -1 **37.** -2.5

39. -4 **41.** -23 **43.** $-\dfrac{11}{10}$ **45.** -8.1 **47.** 16   **49.** 19

51. $\dfrac{9}{4}$ **53.** 14   **55.** -17 **57.** 8   **59.** 2   **61.** $128

63. 23 yd   **65.** 70°   **67.** $95.50   **69.** 1,175   **71.** 3   **73.** -6

75. -23 **77.** -3 **79.** -0.2 **81.** $-\dfrac{15}{4}$ **83.** -9.491

85. -1.0155 **87.** 3.6989   **89.** 120 psi   **91.** 12 V   **93.** $+92$ lb
95. (a) $+7$; (b) -20; (c) $+54$ **97.** Above and Beyond
99. Above and Beyond

1.3
Multiplying and Dividing Real Numbers

< 1.3 Objectives >

1 > Find the product of real numbers

2 > Find the quotient of two real numbers

3 > Use the order of operations to evaluate expressions involving real numbers

When you first considered multiplication, it was thought of as repeated addition. What does our work with the addition of numbers with different signs tell us about multiplication when real numbers are involved?

$$3 \cdot 4 = 4 + 4 + 4 = 12$$

We interpret multiplication as repeated addition to find the product, 12.

Now, consider the product $(3)(-4)$:

$$(3)(-4) = (-4) + (-4) + (-4) = -12$$

Looking at this product suggests the first portion of our rule for multiplying numbers with different signs. The product of a positive number and a negative number is negative.

Property

Multiplying Real Numbers with Different Signs

The product of two numbers with different signs is negative.

To use this rule when multiplying two numbers with different signs, multiply their absolute values and attach a negative sign.

 Example 1 | **Multiplying Numbers with Different Signs**

< Objective 1 >

Multiply.

(a) $(5)(-6) = -30$

The product is negative.

(b) $(-10)(10) = -100$

(c) $(8)(-12) = -96$

(d) $\left(-\dfrac{3}{4}\right)\left(\dfrac{2}{5}\right) = -\dfrac{3}{10}$

 Check Yourself 1

Multiply.

(a) $(-7)(5)$ **(b)** $(-12)(9)$ **(c)** $(-15)(8)$ **(d)** $\left(-\dfrac{4}{7}\right)\left(\dfrac{14}{5}\right)$

The product of two negative numbers is harder to visualize. The following pattern may help you see how we can determine the sign of the product.

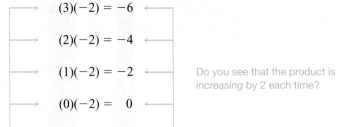

$$(3)(-2) = -6$$

$$(2)(-2) = -4$$

$$(1)(-2) = -2$$

$$(0)(-2) = 0$$

$$(-1)(-2) = 2$$

Do you see that the product is *increasing* by 2 each time?

What should the product $(-2)(-2)$ be? Continuing the pattern shown, we see that

$$(-2)(-2) = 4$$

This suggests that the product of two negative numbers is positive. We can extend our multiplication rule.

Property

Multiplying Real Numbers with the Same Sign

The product of two numbers with the same sign is positive.

 Example 2   **Multiplying Real Numbers with the Same Sign**

RECALL

$(-8)(-5) = (-8) \cdot (-5)$

Multiply.

(a) $9 \cdot 7 = 63$ The product of two positive numbers (same sign, $+$) is positive.

(b) $(-8)(-5) = 40$ The product of two negative numbers (same sign, $-$) is positive.

(c) $\left(-\dfrac{1}{2}\right)\left(-\dfrac{1}{3}\right) = \dfrac{1}{6}$

✓ **Check Yourself 2**

Multiply.

(a) $10 \cdot 12$ **(b)** $(-8)(-9)$ **(c)** $\left(-\dfrac{2}{3}\right)\left(-\dfrac{6}{7}\right)$

Two numbers, 0 and 1, have special properties in multiplication.

Property

Multiplicative Identity Property

The product of 1 and any number is that number. In symbols,

$$a \cdot 1 = 1 \cdot a = a$$

The number 1 is called the **multiplicative identity** for this reason.

Property

Multiplicative Property of Zero

The product of 0 and any number is 0. In symbols,

$$a \cdot 0 = 0 \cdot a = 0$$

Example 3 | **Multiplying Real Numbers Involving 0 and 1**

Find each product.

(a) $(1)(-7) = -7$

(b) $(15)(1) = 15$

(c) $(-7)(0) = 0$

(d) $0 \cdot 12 = 0$

(e) $\left(-\dfrac{4}{5}\right)(0) = 0$

Check Yourself 3

Multiply.

(a) $(-10)(1)$ **(b)** $(0)(-17)$ **(c)** $\left(\dfrac{5}{7}\right)(1)$ **(d)** $(0)\left(\dfrac{3}{4}\right)$

RECALL

$$-\dfrac{2}{3} = \dfrac{-2}{3} = \dfrac{2}{-3}$$

All of these numbers represent the same point on a number line.

Before we continue, consider the following equivalent fractions:

$$-\dfrac{1}{a} = \dfrac{-1}{a} = \dfrac{1}{-a}$$

Any of these forms can occur in the course of simplifying an expression. The first form is generally preferred.

To complete our discussion of the properties of multiplication, we state the following.

Property

Multiplicative Inverse Property

For any nonzero number a, there is a number $\dfrac{1}{a}$ such that

$$a \cdot \dfrac{1}{a} = 1$$

$\dfrac{1}{a}$ is called the multiplicative inverse, or the reciprocal, of a. The product of any nonzero number and its reciprocal is 1.

Example 4 illustrates this property.

Example 4 | **Multiplying Reciprocals**

(a) $3 \cdot \dfrac{1}{3} = 1$ The reciprocal of 3 is $\dfrac{1}{3}$.

(b) $-5\left(-\dfrac{1}{5}\right) = 1$ The reciprocal of -5 is $\dfrac{1}{-5}$ or $-\dfrac{1}{5}$.

(c) $\dfrac{2}{3} \cdot \dfrac{3}{2} = 1$ The reciprocal of $\dfrac{2}{3}$ is $\dfrac{1}{\frac{2}{3}}$, or $\dfrac{3}{2}$.

Check Yourself 4

Find the multiplicative inverse (or the reciprocal) of each of the following numbers.

(a) 6   **(b)** -4 **(c)** $\dfrac{1}{4}$ **(d)** $-\dfrac{3}{5}$

You know from your work in arithmetic that multiplication and division are related operations. We can use that fact, and our work in the earlier part of this section, to determine rules for the division of numbers with different signs. Every equation involving division can be stated as an equivalent equation involving multiplication. For instance,

$$\frac{15}{5} = 3 \qquad \text{can be restated as} \qquad 15 = 5 \cdot 3$$

$$\frac{-24}{6} = -4 \qquad \text{can be restated as} \qquad -24 = (6)(-4)$$

$$\frac{-30}{-5} = 6 \qquad \text{can be restated as} \qquad -30 = (-5)(6)$$

These examples illustrate that because the two operations are related, the rules of signs that we stated in the earlier part of this section for multiplication are also true for division.

Property

Dividing Real Numbers

1. The quotient of two numbers with different signs is negative.

2. The quotient of two numbers with the same sign is positive.

Again, the rules are easy to use. To divide two numbers with different signs, divide their absolute values. Then attach the proper sign according to the rules stated in the box.

Example 5   **Dividing Real Numbers**

< Objective 2 >

Divide.

(a) Positive ⟶ $\dfrac{28}{7} = 4$ ⟵ Positive
Positive ⟶

(b) Negative ⟶ $\dfrac{-36}{-4} = 9$ ⟵ Positive
Negative ⟶

(c) Negative ⟶ $\dfrac{-42}{7} = -6$ ⟵ Negative
Positive ⟶

(d) Positive ⟶ $\dfrac{75}{-3} = -25$ ⟵ Negative
Negative ⟶

(e) Positive ⟶ $\dfrac{15.2}{-3.8} = -4$ ⟵ Negative
Negative ⟶

Check Yourself 5

Divide.

(a) $\dfrac{-55}{11}$ **(b)** $\dfrac{80}{20}$ **(c)** $\dfrac{-48}{-8}$ **(d)** $\dfrac{144}{-12}$ **(e)** $\dfrac{-13.5}{-2.7}$

You should be careful when 0 is involved in a division problem. Remember that 0 divided by any nonzero number is just 0. Recall that

$$\frac{0}{-7} = 0 \qquad \text{because} \qquad 0 = (-7)(0)$$

However, if zero is the *divisor,* we have a special problem. Consider

$$\frac{9}{0} = ?$$

This means that $9 = 0 \cdot ?$.

Can 0 times a number ever be 9? No, so there is no solution.

Because $\frac{9}{0}$ cannot be replaced by any number, we agree that *division by 0 is not allowed.*

Property

Division by Zero	Division by 0 is undefined.

 Example 6　　　**Dividing Numbers Involving Zero**

NOTE

The expression $\frac{0}{0}$ is called an **indeterminate form.** You will learn more about this in later math courses.

Divide, if possible.

(a) $\frac{7}{0}$ is undefined.

(b) $\frac{-9}{0}$ is undefined.

(c) $\frac{0}{5} = 0$

(d) $\frac{0}{-8} = 0$

 Check Yourself 6

Divide if possible.

(a) $\frac{0}{3}$ 　　　　**(b)** $\frac{5}{0}$ 　　　　**(c)** $\frac{-7}{0}$ 　　　　**(d)** $\frac{0}{-9}$

You should remember that the fraction bar serves as a *grouping symbol.* This means that all operations in the numerator and denominator should be performed separately. Then the division is done as the last step. Example 7 illustrates this procedure.

 Example 7　　　**Operations with Grouping Symbols**

< Objective 3 >

Evaluate each expression.

(a) $\dfrac{(-6)(-7)}{3} = \dfrac{42}{3} = 14$ 　　　Multiply in the numerator, and then divide.

(b) $\dfrac{3 + (-12)}{3} = \dfrac{-9}{3} = -3$ Add in the numerator, and then divide.

(c) $\dfrac{-4 + (2)(-6)}{-6 - 2} = \dfrac{-4 + (-12)}{-6 - 2}$ Multiply in the numerator. Then add in the numerator and subtract in the denominator.

$\qquad\qquad = \dfrac{-16}{-8} = 2$ Divide as the last step.

 Check Yourself 7

Evaluate each expression.

(a) $\dfrac{-4 + (-8)}{6}$ **(b)** $\dfrac{3 - (2)(-6)}{-5}$ **(c)** $\dfrac{(-2)(-4) - (-6)(-5)}{(-4)(11)}$

Evaluating fractions with a calculator poses a special problem. Example 8 illustrates this problem.

| **Example 8** | **Using a Calculator to Divide** |

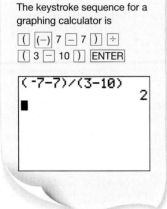 > Calculator

Use your scientific calculator to evaluate each fraction.

(a) $\dfrac{4}{2 - 3}$

As you can see, the correct answer should be -4. To get this answer with your calculator, you must place the denominator in parentheses. The keystroke sequence is

4 $\div$ (2 $-$ 3) $=$

(b) $\dfrac{-7 - 7}{3 - 10}$

NOTE

The keystroke sequence for a graphing calculator is

((−) 7 − 7) ÷
(3 − 10) ENTER

```
(-7-7)/(3-10)
                2
■
```

In this problem, the correct answer is 2. You can get this answer with your calculator by placing both the numerator and the denominator in their own sets of parentheses. The keystroke sequence on a scientific calculator is

(7 +/− − 7) ÷ (3 − 10) =

When evaluating a fraction with a calculator, it is safest to use parentheses in both the numerator and the denominator.

 Check Yourself 8

Evaluate using your calculator.

(a) $\dfrac{-8}{5 - 7}$ **(b)** $\dfrac{-3 - 2}{-13 + 23}$

The order of operations remains the same when performing computations involving negative numbers. You must remain vigilant, though, with any negative signs.

Example 9 | **Order of Operations**

RECALL

$7(-9 + 12)$ means
$7 \cdot (-9 + 12)$.

NOTE

$(-5)^2 = (-5)(-5) = 25$ but
$-5^2 = -25$. The power
applies *only* to the 5 in the
latter expression.

Evaluate each expression.

(a) $7(-9 + 12)$ Evaluate inside the parentheses first.

    $= 7(3) = 21$

(b) $(-8)(-7) - 40$ Multiply first, then subtract.

    $= 56 - 40 = 16$

(c) $(-5)^2 - 3$ Evaluate the power first.

    $= (-5)(-5) - 3$

    $= 25 - 3 = 22$

(d) $-5^2 - 3$

    $= -25 - 3$

    $= -28$

Check Yourself 9

Evaluate each expression.

(a) $8(-9 + 7)$ (b) $(-3)(-5) + 7$
(c) $(-4)^2 - (-4)$ (d) $-4^2 - (-4)$

Many students have difficulty applying the distributive property when negative numbers are involved. Just remember that the sign of a number "travels" with that number.

Example 10 | **Applying the Distributive Property with Negative Numbers**

RECALL

We usually enclose negative
numbers in parentheses in
the middle of an expression
to avoid careless errors.

RECALL

We use brackets rather than
nesting parentheses to avoid
careless errors.

Evaluate each expression.

(a) $-7(3 + 6) = -7 \cdot 3 + (-7) \cdot 6$ Apply the distributive property.

                     $= -21 + (-42)$ Multiply first, then add.

                     $= -63$

(b) $-3(5 - 6) = -3[5 + (-6)]$ First, change the subtraction to addition.

                    $= -3 \cdot 5 + (-3)(-6)$ Distribute the -3.

                    $= -15 + 18$ Multiply first, then add.

                    $= 3$

(c) $5(-2 - 6) = 5[-2 + (-6)]$

                   $= 5 \cdot (-2) + 5 \cdot (-6)$

                   $= -10 + (-30)$ The sum of two negative numbers is negative.

                   $= -40$

Check Yourself 10

Evaluate each expression.

(a) $-2(-3 + 5)$ (b) $4(-3 + 6)$ (c) $-7(-3 - 8)$

Another thing to keep in mind when working with negative signs is the way in which you should evaluate multiple negative signs. Our approach takes into account two ways of looking at positive and negative numbers.

First, a negative sign indicates the opposite of the number that follows. For instance, we have already said that the opposite of 5 is -5, whereas the opposite of -5 is 5. This last instance can be translated as $-(-5) = 5$.

Second, any number must correlate to some point on the number line. That is, any nonzero number is either positive or negative. No matter how many negative signs a quantity has, you can always simplify it so that it is represented by a positive or a negative number.

  Example 11 | **Simplifying Negative Signs**

NOTES

You should see a pattern emerge. An even number of negative signs gives a positive number, whereas an odd number of negative signs produces a negative number.

In this text, we generally choose to write negative fractions with the negative sign outside the fraction, such as $-\dfrac{1}{2}$.

Simplify each expression.

(a) $-(-(-(-4)))$

The opposite of -4 is 4, so $-(-4) = 4$.
    The opposite of 4 is -4, so $-(-(-4)) = -4$. The opposite of this last number, -4, is 4, so

$$-(-(-(-4))) = 4$$

(b) $-\dfrac{-3}{4}$

This is the opposite of $\dfrac{-3}{4}$, which is $\dfrac{3}{4}$, a positive number.

 Check Yourself 11

Simplify each expression.

(a) $-(-(-(-(-(-12)))))$ **(b)** $-\dfrac{-2}{-3}$

  Example 12 | **An Application of Multiplying and Dividing Real Numbers**

Three partners own stock worth $4,680. One partner sells it for $3,678. How much did each partner lose?
    First find the total loss: $4,680 - \$3,678 = \$1,002$
Then divide the total loss by 3: $\dfrac{\$1,002}{3} = \$334$

Each person lost $334.

Check Yourself 12

Sal and Vinnie invested $8,500 in a business. Ten years later they sold the business for $22,000. How much profit did each make?

We conclude this section with a more detailed explanation of the reason the product of two negative numbers is positive.

Property

The Product of Two Negative Numbers

From our earlier work, we know that the sum of a number and its opposite is 0:

$$5 + (-5) = 0$$

Multiply both sides of the equation by -3:

$$(-3)[5 + (-5)] = (-3)(0)$$

Because the product of 0 and any number is 0, on the right we have 0.

$$(-3)[5 + (-5)] = 0$$

We use the distributive property on the left.

$$(-3)(5) + (-3)(-5) = 0$$

We know that $(-3)(5) = -15$, so the equation becomes

$$-15 + (-3)(-5) = 0$$

We now have a statement of the form

$$-15 + \square = 0$$

in which $\square$ is the value of $(-3)(-5)$. We also know that $\square$ is the number that must be added to -15 to get 0, so $\square$ is the opposite of -15, or 15. This means that

$$(-3)(-5) = 15 \qquad \text{The product is positive!}$$

It doesn't matter what numbers we use in this argument. The resulting product of two negative numbers will always be positive.

Check Yourself ANSWERS

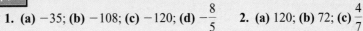

1. (a) -35; (b) -108; (c) -120; (d) $-\dfrac{8}{5}$ 2. (a) 120; (b) 72; (c) $\dfrac{4}{7}$

3. (a) -10; (b) 0; (c) $\dfrac{5}{7}$; (d) 0   4. (a) $\dfrac{1}{6}$; (b) $-\dfrac{1}{4}$; (c) 4; (d) $-\dfrac{5}{3}$

5. (a) -5; (b) 4; (c) 6; (d) -12; (e) 5   6. (a) 0; (b) undefined;

(c) undefined; (d) 0   7. (a) -2; (b) -3; (c) $\dfrac{1}{2}$ 8. (a) 4; (b) -0.5

9. (a) -16; (b) 22; (c) 20; (d) -12 10. (a) -4; (b) 12; (c) 77

11. (a) -12; (b) $-\dfrac{2}{3}$ 12. $6,750

Reading Your Text

The following fill-in-the-blank exercises are designed to ensure that you understand some of the key vocabulary used in this section.

SECTION 1.3

(a) The product of two numbers with different signs is always _____.

(b) The product of two numbers with the same sign is always _____.

(c) The number _____ is called the multiplicative identity.

(d) Division by _____ is undefined.

Name _____

Section _____ Date _____

Answers

1. _____ 2. _____

3. _____ 4. _____

5. _____ 6. _____

7. _____ 8. _____

9. _____ 10. _____

11. _____ 12. _____

13. _____ 14. _____

15. _____ 16. _____

17. _____ 18. _____

19. _____ 20. _____

21. _____ 22. _____

23. _____ 24. _____

25. _____ 26. _____

27. _____ 28. _____

29. _____ 30. _____

31. _____ 32. _____

33. _____ 34. _____

35. _____ 36. _____

< Objective 1 >

Multiply.

1. $4 \cdot 10$ 　　　　　　**2.** $3 \cdot 14$

3. $(-4)(10)$ 　　　　　**4.** $(3)(-14)$

5. $(-4)(-10)$ 　　　　**6.** $(-3)(-14)$

7. $(-13)(5)$ 　　　　　**8.** $(11)(-9)$

9. $(-4)(-17)$ 　　　　**10.** $(-23)(-8)$

11. $4 \cdot \left(-\dfrac{3}{2}\right)$ > Videos 　　**12.** $(-9)\left(-\dfrac{2}{3}\right)$

13. $\left(-\dfrac{1}{4}\right) \cdot (-8)$ 　　　**14.** $\left(-\dfrac{5}{3}\right) \cdot (6)$

15. $\left(-\dfrac{2}{3}\right)\left(\dfrac{3}{5}\right)$ 　　　**16.** $\left(\dfrac{5}{8}\right)\left(-\dfrac{2}{3}\right)$

17. $\left(-\dfrac{1}{2}\right)\left(-\dfrac{10}{3}\right)$ > Videos 　**18.** $\left(-\dfrac{7}{10}\right)\left(-\dfrac{5}{8}\right)$

19. $3.25 \cdot (-4)$ 　　　**20.** $(5.4)(-5)$

21. $(-1.1)(-1.2)$ 　　　**22.** $(0.8)(-3.5)$

23. $0 \cdot (-18)$ 　　　　**24.** $(-5)(0)$

25. $\left(-\dfrac{11}{12}\right)(0)$ 　　　**26.** $(0)(-2.37)$

27. $\left(-\dfrac{1}{2}\right)(2)$ 　　　**28.** $\left(-\dfrac{1}{3}\right)(-3)$

29. $\left(-\dfrac{3}{2}\right)\left(-\dfrac{2}{3}\right)$ 　　**30.** $\left(-\dfrac{4}{7}\right)\left(\dfrac{7}{4}\right)$

< Objective 2 >

Divide.

31. $\dfrac{70}{14}$ 　　　　　　**32.** $48 \div 6$

33. $(-35) \div (-7)$ 　　　**34.** $\dfrac{-48}{-12}$

35. $\dfrac{50}{-5}$ 　　　　　**36.** $\dfrac{-60}{15}$ > Videos

37. $\dfrac{-125}{5}$

38. $\dfrac{24}{-8}$

39. $\dfrac{-11}{-1}$

40. $\dfrac{-13}{1}$

41. $\dfrac{32}{-1}$

42. $\dfrac{-1}{-8}$

43. $\dfrac{0}{-8}$

44. $\dfrac{-10}{0}$

45. $\dfrac{-14}{0}$

46. $\dfrac{0}{-2}$

< Objective 3 >

Evaluate each expression.

47. $\dfrac{(-6)(-3)}{2}$

48. $\dfrac{(-9)(5)}{-3}$

49. $\dfrac{(-8)(2)}{-4}$

50. $\dfrac{(7)(-8)}{-14}$

51. $\dfrac{24}{-4-8}$

52. $\dfrac{-36}{-7+3}$

53. $\dfrac{55-19}{-12-6}$

54. $\dfrac{-11-7}{-14+8}$

55. $\dfrac{5-7}{4-4}$

56. $\dfrac{-3-(-3)}{6-10}$

57. $5(7-2)$

58. $5(2-7)$

59. $-3(-2-5)$

60. $-2[-7-(-3)]$

61. $(-2)(3)-5$

62. $(8)(-6)-27$

63. $(-5)(-2)-12$

64. $(-7)(-3)-25$

65. $-3+(-2)(4)$

66. $-5-(-5)(4)$

67. $12-(-3)(-4)$

68. $20+(-4)(-5)$

69. $(-8)^2-5^2$

70. $(-8)^2-(-4)^2$

71. $-8^2-(-5)^2$

72. -8^2-4^2 > Videos

73. $-(-(-(-(-3))))$

74. $-(-(-(-3.45)))$

75. $\dfrac{-(-2)}{-(-8)}$

76. $\dfrac{-3}{-(-(-4))}$

Answers

37. _____ **38.** _____

39. _____ **40.** _____

41. _____ **42.** _____

43. _____ **44.** _____

45. _____ **46.** _____

47. _____ **48.** _____

49. _____ **50.** _____

51. _____ **52.** _____

53. _____ **54.** _____

55. _____ **56.** _____

57. _____ **58.** _____

59. _____ **60.** _____

61. _____ **62.** _____

63. _____ **64.** _____

65. _____ **66.** _____

67. _____ **68.** _____

69. _____ **70.** _____

71. _____ **72.** _____

73. _____ **74.** _____

75. _____ **76.** _____

Answers

77. _____

78. _____

79. _____

80. _____

81. _____

82. _____

83. _____

84. _____

85. _____

86. _____

87. _____

88. _____

89. _____

90. _____

91. _____

92. _____

93. _____

94. _____

Solve each application.

77. **SCIENCE AND MEDICINE** The temperature is $-6°F$ at 5:00 in the evening. If the temperature drops $2°F$ every hour, what is the temperature at 1:00 A.M.?

78. **SCIENCE AND MEDICINE** A woman lost 42 pounds (lb) while dieting. If she lost 3 lb each week, how long has she been dieting?

79. **BUSINESS AND FINANCE** Patrick worked all day mowing lawns and was paid $9 per hour. If he had $125 at the end of a 9-h day, how much did he have before he started working?

80. **BUSINESS AND FINANCE** Suppose that you and your two brothers bought equal shares of an investment for a total of $20,000 and sold it later for $16,232. How much did each person lose?

81. **SCIENCE AND MEDICINE** Suppose that the temperature outside is dropping at a constant rate. At noon, the temperature is $70°F$ and it drops to $58°F$ at 5:00 P.M. How much did the temperature change each hour?

82. **SCIENCE AND MEDICINE** A chemist has 84 ounces (oz) of a solution. He pours the solution into test tubes. Each test tube holds $\frac{2}{3}$ oz. How many test tubes can he fill?

| Basic Skills | **Challenge Yourself** | Calculator/Computer | Career Applications | Above and Beyond |

Complete each statement with **never, sometimes,** *or* **always.**

83. A product made up of an odd number of negative factors is _____ negative.

84. A product of an even number of negative factors is _____ negative.

85. The quotient $\dfrac{x}{y}$ is _____ positive.

86. The quotient $\dfrac{x}{y}$ is _____ negative.

Evaluate each expression.

87. $4 \cdot 8 \div 2 - 5^2$

88. $36 \div 4 \cdot 3 - (-25)$

89. $-8 + 14 - 2 \cdot 4 - 3$

90. $(-3)^3 - (-8)(-2)$

91. $8 + [2(-3) + 3]^2$

92. $-8^2 - 5^2 + 8 \div (4 \cdot 2)$

93. $\dfrac{-\dfrac{3}{8}}{\dfrac{3}{4}}$

94. $\left(-\dfrac{5}{12}\right) \div \left(-\dfrac{3}{16}\right)$

95. $\left(\dfrac{7}{4}\right) \div \left(-\dfrac{3}{2}\right)$

96. $\dfrac{-\dfrac{1}{2}}{-\dfrac{3}{4}}$

97. $\left(-1\dfrac{1}{2}\right)\left(3\dfrac{1}{3}\right)$

98. $\left(-2\dfrac{1}{2}\right)\left(-3\dfrac{3}{4}\right)$

99. $\left(-5\dfrac{1}{4}\right) \div \left(-2\dfrac{1}{2}\right)$ > Videos

100. $\left(1\dfrac{1}{3}\right) \div \left(-6\dfrac{2}{3}\right)$

| Basic Skills | Challenge Yourself | **Calculator/Computer** | Career Applications | Above and Beyond |

Use a calculator to evaluate each expression to the nearest thousandth.

101. $\dfrac{7}{4-5}$

102. $\dfrac{-8}{-4+2}$

103. $\dfrac{-6-9}{-4+1}$

104. $\dfrac{-10+4}{-7+10}$

105. $(-1.23) \cdot (3.4)$ 23.6504 ÷ -10546

   -22.61 + 1.0404 ÷ -105.6

106. $\dfrac{(3.55)(-12.12)}{(-6.4)}$

107. $3.4 - 5.1^2 + (-1.02)^2 \div 22 \cdot (-4.8)$ 4.22396

108. $-14.6 - \dfrac{3-4}{3} + 2(5+6)^2 - (1.1)^3$

| Basic Skills | Challenge Yourself | Calculator/Computer | **Career Applications** | Above and Beyond |

109. MANUFACTURING TECHNOLOGY Companies occasionally sell products at a loss in order to draw in customers or as a reward to good customers. The theory is that customers will buy other products along with the discounted product and the net result will be a profit.

    Beguhn Industries sells five different products. On product A, they make $18 each; on product B, they lose $4 each; product C makes $11 each; product D makes $38 each; and product E loses $15 each. During the previous month, Beguhn Industries sold 127 units of product A, 273 units of product B, 201 units of product C, 377 units of product D, and 43 units of product E.

    Calculate the profit or loss for the month.

110. MECHANICAL ENGINEERING The bending moment created by a center support on a steel beam is approximated by the formula $-\dfrac{1}{4}PL^3$, in which P is the load on each side of the center support and L is the length of the beam on each side of the center support (assuming a symmetrical beam and load).

    If the total length of the beam is 24 ft (12 ft on each side of the center) and the total load is 4,124 lb (2,062 lb on each side of the center), what is the bending moment (in ft-lb³) at the center support?

Answers

95.

96.

97.

98.

99.

100.

101.

102.

103.

104.

105.

106.

107.

108.

109.

110.

Answers

111. _____

111. Some animal ecologists in Minnesota are planning to reintroduce a group of animals into a wilderness area. The animals, mammals on the endangered species list, will be released into an area where they once prospered and where there is an abundant food supply. But, the animals will face predators. The ecologists expect that the number of mammals will grow about 25 percent each year but that 30 of the animals will die from attacks by predators and hunters.

    The ecologists need to decide how many animals they should release to establish a stable population. Work with other students to try several beginning populations and follow the numbers through 8 years. Is there a number of animals that will lead to a stable population? Write a letter to the editor of your local newspaper explaining how to decide what number of animals to release. Include a formula for the number of animals next year based on the number this year. Begin by filling out this table to track the number of animals living each year after the release:

No. Initially Released	Year							
	1	2	3	4	5	6	7	8
20	+___ −___ =_____							
100	+___ −___ =_____							
200	+___ −___ =_____							

Answers

1. 40   **3.** -40 **5.** 40   **7.** -65 **9.** 68   **11.** -6 **13.** 2

15. $-\dfrac{2}{5}$ **17.** $\dfrac{5}{3}$ **19.** -13 **21.** 1.32   **23.** 0   **25.** 0

27. -1 **29.** 1   **31.** 5   **33.** 5   **35.** -10 **37.** -25 **39.** 11

41. -32 **43.** 0   **45.** Undefined   **47.** 9   **49.** 4   **51.** -2

53. -2 **55.** Undefined   **57.** 25   **59.** 21   **61.** -11 **63.** -2

65. -11 **67.** 0   **69.** 39   **71.** -89 **73.** -3 **75.** $\dfrac{1}{4}$

77. $-22°F$ **79.** \$44   **81.** 2.4°F   **83.** always   **85.** sometimes

87. -9 **89.** -5 **91.** 17   **93.** $-\dfrac{1}{2}$ **95.** $-\dfrac{7}{6}$ **97.** -5

99. $2\dfrac{1}{10}$ **101.** -7 **103.** 5   **105.** -4.182 **107.** -22.837

109. $+\$17{,}086$ **111.** Above and Beyond

1.4

From Arithmetic to Algebra

< 1.4 Objectives >

1 > Use the symbols and language of algebra

2 > Identify algebraic expressions

In arithmetic, you learned how to do calculations with numbers using the basic operations of addition, subtraction, multiplication, and division.

In algebra, we still use numbers and the same four operations. However, we also use letters to represent numbers. Letters such as x, y, L, and W are called **variables** when they represent numerical values.

Here we see two rectangles whose lengths and widths are labeled with numbers.

If we want to represent the length and width of *any* rectangle, we can use the variables L and W.

NOTE

In arithmetic:
+ denotes addition;
− denotes subtraction;
× denotes multiplication;
÷ denotes division.

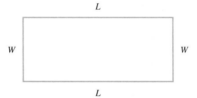

You are familiar with the four symbols $(+, -, \times, \div)$ used to indicate the fundamental operations of arithmetic.

To see how these operations are indicated in algebra, we begin with addition.

Definition

Addition	$x + y$ means the *sum* of x and y or x *plus* y.

 Example 1 | **Writing Expressions That Indicate Addition**

< Objective 1 >

(a) *The sum of a and* 3 is written as $a + 3$.

(b) *L plus W* is written as $L + W$.

(c) 5 *more than m* is written as $m + 5$.

(d) *x increased by* 7 is written as $x + 7$.

(e) 15 *added to x* is written as $x + 15$.

 Check Yourself 1

Write, using symbols.

(a) The sum of y and 4   (b) a plus b
(c) 3 more than x (d) n increased by 6

Similarly, we use a minus sign to indicate subtraction.

Definition

Subtraction	$x - y$ means the *difference* of x and y or x *minus* y.

Example 2 | **Writing Expressions That Indicate Subtraction**

> **CAUTION**

"x minus y," "the difference of x and y," "x decreased by y," and "x take away y" are all written in the same order as the instructions are given, $x - y$.

However, we reverse the order that the quantities are given when writing "x less than y" and "x subtracted from y." These two phrases are translated as $y - x$.

(a) *r minus s* is written as $r - s$.

(b) *The difference of m and* 5 is written as $m - 5$.

(c) *x decreased by* 8 is written as $x - 8$.

(d) 4 *less than a* is written as $a - 4$.

(e) 12 *subtracted from y* is written as $y - 12$.

(f) 7 take away y is written as $7 - y$.

Check Yourself 2

Write, using symbols.

(a) w minus z **(b)** The difference of a and 7
(c) y decreased by 3 **(d)** 5 less than b
(e) m subtracted from 6 **(f)** 4 take away x

You have seen that the operations of addition and subtraction are written exactly the same way in algebra as in arithmetic. This is not true in multiplication because the sign $\times$ looks like the letter x, so we use other symbols to show multiplication to avoid any confusion. Here are some ways to write multiplication.

Definition

Multiplication	A centered dot	$x \cdot y$	All these expressions indicate the
	Parentheses	$(x)(y)$	*product* of x and y or x *times* y. x and y
	Writing the letters next to each other	xy	are called the **factors** of the product xy.

When no operation is shown, the operation is multiplication, so that $2x$ means the product of 2 and x.

Example 3 | **Writing Expressions That Indicate Multiplication**

(a) The product of 5 and a is written as $5 \cdot a$, $(5)(a)$, or $5a$. The last expression, $5a$, is the shortest and the most common way of writing the product.

(b) 3 times 7 can be written as $3 \cdot 7$ or $(3)(7)$.

(c) Twice z is written as $2z$.

(d) The product of 2, s, and t is written as $2st$.

(e) 4 more than the product of 6 and x is written as $6x + 4$.

Check Yourself 3

Write, using symbols.

(a) m times n **(b)** The product of h and b
(c) The product of 8 and 9 **(d)** The product of 5, w, and y
(e) 3 more than the product of 8 and a

Before we move on to division, we define the ways that we can combine the symbols we have learned so far.

Definition

Expression

An **expression** is a meaningful collection of numbers, variables, and symbols of operation.

 Example 4   **Identifying Expressions**

< Objective 2 >

NOTE

Not every collection of symbols is an expression.

(a) $2m + 3$ is an expression. It means that we multiply 2 and m, and then add 3.

(b) $x + \cdot + 3$ is not an expression. The three operations in a row have no meaning.

(c) $y = 2x - 1$ is not an expression. The equal sign is not an operation sign.

(d) $3a + 5b - 4c$ is an expression. Its meaning is clear.

 Check Yourself 4

Identify which are expressions and which are not.

(a) $7 - \cdot x$ (b) $6 + y = 9$
(c) $a + b - c$ (d) $3x - 5yz$

To write more complicated products in algebra, we need some "punctuation marks." Parentheses ( ) mean that an expression is to be thought of as a single quantity. Brackets [ ] are used in exactly the same way as parentheses in algebra. Example 5 shows the use of these signs of grouping.

 Example 5   **Expressions with More Than One Operation**

NOTES

$3(a + b)$ can be read as "3 times the quantity a plus b."

In part (b), no parentheses are needed because the 3 multiplies *only* the a.

(a) 3 times the sum of a and b is written as

$$3(a + b)$$

The sum of a and b is a single quantity, so it is enclosed in parentheses.

(b) The sum of 3 times a and b is written as $3a + b$.

(c) 2 times the difference of m and n is written as $2(m - n)$.

(d) The product of s plus t and s minus t is written as $(s + t)(s - t)$.

(e) The product of b and 3 less than b is written as $b(b - 3)$.

 Check Yourself 5

Write, using symbols.

(a) Twice the sum of p and q
(b) The sum of twice p and q
(c) The product of a and the quantity $b - c$
(d) The product of x plus 2 and x minus 2
(e) The product of x and 4 more than x

NOTE

In algebra, the fraction form is usually used to indicate division.

Now we look at the operation of division. In arithmetic, we use the division sign $\div$, the long division symbol $\overline{)}$, and fraction notation. For example, to indicate the quotient when 9 is divided by 3, we may write

$$9 \div 3 \quad \text{or} \quad 3\overline{)9} \quad \text{or} \quad \frac{9}{3}$$

Definition

Division	$\dfrac{x}{y}$ means *x divided by y* or *the quotient of x and y.*

 Example 6   Writing Expressions That Indicate Division

RECALL

The fraction bar is a grouping symbol.

(a) *m* divided by 3 is written as $\dfrac{m}{3}$.

(b) The quotient when *a* plus *b* is divided by 5 is written as $\dfrac{a + b}{5}$.

(c) The sum *p* plus *q* divided by the difference *p* minus *q* is written as $\dfrac{p + q}{p - q}$.

Check Yourself 6

Write, using symbols.

(a) *r* divided by *s*
(b) The quotient when *x* minus *y* is divided by 7
(c) The difference *a* minus 2 divided by the sum *a* plus 2

We can use many different letters to represent variables. In Example 6, the letters *m*, *a*, *b*, *p*, and *q* represented different variables. We often choose a letter that reminds us of what it represents, for example, *L* for *length* and *W* for *width*.

 Example 7   Writing Geometric Expressions

(a) *Length* times *width* is written $L \cdot W$.
(b) One-half of the *base* times the *height* is written $\dfrac{1}{2} b \cdot h$ or $\dfrac{1}{2}bh$.
(c) *Length* times *width* times *height* is written *LWH*.
(d) Pi (π) times *diameter* is written πd.

Check Yourself 7

Write each geometric expression, using symbols.

(a) Two times *length* plus two times *width*
(b) Two times pi (π) times *radius*

Algebra can be used to model a variety of applications, such as the one shown in Example 8.

 Example 8   Modeling Applications with Algebra

NOTE

We were asked to describe her pay given that her hours may vary.

Carla earns $10.25 per hour in her job. Write an expression that describes her weekly gross pay in terms of the number of hours she works.

We represent the number of hours she works in a week by the variable *h*. Carla's pay is figured by taking the product of her hourly wage and the number of hours she works.

So, the expression

10.25*h*

describes Carla's weekly gross pay.

NOTE

The words "twice" and "doubled" indicate that you should multiply by 2.

Check Yourself 8

The specifications for an engine cylinder call for the stroke length to be two more than twice the diameter of the cylinder. Write an expression for the stroke length of a cylinder based on its diameter.

We close this section by listing many of the common words used to indicate arithmetic operations.

Summary: Words Indicating Operations

The operations listed are usually indicated by the words shown.

Addition (+)	Plus, and, more than, increased by, sum
Subtraction (−)	Minus, from, less than, decreased by, difference, take away
Multiplication (·)	Times, of, by, product
Division (÷)	Divided, into, per, quotient

Check Yourself ANSWERS

1. (a) $y + 4$; **(b)** $a + b$; **(c)** $x + 3$; **(d)** $n + 6$ **2. (a)** $w - z$; **(b)** $a - 7$;
(c) $y - 3$; **(d)** $b - 5$; **(e)** $6 - m$; **(f)** $4 - x$ **3. (a)** mn; **(b)** hb;
(c) $8 \cdot 9$ or $(8)(9)$; **(d)** $5wy$; **(e)** $8a + 3$ **4. (a)** Not an expression;
(b) not an expression; **(c)** an expression; **(d)** an expression
5. (a) $2(p + q)$; **(b)** $2p + q$; **(c)** $a(b - c)$; **(d)** $(x + 2)(x - 2)$; **(e)** $x(x + 4)$
6. (a) $\dfrac{r}{s}$; **(b)** $\dfrac{x - y}{7}$; **(c)** $\dfrac{a - 2}{a + 2}$ **7. (a)** $2L + 2W$; **(b)** $2\pi r$ **8.** $2d + 2$

Reading Your Text

The following fill-in-the-blank exercises are designed to ensure that you understand some of the key vocabulary used in this section.

SECTION 1.4

(a) In algebra, we often use letters, called _____, to represent numerical values that can vary depending on the application.

(b) $x - y$ means the _____ of x and y.

(c) $x \cdot y$, $(x)(y)$, and xy are all ways of indicating _____ in algebra.

(d) An _____ is a meaningful collection of numbers, variables, and symbols of operation.

Name _____

Section _____ Date _____

Answers

1. _____	2. _____
3. _____	4. _____
5. _____	6. _____
7. _____	8. _____
9. _____	10. _____
11. _____	12. _____
13. _____	14. _____
15. _____	16. _____
17. _____	18. _____
19. _____	20. _____
21. _____	22. _____
23. _____	24. _____
25. _____	26. _____
27. _____	28. _____

Basic Skills | Challenge Yourself | Calculator/Computer | Career Applications | Above and Beyond

< Objective 1 >

Write each phrase, using symbols.

1. The sum of c and d

2. a plus 7

3. w plus z

4. The sum of m and n

5. x increased by 5

6. 4 more than c

7. 10 more than y

8. m increased by 4

9. b minus a

10. 5 less than w

11. b decreased by 4

12. r minus 3

13. 6 less than r

14. x decreased by 3

15. w times z

16. The product of 3 and c

17. The product of 5 and t

18. 8 times a

19. The product of 8, m, and n

20. The product of 7, r, and s

21. The product of 3 and the quantity p plus q

22. The product of 5 and the sum of a and b

23. Twice the sum of x and y

24. 7 times the sum of m and n

25. The sum of twice x and y

26. The sum of 3 times m and n

27. Twice the difference of x and y

28. 3 times the difference of a and c

29. The quantity a plus b times the quantity a minus b

30. The product of x plus y and x minus y

31. The product of m and 3 more than m

32. The product of a and 7 less than a > Videos

33. x divided by 5

34. The quotient when b is divided by 8

35. The result of a minus b, divided by 9

36. The difference x minus y, divided by 9

37. The sum of p and q, divided by 4

38. The sum of a and 5, divided by 9

39. The sum of a and 3, divided by the difference of a and 3

40. The sum of m and n, divided by the difference of m and n

< Objective 2 >

Identify which are expressions and which are not.

41. $2(x + 5)$

42. $4 - (x + 3)$

43. $m \div + 4$

44. $6 + a = 7$ > Videos

45. $y(x + 3)$

46. $8 = 4b$

47. $2a + 5b$

48. $4x + \cdot 7$

49. Social Science Earth's population has doubled in the last 40 years. If we let x represent Earth's population 40 years ago, what is the population today?

50. Science and Medicine It is estimated that the earth is losing 4,000 species of plants and animals every year. If S represents the number of species living last year, how many species are on Earth this year?

51. Business and Finance The simple interest (I) earned when a principal (P) is invested at a rate (r) for a time (t) is calculated by multiplying the principal times the rate times the time. Write an expression for the interest earned.

52. Science and Medicine The kinetic energy of a particle of mass m is found by taking one-half the product of the mass and the square of the velocity v. Write an expression for the kinetic energy of a particle.

Basic Skills | **Challenge Yourself** | Calculator/Computer | Career Applications | Above and Beyond

Match each phrase with the proper expression.

53. 8 decreased by x **(a)** $x - 8$

54. 8 less than x **(b)** $8 - x$

55. The difference between 8 and x

56. 8 from x

Answers

29. _____

30. _____

31. _____   32. _____

33. _____   34. _____

35. _____   36. _____

37. _____   38. _____

39. _____   40. _____

41. _____

42. _____

43. _____

44. _____

45. _____

46. _____

47. _____

48. _____

49. _____   50. _____

51. _____   52. _____

53. _____   54. _____

55. _____   56. _____

Answers

57. _____   58. _____

59. _____   60. _____

61. _____   62. _____

63. _____   64. _____

65. _____   66. _____

67. _____

68. _____

69. _____

70. _____

71. _____

72. _____

73. _____

74. _____

75. _____

76. _____

77. _____

78. _____

79. _____

80. _____

81. _____

Write each phrase, using symbols. Use x to represent the variable in each case.

57. 5 more than a number

58. A number increased by 8

59. 7 less than a number

60. A number decreased by 10

61. 9 times a number

62. Twice a number

63. 6 more than 3 times a number

64. 5 times a number, decreased by 10

65. Twice the sum of a number and 5

66. 3 times the difference of a number and 4 > Videos

67. The product of 2 more than a number and 2 less than that same number

68. The product of 5 less than a number and 5 more than that same number

69. The quotient of a number and 7

70. A number divided by 3

71. The sum of a number and 5, divided by 8

72. The quotient when 7 less than a number is divided by 3

73. 6 more than a number divided by 6 less than that same number > Videos

74. The quotient when 3 more than a number is divided by 3 less than that same number

Write each geometric expression using the given symbols.

75. Four times the length of a side (s)

76. $\dfrac{4}{3}$ times π times the cube of the radius (r)

77. The radius (r) squared times the height (h) times π

78. Twice the length (L) plus twice the width (W)

79. One-half the product of the height (h) and the sum of two unequal sides (b_1 and b_2) > Videos

80. Six times the length of a side (s) squared

Basic Skills | Challenge Yourself | Calculator/Computer | **Career Applications** | Above and Beyond
▲

81. **ALLIED HEALTH** The standard dosage given to a patient is equal to the product of the desired dose D and the available quantity Q divided by the available dose H. Write an expression for the standard dosage.

82. **INFORMATION TECHNOLOGY** Mindy is the manager of the help desk at a large cable company. She notices that, on average, her staff can handle 50 calls/hr. Last week, during a thunderstorm, the call volume increased from 65 calls/hr to 150 calls/hr. To figure out the average number of customers in the system, she needs to take the quotient of the average rate of customer arrivals (the call volume) a and the average rate at which customers are served h minus the average rate of customer arrivals a. Write an expression for the average number of customers in the system.

83. **CONSTRUCTION TECHNOLOGY** K Jones Manufacturing produces hex bolts and carriage bolts. They sold 284 more hex bolts than carriage bolts last month. Write an expression that describes the number of carriage bolts they sold last month.

84. **ELECTRICAL ENGINEERING (ADVANCED)** Electrical power P is the product of voltage V and current I. Express this relationship algebraically.

Basic Skills	Challenge Yourself	Calculator/Computer	Career Applications	**Above and Beyond**

Translate each of the given algebraic expressions into words. Exchange papers with another student to edit each other's writing. Be sure the meaning in English is the same as in algebra. Note: Each expression is not a complete sentence, so your English does not have to be a complete sentence, either. Here is an example.

Algebra: $2(x - 1)$

English (some possible answers):

One less than a number is doubled

A number decreased by one, and then multiplied by two

85. $n + 3$

86. $\dfrac{x + 2}{5}$

87. $3(5 + a)$

88. $3 - 4n$

89. $\dfrac{x + 6}{x - 1}$

90. $\dfrac{x^2 - 1}{(x - 1)^2}$

Answers

82. _____

83. _____

84. _____

85. _____

86. _____

87. _____

88. _____

89. _____

90. _____

Answers

1. $c + d$ 3. $w + z$ 5. $x + 5$ 7. $y + 10$ 9. $b - a$
11. $b - 4$ 13. $r - 6$ 15. wz 17. $5t$ 19. $8mn$
21. $3(p + q)$ 23. $2(x + y)$ 25. $2x + y$ 27. $2(x \ y)$
29. $(a + b)(a - b)$ 31. $m(m + 3)$ 33. $\dfrac{x}{5}$ 35. $\dfrac{a - b}{9}$
37. $\dfrac{p + q}{4}$ 39. $\dfrac{a + 3}{a - 3}$ 41. Expression   43. Not an expression
45. Expression   47. Expression   49. $2x$ 51. Prt 53. (b)
55. (b)   57. $x + 5$ 59. $x - 7$ 61. $9x$ 63. $3x + 6$
65. $2(x + 5)$ 67. $(x + 2)(x - 2)$ 69. $\dfrac{x}{7}$ 71. $\dfrac{x + 5}{8}$
73. $\dfrac{x + 6}{x - 6}$ 75. $4s$ 77. $\pi r^2 h$ 79. $\dfrac{1}{2}h(b_1 + b_2)$ 81. $\dfrac{DQ}{H}$
83. $H - 284$ 85. Above and Beyond   87. Above and Beyond
89. Above and Beyond

1.5

Evaluating Algebraic Expressions

< 1.5 Objectives >

1 > Evaluate algebraic expressions given any real-number value for the variables

2 > Use a calculator to evaluate algebraic expressions

When using algebra to solve problems, we often want to find the value of an algebraic expression, given particular values for the variables. Finding the value of an expression is called *evaluating the expression* and uses the following steps.

Step by Step

To Evaluate an Algebraic Expression	Step 1	Replace each variable by its given number value.
	Step 2	Do the necessary arithmetic operations, following the rules for order of operations.

Example 1   Evaluating Algebraic Expressions

< Objective 1 >

Suppose that $a = 5$ and $b = 7$.

(a) To evaluate $a + b$, we replace a with 5 and b with 7.

$a + b = (5) + (7) = 12$

(b) To evaluate $3ab$, we again replace a with 5 and b with 7.

$3ab = 3 \cdot (5) \cdot (7) = 105$

NOTE

We use parentheses when we make the initial substitution. This helps us to avoid careless errors.

Check Yourself 1

If $x = 6$ and $y = 7$, evaluate.

(a) $y - x$ **(b)** $5xy$

Some algebraic expressions require us to follow the rules for the order of operations.

Example 2   Evaluating Algebraic Expressions

Evaluate each expression if $a = 2$, $b = 3$, $c = 4$, and $d = 5$.

(a) $5a + 7b = 5(2) + 7(3)$ Multiply first.

$= 10 + 21 = 31$ Then add.

(b) $3c^2 = 3(4)^2$ Evaluate the power.

$= 3 \cdot 16 = 48$ Then multiply.

> **CAUTION**

This is different from
$(3c)^2 = (3 \cdot 4)^2$
$= 12^2 = 144$

(c) $7(c + d) = 7[(4) + (5)]$ Add inside the brackets.

$= 7 \cdot 9 = 63$

(d) $5a^4 - 2d^2 = 5(2)^4 - 2(5)^2$ Evaluate the powers.

$= 5 \cdot 16 - 2 \cdot 25$ Multiply.

$= 80 - 50 = 30$ Subtract.

Check Yourself 2

If $x = 3$, $y = 2$, $z = 4$, and $w = 5$, evaluate each expression.

(a) $4x^2 + 2$ (b) $5(z + w)$ (c) $7(z^2 - y^2)$

To evaluate an algebraic expression when a fraction bar is used, do the following: Start by doing all the work in the numerator, then do all the work in the denominator. Divide the numerator by the denominator as the last step.

 Example 3   **Evaluating Algebraic Expressions**

If $p = -2$, $q = 3$, and $r = 4$, evaluate:

(a) $\dfrac{8p}{r}$

Replace p with -2 and r with 4.

RECALL

Again, the fraction bar is a grouping symbol, like parentheses. Work first in the numerator and then in the denominator.

$$\dfrac{8p}{r} = \dfrac{8(-2)}{(4)} = \dfrac{-16}{4} = -4 \qquad \text{Divide as the last step.}$$

(b) $\dfrac{7q + r}{p + q} = \dfrac{7(3) + (4)}{(-2) + (3)}$ Now evaluate the top and bottom separately.

$$= \dfrac{21 + 4}{(-2) + (3)}$$

$$= \dfrac{25}{1} = 25$$

Check Yourself 3

Evaluate each expression if $c = -5$, $d = 8$, and $e = 3$.

(a) $\dfrac{6c}{e}$ (b) $\dfrac{4d + e}{c}$ (c) $\dfrac{10d - e}{d + e}$

Often, you will use a calculator or computer to evaluate an algebraic expression. We demonstrate how to do this in Example 4.

 Example 4   **Using a Calculator to Evaluate an Expression**

< Objective 2 >

Use a calculator to evaluate each expression.

(a) $\dfrac{4x + y}{z}$ if $x = 2$, $y = 1$, and $z = 3$.

Begin by making each of the substitutions.

$$\dfrac{4x + y}{z} = \dfrac{4(2) + (1)}{3}$$

Then, enter the numerical expression into a calculator.

(4 × 2 + 1) ÷ 3 ENTER   Remember to enclose the entire numerator in parentheses.

The display should read 3.

(b) $\dfrac{7x - y}{3z - x}$ if $x = 2$, $y = 6$, and $z = -2$.

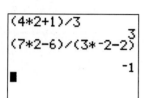

Again, we begin by substituting:

$$\frac{7x - y}{3z - x} = \frac{7(2) - (6)}{3(-2) - 2}$$

Then, we enter the expression into a calculator.

$$\boxed{(}\ 7\ \boxed{\times}\ 2\ \boxed{-}\ 6\ \boxed{)}\ \boxed{\div}\ \boxed{(}\ 3\ \boxed{\times}\ \boxed{(-)}\ 2\ \boxed{-}\ 2\ \boxed{)}\ \boxed{\text{ENTER}}$$

The display should read -1.

 Check Yourself 4

Use a calculator to evaluate each expression if $x = 2$, $y = -6$, and $z = 5$.

(a) $\dfrac{2x + y}{z}$ (b) $\dfrac{4y - 2z}{3x}$

It is important to remember that a calculator follows the correct order of operations when evaluating an expression. For example, if we omit the parentheses in Example 4(b) and enter

$$7\ \boxed{\times}\ 2\ \boxed{-}\ 6\ \boxed{\div}\ 3\ \boxed{\times}\ \boxed{(-)}\ 2\ \boxed{-}\ 2\ \boxed{\text{ENTER}}$$

the calculator will interpret our input as $7 \cdot 2 - \dfrac{6}{3} \cdot (-2) - 2$, which is not what we wanted.

Whether working with a calculator or pencil and paper, you must remember to take care both with signs and with the order of operations.

| **Example 5** | **Evaluating Expressions** |

RECALL

The rules for the order of operations call for us to multiply first, and then add.

Evaluate $5a + 4b$ if $a = -2$ and $b = 3$.

Replace a with -2 and b with 3.

$$\begin{aligned} 5a + 4b &= 5(-2) + 4(3) \\ &= -10 + 12 \\ &= 2 \end{aligned}$$

 Check Yourself 5

Evaluate $3x + 5y$ if $x = -2$ and $y = -5$.

We follow the same rules no matter how many variables are in the expression.

| **Example 6** | **Evaluating Expressions** |

 > CAUTION

When a squared variable is replaced by a negative number, square the negative.

$(-5)^2 = (-5)(-5) = 25$

The exponent applies to -5!

$-5^2 = -(5 \cdot 5) = -25$

The exponent applies only to 5!

Evaluate each expression if $a = -4$, $b = 2$, $c = -5$, and $d = 6$.

This becomes $-(-20)$, or $+20$.

(a) $\begin{aligned} 7a - 4c &= 7(-4) - 4(-5) \\ &= -28 + 20 \\ &= -8 \end{aligned}$

Evaluate the exponent or power first, and then multiply by 7.

(b) $\begin{aligned} 7c^2 &= 7(-5)^2 = 7 \cdot 25 \\ &= 175 \end{aligned}$

(c) $b^2 - 4ac = (2)^2 - 4(-4)(-5)$

$$= 4 - 4(-4)(-5)$$

$$= 4 - 80$$

$$= -76$$

(d) $b(a + d) = (2)[(-4) + (6)]$

$$= 2(2) \quad \text{———— Add inside the brackets first.}$$

$$= 4$$

Check Yourself 6

Evaluate if $p = -4$, $q = 3$, and $r = -2$.

(a) $5p - 3r$ **(b)** $2p^2 + q$ **(c)** $p(q + r)$
(d) $-q^2$ **(e)** $(-q)^2$

If an expression involves a fraction, remember that the fraction bar is a grouping symbol. This means that you should do the required operations first in the numerator and then in the denominator. Divide as the last step.

Example 7   **Evaluating Expressions**

Evaluate each expression if $x = 4$, $y = -5$, $z = 2$, and $w = -3$.

(a) $\dfrac{z - 2y}{x} = \dfrac{(2) - 2(-5)}{(4)} = \dfrac{2 + 10}{4}$

$$= \dfrac{12}{4} = 3$$

(b) $\dfrac{3x - w}{2x + w} = \dfrac{3(4) - (-3)}{2(4) + (-3)} = \dfrac{12 + 3}{8 + (-3)}$

$$= \dfrac{15}{5} = 3$$

Check Yourself 7

Evaluate if $m = -6$, $n = 4$, and $p = -3$.

(a) $\dfrac{m + 3n}{p}$ **(b)** $\dfrac{4m + n}{m + 4n}$

Example 8   **A Business and Finance Application**

NOTE

The principal is the amount invested.
   The growth rate is usually given as a percentage.

The simple interest earned on a principal P at a growth rate r for time t, in years, is given by the product Prt.

Find the simple interest earned on a $6,000 investment if the growth rate is 0.03 and the principal is invested for 2 years.

We substitute the known variable values and compute.

$$Prt = (6,000)(0.03)(2) = 360$$

The investment earns $360 in simple interest over a 2-year period.

Check Yourself 8

In most of the world, temperature is given using a Celsius scale. In the U.S., though, we generally use the Fahrenheit scale. The formula to convert temperatures from Fahrenheit to Celsius is

$$\frac{5}{9}(F - 32)$$

If the temperature is reported to be 41°F, what is the Celsius equivalent?

We provide the following chart as a reference guide for entering expressions into a calculator.

	Algebraic Notation	Calculator Notation
Addition	$6 + 2$	6 $\boxed{+}$ 2
Subtraction	$4 - 8$	4 $\boxed{-}$ 8
Multiplication	$(3)(-5)$	3 $\boxed{\times}$ $\boxed{(-)}$ 5 or 3 $\boxed{\times}$ 5 $\boxed{+/-}$
Division	$\dfrac{8}{6}$	8 $\boxed{\div}$ 6
Exponential	3^4	3 $\boxed{\wedge}$ 4 or 3 $\boxed{y^x}$ 4
	$(-3)^4$	$\boxed{(}$ $\boxed{(-)}$ 3 $\boxed{)}$ $\boxed{\wedge}$ 4 or $\boxed{(}$ 3 $\boxed{+/-}$ $\boxed{)}$ $\boxed{y^x}$ 4

Check Yourself ANSWERS

1. **(a)** 1; **(b)** 210   2. **(a)** 38; **(b)** 45; **(c)** 84   3. **(a)** -10; **(b)** -7; **(c)** 7

4. **(a)** $-\dfrac{2}{5}$; **(b)** $-\dfrac{17}{3}$ 5. -31 6. **(a)** -14; **(b)** 35; **(c)** -4; **(d)** -9; **(e)** 9

7. **(a)** -2; **(b)** -2 8. 5°C

Graphing Calculator Option

Using the Memory Feature to Evaluate Expressions

The memory features of a graphing calculator are a great aid when you need to evaluate several expressions, using the same variables and the same values for those variables.

Your graphing calculator can store variable values for many different variables in different memory spaces. Using these memory spaces saves a great deal of time when evaluating expressions.

Evaluate each expression if $a = 4.6$, $b = -\dfrac{2}{3}$, and $c = 8$. Round your results to the nearest hundredth.

(a) $a + \dfrac{b}{ac}$ **(b)** $b - b^2 + 3(a - c)$

(c) $bc - a^2 - \dfrac{ab}{c}$ **(d)** $a^2b^3c - ab^4c^2$

Begin by entering each variable's value into a calculator memory space. When possible, use the memory space that has the same name as the variable you are saving.

Step 1 Type the value associated with one variable.

Step 2 Press the store key, $\boxed{\text{STO}\blacktriangleright}$, the green alphabet key to access the memory names, $\boxed{\text{ALPHA}}$, and the key indicating which memory space you want to use.

Note: By pressing $\boxed{\text{ALPHA}}$, you are accessing the green letters above selected keys. These letters name the variable spaces.

Step 3 Press $\boxed{\text{ENTER}}$.

Step 4 Repeat until every variable value has been stored in an individual memory space.

In the example above, we store 4.6 in **Memory A**, $-\dfrac{2}{3}$ in **Memory B**, and 8 in **Memory C**.

```
4.6→A
          4.6
■
```
Memory A is with the $\boxed{\text{MATH}}$ key.

```
4.6→A
          4.6
-2/3→B
   -.6666666667
■
```
Memory B is with the $\boxed{\text{APPS}}$ key.
Divide to form a fraction.

```
4.6→A
          4.6
-2/3→B
   -.6666666667
8→C
            8
```
Memory C is with the $\boxed{\text{PRGM}}$ key.

You can use the variables in the memory spaces rather than type in the numbers. Access the memory spaces by pressing the $\boxed{\text{ALPHA}}$ before pressing the key associated with the memory space. This will save time and make careless errors much less likely.

(a) $a + \dfrac{b}{ac}$ The keystrokes are $\boxed{\text{ALPHA}}$ **Memory A**
with $\boxed{\text{MATH}}$: $\boxed{+}$ $\boxed{\text{ALPHA}}$ **Memory B**
with $\boxed{\text{APPS}}$: $\boxed{\div}$ $\boxed{(}$ **A C** $\boxed{)}$ $\boxed{\text{ENTER}}$.

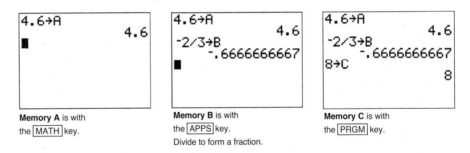

```
A+B/(AC)
      4.581884058
```

$a + \dfrac{b}{ac} = 4.58$, to the nearest hundredth.

Note: Because the fraction bar is a grouping symbol, you must remember to enclose the denominator in parentheses.

(b) $b - b^2 + 3(a - c)$ **(c)** $bc - a^2 - \dfrac{ab}{c}$

```
A+B/(AC)
      4.581884058
B-B²+3(A-C)
      -11.31111111
```

```
A+B/(AC)
      4.581884058
B-B²+3(A-C)
      -11.31111111
BC-A²-(AB)/C
         -26.11
■
```

$b - b^2 + 3(a - c) = -11.31$ $bc - a^2 - \dfrac{ab}{c} = -26.11$

Use $\boxed{x^2}$ to square a value.

(d) $a^2b^3c - ab^4c^2$

```
          4.581884058
B-B²+3(A-C)
         -11.31111111
BC-A²-(AB)/C
              -26.11
A²B^3C-AB^4C²
         -108.3101235
```

$a^2b^3c - ab^4c^2 = -108.31$

Use the caret key, $\boxed{^}$, for general exponents.

Graphing Calculator Check

Evaluate each expression if $x = -8.3$, $y = \dfrac{5}{4}$, and $z = -6$. Round your results to the nearest hundredth.

(a) $\dfrac{xy}{z} - xz$

(b) $5(z - y) + \dfrac{x}{x - z}$

(c) $x^2y^5z - (x + y)^2$

(d) $\dfrac{-2(x + z)^2}{y^3z}$

ANSWERS

(a) -48.07 **(b)** -32.64 **(c)** $-1{,}311.12$ **(d)** 34.90

Note: Throughout this text, we will provide additional graphing-calculator material offset from the exposition. This material is optional. We will not assume that students have learned this, but we feel that students using a graphing calculator will benefit from these materials.

The images and key commands are from the TI-84 Plus model from Texas Instruments. Most calculator models are fairly similar in how they handle memory. If you have a different model, consult your instructor or the instruction manual.

Reading Your Text

The following fill-in-the-blank exercises are designed to ensure that you understand some of the key vocabulary used in this section.

SECTION 1.5

(a) To evaluate an algebraic expression, first replace each _____ by its given numerical value.

(b) Finding the value of an expression given values for the variables is called _____ the expression.

(c) To evaluate an algebraic expression, you must follow the rules for the order of _____.

(d) The amount borrowed or invested in a finance application is known as the _____.

1.5 exercises

< Objective 1 >

Evaluate each expression if $a = -2$, $b = 5$, $c = -4$, *and* $d = 6$.

1. $3c - 2b$

2. $4c - 2b$

3. $8b + 2c$

4. $7a - 2c$

5. $-b^2 + b$

6. $(-b)^2 + b$

7. $3a^2$

8. $6c^2$

9. $c^2 - 2d$

10. $3b^2 + 4c$

11. $2a^2 + 3b^2$

12. $4b^2 - 2c^2$

13. $2(a + b)$

14. $5(b - c)$

15. $-4(2c - a)$

16. $6(3c - d)$

17. $a(b + 3c)$

18. $c(3a - d)$

19. $\dfrac{6d}{c}$

20. $\dfrac{8c}{2a}$

21. $\dfrac{3d + 2c}{b}$

22. $\dfrac{2b + 3d}{2a}$

23. $\dfrac{2b - 3a}{c + 2d}$ > Videos

24. $\dfrac{3d - 2b}{5a + d}$

25. $d^2 - b^2$

26. $c^2 - a^2$

27. $(d - b)^2$

28. $(c - a)^2$

29. $(d - b)(d + b)$

30. $(c - a)(c + a)$

31. $d^3 - b^3$

32. $c^3 + a^3$

Name _____

Section _____ Date _____

Answers

1.	2.
3.	4.
5.	6.
7.	8.
9.	10.
11.	12.
13.	14.
15.	16.
17.	18.
19.	20.
21.	22.
23.	24.
25.	26.
27.	28.
29.	30.
31.	32.

Answers

33. _____ 34. _____

35. _____

36. _____

37. _____

38. _____

39. _____

40. _____

41. _____

42. _____

43. _____

44. _____

45. _____

46. _____

47. _____

48. _____

49. _____

50. _____

51. _____

52. _____

53. _____

54. _____

33. $(d - b)^3$

34. $(c + a)^3$

35. $(d - b)(d^2 + db + b^2)$

36. $(c + a)(c^2 - ac + a^2)$

 > Videos

37. $-(b + a)^2$

38. $(d - a)^2$

39. $3a - 2b + \dfrac{2d}{c}$

40. $4b + 5d - \dfrac{c}{a}$

41. $a^2 + 2ad + d^2$

42. $b^2 - 2bc + c^2$

Evaluate each expression if $x = -3$, $y = 5$, and $z = \dfrac{2}{3}$.

43. $x^2 - y$

44. $\dfrac{y - x}{z}$

45. $z - y^2$

46. $z - \dfrac{z + x}{y - x}$

Evaluate each expression if $m = 4$, $n = -\dfrac{3}{2}$, and $p = \dfrac{2}{3}$.

47. $mn - np + m^2$

48. $n^2 + 2np + p^2$  > Videos

49. $\dfrac{mn}{np}$

50. $-\dfrac{np}{mn}$

Solve each application.

51. **SCIENCE AND MEDICINE** The formula for the total resistance in a parallel circuit is given by the formula $R_T = \dfrac{R_1 R_2}{R_1 + R_2}$. Find the total resistance if $R_1 = 6$ ohms (Ω) and $R_2 = 10\ \Omega$.

52. **GEOMETRY** The formula for the area of a triangle is given by $A = \dfrac{1}{2}bh$. Find the area of a triangle if $b = 4$ cm and $h = 8$ cm.

53. **GEOMETRY** The perimeter of a rectangle of length L and width W is given by the formula $P = 2L + 2W$. Find the perimeter when $L = 10$ in. and $W = 5$ in.

54. **BUSINESS AND FINANCE** The simple interest I on a principal of P dollars at interest rate r for time t, in years, is given by $I = Prt$. Find the simple interest on a principal of $6,000 at 3% for 2 years. (Hint: 3% = 0.03)  > Videos

55. BUSINESS AND FINANCE Use the simple interest formula to find the total interest earned if the principal were $1,875 and the rate of interest were 4% for 2 years.

56. BUSINESS AND FINANCE Use the simple interest formula to find the total interest earned if $5,000 earns 2% interest for 3 years.

57. SCIENCE AND MEDICINE A formula that relates Celsius and Fahrenheit temperature is $F = \frac{9}{5}C + 32$. If the current temperature is $-10°C$, what is the Fahrenheit temperature?

58. GEOMETRY If the area of a circle whose radius is r is given by $A = \pi r^2$, in which $\pi \approx 3.14$, find the area when $r = 3$ meters (m).

Basic Skills	**Challenge Yourself**	Calculator/Computer	Career Applications	Above and Beyond

In each exercise, decide whether the given values for the variables make the statement **true** *or* **false.**

59. $x - 7 = 2y + 5;\quad x = 22, y = 5$

60. $3(x - y) = 6;\quad x = 5, y = -3$

61. $2(x + y) = 2x + y;\quad x = -4, y = -2$ ▶ Videos

62. $x^2 - y^2 = x - y;\quad x = 4, y = -3$

Basic Skills	Challenge Yourself	**Calculator/Computer**	Career Applications	Above and Beyond

< Objective 2 >

Use your calculator to evaluate each expression if $x = -2.34$, $y = -3.14$, *and* $z = 4.12$. *Round your results to the nearest tenth.*

63. $x + yz$ **64.** $y - 2z$

65. $x^2 - z^2$ **66.** $x^2 + y^2$

67. $\dfrac{xy}{z - x}$ **68.** $\dfrac{y^2}{zy}$

69. $\dfrac{2x + y}{2x + z}$ **70.** $\dfrac{x^2 y^2}{xz}$

Answers

55.

56.

57.

58.

59.

60.

61.

62.

63.

64.

65.

66.

67.

68.

69.

70.

71. _____

72. _____

73. _____

74. _____

75. _____

76. _____

77. _____

78. _____

79. _____

80. _____

81. _____

82. _____

83. _____

Use your calculator to evaluate each expression if $m = 232$, $n = -487$, and $p = 58$. Round your results to the nearest tenth.

71. $m + np^2$

72. $p - (m + 2n)$

73. $(p + n)^2 - m^2$

74. $\dfrac{pm - 2n}{n - 2m}$

75. $\dfrac{n^2 - p^2}{p^2 - m^2}$

76. $m^2 + (-n)^2 + (-p^2)$

> Basic Skills | Challenge Yourself | Calculator/Computer | **Career Applications** | Above and Beyond
> ▲

77. **ALLIED HEALTH** The concentration, in micrograms per milliliter (mcg/mL), of an antihistamine in a patient's bloodstream can be approximated using the expression $-2t^2 + 13t + 1$, in which t is the number of hours since the drug was administered. Approximate the concentration of the antihistamine 1 hour after being administered.

78. **ALLIED HEALTH** Use the expression given in exercise 77 to approximate the concentration of the antihistamine 3 hours after being administered.

79. **ELECTRICAL ENGINEERING** Evaluate $\dfrac{rT}{5,252}$ for $r = 1,180$ and $T = 3$ (round to the nearest thousandth).

80. **MECHANICAL ENGINEERING** The kinetic energy (in joules) of a particle is given by $\dfrac{1}{2} mv^2$. Find the kinetic energy of a particle if its mass is 60 kg and its velocity is 6 m/s.

> Basic Skills | Challenge Yourself | Calculator/Computer | Career Applications | **Above and Beyond**
> ▲

81. Write an English interpretation of each algebraic expression.

  (a) $(2x^2 - y)^3$　　　**(b)** $3n - \dfrac{n - 1}{2}$　　　**(c)** $(2n + 3)(n - 4)$

82. Is it true that $a^n + b^n = (a + b)^n$? Try a few numbers and decide whether this is true for all numbers, for some numbers, or never true. Write an explanation of your findings and give examples.

83. Enjoyment of patterns in art, music, and language is common to all cultures, and many cultures also delight in and draw spiritual significance from patterns in numbers. One such set of patterns is that of the "magic" square. One of these squares appears in a famous etching by Albrecht Dürer, who lived from 1471 to 1528 in Europe. He was one of the first artists in Europe to use geometry to give perspective, a feeling of three dimensions, in his work.

The magic square in his work is this one:

16	3	2	13
5	10	11	8
9	6	7	12
4	15	14	1

34

Why is this square "magic"? It is magic because every row, every column, and both diagonals add to the same number. In this square there are sixteen spaces for the numbers 1 through 16.

Part 1: What number does each row and column add to? 34

Write the square that you obtain by adding −17 to each number. Is this still a magic square? If so, what number does each column and row add to? If you add 5 to each number in the original magic square, do you still have a magic square? You have been studying the operations of addition, multiplication, subtraction, and division with integers and with rational numbers. What operations can you perform on this magic square and still have a magic square? Try to find something that will not work. Use algebra to help you decide what will work and what won't. Write a description of your work and explain your conclusions.

Part 2: Here is the oldest published magic square. It is from China, about 250 B.C.E. Legend has it that it was brought from the River Lo by a turtle to the Emperor Yii, who was a hydraulic engineer.

4	9	2
3	5	7
8	1	6

Check to make sure that this is a magic square. Work together to decide what operation might be done to every number in the magic square to make the sum of each row, column, and diagonal the *opposite* of what it is now. What would you do to every number to cause the sum of each row, column, and diagonal to equal zero?

Answers

1. −22	**3.** 32	**5.** −20	**7.** 12	**9.** 4	**11.** 83	**13.** 6
15. 24	**17.** 14	**19.** −9	**21.** 2	**23.** 2	**25.** 11	**27.** 1
29. 11	**31.** 91	**33.** 1	**35.** 91	**37.** −9	**39.** −19	

41. 16   **43.** 4   **45.** $-\dfrac{73}{3}$ **47.** 11   **49.** 6   **51.** 3.75 Ω

53. 30 in.   **55.** $150   **57.** 14°F   **59.** True   **61.** False

63. −15.3   **65.** −11.5   **67.** 1.1   **69.** 14   **71.** −1,638,036

73. 130,217   **75.** −4.6   **77.** 12 mcg/mL   **79.** 0.674

81. Above and Beyond   **83.** Above and Beyond

1.6 Adding and Subtracting Terms

< 1.6 Objectives >

1 > Identify terms and like terms

2 > Combine like terms

To find the perimeter of (or the distance around) a rectangle, we add 2 times the length and 2 times the width. In the language of algebra, this can be written as

L

W [rectangle] W Perimeter $= 2L + 2W$

L

We call $2L + 2W$ an **algebraic expression,** or more simply an **expression.** Recall from Section 1.5 that an expression allows us to write a mathematical idea in symbols. It can be thought of as a meaningful collection of letters, numbers, and operation signs.

Some expressions are

$$5x^2 \qquad 3a + 2b \qquad 4x^3 - 2y + 1$$

Addition and subtraction signs break expressions into smaller parts called *terms.*

Definition	
Term	A **term** can be written as a number, or the product of a number and one or more variables, raised to a whole-number power.

In an expression, each sign ($+$ or $-$) is a part of the term that follows the sign.

Example 1	Identifying Terms

< Objective 1 >

(a) $5x^2$ has one term.

(b) $3a + 2b$ has two terms: $3a$ and $2b$.
    Term Term

NOTE

Each term "owns" the sign that precedes it.

(c) $4x^3 - 2y + 1$ has three terms: $4x^3$, $-2y$, and 1.
    Term Term Term

(d) $x - y$ has two terms: x and $-y$.

Check Yourself 1

List the terms of each expression.

(a) $2b^4$ **(b)** $5m + 3n$ **(c)** $2s^2 - 3t - 6$

NOTE

We usually use *coefficient* instead of "numerical coefficient."

Note that a term in an expression may have any number of factors. For instance, $5xy$ is a term. It has factors of 5, x, and y. The number factor of a term is called the **numerical coefficient.** So for the term $5xy$, the numerical coefficient is 5.

| **Example 2** | **Identifying the Numerical Coefficient**

(a) $4a$ has the numerical coefficient 4.

(b) $6a^3b^4c^2$ has the numerical coefficient 6.

(c) $-7m^2n^3$ has the numerical coefficient -7.

(d) Because $x = 1 \cdot x$, the numerical coefficient of x is understood to be 1.

Check Yourself 2

Give the numerical coefficient for each term.

(a) $8a^2b$ (b) $-5m^3n^4$ (c) y

If terms contain exactly the *same letters* (or variables) raised to the *same powers,* they are called **like terms.**

 | **Example 3** | **Identifying Like Terms**

(a) These are like terms.

$6a$ and $7a$
$5b^2$ and b^2
$10x^2y^3z$ and $-6x^2y^3z$
$-3m^2$ and m^2

Each pair of terms has the same letters, with each letter raised to the same power—the numerical coefficients can be any number.

(b) These are *not* like terms.

Different letters

$6a$ and $7b$

Different exponents

$5b^2$ and $5b^3$

Different exponents

$3x^2y$ and $4xy^2$

Check Yourself 3

Circle the like terms.

$5a^2b$ ab^2 a^2b $-3a^2$ $4ab$ $3b^2$ $-7a^2b$

Like terms of an expression can always be combined into a single term.

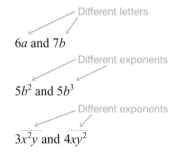

Rather than having to write out all those x's, try

$2x + 5x = (2 + 5)x = 7x$

In the same way,

$9b + 6b = (9 + 6)b = 15b$

and $10a - 4a = (10 - 4)a = 6a$

This leads us to the rule for combining like terms.

RECALL

We use the distributive property from Section 1.1.

Step by Step

Combining Like Terms	To combine like terms, use the following steps.
	Step 1 Add or subtract the numerical coefficients.
	Step 2 Attach the common variables.

Combining like terms is one step we take when *simplifying* an expression.

 Example 4 **Combining Like Terms**

< Objective 2 >

 > CAUTION

Do not change the exponents when combining like terms.

Combine like terms.

(a) $8m + 5m = (8 + 5)m = 13m$

(b) $5pq^3 - 4pq^3 = (5 - 4)pq^3 = 1pq^3 = pq^3$

(c) $7a^3b^2 - 7a^3b^2 = (7 - 7)a^3b^2 = 0a^3b^2 = 0$

Check Yourself 4

Combine like terms.

(a) $6b + 8b$ **(b)** $12x^2 - 3x^2$

(c) $8xy^3 - 7xy^3$ **(d)** $9a^2b^4 - 9a^2b^4$

The idea is the same when expressions involve more than two terms.

 Example 5 **Combining Like Terms**

NOTE

The distributive property can be used with any number of like terms.

NOTE

With practice, you will do this mentally instead of writing out all of these steps.

Combine like terms.

(a) $5ab - 2ab + 3ab$

$= (5 - 2 + 3)ab = 6ab$

Only like terms can be combined.

(b)  $8x - 2x + 5y$

$= (8 - 2)x + 5y$

$= 6x + 5y$

Like terms   Like terms

(c) $5m + 8n \quad\quad + 4m - 3n$

$= (5m + 4m) + (8n - 3n)$

$= \quad 9m \quad + \quad 5n$

Here we have used both the associative and commutative properties.

(d) $4x^2 + 2x - 3x^2 + x$

$= (4x^2 - 3x^2) + (2x + x)$

$= x^2 + 3x$

As these examples illustrate, combining like terms often means changing the grouping and the order in which the terms are written. Again, all this is possible because of the properties of addition that we introduced in Section 1.1.

Check Yourself 5

Combine like terms.

(a) $4m^2 - 3m^2 + 8m^2$ **(b)** $9ab + 3a - 5ab$ **(c)** $4p + 7q + 5p - 3q$

Let us now look at a business and finance application of this section's content.

| Example 6 | A Business and Finance Application |

NOTE

A business can compute the profit it earns on an item by subtracting the costs associated with the item from the revenue earned by the item.

S-Bar Electronics, Inc., sells a certain server for \$1,410. It pays the manufacturer \$849 for each server and there are fixed costs of \$4,500 per week associated with the servers.

Let x be the number of servers bought and sold during the week.

Then, the revenue earned by S-Bar Electronics, Inc., from these servers can be modeled by the formula

$$R = 1,410x$$

The cost can be modeled with the formula

$$C = 849x + 4,500$$

Therefore, the profit can be modeled by the difference between the revenue and the cost.

$$P = 1,410x - (849x + 4,500)$$
$$= 1,410x - 849x - 4,500$$

Simplify the given profit formula.

The like terms are $1,410x$ and $-849x$. We combine these to give a simplified formula

$$P = 561x - 4,500$$

NOTE

A negative profit would mean the company suffered a loss.

Check Yourself 6

S-Bar Electronics, Inc., also sells 19-in. flat-screen monitors for \$799 each. The monitors cost them \$489 each. Additionally, there are weekly fixed costs of \$3,150 associated with the sale of the monitors. We can model the profit earned on the sale of y monitors with the formula

$$P = 799y - 489y - 3,150$$

Simplify the profit formula.

Check Yourself ANSWERS

1. **(a)** $2b^4$; **(b)** $5m, 3n$; **(c)** $2s^2, -3t, -6$ 2. **(a)** 8; **(b)** -5; **(c)** 1
3. The like terms are $5a^2b, a^2b$, and $-7a^2b$ 4. **(a)** $14b$; **(b)** $9x^2$; **(c)** xy^3; **(d)** 0
5. **(a)** $9m^2$; **(b)** $4ab + 3a$; **(c)** $9p + 4q$ 6. $310y - 3,150$

Reading Your Text

The following fill-in-the-blank exercises are designed to ensure that you understand some of the key vocabulary used in this section.

SECTION 1.6

(a) The product of a number and a variable is called a _____.

(b) The number factor of a term is called the _____.

(c) If a variable appears without an exponent, it is understood to be raised to the _____ power.

(d) If a variable appears without a coefficient, it is understood that the coefficient is _____.

Name _____

Section _____ Date _____

Answers

1. _____	2. _____
3. _____	4. _____
5. _____	6. _____
7. _____	8. _____
9. _____	
10. _____	
11. _____	12. _____
13. _____	14. _____
15. _____	16. _____
17. _____	18. _____
19. _____	20. _____
21. _____	22. _____
23. _____	24. _____
25. _____	26. _____
27. _____	
28. _____	
29. _____	
30. _____	
31. _____	
32. _____	
33. _____	

< Objective 1 >

List the terms of each expression.

1. $5a + 2$

2. $7a - 4b$

3. $4x^3$

4. $3x^2$

5. $3x^2 + 3x - 7$

6. $2a^3 - a^2 + a$

Circle the like terms in each group of terms.

7. $5ab, 3b, 3a, 4ab$

8. $9m^2, 8mn, 5m^2, 7m$

9. $4xy^2, 2x^2y, 5x^2, -3x^2y, 5y, 6x^2y$ > Videos

10. $8a^2b, 4a^2, 3ab^2, -5a^2b, 3ab, 5a^2b$

< Objective 2 >

Combine the like terms.

11. $4m + 6m$

12. $6a^2 + 8a^2$

13. $7b^3 + 10b^3$

14. $7rs + 13rs$

15. $21xyz + 7xyz$

16. $-3mn^2 + 9mn^2$

17. $9z^2 - 3z^2$

18. $7m - 6m$

19. $9a^5 - 9a^5$

20. $13xy - 9xy$

21. $19n^2 - 18n^2$

22. $7cd - 7cd$

23. $21p^2q - 6p^2q$

24. $17r^3s^2 - 8r^3s^2$

25. $5x^2 - 3x^2 + 9x^2$

26. $13uv + uv - 12uv$

27. $11b - 9a - 6b$

28. $5m^2 - 3m + 6m^2$

29. $7x + 5y - 4x - 4y$

30. $6a^2 + 11a + 7a^2 - 9a$

31. $4a + 7b + 3 - 2a + 3b - 2$

32. $5p^2 + 2p + 8 + 4p^2 + 5p - 6$

 > Videos

Solve each application.

33. GEOMETRY Provide a simplified expression
for the perimeter of the rectangle shown.

$2x^2 - x + 1$ cm

$3x - 2$ cm

34. **GEOMETRY** Provide a simplified expression for the perimeter of the triangle shown.

x ft $3(x + 1)$ ft

$2x^2 - 5x + 1$ ft

35. **GEOMETRY** A rectangle has sides that measure $8x + 9$ in. and $6x - 7$ in. Provide a simplified expression for its perimeter.

36. **GEOMETRY** A triangle has sides measuring $3x + 7$ mm, $4x - 9$ mm, and $5x + 6$ mm. Find the simplified expression that represents its perimeter.

37. **BUSINESS AND FINANCE** The cost of producing x units of an item is $C = 150 + 25x$. The revenue from selling x units is $R = 90x - x^2$. The profit is given by the revenue minus the cost. Find the simplified expression that represents the profit.

38. **BUSINESS AND FINANCE** The revenue from selling y units is $R = 3y^2 - 2y + 5$ and the cost of producing y units is $C = y^2 + y - 3$. Find the simplified expression that represents the profit.

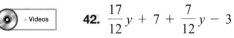

| Basic Skills | **Challenge Yourself** | Calculator/Computer | Career Applications | Above and Beyond |

Simplify each expression by combining like terms.

39. $\dfrac{2}{3}m + 3 + \dfrac{4}{3}m$

40. $\dfrac{a}{5} - 2 + \dfrac{4a}{5}$

41. $\dfrac{13x}{5} + 2 - \dfrac{3x}{5} + 5$ > Videos

42. $\dfrac{17}{12}y + 7 + \dfrac{7}{12}y - 3$

43. $2.3a + 7 + 4.7a + 3$

44. $5.8m + 4 - 2.8m + 11$

Rewrite each statement as an algebraic expression. Simplify each expression, if possible.

45. Find the sum of $5a^4$ and $8a^4$.

46. Find the sum of $9p^2$ and $12p^2$.

47. Find the difference between $15a^3$ and $12a^3$.

48. Subtract $5m^3$ from $18m^3$.

49. Subtract $3mn^2$ from the sum of $9mn^2$ and $5mn^2$.   > Videos

50. Find the difference between the sum of $6x^2y$ and $12x^2y$, and $4x^2y$.

Answers

34.
35.
36.
37.
38.
39.
40.
41.
42.
43.
44.
45.
46.
47.
48.
49.
50.

Use the distributive property to remove the parentheses in each expression. Then, simplify each expression by combining like terms.

51. $2(3x + 2) + 4$

52. $3(4z + 5) - 9$

53. $5(6a - 2) + 12a$

54. $7(4w - 3) - 25w$

55. $4s + 2(s + 4) + 4$ › Videos

56. $5p + 4(p + 3) - 8$

Basic Skills	Challenge Yourself	Calculator/Computer	**Career Applications**	Above and Beyond

57. ALLIED HEALTH The ideal body weight, in pounds, for a woman can be approximated by substituting her height, in inches, into the formula $105 + 5(h - 60)$. Use the distributive property to simplify the expression.

58. ALLIED HEALTH Use exercise 57 to approximate the ideal body weight for a woman who stands 5 ft 4 in. tall.

59. MECHANICAL ENGINEERING A primary beam can support a load of $54p$. A second beam is added that can support a load of $32p$. What is the total load that the two beams can support?

60. MECHANICAL ENGINEERING Two objects are spinning on the same axis. The moment of inertia of the first object is $\frac{6^3}{12}b$. The moment of inertia of the second object is given by $\frac{30^3}{36}b$. The total moment of inertia is given by the sum of the moments of inertia of the two objects. Write a simplified expression for the total moment of inertia for the two objects described.

Basic Skills	Challenge Yourself	Calculator/Computer	Career Applications	**Above and Beyond**

61. Write a paragraph explaining the difference between n^2 and $2n$.

62. Complete the explanation: "x^3 and $3x$ are not the same because . . . ."

63. Complete the statement: "$x + 2$ and $2x$ are different because . . . ."

64. Write an English phrase for each given algebraic expression:

   (a) $2x^3 + 5x$ **(b)** $(2x + 5)^3$ **(c)** $6(n + 4)^2$

65. Work with another student to complete this exercise. Place $>$, $<$, or $=$ in the blank in these statements.

   1^2____2^1
   2^3____3^2
   3^4____4^3
   4^5____5^4

What happens as the table of numbers is extended? Try more examples.

What sign seems to occur the most in your table? $>$, $<$, or $=$?

Write an algebraic statement for the pattern of numbers in this table. Do you think this is a pattern that continues? Add more lines to the table and extend the pattern to the general case by writing the pattern in algebraic notation. Write a short paragraph stating your conjecture.

Answers

51.

52.

53.

54.

55.

56.

57.

58.

59.

60.

61.

62.

63.

64.

65.

66. Work with other students on this exercise.

Part 1: Evaluate the three expressions $\dfrac{n^2 - 1}{2}$, n, $\dfrac{n^2 + 1}{2}$ using odd values of n: 1, 3, 5, 7, and so on. Make a chart like the one below and complete it.

n	$a = \dfrac{n^2 - 1}{2}$	$b = n$	$c = \dfrac{n^2 + 1}{2}$	a^2	b^2	c^2
1						
3						
5						
7						
9						
11						
13						

Part 2: The numbers a, b, and c that you get in each row have a surprising relationship to each other. Complete the last three columns and work together to discover this relationship. You may want to find out more about the history of this famous number pattern.

Answers

1. $5a, 2$ **3.** $4x^3$ **5.** $3x^2, 3x, -7$ **7.** $5ab, 4ab$
9. $2x^2y, -3x^2y, 6x^2y$ **11.** $10m$ **13.** $17b^3$ **15.** $28xyz$ **17.** $6z^2$
19. 0 **21.** n^2 **23.** $15p^2q$ **25.** $11x^2$ **27.** $-9a + 5b$
29. $3x + y$ **31.** $2a + 10b + 1$ **33.** $4x^2 + 4x - 2$ cm   **35.** $28x + 4$ in.
37. $-x^2 + 65x - 150$ **39.** $2m + 3$ **41.** $2x + 7$ **43.** $7a + 10$
45. $13a^4$ **47.** $3a^3$ **49.** $11mn^2$ **51.** $6x + 8$ **53.** $42a - 10$
55. $6s + 12$ **57.** $5h - 195$ **59.** $86p$ **61.** Above and Beyond
63. Above and Beyond   **65.** Above and Beyond

1.7

< 1.7 Objectives >

Multiplying and Dividing Terms

1 > Find the product of algebraic terms

2 > Find the quotient of algebraic terms

Now we consider exponential notation. Remember that the exponent tells us how many times the base is to be used as a factor.

Exponent
$$2^5 = 2 \cdot 2 \cdot 2 \cdot 2 \cdot 2 = 32$$
Base — The fifth power of 2

The notation can also be used when working with letters or variables.

$$x^4 = \underbrace{x \cdot x \cdot x \cdot x}_{4 \text{ factors}}$$

Now look at the product $x^2 \cdot x^3$.

$$x^2 \cdot x^3 = (x \cdot x)(x \cdot x \cdot x) = \underbrace{x \cdot x \cdot x \cdot x \cdot x}_{} = x^5$$

2 factors + 3 factors = 5 factors

So

$$x^2 \cdot x^3 = x^{2+3} = x^5$$

You should recall from the previous section that in order to combine a pair of terms into a single term, we must have like terms. For instance, we cannot combine the sum $x^2 + x^3$ into a single term.

On the other hand, when we multiply a pair of unlike terms, as above, their product is a single term.

This leads us to the following property of exponents.

NOTES

In general,

$$x^m = \underbrace{x \cdot x \cdot \cdots \cdot x}_{m \text{ factors}}$$

in which m is a natural number. **Natural numbers** are the numbers we use for counting: 1, 2, 3, and so on.

Exponents are also called **powers.**

NOTE

The exponent of x^5 is the *sum* of the exponents in x^2 and x^3.

Property

The Product Property of Exponents

For any integers m and n and any real number a,

$$a^m \cdot a^n = a^{m+n}$$

In words, to multiply expressions with the same base, keep the base and add the exponents.

Example 1 | **Using the Product Property of Exponents**

< Objective 1 >

> CAUTION

In part (c), the product is *not* 9^6. The base does not change.

(a) $a^5 \cdot a^7 = a^{5+7} = a^{12}$

(b) $x \cdot x^8 = x^1 \cdot x^8 = x^{1+8} = x^9$ $x = x^1$

(c) $3^2 \cdot 3^4 = 3^{2+4} = 3^6$

(d) $y^2 \cdot y^3 \cdot y^5 = y^{2+3+5} = y^{10}$

(e) $x^3 \cdot y^4$ *cannot* be simplified. The bases are not the same.

Check Yourself 1

Multiply. Write your answer in exponential form.

(a) $b^6 \cdot b^8$ (b) $y^7 \cdot y$ (c) $2^3 \cdot 2^4$ (d) $a^2 \cdot a^4 \cdot a^3$

NOTE

Although it has several factors, this is still a single term.

Suppose that numerical coefficients are involved in a product. To find the product, multiply the coefficients and then use the product property of exponents to combine the variables.

$$2x^3 \cdot 3x^5 = (2 \cdot 3)(x^3 \cdot x^5) \quad \text{Multiply the coefficients.}$$
$$= 6x^{3+5} \quad \text{Add the exponents.}$$
$$= 6x^8$$

You may have noticed that we have again changed the order and grouping. This uses the commutative and associative properties that we introduced in Section 1.1.

Example 2 | **Using the Product Property of Exponents**

NOTE

We have written out all the steps. With practice, you can do the multiplication mentally.

Multiply.

(a) $5a^4 \cdot 7a^6 = (5 \cdot 7)(a^4 \cdot a^6) = 35a^{10}$

(b) $y^2 \cdot 3y^3 \cdot 6y^4 = (1 \cdot 3 \cdot 6)(y^2 \cdot y^3 \cdot y^4) = 18y^9$

(c) $2x^2y^3 \cdot 3x^5y^2 = (2 \cdot 3)(x^2 \cdot x^5)(y^3 \cdot y^2) = 6x^7y^5$

Check Yourself 2

Multiply.

(a) $4x^3 \cdot 7x^5$ (b) $3a^2 \cdot 2a^4 \cdot 2a^5$ (c) $3m^2n^4 \cdot 5m^3n$

What about dividing expressions when exponents are involved? For instance, what if we want to divide x^5 by x^2? We can use the following approach to division:

$$\frac{x^5}{x^2} = \frac{\overbrace{x \cdot x \cdot x \cdot x \cdot x}^{5 \text{ factors}}}{\underbrace{x \cdot x}_{2 \text{ factors}}} = \frac{x \cdot x \cdot x \cdot x \cdot x}{x \cdot x}$$

We can divide by 2 factors of x.

NOTE

The exponent of x^3 is the *difference* of the exponents in x^5 and x^2.

$$= \overbrace{x \cdot x \cdot x}^{3 \text{ factors}} = x^3$$

So

$$\frac{x^5}{x^2} = x^{5-2} = x^3$$

This leads us to a second property of exponents.

Property

The Quotient Property of Exponents

For any integers m and n, and any nonzero number a,

$$\frac{a^m}{a^n} = a^{m-n}$$

In words, to divide expressions with the same base, keep the base and subtract the exponents.

Example 3 | **Using the Quotient Property of Exponents**

< Objective 2 >

Divide the following.

(a) $\dfrac{y^7}{y^3} = y^{7-3} = y^4$

(b) $\dfrac{m^6}{m} = \dfrac{m^6}{m^1} = m^{6-1} = m^5$

> Apply the quotient property to each variable separately.

> **RECALL**
>
> We can write $\dfrac{a^3b^5}{a^2b^2}$ as $\dfrac{a^3}{a^2} \cdot \dfrac{b^5}{b^2}$ because this is how we multiply fractions.

(c) $\dfrac{a^3b^5}{a^2b^2} = a^{3-2} \cdot b^{5-2} = ab^3$

✓ **Check Yourself 3**

Divide.

(a) $\dfrac{m^9}{m^6}$ **(b)** $\dfrac{a^8}{a}$ **(c)** $\dfrac{a^3b^5}{a^2}$ **(d)** $\dfrac{r^5s^6}{r^3s^2}$

If numerical coefficients are involved, just divide the coefficients and then use the quotient property of exponents to divide the variables, as shown in Example 4.

Example 4 | **Using the Quotient Property of Exponents**

Divide the following.

Subtract the exponents.

(a) $\dfrac{6x^5}{3x^2} = 2x^{5-2} = 2x^3$

6 divided by 3

20 divided by 5

(b) $\dfrac{20a^7b^5}{5a^3b^4} = 4a^{7-3} \cdot b^{5-4}$

Again apply the quotient property to each variable separately.

$= 4a^4b$

✓ **Check Yourself 4**

Divide.

(a) $\dfrac{4x^3}{2x}$ **(b)** $\dfrac{20a^6}{5a^2}$ **(c)** $\dfrac{24x^5y^3}{4x^2y^2}$

✓ **Check Yourself ANSWERS**

1. (a) b^{14}; **(b)** y^8; **(c)** 2^7; **(d)** a^9 **2. (a)** $28x^8$; **(b)** $12a^{11}$; **(c)** $15m^5n^5$

3. (a) m^3; **(b)** a^7; **(c)** ab^5; **(d)** r^2s^4 **4. (a)** $2x^2$; **(b)** $4a^4$; **(c)** $6x^3y$

Reading Your Text

The following fill-in-the-blank exercises are designed to ensure that you understand some of the key vocabulary used in this section.

SECTION 1.7

(a) When multiplying expressions with the same base, _____ the exponents.

(b) When multiplying expressions with the same base, the _____ does not change.

(c) When multiplying expressions with the same base, _____ the coefficients.

(d) To divide expressions with the same base, keep the base and _____ the exponents.

1.7 exercises

Name _____

Section _____ Date _____

Answers

1. _____ 2. _____

3. _____ 4. _____

5. _____ 6. _____

7. _____ 8. _____

9. _____ 10. _____

11. _____ 12. _____

13. _____ 14. _____

15. _____ 16. _____

17. _____ 18. _____

19. _____ 20. _____

21. _____ 22. _____

23. _____ 24. _____

25. _____ 26. _____

27. _____ 28. _____

29. _____ 30. _____

31. _____ 32. _____

33. _____ 34. _____

35. _____ 36. _____

37. _____ 38. _____

Basic Skills | Challenge Yourself | Calculator/Computer | Career Applications | Above and Beyond

< Objective 1 >

Multiply.

1. $x^5 \cdot x^7$

2. $b^2 \cdot b^4$

3. $3^2 \cdot 3^6$

4. $y^6 \cdot y^4$

5. $a^9 \cdot a$

6. $3^4 \cdot 3^5$

7. $z^{10} \cdot z^3$

8. $x^6 \cdot x^3$

9. $p^5 \cdot p^7$

10. $s^6 \cdot s^9$

11. $x^3y \cdot x^2y^4$

12. $m^2n^3 \cdot mn^4$

13. $w^3 \cdot w^4 \cdot w^2$

14. $x^5 \cdot x^4 \cdot x^6$

15. $m^3 \cdot m^2 \cdot m^4$

16. $r^3 \cdot r \cdot r^5$ > Videos

17. $a^3b \cdot a^2b^2 \cdot ab^3$

18. $w^2z^3 \cdot wz \cdot w^3z^4$

19. $p^2q \cdot p^3q^5 \cdot pq^4$

20. $c^3d \cdot c^4d^2 \cdot cd^5$

21. $2a^5 \cdot 3a^2$

22. $5x^3 \cdot 3x^2$

23. $x^2 \cdot 3x^5$

24. $2m^4 \cdot 6m^7$

25. $5m^3n^2 \cdot 4mn^3$

26. $7x^2y^5 \cdot 6xy^4$

27. $4x^5y \cdot 3xy^2$

28. $5a^3b \cdot 10ab^4$

29. $2a^2 \cdot a^3 \cdot 3a^7$

30. $2x^3 \cdot 3x^4 \cdot x^5$

31. $3c^2d \cdot 4cd^3 \cdot 2c^5d$

32. $5p^2q \cdot p^3q^2 \cdot 3pq^3$

33. $5m^2 \cdot m^3 \cdot 2m \cdot 3m^4$

34. $3a^3 \cdot 2a \cdot a^4 \cdot 2a^5$

35. $2r^3s \cdot rs^2 \cdot 3r^2s \cdot 5rs$ > Videos

36. $6a^2b \cdot ab \cdot 3ab^3 \cdot 2a^2b$

< Objective 2 >

Divide.

37. $\dfrac{a^{10}}{a^7}$

38. $\dfrac{m^8}{m^2}$

39. $\dfrac{y^{10}}{y^4}$

40. $\dfrac{b^9}{b^4}$

41. $\dfrac{p^{15}}{p^{10}}$

42. $\dfrac{s^{15}}{s^9}$

43. $\dfrac{x^5 y^3}{x^2 y^2}$ > Videos

44. $\dfrac{s^5 t^4}{s^3 t^2}$

45. $\dfrac{10m^6}{5m^4}$

46. $\dfrac{8x^5}{4x}$

47. $\dfrac{24a^7}{6a^4}$

48. $\dfrac{25x^9}{5x^8}$

49. $\dfrac{26m^8 n}{13m^6}$

50. $\dfrac{30a^4 b^5}{6b^4}$

51. $\dfrac{35w^4 z^6}{5w^2 z}$

52. $\dfrac{48p^6 q^7}{8p^4 q}$

53. $\dfrac{48x^4 y^5 z^9}{24x^2 y^3 z^6}$ > Videos

54. $\dfrac{25a^5 b^4 c^3}{5a^4 b c^2}$

| Basic Skills | **Challenge Yourself** | Calculator/Computer | Career Applications | Above and Beyond |

Simplify each expression, if possible.

55. $3a^4 b^3 \cdot 2a^2 b^4$

56. $2xy^3 \cdot 3xy^2$

57. $2a^3 b + 3a^2 b$

58. $2xy^3 + 3xy^2$

59. $2x^2 y^3 \cdot 3x^2 y^3$

60. $5a^3 b^2 \cdot 10a^3 b^2$

61. $2x^3 y^2 + 3x^3 y^2$

62. $5a^3 b^2 + 10a^3 b^2$

63. $\dfrac{8a^2 b \cdot 6a^2 b}{2ab}$

64. $\dfrac{6x^2 y^3 \cdot 9x^2 y^3}{3x^2 y^2}$ > Videos

65. $\dfrac{8a^2 b + 6a^2 b}{2ab}$

66. $\dfrac{6x^2 y^3 + 9x^2 y^3}{3x^2 y^2}$

| Basic Skills | Challenge Yourself | Calculator/Computer | Career Applications | **Above and Beyond** |

67. Complete each statement:

    (a) a^n is negative when _____ because _____.

    (b) a^n is positive when _____ because _____.

              (give all possibilities)

Answers

39. _____	40. _____
41. _____	42. _____
43. _____	44. _____
45. _____	46. _____
47. _____	48. _____
49. _____	50. _____
51. _____	52. _____
53. _____	54. _____
55. _____	56. _____
57. _____	
58. _____	
59. _____	60. _____
61. _____	62. _____
63. _____	64. _____
65. _____	66. _____
67. _____	

Answers

68. _____

69. _____

68. "Earn Big Bucks!" reads an ad for a job. "You will be paid 1 cent for the first day and 2 cents for the second day, 4 cents for the third day, 8 cents for the fourth day, and so on, doubling each day. Apply now!" What kind of deal is this—where is the big money offered in the headline? The fine print at the bottom of the ad says: "Highly qualified people may be paid \$1,000,000 for the first 28 working days if they choose." Well, *that* does sound like big bucks! Work with other students to decide which method of payment is better and how much better. You may want to make a table and try to find a formula for the first offer.

69. An oil spill from a tanker in pristine Prince William Sound in Alaska begins in a circular shape only 2 ft across. The area of the circle is $A = \pi r^2$. Make a table to decide what happens to the area if the diameter is doubling each hour. How large will the spill be in 24 h? (Hint: The radius is one-half the diameter.)

2 ft

Answers

1. x^{12} **3.** 3^8 **5.** a^{10} **7.** z^{13} **9.** p^{12} **11.** $x^5 y^5$ **13.** w^9
15. m^9 **17.** $a^6 b^6$ **19.** $p^6 q^{10}$ **21.** $6a^7$ **23.** $3x^7$ **25.** $20m^4 n^5$
27. $12x^6 y^3$ **29.** $6a^{12}$ **31.** $24c^8 d^5$ **33.** $30m^{10}$ **35.** $30r^7 s^5$
37. a^3 **39.** y^6 **41.** p^5 **43.** $x^3 y$ **45.** $2m^2$ **47.** $4a^3$
49. $2m^2 n$ **51.** $7w^2 z^5$ **53.** $2x^2 y^2 z^3$ **55.** $6a^6 b^7$
57. Cannot simplify   **59.** $6x^4 y^6$ **61.** $5x^3 y^2$ **63.** $24a^3 b$
65. $7a$ **67.** Above and Beyond   **69.** Above and Beyond

Definition/Procedure	Example	Reference
Properties of Real Numbers		Section 1.1
The Commutative Properties		
If a and b are any numbers, **1.** $a + b = b + a$ **2.** $a \cdot b = b \cdot a$	$3 + 8 = 8 + 3$ $2 \cdot 5 = 5 \cdot 2$	*p. 3*
The Associative Properties		
If a, b, and c are any numbers, **1.** $a + (b + c) = (a + b) + c$ **2.** $a \cdot (b \cdot c) = (a \cdot b) \cdot c$	$3 + (7 + 12) = (3 + 7) + 12$ $2 \cdot (5 \cdot 12) = (2 \cdot 5) \cdot 12$	*p. 4*
The Distributive Property		
If a, b, and c are any numbers, $a(b + c) = a \cdot b + a \cdot c$	$6 \cdot (8 + 15) = 6 \cdot 8 + 6 \cdot 15$	*p. 5*
Adding and Subtracting Real Numbers		Section 1.2
Addition		
1. If two numbers have the same sign, add their absolute values. Give the sum the sign of the original numbers.	$9 + 7 = 16$ $(-9) + (-7) = -16$	*p. 12*
2. If two numbers have different signs, subtract their absolute values, the smaller from the larger. Give the result the sign of the number with the larger absolute value.	$15 + (-10) = 5$ $(-12) + 9 = -3$	*p. 13*
Subtraction		

1. Rewrite the subtraction problem as an addition problem by:
  a. Changing the subtraction to addition
  b. Replacing the number being subtracted with its opposite
2. Add the resulting signed numbers as before.
	$16 - 8 = 16 + (-8)$
	$= 8$
	$8 - 15 = 8 + (-15)$
	$= -7$
	$-9 - (-7) = -9 + 7$
	$= -2$
		p. 15

Continued

Definition/Procedure	Example	Reference

Multiplying and Dividing Real Numbers

Section 1.3

Multiplication

Multiply the absolute values of the two numbers.
1. If the numbers have different signs, the product is negative.
2. If the numbers have the same sign, the product is positive.

$5(-7) = -35$
$(-10)(9) = -90$
$8 \cdot 7 = 56$
$(-9)(-8) = 72$

p. 25

p. 26

Division

Divide the absolute values of the two numbers.
1. If the numbers have different signs, the quotient is negative.
2. If the numbers have the same sign, the quotient is positive.

$$\frac{-32}{4} = -8$$

$$\frac{75}{-5} = -15$$

$$\frac{20}{5} = 4$$

$$\frac{-18}{-9} = 2$$

p. 28

From Arithmetic to Algebra

Section 1.4

Addition

$x + y$ means the *sum* of x and y or x *plus* y. Some other words indicating addition are "more than" and "increased by."

The sum of x and 5 is $x + 5$.
7 more than a is $a + 7$.
b increased by 3 is $b + 3$.

p. 39

Subtraction

$x - y$ means the *difference* of x and y or x *minus* y. Some other words indicating subtraction are "less than" and "decreased by."

The difference of x and 3 is $x - 3$.
5 less than p is $p - 5$.
a decreased by 4 is $a - 4$.

p. 40

Multiplication

$\left.\begin{matrix} x \cdot y \\ (x)(y) \\ xy \end{matrix}\right\}$ All these mean the *product* of x and y or x *times* y.

The product of m and n is mn.
The product of 2 and the sum of a and b is $2(a + b)$.

p. 40

Definition/Procedure	Example	Reference

Expressions

An expression is a meaningful collection of numbers, variables, and signs of operation.

$3x + y$ is an expression.
$3x = y$ is not an expression.

p. 41

Division

$\dfrac{x}{y}$ means *x divided* by *y* or the *quotient* when *x* is divided by *y*.

n divided by 5 is $\dfrac{n}{5}$.

The sum of *a* and *b*, divided by 3, is $\dfrac{a + b}{3}$.

p. 42

Evaluating Algebraic Expressions

Section 1.5

Evaluating Algebraic Expressions

To evaluate an algebraic expression:
1. Replace each variable or letter with its number value.
2. Do the necessary arithmetic, following the rules for the order of operations.

Evaluate $2x + 3y$ if $x = 5$ and $y = -2$.

$2x + 3y$

$= 2(5) + (3)(-2)$
$= 10 - 6 = 4$

p. 48

Adding and Subtracting Terms

Section 1.6

Term

A term can be written as a number or the product of a number and one or more variables.

p. 60

Combining Like Terms

To combine like terms:
1. Add or subtract the numerical coefficients (the numbers multiplying the variables).
2. Attach the common variables.

$5x + 2x = 7x$

$5 + 2$

$8a - 5a = 3a$

$8 - 5$

p. 62

Continued

Definition/Procedure	Example	Reference
Multiplying and Dividing Terms		Section 1.7
The Product Property of Exponents		
$a^m \cdot a^n = a^{m+n}$	$x^7 \cdot x^3 = x^{7+3} = x^{10}$	p. 68
The Quotient Property of Exponents		
$\dfrac{a^m}{a^n} = a^{m-n}$	$\dfrac{y^7}{y^3} = y^{7-3} = y^4$	p. 69

This summary exercise set is provided to give you practice with each of the objectives of this chapter. Each exercise is keyed to the appropriate chapter section. When you are finished, you can check your answers to the odd-numbered exercises in the back of the text. If you have difficulty with any of these questions, go back and reread the examples from that section. The answers to the even-numbered exercises appear in the *Instructor's Solutions Manual.* Your instructor will give you guidelines on how best to use these exercises in your instructional setting.

1.1 *Identify the property that is illustrated by each statement.*

1. $5 + (7 + 12) = (5 + 7) + 12$

2. $2(8 + 3) = 2 \cdot 8 + 2 \cdot 3$

3. $4 \cdot (5 \cdot 3) = (4 \cdot 5) \cdot 3$

4. $4 \cdot 7 = 7 \cdot 4$

Verify that each statement is true by evaluating each side of the equation separately and comparing the results.

5. $8(5 + 4) = 8 \cdot 5 + 8 \cdot 4$

6. $2(3 + 7) = 2 \cdot 3 + 2 \cdot 7$

7. $(7 + 9) + 4 = 7 + (9 + 4)$

8. $(2 + 3) + 6 = 2 + (3 + 6)$

9. $(8 \cdot 2) \cdot 5 = 8(2 \cdot 5)$

10. $(3 \cdot 7) \cdot 2 = 3 \cdot (7 \cdot 2)$

Use the distributive law to remove the parentheses.

11. $3(7 + 4)$

12. $4(2 + 6)$

13. $\dfrac{1}{2}(5 + 8)$

14. $0.05(1.35 + 8.1)$

1.2 *Add.*

15. $-3 + (-8)$

16. $10 + (-4)$

17. $6 + (-6)$

18. $-16 + (-16)$

19. $-18 + 0$

20. $\dfrac{3}{8} + \left(-\dfrac{11}{8}\right)$

21. $5.7 + (-9.7)$

22. $-18 + 7 + (-3)$

Subtract.

23. $8 - 13$

24. $-7 - 10$

25. $10 - (-7)$

26. $-5 - (-1)$

27. $-9 - (-9)$

28. $0 - (-2)$

29. $-\dfrac{5}{4} - \left(-\dfrac{17}{4}\right)$

30. $7.9 - (-8.1)$

Use a calculator to perform the indicated operations.

31. $489 + (-332)$

32. $1,024 - (-3,206)$

33. $-234 + (-321) - (-459)$

34. $981 - 1,854 - (-321)$

35. $4.56 + (-0.32)$

36. $-32.14 - 2.56$

37. $-3.112 - (-0.1) + 5.06$

38. $10.01 - 12.566 + 2$

39. $13 - (-12.5) + 4\dfrac{1}{4}$

40. $3\dfrac{1}{8} - 6.19 + (-8)$

1.3 *Multiply.*

41. $(10)(-7)$

42. $(-8)(-5)$

43. $(-3)(-15)$

44. $(1)(-15)$

45. $(0)(-8)$

46. $\left(\dfrac{2}{3}\right)\left(-\dfrac{3}{2}\right)$

47. $(-4)\left(\dfrac{3}{8}\right)$

48. $\left(-\dfrac{5}{4}\right)(-1)$

Divide.

49. $\dfrac{80}{16}$

50. $\dfrac{-63}{7}$

51. $\dfrac{-81}{-9}$

52. $\dfrac{0}{-5}$

53. $\dfrac{32}{-8}$

54. $\dfrac{-7}{0}$

Perform the indicated operations.

55. $\dfrac{-8 + 6}{-8 - (-10)}$

56. $\dfrac{-6 - 1}{5 - (-2)}$

57. $\dfrac{25 - 4}{-5 - (-2)}$

58. $\dfrac{3 - (-6)}{-4 + 2}$

1.4 *Write, using symbols.*

59. 5 more than y

60. c decreased by 10

61. The product of 8 and a

62. The quotient when y is divided by 3

63. 5 times the product of m and n

64. The product of a and 5 less than a

65. 3 more than the product of 17 and x

66. The quotient when a plus 2 is divided by a minus 2

Identify which are expressions and which are not.

67. $4(x + 3)$

68. $7 \div \cdot 8$

69. $y + 5 = 9$

70. $11 + 2(3x - 9)$

1.5 *Evaluate each expression.*

71. $18 - 3 \cdot 5$

72. $(18 - 3) \cdot 5$

73. $5 \cdot 4^2$

74. $(5 \cdot 4)^2$

75. $5 \cdot 3^2 - 4$

76. $5(3^2 - 4)$

77. $5(4 - 2)^2$

78. $5 \cdot 4 - 2^2$

79. $(5 \cdot 4 - 2)^2$

80. $3(5 - 2)^2$

81. $3 \cdot 5 - 2^2$

82. $(3 \cdot 5 - 2)^2$

Evaluate each expression if $x = -3$, $y = 6$, $z = -4$, and $w = 2$.

83. $3x + w$

84. $5y - 4z$

85. $x + y - 3z$

86. $5z^2$

87. $3x^2 - 2w^2$

88. $3x^3$

89. $5(x^2 - w^2)$

90. $\dfrac{6z}{2w}$

91. $\dfrac{2x - 4z}{y - z}$

92. $\dfrac{3x - y}{w - x}$

93. $\dfrac{x(y^2 - z^2)}{(y + z)(y - z)}$

94. $\dfrac{y(x - w)^2}{x^2 - 2xw + w^2}$

1.6 *List the terms of each expression.*

95. $4a^3 - 3a^2$

96. $5x^2 - 7x + 3$

Circle like terms.

97. $5m^2, -3m, -4m^2, 5m^3, m^2$

98. $4ab^2, 3b^2, -5a, ab^2, 7a^2, -3ab^2, 4a^2b$

Combine like terms.

99. $5c + 7c$

100. $2x + 5x$

101. $4a - 2a$

102. $6c - 3c$

103. $9xy - 6xy$

104. $5ab^2 + 2ab^2$

105. $7a + 3b + 12a - 2b$

106. $6x - 2x + 5y - 3x$

107. $5x^3 + 17x^2 - 2x^3 - 8x^2$

108. $3a^3 + 5a^2 + 4a - 2a^3 - 3a^2 - a$

109. Subtract $4a^3$ from the sum of $2a^3$ and $12a^3$.

110. Subtract the sum of $3x^2$ and $5x^2$ from $15x^2$.

1.7 *Simplify.*

111. $\dfrac{x^{10}}{x^3}$

112. $\dfrac{a^5}{a^4}$

113. $\dfrac{x^2 \cdot x^3}{x^4}$

114. $\dfrac{m^2 \cdot m^3 \cdot m^4}{m^5}$

115. $\dfrac{18p^7}{9p^5}$

116. $\dfrac{24x^{17}}{8x^{13}}$

117. $\dfrac{30m^7n^5}{6m^2n^3}$

118. $\dfrac{108x^9y^4}{9xy^4}$

119. $\dfrac{48p^5q^3}{6p^3q}$

120. $\dfrac{52a^5b^3c^5}{13a^4c}$

121. $(4x^3)(5x^4)$

122. $(3x)^2(4xy)$

123. $(8x^2y^3)(3x^3y^2)$

124. $(-2x^3y^3)(-5xy)$

125. $(6x^4)(2x^2y)$

Write an algebraic expression to model each application.

126. CONSTRUCTION If x ft are cut off the end of a board that is 23 ft long, how much is left?

127. BUSINESS AND FINANCE Joan has 25 nickels and dimes in her pocket. If x of these are dimes, how many of the coins are nickels?

128. SOCIAL SCIENCE Sam is 5 years older than Angela. If Angela is x years old now, how old is Sam?

129. BUSINESS AND FINANCE Margaret has $5 more than twice as much money as Gerry. Write an expression for the amount of money that Margaret has.

130. GEOMETRY The length of a rectangle is 4 m more than the width. Write an expression for the length of the rectangle.

131. NUMBER PROBLEM A number is 7 less than 6 times the number n. Write an expression for the number.

132. CONSTRUCTION A 25-ft plank is cut into two pieces. Write expressions for the length of each piece.

133. BUSINESS AND FINANCE Bernie has x dimes and q quarters in his pocket. Write an expression for the amount of money that Bernie has in his pocket.

The purpose of this self-test is to help you assess your progress so that you can find concepts that you need to review before the next exam. Allow yourself about an hour to take this test. At the end of that hour, check your answers against those given in the back of this text. If you miss any, go back to the appropriate section to reread the examples until you have mastered that particular concept.

Name _____

Section _____ Date _____

Answers

Evaluate each expression.

1. $-8 + (-5)$

2. $6 + (-9)$

3. $(-9) + (-12)$

4. $-\dfrac{5}{3} + \dfrac{8}{3}$

5. $9 - 15$

6. $-10 - 11$

7. $5 - (-4)$

8. $-7 - (-7)$

9. $(8)(-5)$

10. $(-9)(-7)$

11. $(4.5)(-6)$

12. $(6)(-4)$

13. $\dfrac{-100}{4}$

14. $\dfrac{-36 + 9}{-9}$

15. $\dfrac{(-15)(-3)}{-9}$

16. $\dfrac{9}{0}$

17. $29 - 3 \cdot 4$

18. $4 \cdot 5^2 - 35$

19. $4(2 + 4)^2$

20. $\dfrac{16}{-4} + (-5)$

Simplify each expression.

21. $9a + 4a$

22. $10x + 8y + 9x - 3y$

23. $a^5 \cdot a^9$

24. $2x^3y^2 \cdot 4x^4y$

25. $\dfrac{9x^9}{3x^3}$

26. $\dfrac{20a^3b^5}{5a^2b^2}$

27. $\dfrac{x^{10}x^5}{x^6}$

28. Subtract $9a^2$ from the sum of $12a^2$ and $5a^2$.

Translate each phrase into an algebraic expression.

29. 5 less than a

30. The product of 6 and m

31. 4 times the sum of m and n

32. The quotient when the sum of a and b is divided by 3

1. _____	2. _____
3. _____	4. _____
5. _____	6. _____
7. _____	8. _____
9. _____	10. _____
11. _____	12. _____
13. _____	14. _____
15. _____	16. _____
17. _____	18. _____
19. _____	20. _____
21. _____	22. _____
23. _____	24. _____
25. _____	26. _____
27. _____	28. _____
29. _____	30. _____
31. _____	32. _____

Answers

33. Evaluate $\dfrac{9x^2y}{3z}$ if $x = 2, y = -1,$ and $z = 3.$

Identify the property illustrated by each equation.

34. $6 \cdot 7 = 7 \cdot 6$

35. $2(6 + 7) = 2 \cdot 6 + 2 \cdot 7$

36. $4 + (3 + 7) = (4 + 3) + 7$

Use the distributive property to simplify each expression.

37. $3(5 + 2)$ **38.** $4(5x + 3)$

Determine whether each "collection" is an expression or not.

39. $5x + 6 = 4$ **40.** $4 + (6 + x)$

41. **SOCIAL SCIENCE**   Tom is 8 years younger than twice Moira's age. Let x represent Moira's age and write an expression for Tom's age.

42. **GEOMETRY**   The length of a rectangle is 4 more than twice its width. Write an expression for the length of the rectangle.

34. _____

35. _____

36. _____

37. _____

38. _____

39. _____

40. _____

41. _____

42. _____

Activity 1 ::
An Introduction to Searching

Each activity in this text is designed to either enhance your understanding of the topics of the preceding chapter or provide you with a mathematical extension of those topics, or both. The activities can be undertaken by one student, but they are better suited for a small group project. Occasionally it is only through discussion that different facets of the activity become apparent.

There are many resources available to help you when you have difficulty with your math work. Your instructor can answer many of your questions, but there are other resources to help you learn, as well.

Studying with friends and classmates is a great way to learn math. Your school may have a "math lab" where instructors or peers provide tutoring services. This text provides examples and exercises to help you learn and understand new concepts.

Another place to go for help is the **Internet.** There are many math tutorials on the Web. This activity is designed to introduce you to searching the Web and evaluating what you find there.

If you are new to computers or the Internet, your instructor or a classmate can help you get started. You will need to access the Internet through one of the many **Web browsers** such as Microsoft's Internet Explorer, Mozilla Firefox, Netscape Navigator, AOL's browser, or Opera.

First, you need to connect to the Internet. Then, you need to access a page containing a **search engine.** Many *default* home pages contain a *search* field. If yours does not, several of the more popular search engines are at these sites:

http://www.ask.com
http://www.dogpile.com
http://www.google.com
http://www.yahoo.com

Access one of these search engines or use one from another site as you work through this activity.

1. Type the word *integers* in the search field. You should see a long list of websites related to your search.

2. Look at the page titles and descriptions. Find a page that has an introduction to integers and click on that link.

3. Write two or three sentences describing the layout of the Web page. Is it "user friendly"? Are the topics presented in an easy-to-find and useful way? Are the colors and images helpful?

4. Choose a topic such as integer multiplication or even some math game. Describe the instruction that the website has for the topic. In what format is the information given? Is there an interactive component to the instruction?

5. Does the website offer free tutoring services? If so, try to get some help with a homework problem. Briefly evaluate the tutoring services.

6. Chapter 4 in this text introduces you to *systems of equations*. Are there activities or links on the website related to systems of equations? Do they appear to be helpful to a student having difficulty with this topic?

7. Return to your search engine. Find a second math Web page by typing "systems of equations" (including the quotation marks) into the search field. Choose a page that offers instruction, tutoring, and activities related to systems of equations. Save the link for this page—this is called a bookmark, favorite, or preference, depending on your browser. If you find yourself struggling with systems of equations in Chapter 4, try using this page to get some additional help.

CHAPTER

2

> chapter 2 > Make the Connection

INTRODUCTION

Every year, millions of people travel to other countries for business and pleasure. When traveling to another country, you need to consider many things, such as passports and visas, immunizations, local sights, restaurants and hotels, and language.

Another consideration when traveling internationally is currency. Nearly every country has its own money. For example, the Japanese currency is the yen (¥), Europeans use the euro (€), and Canadians use Canadian dollars (CAN$), whereas the United States of America uses the US$.

When visiting another country, you need to acquire the local currency. Many sources publish exchange rates for currency on a daily basis. For instance, on May 26, 2009, *Yahoo!Finance* listed the US$ to CAN$ exchange rate as 1.1155. We can use this to construct an equation to determine the amount of Canadian dollars that one receives for U.S. dollars.

$$C = 1.1155U$$

in which U represents the amount of US$ to be exchanged and C represents the amount of CAN$ to be received.

The equation is an ancient tool used to solve problems and describe numerical relationships accurately and clearly. In this chapter, you will learn methods to solve linear equations and practice writing equations to model real-world problems.

Equations and Inequalities

CHAPTER 2 OUTLINE

87

Name _____

Section _____ Date _____

Answers

1. _____

2. _____

3. _____

4. _____

5. _____

6. _____

7. _____

8. _____

9. _____

10. _____

11. _____

12. _____

This prerequisite test provides some exercises requiring skills that you will need to be successful in the coming chapter. The answers for these exercises can be found in the back of this text. This prerequisite test can help you identify topics that you will need to review before beginning the chapter.

Use the distributive property to remove the parentheses in each expression.

1. $4(2x + 3)$ **2.** $-2(3x - 8)$

Find the reciprocal of each number.

3. -10 **4.** $-\dfrac{3}{4}$

Evaluate as indicated.

5. $\left(-\dfrac{3}{5}\right) \times \left(-\dfrac{5}{3}\right)$ **6.** $(-6)\left(-\dfrac{1}{6}\right)$

7. -7^2 **8.** $(-7)^2$

Simplify each expression.

9. $3x^2 - 5x + x^2 + 2x$ **10.** $8x + 2y - 7x$

11. **BUSINESS AND FINANCE** An auto body shop sells 12 sets of windshield wipers at $19.95 each. How much revenue did it earn from the sales of wiper blades?

12. **BUSINESS AND FINANCE** An auto body shop charges $19.95 for a set of windshield wipers after applying a 25% markup to the wholesale price. What was the wholesale price of the wiper blades?

2.1 Solving Equations by the Addition Property

< 2.1 Objectives >

1 > Determine whether a given number is a solution for an equation

2 > Identify expressions and equations

3 > Use the addition property to solve an equation

4 > Use the distributive property in solving equations

> **Tips for Student Success**

Don't procrastinate!

1. Do your math homework while you are still fresh. If you wait until too late at night, your tired mind will have much more difficulty understanding the concepts.

2. Do your homework the day it is assigned. The more recent the explanation, the easier it is to recall.

3. When you finish your homework, try reading through the next section one time. This will give you a sense of direction when you next hear the material. This works in a lecture or lab setting.

Remember that, in a typical math class, you are expected to do two or three hours of homework for each weekly class hour. This means two or three hours per night. Schedule the time and stick to your schedule.

In this chapter we work with one of the most important tools of mathematics, the equation. The ability to recognize and solve various types of equations is probably the most useful algebraic skill you will learn. We will continue to build upon the methods of this chapter throughout the text. To begin, we define the word *equation*.

Definition

Equation

An **equation** is a mathematical statement that two expressions are equal.

Some examples are $3 + 4 = 7$, $x + 3 = 5$, and $P = 2L + 2W$.

As you can see, an equal sign ($=$) separates the two expressions. These expressions are usually called the *left side* and the *right side* of the equation.

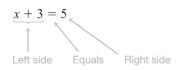

$$x + 3 = 5$$

Left side   Equals   Right side

Just as the balance scale may be in balance or out of balance, an equation may be either true or false. For instance, $3 + 4 = 7$ is true because both sides name the same number. What about an equation such as $x + 3 = 5$ that has a letter or variable on one

NOTE

An equation such as
$x + 3 = 5$
is called a **conditional equation** because it can be either true or false, depending on the value given to the variable.

side? Any number can replace x in the equation. However, only one number will make this equation a true statement.

$x + 3 = 5$

$$\text{If } x = 2 \begin{cases} 1 & (1) + 3 = 5 \text{ is false} \\ 2 & (2) + 3 = 5 \text{ is true} \\ 3 & (3) + 3 = 5 \text{ is false} \end{cases}$$

The number 2 is called the **solution** (or *root*) of the equation $x + 3 = 5$ because substituting 2 for x gives a true statement.

Definition

Solution

A **solution** for an equation is any value for the variable that makes the equation a true statement.

Example 1   Verifying a Solution

< Objective 1 >

RECALL

The rules for order of operations require that we multiply first; then add or subtract.

(a) Is 3 a solution for the equation $2x + 4 = 10$?

To find out, replace x with 3 and evaluate $2x + 4$ on the left.

Left side		Right side
$2(3) + 4$	$\overset{?}{=}$	10
$6 + 4$	$\overset{?}{=}$	10
10	$=$	10

Because $10 = 10$ is a true statement, 3 is a solution of the equation.

(b) Is 5 a solution of the equation $3x - 2 = 2x + 1$?

To find out, replace x with 5 and evaluate each side separately.

Left side		Right side
$3(5) - 2$	$\overset{?}{=}$	$2(5) + 1$
$15 - 2$	$\overset{?}{=}$	$10 + 1$
13	$\neq$	11

Because the two sides do not name the same number, we do not have a true statement, and 5 is not a solution.

Check Yourself 1

For the equation

$2x - 1 = x + 5$

(a) Is 4 a solution?   (b) Is 6 a solution?

$2(4) - 1 = 4 + 5$
$8 - 1 = 9$
$7 = 9$

$2(6) - 1 = 6 + 5$
$12 - 1 \quad 11$
$11 = 11$

NOTE

$x^2 = 9$ is an example of a **quadratic equation.** We consider such equations in Chapter 4 and then again in Chapter 10.

You may be wondering whether an equation can have more than one solution. It certainly can. For instance,

$x^2 = 9$

has two solutions. They are 3 and -3 because

$3^2 = 9$ and $(-3)^2 = 9$

In this chapter, however, we work with *linear equations in one variable*. These are equations that can be put into the form

$$ax + b = 0$$

in which the variable is x, a and b are any numbers, and a is not equal to 0. In a linear equation, the variable can appear only to the first power. No other power (x^2, x^3, and so on) can appear. Linear equations are also called **first-degree equations.** The degree of an equation in one variable is the highest power to which the variable appears.

Property

Linear Equations

Linear equations in one variable are equations that can be written in the form

$$ax + b = 0 \qquad a \neq 0$$

Every such equation has exactly one solution.

| **Example** 2 | Identifying Expressions and Equations |

< Objective 2 >

Label each statement as an expression, a linear equation, or an equation that is not linear.

(a) $4x + 5$ is an expression.

(b) $2x + 8 = 0$ is a linear equation.

(c) $3x^2 - 9 = 0$ is an equation that is not linear.

(d) $5x = 15$ is a linear equation.

(e) $5 - \dfrac{7}{x} = 4x$ is an equation that is not linear.

NOTE

In part (e) we see that an equation that includes a variable in a denominator is not a linear equation.

Check Yourself 2

Label each as an expression, a linear equation, or an equation that is not linear.

(a) $2x^2 = 8$ **(b)** $2x - 3 = 0$ **(c)** $5x - 10$

(d) $2x + 1 = 7$ **(e)** $\dfrac{3}{x} - 4 = x$

It is not difficult to find the solution for an equation such as $x + 3 = 8$ by guessing the answer to the question "What plus 3 is 8?" Here the answer to the question is 5, which is also the solution for the equation. But for more complicated equations we need something more than guesswork. A better method is to transform the given equation to an *equivalent equation* whose solution can be found by inspection.

Definition

Equivalent Equations

Equations that have exactly the same solution(s) are called **equivalent equations.**

NOTE

In some cases we write the equation in the form

$$\square = x$$

The number is the solution when the equation has the variable isolated on either side.

These are equivalent equations.

$$2x + 3 = 5 \qquad 2x = 2 \qquad \text{and} \qquad x = 1$$

They all have the same solution, 1. We say that a linear equation is *solved* when it is transformed to an equivalent equation of the form

$$x = \square$$

↑ The variable is alone on the left side.   The right side is some number, the solution.

The *addition property of equality* is the first property you need to transform an equation to an equivalent form.

The Addition Property of Equality

If $a = b$

then $a + c = b + c$

In words, adding the same quantity to both sides of an equation gives an equivalent equation.

RECALL

An equation is a statement that the two sides are equal. Adding the same quantity to both sides does not change the equality or "balance."

Recall that we said that a true equation was like a scale in balance.

The addition property is equivalent to adding the same weight to both sides of the scale. It remains in balance.

NOTE

This scale represents
$a + c = b + c$

 Example 3   **Using the Addition Property to Solve an Equation**

< Objective 3 >

Solve.

$x - 3 = 9$

Remember that our goal is to isolate x on one side of the equation. Because 3 is being subtracted from x, we can add 3 to remove it. We must use the addition property to add 3 to both sides of the equation.

NOTE

To check, replace x with 12 in the original equation:

$x - 3 \overset{?}{=} 9$

$(12) - 3 \overset{?}{=} 9$

$9 = 9$

Because we have a true statement, 12 is the solution.

$$
\begin{array}{rl}
x - 3 = & 9 \\
+3 \quad & +3 \\
\hline
x \quad = & 12
\end{array}
$$

Adding 3 "undoes" the subtraction and leaves x alone on the left.

Because 12 is the solution for the equivalent equation $x = 12$, it is the solution for our original equation.

Check Yourself 3

Solve and check.

$x - 5 = 4$

The addition property also allows us to add a negative number to both sides of an equation. This is really the same as subtracting the same quantity from both sides.

Example 4   Using the Addition Property to Solve an Equation

Solve.

$x + 5 = 9$

In this case, 5 is *added* to x on the left. We can use the addition property to add a -5 to both sides. Because $5 + (-5) = 0$, this "undoes" the addition and leaves the variable x alone on one side of the equation.

$$
\begin{array}{rcr}
x + 5 = & & 9 \\
-5 & & -5 \\
\hline
x & = & 4
\end{array}
$$

The solution is 4. To check, replace x with 4:

$(4) + 5 = 9$ (True)

RECALL

Earlier, we stated that we could write an equation in the equivalent forms $x = \square$ or $\square = x$, in which $\square$ represents some number. Suppose we have an equation like

$12 = x + 7$

Adding -7 isolates x *on the right:*

$$
\begin{array}{r}
12 = x + 7 \\
-7 \quad\quad -7 \\
\hline
5 = x
\end{array}
$$

The solution is 5.

✓ **Check Yourself 4**

Solve and check.

$x + 6 = 13$

What if the equation has a variable term on both sides? We have to use the addition property to add or subtract a term involving the variable to get the desired result.

Example 5   Using the Addition Property to Solve an Equation

Solve.

$5x = 4x + 7$

We start by subtracting $4x$ from both sides of the equation. Do you see why? Remember that an equation is solved when we have an equivalent equation of the form $x = \square$.

$$
\begin{array}{rcr}
5x = & 4x + 7 \\
-4x & -4x \\
\hline
x = & 7
\end{array}
$$

Subtracting $4x$ from both sides *removes* $4x$ from the right.

To check: Because 7 is a solution for the equivalent equation $x = 7$, it should be a solution for the original equation. To find out, replace x with 7.

$5(7) \overset{?}{=} 4(7) + 7$

$35 \overset{?}{=} 28 + 7$

$35 = 35$ (True)

RECALL

Subtracting $4x$ is the same as adding $-4x$.

✓ **Check Yourself 5**

Solve and check.

$7x = 6x + 3$

You may have to apply the addition property more than once to solve an equation. Look at Example 6.

Example 6   Using the Addition Property to Solve an Equation

Solve.

$7x - 8 = 6x$

NOTE

We could add 8 to both sides, and then subtract $6x$. However, we find it easiest to bring the variable terms to one side first, and then work with the constant (or numerical) terms.

We want all variables on *one* side of the equation. If we choose the left, we subtract $6x$ from both sides of the equation. This removes the $6x$ from the right:

$$
\begin{array}{rcr}
7x - 8 = & & 6x \\
-6x & & -6x \\
\hline
x - 8 = & & 0
\end{array}
$$

We want the variable alone, so we add 8 to both sides. This isolates x on the left.

$$
\begin{array}{rcr}
x - 8 = & & 0 \\
+ 8 & & +8 \\
\hline
x \quad\;\; = & & 8
\end{array}
$$

The solution is 8. We leave it to you to check this result.

 Check Yourself 6

Solve and check.

$9x + 3 = 8x$

Often an equation has more than one variable term *and* more than one number. You have to apply the addition property twice to solve these equations.

 Example 7   **Using the Addition Property to Solve an Equation**

Solve.

$5x - 7 = 4x + 3$

We would like the variable terms on the left, so we start by subtracting $4x$ from both sides of the equation:

$$
\begin{array}{rcr}
5x - 7 = & & 4x + 3 \\
-4x & & -4x \\
\hline
x - 7 = & & 3
\end{array}
$$

NOTE

You could just as easily have added 7 to both sides and *then* subtracted $4x$. The result would be the same. In fact, some students prefer to combine the two steps.

Now, to isolate the variable, we add 7 to both sides.

$$
\begin{array}{rcr}
x - 7 = & & 3 \\
+ 7 & & +7 \\
\hline
x \quad\;\; = & & 10
\end{array}
$$

The solution is 10. To check, replace x with 10 in the original equation:

$$5(10) - 7 \stackrel{?}{=} 4(10) + 3$$
$$43 = 43 \quad\text{(True)}$$

RECALL

Combining like terms is one of the steps we take when simplifying an expression.

 Check Yourself 7

Solve and check.

(a) $4x - 5 = 3x + 2$ (b) $6x + 2 = 5x - 4$

In solving an equation, you should always simplify each side as much as possible before using the addition property.

 Example 8   **Simplifying an Equation**

Solve $5 + 8x - 2 = 2x - 3 + 5x$.

We begin by identifying like terms on each side of the equation.

Like terms   Like terms

$$5 + 8x - 2 = 2x - 3 + 5x$$

Because like terms appear on both sides of the equation, we start by combining the numbers on the left (5 and -2). Then we combine the like terms ($2x$ and $5x$) on the right. We have

$3 + 8x = 7x - 3$

Now we can apply the addition property, as before.

$$
\begin{array}{rcl}
3 + 8x = & 7x - 3 & \\
\underline{-7x = -7x} & & \text{Subtract } 7x. \\
3 + x = & -3 & \\
\underline{-3 -3} & & \text{Subtract 3 to isolate } x. \\
x = & -6 &
\end{array}
$$

The solution is -6. To check, always return to the original equation. That catches any possible errors in simplifying. Replacing x with -6 gives

$$5 + 8(-6) - 2 \overset{?}{=} 2(-6) - 3 + 5(-6)$$
$$5 - 48 - 2 \overset{?}{=} -12 - 3 - 30$$
$$-45 = -45 \quad \text{(True)}$$

Check Yourself 8

Solve and check.

(a) $3 + 6x + 4 = 8x - 3 - 3x$ (b) $5x + 21 + 3x = 20 + 7x - 2$

We may have to apply some of the properties discussed in Section 1.1 in solving equations. Example 9 illustrates the use of the distributive property to clear an equation of parentheses.

Example 9	Using the Distributive Property and Solving Equations

< Objective 4 >

NOTE

$2(3x + 4)$

$= 2(3x) + 2(4)$

$= 6x + 8$

Solve.

$2(3x + 4) = 5x - 6$

Applying the distributive property on the left gives

$6x + 8 = 5x - 6$

We can then proceed as before:

$$
\begin{array}{rcl}
6x + 8 = & 5x - 6 & \\
\underline{-5x -5x} & & \text{Subtract } 5x. \\
x + 8 = & -6 & \\
\underline{-8 -8} & & \text{Subtract 8.} \\
x = & -14 &
\end{array}
$$

The solution is -14. We leave it to you to check this result.

Remember: Always return to the original equation to check.

Check Yourself 9

Solve and check each equation.

(a) $4(5x - 2) = 19x + 4$ (b) $3(5x + 1) = 2(7x - 3) - 4$

Given an expression such as

$-2(x - 5)$

the distributive property can be used to create the equivalent expression

$-2x + 10$

The distribution of a negative number is shown in Example 10.

Example 10 | **Distributing a Negative Number**

Solve each equation.

(a) $-2(x - 5) = -3x + 2$

$$-2x + 10 = -3x + 2$$ Distribute -2 to remove the parentheses.

$$\underline{+3x \qquad\qquad +3x}$$ Add $3x$ to bring the variable terms to the same side.

$$x + 10 = \qquad 2$$

$$\underline{\quad -10 = \quad -10}$$ Subtract 10 to isolate the variable.

$$x \quad = \quad -8$$

(b) $-3(3x + 5) = -5(2x - 2)$

$$-9x - 15 = -5(2x - 2)$$ Distribute -3.

$$-9x - 15 = -10x + 10$$ Distribute -5.

$$\underline{+10x \qquad\qquad +10x}$$ Add $10x$.

$$x - 15 = \qquad 10$$

$$\underline{\quad +15 \qquad\qquad +15}$$ Add 15.

$$x \quad = \quad 25$$ The solution is 25.

RECALL

Return to the original equation to check your solution.

Check

$$-3[3(25) + 5] \overset{?}{=} -5[2(25) - 2]$$

$$-3(75 + 5) \overset{?}{=} -5(50 - 2)$$ Follow the order of operations.

$$-3(80) \overset{?}{=} -5(48)$$

$$-240 = -240$$ True

 Check Yourself 10

Solve each equation.

(a) $-2(x - 3) = -x + 5$ **(b)** $-4(2x - 1) = -3(3x + 2)$

When parentheses are preceded only by a negative, or by the minus sign, we say that we have a silent -1. Example 11 illustrates this case.

    Example 11 | **Distributing a Silent −1**

Solve.

$$-(2x + 3) = -3x + 7$$

$$-1(2x + 3) = -3x + 7$$

$$(-1)(2x) + (-1)(3) = -3x + 7$$ Distribute the -1.

$$-2x - 3 = -3x + 7$$

$$\underline{+3x \qquad\qquad +3x}$$ Add $3x$.

$$x - 3 = \qquad 7$$

$$\underline{\quad +3 \qquad\qquad +3}$$ Add 3.

$$x \quad = \quad 10$$

 Check Yourself 11

Solve and check.

$-(3x + 2) = -2x - 6$

Of course, there are many applications that require us to use the addition property to solve an equation. Consider the consumer application in the next example.

| Example 12 | A Consumer Application |

NOTE

Applications should always be answered with a full sentence.

An appliance store is having a sale on washers and dryers. They are charging $999 for a washer and dryer combination. If the washer sells for $649, how much is a customer paying for the dryer as part of the combination?

Let d be the cost of the dryer and solve the equation $d + 649 = 999$ to answer the question.

$$
\begin{array}{rl}
d + 649 = & 999 \\
-649 & -649 \\
\hline
d \quad\quad = & 350
\end{array}
$$

Subtract 649 from both sides.

The dryer adds $350 to the price.

Check Yourself 12

Of 18,540 votes cast in the school board election, 11,320 went to Carla. How many votes did her opponent Marco receive? Who won the election?

Let m be the number of votes Marco received and solve the equation $11{,}320 + m = 18{,}540$ in order to answer the questions.

Check Yourself ANSWERS

1. (a) 4 is not a solution; (b) 6 is a solution
2. (a) An equation that is not linear; (b) linear equation; (c) expression; (d) linear equation; (e) an equation that is not linear   3. 9   4. 7
5. 3   6. −3   7. (a) 7; (b) −6   8. (a) −10; (b) −3   9. (a) 12; (b) −13   10. (a) 1; (b) −10   11. 4
12. Marco received 7,220 votes; Carla won the election.

Reading Your Text

The following fill-in-the-blank exercises are designed to ensure that you understand some of the key vocabulary used in this section.

SECTION 2.1

(a) An _____ is a mathematical statement that two expressions are equal.

(b) A _____ for an equation is any value for the variable that makes the equation a true statement.

(c) Linear equations in one variable have exactly _____ solution.

(d) Equivalent equations have exactly the same _____.

Name _____

Section _____ Date _____

Answers

1. _____ 2. _____
3. _____ 4. _____
5. _____ 6. _____
7. _____ 8. _____
9. _____ 10. _____
11. _____ 12. _____
13. _____ 14. _____
15. _____ 16. _____
17. _____ 18. _____
19. _____ 20. _____
21. _____ 22. _____
23. _____
24. _____
25. _____
26. _____
27. _____
28. _____
29. _____
30. _____
31. _____ 32. _____
33. _____ 34. _____
35. _____ 36. _____

Basic Skills | Challenge Yourself | Calculator/Computer | Career Applications | Above and Beyond

< Objective 1 >

Is the number shown in parentheses a solution for the given equation?

1. $x + 7 = 12$ (5)

2. $x + 2 = 11$ (8)

3. $x - 15 = 6$ (−21)

4. $x - 11 = 5$ (16)

5. $5 - x = 2$ (4)

6. $10 - x = 7$ (3)

7. $8 - x = 5$ (−3)

8. $5 - x = 6$ (−3)

9. $3x + 4 = 13$ (8)

10. $5x + 6 = 31$ (5)

11. $4x - 5 = 7$ (2)

12. $4x - 3 = 9$ (3)

13. $7 - 3x = 10$ (−1)

14. $4 - 5x = 9$ (−2)

15. $4x - 5 = 2x + 3$ (4)

16. $5x + 4 = 2x + 10$ (4)

17. $x + 3 + 2x = 5 + x + 8$ (5)   > Videos

18. $5x - 3 + 2x = 3 + x - 12$ (−2)

19. $\dfrac{2}{3}x = 9$ (15)

20. $\dfrac{3}{5}x = 24$ (40)

21. $\dfrac{3}{5}x + 5 = 11$ (10)

22. $\dfrac{2}{3}x + 8 = -12$ (−6)

< Objective 2 >

Label each as an expression, a linear equation, or an equation that is not linear.

23. $2x + 1 = 9$

24. $7x + 14$

25. $2x - 8$ > Videos

26. $5x - 3 = 12$

27. $2x^2 - 8 = 0$

28. $x + 5 = 13$

29. $2x - 8 = 3$

30. $\dfrac{2}{x} - 4 = 3x$

< Objectives 3–4 >

Solve and check each equation.

31. $x + 9 = 11$

32. $x - 4 = 6$

33. $x - 5 = -9$

34. $x + 11 = 15$

35. $x - 8 = -10$

36. $x + 5 = 2$

37. $x + 4 = -3$ **38.** $x - 6 = -5$

39. $17 = x + 11$ **40.** $x + 7 = 0$

41. $4x = 3x + 4$ **42.** $7x = 6x - 8$

43. $9x = 8x - 12$ **44.** $9x = 8x + 5$

45. $6x + 3 = 5x$ **46.** $12x - 6 = 11x$

47. $7x - 5 = 6x$ **48.** $9x - 7 = 8x$

49. $2x + 3 = x + 5$ > Videos   **50.** $5x - 6 = 4x + 2$

| Basic Skills | **Challenge Yourself** | Calculator/Computer | Career Applications | Above and Beyond |

51. CRAFTS Jeremiah had found 50 bones for a Halloween costume. In order to complete his 62-bone costume, how many more does he need?
  Let b be the number of bones he needs and use the equation $b + 50 = 62$ to solve the problem.

52. BUSINESS AND FINANCE Four hundred tickets were sold to the opening of an art exhibit. General admission tickets cost $5.50, whereas students were required to pay only $4.50 for tickets. If total ticket sales were $1,950, how many of each type of ticket were sold?
  Let x be the number of general admission tickets sold and $400 - x$ be the number of student tickets sold. Use the equation $5.5x + 4.5(400 - x) = 1,950$ to solve the problem.

53. BUSINESS AND FINANCE A shop pays $2.25 for each copy of a magazine and sells the magazines for $3.25 each. If the fixed costs associated with the sale of these magazines are $50 per month, how many must the shop sell in order to realize $175 in profit from the magazines?
  Let m be the number of magazines they must sell and use the equation $3.25m - 2.25m - 50 = 175$ to solve the problem.

54. NUMBER PROBLEM The sum of a number and 15 is 22. Find the number.
  Let x be the number and solve the equation $x + 15 = 22$ to find the number.

55. Which equation is equivalent to $5x - 7 = 4x - 12$?

(a) $9x = 19$ (b) $x + 7 = -12$
(c) $x = -18$ (d) $4x - 5 = 8$

56. Which equation is equivalent to $12x - 6 = 8x + 14$?

(a) $4x - 6 = 14$ (b) $x = 20$
(c) $20x = 20$ (d) $4x = 8$

Answers

37. _____ 38. _____

39. _____ 40. _____

41. _____

42. _____

43. _____

44. _____

45. _____

46. _____

47. _____

48. _____

49. _____

50. _____

51. _____

52. _____

53. _____

54. _____

55. _____

56. _____

Answers

57. _____

58. _____

59. _____

60. _____

61. _____

62. _____

63. _____

64. _____

65. _____

66. _____

67. _____

68. _____

69. _____

70. _____

71. _____

72. _____

73. _____

74. _____

75. _____

76. _____

77. _____

78. _____

57. Which equation is equivalent to $7x + 5 = 12x - 10$?

   (a) $5x = -15$ **(b)** $7x - 5 = 12x$

   (c) $-5 = 5x$ **(d)** $7x + 15 = 12x$

58. Which equation is equivalent to $8x + 5 = 9x - 4$?

   (a) $17x = -9$ **(b)** $x = -9$

   (c) $8x + 9 = 9x$ **(d)** $9 = 17x$

Determine whether each statement is **true** *or* **false.**

59. Every linear equation with one variable has no more than one solution.

60. Isolating the variable on the right side of an equation results in a negative solution.

Solve and check each equation.

61. $4x - \dfrac{3}{5} = 3x + \dfrac{1}{10}$ **62.** $5\left(x - \dfrac{3}{4}\right) = 4x + \dfrac{3}{8}$

63. $\dfrac{7}{8}(x - 2) = \dfrac{3}{4} - \dfrac{1}{8}x$ ▷ Videos   **64.** $\dfrac{5}{6}(3x - 2) = \dfrac{3}{2}(x + 1)$

65. $3x - 0.54 = 2(x - 0.15)$ **66.** $7x + 0.125 = 6x - 0.289$

67. $6x + 3(x - 0.2789) = 4(2x + 0.3912)$

68. $9x - 2(3x - 0.124) = 2x + 0.965$

69. $5x - 7 + 6x - 9 - x = 2x - 8 + 7x$

70. $5x + 8 + 3x - x + 5 = 6x - 3$

71. $5x - (0.345 - x) = 5x + 0.8713$ **72.** $-3(0.234 - x) = 2(x + 0.974)$

73. $3(7x + 2) = 5(4x + 1) + 17$ **74.** $5(5x + 3) = 3(8x - 2) + 4$

                          ▷ Videos

75. $\dfrac{5}{4}x - 1 = \dfrac{1}{4}x + 7$ **76.** $\dfrac{7x}{5} + 3 = \dfrac{2x}{5} - 8$

77. $\dfrac{9x}{2} - \dfrac{3}{4} = \dfrac{7x}{2} + \dfrac{5}{4}$ **78.** $\dfrac{11}{3}x + \dfrac{1}{6} = \dfrac{8}{3}x + \dfrac{19}{6}$

Basic Skills | Challenge Yourself | Calculator/Computer | Career Applications | **Above and Beyond**

An algebraic equation is a complete sentence. It has a subject and a predicate. For example, the equation $x + 2 = 5$ can be written in English as "two more than a number is five," or "a number added to two is five."

Write an English version of each equation. Be sure that you write complete sentences and that your sentences express the same idea as the equations. Exchange sentences with another student and see whether each other's sentences result in the same equation.

79. $2x - 5 = x + 1$

80. $2(x + 2) = 14$

81. $n + 5 = \frac{n}{2} - 6$

82. $7 - 3a = 5 + a$

83. Complete the sentence in your own words. "The difference between $3(x - 1) + 4 - 2x$ and $3(x - 1) + 4 = 2x$ is. . . ."

84. "Surprising Results!" Work with other students to try this experiment. Each person should do the six steps mentally, not telling anyone else what their calculations are:

(a) Think of a number.
(b) Add 7.
(c) Multiply by 3.
(d) Add 3 more than the original number.
(e) Divide by 4.
(f) Subtract the original number.

What number do you end up with? Compare your answer with everyone else's. Does everyone have the same answer? Make sure that everyone followed the directions accurately. How do you explain the results? Algebra makes the explanation clear. Work together to do the problem again, using a variable for the number. Make up another series of computations that yields "surprising results."

Answers

79. ______

80. ______

81. ______

82. ______

83. ______

84. ______

Answers

1. Yes **3.** No **5.** No **7.** No **9.** No **11.** No **13.** Yes **15.** Yes **17.** Yes **19.** No **21.** Yes **23.** Linear equation **25.** Expression **27.** An equation that is not linear **29.** Linear equation **31.** 2 **33.** −4 **35.** −2 **37.** −7 **39.** 6 **41.** 4 **43.** −12 **45.** −3 **47.** 5 **49.** 2 **51.** 12 **53.** 225 **55.** (b) **57.** (d) **59.** True **61.** $\frac{7}{10}$ **63.** $\frac{5}{2}$ **65.** 0.24 **67.** 2.4015 **69.** 8 **71.** 1.2163 **73.** 16 **75.** 8 **77.** 2 **79.** Above and Beyond **81.** Above and Beyond **83.** Above and Beyond

Solving Equations by the Multiplication Property

< 2.2 Objectives >

1 > Use the multiplication property to solve equations

2 > Solve an application involving the multiplication property

Consider a different type of equation. For instance, what if we want to solve the equation

$6x = 18$

The addition property does not help, so we need a second property for solving such equations.

Property

| The Multiplication Property of Equality | If $a = b$ then $ac = bc$ with $c \neq 0$

 In words, multiplying both sides of an equation by the same nonzero number produces an equivalent equation. |

Again, we return to the image of the balance scale. We start with the assumption that a and b have the same weight.

RECALL

As long as you do the *same* thing to *both* sides of the equation, the "balance" is maintained.

The multiplication property tells us that the scale will be in balance as long as we have the same number of "a weights" as we have of "b weights."

NOTE

The scale represents the equation $5a = 5b$.

We work through some examples, using this second rule.

| Example 1 | Solving Equations Using the Multiplication Property |

< Objective 1 >

Solve.

$6x = 18$

NOTE

$$\frac{1}{6}(6x) = \left(\frac{1}{6} \cdot 6\right)x$$

$$= 1 \cdot x \text{ or } x$$

We now have x alone on the left, which was our goal.

Here the variable x is multiplied by 6. So we apply the multiplication property and multiply both sides by $\frac{1}{6}$. Keep in mind that we want an equation of the form

$$x = \square$$

$$\frac{1}{6}(6x) = \frac{1}{6}(18)$$

We can now simplify.

$$1 \cdot x = 3 \qquad \text{or} \qquad x = 3$$

The solution is 3. To check, replace x with 3:

$$6(3) \overset{?}{=} 18$$

$$18 = 18 \qquad \text{(True)}$$

  Check Yourself 1

Solve and check.

$8x = 32$

In Example 1, we solved the equation by multiplying both sides by the reciprocal of the coefficient of the variable.

Example 2 illustrates a slightly different approach to solving an equation by using the multiplication property.

| Example 2 | Solving Equations Using the Multiplication Property |

Solve.

$$5x = -35$$

NOTE

Because division is defined in terms of multiplication, we can also divide both sides of an equation by the same nonzero number.

The variable x is multiplied by 5. We *divide* both sides by 5 to "undo" that multiplication:

$$\frac{5x}{5} = \frac{-35}{5} \qquad \text{This is the same as multiplying by } \frac{1}{5}.$$

$$x = -7 \qquad \begin{array}{l}\text{Note that the right side} \\ \text{simplifies to } -7. \text{ Be careful} \\ \text{with the rules for signs.}\end{array}$$

We leave it to you to check the solution.

  Check Yourself 2

Solve and check.

$7x = -42$

| Example 3 | Equations with Negative Coefficients |

Solve.

$$-9x = 54$$

RECALL

Dividing by -9 and multiplying by $-\frac{1}{9}$ produce the same result—they are the same operation.

In this case, x is multiplied by -9, so we divide both sides by -9 to isolate x on the left:

$$\frac{-9x}{-9} = \frac{54}{-9}$$

$$x = -6$$

The solution is -6. To check:

$$(-9)(-6) \overset{?}{=} 54$$

$$54 = 54 \quad \text{(True)}$$

 Check Yourself 3

Solve and check.

$$-10x = -60$$

Example 4 illustrates the use of the multiplication property when fractions appear in an equation.

Example 4   **Solving Equations That Contain Fractions**

RECALL

$$\frac{x}{3} = \frac{1}{3}x$$

(a) Solve.

$$\frac{x}{3} = 6$$

Here x is *divided* by 3. We use multiplication to isolate x.

$$3\left(\frac{x}{3}\right) = 3 \cdot 6 \qquad \text{This leaves } x \text{ alone on the left because}$$

$$x = 18 \qquad 3\left(\frac{x}{3}\right) = \frac{3}{1} \cdot \frac{x}{3} = \frac{x}{1} = x$$

To check:

$$\left(\frac{18}{3}\right) \overset{?}{=} 6$$

$$6 = 6 \quad \text{(True)}$$

RECALL

$$\frac{x}{5} = \frac{1}{5}x$$

(b) Solve.

$$\frac{x}{5} = -9$$

$$5\left(\frac{x}{5}\right) = 5(-9) \qquad \text{Because } x \text{ is divided by 5, multiply both sides by 5.}$$

$$x = -45$$

The solution is -45. To check, we replace x with -45:

$$\left(\frac{-45}{5}\right) \overset{?}{=} -9$$

$$-9 = -9 \quad \text{(True)}$$

The solution is verified.

 Check Yourself 4

Solve and check.

(a) $\dfrac{x}{7} = 3$ **(b)** $\dfrac{x}{4} = -8$

When the variable is multiplied by a fraction that has a numerator other than 1, there are two approaches to finding the solution.

Example 5 | Solving Equations Using Reciprocals

Solve.

$$\frac{3}{5}x = 9$$

One approach is to multiply by 5 as the first step.

$$5\left(\frac{3}{5}x\right) = 5 \cdot 9$$

$$3x = 45$$

Now we divide by 3.

$$\frac{3x}{3} = \frac{45}{3}$$

$$x = 15$$

To check:

$$\frac{3}{5}(15) \stackrel{?}{=} 9$$

$$9 = 9 \qquad \text{(True)}$$

A second approach combines the multiplication and division steps and is generally more efficient. We multiply by $\frac{5}{3}$.

$$\frac{5}{3}\left(\frac{3}{5}x\right) = \frac{5}{3} \cdot 9$$

$$x = \frac{5}{3} \cdot \frac{9}{1} = 15$$

So $x = 15$, as before.

RECALL

$\frac{5}{3}$ is the *reciprocal* of $\frac{3}{5}$, and the product of a number and its reciprocal is just 1! So

$$\left(\frac{5}{3}\right)\left(\frac{3}{5}\right) = 1$$

Check Yourself 5

Solve and check.

$$\frac{2}{3}x = 18$$

You may have to simplify an equation before applying the methods of this section. Example 6 illustrates this procedure.

Example 6 | Simplifying an Equation

Solve and check.

$$3x + 5x = 40$$

Using the distributive property, we can combine the like terms on the left to write

$$8x = 40$$

We can now proceed as before.

$$\frac{8x}{8} = \frac{40}{8} \qquad \text{Divide by 8.}$$

$$x = 5$$

RECALL

$3x + 5x = (3 + 5)x$
$= 8x$

The solution is 5. To check, we return to the original equation. Substituting 5 for x yields

$$3(5) + 5(5) \stackrel{?}{=} 40$$
$$15 + 25 \stackrel{?}{=} 40$$
$$40 = 40 \quad \text{(True)}$$

Check Yourself 6

Solve and check.

$$7x + 4x = -66$$

As with the addition property, there are many applications that require us to use the multiplication property.

    Example 7   **An Application Involving the Multiplication Property**

< Objective 2 >

RECALL

You should always use a sentence to give the answer to an application.

chapter 2 > Make the Connection

NOTE

The yen (¥) is the monetary unit of Japan.

On her first day on the job in a photography lab, Samantha processed all of the film given to her. The following day, her boss gave her four times as much film to process. Over the two days, she processed 60 rolls of film. How many rolls did she process on the first day?

Let x be the number of rolls Samantha processed on her first day and solve the equation $x + 4x = 60$ to answer the question.

$$x + 4x = 60$$
$$5x = 60 \quad \text{Combine like terms first.}$$
$$\frac{1}{5}(5x) = \frac{1}{5}(60) \quad \text{Multiply by } \frac{1}{5}, \text{ to isolate the variable.}$$
$$x = 12$$

Samantha processed 12 rolls of film on her first day.

Check Yourself 7

On a recent trip to Japan, Marilyn exchanged $1,200 and received 139,812 yen. What exchange rate did she receive?
Let x be the exchange rate and solve the equation
$1,200x = 139,812$ to answer the question (to the nearest hundredth).

Check Yourself ANSWERS

1. 4   **2.** −6   **3.** 6   **4. (a)** 21; **(b)** −32   **5.** 27   **6.** −6
7. She received 116.51 yen for each dollar.

Reading Your Text

The following fill-in-the-blank exercises are designed to ensure that you understand some of the key vocabulary used in this section.

SECTION 2.2

(a) Multiplying both sides of an equation by the same nonzero number yields an _____ equation.

(b) Division is defined in terms of _____.

(c) Dividing by 5 is the same as _____ by $\frac{1}{5}$.

(d) The product of a nonzero number and its _____ is 1.

Basic Skills | Challenge Yourself | Calculator/Computer | Career Applications | Above and Beyond

< Objective 1 >

Solve and check.

1. $5x = 20$

2. $6x = 30$

3. $8x = 48$

4. $6x = -42$

5. $77 = 11x$

6. $66 = 6x$

7. $4x = -16$

8. $-3x = 27$

9. $-9x = 72$ > Videos

10. $10x = -100$

11. $6x = -54$

12. $-7x = 49$

13. $-5x = -15$

14. $52 = -4x$

15. $-42 = 6x$

16. $-7x = -35$

17. $-6x = -54$

18. $-7x = -42$

19. $\dfrac{x}{2} = 4$

20. $\dfrac{x}{3} = 2$

21. $\dfrac{x}{5} = 3$

22. $\dfrac{x}{8} = 5$

23. $5 = \dfrac{x}{8}$

24. $6 = \dfrac{x}{3}$

25. $\dfrac{x}{5} = -4$

26. $\dfrac{x}{7} = -5$

27. $-\dfrac{x}{3} = 8$ > Videos

28. $-\dfrac{x}{6} = -2$

29. $\dfrac{2}{3}x = 0.9$

30. $\dfrac{3}{7}x = 15$

31. $\dfrac{3}{4}x = -15$

32. $\dfrac{3}{5}x = 10 - \dfrac{6}{5}$

33. $-\dfrac{6}{5}x = -18$

34. $5x + 4x = 36$

35. $16x - 9x = -16.1$ > Videos

36. $4x - 2x + 7x = 36$

Boost *your* **GRADE** at
ALEKS.com!

ALEKS®

- Practice Problems
- Self-Tests
- NetTutor
- e-Professors
- Videos

Name _____

Section _____ Date _____

Answers

1. _____	2. _____
3. _____	4. _____
5. _____	6. _____
7. _____	8. _____
9. _____	10. _____
11. _____	12. _____
13. _____	14. _____
15. _____	16. _____
17. _____	18. _____
19. _____	20. _____
21. _____	22. _____
23. _____	24. _____
25. _____	26. _____
27. _____	28. _____
29. _____	30. _____
31. _____	32. _____
33. _____	34. _____
35. _____	36. _____

Answers

37. _____

38. _____

39. _____

40. _____

41. _____

42. _____

43. _____

44. _____

45. _____

46. _____

47. _____

48. _____

49. _____

50. _____

51. _____

52. _____

53. _____

54. _____

37. BUSINESS AND FINANCE Returning from Mexico City, Sung-A exchanged her remaining 450 pesos for $41.70. What exchange rate did she receive?

Use the equation $450x = 41.70$ to solve this problem (round to the nearest thousandth).

38. BUSINESS AND FINANCE Upon arrival in Portugal, Nicolas exchanged $500 and received 417.35 euros (€). What exchange rate did he receive?

Use the equation $500x = 417.35$ to solve this problem (round to the nearest hundredth).

39. SCIENCE AND TECHNOLOGY On Tuesday, there were twice as many patients in the clinic as on Monday. Over the 2-day period, 48 patients were treated. How many patients were treated on Monday?

Let p be the number of patients who came in on Monday and use the equation $p + 2p = 48$ to answer the question.

40. NUMBER PROBLEM Two-thirds of a number is 46. Find the number.

Use the equation $\frac{2}{3}x = 46$ to solve the problem.

| Basic Skills | **Challenge Yourself** | Calculator/Computer | Career Applications | Above and Beyond |

Certain equations involving decimals can be solved by the methods of this section. For instance, to solve $2.3x = 6.9$, we use the multiplication property to divide both sides of the equation by 2.3. This isolates x on the left, as desired. Use this idea to solve each equation.

41. $3.2x = 12.8$ **42.** $5.1x = -15.3$

43. $-4.5x = 13.5$ **44.** $-8.2x = -32.8$

45. $1.3x + 2.8x = 12.3$ **46.** $2.7x + 5.4x = -16.2$

47. $9.3x - 6.2x = 12.4$ **48.** $12.5x - 7.2x = -21.2$

| Basic Skills | Challenge Yourself | **Calculator/Computer** | Career Applications | Above and Beyond |

Use your calculator to solve each equation. Round your answers to the nearest hundredth.

49. $230x = 157$ **50.** $31x = -15$

51. $-29x = 432$ **52.** $-141x = -3,467$

53. $23.12x = 94.6$ **54.** $46.1x = -1$

Answers

55. INFORMATION TECHNOLOGY A 50-GB-capacity hard drive contains 30 GB of used space. What percent of the hard drive is full?

56. INFORMATION TECHNOLOGY A compression program reduces the size of files and folders by 36%. If a folder contains 17.5 MB, how large will it be after it is compressed?

57. AUTOMOTIVE TECHNOLOGY It is estimated that 8% of rebuilt alternators do not last through the 90-day warranty period. If a parts store had 6 bad alternators returned during the year, how many did they sell?

58. AGRICULTURAL TECHNOLOGY A farmer sold 2,200 bushels of barley on the futures market. Due to a poor harvest, he was able to make only 94% of his bid. How many bushels did he actually harvest?

59. Describe the difference between the multiplication property and the addition property for solving equations. Give examples of when to use one property or the other.

60. Describe when you should add a quantity to or subtract a quantity from both sides of an equation as opposed to when you should multiply or divide both sides by the same quantity.

BUSINESS AND FINANCE Motors, Windings, and More! sells every motor, regardless of type, for $2.50. This vendor also has a deal in which customers can choose whether to receive a markdown or free shipping. Shipping costs are $1.00 per item. If you do not choose the free shipping option, you can deduct 17.5% from your total order (but not the cost of shipping).

61. If you buy six motors, calculate the total cost for each of the two options. Which option is cheaper?

62. Is one option *always* cheaper than the other? Justify your result.

Answers

55.			
56.			
57.			
58.			
59.			
60.			
61.			
62.			

1. 4 **3.** 6 **5.** 7 **7.** −4 **9.** −8 **11.** −9 **13.** 3
15. −7 **17.** 9 **19.** 8 **21.** 15 **23.** 40 **25.** −20
27. −24 **29.** 1.35 **31.** −20 **33.** 15 **35.** −2.3
37. 0.093 dollar for each peso **39.** 16 **41.** 4 **43.** −3 **45.** 3
47. 4 **49.** 0.68 **51.** −14.90 **53.** 4.09 **55.** 60% **57.** 75
59. Above and Beyond **61.** Above and Beyond

Combining the Rules to Solve Equations

< 2.3 Objectives >

1 > Combine the addition and multiplication properties to solve an equation

2 > Solve equations containing parentheses

3 > Solve equations containing fractions

4 > Recognize identities and contradictions

In each example so far, we used either the addition property or the multiplication property to solve an equation. Often, finding a solution requires that we use both properties.

Example 1	Solving Equations

< Objective 1 >

Solve each equation.

(a) $4x - 5 = 7$

Here x is *multiplied* by 4. The result, $4x$, then has 5 subtracted from it (or -5 added to it) on the left side of the equation. These two operations mean that both properties must be applied to solve the equation.

Because there is only one variable term, we start by adding 5 to both sides:
The first step is to *isolate* the variable term, $4x$, on one side of the equation.

> CAUTION

Use the addition property before applying the multiplication property. That is, do not divide by 4 until after you have added 5!

$$
\begin{array}{rl}
4x - 5 = & 7 \\
\underline{+5} & \underline{+5} \\
4x \quad = & 12
\end{array}
$$

The first step is to *isolate* the variable term, $4x$, on one side of the equation.

We now divide both sides by 4:

$$\frac{4x}{4} = \frac{12}{4} \qquad \text{Next, } \textit{isolate} \text{ the variable } x.$$

$$x = 3$$

The solution is 3. To check, replace x with 3 in the original equation. Be careful to follow the rules for the order of operations.

$$4(3) - 5 \stackrel{?}{=} 7$$

$$12 - 5 \stackrel{?}{=} 7$$

$$7 = 7 \qquad \text{(True)}$$

(b) $3x + 8 = -4$

$$
\begin{array}{rl}
3x + 8 = & -4 \\
\underline{-8} & \underline{-8} \qquad \text{Subtract 8 from both sides.} \\
3x \quad = & -12
\end{array}
$$

NOTES

Isolate the variable term, 3*x*.

Isolate the variable.

Now divide both sides by 3 to isolate x.

$$\frac{3x}{3} = \frac{-12}{3}$$

$$x = -4$$

The solution is -4. We leave it to you to check this result.

 Check Yourself 1

Solve and check.

(a) $6x + 9 = -15$ **(b)** $5x - 8 = 7$

The variable may appear in any position in an equation. Just apply the rules carefully as you try to write an equivalent equation, and you will find the solution.

Example 2   Solving Equations

Solve.

$$
\begin{array}{rcl}
3 - 2x &=& 9 \\
-3 && -3 \\
\hline
-2x &=& 6
\end{array}
$$
First, subtract 3 from both sides.

NOTE

$\dfrac{-2}{-2} = 1$, so we divide by -2 to isolate x.

Now divide both sides by -2. This leaves x alone on the left.

$$\frac{-2x}{-2} = \frac{6}{-2}$$

$$x = -3$$

The solution is -3. We leave it to you to check this result.

 Check Yourself 2

Solve and check.

$10 - 3x = 1$

You may also have to combine multiplication with addition or subtraction to solve an equation. Consider Example 3.

Example 3   Solving Equations

Solve each equation.

(a) $\dfrac{x}{5} - 3 = 4$

RECALL

A *variable term* is a term that has a variable as a factor.

To get the variable term $\dfrac{x}{5}$ alone, we first add 3 to both sides.

$$
\begin{array}{rcl}
\dfrac{x}{5} - 3 &=& 4 \\[2mm]
+ 3 && +3 \\
\hline
\dfrac{x}{5} &=& 7
\end{array}
$$

To undo the division, multiply both sides of the equation by 5.

$$5\left(\frac{x}{5}\right) = 5 \cdot 7$$

$$x = 35$$

The solution is 35. Return to the original equation to check the result.

$$\frac{(35)}{5} - 3 \stackrel{?}{=} 4$$

$$7 - 3 \stackrel{?}{=} 4$$

$$4 = 4 \qquad \text{(True)}$$

(b) $\frac{2}{3}x + 5 = 13$

$$\underline{\quad -5 \quad -5 \quad} \qquad \text{First, subtract 5 from both sides.}$$

$$\frac{2}{3}x \qquad = 8$$

Now multiply both sides by $\frac{3}{2}$, the reciprocal of $\frac{2}{3}$.

$$\frac{3}{2}\left(\frac{2}{3}x\right) = \left(\frac{3}{2}\right)8$$

or

$$x = 12$$

The solution is 12. We leave it to you to check this result.

Check Yourself 3

Solve and check.

(a) $\frac{x}{6} + 5 = 3$ **(b)** $\frac{3}{4}x - 8 = 10$

In Section 2.1, you learned how to solve certain equations when the variable appeared on both sides. Example 4 shows you how to extend that work when using the multiplication and addition properties of equality.

| ▶ | **Example 4** | **Solving an Equation** |

Solve.

$$6x - 4 = 3x - 2$$

We begin by bringing all the variable terms to one side. To do this, we subtract $3x$ from both sides of the equation. This removes the variable term from the right side.

$$\begin{array}{r} 6x - 4 = \quad 3x - 2 \\ \underline{-3x \qquad\quad -3x} \\ 3x - 4 = \qquad -2 \end{array}$$

We now isolate the variable term by adding 4 to both sides.

$$\begin{array}{r} 3x - 4 = -2 \\ \underline{+4 \quad +4} \\ 3x \quad = \quad 2 \end{array}$$

NOTE

The basic idea is to use the two properties to form an equivalent equation with the x isolated. Here we subtracted $3x$ and then added 4. You can do these steps in either order. Try it for yourself the other way. In either case, the multiplication property is then used as the *last step* in finding the solution.

Finally, divide by 3.

$$\frac{3x}{3} = \frac{2}{3}$$

$$x = \frac{2}{3}$$

Check:

$$6\left(\frac{2}{3}\right) - 4 \overset{?}{=} 3\left(\frac{2}{3}\right) - 2$$

$$4 - 4 \overset{?}{=} 2 - 2$$

$$0 = 0 \quad \text{(True)}$$

Check Yourself 4

Solve and check.

$7x - 5 = 3x + 5$

Next, we look at two approaches to solving equations in which the coefficient on the right side is greater than the coefficient on the left side.

Example 5   Solving an Equation (Two Methods)

Solve $4x - 8 = 7x + 7$.

Method 1

$$\begin{array}{rcl} 4x - 8 = & 7x + 7 \\ -7x & -7x \end{array}$$ Bring the variable terms to the same (left) side.

$$\begin{array}{rcl} -3x - 8 = & +7 \\ +8 & +8 \end{array}$$ Isolate the variable term.

$$-3x \quad = \quad 15$$

$$\frac{-3x}{-3} = \frac{15}{-3}$$ Isolate the variable.

$$x = -5$$

We let you check this result.

To avoid a negative coefficient (in this example, -3), some students prefer a different approach.

This time we work toward having the number on the *left* and the x term on the *right,* or $\square = x$.

NOTE

It is usually easier to isolate the variable term on the side that results in a positive coefficient.

Method 2

$$\begin{array}{rcl} 4x - 8 = & 7x + 7 \\ -4x & -4x \end{array}$$ Bring the variable terms to the same (right) side.

$$\begin{array}{rcl} - 8 = & 3x + 7 \\ - 7 & - 7 \end{array}$$ Isolate the variable term.

$$-15 = \quad 3x$$

$$\frac{-15}{3} = \frac{3x}{3}$$ Isolate the variable.

$$-5 = x$$

Because $-5 = x$ and $x = -5$ are equivalent equations, it really makes no difference; the solution is still -5! You can use whichever approach you prefer.

Check Yourself 5

Solve $5x + 3 = 9x - 21$ by finding equivalent equations of the form $x = \boxed{}$ and $\boxed{} = x$ to compare the two methods of finding the solution.

It may also be necessary to remove grouping symbols to solve an equation.

Example 6   Solving Equations That Contain Parentheses

< Objective 2 >

NOTE

$5(x - 3)$
$= 5(x) - 5(3)$
$= 5x - 15$

Solve.

$5(x - 3) - 2x = x + 7$	Apply the distributive property.
$5x - 15 - 2x = x + 7$	Combine like terms.
$3x - 15 = x + 7$	

We now have an equation that we can solve by the usual methods. First, bring the variable terms to one side, then isolate the variable term, and finally, isolate the variable.

$$
\begin{array}{rl}
3x - 15 = x + 7 & \\
\underline{-x \qquad\quad -x} & \text{Subtract } x \text{ to bring the variable terms to the same side.} \\
2x - 15 = \qquad 7 & \\
\underline{+ 15 \qquad + 15} & \text{Add 15 to isolate the variable term.} \\
\dfrac{2x}{2} = \dfrac{22}{2} & \text{Divide by 2 to isolate the variable.} \\
x = 11 &
\end{array}
$$

The solution is 11. To check, substitute 11 for x in the original equation. Again note the use of our rules for the order of operations.

$5[(11) - 3] - 2(11) \stackrel{?}{=} (11) + 7$	Simplify terms in parentheses.
$5 \cdot 8 - 2 \cdot 11 \stackrel{?}{=} 11 + 7$	Multiply.
$40 - 22 \stackrel{?}{=} 11 + 7$	Add and subtract.
$18 = 18$	A true statement

Check Yourself 6

Solve and check.

$7(x + 5) - 3x = x - 7$

We now look at equations that contain fractions with different denominators. To solve an equation involving fractions, the first step is to multiply both sides of the equation by the **least common multiple (LCM)** of all denominators in the equation. Recall that the **LCM** of a set of numbers is the *smallest* number into which all the numbers divide evenly.

Example 7   Solving an Equation That Contains Fractions

< Objective 3 >

Solve.

$$\frac{x}{2} - \frac{2}{3} = \frac{5}{6}$$

First, multiply each side by 6, the LCM of 2, 3, and 6.

$$6\left(\frac{x}{2} - \frac{2}{3}\right) = 6\left(\frac{5}{6}\right) \qquad \text{Apply the distributive property.}$$

$$6\left(\frac{x}{2}\right) - 6\left(\frac{2}{3}\right) = 6\left(\frac{5}{6}\right)$$ Simplify.

$$3x - 4 = 5$$

Next, isolate the variable term on the left side.

$$3x = 9$$

$$x = 3$$

The solution can be checked by returning to the original equation.

 Check Yourself 7

Solve and check.

$$\frac{x}{4} - \frac{4}{5} = \frac{19}{20}$$

 Example 8   **Solving an Equation That Contains Fractions**

Solve.

$$\frac{2x - 1}{5} + 1 = \frac{x}{2}$$

First multiply each side by 10, the LCM of 5 and 2.

NOTE

You must remember to distribute because you are multiplying the *entire* left side by 10.

$$10\left(\frac{2x - 1}{5} + 1\right) = 10\left(\frac{x}{2}\right)$$ Apply the distributive property on the left and simplify.

$$10\left(\frac{2x - 1}{5}\right) + 10(1) = 10\left(\frac{x}{2}\right)$$

$$2(2x - 1) + 10 = 5x$$

$$4x - 2 + 10 = 5x$$

$$\frac{20}{50}$$

$$4x + 8 = 5x$$ Next, isolate x on the right side.

$$8 = x$$ The solution to the original equation is 8.

 Check Yourself 8

Solve and check.

$$\frac{3x + 1}{4} - 2 = \frac{x + 1}{3}$$

An equation that is true for any value of x is called an **identity**.

 Example 9   **Solving an Equation**

< Objective 4 >

NOTE

We could ask the question "For what values of x does $-6 = -6$?"

Solve the equation $2(x - 3) = 2x - 6$.

$$
\begin{array}{rcl}
2(x - 3) = & 2x - 6 \\
2x - 6 = & 2x - 6 \\
-2x & -2x \\
\hline
-6 = & -6
\end{array}
$$

The statement $-6 = -6$ is true for any value of x. The original equation is an identity. This means that all real numbers are solutions.

 Check Yourself 9

Solve the equation $3(x - 4) - 2x = x - 12$.

There are also equations for which there are no solutions. We call such equations **contradictions.**

 Example 10 | **Solving an Equation**

Solve the equation $3(2x - 5) - 4x = 2x + 1$.

$$
\begin{aligned}
3(2x - 5) - 4x &= 2x + 1 \\
6x - 15 - 4x &= 2x + 1 \\
2x - 15 &= 2x + 1 \\
\underline{-2x \qquad\quad -2x} & \\
-15 &= 1
\end{aligned}
$$

NOTE

We could ask the question "For what values of x does $-15 = 1$?"

These two numbers are never equal. The original equation has no solutions.

 Check Yourself 10

Solve the equation $2(x - 5) + x = 3x - 3$.

A series of steps to solve a problem is called an **algorithm.** The following algorithm can be used to solve a linear equation.

Step by Step

Solving Linear Equations

If no variable remains after step 3, determine whether the equation is an identity or a contradiction.

Step 1	Use the distributive property to remove any grouping symbols.
Step 2	Combine like terms on each side of the equation.
Step 3	Add or subtract variable terms to bring the variable term to one side of the equation.
Step 4	Add or subtract numbers to isolate the variable term.
Step 5	Multiply by the reciprocal of the coefficient to isolate the variable.
Step 6	Check your result.

 Check Yourself ANSWERS

1. (a) -4; (b) 3 2. 3 3. (a) -12; (b) 24 4. $\dfrac{5}{2}$ 5. 6 6. -14

7. 7 8. 5 9. The equation is an identity, so x can be any real number.

10. There are no solutions.

Reading Your Text

The following fill-in-the-blank exercises are designed to ensure that you understand some of the key vocabulary used in this section.

SECTION 2.3

(a) The first goal for solving an equation is to _____ the variable term on one side of the equation.

(b) Apply the _____ property before applying the multiplication property.

(c) Always return to the _____ equation to check your result.

(d) It is usually easiest to clear the _____ by multiplying both sides by the LCM of the denominators when solving an equation with unlike fractions.

< Objectives 1–3 >

Solve and check.

1. $3x + 2 = 14$

2. $3x - 1 = 17$

3. $3x - 2 = 7$

4. $7x + 9 = 37$

5. $4x + 7 = 35$

6. $7x - 8 = 13$

7. $2x + 9 = 5$

8. $6x + 25 = -5$

9. $4 - 7x = 18$

10. $8 - 5x = -7$

11. $5 - 3x = 11$

12. $5 - 4x = 25$ > Videos

13. $\dfrac{x}{2} + 1 = 5$

14. $\dfrac{x}{5} - 3 = 2$

15. $\dfrac{x}{5} - 3 = 4$

16. $\dfrac{x}{5} + 3 = 8$

17. $\dfrac{2}{3}x + 5 = 17$

18. $\dfrac{3}{4}x - 5 = 4$

19. $\dfrac{3}{4}x - 2 = 16$

20. $\dfrac{5}{7}x + 4 = 14$

21. $5x = 2x + 9$

22. $7x = 18 - 2x$

23. $3x = 10 - 2x$

24. $11x = 7x + 20$

25. $9x + 2 = 3x + 38$

26. $8x - 3 = 4x + 17$

27. $4x - 8 = x - 14$

28. $6x - 5 = 3x - 29$

Answers

1. _____   2. _____

3. _____   4. _____

5. _____   6. _____

7. _____   8. _____

9. _____   10. _____

11. _____   12. _____

13. _____   14. _____

15. _____   16. _____

17. _____   18. _____

19. _____   20. _____

21. _____   22. _____

23. _____   24. _____

25. _____   26. _____

27. _____   28. _____

Answers

29. _____

30. _____

31. _____

32. _____

33. _____

34. _____

35. _____

36. _____

37. _____

38. _____

39. _____

40. _____

41. _____

42. _____

43. _____

44. _____

45. _____

46. _____

47. _____

48. _____

49. _____

50. _____

51. _____

52. _____

53. _____

54. _____

29. $5x + 7 = 2x - 3$

30. $9x + 7 = 5x - 3$

31. $7x - 3 = 9x + 5$

32. $5x - 2 = 8x - 11$

33. $5x + 4 = 7x - 8$

34. $2x + 23 = 6x - 5$

35. $2x - 3 + 5x = 7 + 4x + 2$

36. $8x - 7 - 2x = 2 + 4x - 5$

37. $6x + 7 - 4x = 8 + 7x - 26$

> Videos

38. $7x - 2 - 3x = 5 + 8x + 13$

39. $9x - 2 + 7x + 13 = 10x - 13$

40. $5x + 3 + 6x - 11 = 8x + 25$

41. $2(x + 3) = 8$

42. $-3(x - 1) = 4(x + 2) + 2$

> Videos

43. $7(2x - 1) - 5x = x + 25$

44. $9(3x + 2) - 10x = 12x - 7$

< Objective 4 >

45. $5(x + 1) - 4x = x - 5$

> Videos

46. $-4(2x - 3) = -8x + 5$

47. $6x - 4x + 1 = 12 + 2x - 11$

48. $-2x + 5x - 9 = 3(x - 4) - 5$

49. $-4(x + 2) - 11 = 2(-2x - 3) - 13$

50. $4(-x - 2) + 5 = -2(2x + 7)$

Basic Skills | **Challenge Yourself** | Calculator/Computer | Career Applications | Above and Beyond
▲

Find the length of each side of the figure for the given perimeter.

51.

$P = 24$ in.

52.

$P = 32$ cm

53.

$P = 90$ in.

54.

$P = 34$ cm

Solve each equation and check your solution.

55. $3x + 2(4x - 3) = 6x - 9$

56. $7x + 3(2x + 5) = 10x + 17$

57. $\dfrac{8}{3}x - 3 = \dfrac{2}{3}x + 15$ ▸ Videos

58. $\dfrac{12x}{5} + 7 = 31 - \dfrac{3x}{5}$

59. $\dfrac{2x}{5} - \dfrac{x}{3} = \dfrac{7}{15}$

60. $\dfrac{2}{7}x - \dfrac{3}{5}x = \dfrac{6}{35}$

61. $5.3x - 7 = 2.3x + 5$

62. $9.8x + 2 = 3.8x + 20$

63. $\dfrac{5x - 3}{4} - 2 = \dfrac{x}{3}$

64. $\dfrac{6x - 1}{5} - \dfrac{2x}{3} = 3$

65. $3 - (x - 2) = 2x + 1$

66. $4x + 2(3 - 2x) = 4 - 3(2x + 5)$

67. $2(1 - 3x) + 2(5x + 4) = 3 - (4x - 1)$

68. $11x + 5(3 - 2x) = -2(-3x + 2)$

69. $\dfrac{3x + 1}{5} + \dfrac{2x - 2}{3} = x$

70. $\dfrac{1 - x}{4} - \dfrac{2x + 3}{2} = \dfrac{3}{4}$

71. $3x + \dfrac{2x - 3}{3} = \dfrac{3(4x + 1)}{3}$

72. $\dfrac{-3(x - 1)}{2} = \dfrac{2x - 3}{3}$

Basic Skills | Challenge Yourself | **Calculator/Computer** | Career Applications | Above and Beyond
▲

Use your calculator to solve each equation. Round your answers to the nearest hundredth.

73. $230x - 52 = 191$

74. $321 - 45x = 1,021x + 658$

75. $360 - 29(2x + 1) = 2,464$

76. $81(x + 26) = 35(86 - 4x)$

Answers

55. _____

56. _____

57. _____

58. _____

59. _____

60. _____

61. _____

62. _____

63. _____

64. _____

65. _____

66. _____

67. _____

68. _____

69. _____

70. _____

71. _____

72. _____

73. _____

74. _____

75. _____

76. _____

Answers

77. _____

78. _____

79. _____

80. _____

81. _____

82. _____

83. _____

84. _____

77. $23.12x - 34.2 = 34.06$

78. $46.1x + 5.78 = x - 12$

79. $3.2(0.5x - 5.1) = -6.4(9.7x + 15.8)$

80. $x - 11.304(2 - 1.8x) = 2.4x + 3.7$

Basic Skills	Challenge Yourself	Calculator/Computer	**Career Applications**	Above and Beyond
			▲	

81. **AGRICULTURAL TECHNOLOGY** The estimated yield Y of a field of corn (in bushels per acre) can be found by multiplying the rainfall r, in inches, during the growing season by 16 and then subtracting 15. This relationship can be modeled by the formula

$$Y = 16r - 15$$

If a farmer wants a yield of 159 bushels per acre, then we can write the equation shown to determine the amount of rainfall required.

$$159 = 16r - 15$$

How much rainfall is necessary to achieve a yield of 159 bushels of corn per acre?

82. **CONSTRUCTION TECHNOLOGY** The number of studs s required to build a wall (with studs spaced 16 inches on center) is equal to one more than $\frac{3}{4}$ times the length w of the wall, in feet. We model this with the formula

$$s = \frac{3}{4}w + 1$$

If a contractor uses 22 studs to build a wall, how long is the wall?

83. **ALLIED HEALTH** The internal diameter D (in mm) of an endotracheal tube for a child is calculated using the formula

$$D = \frac{t + 16}{4}$$

in which t is the child's age (in years).

How old is a child who requires an endotracheal tube with an internal diameter of 7 mm?

84. **MECHANICAL ENGINEERING** The number of BTUs required to heat a house is $2\frac{3}{4}$ times the volume of the air in the house (in cubic feet). What is the maximum air volume that can be heated with a 90,000-BTU furnace?

Basic Skills | Challenge Yourself | Calculator/Computer | Career Applications | **Above and Beyond**

▲

85. Create an equation of the form $ax + b = c$ that has 2 as a solution.

86. Create an equation of the form $ax + b = c$ that has 7 as a solution.

87. The equation $3x = 3x + 5$ has no solution, whereas the equation $7x + 8 = 8$ has zero as a solution. Explain the difference between a solution of zero and no solution.

88. Construct an equation for which every real number is a solution.

85. _____

86. _____

87. _____

88. _____

Answers

1. 4   **3.** 3   **5.** 7   **7.** −2   **9.** −2   **11.** −2   **13.** 8
15. 35   **17.** 18   **19.** 24   **21.** 3   **23.** 2   **25.** 6   **27.** −2
29. $-\dfrac{10}{3}$ **31.** −4   **33.** 6   **35.** 4   **37.** 5   **39.** −4   **41.** 1
43. 4   **45.** No solution   **47.** Identity   **49.** Identity
51. 6 in., 8 in., 10 in.   **53.** 12 in., 19 in., 29 in., 30 in.   **55.** $-\dfrac{3}{5}$
57. 9   **59.** 7   **61.** 4   **63.** 3   **65.** $\dfrac{4}{3}$ **67.** $-\dfrac{3}{4}$ **69.** $\dfrac{7}{4}$
71. −6   **73.** 1.06   **75.** −36.78   **77.** 2.95   **79.** −1.33   **81.** $10\dfrac{7}{8}$ in.
83. 12 yr old   **85.** Above and Beyond   **87.** Above and Beyond

2.4

Formulas and Problem Solving

< 2.4 Objectives >

1 > Solve a literal equation for one of its variables

2 > Solve an application involving a literal equation

3 > Translate a word phrase to an expression or an equation

4 > Use an equation to solve an application

Formulas are extremely useful tools in any field in which mathematics is applied. Formulas are simply equations that express a relationship between more than one letter or variable. You are no doubt familiar with all kinds of formulas, such as

$$A = \frac{1}{2} bh \qquad \text{The area of a triangle}$$

$$I = Prt \qquad \text{Interest}$$

$$V = \pi r^2 h \qquad \text{The volume of a cylinder}$$

A formula is also called a **literal equation** because it involves several letters or variables. For instance, our first formula or literal equation, $A = \frac{1}{2} bh$, involves the three variables A (for area), b (for base), and h (for height).

Unfortunately, formulas are not always given in the form needed to solve a particular problem. In such cases, we use algebra to change the formula to a more useful equivalent equation solved for a particular variable. The steps used in the process are very similar to those you used in solving linear equations. Consider an example.

| ⓘ | **Example 1** | Solving a Literal Equation for a Variable |

< Objective 1 >

Suppose that we know the area A and the base b of a triangle and want to find its height h.

We are given

$$A = \frac{1}{2} bh$$

We need to find an equivalent equation with h, the unknown, by itself on one side. We can think of $\frac{1}{2} b$ as the **coefficient** of h. We can remove the two *factors* of that coefficient, $\frac{1}{2}$ and b, separately.

NOTE

$$2\left(\frac{1}{2} bh\right) = \left(2 \cdot \frac{1}{2}\right)(bh)$$
$$= 1(bh)$$
$$= bh$$

$$2A = 2\left(\frac{1}{2} bh\right) \qquad \text{Multiply both sides by 2 to clear the equation of fractions.}$$

or

$$2A = bh$$

$$\frac{2A}{b} = \frac{bh}{b} \qquad \text{Divide by } b \text{ to isolate } h.$$

$$\frac{2A}{b} = h$$

or

$$h = \frac{2A}{b}$$　　　Reverse the sides to write h on the left.

We now have the height h in terms of the area A and the base b. This is called **solving the equation for h** and means that we are rewriting the formula as an equivalent equation of the form

$$h = \square.$$

NOTE

Here, $\square$ means an expression containing all the numbers or variables *other* than h.

 Check Yourself 1

Solve $V = \frac{1}{3}Bh$ for h.

You have already learned the methods needed to solve most literal equations or formulas for some specified variable. As Example 1 illustrates, the rules you learned in Sections 2.1 and 2.2 are applied in exactly the same way as they were applied to equations with one variable.

You may have to apply both the addition and the multiplication properties when solving a formula for a specified variable. Example 2 illustrates this process.

| **Example 2** | **Solving a Literal Equation** |

(a) Solve $y = mx + b$ for x.

Remember that we want to end up with x alone on one side of the equation. Start by subtracting b from both sides to "undo" the addition on the right.

$$\begin{array}{rl} y & = mx + b \\ -b & -b \\ \hline y - b & = mx \end{array}$$

If we now divide both sides by m, then x will be alone on the right-hand side.

$$\frac{y - b}{m} = \frac{mx}{m}$$

$$\frac{y - b}{m} = x$$

or

$$x = \frac{y - b}{m}$$

(b) Solve $3x + 2y = 12$ for y.

Begin by isolating the y term.

$$\begin{array}{rl} 3x + 2y = & 12 \\ -3x & -3x \\ \hline 2y = & -3x + 12 \end{array}$$

Then, isolate y by dividing by its coefficient.

$$\frac{2y}{2} = \frac{-3x + 12}{2}$$

$$y = \frac{-3x + 12}{2}$$

RECALL

Dividing by 2 is the same as multiplying by $\frac{1}{2}$.

Often, in a situation like this, we use the distributive property to separate the terms on the right-hand side of the equation.

$$y = \frac{-3x + 12}{2}$$

$$= \frac{-3x}{2} + \frac{12}{2}$$

$$= \frac{-3}{2}x + 6 = -\frac{3}{2}x + 6$$

NOTE

v and v_0 represent distinct quantities.

Check Yourself 2

(a) Solve $v = v_0 + gt$ for t.   **(b)** Solve $4x - 3y = 8$ for x.

Here is a summary of the steps illustrated by our examples.

Step by Step

Solving a Formula or Literal Equation		
Step 1	If necessary, multiply both sides of the equation by the LCD to clear it of fractions.
Step 2	Add or subtract the same term on each side of the equation so that all terms involving the variable that you are solving for are on one side of the equation and all other terms are on the other side.
Step 3	Divide both sides of the equation by the coefficient of the variable that you are solving for.

Look at one more example using these steps.

 Example 3   **Solving a Literal Equation Involving Money**

Solve $A = P + Prt$ for r.

NOTE

This is a formula for the *amount* of money in an account after interest has been earned.

$$\begin{aligned} A &= P + Prt \\ -P & \quad -P \\ \hline A - P &= \quad Prt \end{aligned}$$

Subtracting P from both sides leaves the term involving r alone on the right.

$$\frac{A - P}{Pt} = \frac{Prt}{Pt}$$

Dividing both sides by Pt isolates r on the right.

$$\frac{A - P}{Pt} = r$$

or

$$r = \frac{A - P}{Pt}$$

Check Yourself 3

Solve $2x + 3y = 6$ for y.

Now look at an application of solving a literal equation.

| Example 4 | Using a Literal Equation |

< Objective 2 >

Suppose that the amount in an account, 3 years after a principal of $5,000 was invested, is $6,050. What was the interest rate?

From Example 3,

$$A = P + Prt$$

in which A is the amount in the account, P is the principal, r is the interest rate, and t is the time that the money has been invested. By the result of Example 3 we have

$$r = \frac{A - P}{Pt}$$

and we can substitute the known values into this equation.

NOTE

Do you see the advantage of having our equation solved for the desired variable?

$$r = \frac{(6,050) - (5,000)}{(5,000)(3)}$$

$$= \frac{1,050}{15,000} = 0.07 = 7\%$$ The interest rate was 7%.

 Check Yourself 4

Suppose that the amount in an account, 4 years after a principal of $3,000 was invested, is $3,480. What was the interest rate?

The main reason for learning how to set up and solve algebraic equations is so that we can use them to solve word problems and applications. In fact, algebraic equations were *invented* to make solving word problems much easier. The first word problems that we know about are over 4,000 years old. They were literally "written in stone," on Babylonian tablets, about 500 years before the first algebraic equation made its appearance.

Before algebra, people solved word problems primarily by "guess-and-check," which is a method of finding unknown numbers by using trial and error in a logical way. Example 5 shows how to solve a word problem using this method. We sometimes refer to this method as *inspection*.

| Example 5 | Solving a Word Problem by Substitution |

The sum of two consecutive integers is 37. Find the two integers.

If the two integers were 20 and 21, their sum would be 41, which is more than 37, so the integers must be smaller. If the integers were 15 and 16, the sum would be 31. More trials yield that the sum of 18 and 19 is 37.

Check Yourself 5

The sum of two consecutive integers is 91. Find the two integers.

Most word problems are not so easily solved by the guess-and-check method. For more complicated word problems, we use a five-step procedure. This step-by-step approach will, with practice, allow you to organize your work. Organization is the key to solving word problems. Here are the five steps.

Step by Step

To Solve Word Problems		
	Step 1	Read the problem carefully. Then reread it to decide what you are asked to find.
	Step 2	Choose a letter to represent one of the unknowns in the problem. Then represent all other unknowns of the problem with expressions that use the same letter.
	Step 3	Translate the problem to the language of algebra to form an equation.
	Step 4	Solve the equation.
	Step 5	Answer the question—include units in your answer, when appropriate, and check your solution by returning to the original problem.

RECALL

We discussed these translations in Section 1.4. You might find it helpful to review that section before going on.

The third step is usually the hardest part. We must translate words to the language of algebra. Before we look at a complete example, the following table may help you review that translation step.

Translating Words to Algebra

Words	Algebra
The sum of x and y	$x + y$
3 plus a	$3 + a$ or $a + 3$
5 more than m	$m + 5$
b increased by 7	$b + 7$
The difference between x and y	$x - y$
4 less than a	$a - 4$
s decreased by 8	$s - 8$
The product of x and y	$x \cdot y$ or xy
5 times a	$5 \cdot a$ or $5a$
Twice m	$2m$
The quotient of x and y	$\dfrac{x}{y}$
a divided by 6	$\dfrac{a}{6}$
One-half of b	$\dfrac{b}{2}$ or $\dfrac{1}{2}b$

Here are some typical examples of translating phrases to algebra to help you review.

(▶)   **Example 6**   **Translating Statements**

< Objective 3 >

Translate each statement to an algebraic expression.

(a) The sum of a and twice b $a + 2b$

Sum   Twice b

(b) 5 times m increased by 1 $5m + 1$

5 times m Increased by 1

(c) 5 less than 3 times x $\qquad$ $3x - 5$

3 times x $\qquad$ 5 less than

(d) The product of x and y, divided by 3 $\qquad$ $\dfrac{xy}{3}$

The product of x and y

Divided by 3

Check Yourself 6

Translate to algebra.

(a) 2 more than twice x $\qquad\qquad$ **(b)** 4 less than 5 times n
(c) The product of twice a and b $\qquad$ **(d)** The sum of s and t, divided by 5

Now we work through a complete example. Although this problem could be solved by substitution, it is presented here to help you practice the five-step approach.

▶ **Example 7** $\qquad$ **Solving an Application**

< Objective 4 >

The sum of a number and 5 is 17. What is the number?

Step 1 *Read carefully.* You must find the unknown number.

> **NOTE**
>
> The word *is* usually translates into an equal sign, =.

Step 2 *Choose letters or variables.* Let x represent the unknown number. There are no other unknowns.

Step 3 *Translate.*

The sum of

$x + 5 = 17$

is

Step 4 *Solve.*

> **NOTE**
>
> Always return to the *original problem* to check your result and *not* to the equation of step 3. This prevents many errors!

$$
\begin{array}{r}
x + 5 = 17 \\
\underline{-5 \quad -5} \quad \text{Subtract 5.} \\
x = 12
\end{array}
$$

Step 5 The number is 12.

Check. Is the sum of 12 and 5 equal to 17? Yes $(12 + 5 = 17)$.

Check Yourself 7

The sum of a number and 8 is 35. What is the number?

Property

Consecutive Integers

Consecutive integers are integers that follow one another, such as 10, 11, and 12. To represent them in algebra:

If x is an integer, then $x + 1$ is the next consecutive integer, $x + 2$ is the one after that, and so on.

We need this idea in Example 8.

Example 8   Solving an Application

The sum of two consecutive integers is 41. What are the two integers?

Step 1   We want to find the two consecutive integers.

Step 2   Let x be the first integer. Then $x + 1$ must be the next.

Step 3

The first   The second
integer   integer

$$x + (x + 1) = 41$$

The sum   Is

Step 4

$$x + x + 1 = 41$$
$$2x + 1 = 41$$
$$2x = 40$$
$$x = 20$$

Step 5   The first integer (x) is 20, and the next integer ($x + 1$) is 21.
The sum of the two integers 20 and 21 is 41.

 Check Yourself 8

The sum of three consecutive integers is 51. What are the three integers?

Sometimes algebra is used to reconstruct missing information. Example 9 does just that with some election information.

 Example 9   Solving an Application

There were 55 more yes votes than no votes on an election measure. If 735 votes were cast in all, how many yes votes were there? How many no votes?

RECALL

Read the problem carefully. What do you need to find? Assign letters to the unknown or unknowns. Write an equation.

NOTES

What do you need to find?
Assign letters to the unknowns.

Step 1   We want to find the number of yes votes and the number of no votes.

Step 2   Let x be the number of no votes. Then

$$x + 55$$

55 more than x

is the number of yes votes.

Step 3

$$x + x + 55 = 735$$

No votes   Yes votes

Step 4

$$x + x + 55 = 735$$
$$2x + 55 = 735$$
$$2x = 680$$
$$x = 340$$

Step 5   No votes $(x) = 340$
Yes votes $(x + 55) = 395$

340 no votes plus 395 yes votes equals 735 total votes. The solution checks.

Check Yourself 9

Francine earns $120 per month more than Rob. If they earn a total of $2,680 per month, what are their monthly salaries?

Similar methods allow you to solve a variety of word problems. Example 10 includes three unknown quantities but uses the same basic solution steps.

   Example 10   Solving an Application

Juan worked twice as many hours as Jerry. Marcia worked 3 more hours than Jerry. If they worked a total of 31 hours, find out how many hours each worked.

Step 1   We want to find the hours each worked, so there are three unknowns.

Step 2   Let x be the hours that Jerry worked.

NOTE

There are other choices for x, but choosing the smallest quantity usually gives the easiest equation to write and solve.

Twice Jerry's hours

Then $2x$ is Juan's hours worked

3 more hours than Jerry worked

and $x + 3$ is Marcia's hours.

Step 3

Jerry   Juan   Marcia

$$x + 2x + (x + 3) = 31$$

Sum of their hours

Step 4

$$x + 2x + x + 3 = 31$$
$$4x + 3 = 31$$
$$4x = 28$$
$$x = 7$$

Step 5   Jerry's hours (x) $= 7$
Juan's hours ($2x$) $= 14$
Marcia's hours ($x + 3$) $= 10$

The sum of their hours ($7 + 14 + 10$) is 31, and the solution is verified.

 Check Yourself 10

Paul jogged half as many miles (mi) as Lucy and 7 less than Isaac. If the three ran a total of 23 mi, how far did each person run?

 Check Yourself ANSWERS

1. $h = \dfrac{3V}{B}$ **2. (a)** $t = \dfrac{v - v_0}{g}$; **(b)** $x = \dfrac{3}{4}y + 2$

3. $y = \dfrac{6 - 2x}{3}$ or $y = -\dfrac{2}{3}x + 2$ **4.** 4%   **5.** 45 and 46

6. (a) $2x + 2$; **(b)** $5n - 4$; **(c)** $2ab$; **(d)** $\dfrac{s + t}{5}$

7. The equation is $x + 8 = 35$. The number is 27.

8. The equation is $x + x + 1 + x + 2 = 51$. The integers are 16, 17, and 18.

9. The equation is $x + x + 120 = 2{,}680$. Rob's salary is \$1,280 and Francine's is \$1,400.   **10.** Paul: 4 mi; Lucy: 8 mi; Isaac: 11 mi

Reading Your Text

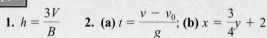

The following fill-in-the-blank exercises are designed to ensure that you understand some of the key vocabulary used in this section.

SECTION 2.4

(a) A _____ is also called a literal equation because it involves several letters or variables.

(b) A _____ is the factor by which a variable is multiplied.

(c) When translating a sentence into algebra, the word "is" usually indicates _____.

(d) Always return to the _____ equation or statement when checking your result.

Basic Skills | Challenge Yourself | Calculator/Computer | Career Applications | Above and Beyond

< Objective 1 >

Solve each literal equation for the indicated variable.

1. $P = 4s$ (for s) Perimeter of a square

2. $V = Bh$ (for B) Volume of a prism

3. $E = IR$ (for R) Voltage in an electric circuit

4. $I = Prt$ (for r) Simple interest

5. $V = LWH$ (for H) Volume of a rectangular solid

6. $V = \pi r^2 h$ (for h) Volume of a cylinder

7. $A + B + C = 180$ (for B) Measure of angles in a triangle

8. $P = I^2 R$ (for R) Power in an electric circuit

9. $ax + b = 0$ (for x) Linear equation in one variable

10. $y = mx + b$ (for m) Slope-intercept form for a line

 > Videos

11. $s = \dfrac{1}{2}gt^2$ (for g) Distance

12. $K = \dfrac{1}{2}mv^2$ (for m) Energy

13. $x + 5y = 15$ (for y) Linear equation in two variables

14. $2x + 3y = 6$ (for x) Linear equation in two variables

15. $P = 2L + 2W$ (for L) Perimeter of a rectangle

16. $ax + by = c$ (for y) Linear equation in two variables

17. $V = \dfrac{KT}{P}$ (for T) Volume of a gas

18. $V = \dfrac{1}{3}\pi r^2 h$ (for h) Volume of a cone

19. $x = \dfrac{a + b}{2}$ (for b) Mean of two numbers

Boost *your* GRADE at ALEKS.com!

ALEKS®

• Practice Problems   • e-Professors
• Self-Tests   • Videos
• NetTutor

Name _____

Section _____ Date _____

Answers

1. _____   2. _____

3. _____   4. _____

5. _____   6. _____

7. _____

8. _____   9. _____

10. _____   11. _____

12. _____

13. _____

14. _____

15. _____

16. _____

17. _____

18. _____

19. _____

Answers

20. _____

21. _____

22. _____

23. _____

24. _____

25. _____

26. _____

27. _____

28. _____

29. _____

30. _____

31. _____

32. _____

33. _____

20. $D = \dfrac{C - s}{n}$ (for s)   Depreciation

21. $F = \dfrac{9}{5}C + 32$ (for C)   Celsius/Fahrenheit

22. $A = P + Prt$ (for t)   Amount at simple interest

23. $S = 2\pi r^2 + 2\pi rh$ (for h)   Total surface area of a cylinder

24. $A = \dfrac{1}{2}h(B + b)$ (for b)   Area of a trapezoid   > Videos

< Objective 2 >

25. GEOMETRY A rectangular solid has a base with length 8 cm and width 5 cm. If the volume of the solid is 120 cm³, find the height of the solid. (See exercise 5.)  > Videos

26. GEOMETRY A cylinder has a radius of 4 in. If the volume of the cylinder is 48π in.³, what is the height of the cylinder? (See exercise 6.)

27. BUSINESS AND FINANCE A principal of $3,000 was invested in a savings account for 3 years. If the interest earned for the period was $450, what was the interest rate? (See exercise 4.)

28. GEOMETRY If the perimeter of a rectangle is 60 ft and the width is 12 ft, find its length. (See exercise 15.)

29. SCIENCE AND MEDICINE The high temperature in New York for a particular day was reported at 77°F. How would the same temperature have been given in degrees Celsius? (See exercise 21.)

30. CRAFTS Rose's garden is in the shape of a trapezoid. If the height of the trapezoid is 16 m, one base is 20 m, and the area is 224 m², find the length of the other base. (See exercise 24.)

$A = 224\ m^2$

16 m

20 m

< Objective 3 >

Translate each statement to an algebraic equation. Let x represent the number in each case.

31. 3 more than a number is 7.

32. 5 less than a number is 12.

33. 7 less than 3 times a number is twice that same number.   > Videos

34. 4 more than 5 times a number is 6 times that same number.

35. 2 times the sum of a number and 5 is 18 more than that same number.

36. 3 times the sum of a number and 7 is 4 times that same number.

37. 3 more than twice a number is 7.

38. 5 less than 3 times a number is 25.

39. 7 less than 4 times a number is 41.

40. 10 more than twice a number is 44.

41. 5 more than two-thirds of a number is 21.

42. 3 less than three-fourths of a number is 24.

43. 3 times a number is 12 more than that number.

44. 5 times a number is 8 less than that number.

< Objective 4 >

Solve each word problem. Be sure to label the unknowns and to show the equation you use for the solution.

45. NUMBER PROBLEM The sum of a number and 7 is 33. What is the number?

46. NUMBER PROBLEM The sum of a number and 15 is 22. What is the number?

47. NUMBER PROBLEM The sum of a number and −15 is 7. What is the number?

48. NUMBER PROBLEM The sum of a number and −8 is 17. What is the number?

49. SOCIAL SCIENCE In an election, the winning candidate has 1,840 votes. If the total number of votes cast was 3,260, how many votes did the losing candidate receive?

50. BUSINESS AND FINANCE Mike and Stefanie work at the same company and make a total of $2,760 per month. If Stefanie makes $1,400 per month, how much does Mike earn every month?

51. NUMBER PROBLEM The sum of twice a number and 5 is 35. What is the number?

52. NUMBER PROBLEM 3 times a number, increased by 8, is 50. Find the number.

53. NUMBER PROBLEM 5 times a number, minus 12, is 78. Find the number.

Answers

34. _____

35. _____

36. _____

37. _____

38. _____

39. _____

40. _____

41. _____

42. _____

43. _____

44. _____

45. _____

46. _____

47. _____

48. _____

49. _____

50. _____

51. _____

52. _____

53. _____

Answers

54. _____

55. _____

56. _____

57. _____

58. _____

59. _____

60. _____

61. _____

62. _____

63. _____

64. _____

65. _____

66. _____

67. _____

68. _____

69. _____

54. **NUMBER PROBLEM** 4 times a number, decreased by 20, is 44. What is the number?

55. **NUMBER PROBLEM** The sum of two consecutive integers is 47. Find the two integers.

56. **NUMBER PROBLEM** The sum of two consecutive integers is 145. Find the two integers.

57. **NUMBER PROBLEM** The sum of three consecutive integers is 63. What are the three integers?

58. **NUMBER PROBLEM** If the sum of three consecutive integers is 93, find the three integers. ⊙ › Videos

59. **NUMBER PROBLEM** The sum of two consecutive even integers is 66. What are the two integers? (*Hint:* Consecutive even integers such as 10, 12, and 14 can be represented by $x, x + 2, x + 4$, and so on.)

60. **NUMBER PROBLEM** If the sum of two consecutive even integers is 114, find the two integers.

61. **NUMBER PROBLEM** If the sum of two consecutive odd integers is 52, what are the two integers? (*Hint:* Consecutive odd integers such as 21, 23, and 25 can be represented by $x, x + 2, x + 4$, and so on.)

62. **NUMBER PROBLEM** The sum of two consecutive odd integers is 88. Find the two integers.

63. **NUMBER PROBLEM** The sum of three consecutive odd integers is 63. What are the three integers?

64. **NUMBER PROBLEM** The sum of three consecutive even integers is 126. What are the three integers?

65. **NUMBER PROBLEM** The sum of four consecutive integers is 86. What are the four integers?

66. **NUMBER PROBLEM** The sum of four consecutive integers is 62. What are the four integers?

67. **NUMBER PROBLEM** 4 times an integer is 9 more than 3 times the next consecutive integer. What are the two integers?

68. **NUMBER PROBLEM** 4 times an integer is 30 less than 5 times the next consecutive even integer. Find the two integers.

69. **SOCIAL SCIENCE** In an election, the winning candidate had 160 more votes than the loser. If the total number of votes cast was 3,260, how many votes did each candidate receive?

70. **BUSINESS AND FINANCE** Jody earns $140 more per month than Frank. If their monthly salaries total $2,760, what amount does each earn?

71. **BUSINESS AND FINANCE** A washer-dryer combination costs $650. If the washer costs $70 more than the dryer, what does each appliance cost?

72. **CRAFTS** Yuri has a board that is 98 in. long. He wishes to cut the board into two pieces so that one piece will be 10 in. longer than the other. What should the length of each piece be?

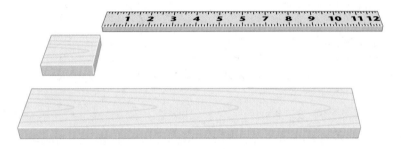

73. **SOCIAL SCIENCE** Yan Ling is 1 year less than twice as old as his sister. If the sum of their ages is 14 years, how old is Yan Ling?

74. **SOCIAL SCIENCE** Diane is twice as old as her brother Dan. If the sum of their ages is 27 years, how old are Diane and her brother?

75. **SOCIAL SCIENCE** Maritza is 3 years less than 4 times as old as her daughter. If the sum of their ages is 37, how old is Maritza?

76. **SOCIAL SCIENCE** Mrs. Jackson is 2 years more than 3 times as old as her son. If the difference between their ages is 22 years, how old is Mrs. Jackson?

77. **BUSINESS AND FINANCE** On her vacation in Europe, Jovita's expenses for food and lodging were $60 less than twice as much as her airfare. If she spent $2,400 in all, what was her airfare?

78. **BUSINESS AND FINANCE** Rachel earns $6,000 less than twice as much as Tom. If their two incomes total $48,000, how much does each earn?

79. **STATISTICS** There are 99 students registered in three sections of algebra. There are twice as many students in the 10 A.M. section as the 8 A.M. section and 7 more students at 12 P.M. than at 8 A.M. How many students are in each section?

80. **BUSINESS AND FINANCE** The Randolphs used 12 more gal of fuel oil in October than in September and twice as much oil in November as in September. If they used 132 gal for the 3 months, how much was used during each month?

Answers

70. _____

71. _____

72. _____

73. _____

74. _____

75. _____

76. _____

77. _____

78. _____

79. _____

80. _____

Answers

81. _____

82. _____

83. _____

84. _____

85. _____

86. _____

Basic Skills | Challenge Yourself | Calculator/Computer | **Career Applications** | Above and Beyond
▲

81. MECHANICAL ENGINEERING A motor's horsepower (hp) is approximated by the equation

$$hp = \frac{6.2832 \cdot T \cdot (rpm)}{33,000}$$

in which T is the torque of the motor and (rpm) is its revolutions per minute.
    Find the rpm required to produce 240 hp in a motor that produces 380 foot-pounds of torque (nearest hundredth).

82. MECHANICAL ENGINEERING In a planetary gear, the size and number of teeth must satisfy the equation

$$Cx = By(F - 1)$$

    Calculate the number of teeth y needed if $C = 9$ in., $x = 14$ teeth, $B = 2$ in., and $F = 8$.

83. ELECTRICAL ENGINEERING Power dissipation, in watts, is given by the quotient of the square of the voltage and the resistance.

  (a) Express the given relationship with a formula.

  (b) Determine the power dissipation when 13.2 volts pass through a 220-Ω resistor (nearest thousandth).

84. INFORMATION TECHNOLOGY The total distance around a circular ring network in a metropolitan area is 100 mi. What is the diameter of the ring network (three decimal places)?

85. ALLIED HEALTH A patient enters treatment with an abdominal tumor weighing 32 g. Each day, chemotherapy reduces the size of the tumor by 2.33 g. Therefore, a formula to describe the mass m of the tumor after t days of treatment is

$$m = 32 - 2.33t$$

  (a) How much does the tumor weigh after one week of treatment?

  (b) When will the tumor weigh less than 10 g?

  (c) How many days of chemotherapy are required to eliminate the tumor?

86. ALLIED HEALTH Yohimbine is used to reverse the effects of xylazine in deer. The recommended dose is 0.125 mg per kilogram of a deer's weight.

  (a) Write a formula that expresses the required dosage level d for a deer of weight w.

  (b) How much yohimbine should be administered to a 15-kg fawn?

  (c) What size deer requires a 5.0-mg dosage?

ELECTRONICS TECHNOLOGY *Temperature sensors output voltage at a certain temperature. The output voltage varies with respect to temperature. For a particular sensor, the output voltage V for a given Celsius temperature C is given by*

$$V = 0.28C + 2.2$$

87. Determine the output voltage at 0°C.

88. Determine the output voltage at 22°C.

89. Determine the temperature if the sensor outputs 14.8 V.

90. At what temperature is there no voltage output (two decimal places)?

Answers

87. _____

88. _____

89. _____

90. _____

91. _____

92. _____

93. _____

Basic Skills | Challenge Yourself | Calculator/Computer | Career Applications | **Above and Beyond**
▲

91. "I make $2.50 an hour more in my new job." If x = the amount I used to make per hour and y = the amount I now make, which equation(s) below say the same thing as the statement above? Explain your choice(s) by translating the equation into English and comparing with the original statement.

(a) $x + y = 2.50$ (b) $x - y = 2.50$

(c) $x + 2.50 = y$ (d) $2.50 + y = x$

(e) $y - x = 2.50$ (f) $2.50 - x = y$

92. "The river rose 4 feet above flood stage last night." If a = the river's height at flood stage and b = the river's height now (the morning after), which equation(s) below say the same thing as the statement? Explain your choice(s) by translating the equations into English and comparing with the original statement.

(a) $a + b = 4$ (b) $b - 4 = a$

(c) $a - 4 = b$ (d) $a + 4 = b$

(e) $b + 4 = b$ (f) $b - a = 4$

93. Maxine lives in Pittsburgh, Pennsylvania, and pays 8.33 cents per kilowatt hour (kWh) for electricity. During the 6 months of cold winter weather, her household uses about 1,500 kWh of electric power per month. During the two hottest summer months, the usage is also high because the family uses electricity to run an air conditioner. During these summer months, the usage is 1,200 kWh per month; the rest of the year, usage averages 900 kWh per month.

(a) Write an expression for the total yearly electric bill.

(b) Maxine is considering spending $2,000 for more insulation for her home so that it is less expensive to heat and to cool. The insulation company claims that "with proper installation the insulation will reduce your heating and cooling bills by 25 percent." If Maxine invests the money in insulation, how long will it take her to get her money back by saving on her electric bill? Write to her about what information she needs to answer this question. Give her your opinion about how long it will take to save $2,000 on heating and cooling bills, and explain your reasoning. What is your advice to Maxine?

Answers

1. $s = \dfrac{P}{4}$ **3.** $R = \dfrac{E}{I}$ **5.** $H = \dfrac{V}{LW}$ **7.** $B = 180 - A - C$

9. $x = -\dfrac{b}{a}$ **11.** $g = \dfrac{2s}{t^2}$ **13.** $y = \dfrac{15 - x}{5}$ or $y = -\dfrac{1}{5}x + 3$

15. $L = \dfrac{P - 2W}{2}$ or $L = \dfrac{P}{2} - W$ **17.** $T = \dfrac{PV}{K}$ **19.** $b = 2x - a$

21. $C = \dfrac{5}{9}(F - 32)$ or $C = \dfrac{5(F - 32)}{9}$

23. $h = \dfrac{S - 2\pi r^2}{2\pi r}$ or $h = \dfrac{S}{2\pi r} - r$ **25.** 3 cm   **27.** 5%   **29.** 25°C

31. $x + 3 = 7$ **33.** $3x - 7 = 2x$ **35.** $2(x + 5) = x + 18$

37. $2x + 3 = 7$ **39.** $4x - 7 = 41$ **41.** $\dfrac{2}{3}x + 5 = 21$

43. $3x = x + 12$ **45.** $x + 7 = 33; 26$ **47.** $x - 15 = 7; 22$
49. $x + 1,840 = 3,260; 1,420$ **51.** $2x + 5 = 35; 15$ **53.** $5x - 12 = 78; 18$
55. $x + x + 1 = 47; 23, 24$ **57.** $x + x + 1 + x + 2 = 63; 20, 21, 22$
59. $x + x + 2 = 66; 32, 34$ **61.** $x + x + 2 = 52; 25, 27$
63. $x + x + 2 + x + 4 = 63; 19, 21, 23$
65. $x + x + 1 + x + 2 + x + 3 = 86; 20, 21, 22, 23$
67. $4x = 3(x + 1) + 9; 12, 13$ **69.** $x + x + 160 = 3,260; 1,550, 1,710$
71. $x + x + 70 = 650$; Washer, \$360; dryer, \$290
73. $x + 2x - 1 = 14$; 9 years old   **75.** $x + 4x - 3 = 37$; 29 years old
77. $x + 2x - 60 = 2,400$; \$820
79. $x + 2x + x + 7 = 99$; 8 A.M.: 23, 10 A.M.: 46, 12 P.M.: 30

81. 3,317.12 rpm   **83.** (a) $D = \dfrac{V^2}{R}$; (b) 0.792

85. (a) 15.69 g; (b) 10 days; (c) 14 days   **87.** 2.2 V   **89.** 45°C
91. Above and Beyond   **93.** Above and Beyond

2.5

Applications of Linear Equations

< 2.5 Objectives >

1 > Set up and solve an application

2 > Solve geometry problems

3 > Solve mixture problems

4 > Solve motion problems

5 > Identify the elements of a percent problem

6 > Solve applications involving percents

We now have all the tools needed to solve problems that can be modeled by linear equations. Before moving to real-world applications, we look at a number problem to review the five-step process for solving word problems outlined in the previous section.

| Example 1 | Solving an Application—The Five-Step Process |

< Objective 1 >

One number is 5 more than a second number. The sum of the smaller number multiplied by 3 and the larger number times 4 is 104. Find the two numbers.

Step 1 What are you asked to find? You must find the two numbers.

Step 2 Represent the unknowns. Let x be the smaller number. Then

$$x + 5$$

is the larger number.

NOTES

In step 2, "5 more than" x translates to $x + 5$.

The parentheses are *essential* in writing the correct equation.

Step 3 Write an equation.

$$3x + 4(x + 5) = 104$$

3 times   Plus   4 times
the smaller   the larger

Step 4 Solve the equation.

$$3x + 4(x + 5) = 104$$
$$3x + 4x + 20 = 104$$
$$7x + 20 = 104$$
$$7x = 84$$
$$x = 12$$

Step 5 The smaller number (x) is 12, and the larger number ($x + 5$) is 17.

Check the solution:

$$3 \cdot (12) + 4 \cdot [(12) + 5] = 104 \qquad \text{(True)}$$

 Check Yourself 1

One number is 4 more than another. If 6 times the smaller minus 4 times the larger is 4, what are the two numbers?

The solutions for many problems from geometry will also yield equations involving parentheses. Consider Example 2.

Example 2   **Solving a Geometry Application**

< Objective 2 >

The length of a rectangle is 1 cm less than 3 times the width. If the perimeter is 54 cm, find the dimensions of the rectangle.

Step 1   You want to find the dimensions (the width and length).

Step 2   Let x be the width.

Then $3x - 1$ is the length.

3 times   1 less than
the width

Step 3   To write an equation, we use this formula for the perimeter of a rectangle.

$$P = 2W + 2L$$

So

$$2x + \underbrace{2(3x - 1)} = 54$$

Twice the   Twice the   Perimeter
width   length

NOTE

When working with geometric figures, you should always draw a sketch of the problem, including the labels assigned in step 2.

Length $3x - 1$

Width
x

Step 4   Solve the equation.

$$2x + 2(3x - 1) = 54$$
$$2x + 6x - 2 = 54$$
$$8x = 56$$
$$x = 7$$

RECALL

Be sure to return to the original statement of the problem when checking your result.

Step 5   The width x is 7 cm, and the length, $3x - 1$, is 20 cm. We leave the check to you.

 Check Yourself 2

The length of a rectangle is 5 in. more than twice the width. If the perimeter of the rectangle is 76 in., what are the dimensions of the rectangle?

Often, we need parentheses to set up a *mixture problem*. Mixture problems involve combining things that have a different value, rate, or strength, as shown in Example 3.

 Example 3   **Solving a Mixture Problem**

< Objective 3 >

Four hundred tickets were sold for a school play. General admission tickets were $4, and student tickets were $3. If the total ticket sales were $1,350, how many of each type of ticket were sold?

Step 1   You want to find the number of each type of ticket sold.

Step 2   Let x be the number of general admission tickets.

Then $\underbrace{400 - x}$ student tickets were sold.

400 tickets were
sold in all.

NOTE

We subtract x, the number of general admission tickets, from 400, the total number of tickets, to find the number of student tickets.

Step 3 The revenue from each kind of ticket is found by multiplying the price of the ticket by the number sold.

General admission tickets: $4x$ $4 for each of the x tickets

Student tickets: $3(400 - x)$ $3 for each of the $400 - x$ tickets

So to form an equation, we have

$$4x + 3(400 - x) = 1,350$$

Revenue from general admission tickets   Revenue from student tickets   Total revenue

Step 4 Solve the equation.

$$4x + 3(400 - x) = 1,350$$
$$4x + 1,200 - 3x = 1,350$$
$$x + 1,200 = 1,350$$
$$x = 150$$

Step 5 So 150 general admission and $400 - 150$ or 250 student tickets were sold. We leave the check to you.

Check Yourself 3

Beth bought 40¢ stamps and 3¢ stamps at the post office. If she purchased 92 stamps at a total cost of $22, how many of each kind did she buy?

The next group of applications that we look at are *motion problems*. They involve a distance traveled, a rate or speed, and time. To solve motion problems, we need a relationship among these three quantities.

Suppose you travel at a rate of 50 mi/h on a highway for 6 h. How far (what distance) will you have gone? To find the distance, you multiply:

$$(50 \text{ mi/h})(6 \text{ h}) = 300 \text{ mi}$$

Speed or rate   Time   Distance

> CAUTION

Make your units consistent. If a rate is given in *miles per hour,* then the time must be given in *hours* and the distance in *miles.*

Property

Relationship for Motion Problems

In general, if r is a rate, t is the time, and d is the distance traveled, then

$$d = r \cdot t$$

This is the key relationship, and it will be used in all motion problems. We apply this relationship in Example 4.

Example 4   Solving a Motion Problem

< Objective 4 >

On Friday morning Ricardo drove from his house to the beach in 4 h. In coming back on Sunday afternoon, heavy traffic slowed his speed by 10 mi/h, and the trip took 5 h. What was his average speed (rate) in each direction?

Step 1 We want the speed or rate in each direction.

Step 2   Let x be Ricardo's speed to the beach. Then $x - 10$ is his return speed.
It is always a good idea to sketch the given information in a motion problem. Here we would have

Going $\xrightarrow{\hspace{3cm}}$ x mi/h for 4 h

Returning $\xleftarrow{\hspace{3cm}}$ $(x - 10)$ mi/h for 5 h

Step 3   Because we know that the distance is the same each way, we can write an equation, using the fact that the product of the rate and the time each way must be the same.

So

$$4x = 5(x - 10)$$

Time · rate (going)   Time · rate (returning)

NOTE

Distance (going) = distance (returning)

or

Time · rate (going) = time · rate (returning)

An alternate method is to use a chart, which can help summarize the given information. We begin by filling in the information given in the problem.

	Distance	Rate	Time
Going		x	4
Returning		$x - 10$	5

Now we fill in the missing information. Here we use the fact that $d = rt$ to complete the chart.

	Distance	Rate	Time
Going	$4x$	x	4
Returning	$5(x - 10)$	$x - 10$	5

From here we set the two distances equal to each other and solve as before.

Step 4   Solve.

$$4x = 5(x - 10)$$
$$4x = 5x - 50$$
$$-x = -50$$
$$x = 50 \text{ mi/h}$$

NOTE

x was his rate going, $x - 10$ was his rate returning.

Step 5   So Ricardo's rate going to the beach was 50 mi/h, and his rate returning was 40 mi/h.

To check, you should verify that the product of the time and the rate is the same in each direction.

 Check Yourself 4

A plane made a flight (with the wind) between two towns in 2 h. Returning against the wind, the plane's speed was 60 mi/h slower, and the flight took 3 h. What was the plane's speed in each direction?

Example 5 illustrates another way of using the distance relationship.

Example 5	Solving a Motion Problem

Katy leaves Las Vegas for Los Angeles at 10 A.M., driving at 50 mi/h. At 11 A.M. Jensen leaves Los Angeles for Las Vegas, driving at 55 mi/h along the same route. If the cities are 260 mi apart, at what time will Katy and Jensen meet?

Step 1   Find the time that Katy travels until they meet.

Step 2   Let x be Katy's time.

Then $x - 1$ is Jensen's time.

Jensen left 1 h later.

Again, you should draw a sketch of the given information.

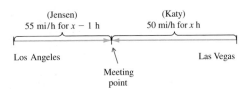

(Jensen) 55 mi/h for $x - 1$ h   (Katy) 50 mi/h for x h

Los Angeles   Meeting point   Las Vegas

Step 3   To write an equation, we again need the relationship $d = rt$. From this equation, we can write

Katy's distance $= 50x$

Jensen's distance $= 55(x - 1)$

As before, we can use a chart to solve.

	Distance	Rate	Time
Katy	$50x$	50	x
Jensen	$55(x - 1)$	55	$x - 1$

From the original problem, the sum of the distances is 260 mi, so

$$50x + 55(x - 1) = 260$$

Step 4

$$50x + 55(x - 1) = 260$$
$$50x + 55x - 55 = 260$$
$$105x - 55 = 260$$
$$105x = 315$$
$$x = 3 \text{ h}$$

NOTE

Be sure to answer the question asked in the problem.

Step 5   Finally, because Katy left at 10 A.M., the two will meet at 1 P.M. We leave the check of this result to you.

Check Yourself 5

At noon a jogger leaves one point, running at 8 mi/h. One hour later a bicyclist leaves the same point, traveling at 20 mi/h in the opposite direction. At what time will they be 36 mi apart?

The final type of problem we look at in this section involves percents. Percents come up in more applications than nearly any other type of problem, so it is important that you become comfortable modeling and solving percent problems.

Every complete percent statement has three parts that need to be identified. We call these parts the *base,* the *rate,* and the *amount.* Here are definitions for each of these terms.

Definition	
Base, Amount, and Rate	The **base** is the whole in a problem. It is the standard used for comparison. The **amount** is the part of the whole being compared to the base. The **rate** is the ratio of the amount to the base. It is usually written as a percent.

The next examples provide some practice in determining the parts of a percent problem.

 Example 6   Identifying the Parts of a Percent Problem

< Objective 5 >

In each case, identify the base, the amount, and the rate.

(a) 50% of 480 is 240.

The base in this problem is 480.

The amount is 240. This is being compared to the base.

The rate is 50%. It is the percent.

(b) Delia borrows $10,000 for 1 year at 11.49% interest. How much interest will she pay?

The base is the beginning amount, $10,000.

In this case, the amount is the interest she will pay. The amount is *unknown*.

The rate is given by the percent, 11.49%.

NOTES

The base is usually the quantity we begin with.

We will solve this type of problem for the unknown amount.

 Check Yourself 6

Identify the base, the amount, and the rate in each case.

(a) 150 is 25% of what number?

(b) Steffen earned $120 in interest from a CD account that paid 8% interest when he invested $1,500 for one year.

As we said, every percent problem consists of these three parts: base, amount, and rate. In nearly every such problem, one of these parts is unknown. Solving a percent problem is a matter of identifying and finding the missing part. To do this, we use the *percent relationship*.

Property	
The Percent Relationship	In a percent statement, the amount is equal to the product of the rate and the base. We can write this as a formula with B equal to the base, A the amount, and R the rate. $A = R \cdot B$

NOTE

To solve problems involving percents, we write the rate as a decimal or fraction.

Now we are ready to solve percent problems. We begin with some straightforward ones and work our way to more involved applications. In all cases, your first step should be to identify the parts of the percent relationship.

 Example 7   Solving Percent Problems

< Objective 6 >

(a) 84 is 5% of what number?

5% is the rate and 84 is the amount. The base is unknown.

We substitute these values into the percent-relationship equation and solve.

$$\underbrace{(84)}_{A=\text{Amount}} = \underbrace{(0.05)}_{R=\text{Rate}} \cdot \underbrace{B}_{\substack{B=\text{Unknown}\\\text{Base}}}$$ Write the rate as a decimal.

$$\frac{84}{0.05} = B$$ Divide by 0.05 to isolate the variable.

$$1{,}680 = B$$

Answer the question using a sentence: 84 is 5% of 1,680.

(b) Delia borrows $10,000 for 1 year at 11.49% interest. How much interest will she pay?

From Example 6(b), we know that the missing element is the amount.

$$A = R \cdot B$$
$$= (0.1149) \cdot (10{,}000)$$
$$= 1{,}149$$

Delia's interest payment comes to $1,149 after one year.

Check Yourself 7

Solve each problem.

(a) 32 is what percent of 128?

(b) If you invest $5,000 for one year at $8\frac{1}{2}\%$, how much interest will you earn?

We conclude this section with some more involved percent applications.

   Example 8   Solving Percent Applications

(a) A state adds a 7.25% sales tax to the price of most goods. If a 30-GB iPod is listed for $299, how much will it cost after the sales tax has been added?

This problem is similar to the application in Example 7(b), in that we are missing the amount. There is the further complication that we need to add the sales tax to the original price.

If we use the price, including tax, as the unknown amount, then the rate is

$$R = 107.25\% = 1.0725$$

The base is the list price, $B = \$299$.
As before, we use the percent relationship to solve the problem.

$$A = R \cdot B$$
$$= (1.0725) \cdot (299)$$
$$= 320.6775$$

Because our answer refers to money, we round to two decimal places. The iPod sells for $320.68, after the sales tax has been included.

(b) A store sells a certain Kicker amplifier model for a car stereo system for $249.95. If the store pays $199.95 for the amplifier, what is its markup percentage for the item (to the nearest whole percent)?

The base is given by the wholesale price, $B = \$199.95$.

In this case, though, the amount is not the selling price, but rather, the difference between the selling price and the wholesale price.

$$A = 249.95 - 199.95 = 50$$

Therefore, in this problem we are missing the rate.
Once we have the amount, we can use the percent relationship, as before.

$$A = R \cdot B$$

$$(50) = R \cdot (199.95) \qquad \text{Isolate the variable.}$$

$$\frac{50}{199.95} = R$$

$$0.250 = R$$

The store marked up the amplifier by 25%.

Check Yourself 8

(a) In order to make room for the new fall line of merchandise, a proprietor offers to discount all existing stock by 15%. How much would you pay for a Fendi handbag that the store usually sells for $229?

(b) A grocery store adds a 30% markup to the wholesale price of an item to determine the selling price. If the store sells a half-gallon container of orange juice for $2.99, what is the wholesale price of the orange juice?

Check Yourself ANSWERS

1. The numbers are 10 and 14.   **2.** The width is 11 in. and the length is 27 in.
3. Beth bought fifty-two 40¢ stamps and forty 3¢ stamps.   **4.** The plane flew at a rate of 180 mi/h with the wind and 120 mi/h against the wind.
5. At 2 P.M. the jogger and the bicyclist will be 36 mi apart.
6. (a) B = unknown, $A = 150$, $R = 25\%$; **(b)** $B = \$1,500$, $A = \$120$, $R = 8\%$
7. (a) 25%; **(b)** $425   **8. (a)** $194.65; **(b)** $2.30

Reading Your Text

The following fill-in-the-blank exercises are designed to ensure that you understand some of the key vocabulary used in this section.

SECTION 2.5

(a) Always try to draw a sketch of the figures when solving _____ applications.

(b) _____ problems involve combining things that have a different value, rate, or strength.

(c) In a percent problem, the rate is the ratio of the _____ to the base.

(d) To solve a percent problem, begin by _____ the parts of the percent relationship.

RECALL

To round to the nearest whole percent (two decimal places), we need to divide to a third decimal place.

2.5 exercises

< Objective 1 >

Solve each word problem. Be sure to show the equation you use for the solution.

1. **NUMBER PROBLEM** One number is 8 more than another. If the sum of the smaller number and twice the larger number is 46, find the two numbers.

2. **NUMBER PROBLEM** One number is 3 less than another. If 4 times the smaller number minus 3 times the larger number is 4, find the two numbers.

3. **NUMBER PROBLEM** One number is 7 less than another. If 4 times the smaller number plus 2 times the larger number is 62, find the two numbers.

4. **NUMBER PROBLEM** One number is 10 more than another. If the sum of twice the smaller number and 3 times the larger number is 55, find the two numbers.

 > Videos

5. **NUMBER PROBLEM** Find two consecutive integers such that the sum of twice the first integer and 3 times the second integer is 28. (*Hint:* If x represents the first integer, $x + 1$ represents the next consecutive integer.)

6. **NUMBER PROBLEM** Find two consecutive odd integers such that 3 times the first integer is 5 more than twice the second. (*Hint:* If x represents the first integer, $x + 2$ represents the next consecutive odd integer.)

< Objective 2 >

7. **GEOMETRY** The length of a rectangle is 1 in. more than twice its width. If the perimeter of the rectangle is 74 in., find the dimensions of the rectangle.

8. **GEOMETRY** The length of a rectangle is 5 cm less than 3 times its width. If the perimeter of the rectangle is 46 cm, find the dimensions of the rectangle.   > Videos

9. **GEOMETRY** The length of a rectangular garden is 4 m more than 3 times its width. The perimeter of the garden is 56 m. What are the dimensions of the garden?

10. **GEOMETRY** The length of a rectangular playing field is 5 ft less than twice its width. If the perimeter of the playing field is 230 ft, find the length and width of the field.

11. **GEOMETRY** The base of an isosceles triangle is 3 cm less than the length of the equal sides. If the perimeter of the triangle is 36 cm, find the length of each of the sides.

12. **GEOMETRY** The length of one of the equal legs of an isosceles triangle is 3 in. less than twice the length of the base. If the perimeter is 29 in., find the length of each of the sides.

< Objective 3 >

13. **BUSINESS AND FINANCE** Tickets for a play cost $8 for the main floor and $6 in the balcony. If the total receipts from 500 tickets were $3,600, how many of each type of ticket were sold?

14. **BUSINESS AND FINANCE** Tickets for a basketball tournament were $6 for students and $9 for nonstudents. Total sales were $10,500, and 250 more student tickets were sold than nonstudent tickets. How many of each type of ticket were sold?   > Videos

Boost *your* GRADE at ALEKS.com!

ALEKS®

- Practice Problems   • e-Professors
- Self-Tests   • Videos
- NetTutor

Name _____

Section _____ Date _____

Answers

1. _____

2. _____

3. _____

4. _____

5. _____

6. _____

7. _____

8. _____

9. _____

10. _____

11. _____

12. _____

13. _____

14. _____

Answers

15. _____

16. _____

17. _____

18. _____

19. _____

20. _____

21. _____

22. _____

23. _____

24. _____

25. _____

15. **BUSINESS AND FINANCE** Maria bought 50 stamps at the post office in 27¢ and 42¢ denominations. If she paid $18 for the stamps, how many of each denomination did she buy?

16. **BUSINESS AND FINANCE** A bank teller had a total of 125 $10 bills and $20 bills to start the day. If the value of the bills was $1,650, how many of each denomination did he have?

17. **BUSINESS AND FINANCE** Tickets for a train excursion were $120 for a sleeping room, $80 for a berth, and $50 for a coach seat. The total ticket sales were $8,600. If there were 20 more berth tickets sold than sleeping room tickets and 3 times as many coach tickets as sleeping room tickets, how many of each type of ticket were sold?

18. **BUSINESS AND FINANCE** Admission for a college baseball game is $6 for box seats, $5 for the grandstand, and $3 for the bleachers. The total receipts for one evening were $9,000. There were 100 more grandstand tickets sold than box seat tickets. Twice as many bleacher tickets were sold as box seat tickets. How many tickets of each type were sold?

< Objective 4 >

19. **SCIENCE AND MEDICINE** Patrick drove 3 h to attend a meeting. On the return trip, his speed was 10 mi/h less and the trip took 4 h. What was his speed each way?

20. **SCIENCE AND MEDICINE** A bicyclist rode into the country for 5 h. In returning, her speed was 5 mi/h faster and the trip took 4 h. What was her speed each way?

21. **SCIENCE AND MEDICINE** A car leaves a city and goes north at a rate of 50 mi/h at 2 P.M. One hour later a second car leaves, traveling south at a rate of 40 mi/h. At what time will the two cars be 320 mi apart?

> Videos

22. **SCIENCE AND MEDICINE** A bus leaves a station at 1 P.M., traveling west at an average rate of 44 mi/h. One hour later a second bus leaves the same station, traveling east at a rate of 48 mi/h. At what time will the two buses be 274 mi apart?

23. **SCIENCE AND MEDICINE** At 8:00 A.M., Catherine leaves on a trip at 45 mi/h. One hour later, Max decides to join her and leaves along the same route, traveling at 54 mi/h. When will Max catch up with Catherine?

24. **SCIENCE AND MEDICINE** Martina leaves home at 9 A.M., bicycling at a rate of 24 mi/h. Two hours later, John leaves, driving at the rate of 48 mi/h. At what time will John catch up with Martina?

25. **SCIENCE AND MEDICINE** Mika leaves Boston for Baltimore at 10:00 A.M., traveling at 45 mi/h. One hour later, Hiroko leaves Baltimore for Boston on the same route, traveling at 50 mi/h. If the two cities are 425 mi apart, when will Mika and Hiroko meet?

26. SCIENCE AND MEDICINE A train leaves town A for town B, traveling at 35 mi/h. At the same time, a second train leaves town B for town A at 45 mi/h. If the two towns are 320 mi apart, how long will it take for the two trains to meet?

27. BUSINESS AND FINANCE There are a total of 500 Douglas fir and hemlock trees in a section of forest bought by Hoodoo Logging Co. The company paid an average of $250 for each Douglas fir and $300 for each hemlock. If the company paid $132,000 for the trees, how many of each kind did the company buy?

28. BUSINESS AND FINANCE There are 850 Douglas fir and ponderosa pine trees in a section of forest bought by Sawz Logging Co. The company paid an average of $300 for each Douglas fir and $225 for each ponderosa pine. If the company paid $217,500 for the trees, how many of each kind did the company buy?

< Objective 5 >

Identify the indicated quantity in each statement.

29. The rate in the statement "23% of 400 is 92."

30. The base in the statement "40% of 600 is 240."

31. The amount in the statement "200 is 40% of 500."

32. The rate in the statement "480 is 60% of 800."

33. The base in the statement "16% of 350 is 56."

34. The amount in the statement "150 is 75% of 200."

Identify the rate, the base, and the amount in each application. Do not solve the applications at this point.

35. BUSINESS AND FINANCE Jan has a 5% commission rate on all her sales. If she sells $40,000 worth of merchandise in 1 month, what commission will she earn? > Videos

36. BUSINESS AND FINANCE 22% of Shirley's monthly salary is deducted for withholding. If those deductions total $209, what is her salary?

37. SCIENCE AND MEDICINE In a chemistry class of 30 students, 5 received a grade of A. What percent of the students received A's?

38. BUSINESS AND FINANCE A can of mixed nuts contains 80% peanuts. If the can holds 16 oz, how many ounces of peanuts does it contain?

Answers

26. _____

27. _____

28. _____

29. _____

30. _____

31. _____

32. _____

33. _____

34. _____

35. _____

36. _____

37. _____

38. _____

Answers

39. _____

40. _____

41. _____

42. _____

43. _____

44. _____

45. _____

46. _____

47. _____

48. _____

49. _____

50. _____

51. _____

52. _____

39. **STATISTICS** A college had 9,000 students at the start of a school year. If there is an enrollment increase of 6% by the beginning of the next year, how many additional students will there be?

40. **BUSINESS AND FINANCE** Paul invested $5,000 in a time deposit. What interest will he earn for 1 year if the interest rate is 6.5%?

< Objective 6 >

Solve each application.

41. **BUSINESS AND FINANCE** What interest will you pay on a $3,400 loan for 1 year if the interest rate is 12%?

42. **SCIENCE AND MEDICINE** A chemist has 300 milliliters (mL) of solution that is 18% acid. How many milliliters of acid are in the solution?

43. **BUSINESS AND FINANCE** Roberto has 26% of his pay withheld for deductions. If he earns $550 per week, what amount is withheld?

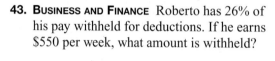

44. **BUSINESS AND FINANCE** A real estate agent's commission rate is 6%. What will the amount of the commission be on the sale for a $185,000 home?

45. **BUSINESS AND FINANCE** If a salesman is paid a $140 commission on the sale of a $2,800 sailboat, what is his commission rate?

46. **BUSINESS AND FINANCE** Ms. Jordan has been given a loan of $2,500 for 1 year. If the interest charged is $275, what is the interest rate on the loan?

47. **BUSINESS AND FINANCE** Joan was charged $18 interest for 1 month on a $1,200 credit card balance. What was the monthly interest rate?

48. **SCIENCE AND MEDICINE** There are 117 grams (g) of acid in 900 g of a solution of acid and water. What percent of the solution is acid?

49. **STATISTICS** On a test, Alice had 80% of the problems right. If she had 20 problems correct, how many questions were on the test?

50. **BUSINESS AND FINANCE** A state sales tax rate is 3.5%. If the tax on a purchase is $7, what was the amount of the purchase?

51. **BUSINESS AND FINANCE** If a house sells for $125,000 and the commission rate is $6\frac{1}{2}$%, how much will the salesperson make for the sale?

52. **STATISTICS** Marla needs 70% on a final test to receive a C for a course. If the exam has 120 questions, how many questions must she answer correctly?

> Videos

53. **SOCIAL SCIENCE** A study has shown that 102 of the 1,200 people in the workforce of a small town are unemployed. What is the town's unemployment rate?

54. **STATISTICS** A survey of 400 people found that 66 were left-handed. What percent of those surveyed were left-handed?

55. **STATISTICS** Of 60 people who start a training program, 45 complete the course. What is the dropout rate?

56. **BUSINESS AND FINANCE** In a shipment of 250 parts, 40 are found to be defective. What percent of the parts are faulty?

57. **STATISTICS** In a recent survey, 65% of those responding were in favor of a freeway improvement project. If 780 people were in favor of the project, how many people responded to the survey?

58. **STATISTICS** A college finds that 42% of the students taking a foreign language are enrolled in Spanish. If 1,512 students are taking Spanish, how many foreign language students are there?

59. **BUSINESS AND FINANCE** An appliance dealer marks up refrigerators 22% (based on cost). If the cost of one model was $600, what will its selling price be?

60. **STATISTICS** A school had 900 students at the start of a school year. If there is an enrollment increase of 7% by the beginning of the next year, what is the new enrollment?

61. **BUSINESS AND FINANCE** A home lot purchased for $125,000 increased in value by 25% over 3 years. What was the lot's value at the end of the period?

62. **BUSINESS AND FINANCE** New cars depreciate an average of 28% in their first year of use. What would an $18,000 car be worth after 1 year?

63. **STATISTICS** A school's enrollment was up from 950 students in 1 year to 1,064 students in the next. What was the rate of increase?

64. **BUSINESS AND FINANCE** Under a new contract, the salary for a position increases from $31,000 to $33,635. What rate of increase does this represent?

65. **BUSINESS AND FINANCE** The price of a new van has increased $4,830, which amounts to a 14% increase. What was the price of the van before the increase?

66. **BUSINESS AND FINANCE** A television set is marked down $75, for a sale. If this is a 12.5% decrease from the original price, what was the selling price before the sale?

67. **STATISTICS** A company had 66 fewer employees in July 2005 than in July 2004. If this represents a 5.5% decrease, how many employees did the company have in July 2004?

Answers

53. _____

54. _____

55. _____

56. _____

57. _____

58. _____

59. _____

60. _____

61. _____

62. _____

63. _____

64. _____

65. _____

66. _____

67. _____

68. BUSINESS AND FINANCE Carlotta received a monthly raise of $162.50. If this represented a 6.5% increase, what was her monthly salary before the raise?

69. BUSINESS AND FINANCE A pair of shorts, advertised for $48.75, is being sold at 25% off the original price. What was the original price?

70. BUSINESS AND FINANCE If the total bill at a restaurant, including a 15% tip, is $65.32, what was the cost of the meal alone?

The chart below gives U.S.-Mexico trade data from 2000 to 2005. Use this information for exercises 71–74.

U.S. Trade with Mexico, 2000 to 2005 (in millions of dollars)

Year	Exports	Imports	Trade Balance
2000	$111,349	$135,926	−$24,577
2001	101,297	131,338	−30,041
2002	97,470	134,616	−37,146
2003	97,412	138,060	−40,648
2004	110,835	155,902	−45,067
2005	120,049	170,198	−50,149

Source: U.S. Census Bureau, Foreign Trade Division.

71. What was the percent increase (to the nearest whole percent) of exports from 2000 to 2005?

72. What was the percent increase (to the nearest whole percent) of imports from 2000 to 2005?

73. By what percent (to the nearest whole percent) did imports exceed exports in 2000? 2005?

74. By what percent (to the nearest whole percent) did the trade imbalance increase between 2000 and 2005?

75. STATISTICS In 1990, there were an estimated 145.0 million passenger cars registered in the United States. The total number of vehicles registered in the United States for 1990 was estimated at 194.5 million. What percent of the vehicles registered were passenger cars (to the nearest tenth)?

76. STATISTICS Gasoline accounts for 85% of the motor fuel consumed in the United States every day. If 8,882 thousand barrels (bbl) of motor fuel are consumed each day, how much gasoline is consumed each day in the United States (to the nearest gallon)?

77. STATISTICS In 1999, transportation accounted for 63% of U.S. petroleum consumption. Assuming that same rate applies now, and 10.85 million bbl of petroleum are used each day for transportation in the United States, what is the total daily petroleum consumption by all sources in the United States (to the nearest hundredth)?

78. **STATISTICS** Each year, 540 million metric tons (t) of carbon dioxide are added to the atmosphere by the United States. Burning gasoline and other transportation fuels is responsible for 35% of the carbon dioxide emissions in the United States. How much carbon dioxide is emitted each year by the burning of transportation fuels in the United States?

Answers

78. _____

79. _____

80. _____

81. _____

82. _____

Basic Skills	Challenge Yourself	Calculator/Computer	Career Applications	**Above and Beyond**

79. There is a universally accepted "order of operations" used to simplify expressions. Explain how the order of operations is used in solving equations. Be sure to use complete sentences.

80. A common mistake when solving equations is

The equation: $2(x - 2) = x + 3$

First step in solving: $2x - 2 = x + 3$

Write a clear explanation of what error has been made. What could be done to avoid this error?

81. Another common mistake is shown in the equation below.

The equation: $6x - (x + 3) = 5 + 2x$

First step in solving: $6x - x + 3 = 5 + 2x$

Write a clear explanation of what error has been made and what could be done to avoid the mistake.

82. Write an algebraic equation for the English statement "Subtract 5 from the sum of x and 7 times 3 and the result is 20." Compare your equation with those of other students. Did you all write the same equation? Are all the equations correct even though they don't look alike? Do all the equations have the same solution? What is wrong? The English statement is *ambiguous*. Write another English statement that leads correctly to more than one algebraic equation. Exchange with another student and see whether the other student thinks the statement is ambiguous. Notice that the algebra is *not* ambiguous!

Answers

1. 10, 18 **3.** 8, 15 **5.** 5, 6 **7.** 12 in., 25 in. **9.** 6 m, 22 m

11. 13-cm legs, 10-cm base **13.** 200 $6 tickets, 300 $8 tickets

15. 20 27¢ stamps, 30 42¢ stamps **17.** 60 coach, 40 berth, 20 sleeping room

19. 40 mi/h, 30 mi/h **21.** 6 P.M. **23.** 2 P.M. **25.** 3 P.M.

27. 360 Douglas firs, 140 hemlocks **29.** 23% **31.** 200 **33.** 350

35. $R = 5\%, B = \$40{,}000, A = $ unknown **37.** $R = $ unknown, $B = 30, A = 5$

39. $R = 6\%, B = 9{,}000, A = $ unknown **41.** $408 **43.** $143 **45.** 5%

47. 1.5% **49.** 25 questions **51.** $8,125 **53.** 8.5% **55.** 25%

57. 1,200 people **59.** $732 **61.** $156,250 **63.** 12%

65. $34,500 **67.** 1,200 employees **69.** $65 **71.** 8%

73. 22%; 42% **75.** 74.6% **77.** 17.22 million bbl

79. Above and Beyond **81.** Above and Beyond

2.6

Inequalities—An Introduction

< 2.6 Objectives >

1 > Use inequality notation

2 > Graph the solution set of an inequality

3 > Solve an inequality and graph the solution set

4 > Solve an application using inequalities

As we pointed out earlier, an equation is a statement that two expressions are equal. In algebra, an **inequality** is a statement that one expression is less than or greater than another. We show two of the inequality symbols in Example 1.

| ▶ Example 1 | Reading the Inequality Symbol |

< Objective 1 >

NOTE

To help you remember, the "arrowhead" always points toward the smaller quantity.

$5 < 8$ is an inequality read "5 is less than 8."

$9 > 6$ is an inequality read "9 is greater than 6."

Check Yourself 1

Fill in the blanks using the symbols < and >.

(a) 12 ___>___ 8   (b) 20 ___<___ 25

Like an equation, an inequality can be represented by a balance scale. Note that, in each case, the inequality arrow points to the side that is "lighter."

$2x < 4x - 3$

NOTE

The $2x$ side is less than the $4x - 3$ side, so it is "lighter."

$5x - 6 > 9$

Just as was the case with equations, inequalities that involve variables may be either true or false depending on the value that we give to the variable. For instance, consider the inequality

$x < 6$

The Streeter/Hutchison Series in Mathematics   Beginning Algebra

$$\text{If } x = \begin{cases} 3 & 3 < 6 \text{ is true} \\ 5 & 5 < 6 \text{ is true} \\ -10 & -10 < 6 \text{ is true} \\ 8 & 8 < 6 \text{ is false} \end{cases}$$

Therefore, 3, 5, and -10 are *solutions* for the inequality $x < 6$; they make the inequality a true statement. You should see that 8 is *not* a solution. We call the set of all solutions the **solution set** for the inequality. Of course, there are many possible solutions.

Because there are so many solutions (an infinite number, in fact), we certainly do not want to try to list them all! A convenient way to show the solution set of an inequality is with a number line.

Example 2   Solving Inequalities

< Objective 2 >

NOTE

The colored arrow indicates the direction of the *solution set.*

To graph the solution set for the inequality $x < 6$, we want to include all real numbers that are "less than" 6. This means all numbers *to the left* of 6 on the number line. We start at 6 and draw an arrow extending left, as shown:

Note: The **open circle** at 6 means that we do not include 6 in the solution set (6 is not less than itself). The colored arrow shows all the numbers in the solution set, with the arrowhead indicating that the solution set continues indefinitely to the left.

✓ **Check Yourself 2**

Graph the solution set of $x < -2$.

Two other symbols are used in writing inequalities. They are used with inequalities such as

$$x \geq 5 \qquad \text{and} \qquad x \leq 2$$

Here $x \geq 5$ is really a combination of the two statements $x > 5$ and $x = 5$. It is read "x is greater than or equal to 5." The solution set includes 5 in this case.

The inequality $x \leq 2$ combines the statements $x < 2$ and $x = 2$. It is read "x is less than or equal to 2."

Example 3   Graphing Inequalities

NOTE

Here the filled-in circle means that we include 5 in the solution set. This is often called a **closed** circle.

The solution set for $x \geq 5$ is graphed as follows.

✓ **Check Yourself 3**

Graph the solution sets.

(a) $x \leq -4$ (b) $x \geq 3$

NOTE

Equivalent inequalities have exactly the same solution sets.

You have learned how to graph the solution sets of some simple inequalities, such as $x < 8$ or $x \geq 10$. Now we look at more complicated inequalities, such as

$$2x - 3 < x + 4$$

This is called a **linear inequality in one variable.** Only one variable is involved in the inequality, and it appears only to the first power. Fortunately, the methods used to solve this type of inequality are very similar to those we used earlier in this chapter to solve linear equations in one variable. Here is our first property for inequalities.

The Streeter/Hutchison Series in Mathematics   Beginning Algebra

Property

| The Addition Property of Inequality | If $a < b$ then $a + c < b + c$
In words, adding the same quantity to both sides of an inequality gives an equivalent inequality. |

NOTES

Because $a < b$, the scale shows b to be heavier.

The second scale represents $a + c < b + c$

Again, we can use the idea of a balance scale to see the significance of this property. If we add the same weight to both sides of an unbalanced scale, it stays unbalanced.

Example 4   **Solving Inequalities**

< Objective 3 >

NOTE

The inequality is solved when an equivalent inequality has the form
$x < \square$ or $x > \square$

Solve and graph the solution set for $x - 8 < 7$.

To solve $x - 8 < 7$, add 8 to both sides of the inequality by the addition property.

$$\begin{array}{rl} x - 8 < & 7 \\ \underline{+8 \quad +8} & \\ x \quad\quad < & 15 \end{array}$$ (The inequality is solved.)

The graph of the solution set is

 Check Yourself 4

Solve and graph the solution set.

$x - 9 > -3$

As with equations, the addition property allows us to subtract the same quantity from both sides of an inequality.

Example 5   **Solving Inequalities**

NOTE

We subtracted $3x$ and then added 2 to both sides. If these steps are done in the reverse order, the result is the same.

Solve and graph the solution set for $4x - 2 \geq 3x + 5$.

First, we subtract $3x$ from both sides of the inequality.

$$\begin{array}{rl} 4x - 2 \geq & 3x + 5 \\ \underline{-3x \quad\quad\quad -3x} & \\ x - 2 \geq & 5 \\ \underline{+2 \quad\quad +2} & \\ x \quad\quad \geq & 7 \end{array}$$ Subtract $3x$ from both sides.

    Now we add 2 to both sides.

The graph of the solution set is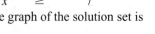

Check Yourself 5

Solve and graph the solution set.

$7x - 8 \leq 6x + 2$

We also need a rule for multiplying on both sides of an inequality. Here we have to be a bit careful. There is a difference between the multiplication property for inequalities and that for equations. Look at the following:

$2 < 7$ (A true inequality)

Multiply both sides by 3.

$2 < 7$

$3 \cdot 2 < 3 \cdot 7$

$6 < 21$ (A true inequality)

Now we multiply both sides of the original inequality by -3.

$2 < 7$

$(-3)(2) < (-3)(7)$

$-6 < -21$ (*Not* a true inequality)

But,

$2 < 7$

$(-3)(2) > (-3)(7)$ Change the direction of the inequality: $<$ becomes $>$. (This is now a true inequality.)

$-6 > -21$

This suggests that multiplying both sides of an inequality by a negative number changes the direction of the inequality.

We can state the following general property.

Property

The Multiplication Property of Inequality

> If $a < b$ then $ac < bc$ if $c > 0$
>
>   and $ac > bc$ if $c < 0$

In words, multiplying both sides of an inequality by the same *positive* number gives an equivalent inequality.

When both sides of an inequality are multiplied by the same *negative* number, it is necessary to *reverse the direction* of the inequality to give an equivalent inequality.

NOTE

Because division is defined in terms of multiplication, this rule applies to division, as well.

 Example 6   Solving and Graphing Inequalities

(a) Solve and graph the solution set for $5x < 30$.

Multiplying both sides of the inequality by $\dfrac{1}{5}$ gives

$$\frac{1}{5}(5x) < \frac{1}{5}(30)$$

Simplifying, we have

$x < 6$

The graph of the solution set is

(b) Solve and graph the solution set for $-4x \geq 28$.

In this case we want to multiply both sides of the inequality by $-\dfrac{1}{4}$ to leave x alone on the left.

$$\left(-\frac{1}{4}\right)(-4x) \leq \left(-\frac{1}{4}\right)(28)$$

Reverse the direction of the inequality because you are multiplying by a negative number!

or $x \leq -7$

The graph of the solution set is

Check Yourself 6

Solve and graph the solution sets.

(a) $7x > 35$ **(b)** $-8x \leq 48$

Example 7 illustrates the use of the multiplication property when fractions are involved in an inequality.

Example 7   **Solving and Graphing Inequalities**

(a) Solve and graph the solution set for

$$\frac{x}{4} > 3$$

Here we multiply both sides of the inequality by 4. This isolates x on the left.

$$4\left(\frac{x}{4}\right) > 4(3)$$

$$x > 12$$

The graph of the solution set is

(b) Solve and graph the solution set for

$$-\frac{x}{6} \geq -3$$

NOTE

We reverse the direction of the inequality because we are multiplying by a negative number.

In this case, we multiply both sides of the inequality by -6:

$$(-6)\left(-\frac{x}{6}\right) \leq (-6)(-3)$$

$$x \leq 18$$

The graph of the solution set is

Check Yourself 7

Solve and graph the solution sets.

(a) $\dfrac{x}{5} \leq 4$ **(b)** $-\dfrac{x}{3} < -7$

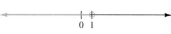

Example 8 | **Solving and Graphing Inequalities**

(a) Solve and graph the solution set for $5x - 3 < 2x$.

$$\begin{array}{rcl} 5x - 3 &<& 2x \\ -2x && -2x \\ \hline 3x - 3 &<& 0 \\ +3 && +3 \\ \hline 3x &<& 3 \end{array}$$

Bring the variable terms to the same (left) side.

Isolate the variable term.

Next, *divide* both sides by 3.

$$\frac{3x}{3} < \frac{3}{3}$$

$$x < 1$$

The graph of the solution set is

$$\quad 0 \;\; 1$$

(b) Solve and graph the solution set for $2 - 5x < 7$.

$$\begin{array}{rcl} 2 - 5x &<& 7 \\ -2 && -2 \\ \hline -5x &<& 5 \end{array}$$

Add -2.

$$\frac{-5x}{-5} > \frac{5}{-5}$$

Divide by -5. Be sure to reverse the direction of the inequality.

or $x > -1$

The graph is

$$\quad -1 \qquad 0$$

Check Yourself 8

Solve and graph the solution sets.

(a) $4x + 9 \geq x$ **(b)** $5 - 6x < 41$

As with equations, we collect all variable terms on one side and all constant terms on the other.

Example 9 | **Solving and Graphing Inequalities**

Solve and graph the solution set for $5x - 5 \geq 3x + 4$.

$$\begin{array}{rcl} 5x - 5 &\geq& 3x + 4 \\ -3x && -3x \\ \hline 2x - 5 &\geq& 4 \\ +5 && +5 \\ \hline 2x &\geq& 9 \end{array}$$

Bring the variable terms to the same (left) side.

Isolate the variable term.

$$\frac{2x}{2} \geq \frac{9}{2}$$

Isolate the variable.

$$x \geq \frac{9}{2}$$

The graph of the solution set is

$$\quad 0 \qquad \frac{9}{2}$$

Check Yourself 9

Solve and graph the solution set.

$8x + 3 < 4x - 13$

Be especially careful when negative coefficients occur in the process of solving.

 Example 10   **Solving and Graphing Inequalities**

Solve and graph the solution set for $2x + 4 < 5x - 2$.

$$
\begin{array}{rll}
2x + 4 <\quad 5x - 2 & \\
\underline{-5x \qquad\quad -5x} & \qquad \text{Bring the variable terms to the same (left) side.}\\
-3x + 4 <\qquad -2 & \\
\underline{\quad\; -4 \qquad\quad -4} & \qquad \text{Isolate the variable term.}\\
-3x \quad < \qquad -6 & \\
\dfrac{-3x}{-3} \;>\; \dfrac{-6}{-3} & \qquad \text{Isolate the variable. Be sure to reverse the direction of the}\\
& \qquad \text{inequality when you divide by a negative number.}\\
x > 2 &
\end{array}
$$

The graph of the solution set is

Check Yourself 10

Solve and graph the solution set.

$5x + 12 \geq 10x - 8$

Solving inequalities may also require the distributive property.

 Example 11   **Solving and Graphing Inequalities**

Solve and graph the solution set for

$5(x - 2) \geq -8$

Applying the distributive property on the left yields

$5x - 10 \geq -8$

Solving as before yields

$$
\begin{array}{rll}
5x - 10 \geq -\ 8 & \\
\underline{+ 10 \quad +10} & \qquad \text{Add 10.}\\
5x \qquad \geq \quad 2 & \\
\text{or} \quad x \geq \dfrac{2}{5} & \qquad \text{Divide by 5.}
\end{array}
$$

The graph of the solution set is

Check Yourself 11

Solve and graph the solution set.

$4(x + 3) < 9$

Some applications are solved by using an inequality instead of an equation. Example 12 illustrates such an application.

| Example 12 | Solving an Inequality Application |

< Objective 4 >

Mohammed needs a mean score of 92 or higher on four tests to get an A. So far his scores are 94, 89, and 88. What scores on the fourth test will get him an A?

NOTE

The *mean* of a data set is its arithmetic average.

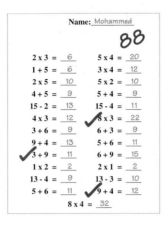

NOTES

What do you need to find?

Assign a letter to the unknown.

Write an inequality.

Solve the inequality.

Step 1   We are looking for the scores that will, when combined with the other scores, give Mohammed an A.

Step 2   Let x represent a fourth-test score that will get him an A.

Step 3   The inequality will have the mean on the left side, which must be greater than or equal to the 92 on the right.

$$\frac{94 + 89 + 88 + x}{4} \geq 92$$

Step 4   First, multiply both sides by 4:

$$94 + 89 + 88 + x \geq 368$$

Then add the test scores:

$$183 + 88 + x \geq 368$$
$$271 + x \geq 368$$

Subtracting 271 from both sides,

$x \geq 97$ Mohammed needs to score 97 or higher to earn an A.

Step 5   To check the solution, we find the mean of the four test scores, 94, 89, 88, and 97.

$$\frac{94 + 89 + 88 + (97)}{4} = \frac{368}{4} = 92$$

Check Yourself 12

Felicia needs a mean score of at least 75 on five tests to get a passing grade in her health class. On her first four tests she has scores of 68, 79, 71, and 70. What scores on the fifth test will give her a passing grade?

The following outline (or algorithm) summarizes our work in this section.

Step by Step

Solving Linear Inequalities

Step 1 Perform operations, as needed, to write an equivalent inequality without any grouping symbols, and combine any like terms appearing on either side of the inequality.

Step 2 Apply the addition property to write an equivalent inequality with the variable term on one side of the inequality and the number on the other.

Step 3 Apply the multiplication property to write an equivalent inequality with the variable isolated on one side of the inequality. Be sure to reverse the direction of the inequality if you multiply or divide by a negative number. The set of solutions derived in step 3 can then be graphed on a number line.

Check Yourself ANSWERS

1. (a) $>$; (b) $<$

2.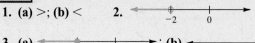

3. (a) [graph: -4, 0] ; (b) [graph: 0, 3]

4. $x > 6$ [graph: 0, 6]   **5.** $x \leq 10$ [graph: 0, 10]

6. (a) $x > 5$ [graph: 0, 5] ; (b) $x \geq -6$ [graph: -6, 0]

7. (a) $x \leq 20$ [graph: 0, 20] ; (b) $x > 21$ [graph: 0, 21]

8. (a) $x \geq -3$ [graph: -3, 0] ; (b) $x > -6$ [graph: -6, 0]

9. $x < -4$ [graph: -4, 0]   **10.** $x \leq 4$ [graph: 0, 4]

11. $x < -\dfrac{3}{4}$ [graph: $-\frac{3}{4}$, 0]   **12.** 87 or greater

Reading Your Text

The following fill-in-the-blank exercises are designed to ensure that you understand some of the key vocabulary used in this section.

SECTION 2.6

(a) A statement that one expression is less than another is an _____.

(b) In an inequality, the "arrowhead" always points to the _____ quantity.

(c) A filled-in or closed circle on a number line indicates that the number is part of the _____ set.

(d) When multiplying both sides of an inequality by a _____ number, remember to switch the direction of the inequality symbol.

< Objective 1 >

Complete the statements, using the symbol < or >.

1. 9 _____ 6

2. 9 _____ 8

3. 7 _____ −2

4. 0 _____ −5

5. 0 _____ 4

6. −12 _____ −7

7. −2 _____ −5 > Videos

8. −4 _____ −11

Write each inequality in words.

9. $x < 3$

10. $x \leq -5$

11. $x \geq -4$

12. $x < -2$

13. $-5 \leq x$

14. $2 < x$

< Objective 2 >

Graph the solution set of each inequality.

15. $x > 2$

16. $x < -3$

17. $x < 10$

18. $x > 4$

19. $x > 1$

20. $x < -2$

21. $x < 8$

22. $x > 5$

23. $x > -7$

24. $x < -4$

Boost *your* GRADE at
ALEKS.com!

ALEKS®

- Practice Problems
- Self-Tests
- NetTutor
- e-Professors
- Videos

Name _____

Section _____ Date _____

Answers

1. _____   2. _____
3. _____   4. _____
5. _____   6. _____
7. _____   8. _____
9. _____
10. _____
11. _____
12. _____
13. _____
14. _____
15. _____
16. _____
17. _____
18. _____
19. _____
20. _____
21. _____
22. _____
23. _____
24. _____

Answers

25. _____

26. _____

27. _____

28. _____

29. _____ 30. _____

31. _____ 32. _____

33. _____ 34. _____

35. _____ 36. _____

37. _____

38. _____

39. _____

40. _____

41. _____

42. _____

43. _____

44. _____

45. _____

46. _____

47. _____

48. _____

---

25. $x \geq 11$

$\longleftarrow\!\!\!\longrightarrow$

26. $x \geq 0$

 > Videos

$\longleftarrow\!\!\!\longrightarrow$

27. $x < 0$

$\longleftarrow\!\!\!\longrightarrow$

28. $x \leq -3$

$\longleftarrow\!\!\!\longrightarrow$

< Objective 3 >

Solve and graph the solution set of each inequality.

29. $x + 9 < 22$

$\longleftarrow\!\!\!\longrightarrow$

30. $x + 5 \leq 4$

$\longleftarrow\!\!\!\longrightarrow$

31. $x + 8 \geq 10$

$\longleftarrow\!\!\!\longrightarrow$

32. $x - 14 > -17$

$\longleftarrow\!\!\!\longrightarrow$

33. $5x < 4x + 7$

$\longleftarrow\!\!\!\longrightarrow$

34. $3x \geq 2x - 4$

$\longleftarrow\!\!\!\longrightarrow$

35. $6x - 8 \leq 5x$

$\longleftarrow\!\!\!\longrightarrow$

36. $3x + 2 > 2x$

$\longleftarrow\!\!\!\longrightarrow$

37. $6x + 5 \geq 5x + 19$

$\longleftarrow\!\!\!\longrightarrow$

38. $5x + 2 \leq 4x - 6$

$\longleftarrow\!\!\!\longrightarrow$

39. $7x + 5 < 6x - 4$

$\longleftarrow\!\!\!\longrightarrow$

40. $8x - 7 > 7x + 3$

$\longleftarrow\!\!\!\longrightarrow$

41. $4x \leq 12$

$\longleftarrow\!\!\!\longrightarrow$

42. $5x > 20$

$\longleftarrow\!\!\!\longrightarrow$

43. $5x > -35$

$\longleftarrow\!\!\!\longrightarrow$

44. $8x \leq -24$

$\longleftarrow\!\!\!\longrightarrow$

45. $-6x \geq 18$

$\longleftarrow\!\!\!\longrightarrow$

46. $-9x < 45$

$\longleftarrow\!\!\!\longrightarrow$

47. $-12x < -72$

$\longleftarrow\!\!\!\longrightarrow$

48. $-12x \geq -48$

$\longleftarrow\!\!\!\longrightarrow$

49. $\dfrac{x}{4} > 5$

50. $\dfrac{x}{3} \leq -3$

51. $-\dfrac{x}{2} \geq -3$ 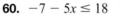 > Videos

52. $-\dfrac{x}{4} < 5$

53. $\dfrac{2x}{3} < 6$

54. $\dfrac{3x}{4} \geq -9$

55. $6x > 3x + 12$

56. $4x \leq x - 9$

57. $5x - 2 > 3x$

58. $7x + 3 \geq 2x$

59. $3 - 2x > 5$

60. $-7 - 5x \leq 18$

61. $2x \geq 5x + 18$

62. $3x < 7x - 28$

63. $5x - 3 \leq 3x + 15$

64. $8x + 7 > 5x + 34$

65. $11x + 8 > 4x - 6$

66. $10x - 5 \leq 8x - 25$

67. $7x - 5 < 3x + 2$

68. $5x - 2 \geq 2x - 7$

69. $5x + 7 > 8x - 17$

70. $4x - 3 \leq 9x + 27$

49.	50.
51.	52.
53.	54.

55.

56.

57.

58.

59.

60.

61.

62.

63.

64.

65.

66.

67.

68.

69.

70.

The Streeter/Hutchison Series in Mathematics   Beginning Algebra

Answers

71. _____

72. _____

73. _____

74. _____

75. _____

76. _____

77. _____

78. _____

79. _____

80. _____

81. _____

82. _____

71. $3x - 2 \leq 5x + 3$

⟵————————————⟶

72. $2x + 3 > 8x - 2$

⟵————————————⟶

73. $4(x + 7) \leq 2x + 31$

⟵————————————⟶

74. $7(x - 3) > 5x - 14$

⟵————————————⟶

75. $2(x - 7) > 5x - 12$

⟵————————————⟶

76. $3(x + 4) \leq 7x + 7$

⟵————————————⟶

< Objective 4 >

77. SOCIAL SCIENCE There are fewer than 1,000 wild giant pandas left in the bamboo forests of China. Write an inequality expressing this relationship.

78. SCIENCE AND MEDICINE Let C represent the amount of Canadian forest and M represent the amount of Mexican forest. Write an inequality showing the relationship of the forests of Mexico and Canada if Canada contains at least 9 times as much forest as Mexico.

79. STATISTICS To pass a course with a grade of B or better, Liza must have an average of 80 or more. Her grades on three tests are 72, 81, and 79. Write an inequality representing the score that Liza must get on the fourth test to obtain a B average or better for the course.

80. STATISTICS Sam must have an average of 70 or more in his summer course to obtain a grade of C. His first three test grades were 75, 63, and 68. Write an inequality representing the score that Sam must get on the last test to get a C grade. ⦿ > Videos

81. BUSINESS AND FINANCE Juanita is a salesperson for a manufacturing company. She may choose to receive $500 or 5% commission on her sales as payment for her work. How much does she need to sell to make the 5% offer a better deal?

82. BUSINESS AND FINANCE The cost for a long-distance telephone call is $0.36 for the first minute and $0.21 for each additional minute or portion thereof. Write an inequality representing the number of minutes a person could talk without exceeding $3.

83. GEOMETRY The perimeter of a rectangle is to be no greater than 250 cm and the length must be 105 cm. Find the maximum width of the rectangle.

84. STATISTICS Sarah bowled 136 and 189 in her first two games. What must she bowl in her third game to have an average of at least 170?

Basic Skills | **Challenge Yourself** | Calculator/Computer | Career Applications | Above and Beyond
▲

Translate each statement into an inequality. Let x represent the number in each case.

85. 6 more than a number is greater than 5.

86. 3 less than a number is less than or equal to 5.

87. 4 less than twice a number is less than or equal to 7.

88. 10 more than a number is greater than negative 2.

89. 4 times a number, decreased by 15, is greater than that number.

90. 2 times a number, increased by 28, is less than or equal to 6 times that number.

Match each inequality on the right with a statement on the left.

91. x is nonnegative   **(a)** $x \geq 0$

92. x is negative   **(b)** $x \geq 5$

93. x is no more than 5   **(c)** $x \leq 5$

94. x is positive   **(d)** $x > 0$

95. x is at least 5   **(e)** $x < 5$

96. x is less than 5   **(f)** $x < 0$

Basic Skills | Challenge Yourself | Calculator/Computer | Career Applications | **Above and Beyond**
▲

---

97. You are the office manager for a small company. You need to acquire a new copier for the office. You find a suitable one that leases for $250 a month from the copy machine company. It costs 2.5¢ per copy to run the machine. You purchase paper for $3.50 a ream (500 sheets). If your copying budget is no more than $950 per month, is this machine a good choice? Write a brief recommendation to the purchasing department. Use equations and inequalities to explain your recommendation.

Answers

83. _____

84. _____

85. _____

86. _____

87. _____

88. _____

89. _____

90. _____

91. _____

92. _____

93. _____

94. _____

95. _____

96. _____

97. _____

Answers

98. Your aunt calls to ask for your help in making a decision about buying a new refrigerator. She says that she found two that seem to fit her needs, and both are supposed to last at least 14 years, according to *Consumer Reports*. The initial cost for one refrigerator is $712, but it uses only 88 kilowatt-hours (kWh) per month. The other refrigerator costs $519 and uses an estimated 100 kWh per month. You do not know the price of electricity per kilowatt-hour where your aunt lives, so you will have to decide what prices in cents per kilowatt-hour will make the first refrigerator cheaper to run during its 14 years of expected usefulness. Write your aunt a letter explaining what you did to calculate this cost, and tell her to make her decision based on how the kilowatt-hour rate she has to pay in her area compares with your estimation.

Answers

1. $9 > 6$ **3.** $7 > -2$ **5.** $0 < 4$ **7.** $-2 > -5$ **9.** x is less than 3.

11. x is greater than or equal to -4. **13.** -5 is less than or equal to x.

15. **17.**

19. **21.**

23. **25.**

27. **29.** $x < 13$

31. $x \geq 2$ **33.** $x < 7$

35. $x \leq 8$ **37.** $x \geq 14$

39. $x < -9$ **41.** $x \leq 3$

43. $x > -7$ **45.** $x \leq -3$

47. $x > 6$ **49.** $x > 20$

51. $x \leq 6$ **53.** $x < 9$

55. $x > 4$ **57.** $x > 1$

59. $x < -1$ **61.** $x \leq -6$

63. $x \leq 9$ **65.** $x > -2$

67. $x < \dfrac{7}{4}$ **69.** $x < 8$

71. $x \geq -\dfrac{5}{2}$ **73.** $x \leq \dfrac{3}{2}$

75. $x < -\dfrac{2}{3}$ **77.** $P < 1,000$ **79.** $x \geq 88$

81. More than $10,000 **83.** 20 cm **85.** $x + 6 > 5$ **87.** $2x - 4 \leq 7$

89. $4x - 15 > x$ **91.** (a) **93.** (c) **95.** (b) **97.** Above and Beyond

Definition/Procedure	Example	Reference
Solving Equations by the Addition Property		Section 2.1
Equation		
A mathematical statement that two expressions are equal	$2x - 3 = 5$ is an equation.	p. 89
Solution		
A value for a variable that makes an equation a true statement	4 is a solution for the above equation because $2(4) - 3 = 5$.	p. 90
Equivalent Equations		
Equations that have exactly the same solutions	$2x - 3 = 5$ and $x = 4$ are equivalent equations.	p. 91
The Addition Property of Equality		
If $a = b$, then $a + c = b + c$.	If $2x - 3 = 7$, then $2x - 3 + 3 = 7 + 3$.	p. 92
Solving Equations by the Multiplication Property		Section 2.2
The Multiplication Property of Equality		
If $a = b$, then $ac = bc$ with $c \neq 0$.	If $\dfrac{1}{2}x = 7$, then $2\left(\dfrac{1}{2}x\right) = 2(7)$.	p. 102
Combining the Rules to Solve Equations		Section 2.3
Solving Linear Equations		
The steps of solving a linear equation are as follows: **1.** Use the distributive property to remove any grouping symbols. Then simplify by combining like terms. **2.** Add or subtract the same term on each side of the equation until the variable term is on one side and a number is on the other. **3.** Multiply or divide both sides of the equation by the same nonzero number so that the variable is alone on one side of the equation. **4.** Check the solution in the original equation.	Solve: $3(x - 2) + 4x = 3x + 14$ $3x - 6 + 4x = 3x + 14$ $7x - 6 = 3x + 14$ $\underline{-3x \qquad\qquad -3x}$ $4x - 6 = \qquad 14$ $\underline{+6 \qquad\qquad + 6}$ $4x = \qquad 20$ $\dfrac{4x}{4} = \dfrac{20}{4}$ $x = 5$	p. 116

Continued

Definition/Procedure	Example	Reference
Formulas and Problem Solving		Section 2.4
Literal Equation		
An equation that involves more than one letter or variable	$a = \dfrac{2b + c}{3}$	*p. 122*
Solving Literal Equations		*p. 124*
1. Multiply both sides of the equation by the same term to clear it of fractions. 2. Add or subtract the same term on both sides of the equation so that all terms containing the variable you are solving for are on one side. 3. Divide both sides by the coefficient of the variable that you are solving for.	Solve for b. $a = \dfrac{2b + c}{3}$ $3a = \left(\dfrac{2b + c}{3}\right)3$ $3a = 2b + c$ $3a - c = 2b$ $\dfrac{3a - c}{2} = b$	
Applications of Linear Equations		Section 2.5
The base is the whole in a percent statement.	14 is 25% of 56. 56 is the base.	*p. 144*
The amount is the part being compared to the base.	14 is the amount.	*p. 144*
The rate is the ratio of the amount to the base.	25% is the rate.	*p. 144*
The **percent relationship** is given by $A = R \cdot B$ Amount = Rate · Base	$14 = 0.25 \cdot 56$ A = Amount R = Rate in decimal form B = Base	*p. 144*
Inequalities—An Introduction		Section 2.6
Inequality		
A statement that one quantity is less than (or greater than) another. Four symbols are used: $a < b$ $a > b$ $a \le b$ $a \ge b$ a is less than b a is greater than b a is less than or equal to b a is greater than or equal to b	$-4 < -1$ $x^2 + 1 \ge x + 1$	*p. 154*

Definition/Procedure	Example	Reference

Graphing Inequalities

To graph $x < a$, we use an open circle and an arrow pointing left.

The heavy arrow indicates all numbers less than (or to the left of) a.

The open circle means a is not included in the solution set.

Graph $x < 3$.

p. 155

To graph $x \geq b$, we use a closed circle and an arrow pointing right.

The closed circle means that in this case b is included in the solution set.

Graph $x \geq -1$.

p. 155

Solving Inequalities

An inequality is "solved" when it is in the form $x < \square$ or $x > \square$.

Proceed as in solving equations by using the following properties.

1. If $a < b$, then $a + c < b + c$.

Adding (or subtracting) the same quantity to each side of an inequality gives an equivalent inequality.

2. If $a < b$, then $ac < bc$ when $c > 0$ and $ac > bc$ when $c < 0$.

Multiplying both sides of an inequality by the same *positive number* gives an equivalent inequality. When both sides of an inequality are multiplied by the same *negative number, you must reverse the direction* of the inequality to give an equivalent inequality.

$$2x - 3 > 5x + 6$$
$$\underline{\quad +3 \qquad\quad +3\quad}$$
$$2x \quad > \quad 5x + 9$$
$$\underline{-5x \qquad\quad -5x\quad}$$
$$-3x \quad > \quad 9$$
$$\frac{-3x}{-3} < \frac{9}{-3}$$
$$x < -3$$

p. 156

This summary exercise set is provided to give you practice with each of the objectives of this chapter. Each exercise is keyed to the appropriate chapter section. When you are finished, you can check your answers to the odd-numbered exercises in the back of the text. If you have difficulty with any of these questions, go back and reread the examples from that section. The answers to the even-numbered exercises appear in the *Instructor's Solutions Manual.* Your instructor will give you guidelines on how best to use these exercises in your instructional setting.

2.1 *Tell whether the number shown in parentheses is a solution for the given equation.*

1. $7x + 2 = 16$ (2)

2. $5x - 8 = 3x + 2$ (4)

3. $7x - 2 = 2x + 8$ (2)

4. $4x + 3 = 2x - 11$ (−7)

5. $x + 5 + 3x = 2 + x + 23$ (6)

6. $\dfrac{2}{3}x - 2 = 10$ (21)

2.1–2.3 *Solve each equation and check your results.*

7. $x + 5 = 7$

8. $x - 9 = 3$

9. $7 + 6x = 5x$

10. $3x - 9 = 2x$

11. $5x - 3 = 4x + 2$

12. $9x + 2 = 8x - 7$

13. $7x - 5 = 6x - 4$

14. $3 + 4x - 1 = x - 7 + 2x$

15. $4(2x + 3) = 7x + 5$

16. $5(5x - 3) = 6(4x + 1)$

17. $6x = 42$

18. $7x = -28$

19. $-6x = 24$

20. $-9x = -63$

21. $\dfrac{x}{8} = 4$

22. $-\dfrac{x}{3} = -5$

23. $\dfrac{2}{3}x = 18$

24. $\dfrac{3}{4}x = 24$

25. $5x - 3 = 12$

26. $4x + 3 = -13$

27. $7x + 8 = 3x$

28. $3 - 5x = -17$

29. $3x - 7 = x$

30. $2 - 4x = 5$

31. $\dfrac{x}{3} - 5 = 1$

32. $\dfrac{3}{4}x - 2 = 7$

33. $6x - 5 = 3x + 13$

34. $3x + 7 = x - 9$

35. $7x + 4 = 2x + 6$

36. $9x - 8 = 7x - 3$

37. $2x + 7 = 4x - 5$

38. $3x - 15 = 7x - 10$

39. $\dfrac{10}{3}x - 5 = \dfrac{4}{3}x + 7$

40. $\dfrac{11}{4}x - 15 = 5 - \dfrac{5}{4}x$

41. $3.7x + 8 = 1.7x + 16$

42. $5.4x - 3 = 8.4x + 9$

43. $3x - 2 + 5x = 7 + 2x + 21$

44. $8x + 3 - 2x + 5 = 3 - 4x$

45. $5(3x - 1) - 6x = 3x - 2$

2.4 *Solve for the indicated variable.*

46. $V = LWH$ (for L)

47. $P = 2L + 2W$ (for L)

48. $ax + by = c$ (for y)

49. $A = \dfrac{1}{2}bh$ (for h)

50. $A = P + Prt$ (for t)

51. $m = \dfrac{n - p}{q}$ (for n)

2.4–2.5 *Solve each word problem. Be sure to label the unknowns and to show the equation you used.*

52. NUMBER PROBLEM The sum of 3 times a number and 7 is 25. What is the number?

53. NUMBER PROBLEM 5 times a number, decreased by 8, is 32. Find the number.

54. NUMBER PROBLEM If the sum of two consecutive integers is 85, find the two integers.

55. NUMBER PROBLEM The sum of three consecutive odd integers is 57. What are the three integers?

56. BUSINESS AND FINANCE Rafael earns $35 more per week than Andrew. If their weekly salaries total $715, what amount does each earn?

57. NUMBER PROBLEM Larry is 2 years older than Susan, and Nathan is twice as old as Susan. If the sum of their ages is 30 years, find each of their ages.

58. BUSINESS AND FINANCE Joan works on a 4% commission basis. She sold $45,000 in merchandise during 1 month. What was the amount of her commission?

59. BUSINESS AND FINANCE David buys a dishwasher that is marked down $77 from its original price of $350. What is the discount rate?

60. SCIENCE AND MEDICINE A chemist prepares a 400-milliliter (400-mL) acid-water solution. If the solution contains 30 mL of acid, what percent of the solution is acid?

61. BUSINESS AND FINANCE The price of a new compact car has increased $819 over the previous year. If this amounts to a 4.5% increase, what was the price of the car before the increase?

62. BUSINESS AND FINANCE A store advertises, "Buy the red-tagged items at 25% off their listed price." If you buy a coat marked $136, what will you pay for the coat during the sale?

63. BUSINESS AND FINANCE Tom has 6% of his salary deducted for a retirement plan. If that deduction is $168, what is his monthly salary?

64. STATISTICS A college finds that 35% of its science students take biology. If there are 252 biology students, how many science students are there altogether?

65. BUSINESS AND FINANCE A company finds that its advertising costs increased from $72,000 to $76,680 in 1 year. What was the rate of increase?

66. BUSINESS AND FINANCE A savings bank offers 3.25% on 1-year time deposits. If you place $900 in an account, how much will you have at the end of the year?

67. BUSINESS AND FINANCE Maria's company offers her a 4% pay raise. This will amount to a $126 per month increase in her salary. What is her monthly salary before and after the raise?

68. STATISTICS A computer has 8 gigabytes (GB) of storage space. Arlene is going to add 16 GB of storage space. By what percent will the available storage space be increased?

69. STATISTICS A virus scanning program is checking every file for viruses. It has completed 30% of the files in 150 s. How long should it take to check all the files?

70. BUSINESS AND FINANCE If the total bill at a restaurant for 10 people is $572.89, including an 18% tip, what was the cost of the food?

71. BUSINESS AND FINANCE A pair of running shoes is advertised at 30% off the original price for $80.15. What was the original price?

2.6 Solve and graph the solution set for each inequality.

72. $x - 4 \le 7$

73. $x + 3 > -2$

74. $5x > 4x - 3$

75. $4x \ge -12$

76. $-12x < 36$

77. $-\dfrac{x}{5} \ge 3$

78. $2x \le 8x - 3$

79. $2x + 3 \ge 9$

80. $4 - 3x > 8$

81. $5x - 2 \le 4x + 5$

82. $7x + 13 \ge 3x + 19$

83. $4x - 2 < 7x + 16$

self-test 2

The purpose of this self-test is to help you assess your progress so that you can find concepts that you need to review before the next exam. Allow yourself about an hour to take this test. At the end of that hour, check your answers against those given in the back of this text. If you miss any, go back to the appropriate section to reread the examples until you have mastered that particular concept.

Tell whether the number shown in parentheses is a solution for the given equation.

1. $7x - 3 = 25$ (5)

2. $8x - 3 = 5x + 9$ (4)

Solve each equation and check your results.

3. $x - 7 = 4$

4. $7x - 12 = 6x$

5. $9x - 2 = 8x + 5$

6. $7x = 49$

7. $\dfrac{1}{4}x = -3$

8. $\dfrac{4}{5}x = 20$

9. $7x - 5 = 16$

10. $10 - 3x = -2$

11. $7x - 3 = 4x - 5$

12. $\dfrac{3x}{2} - 5 = 4x + \dfrac{5}{8}$

Solve for the indicated variable.

13. $C = 2\pi r$ (for r)

14. $V = \dfrac{1}{3}Bh$ (for h)

15. $3x + 2y = 6$ (for y)

Solve and graph the solution sets for each inequality.

16. $x - 5 \le 9$

17. $5 - 3x > 17$

18. $5x + 13 \ge 2x + 17$

19. $2x - 3 < 7x + 2$

Name _____

Section _____ **Date** _____

Answers

1. _____
2. _____
3. _____
4. _____
5. _____
6. _____
7. _____
8. _____
9. _____
10. _____
11. _____
12. _____
13. _____
14. _____
15. _____
16. _____
17. _____
18. _____
19. _____

Answers

20. _____

21. _____

22. _____

23. _____

24. _____

25. _____

Solve each application.

20. NUMBER PROBLEM 5 times a number, decreased by 7, is 28. What is the number?

21. NUMBER PROBLEM The sum of three consecutive integers is 66. Find the three integers.

22. NUMBER PROBLEM Jan is twice as old as Juwan, and Rick is 5 years older than Jan. If the sum of their ages is 35 years, find each of their ages.

23. GEOMETRY The perimeter of a rectangle is 62 in. If the length of the rectangle is 1 in. more than twice its width, what are the dimensions of the rectangle?

24. BUSINESS AND FINANCE Mrs. Moore made a $450 commission on the sale of a $9,000 pickup truck. What was her commission rate?

25. BUSINESS AND FINANCE Cynthia makes a 5% commission on all her sales. She earned $1,750 in commissions during 1 month. What were her gross sales for the month?

The Streeter/Hutchison Series in Mathematics   Beginning Algebra

Activity 2 ::
Monetary Conversions

Each activity in this text is designed to either enhance your understanding of the topics of the preceding chapter, provide you with a mathematical extension of those topics, or both. The activities can be undertaken by one student, but they are better suited for a small-group project. Occasionally it is only through discussion that different facets of the activity become apparent.

In the opener to this chapter, we discussed international travel and using exchange rates to acquire local currency. In this activity, we use these exchange rates to explore the idea of variables. You should recall that a **variable** is a symbol used to represent an unknown quantity or a quantity that varies.

Currency exchange rates are published on a daily basis by many sources such as *Yahoo!Finance* and the *Wall Street Journal*. For instance, on May 20, 2006, the exchange rate for trading US$ for CAN$ was 1.1191. This means that US$1 is equivalent to CAN$1.1191. That is, if you exchanged $100 of U.S. money, you would have received $111.91 in Canadian dollars. We compute this as follows:

CAN$ = Exchange rate × US$

Activity

I.

1. Choose a country that you would like to visit. Use a search engine to find the exchange rate between US$ and the currency of your chosen country.
2. If you are visiting for only a short time, you may not need too much money. Determine how much of the local currency you will receive in exchange for US$250.
3. If you stay for an extended period, you will need more money. How much would you receive in exchange for US$900?

In part I, we treated the amount (US$) as a *variable.* This quantity varied depending upon our needs. If we visit Canada and let x = the amount exchanged in US$ and y = the amount received in CAN$, then, using the exchange rate previously given, we have the equation

$$y = 1.1191x$$

You may ask, "Isn't the amount of Canadian money received (y) a variable, too?" The answer to this question is yes; in fact, all three quantities are variables. According to *Yahoo!Finance,* the exchange rate for US-CAN currency was 1.372 on December 14, 2001. The exchange rate varies on a daily basis. If we let r = the exchange rate, then we can write our equation as

$$y = rx$$

II.

1. Consider the country you chose to visit in part I. Find the exchange rate for another date and repeat steps I.2 and I.3 for this other exchange rate.
2. Choose another nation that you would like to visit. Repeat the steps in part I for this country.

Data Set

Currency	US$	Yen (¥)	Euro (€)	CAN$	U.K. (£)	Aust$
1 US$	1	111.705	0.7833	1.1191	0.5327	1.3181
1 Yen (¥)	0.008952	1	0.007012	0.010018	0.004769	0.0118
1 Euro (€)	1.2766	142.6026	1	1.4286	0.6801	1.6827
1 CAN$	0.8936	99.8213	0.7	1	0.476	1.1779
1 U.K. (£)	1.8772	209.6924	1.4705	2.1007	1	2.4744
1 Aust$	0.7586	84.745	0.5943	0.849	0.4041	1

Source: Yahoo!Finance; 5/20/06.

I.1 We chose to visit Canada and will use the 5/20/06 exchange rate of 1.1191 from the sample data set.

I.2 Exchange rate × US$ = CAN$

$(1.1191) \cdot (US\$250) = CAN\$279.775$

We would receive $279.78 in Canadian dollars for $250 in U.S. money (round Canadian money to two decimal places).

I.3 $(1.1191) \cdot (US\$900) = CAN\$1,007.19$

II.1 Had we visited Canada on 12/14/01, we would have received an exchange rate of 1.372.

$(1.372) \cdot (US\$250) = CAN\$343$

$(1.372) \cdot (US\$900) = CAN\$1,234.80$

II.2 We choose to visit Japan. The 5/20/06 exchange rate was 111.705 Yen (¥) for each US$.

$(111.705) \cdot (US\$250) = ¥27,926.25$

$(111.705) \cdot (US\$900) = ¥100,534.5$

We would receive 27,926 yen for US$250, and 100,535 yen for US$900.

The following exercises are presented to help you review concepts from earlier chapters. This is meant as review material and not as a comprehensive exam. The answers are presented in the back of the text. Beside each answer is a section reference for the concept. If you have difficulty with any of these exercises, be certain to at least read through the summary related to that section.

Name _____

Section _____ Date _____

Answers

Perform the indicated operations.

1. $8 + (-4)$

2. $-7 + (-5)$

3. $6 - (-2)$

4. $-4 - (-7)$

5. $(-6)(3)$

6. $(-11)(-4)$

7. $20 \div (-4)$

8. $(-50) \div (-5)$

9. $0 \div (-26)$

10. $15 \div 0$

1. _____	2. _____
3. _____	4. _____
5. _____	6. _____
7. _____	8. _____
9. _____	10. _____
11. _____	12. _____
13. _____	14. _____

Evaluate the expressions if $x = 5$, $y = 2$, $z = -3$, and $w = -4$.

11. $2xy$

12. $2x + 7z$

13. $3z^2$

14. $4(x + 3w)$

15. $\dfrac{2w}{y}$

16. $\dfrac{2x - w}{2y - z}$

15. _____

16. _____

17. _____

18. _____

19. _____

Simplify each expression.

17. $14x^2y - 11x^2y$

18. $2x^3(3x - 5y)$

19. $\dfrac{x^2y - 2xy^2 + 3xy}{xy}$

20. $10x^2 + 5x + 2x^2 - 2x$

20. _____

21. _____

22. _____

23. _____

Solve each equation and check your results.

21. $9x - 5 = 8x$

22. $-\dfrac{3}{4}x = 18$

23. $6x - 8 = 2x - 3$

24. _____

24. $2x + 3 = 7x + 5$

25. $\dfrac{4}{3}x - 6 = 4 - \dfrac{2}{3}x$

25. _____

Answers

26. _____

27. _____

28. _____

29. _____

30. _____

31. _____

32. _____

33. _____

34. _____

35. _____

36. _____

37. _____

38. _____

39. _____

40. _____

Solve each equation for the indicated variable.

26. $I = Prt$ (for r)   **27.** $A = \dfrac{1}{2}bh$ (for h)   **28.** $ax + by = c$ (for y)

Solve and graph the solution sets for each inequality.

29. $3x - 5 < 4$

30. $7 - 2x \geq 10$

31. $7x - 2 > 4x + 10$

32. $2x + 5 \leq 8x - 3$

Solve each word problem. Be sure to show the equation used for the solution.

33. **NUMBER PROBLEM** If 4 times a number decreased by 7 is 45, find that number.

34. **NUMBER PROBLEM** The sum of two consecutive integers is 85. What are those two integers?

35. **NUMBER PROBLEM** If 3 times an integer is 12 more than the next consecutive odd integer, what is that integer?

36. **BUSINESS AND FINANCE** Michelle earns $120 more per week than Dmitri. If their weekly salaries total $720, how much does Michelle earn?

37. **GEOMETRY** The length of a rectangle is 2 cm more than 3 times its width. If the perimeter of the rectangle is 44 cm, what are the dimensions of the rectangle?

38. **GEOMETRY** One side of a triangle is 5 in. longer than the shortest side. The third side is twice the length of the shortest side. If the triangle perimeter is 37 in., find the length of each leg.

39. **BUSINESS AND FINANCE** Jesse paid $1,562.50 in state income tax last year. If his salary was $62,500, what was the rate of tax?

40. **BUSINESS AND FINANCE** A car is marked down from $31,500 to $29,137.50. What was the discount rate?

CHAPTER

3

INTRODUCTION

Polynomials are used in many disciplines and industries to model applications and solve problems. For example, aerospace engineers use complex formulas to plan and guide space shuttle flights, and telecommunications engineers use them to improve digital signal processing. Equations expressing relationships among variables play a significant role in building construction, estimating electrical power generation needs and consumption, astronomy, medicine and pharmacological measurements, determining manufacturing costs, and projecting retail revenue.

The field of personal investments and savings presents an opportunity to estimate the future value of savings accounts, Individual Retirement Accounts, and other investment products. In the chapter activity we explore the power of compound interest.

Polynomials

CHAPTER 3 OUTLINE

3 prerequisite test

Name _____

Section _____ Date _____

This prerequisite test provides some exercises requiring skills that you will need to be successful in the coming chapter. The answers for these exercises can be found in the back of this text. This prerequisite test can help you identify topics that you will need to review before beginning the chapter.

Answers

1. _____

2. _____

3. _____

4. _____

5. _____

6. _____

7. _____

8. _____

9. _____

10. _____

11. _____

12. _____

Evaluate each expression.

1. 5^4

2. $2 \cdot 6^3$

3. -3^4

4. $(-3)^4$

5. 2.3×10^5

6. $\dfrac{2.3}{10^5}$

Simplify each expression.

7. $5x - 2(3x - 4)$

8. $2x + 5y - y$

Evaluate each expression.

9. $7x^2 + 4x - 3$ for $x = -1$

10. $4x^2 - 3xy + y^2$ for $x = 3$ and $y = -2$

Solve each application.

11. NUMBER PROBLEM Find two consecutive odd integers such that 3 times the first integer is 5 more than twice the second integer.

12. ELECTRICAL ENGINEERING Resistance (in ohms, Ω) is given by the formula

$$R = \frac{V^2}{D}$$

in which D is the power dissipation (in watts) and V is the voltage.
    Determine the power dissipation when 13.2 volts pass through a 220-Ω resistor.

3.1

Exponents and Polynomials

< 3.1 Objectives >

1 > Use the properties of exponents to simplify expressions

2 > Identify types of polynomials

3 > Find the degree of a polynomial

4 > Write a polynomial in descending order

5 > Evaluate a polynomial

 Tips for Student Success

Preparing for a Test

Preparing for a test begins on the first day of class. Everything you do in class and at home is part of that preparation. In fact, if you attend class every day, take good notes, and keep up with the homework, then you will already be prepared and not need to "cram" for your exam.

Instead of cramming, here are a few things to focus on in the days before a scheduled test.

1. Study for your exam, but finish studying 24 hours before the test. Make certain to get some rest before taking a test.

2. Study for an exam by going over homework and class notes. Write down all of the problem types, formulas, and definitions that you think might give you trouble on the test.

3. The last item before you finish studying is to take the notes you made in step 2 and transfer the most important ideas to a 3×5 (index) card. You should complete this step a full 24 hours before your exam.

4. One hour before your exam, review the information on the 3×5 card you made in step 3. You will be surprised at how much you remember about each concept.

5. The biggest obstacle for many students is believing that they can be successful on the test. You can overcome this obstacle easily enough. If you have been completing the homework and keeping up with the classwork, then you should perform quite well on the test. Truly anxious students are often surprised to score well on an exam. These students attribute a good test score to blind luck when it is not luck at all. This is the first sign that they "get it." Enjoy the success!

Recall that exponential notation indicates repeated multiplication; the exponent or power tells us how many times the base is to be used as a factor.

Exponent or Power
↓

$$3^5 = \underbrace{3 \cdot 3 \cdot 3 \cdot 3 \cdot 3}_{5 \text{ factors}} = 243$$

↑
Base

183

In order to effectively use exponential notation, we need to understand how to evaluate and simplify expressions that contain exponents. To do this, we need to understand some properties associated with exponents.

$$2^3 \cdot 2^2 = 8 \cdot 4 \qquad \text{\small $2^3 = 8; 2^2 = 4$}$$
$$= 32$$

Another way to look at this same product is to expand each exponential expression.

$$2^3 \cdot 2^2 = (2 \cdot 2 \cdot 2) \cdot (2 \cdot 2)$$
$$= 2 \cdot 2 \cdot 2 \cdot 2 \cdot 2 \qquad \text{\small We can remove the parentheses.}$$
$$= 2^5 \qquad \text{\small There are 5 factors (of 2).}$$

NOTE

$2^5 = 32$

Now consider what happens when we replace 2 by a variable.

$$a^3 \cdot a^2 = \underbrace{(a \cdot a \cdot a)}_{a^3} \underbrace{(a \cdot a)}_{a^2}$$
$$= a \cdot a \cdot a \cdot a \cdot a \qquad \text{\small Five factors.}$$
$$= a^5$$

You should see that the result, a^5, can be found by simply adding the exponents because this gives the number of times the base appears as a factor in the final product.

$$a^3 \cdot a^2 = a^{3+2} \qquad \text{\small Add the exponents.}$$
$$= a^5$$

 > CAUTION

The base must be the same in both factors. We cannot combine $a^2 \cdot b^3$ any further.

We can now state our first property, the **product property of exponents,** for the general case.

Property
Product Property of Exponents

For any real number a and positive integers m and n,

$$a^m \cdot a^n = a^{m+n}$$

In words, the product of two terms with the same base is the base taken to the power that is the sum of the exponents.

For example, $2^5 \cdot 2^7 = 2^{5+7} = 2^{12}$

Here is an example illustrating the product property of exponents.

Example 1   Using the Product Property of Exponents

NOTE

In every case, the base stays the same.

Write each expression as a single base to a power.

(a) $b^4 \cdot b^6 = b^{4+6}$ Add the exponents.
$$= b^{10}$$

(b) $(-2)^5 (-2)^4 = (-2)^{5+4}$
$$= (-2)^9$$

RECALL

If a factor has no exponent, it is understood to be to the first power (the exponent is one).

(c) $10^7 \cdot 10^{11} = 10^{7+11}$ The base does not change; we are already multiplying the base by adding the exponents.
$$= 10^{18}$$

(d) $x^5 \cdot x = x^{5+1}$ $x = x^1$
$$= x^6$$

Check Yourself 1

Write each expression as a single base to a power.

(a) $x^7 \cdot x^3$ (b) $(-3)^4(-3)^3$ (c) $(x^2y)^3(x^2y)^5$ (d) $y \cdot y^6$

By applying the commutative and associative properties of multiplication, we can simplify products that have coefficients. Consider the following case.

$$2x^3 \cdot 3x^4 = (2 \cdot 3)(x^3 \cdot x^4) \quad \text{We can group the factors any way we want.}$$
$$= 6x^7$$

The next example expands on this idea.

| Example 2 | Using the Properties of Exponents |

RECALL

Multiply the coefficients but add the exponents. With practice, you will not need to write the regrouping step.

Simplify each expression.

(a) $(3x^4)(5x^2) = (3 \cdot 5)(x^4 \cdot x^2) \quad \text{Regroup the factors.}$
$$= 15x^6 \quad \text{Add the exponents.}$$

(b) $(2x^5y)(9x^3y^4) = (2 \cdot 9)(x^5 \cdot x^3)(y \cdot y^4)$
$$= 18x^8y^5$$

Check Yourself 2

Simplify each expression.

(a) $(7x^5)(2x^2)$ (b) $(-2x^3y)(x^2y^2)$

(c) $(-5x^3y^2)(-3x^2y^3)$ (d) $x \cdot x^5 \cdot x^3$

What happens when we divide two exponential expressions with the same base? Consider the following cases.

$$\frac{2^5}{2^2} = \frac{2 \cdot 2 \cdot 2 \cdot 2 \cdot 2}{2 \cdot 2} \quad \text{Expand and simplify.}$$

$$= \frac{2 \cdot 2 \cdot 2}{1}$$

$$= 2^3$$

You should immediately see that the final exponent is the difference between the two exponents: $3 = 5 - 2$.

This is true in the more general case:

$$\frac{a^6}{a^4} = \frac{a \cdot a \cdot a \cdot a \cdot a \cdot a}{a \cdot a \cdot a \cdot a}$$

$$= a^2$$

We can now state our second rule, the **quotient property of exponents.**

Property

Quotient Property of Exponents

For any nonzero real number a and positive integers m and n, with $m > n$,

$$\frac{a^m}{a^n} = a^{m-n}$$

For example, $\dfrac{2^{12}}{2^7} = 2^{12-7} = 2^5$

Example 3 Using the Quotient Properties of Exponents

Simplify each expression.

(a) $\frac{x^{10}}{x^4} = x^{10-4}$ Subtract the exponents.

$= x^6$

(b) $\frac{a^8}{a^7} = a^{8-7}$

$= a$ $a^1 = a$; we do not need to write the exponent.

(c) $\frac{-32a^4b^5}{8a^2b} = \frac{-32}{8} \cdot \frac{a^4}{a^2} \cdot \frac{b^5}{b}$ Use the properties of fractions to regroup the factors.

$= -4a^{4-2}b^{5-1}$ Apply the quotient property to each grouping.

$= -4a^2b^4$

Check Yourself 3

Simplify each expression.

(a) $\frac{y^{12}}{y^5}$ (b) $\frac{x^9}{x}$ (c) $\frac{-45r^8}{-9r^7}$ (d) $\frac{56m^6n^7}{-7mn^3}$

NOTE

This means that the base, x^2, is used as a factor 4 times.

Consider the following:

$(x^2)^4 = x^2 \cdot x^2 \cdot x^2 \cdot x^2 = x^8$

This leads us to our third property for exponents.

Property

Power to a Power Property of Exponents

For any real number a and positive integers m and n,

$(a^m)^n = a^{m \cdot n}$

For example, $(2^3)^2 = 2^{3 \cdot 2} = 2^6$.

We illustrate this property in the next example.

Example 4 Using the Power to a Power Property of Exponents

< Objective 1 >

> CAUTION

Be sure to distinguish between the correct use of the product property and the power to a power property.

$(x^4)^5 = x^{4 \cdot 5} = x^{20}$

but

$x^4 \cdot x^5 = x^{4+5} = x^9$

Simplify each expression.

(a) $(x^4)^5 = x^{4 \cdot 5} = x^{20}$

(b) $(2^3)^4 = 2^{3 \cdot 4} = 2^{12}$ Multiply the exponents.

Check Yourself 4

Simplify each expression.

(a) $(m^5)^6$ (b) $(m^5)(m^6)$ (c) $(3^2)^4$ (d) $(3^2)(3^4)$

The Streeter/Hutchison Series in Mathematics Beginning Algebra
© The McGraw-Hill Companies. All Rights Reserved.

NOTES

Here the base is $3x$.

We apply the commutative and associative properties.

Suppose we have a product raised to a power, such as $(3x)^4$. We know that

$$(3x)^4 = (3x)(3x)(3x)(3x)$$
$$= (3 \cdot 3 \cdot 3 \cdot 3)(x \cdot x \cdot x \cdot x)$$
$$= 3^4 \cdot x^4 = 81x^4$$

Note that the power, here 4, has been applied to each factor, 3 and x. In general, we have:

Property

Product to a Power Property of Exponents

For any real numbers a and b and positive integer m,

$$(ab)^m = a^m b^m$$

For example, $(3x)^3 = 3^3 \cdot x^3 = 27x^3$

The use of this property is shown in Example 5.

Example 5 Using the Product to a Power Property of Exponents

NOTE

$(2x)^5$ and $2x^5$ are different expressions. For $(2x)^5$, the base is $2x$, so we raise each factor to the fifth power. For $2x^5$, the base is x, and so the exponent applies only to x.

Simplify each expression.

(a) $(2x)^5 = 2^5 \cdot x^5 = 32x^5$

(b) $(3ab)^4 = 3^4 \cdot a^4 \cdot b^4 = 81a^4b^4$

(c) $5(-2r)^3 = 5 \cdot (-2)^3 \cdot (r)^3 = 5 \cdot (-8) \cdot r^3 = -40r^3$

Check Yourself 5

Simplify each expression.

(a) $(3y)^4$ **(b)** $(2mn)^6$ **(c)** $3(4x)^2$ **(d)** $6(-2x)^3$

We may have to use more than one property when simplifying an expression involving exponents, as shown in Example 6.

Example 6 Using the Properties of Exponents

NOTE

To help you understand each step of the simplification, we refer to the property being applied. Make a list of the properties now to help you as you work through the remainder of this section and Section 3.2.

Simplify each expression.

(a) $(r^4 s^3)^3 = (r^4)^3 \cdot (s^3)^3$ Product to a power property

$\qquad\qquad = r^{12} s^9$ Power to a power property

(b) $(3x^2)^2 \cdot (2x^3)^3$

$\qquad = 3^2 (x^2)^2 \cdot 2^3 \cdot (x^3)^3$ Product to a power property

$\qquad = 9x^4 \cdot 8x^9$ Power to a power property

$\qquad = 72x^{13}$ Multiply the coefficients and apply the product property.

(c) $\dfrac{(a^3)^5}{a^4} = \dfrac{a^{15}}{a^4}$ Power to a power property

$\qquad = a^{11}$ Quotient property

Check Yourself 6

Simplify each expression.

(a) $(m^5 n^2)^3$ (b) $(2p)^4(-4p^2)^2$ (c) $\dfrac{(s^4)^3}{s^5}$

We have one final exponent property to develop. Suppose we have a quotient raised to a power. Consider the following:

$$\left(\frac{x}{3}\right)^3 = \frac{x}{3} \cdot \frac{x}{3} \cdot \frac{x}{3} = \frac{x \cdot x \cdot x}{3 \cdot 3 \cdot 3} = \frac{x^3}{3^3}$$

Note that the power, here 3, has been applied to the numerator x and to the denominator 3. This gives us our fifth property of exponents.

Property

Quotient to a Power Property of Exponents

For any real numbers a and b, when b is not equal to 0, and positive integer m,

$$\left(\frac{a}{b}\right)^m = \frac{a^m}{b^m}$$

For example,

$$\left(\frac{2}{5}\right)^3 = \frac{2^3}{5^3} = \frac{8}{125}$$

Example 7 illustrates the use of this property. Again note that the other properties may also be applied when simplifying an expression.

Example 7   **Using the Quotient to a Power Property of Exponents**

Simplify each expression.

(a) $\left(\dfrac{3}{4}\right)^3 = \dfrac{3^3}{4^3} = \dfrac{27}{64}$ Quotient to a power property

(b) $\left(\dfrac{x^3}{y^2}\right)^4 = \dfrac{(x^3)^4}{(y^2)^4}$ Quotient to a power property

$\quad\quad = \dfrac{x^{12}}{y^8}$ Power to a power property

(c) $\left(\dfrac{r^2 s^3}{t^4}\right)^2 = \dfrac{(r^2 s^3)^2}{(t^4)^2}$ Quotient to a power property

$\quad\quad = \dfrac{(r^2)^2 (s^3)^2}{(t^4)^2}$ Product to a power property

$\quad\quad = \dfrac{r^4 s^6}{t^8}$ Power to a power property

Check Yourself 7

Simplify each expression.

(a) $\left(\dfrac{2}{3}\right)^4$ (b) $\left(\dfrac{m^3}{n^4}\right)^5$ (c) $\left(\dfrac{a^2 b^3}{c^5}\right)^2$

The following table summarizes the five properties of exponents that were discussed in this section:

Property	General Form	Example
Product	$a^m a^n = a^{m+n}$	$x^2 \cdot x^3 = x^5$
Quotient	$\dfrac{a^m}{a^n} = a^{m-n} \quad (m > n)$	$\dfrac{5^7}{5^3} = 5^4$
Power to a power	$(a^m)^n = a^{mn}$	$(z^5)^4 = z^{20}$
Product to a power	$(ab)^m = a^m b^m$	$(4x)^3 = 4^3 x^3 = 64x^3$
Quotient to a power	$\left(\dfrac{a}{b}\right)^m = \dfrac{a^m}{b^m}$	$\left(\dfrac{2}{3}\right)^3 = \dfrac{2^3}{3^3} = \dfrac{8}{27}$

Our work in this chapter deals with the most common kind of algebraic expression, a *polynomial*. To define a polynomial, we recall our earlier definition of the word *term*.

Definition

Term

A **term** can be written as a number or the product of a number and one or more variables.

This definition indicates that constants, such as the number 3, and single variables, such as x, are terms.

For instance, x^5, $3x$, $-4xy^2$, and 8 are all examples of terms.

You should recall that the number factor of a term is called the **numerical coefficient** or simply the **coefficient.**

In the terms above, 1 is the coefficient of x^5, 3 is the coefficient of $3x$, -4 is the coefficient of $-4xy^2$ because the negative sign is part of the coefficient, and 8 is the coefficient of the term 8.

We combine terms to form expressions called **polynomials.** Polynomials are one of the most common expressions in algebra.

Definition

Polynomial

A **polynomial** is an algebraic expression that can be written as a term or as the sum or difference of terms.
Any variable factors with exponents must be to whole number powers.

Example 8 — Identifying Polynomials

< Objective 2 >

State whether each expression is a polynomial. List the terms of each polynomial and the coefficient of each term.

NOTE

In a polynomial, terms are separated by + and − signs.

(a) $x + 3$ is a polynomial. The terms are x and 3. The coefficients are 1 and 3.

(b) $3x^2 - 2x + 5$, or $3x^2 + (-2x) + 5$, is also a polynomial. Its terms are $3x^2$, $-2x$, and 5. The coefficients are 3, -2, and 5.

(c) $5x^3 + 2 - \dfrac{3}{x}$ is *not* a polynomial because of the division by x in the third term.

Check Yourself 8

Which expressions are polynomials?

(a) $5x^2$ (b) $3y^3 - 2y + \dfrac{5}{y}$ (c) $4x^2 - \dfrac{2}{3}x + 3$

Certain polynomials are given special names because of the number of terms that they have.

Definition	
Monomial, Binomial, and Trinomial	A polynomial with one term is called a **monomial**. The prefix *mono-* means 1.
	A polynomial with two terms is called a **binomial**. The prefix *bi-* means 2.
	A polynomial with three terms is called a **trinomial**. The prefix *tri-* means 3.

We do not use special names for polynomials with more than three terms.

Example 9   Identifying Types of Polynomials

(a) $3x^2y$ is a monomial. It has one term.

(b) $2x^3 + 5x$ is a binomial. It has two terms, $2x^3$ and $5x$.

(c) $5x^2 - 4x + 3$ is a trinomial. Its three terms are $5x^2$, $-4x$, and 3.

Check Yourself 9

Classify each polynomial as a monomial, binomial, or trinomial.

(a) $5x^4 - 2x^3$ (b) $4x^7$ (c) $2x^2 + 5x - 3$

We also classify polynomials by their *degree*. The **degree** of a polynomial that has only one variable is the highest power appearing in any one term.

Example 10   Classifying Polynomials by Their Degree

< Objective 3 >

NOTE

We will see in the next section that $x^0 = 1$.

The highest power

(a) $5x^3 - 3x^2 + 4x$ has degree 3.

The highest power

(b) $4x - 5x^4 + 3x^3 + 2$ has degree 4.

(c) $8x$ has degree 1.   Because $8x = 8x^1$

(d) 7 has degree 0.   The degree of any nonzero constant expression is zero.

Note: Polynomials can have more than one variable, such as $4x^2y^3 + 5xy^2$. The degree is then the highest sum of the powers in any single term (here $2 + 3$, or 5). In general, we will be working with polynomials in a single variable, such as x.

Check Yourself 10

Find the degree of each polynomial.

(a) $6x^5 - 3x^3 - 2$ (b) $5x$ (c) $3x^3 + 2x^6 - 1$ (d) 9

Working with polynomials is much easier if you get used to writing them in **descending order** (sometimes called *descending-exponent form*). This simply means that the term with the highest exponent is written first, then the term with the next highest exponent, and so on.

Example 11   **Writing Polynomials in Descending Order**

< Objective 4 >

The exponents get smaller from left to right.

(a) $5x^7 - 3x^4 + 2x^2$ is in descending order.

(b) $4x^4 + 5x^6 - 3x^5$ is *not* in descending order. The polynomial should be written as

$$5x^6 - 3x^5 + 4x^4$$

The degree of the polynomial is the power of the *first,* or *leading,* term once the polynomial is arranged in descending order.

 Check Yourself 11

Write each polynomial in descending order.

(a) $5x^4 - 4x^5 + 7$ **(b)** $4x^3 + 9x^4 + 6x^8$

A polynomial can represent any number. Its value depends on the value given to the variable.

 Example 12   **Evaluating Polynomials**

< Objective 5 >

RECALL

We use the rules for order of operations to evaluate each polynomial.

> C A U T I O N

Be particularly careful when dealing with powers of negative numbers!

Given the polynomial

$$3x^3 - 2x^2 - 4x + 1$$

(a) Find the value of the polynomial when $x = 2$.

To evaluate the polynomial, substitute 2 for x.

$$3(2)^3 - 2(2)^2 - 4(2) + 1$$
$$= 3(8) - 2(4) - 4(2) + 1$$
$$= 24 - 8 - 8 + 1$$
$$= 9$$

(b) Find the value of the polynomial when $x = -2$.

Now we substitute -2 for x.

$$3(-2)^3 - 2(-2)^2 - 4(-2) + 1$$
$$= 3(-8) - 2(4) - 4(-2) + 1$$
$$= -24 - 8 + 8 + 1$$
$$= -23$$

 Check Yourself 12

Find the value of the polynomial

$$4x^3 - 3x^2 + 2x - 1$$

when

(a) $x = 3$ **(b)** $x = -3$

Polynomials are used in almost every professional field. Many applications are related to predictions and forecasts. In allied health, polynomials can be used to calculate the concentration of a medication in the bloodstream after a given amount of time, as the next example demonstrates.

Example 13   An Allied Health Application

The concentration of digoxin, a medication prescribed for congestive heart failure, in a patient's bloodstream t hours after injection is given by the polynomial

$$-0.0015t^2 + 0.0845t + 0.7170$$

where concentration is measured in nanograms per milliliter (ng/mL). Determine the concentration of digoxin in a patient's bloodstream 19 hours after injection.

We are asked to evaluate the polynomial

$$-0.0015t^2 + 0.0845t + 0.7170$$

for the variable value $t = 19$. We substitute 19 for t in the polynomial.

$$-0.0015(19)^2 + 0.0845(19) + 0.7170$$
$$= -0.0015(361) + 1.6055 + 0.7170$$
$$= -0.5415 + 1.6055 + 0.7170$$
$$= 1.781$$

The concentration is 1.781 nanograms per milliliter.

Check Yourself 13

The concentration of a sedative, in micrograms per milliliter (mcg/mL), in a patient's bloodstream t hours after injection is given by the polynomial $-1.35t^2 + 10.81t + 7.38$. Determine the concentration of the sedative in a patient's bloodstream 3.5 hours after injection. Round to the nearest tenth.

Check Yourself ANSWERS

1. (a) x^{10}; (b) $(-3)^7$; (c) $(x^2y)^8$; (d) y^7 **2.** (a) $14x^7$; (b) $-2x^5y^3$; (c) $15x^5y^5$; (d) x^9

3. (a) y^7; (b) x^8; (c) $5r$; (d) $-8m^5n^4$ **4.** (a) m^{30}; (b) m^{11}; (c) 3^8; (d) 3^6

5. (a) $81y^4$; (b) $64m^6n^6$; (c) $48x^2$; (d) $-48x^3$ **6.** (a) $m^{15}n^6$; (b) $256p^8$; (c) s^7

7. (a) $\dfrac{16}{81}$; (b) $\dfrac{m^{15}}{n^{20}}$; (c) $\dfrac{a^4b^6}{c^{10}}$ **8.** (a) polynomial; (b) not a polynomial;

(c) polynomial   **9.** (a) binomial; (b) monomial; (c) trinomial

10. (a) 5; (b) 1; (c) 6; (d) 0   **11.** (a) $-4x^5 + 5x^4 + 7$; (b) $6x^8 + 9x^4 + 4x^3$

12. (a) 86; (b) -142 **13.** 28.7 mcg/mL

Reading Your Text

The following fill-in-the-blank exercises are designed to ensure that you understand some of the key vocabulary used in this section.

SECTION 3.1

(a) Exponential notation indicates repeated _____.

(b) A _____ can be written as a number or product of a number and one or more variables.

(c) In each term of a polynomial, the number factor is called the numerical _____.

(d) The _____ of a polynomial in one variable is the highest power of the variable that appears in a term.

The Streeter/Hutchison Series in Mathematics   Beginning Algebra

Basic Skills | Challenge Yourself | Calculator/Computer | Career Applications | Above and Beyond

< Objective 1 >

Simplify each expression.

1. $(x^2)^3$

2. $(a^5)^3$

3. $(m^4)^4$

4. $(p^7)^2$

5. $(2^4)^2$

6. $(3^3)^2$

7. $(5^3)^5$

8. $(7^2)^4$

9. $(3x)^3$

10. $(4m)^2$

11. $(2xy)^4$

12. $(5pq)^3$

13. $\left(\dfrac{3}{4}\right)^2$

14. $\left(\dfrac{2}{3}\right)^3$

15. $\left(\dfrac{x}{5}\right)^3$

16. $\left(\dfrac{a}{2}\right)^5$

17. $(2x^2)^4$

18. $(3y^2)^5$

19. $(a^8b^6)^2$

20. $(p^3q^4)^2$

21. $(4x^2y)^3$

22. $(4m^4n^4)^2$ > Videos

23. $(3m^2)^4(-2m^3)^2$

24. $(-2y^4)^3(4y^3)^2$

25. $\dfrac{(x^4)^3}{x^2}$

26. $\dfrac{(m^5)^3}{m^6}$

27. $\dfrac{(s^3)^2(s^2)^3}{(s^5)^2}$

28. $\dfrac{(y^5)^3(y^3)^2}{(y^4)^4}$ > Videos

29. $\left(\dfrac{m^3}{n^2}\right)^3$

30. $\left(\dfrac{a^4}{b^3}\right)^4$

31. $\left(\dfrac{a^3b^2}{c^4}\right)^2$

32. $\left(\dfrac{x^5y^2}{z^4}\right)^3$

Answers

33. _____

34. _____

35. _____

36. _____

37. _____

38. _____

39. _____

40. _____

41. _____

42. _____

43. _____

44. _____

45. _____

46. _____

47. _____

48. _____

49. _____

50. _____

51. _____

52. _____

53. _____

54. _____

55. _____

56. _____

57. _____

58. _____

59. _____

60. _____

< Objective 2 >

Which expressions are polynomials?

33. $7x^3$

34. $5x^3 - \dfrac{3}{x}$

35. 7

36. $4x^3 + x$

37. $\dfrac{3 + x}{x^2}$

38. $5a^2 - 2a + 7$

For each polynomial, list the terms and their coefficients.

39. $2x^2 - 3x$

40. $5x^3 + x$

41. $4x^3 - 3x + 2$ > Videos

42. $7x^2$

Classify each expression as a **monomial, binomial,** *or* **trinomial,** *where possible.*

43. $7x^3 - 3x^2$

44. $4x^7$

45. $7y^2 + 4y + 5$

46. $2x^2 + \dfrac{1}{3}xy + y^2$

47. $2x^4 - 3x^2 + 5x - 2$

48. $x^4 + \dfrac{5}{x} + 7$

49. $6y^8$

50. $4x^4 - 2x^2 + \dfrac{3}{4}x - 7$

51. $x^5 - \dfrac{3}{x^2}$

52. $4x^2 - 9$

< Objectives 3–4 >

Arrange in descending order if necessary, and give the degree of each polynomial.

53. $4x^5 - 3x^2$

54. $5x^2 + 3x^3 + 4$

55. $7x^7 - 5x^9 + 4x^3$

56. $2 + x$

57. $4x$

58. $x^{17} - 3x^4$

59. $5x^2 - 3x^5 + x^6 - 7$ > Videos

60. 5

< Objective 5 >

Evaluate each polynomial for the given values of the variable.

61. $6x + 1$, $x = 1$ and $x = -1$

62. $5x - 5$, $x = 2$ and $x = -2$

63. $x^3 - 2x$, $x = 2$ and $x = -2$

64. $3x^2 + 7$, $x = 3$ and $x = -3$

> Videos

65. $3x^2 + 4x - 2$, $x = 4$ and $x = -4$

66. $2x^2 - 5x + 1$, $x = 2$ and $x = -2$

67. $-x^2 - 2x + 3$, $x = 1$ and $x = -3$

68. $-x^2 - 5x - 6$, $x = -3$ and $x = -2$

Basic Skills | **Challenge Yourself** | Calculator/Computer | Career Applications | Above and Beyond

Complete each statement with **never, sometimes,** *or* **always.**

69. A polynomial is _____ a trinomial.

70. A trinomial is _____ a polynomial.

71. The product of two monomials is _____ a monomial.

72. A term is _____ a binomial.

Determine whether each statement is **always true, sometimes true,** *or* **never true.**

73. A monomial is a polynomial.

74. A binomial is a trinomial.

75. The degree of a trinomial is 3.

76. A trinomial has three terms.

77. A polynomial has four or more terms.

78. A binomial must have two coefficients.

Basic Skills | Challenge Yourself | Calculator/Computer | Career Applications | **Above and Beyond**

Solve each problem.

79. Write x^{12} as a power of x^2.

80. Write y^{15} as a power of y^3.

81. Write a^{16} as a power of a^2.

82. Write m^{20} as a power of m^5.

Answers

61.

62.

63.

64.

65.

66.

67.

68.

69.

70.

71.

72.

73.

74.

75.

76.

77.

78.

79.

80.

81.

82.

83. _____

84. _____

85. _____

86. _____

87. _____

88. _____

89. _____

90. _____

91. _____

92. _____

93. _____

94. _____

95. _____

96. _____

83. Write each expression as a power of 8. (Remember that $8 = 2^3$.)

$2^{12}, 2^{18}, (2^5)^3, (2^7)^6$

84. Write each expression as a power of 9.

$3^8, 3^{14}, (3^5)^8, (3^4)^7$

85. What expression raised to the third power is $-8x^6y^9z^{15}$?

86. What expression raised to the fourth power is $81x^{12}y^8z^{16}$?

The formula $(1 + R)^y = G$ gives us useful information about the growth of a population. Here R is the rate of growth expressed as a decimal, y is the time in years, and G is the growth factor. If a country has a 2% growth rate for 35 years, then its population will double:

$$(1.02)^{35} \approx 2$$

87. SOCIAL SCIENCE

(a) With a 2% growth rate, how many doublings will occur in 105 years? How much larger will the country's population be to the nearest whole number?

(b) The less-developed countries of the world had an average growth rate of 2% in 1986. If their total population was 3.8 billion, what will their population be in 105 years if this rate remains unchanged?

88. SOCIAL SCIENCE The United States has a growth rate of 0.7%. What will be its growth factor after 35 years (to the nearest percent)?

89. Write an explanation of why $(x^3)(x^4)$ is *not* x^{12}.

90. Your algebra study partners are confused. "Why isn't $x^2 \cdot x^3 = 2x^5$?" they ask you. Write an explanation that will convince them.

Capital italic letters such as P and Q are often used to name polynomials. For example, we might write $P(x) = 3x^3 - 5x^2 + 2$ in which $P(x)$ is read "P of x." The notation permits a convenient shorthand. We write P(2), read "P of 2," to indicate the value of the polynomial when $x = 2$. Here

$$P(2) = 3(2)^3 - 5(2)^2 + 2$$
$$= 3 \cdot 8 - 5 \cdot 4 + 2$$
$$= 6$$

Use the preceding information to complete exercises 91–104.

If $P(x) = x^3 - 2x^2 + 5$ and $Q(x) = 2x^2 + 3$, find:

91. $P(1)$ **92.** $P(-1)$

93. $Q(2)$ **94.** $Q(-2)$

95. $P(3)$ **96.** $Q(-3)$

97. $P(0)$

98. $Q(0)$

99. $P(2) + Q(-1)$

100. $P(-2) + Q(3)$

101. $P(3) - Q(-3) \div Q(0)$

102. $Q(-2) \div Q(2) \cdot P(0)$

103. $|Q(4)| - |P(4)|$

104. $\dfrac{P(-1) + Q(0)}{P(0)}$

105. BUSINESS AND FINANCE The cost, in dollars, of typing a term paper is given as 3 times the number of pages plus 20. Use y as the number of pages to be typed and write a polynomial to describe this cost. Find the cost of typing a 50-page paper.

106. BUSINESS AND FINANCE The cost, in dollars, of making suits is described as 20 times the number of suits plus 150. Use s as the number of suits and write a polynomial to describe this cost. Find the cost of making seven suits.

Answers

97. _____

98. _____

99. _____

100. _____

101. _____

102. _____

103. _____

104. _____

105. _____

106. _____

Answers

1. x^6 **3.** m^{16} **5.** 2^8 **7.** 5^{15} **9.** $27x^3$ **11.** $16x^4y^4$

13. $\dfrac{9}{16}$ **15.** $\dfrac{x^3}{125}$ **17.** $16x^8$ **19.** $a^{16}b^{12}$ **21.** $64x^6y^3$

23. $324m^{14}$ **25.** x^{10} **27.** s^2 **29.** $\dfrac{m^9}{n^6}$ **31.** $\dfrac{a^6b^4}{c^8}$ **33.** Polynomial

35. Polynomial **37.** Not a polynomial **39.** $2x^2, -3x; 2, -3$
41. $4x^3, -3x, 2; 4, -3, 2$ **43.** Binomial **45.** Trinomial
47. Not classified **49.** Monomial **51.** Not a polynomial
53. $4x^5 - 3x^2; 5$ **55.** $-5x^9 + 7x^7 + 4x^3; 9$ **57.** $4x; 1$
59. $x^6 - 3x^5 + 5x^2 - 7; 6$ **61.** $7, -5$ **63.** $4, -4$ **65.** $62, 30$
67. $0, 0$ **69.** sometimes **71.** always **73.** Always
75. Sometimes **77.** Sometimes **79.** $(x^2)^6$ **81.** $(a^2)^8$
83. $8^4, 8^6, 8^5, 8^{14}$ **85.** $-2x^2y^3z^5$ **87.** **(a)** Three doublings, 8 times as large; **(b)** 30.4 billion **89.** Above and Beyond **91.** 4 **93.** 11
95. 14 **97.** 5 **99.** 10 **101.** 7 **103.** -2
105. $3y + 20$, 170

3.2

Negative Exponents and Scientific Notation

< 3.2 Objectives >

1 > Evaluate expressions involving a zero or negative exponent

2 > Simplify expressions involving a zero or negative exponent

3 > Write a number in scientific notation

4 > Solve applications involving scientific notation

In Section 3.1, we discussed exponents. We now want to extend our exponent notation to include 0 and negative integers as exponents.

First, what do we do with x^0? It will help to look at a problem that gives us x^0 as a result. What if the numerator and denominator of a fraction have the same base raised to the same power and we extend our division rule? For example,

$$\frac{a^5}{a^5} = a^{5-5} = a^0$$

But from our experience with fractions we know that

$$\frac{a^5}{a^5} = 1$$

By comparing these equations, it seems reasonable to make the following definition:

RECALL

By the quotient property,

$$\frac{a^m}{a^n} = a^{m-n}$$

when $m > n$. Here m and n are *both* 5, so $m = n$.

Definition

Zero Power

For any nonzero number a,

$a^0 = 1$

In words, any expression, except 0, raised to the 0 power is 1.

Example 1 illustrates the use of this definition.

 Example 1   Raising Expressions to the Zero Power

< Objective 1 >

Evaluate each expression. Assume all variables are nonzero.

(a) $5^0 = 1$

(b) $(-27)^0 = 1$ The exponent is applied to -27.

(c) $(x^2y)^0 = 1$

(d) $6x^0 = 6 \cdot 1 = 6$

(e) $-27^0 = -1$ The exponent is applied to 27, but not to the silent -1.

> **CAUTION**

In part (d) the 0 exponent applies only to the x and *not* to the factor 6, because the base is x.

 Check Yourself 1

Evaluate each expression. Assume all variables are nonzero.

(a) 7^0 (b) $(-8)^0$ (c) $(xy^3)^0$ (d) $3x^0$ (e) -5^0

198

© The McGraw-Hill Companies. All Rights Reserved.   The Streeter/Hutchison Series in Mathematics   Beginning Algebra

Before we introduce the next property, we look at some examples that use the properties of Section 3.1.

Example 2	Evaluating Expressions

Evaluate each expression.

(a) $\dfrac{5^6}{5^2}$ From our earlier work, we get $5^{6-2} = 5^4 = 625$.

(b) $\dfrac{5^2}{5^6}$

$$\frac{5^2}{5^6} = \frac{5 \cdot 5}{5 \cdot 5 \cdot 5 \cdot 5 \cdot 5 \cdot 5} = \frac{1}{5^4} = \frac{1}{625}$$

(c) $\dfrac{10^3}{10^9} = \dfrac{10 \cdot 10 \cdot 10}{10 \cdot 10 \cdot 10 \cdot 10 \cdot 10 \cdot 10 \cdot 10 \cdot 10 \cdot 10} = \dfrac{1}{10^6}$ or $\dfrac{1}{1,000,000}$

Check Yourself 2

Evaluate each expression.

(a) $\dfrac{5^9}{5^6}$ **(b)** $\dfrac{5^6}{5^9}$ **(c)** $\dfrac{10^6}{10^{10}}$ **(d)** $\dfrac{x^3}{x^5}$

NOTES

John Wallis (1616–1703), an English mathematician, was the first to fully discuss the meaning of 0 and negative exponents.

Divide the numerator and denominator by the two common x factors.

The quotient property of exponents allows us to define a negative exponent. Suppose that the exponent in the denominator is *greater than* the exponent in the numerator. Consider the expression $\dfrac{x^2}{x^5}$.

Our previous work with fractions tells us that

$$\frac{x^2}{x^5} = \frac{x \cdot x}{x \cdot x \cdot x \cdot x \cdot x} = \frac{1}{x^3}$$

However, if we extend the quotient property to let n be greater than m, we have

$$\frac{x^2}{x^5} = x^{2-5} = x^{-3}$$

Now, by comparing these equations, it seems reasonable to define x^{-3} as $\dfrac{1}{x^3}$.

In general, we have the following results.

Definition
Negative Powers

For any nonzero number a,

$$a^{-1} = \frac{1}{a}$$

For any nonzero number a, and any integer n,

$$a^{-n} = \frac{1}{a^n}$$

This definition tells us that if we have a base a raised to a negative integer power, such as a^{-5}, we may rewrite this as 1 over the base a raised to a positive integer power: $\dfrac{1}{a^5}$.

We work with this in Example 3.

Example 3   **Rewriting Expressions That Contain Negative Exponents**

< Objective 2 >   Rewrite each expression using only positive exponents. Simplify when possible.

Negative exponent in numerator

(a) $x^{-4} = \dfrac{1}{x^4}$

Positive exponent in
denominator

(b) $m^{-7} = \dfrac{1}{m^7}$

(c) $3^{-2} = \dfrac{1}{3^2}$ or $\dfrac{1}{9}$

(d) $\dfrac{1}{10^{-3}} = \dfrac{1}{\left(\dfrac{1}{10}\right)^3} = \dfrac{1 \cdot \left(\dfrac{10}{1}\right)^3}{\left(\dfrac{1}{10}\right)^3 \cdot \left(\dfrac{10}{1}\right)^3} = \left(\dfrac{10}{1}\right)^3 = 1{,}000$

A negative power in the
denominator is equivalent
to a positive power in the
numerator.

So, $\dfrac{1}{x^{-3}} = x^3$

 > CAUTION

$2x^{-3}$ is not the same as $(2x)^{-3}$.

(e) $2x^{-3} = 2 \cdot \dfrac{1}{x^3} = \dfrac{2}{x^3}$

The -3 exponent applies only
to x, because x is the base.

(f) $\left(\dfrac{2}{5}\right)^{-1} = \dfrac{1}{\dfrac{2}{5}} = \dfrac{5}{2}$

(g) $-4x^{-5} = -4 \cdot \dfrac{1}{x^5} = -\dfrac{4}{x^5}$

 Check Yourself 3

Write each expression using only positive exponents.

(a) a^{-10} **(b)** 4^{-3} **(c)** $3x^{-2}$ **(d)** $\left(\dfrac{3}{2}\right)^{-2}$

We can now use negative integers as exponents in our product property for exponents. Consider Example 4.

 Example 4   **Simplifying Expressions Containing Exponents**

RECALL

$a^m \cdot a^n = a^{m+n}$ for *any* integers m and n. So add the exponents.

Rewrite each expression using only positive exponents.

(a) $x^5 x^{-2} = x^{5+(-2)} = x^3$

Note: An alternative approach would be

$x^5 x^{-2} = x^5 \cdot \dfrac{1}{x^2} = \dfrac{x^5}{x^2} = x^3$

(b) $a^7a^{-5} = a^{7+(-5)} = a^2$

(c) $y^5y^{-9} = y^{5+(-9)} = y^{-4} = \dfrac{1}{y^4}$

 Check Yourself 4

Rewrite each expression using only positive exponents.

(a) x^7x^{-2} **(b)** b^3b^{-8}

Example 5 shows that all the properties of exponents introduced in the last section can be extended to expressions with negative exponents.

Example 5	Simplifying Expressions Containing Exponents

Simplify each expression.

(a) $\dfrac{m^{-3}}{m^4} = m^{-3-4}$ Quotient property

$\qquad = m^{-7} = \dfrac{1}{m^7}$

(b) $\dfrac{a^{-2}b^6}{a^5b^{-4}} = a^{-2-5}b^{6-(-4)}$ Apply the quotient property to each variable.

$\qquad = a^{-7}b^{10} = \dfrac{b^{10}}{a^7}$

(c) $(2x^4)^{-3} = \dfrac{1}{(2x^4)^3}$ Definition of a negative exponent

$\qquad = \dfrac{1}{2^3(x^4)^3}$ Product to a power property

$\qquad = \dfrac{1}{8x^{12}}$ Power to a power property

(d) $\dfrac{(y^{-2})^4}{(y^3)^{-2}} = \dfrac{y^{-8}}{y^{-6}}$ Power to a power property

$\qquad = y^{-8+(+6)}$ Quotient property

$\qquad = y^{-2} = \dfrac{1}{y^2}$

NOTE

We can also complete (c) by using the power to a power property first, so

$(2x^4)^{-3} = 2^{-3} \cdot (x^4)^{-3} = 2^{-3}x^{-12}$

$\qquad = \dfrac{1}{2^3x^{12}}$

$\qquad = \dfrac{1}{8x^{12}}$

 Check Yourself 5

Simplify each expression.

(a) $\dfrac{x^5}{x^{-3}}$ **(b)** $\dfrac{m^3n^{-5}}{m^{-2}n^3}$ **(c)** $(3a^3)^{-4}$ **(d)** $\dfrac{(r^3)^{-2}}{(r^{-4})^2}$

Scientific notation is one important use of exponents.

We begin the discussion with a calculator exercise. On most calculators, if you multiply 2.3 times 1,000, the display reads

2300

Multiply by 1,000 a second time and you see

2300000

> Calculator

NOTE

2.3 E09 must equal 2,300,000,000.

NOTE

Consider the following table:

$2.3 = 2.3 \times 10^0$

$23 = 2.3 \times 10^1$

$230 = 2.3 \times 10^2$

$2300 = 2.3 \times 10^3$

$23,000 = 2.3 \times 10^4$

$230,000 = 2.3 \times 10^5$

On most calculators, multiplying by 1,000 a third time results in the display

2.3 09   or   2.3 E09

Multiplying by 1,000 again yields

2.3 12   or   2.3 E12

Can you see what is happening? This is the way calculators display very large numbers. The number on the left is always between 1 and 10, and the number on the right indicates the number of places the decimal point must be moved to the right to put the answer in standard (or decimal) form.

This notation is used frequently in science. It is not uncommon in scientific applications of algebra to find yourself working with very large or very small numbers. Even in the time of Archimedes (287–212 B.C.E.), the study of such numbers was not unusual. Archimedes estimated that the universe was 23,000,000,000,000,000 m in diameter, which is the approximate distance light travels in $2\frac{1}{2}$ years. By comparison, Polaris (the North Star) is actually 680 light-years from Earth. Example 7 looks at the idea of light-years.

In scientific notation, Archimedes' estimate for the diameter of the universe would be

2.3×10^{16} m

If a number is divided by 1,000 again and again, we get a negative exponent on the calculator. In scientific notation, we use positive exponents to write very large numbers, such as the distance of stars. We use negative exponents to write very small numbers, such as the width of an atom.

Definition	
Scientific Notation	Any number written in the form $a \times 10^n$ in which $1 \le a < 10$ and n is an integer, is written in **scientific notation**.

Scientific notation is one of the few places that we still use the multiplication symbol $\times$.

▶	**Example 6**	**Using Scientific Notation**

< Objective 3 >

NOTE

The exponent on 10 shows the *number of places* we must move the decimal point. A positive exponent tells us to move right, and a negative exponent indicates a move to the left.

Write each number in scientific notation.

(a) $120,000. = 1.2 \times 10^5$

5 places

The power is 5.

(b) $88,000,000. = 8.8 \times 10^7$

7 places

The power is 7.

(c) $520,000,000. = 5.2 \times 10^8$

8 places

(d) $4000,000,000. = 4 \times 10^9$

9 places

NOTE

To convert back to standard or decimal form, the process is simply reversed.

(e) $0.0005 = 5 \times 10^{-4}$

4 places

If the decimal point is to be moved to the left, the exponent is negative.

(f) $0.0000000081 = 8.1 \times 10^{-9}$

9 places

Check Yourself 6

Write in scientific notation.
(a) 212,000,000,000,000,000   **(b)** 0.00079
(c) 5,600,000   **(d)** 0.0000007

Example 7   **An Application of Scientific Notation**

< Objective 4 >

(a) Light travels at a speed of 3.0×10^8 meters per second (m/s). There are approximately 3.15×10^7 s in a year. How far does light travel in a year?

We multiply the distance traveled in 1 s by the number of seconds in a year. This yields

NOTE

$9.45 \times 10^{15} \approx 10 \times 10^{15} = 10^{16}$

$$(3.0 \times 10^8)(3.15 \times 10^7) = (3.0 \cdot 3.15)(10^8 \cdot 10^7)$$
$$= 9.45 \times 10^{15}$$

Multiply the coefficients, and add the exponents.

For our purposes we round the distance light travels in 1 year to 10^{16} m. This unit is called a **light-year,** and it is used to measure astronomical distances.

(b) The distance from Earth to the star Spica (in Virgo) is 2.2×10^{18} m. How many light-years is Spica from Earth?

NOTE

We divide the distance (in meters) by the number of meters in 1 light-year.

$$\frac{2.2 \times 10^{18}}{10^{16}} = 2.2 \times 10^{18-16}$$
$$= 2.2 \times 10^2 = 220 \text{ light-years}$$

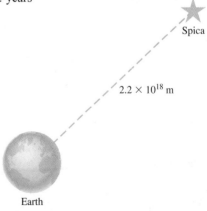

Spica

2.2×10^{18} m

Earth

Check Yourself 7

The farthest object that can be seen with the unaided eye is the Andromeda galaxy. This galaxy is 2.3×10^{22} m from Earth. What is this distance in light-years?

Check Yourself ANSWERS

1. (a) 1; (b) 1; (c) 1; (d) 3; (e) -1 **2.** (a) 125; (b) $\dfrac{1}{125}$; (c) $\dfrac{1}{10,000}$; (d) $\dfrac{1}{x^2}$

3. (a) $\dfrac{1}{a^{10}}$; (b) $\dfrac{1}{4^3}$ or $\dfrac{1}{64}$; (c) $\dfrac{3}{x^2}$; (d) $\dfrac{4}{9}$ **4.** (a) x^5; (b) $\dfrac{1}{b^5}$

5. (a) x^8; (b) $\dfrac{m^5}{n^8}$; (c) $\dfrac{1}{81a^{12}}$; (d) r^2 **6.** (a) 2.12×10^{17}; (b) 7.9×10^{-4};

(c) 5.6×10^6; (d) 7×10^{-7} **7.** 2,300,000 light-years

Reading Your Text

The following fill-in-the-blank exercises are designed to ensure that you understand some of the key vocabulary used in this section.

SECTION 3.2

(a) A nonzero number raised to the zero power is always equal to _____.

(b) A negative exponent in the denominator is equivalent to a _____ exponent in the numerator.

(c) All of the properties of _____ can be extended to terms with negative exponents.

(d) The base a in a number written in scientific notation cannot be greater than or equal to _____.

3.2 exercises

Name _____

Section _____ Date _____

< Objective 1 >

Evaluate (assume any variables are nonzero).

1. 4^0

2. $(-7)^0$

3. $(-29)^0$

4. 75^0

5. $(x^3 y^2)^0$

6. $7m^0$

7. $11x^0$ > Videos

8. $(2a^3 b^7)^0$

9. $(-3p^6 q^8)^0$

10. $-7x^0$

< Objective 2 >

Write each expression using positive exponents; simplify when possible.

11. b^{-8}

12. p^{-12}

13. 3^{-4}

14. 2^{-5}

15. $\left(\dfrac{1}{5}\right)^{-2}$

16. $\left(\dfrac{1}{4}\right)^{-3}$

17. $\dfrac{1}{10^{-4}}$

18. $\dfrac{1}{10^{-5}}$

19. $5x^{-1}$

20. $3a^{-2}$

21. $(5x)^{-1}$

22. $(3a)^{-2}$

23. $-2x^{-5}$

24. $3x^{-4}$

25. $(-2x)^{-5}$ > Videos

26. $(3x)^{-4}$

Answers

1. _____   2. _____

3. _____   4. _____

5. _____   6. _____

7. _____   8. _____

9. _____   10. _____

11. _____   12. _____

13. _____   14. _____

15. _____   16. _____

17. _____   18. _____

19. _____   20. _____

21. _____   22. _____

23. _____   24. _____

25. _____   26. _____

Answers

27. _____   28. _____

29. _____   30. _____

31. _____   32. _____

33. _____

34. _____

35. _____

36. _____

37. _____

38. _____

39. _____

40. _____

41. _____

42. _____

43. _____

44. _____

45. _____

46. _____

47. _____

48. _____

49. _____

50. _____

Simplify each expression and write your answers with only positive exponents.

27. $a^5 a^3$

28. $m^5 m^7$

29. $x^8 x^{-2}$

30. $a^{12} a^{-8}$

31. $x^0 x^5$

32. $r^{-3} r^0$

33. $\dfrac{a^8}{a^5}$

34. $\dfrac{m^9}{m^4}$

35. $\dfrac{x^7}{x^9}$ > Videos

36. $\dfrac{a^3}{a^{10}}$

| Basic Skills | **Challenge Yourself** | Calculator/Computer | Career Applications | Above and Beyond |

Determine whether each statement is **true** *or* **false**.

37. Zero raised to any power is one.

38. One raised to any power is one.

39. When multiplying two terms with the same base, add the exponents to find the power of that base in the product.

40. When multiplying two terms with the same base, multiply the exponents to find the power of that base in the product.

Simplify each expression. Write your answers with positive exponents only.

41. $\dfrac{x^{-4}yz}{x^{-5}yz}$

42. $\dfrac{p^{-6}q^{-3}}{p^{-3}q^{-6}}$

43. $\dfrac{m^5 n^{-3}}{m^{-4} n^5}$

44. $\dfrac{p^{-3}q^{-2}}{p^4 q^{-3}}$ > Videos

45. $(-2a^{-3})^4$

46. $(3x^2)^{-3}$

47. $(x^{-2}y^3)^{-2}$

48. $(-a^5 b^{-3})^{-3}$

49. $\dfrac{(r^{-2})^3}{r^{-4}}$

50. $\dfrac{(y^3)^{-4}}{y^{-6}}$

51. $\dfrac{m^{-2}n^3}{m^2n^4}$

52. $\dfrac{c^2d^{-3}}{c^{-4}d^{-5}}$

53. $\dfrac{r^3s^{-3}}{s^4t^{-2}}$

54. $\dfrac{x^3yz^{-2}}{x^{-2}y^{-3}z^4}$

55. $\dfrac{a^{-5}(b^2)^{-3}c^{-1}}{a(b^{-4})^3c^{-1}}$

56. $\dfrac{x^4y^{-3}z}{(xy^2)^{-2}z^{-1}}$

57. $\dfrac{(p^0q^{-2})^{-3}}{p(q^0)^2(p^{-1}q)^0}$

58. $\dfrac{x^{-1}(x^2y^{-2})^{-3}z^{-2}}{xy^{-3}z^0}$

59. $3(2x^{-2})^{-3}$

60. $2b^{-1}(2b^{-3})^{-2}$

61. $ab^{-2}(a^{-3}b^0)^{-2}$

62. $m^{-1}(m^2n^{-3})^2$

63. $2a^6(3a^{-4})^2$

64. $4x^2y^{-1}(2x^{-2}y^3)^{-2}$

65. $[c(c^{-2}d^0)^{-2}]^3$

66. $[x^2y(x^4y^{-3})^{-1}]^{-2}$

67. $\dfrac{w(w^2)^{-3}}{(w^2)^{-2}}$

68. $\dfrac{(2n^2)^{-3}}{(2n^{-2})^{-4}}$

69. $\dfrac{a^{-5}(a^2)^{-3}}{a(a^{-4})^3}$

70. $\dfrac{y^2(y^2)^{-2}}{(y^3)^{-2}(y^0)^2}$

< Objective 3 >

In exercises 71–74, express each number in scientific notation.

71. SCIENCE AND MEDICINE The distance from Earth to the Sun: 93,000,000 mi.

> Videos

Answers

51. _____

52. _____

53. _____

54. _____

55. _____

56. _____

57. _____

58. _____

59. _____

60. _____

61. _____

62. _____

63. _____

64. _____

65. _____

66. _____

67. _____

68. _____

69. _____   70. _____

71. _____

Answers

72. _____

73. _____

74. _____

75. _____

76. _____

77. _____

78. _____

79. _____

80. _____

81. _____

82. _____

83. _____

84. _____

85. _____

86. _____

87. _____

88. _____

89. _____

90. _____

91. _____

92. _____

93. _____

94. _____

95. _____

96. _____

72. **SCIENCE AND MEDICINE** The diameter of a grain of sand: 0.000021 m.

73. **SCIENCE AND MEDICINE** The diameter of the Sun: 130,000,000,000 cm.

74. **SCIENCE AND MEDICINE** The number of molecules in 22.4 L of a gas: 602,000,000,000,000,000,000,000 (Avogadro's number).

75. **SCIENCE AND MEDICINE** The mass of the Sun is approximately 1.99×10^{30} kg. If this were written in standard or decimal form, how many 0's would follow the second 9's digit?

76. **SCIENCE AND MEDICINE** Archimedes estimated the universe to be 2.3×10^{19} millimeters (mm) in diameter. If this number were written in standard or decimal form, how many 0's would follow the digit 3?

Write each expression in standard notation.

77. 8×10^{-3}

78. 7.5×10^{-6}

79. 2.8×10^{-5}

80. 5.21×10^{-4}

Write each number in scientific notation.

81. 0.0005

82. 0.000003

83. 0.00037

84. 0.000051

Evaluate the expressions using scientific notation, and write your answers in that form.

85. $(4 \times 10^{-3})(2 \times 10^{-5})$

86. $(1.5 \times 10^{-6})(4 \times 10^2)$

87. $\dfrac{9 \times 10^3}{3 \times 10^{-2}}$

88. $\dfrac{7.5 \times 10^{-4}}{1.5 \times 10^2}$

Evaluate each expression. Write your results in scientific notation.

89. $(2 \times 10^5)(4 \times 10^4)$

90. $(2.5 \times 10^7)(3 \times 10^5)$

91. $\dfrac{6 \times 10^9}{3 \times 10^7}$

92. $\dfrac{4.5 \times 10^{12}}{1.5 \times 10^7}$

93. $\dfrac{(3.3 \times 10^{15})(6 \times 10^{15})}{(1.1 \times 10^8)(3 \times 10^6)}$

94. $\dfrac{(6 \times 10^{12})(3.2 \times 10^8)}{(1.6 \times 10^7)(3 \times 10^2)}$

 > Videos

In 1975 the population of Earth was approximately 4 billion and doubling every 35 years. The formula for the population P in year y for this doubling rate is

P (in billions) $= 4 \times 2^{(y-1975)/35}$

95. **SOCIAL SCIENCE** What was the approximate population of Earth in 1960?

96. **SOCIAL SCIENCE** What will Earth's population be in 2025?

The U.S. population in 1990 was approximately 250 million, and the average growth rate for the past 30 years gives a doubling time of 66 years. The formula just given for the United States then becomes

$P \text{ (in millions)} = 250 \times 2^{(y-1990)/66}$

97. SOCIAL SCIENCE What was the approximate population of the United States in 1960?

98. SOCIAL SCIENCE What will the population of the United States be in 2025 if this growth rate continues?

< Objective 4 >

99. SCIENCE AND MEDICINE Megrez, the nearest of the Big Dipper stars, is 6.6×10^{17} m from Earth. Approximately how long does it take light, traveling at $10^{16} \dfrac{\text{m}}{\text{year}}$, to travel from Megrez to Earth?

100. SCIENCE AND MEDICINE Alkaid, the most distant star in the Big Dipper, is 2.1×10^{18} m from Earth. Approximately how long does it take light to travel from Alkaid to Earth?

101. SOCIAL SCIENCE The number of liters of water on Earth is 15,500 followed by 19 zeros. Write this number in scientific notation. Then use the number of liters of water on Earth to find out how much water is available for each person on Earth. The population of Earth is 6 billion.

102. SOCIAL SCIENCE If there are 6×10^9 people on Earth and there is enough freshwater to provide each person with 8.79×10^5 L, how much freshwater is on Earth?

103. SOCIAL SCIENCE The United States uses an average of 2.6×10^6 L of water per person each year. The United States has 3.2×10^8 people. How many liters of water does the United States use each year?

Answers

97.	
98.	
99.	
100.	
101.	
102.	
103.	

Answers

1. 1   **3.** 1   **5.** 1   **7.** 11   **9.** 1   **11.** $\dfrac{1}{b^8}$ **13.** $\dfrac{1}{81}$

15. 25   **17.** 10,000   **19.** $\dfrac{5}{x}$ **21.** $\dfrac{1}{5x}$ **23.** $-\dfrac{2}{x^5}$

25. $-\dfrac{1}{32x^5}$ **27.** a^8 **29.** x^6 **31.** x^5 **33.** a^3 **35.** $\dfrac{1}{x^2}$

37. False   **39.** True   **41.** x **43.** $\dfrac{m^9}{n^8}$ **45.** $\dfrac{16}{a^{12}}$ **47.** $\dfrac{x^4}{y^6}$

49. $\dfrac{1}{r^2}$ **51.** $\dfrac{1}{m^4n}$ **53.** $\dfrac{r^3t^2}{s^7}$ **55.** $\dfrac{b^6}{a^6}$ **57.** $\dfrac{q^6}{p}$ **59.** $\dfrac{3x^6}{8}$

61. $\dfrac{a^7}{b^2}$ **63.** $\dfrac{18}{a^2}$ **65.** c^{15} **67.** $\dfrac{1}{w}$ **69.** 1

71. 9.3×10^7 mi   **73.** 1.3×10^{11} cm   **75.** 28   **77.** 0.008

79. 0.000028   **81.** 5×10^{-4} **83.** 3.7×10^{-4} **85.** 8×10^{-8}

87. 3×10^5 **89.** 8×10^9 **91.** 2×10^2 **93.** 6×10^{16}

95. 2.97 billion   **97.** 182 million   **99.** 66 years

101. 1.55×10^{23} L; 2.58×10^{13} L   **103.** 8.32×10^{14} L

3.3
Adding and Subtracting Polynomials

< 3.3 Objectives >

1 > Add polynomials

2 > Distribute a negative sign over a polynomial

3 > Subtract polynomials

Addition is always a matter of combining like quantities (two apples plus three apples, four books plus five books, and so on). If you keep that basic idea in mind, adding polynomials is easy. It is just a matter of combining like terms.

Suppose that you want to add

$$5x^2 + 3x + 4 \qquad \text{and} \qquad 4x^2 + 5x - 6$$

RECALL

The plus sign between the parentheses indicates addition.

Parentheses are sometimes used when adding, so for the sum of these polynomials, we can write

$$(5x^2 + 3x + 4) + (4x^2 + 5x - 6)$$

Now what about the parentheses? You can use the following rule.

Property

Removing Signs of Grouping Case 1

When finding the sum of two polynomials, if a plus sign (+) or nothing at all appears in front of parentheses, simply remove the parentheses. No other changes are necessary.

NOTES

Remove the parentheses. No other changes are necessary.

We use the associative and commutative properties in reordering and regrouping.

We use the distributive property. For example,
$5x^2 + 4x^2 = (5 + 4)x^2 = 9x^2$

Now let's return to the addition.

$$(5x^2 + 3x + 4) + (4x^2 + 5x - 6)$$
$$= 5x^2 + 3x + 4 + 4x^2 + 5x - 6$$

Like terms   Like terms   Like terms

Collect like terms. (*Remember:* Like terms have the same variables raised to the same power).

$$= (5x^2 + 4x^2) + (3x + 5x) + (4 - 6)$$

Combine like terms for the result:

$$= 9x^2 + 8x - 2$$

As should be clear, much of this work can be done mentally. You can then write the sum directly by locating like terms and combining. Example 1 illustrates this property.

Example 1 | **Combining Like Terms**

< Objective 1 >

Add $3x - 5$ and $2x + 3$.

Write the sum.

$$(3x - 5) + (2x + 3) = 3x - 5 + 2x + 3 = 5x - 2$$

Like terms   Like terms

NOTE

We call this the "horizontal method" because the entire problem is written on one line. $3 + 4 = 7$ is the horizontal method.

$$\begin{array}{r} 3 \\ + 4 \\ \hline 7 \end{array}$$

is the vertical method.

Check Yourself 1

Add $6x^2 + 2x$ and $4x^2 - 7x$.

The same technique is used to find the sum of two trinomials.

Example 2 | **Adding Polynomials Using the Horizontal Method**

Add $4a^2 - 7a + 5$ and $3a^2 + 3a - 4$.

Write the sum.

$$(4a^2 - 7a + 5) + (3a^2 + 3a - 4)$$
$$= 4a^2 - 7a + 5 + 3a^2 + 3a - 4 = 7a^2 - 4a + 1$$

Like terms

Like terms

Like terms

RECALL

Only the like terms are combined in the sum.

Check Yourself 2

Add $5y^2 - 3y + 7$ and $3y^2 - 5y - 7$.

Example 3 | **Adding Polynomials Using the Horizontal Method**

Add $2x^2 + 7x$ and $4x - 6$.

Write the sum.

$$(2x^2 + 7x) + (4x - 6)$$
$$= 2x^2 + 7x + 4x - 6$$

These are the only like terms; $2x^2$ and -6 cannot be combined.

$$= 2x^2 + 11x - 6$$

Check Yourself 3

Add $5m^2 + 8$ and $8m^2 - 3m$.

Writing polynomials in descending order usually makes the work easier.

Example 4 | **Adding Polynomials Using the Horizontal Method**

Add $3x - 2x^2 + 7$ and $5 + 4x^2 - 3x$.

Write the polynomials in descending order and then add.

$(-2x^2 + 3x + 7) + (4x^2 - 3x + 5)$

$= 2x^2 + 12$

Check Yourself 4

Add $8 - 5x^2 + 4x$ and $7x - 8 + 8x^2$.

Subtracting polynomials requires another rule for removing signs of grouping.

Property

Removing Signs of Grouping Case 2

When finding the difference of two polynomials, if a minus sign ($-$) appears in front of a set of parentheses, the parentheses can be removed by changing the sign of each term inside the parentheses.

We illustrate this rule in Example 5.

Example 5 | **Removing Parentheses**

< Objective 2 >

Remove the parentheses in each expression.

NOTE

We are using the distributive property in part (a), because

$-(2x + 3y) = (-1)(2x + 3y)$

$= (-1)(2x) + (-1)(3y)$

$= -2x - 3y$

(a) $-(2x + 3y) = -2x - 3y$ Change each sign to remove the parentheses.

(b) $m - (5n - 3p) = m - 5n + 3p$

Sign changes

(c) $2x - (-3y + z) = 2x + 3y - z$

Sign changes

Check Yourself 5

In each expression, remove the parentheses.

(a) $-(3m + 5n)$ **(b)** $-(5w - 7z)$
(c) $3r - (2s - 5t)$ **(d)** $5a - (-3b - 2c)$

Subtracting polynomials is now a matter of using the previous rule to remove the parentheses and then combining the like terms. Consider Example 6.

Example 6   **Subtracting Polynomials Using the Horizontal Method**

< Objective 3 >

RECALL

The expression following "from" is written first in the problem.

(a) Subtract $5x - 3$ from $8x + 2$.

Write

$(8x + 2) - (5x - 3)$

$= 8x + 2 \underbrace{- 5x + 3}_{\text{Sign changes}}$ Recall that subtracting $5x$ is the same as adding $-5x$.

$= 3x + 5$

(b) Subtract $4x^2 - 8x + 3$ from $8x^2 + 5x - 3$.

Write

$(8x^2 + 5x - 3) - (4x^2 - 8x + 3)$

$= 8x^2 + 5x - 3 \underbrace{- 4x^2 + 8x - 3}_{\text{Sign changes}}$

$= 4x^2 + 13x - 6$

 Check Yourself 6

(a) Subtract $7x + 3$ from $10x - 7$.
(b) Subtract $5x^2 - 3x + 2$ from $8x^2 - 3x - 6$.

Again, writing all polynomials in descending order makes locating and combining like terms much easier. Look at Example 7.

Example 7   **Subtracting Polynomials Using the Horizontal Method**

(a) Subtract $4x^2 - 3x^3 + 5x$ from $8x^3 - 7x + 2x^2$.

Write

$(8x^3 + 2x^2 - 7x) - (-3x^3 + 4x^2 + 5x)$

$= 8x^3 + 2x^2 - 7x \underbrace{+ 3x^3 - 4x^2 - 5x}_{\text{Sign changes}}$

$= 11x^3 - 2x^2 - 12x$

(b) Subtract $8x - 5$ from $-5x + 3x^2$.

Write

$(3x^2 - 5x) - (8x - 5)$

$= 3x^2 \underbrace{- 5x - 8x}_{} + 5$

Only the like terms can be combined.

$= 3x^2 - 13x + 5$

Check Yourself 7

(a) Subtract $7x - 3x^2 + 5$ from $5 - 3x + 4x^2$.
(b) Subtract $3a - 2$ from $5a + 4a^2$.

If you think back to addition and subtraction in arithmetic, you should remember that the work was arranged vertically. That is, the numbers being added or subtracted were placed under one another so that each column represented the same place value. This meant that in adding or subtracting columns you were always dealing with "like quantities."

It is also possible to use a vertical method for adding or subtracting polynomials. First rewrite the polynomials in descending order, and then arrange them one under another, so that each column contains like terms. Then add or subtract in each column.

Example 8   **Adding Using the Vertical Method**

Add $2x^2 - 5x$, $3x^2 + 2$, and $6x - 3$.

Like terms are placed in columns.

$$
\begin{array}{r}
2x^2 - 5x \\
3x^2 \ + 2 \\
6x - 3 \\
\hline
5x^2 + \ x - 1
\end{array}
$$

Check Yourself 8

Add $3x^2 + 5$, $x^2 - 4x$, and $6x + 7$.

Example 9 illustrates subtraction by the vertical method.

Example 9   **Subtracting Using the Vertical Method**

(a) Subtract $5x - 3$ from $8x - 7$.

Write

$$
\begin{array}{r}
8x - 7 \\
(-)\ (5x - 3) \\
\hline
3x - 4
\end{array}
$$

To subtract, change each sign of $5x - 3$ to get $-5x + 3$ and then add.

$$
\begin{array}{r}
8x - 7 \\
-5x + 3 \\
\hline
3x - 4
\end{array}
$$

(b) Subtract $5x^2 - 3x + 4$ from $8x^2 + 5x - 3$.

Write

$$
\begin{array}{r}
8x^2 + 5x - 3 \\
(-)\ (5x^2 - 3x + 4) \\
\hline
3x^2 + 8x - 7
\end{array}
$$

To subtract, change each sign of $5x^2 - 3x + 4$ to get $-5x^2 + 3x - 4$ and then add.

$$
\begin{array}{r}
8x^2 + 5x - 3 \\
-5x^2 + 3x - 4 \\
\hline
3x^2 + 8x - 7
\end{array}
$$

Subtracting using the vertical method takes some practice. Take time to study the method carefully. You will use it in long division in Section 3.5.

Check Yourself 9

Subtract, using the vertical method.

(a) $4x^2 - 3x$ from $8x^2 + 2x$ **(b)** $8x^2 + 4x - 3$ from $9x^2 - 5x + 7$

Check Yourself ANSWERS

1. $10x^2 - 5x$ 2. $8y^2 - 8y$ 3. $13m^2 - 3m + 8$ 4. $3x^2 + 11x$

5. **(a)** $-3m - 5n$; **(b)** $-5w + 7z$; **(c)** $3r - 2s + 5t$; **(d)** $5a + 3b + 2c$

6. **(a)** $3x - 10$; **(b)** $3x^2 - 8$ 7. **(a)** $7x^2 - 10x$; **(b)** $4a^2 + 2a + 2$

8. $4x^2 + 2x + 12$ 9. **(a)** $4x^2 + 5x$; **(b)** $x^2 - 9x + 10$

Reading Your Text

The following fill-in-the-blank exercises are designed to ensure that you understand some of the key vocabulary used in this section.

SECTION 3.3

(a) If a _____ sign appears in front of parentheses, simply remove the parentheses.

(b) If a minus sign appears in front of parentheses, the subtraction can be changed to addition by changing the _____ in front of each term inside the parentheses.

(c) When subtracting polynomials, the expression following the word *from* is written _____ when writing the problem.

(d) When adding or subtracting polynomials, we can only combine _____ terms.

3.3 exercises

Name _____

Section _____ Date _____

Answers

1. _____   2. _____

3. _____   4. _____

5. _____   6. _____

7. _____

8. _____

9. _____

10. _____

11. _____   12. _____

13. _____   14. _____

15. _____

16. _____

17. _____   18. _____

19. _____

20. _____

21. _____   22. _____

23. _____   24. _____

Basic Skills | Challenge Yourself | Calculator/Computer | Career Applications | Above and Beyond

< Objective 1 >

Add.

1. $6a - 5$ and $3a + 9$

2. $9x + 3$ and $3x - 4$

3. $8b^2 - 11b$ and $5b^2 - 7b$

4. $2m^2 + 3m$ and $6m^2 - 8m$

5. $3x^2 - 2x$ and $-5x^2 + 2x$

6. $3p^2 + 5p$ and $-7p^2 - 5p$

7. $2x^2 + 5x - 3$ and $3x^2 - 7x + 4$ > Videos

8. $4d^2 - 8d + 7$ and $5d^2 - 6d - 9$

9. $2b^2 + 8$ and $5b + 8$

10. $4x - 3$ and $3x^2 - 9x$

11. $8y^3 - 5y^2$ and $5y^2 - 2y$

12. $9x^4 - 2x^2$ and $2x^2 + 3$

13. $2a^2 - 4a^3$ and $3a^3 + 2a^2$

14. $9m^3 - 2m$ and $-6m - 4m^3$

15. $4x^2 - 2 + 7x$ and $5 - 8x - 6x^2$

$-6x^2 + 5 - 8x$

16. $5b^3 - 8b + 2b^2$ and $3b^2 - 7b^3 + 5b$

< Objective 2 >

Remove the parentheses in each expression and simplify when possible.

17. $-(2a + 3b)$

18. $-(7x - 4y)$

19. $5a - (2b - 3c)$

20. $7x - (4y + 3z)$

21. $9r - (3r + 5s)$

22. $10m - (3m - 2n)$

23. $5p - (-3p + 2q)$ > Videos

24. $8d - (-7c - 2d)$

< Objective 3 >

Subtract.

25. $x + 4$ from $2x - 3$

26. $x - 2$ from $3x + 5$

27. $3m^2 - 2m$ from $4m^2 - 5m$

28. $9a^2 - 5a$ from $11a^2 - 10a$

29. $6y^2 + 5y$ from $4y^2 + 5y$

30. $9n^2 - 4n$ from $7n^2 - 4n$

31. $x^2 - 4x - 3$ from $3x^2 - 5x - 2$

32. $3x^2 - 2x + 4$ from $5x^2 - 8x - 3$

33. $3a + 7$ from $8a^2 - 9a$

34. $3x^3 + x^2$ from $4x^3 - 5x$

35. $4b^2 - 3b$ from $5b - 2b^2$

36. $7y - 3y^2$ from $3y^2 - 2y$

37. $x^2 - 5 - 8x$ from $3x^2 - 8x + 7$

38. $4x - 2x^2 + 4x^3$ from $4x^3 + x - 3x^2$

> Videos

Perform the indicated operations.

39. Subtract $3b + 2$ from the sum of $4b - 2$ and $5b + 3$.

40. Subtract $5m - 7$ from the sum of $2m - 8$ and $9m - 2$.

41. Subtract $3x^2 + 2x - 1$ from the sum of $x^2 + 5x - 2$ and $2x^2 + 7x - 8$.

42. Subtract $4x^2 - 5x - 3$ from the sum of $x^2 - 3x - 7$ and $2x^2 - 2x + 9$.

43. Subtract $2x^2 - 3x$ from the sum of $4x^2 - 5$ and $2x - 7$.

44. Subtract $5a^2 - 3a$ from the sum of $3a - 3$ and $5a^2 + 5$.

45. Subtract the sum of $3y^2 - 3y$ and $5y^2 + 3y$ from $2y^2 - 8y$. > Videos

46. Subtract the sum of $7r^3 - 4r^2$ and $-3r^3 + 4r^2$ from $2r^3 + 3r^2$.

Add using the vertical method.

47. $2w^2 + 7$, $3w - 5$, and $4w^2 - 5w$

48. $3x^2 - 4x - 2$, $6x - 3$, and $2x^2 + 8$

49. $3x^2 + 3x - 4$, $4x^2 - 3x - 3$, and $2x^2 - x + 7$

50. $5x^2 + 2x - 4$, $x^2 - 2x - 3$, and $2x^2 - 4x - 3$

Answers

25. ___ 26. ___
27. ___ 28. ___
29. ___ 30. ___
31. ___
32. ___
33. ___
34. ___
35. ___ 36. ___
37. ___ 38. ___
39. ___
40. ___
41. ___
42. ___
43. ___
44. ___
45. ___
46. ___
47. ___
48. ___
49. ___
50. ___

Answers

51.

52.

53.

54.

55.

56.

57.

58.

Subtract using the vertical method.

51. $5x^2 - 3x$ from $8x^2 - 9$

52. $7x^2 + 6x$ from $9x^2 - 3$

Basic Skills | **Challenge Yourself** | Calculator/Computer | Career Applications | Above and Beyond

▲

Perform the indicated operations.

53. $[(9x^2 - 3x + 5) - (3x^2 + 2x - 1)] - (x^2 - 2x - 3)$ › Videos

54. $[(5x^2 + 2x - 3) - (-2x^2 + x - 2)] - (2x^2 + 3x - 5)$

Basic Skills | Challenge Yourself | Calculator/Computer | **Career Applications** | Above and Beyond

▲

55. ALLIED HEALTH A patient's arterial oxygen content (CaO_2), as a percentage measurement, is calculated using the formula $CaO_2 = 1.34(Hb)(SaO_2) + 0.003PaO_2$, which is based on a patient's hemoglobin content (Hb), as a percentage measurement, arterial oxygen saturation (SaO_2), a percent expressed as a decimal, and arterial oxygen tension (PaO_2), in millimeters of mercury (mm Hg). Similarly, a patient's end-capillary oxygen content (CcO_2), as a percentage measurement, is calculated using the formula $CcO_2 = 1.34(Hb)(SaO_2) + 0.003P_AO_2$, which is based on the alveolar oxygen tension (P_AO_2), in mm Hg, instead of the arterial oxygen tension. Write a simplified formula for the difference between the end-capillary and arterial oxygen contents.

56. ALLIED HEALTH A diabetic patient's morning (m) and evening (n) blood glucose levels depend on the number of days (t) since the patient's treatment began and can be approximated by the formulas $m = 0.472t^3 - 5.298t^2 + 11.802t + 93.143$ and $n = -1.083t^3 + 11.464t^2 - 29.524t + 117.429$. Write a formula for the difference (d) in morning and evening blood glucose levels based on the number of days since treatment began.

57. MANUFACTURING TECHNOLOGY The shear polynomial for a polymer is

$0.4x^2 - 144x + 318$

After vulcanization of the polymer, the shear factor is increased by

$0.2x^2 - 14x + 144$

Find the shear polynomial for the polymer after vulcanization (add the polynomials).

58. MANUFACTURING TECHNOLOGY The moment of inertia of a square object is given by

$$I = \frac{s^4}{12}$$

The moment of inertia for a circular object is approximately given by

$$I = \frac{3.14s^4}{48}$$

Find the moment of inertia of a square with a circular inlay (add the polynomials).

Basic Skills | Challenge Yourself | Calculator/Computer | Career Applications | **Above and Beyond**
▲

Find values for a, b, c, and d so that each equation is true.

59. $3ax^4 - 5x^3 + x^2 - cx + 2 = 9x^4 - bx^3 + x^2 - 2d$

60. $(4ax^3 - 3bx^2 - 10) - 3(x^3 + 4x^2 - cx - d) = x^2 - 6x + 8$

61. GEOMETRY A rectangle has sides of $8x + 9$ and $6x - 7$. Find the polynomial that represents its perimeter.

$6x - 7$

$8x + 9$

62. GEOMETRY A triangle has sides $3x + 7$, $4x - 9$, and $5x + 6$. Find the polynomial that represents its perimeter.

$5x + 6$ $3x + 7$

$4x - 9$

63. BUSINESS AND FINANCE The cost of producing x units of an item is $C = 150 + 25x$. The revenue for selling x units is $R = 90x - x^2$. The profit is given by the revenue minus the cost. Find the polynomial that represents profit.

64. BUSINESS AND FINANCE The revenue for selling y units is $R = 3y^2 - 2y + 5$ and the cost of producing y units is $C = y^2 + y - 3$. Find the polynomial that represents profit.

59. _____

60. _____

61. _____

62. _____

63. _____

64. _____

Answers

1. $9a + 4$ **3.** $13b^2 - 18b$ **5.** $-2x^2$ **7.** $5x^2 - 2x + 1$
9. $2b^2 + 5b + 16$ **11.** $8y^3 - 2y$ **13.** $-a^3 + 4a^2$
15. $-2x^2 - x + 3$ **17.** $-2a - 3b$ **19.** $5a - 2b + 3c$
21. $6r - 5s$ **23.** $8p - 2q$ **25.** $x - 7$ **27.** $m^2 - 3m$ **29.** $-2y^2$
31. $2x^2 - x + 1$ **33.** $8a^2 - 12a - 7$ **35.** $-6b^2 + 8b$ **37.** $2x^2 + 12$
39. $6b - 1$ **41.** $10x - 9$ **43.** $2x^2 + 5x - 12$ **45.** $-6y^2 - 8y$
47. $6w^2 - 2w + 2$ **49.** $9x^2 - x$ **51.** $3x^2 + 3x - 9$ **53.** $5x^2 - 3x + 9$
55. $CcO_2 - CaO_2 = 0.003(P_AO_2 - PaO_2)$ **57.** $0.6x^2 + 158x + 462$
59. $a = 3, b = 5, c = 0, d = -1$ **61.** $28x + 4$ **63.** $-x^2 + 65x - 150$

3.4

Multiplying Polynomials

< 3.4 Objectives >

1 > Find the product of a monomial and a polynomial
2 > Find the product of two binomials
3 > Find the product of two polynomials
4 > Square a binomial

You have already had some experience in multiplying polynomials. In Section 3.1, we stated the product property of exponents and used that property to find the product of two monomial terms.

Step by Step

| To Find the Product of Monomials | Step 1 | Multiply the coefficients. |
| | Step 2 | Use the product property of exponents to combine the variables. |

Example 1 — **Multiplying Monomials**

< Objective 1 >

RECALL

We use the commutative and associative properties to regroup the factors.

Multiply $3x^2y$ and $2x^3y^5$.

Write

$(3x^2y)(2x^3y^5)$

$= (3 \cdot 2)(x^2 \cdot x^3)(y \cdot y^5)$

↑ Multiply the coefficients. ↑ Add the exponents.

$= 6x^5y^6$

Check Yourself 1

Multiply.

(a) $(5a^2b)(3a^2b^4)$ (b) $(-3xy)(4x^3y^5)$

Our next task is to find the product of a monomial and a polynomial. Here we use the distributive property, which leads us to the following rule for multiplication.

Property

| To Multiply a Polynomial by a Monomial | Use the distributive property to multiply each term of the polynomial by the monomial. |

Example 2   Multiplying a Monomial and a Binomial

NOTES

Distributive property:

$a(b + c) = ab + ac$

With practice you will do this step mentally.

(a) Multiply $2x + 3$ by x.

Write

$x(2x + 3)$

$= x \cdot 2x + x \cdot 3$

$= 2x^2 + 3x$

Multiply x by $2x$ and then by 3 (the terms of the polynomial). That is, "distribute" the multiplication over the sum.

(b) Multiply $2a^3 + 4a$ by $3a^2$.

Write

$3a^2(2a^3 + 4a)$

$= 3a^2 \cdot 2a^3 + 3a^2 \cdot 4a = 6a^5 + 12a^3$

Check Yourself 2

Multiply.

(a) $2y(y^2 + 3y)$ **(b)** $3w^2(2w^3 + 5w)$

The pattern above extends to *any* number of terms.

Example 3   Multiplying a Monomial and a Polynomial

NOTE

We show all the steps of the process. With practice, you will be able to write the product directly and should try to do so.

Multiply the following.

(a) $3x(4x^3 + 5x^2 + 2)$

$= 3x \cdot 4x^3 + 3x \cdot 5x^2 + 3x \cdot 2 = 12x^4 + 15x^3 + 6x$

(b) $5y^2(2y^3 - 4)$

$= 5y^2 \cdot 2y^3 - 5y^2 \cdot 4 = 10y^5 - 20y^2$

(c) $-5c(4c^2 - 8c)$

$= (-5c)(4c^2) - (-5c)(8c) = -20c^3 + 40c^2$

(d) $3c^2d^2(7cd^2 - 5c^2d^3)$

$= 3c^2d^2 \cdot 7cd^2 - 3c^2d^2 \cdot 5c^2d^3 = 21c^3d^4 - 15c^4d^5$

Check Yourself 3

Multiply.

(a) $3(5a^2 + 2a + 7)$ **(b)** $4x^2(8x^3 - 6)$

(c) $-5m(8m^2 - 5m)$ **(d)** $9a^2b(3a^3b - 6a^2b^4)$

Example 4   **Multiplying Binomials**

< Objective 2 >

NOTE

This ensures that each term, *x* and 2, of the first binomial is multiplied by each term, *x* and 3, of the second binomial.

(a) Multiply $x + 2$ by $x + 3$.

We can think of $x + 2$ as a single quantity and apply the distributive property.

$(x + 2)(x + 3)$ Multiply $x + 2$ by x and then by 3.
$= (x + 2)x + (x + 2)3$
$= x \cdot x + 2 \cdot x + x \cdot 3 + 2 \cdot 3$
$= x^2 + 2x + 3x + 6$
$= x^2 + 5x + 6$

(b) Multiply $a - 3$ by $a - 4$. (Think of $a - 3$ as a single quantity and distribute.)

$(a - 3)(a - 4)$
$= (a - 3)a - (a - 3)(4)$
$= a \cdot a - 3 \cdot a - [(a \cdot 4) - (3 \cdot 4)]$
$= a^2 - 3a - (4a - 12)$ The parentheses are needed here
$= a^2 - 3a - 4a + 12$ because a *minus sign* precedes the
$= a^2 - 7a + 12$ binomial.

 Check Yourself 4

Multiply.

(a) $(x + 2)(x + 5)$ **(b)** $(y + 5)(y - 6)$

Fortunately, there is a pattern to this kind of multiplication that allows you to write the product of two binomials without going through all these steps. We call it the **FOIL method** of multiplying. The reason for this name will be clear as we look at the process in more detail.

To multiply $(x + 2)(x + 3)$:

NOTES

Remember this by F!

Remember this by O!

Remember this by I!

Remember this by L!

1. $(x + 2)(x + 3)$
        $x \cdot x$ Find the product of the
                    first terms of the factors.

2. $(x + 2)(x + 3)$
          $x \cdot 3$ Find the product of
                    the *outer* terms.

3. $(x + 2)(x + 3)$
        $2 \cdot x$ Find the product of
                    the *inner* terms.

4. $(x + 2)(x + 3)$
          $2 \cdot 3$ Find the product of
                    the *last* terms.

NOTE

Of course, these are the same four terms found in Example 4(a).

Combining the four steps, we have

$(x + 2)(x + 3)$
$= x^2 + 3x + 2x + 6$
$= x^2 + 5x + 6$

With practice, you can use the FOIL method to write products quickly and easily. Consider Example 5, which illustrates this approach.

Example 5 | **Using the FOIL Method**

Find each product using the FOIL method.

NOTE

It is called FOIL to give you an easy way of remembering the steps: *First, Outer, Inner,* and *Last.*

(a) $(x + 4)(x + 5)$

$$= x^2 + 5x + 4x + 20$$

$$= x^2 + 9x + 20$$

NOTE

When possible, you should combine the outer and inner products mentally and write just the final product.

(b) $(x - 7)(x + 3)$

Combine the outer and inner products as $-4x$.

$$= x^2 - 4x - 21$$

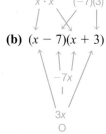

Check Yourself 5

Multiply.

(a) $(x + 6)(x + 7)$ **(b)** $(x + 3)(x - 5)$ **(c)** $(x - 2)(x - 8)$

Using the FOIL method, you can also find the product of binomials with coefficients other than 1 or with more than one variable.

Example 6 | **Using the FOIL Method**

Find each product using the FOIL method.

(a) $(4x - 3)(3x + 2)$

Combine:
$-9x + 8x = -x$

$$= 12x^2 - x - 6$$

$$6x^2 \qquad 35y^2$$

(b) $(3x - 5y)(2x - 7y)$

$-10xy$

$-21xy$

Combine:

$-10xy - 21xy = -31xy$

$$= 6x^2 - 31xy + 35y^2$$

This rule summarizes our work in multiplying binomials.

Step by Step

To Multiply Two Binomials	Step 1	Find the first term of the product of the binomials by multiplying the first terms of the binomials (F).
	Step 2	Find the outer and inner products and add them (O + I) if they are like terms.
	Step 3	Find the last term of the product by multiplying the last terms of the binomials (L).

Check Yourself 6

Multiply.

(a) $(5x + 2)(3x - 7)$ **(b)** $(4a - 3b)(5a - 4b)$

(c) $(3m + 5n)(2m + 3n)$

Sometimes, especially with larger polynomials, it is easier to use the vertical method to find their product. This is the same method you originally learned when multiplying two large integers.

Example 7   **Multiplying Using the Vertical Method**

Use the vertical method to find the product $(3x + 2)(4x - 1)$.

First, we rewrite the multiplication in vertical form.

$$3x + 2$$
$$4x - 1$$

Multiplying the quantity $3x + 2$ by -1 yields

$$\begin{array}{r} 3x + 2 \\ 4x - 1 \\ \hline -3x - 2 \end{array}$$

We maintain the columns of the original binomial when we find the product. We continue with those columns as we multiply by the $4x$ term.

$$\begin{array}{r} 3x + 2 \\ 4x - 1 \\ \hline -3x - 2 \\ 12x^2 + 8x \\ \hline 12x^2 + 5x - 2 \end{array}$$

We write the product as $(3x + 2)(4x - 1) = 12x^2 + 5x - 2$.

Check Yourself 7

Use the vertical method to find the product $(5x - 3)(2x + 1)$.

We use the vertical method again in Example 8. This time, we multiply a binomial and a trinomial. Note that the FOIL method is only used to find the product of two binomials.

| **Example 8** | Using the Vertical Method to Multiply Polynomials |

< Objective 3 >

Multiply $x^2 - 5x + 8$ by $x + 3$.

Step 1

$$
\begin{array}{r}
x^2 - 5x + 8 \\
x + \enclose{circle}{3} \\
\hline
3x^2 - 15x + 24
\end{array}
$$

Multiply each term of $x^2 - 5x + 8$ by 3.

Step 2

$$
\begin{array}{r}
x^2 - 5x + 8 \\
\enclose{circle}{x + 3} \\
\hline
3x^2 - 15x + 24 \\
x^3 - 5x^2 + 8x
\end{array}
$$

Now multiply each term by x.

Note that this line is shifted over so that like terms are in the same columns.

NOTE

Using the vertical method ensures that each term of one factor multiplies each term of the other. That's why it works!

Step 3

$$
\begin{array}{r}
x^2 - 5x + 8 \\
x + 3 \\
\hline
3x^2 - 15x + 24 \\
x^3 - 5x^2 + 8x \\
\hline
x^3 - 2x^2 - 7x + 24
\end{array}
$$

Now combine like terms to write the product.

Check Yourself 8

Multiply $2x^2 - 5x + 3$ by $3x + 4$.

Certain products occur frequently enough in algebra that it is worth learning special formulas for dealing with them. First, look at the **square of a binomial,** which is the product of two equal binomial factors.

$$(x + y)^2 = (x + y)(x + y)$$
$$= x^2 + 2xy + y^2$$
$$(x - y)^2 = (x - y)(x - y)$$
$$= x^2 - 2xy + y^2$$

The patterns above lead us to the following rule.

Step by Step

To Square a Binomial		
	Step 1	Find the first term of the square by squaring the first term of the binomial.
	Step 2	Find the middle term of the square as twice the product of the two terms of the binomial.
	Step 3	Find the last term of the square by squaring the last term of the binomial.

Example 9   **Squaring a Binomial**

< Objective 4 >

  >CAUTION

A very common mistake in squaring binomials is to forget the middle term.

(a) $(x + 3)^2 = x^2 + \underbrace{2 \cdot x \cdot 3} + 3^2$

Square of   Twice the   Square of
first term   product of   the last term
                    the two terms

$= x^2 + 6x + 9$

(b) $(3a + 4b)^2 = (3a)^2 + 2(3a)(4b) + (4b)^2$

$= 9a^2 + 24ab + 16b^2$

(c) $(y - 5)^2 = y^2 + 2 \cdot y \cdot (-5) + (-5)^2$

$= y^2 - 10y + 25$

(d) $(5c - 3d)^2 = (5c)^2 + 2(5c)(-3d) + (-3d)^2$

$= 25c^2 - 30cd + 9d^2$

Again we have shown all the steps. With practice you can write just the square.

  Check Yourself 9

Simplify.

(a) $(2x + 1)^2$ **(b)** $(4x - 3y)^2$

  Example 10   **Squaring a Binomial**

NOTE

You should see that $(2 + 3)^2 \neq 2^2 + 3^2$ because $5^2 \neq 4 + 9$.

Find $(y + 4)^2$.

$(y + 4)^2$ is *not* equal to $y^2 + 4^2$ or $y^2 + 16$

The correct square is

$(y + 4)^2 = y^2 + 8y + 16$

↑

The middle term is twice the product of y and 4.

  Check Yourself 10

Simplify.

(a) $(x + 5)^2$ **(b)** $(3a + 2)^2$
(c) $(y - 7)^2$ **(d)** $(5x - 2y)^2$

A second special product will be very important in Chapter 4, which presents factoring. Suppose the form of a product is

$(x + y)(x - y)$

The two terms differ only in sign.

Let's see what happens when we multiply these two terms.

$(x + y)(x - y)$
$= x^2 \underbrace{- xy + xy} - y^2$

$= 0$

$= x^2 - y^2$

Because the middle term becomes 0, we have the following rule.

Property

Special Product

The product of two binomials that differ only in the sign between the terms is the square of the first term minus the square of the second term.

Here are some examples of this rule.

Example 11   **Finding a Special Product**

Multiply each pair of binomials.

(a) $(x + 5)(x - 5) = x^2 - 5^2$

Square of the first term   Square of the second term

$= x^2 - 25$

RECALL

$(2y)^2 = (2y)(2y)$
$= 4y^2$

(b) $(x + 2y)(x - 2y) = x^2 - (2y)^2$

Square of the first term   Square of the second term

$= x^2 - 4y^2$

(c) $(3m + n)(3m - n) = 9m^2 - n^2$

(d) $(4a - 3b)(4a + 3b) = 16a^2 - 9b^2$

Check Yourself 11

Find the products.

(a) $(a - 6)(a + 6)$ **(b)** $(x - 3y)(x + 3y)$
(c) $(5n + 2p)(5n - 2p)$ **(d)** $(7b - 3c)(7b + 3c)$

When finding the product of three or more factors, it is useful to first look for the pattern in which two binomials differ only in their sign. Finding this product first will make it easier to find the product of all the factors.

Example 12   **Multiplying Polynomials**

(a) $x(x - 3)(x + 3)$ These binomials differ only in the sign.

$= x(x^2 - 9)$

$= x^3 - 9x$

(b) $(x + 1)(x - 5)(x + 5)$ These binomials differ only in the sign.

$= (x + 1)(x^2 - 25)$ With two binomials, use the FOIL method.

$= x^3 + x^2 - 25x - 25$

(c) $(2x - 1)(x + 3)(2x + 1)$ These two binomials differ only in the sign of the second term. We can use the commutative property to rearrange the terms.

$= (x + 3)(2x - 1)(2x + 1)$

$= (x + 3)(4x^2 - 1)$

$= 4x^3 + 12x^2 - x - 3$

Check Yourself 12

Multiply.

(a) $3x(x - 5)(x + 5)$ **(b)** $(x - 4)(2x + 3)(2x - 3)$

(c) $(x - 7)(3x - 1)(x + 7)$

We can use either the horizontal or vertical method to multiply polynomials with any number of terms. The key to multiplying polynomials successfully is to make sure each term in the first polynomial multiplies with every term in the second polynomial. Then, combine like terms and write your result in descending order, if you can.

 Example 13   **Multiplying Polynomials**

NOTE

Although it may seem tedious, you can do this if you are very careful. In each case, we are simply using a pattern to find the product of every pair of monomials.

Because one polynomial has three terms and one has four terms, we are initially finding $3 \times 4 = 12$ products.

Find the product.

$(2x^2 - 3x + 5)(3x^3 + 4x^2 - x - 1)$

$= (2x^2)(3x^3) + (2x^2)(4x^2) + (2x^2)(-x) + (2x^2)(-1) + (-3x)(3x^3) + (-3x)(4x^2)$
$\quad + (-3x)(-x) + (-3x)(-1) + (5)(3x^3) + (5)(4x^2) + (5)(-x) + (5)(-1)$

$= 6x^5 + 8x^4 - 2x^3 - 2x^2 - 9x^4 - 12x^3 + 3x^2 + 3x + 15x^3 + 20x^2 - 5x - 5$

$= 6x^5 - x^4 + x^3 + 21x^2 - 2x - 5$

Check Yourself 13

Find the product.

$(3x^2 + 2x - 5)(x^2 - 2xy + y^2)$

Check Yourself ANSWERS

1. (a) $15a^4b^5$; **(b)** $-12x^4y^6$ **2. (a)** $2y^3 + 6y^2$; **(b)** $6w^5 + 15w^3$
3. (a) $15a^2 + 6a + 21$; **(b)** $32x^5 - 24x^2$; **(c)** $-40m^3 + 25m^2$;
(d) $27a^5b^2 - 54a^4b^5$ **4. (a)** $x^2 + 7x + 10$; **(b)** $y^2 - y - 30$
5. (a) $x^2 + 13x + 42$; **(b)** $x^2 - 2x - 15$; **(c)** $x^2 - 10x + 16$
6. (a) $15x^2 - 29x - 14$; **(b)** $20a^2 - 31ab + 12b^2$; **(c)** $6m^2 + 19mn + 15n^2$
7. $10x^2 - x - 3$ **8.** $6x^3 - 7x^2 - 11x + 12$ **9. (a)** $4x^2 + 4x + 1$;
(b) $16x^2 - 24xy + 9y^2$ **10. (a)** $x^2 + 10x + 25$; **(b)** $9a^2 + 12a + 4$;
(c) $y^2 - 14y + 49$; **(d)** $25x^2 - 20xy + 4y^2$ **11. (a)** $a^2 - 36$; **(b)** $x^2 - 9y^2$;
(c) $25n^2 - 4p^2$; **(d)** $49b^2 - 9c^2$ **12. (a)** $3x^3 - 75x$;
(b) $4x^3 - 16x^2 - 9x + 36$; **(c)** $3x^3 - x^2 - 147x + 49$
13. $3x^4 - 6x^3y + 3x^2y^2 + 2x^3 - 4x^2y + 2xy^2 - 5x^2 + 10xy - 5y^2$

Reading Your Text

The following fill-in-the-blank exercises are designed to ensure that you understand some of the key vocabulary used in this section.

SECTION 3.4

(a) When multiplying monomials, we use the product property of exponents to combine the _____.

(b) The F in FOIL stands for the product of the _____ terms.

(c) The O in FOIL stands for the product of the _____ terms.

(d) The square of a binomial always has exactly _____ terms.

3.4 exercises

Name _____

Section _____ Date _____

Answers

1. _____ 2. _____

3. _____ 4. _____

5. _____ 6. _____

7. _____ 8. _____

9. _____ 10. _____

11. _____ 12. _____

13. _____ 14. _____

15. _____ 16. _____

17. _____

18. _____

19. _____

20. _____

21. _____

22. _____

23. _____

24. _____

< Objectives 1–2 >

Multiply.

1. $(5x^2)(3x^3)$

2. $(7a^5)(4a^6)$

3. $(-2b^2)(14b^8)$

4. $(14y^4)(-4y^6)$

5. $(-10p^6)(-4p^7)$

6. $(-6m^8)(9m^7)$

7. $(4m^5)(-3m)$

8. $(-5r^7)(-3r)$

9. $(4x^3y^2)(8x^2y)$

10. $(-3r^4s^2)(-7r^2s^5)$

11. $(-3m^5n^2)(2m^4n)$

12. $(7a^3b^5)(-6a^4b)$

13. $5(2x + 6)$

14. $4(7b - 5)$

15. $3a(4a + 5)$

16. $5x(2x - 7)$

17. $3s^2(4s^2 - 7s)$

18. $9a^2(3a^3 + 5a)$

19. $2x(4x^2 - 2x + 1)$

20. $5m(4m^3 - 3m^2 + 2)$

21. $3xy(2x^2y + xy^2 + 5xy)$

22. $5ab^2(ab - 3a + 5b)$

23. $6m^2n(3m^2n - 2mn + mn^2)$

24. $8pq^2(2pq - 3p + 5q)$

> Videos

25. $(x + 3)(x + 2)$ **26.** $(a - 3)(a - 7)$

27. $(m - 5)(m - 9)$ **28.** $(b + 7)(b + 5)$

29. $(p - 8)(p + 7)$ **30.** $(x - 10)(x + 9)$

31. $(w + 10)(w + 20)$ **32.** $(s - 12)(s - 8)$

33. $(3x - 5)(x - 8)$ **34.** $(w + 5)(4w - 7)$

35. $(2x - 3)(3x + 4)$ **36.** $(5a + 1)(3a + 7)$

37. $(3a - b)(4a - 9b)$ > Videos   **38.** $(7s - 3t)(3s + 8t)$

39. $(3p - 4q)(7p + 5q)$ **40.** $(5x - 4y)(2x - y)$

41. $(2x + 5y)(3x + 4y)$ **42.** $(4x - 5y)(4x + 3y)$

43. $(x + 5)(x + 5)$ **44.** $(y + 8)(y + 8)$

45. $(y - 9)(y - 9)$ **46.** $(2a + 3)(2a + 3)$

47. $(6m + n)(6m + n)$ **48.** $(7b - c)(7b - c)$

49. $(a - 5)(a + 5)$ **50.** $(x - 7)(x + 7)$

51. $(x - 2y)(x + 2y)$ **52.** $(7x + y)(7x - y)$

53. $(5s + 3t)(5s - 3t)$ **54.** $(9c - 4d)(9c + 4d)$

Answers

25. _____

26. _____

27. _____

28. _____

29. _____

30. _____

31. _____

32. _____

33. _____

34. _____

35. _____

36. _____

37. _____

38. _____

39. _____

40. _____

41. _____

42. _____

43. _____

44. _____

45. _____

46. _____

47. _____

48. _____

49. _____   50. _____

51. _____   52. _____

53. _____

54. _____

Answers

55. _____

56. _____

57. _____

58. _____

59. _____

60. _____

61. _____

62. _____

63. _____

64. _____

65. _____

66. _____

67. _____ 68. _____

69. _____ 70. _____

71. _____ 72. _____

73. _____ 74. _____

75. _____

76. _____

77. _____

78. _____

79. _____ 80. _____

81. _____ 82. _____

55. $(x + 5)^2$

56. $(y + 9)^2$

57. $(2a - 1)^2$

58. $(3x - 2)^2$

59. $(6m + 1)^2$

60. $(7b - 2)^2$

61. $(3x - y)^2$

62. $(5m + n)^2$

63. $(2r + 5s)^2$

64. $(3a - 4b)^2$

65. $\left(x + \dfrac{1}{2}\right)^2$ > Videos

66. $\left(w - \dfrac{1}{4}\right)^2$

67. $(x - 6)(x + 6)$

68. $(y + 8)(y - 8)$

69. $(m + 12)(m - 12)$ > Videos

70. $(w - 10)(w + 10)$

71. $\left(x - \dfrac{1}{2}\right)\left(x + \dfrac{1}{2}\right)$

72. $\left(x + \dfrac{2}{3}\right)\left(x - \dfrac{2}{3}\right)$

73. $(p - 0.4)(p + 0.4)$

74. $(m - 0.6)(m + 0.6)$

75. $(a - 3b)(a + 3b)$

76. $(p + 4q)(p - 4q)$

77. $(4r - s)(4r + s)$

78. $(7x - y)(7x + y)$

Basic Skills | **Challenge Yourself** | Calculator/Computer | Career Applications | Above and Beyond
▲

Label each equation as **true** *or* **false.**

79. $(x + y)^2 = x^2 + y^2$

80. $(x - y)^2 = x^2 - y^2$

81. $(x + y)^2 = x^2 + 2xy + y^2$

82. $(x - y)^2 = x^2 - 2xy + y^2$

83. GEOMETRY The length of a rectangle is given by $(3x + 5)$ cm and the width is given by $(2x - 7)$ cm. Express the area of the rectangle in terms of x.

84. GEOMETRY The base of a triangle measures $(3y + 7)$ in. and the height is $(2y - 3)$ in. Express the area of the triangle in terms of y.

Find each product.

85. $(2x + 5)(3x^2 - 4x + 1)$

86. $(2x^2 - 5)(x^2 + 3x + 4)$

87. $(x^2 + x + 9)(3x^2 - 2x - 5)$

88. $(x + 2)(2x - 1)(x^2 - x + 6)$

Basic Skills | Challenge Yourself | Calculator/Computer | Career Applications | **Above and Beyond**

▲

Note that $(28)(32) = (30 - 2)(30 + 2) = 900 - 4 = 896.$ *Use this pattern to find each product.*

89. $(49)(51)$ **90.** $(27)(33)$

91. $(34)(26)$ **92.** $(98)(102)$

93. $(55)(65)$ **94.** $(64)(56)$ > Videos

95. AGRICULTURE Suppose an orchard is planted with trees in straight rows. If there are $(5x - 4)$ rows with $(5x - 4)$ trees in each row, how many trees are there in the orchard?

96. GEOMETRY A square has sides of length $(3x - 2)$ cm. Express the area of the square as a polynomial.

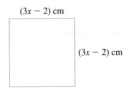

$(3x - 2)$ cm

$(3x - 2)$ cm

97. Complete the following statement: $(a + b)^2$ is not equal to $a^2 + b^2$ because.... But, wait! Isn't $(a + b)^2$ *sometimes* equal to $a^2 + b^2$? What do you think?

98. Is $(a + b)^3$ ever equal to $a^3 + b^3$? Explain.

Answers

83. _____
84. _____
85. _____
86. _____
87. _____
88. _____
89. _____
90. _____
91. _____
92. _____
93. _____
94. _____
95. _____
96. _____
97. _____
98. _____

Answers

99.
_____

100.
_____

99. GEOMETRY Identify the length, width, and area of each square.

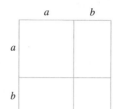

Length = _____

Width = _____

Area = _____

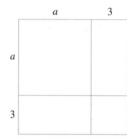

Length = _____

Width = _____

Area = _____

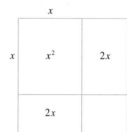

Length = _____

Width = _____

Area = _____

100. GEOMETRY The square shown here is x units on a side. The area is _____.

Draw a picture of what happens when the sides are doubled. The area is _____.

Continue the picture to show what happens when the sides are tripled. The area is _____.

If the sides are quadrupled, the area is _____.

In general, if the sides are multiplied by n, the area is _____.

If each side is increased by 3, the area is increased by _____.

If each side is decreased by 2, the area is decreased by _____.

In general, if each side is increased by n, the area is increased by _____, and if each side is decreased by n, the area is decreased by _____.

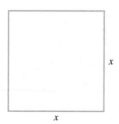

101. **GEOMETRY** Find the volume of a rectangular solid whose length measures $(2x + 4)$, width measures $(x + 2)$, and height measures $(x - 3)$.

102. **GEOMETRY** Neil and Suzanne are building a pool. Their backyard measures $(2x + 3)$ feet by $(2x + 12)$ feet, and the pool will measure $(x + 4)$ feet by $(x + 10)$ feet. If the remainder of the yard will be cement, how many square feet of the backyard will be covered by cement?

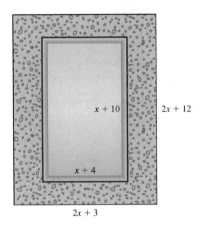

Answers

1. $15x^5$ **3.** $-28b^{10}$ **5.** $40p^{13}$ **7.** $-12m^6$ **9.** $32x^5y^3$

11. $-6m^9n^3$ **13.** $10x + 30$ **15.** $12a^2 + 15a$ **17.** $12s^4 - 21s^3$

19. $8x^3 - 4x^2 + 2x$ **21.** $6x^3y^2 + 3x^2y^3 + 15x^2y^2$

23. $18m^4n^2 - 12m^3n^2 + 6m^3n^3$ **25.** $x^2 + 5x + 6$ **27.** $m^2 - 14m + 45$

29. $p^2 - p - 56$ **31.** $w^2 + 30w + 200$ **33.** $3x^2 - 29x + 40$

35. $6x^2 - x - 12$ **37.** $12a^2 - 31ab + 9b^2$ **39.** $21p^2 - 13pq - 20q^2$

41. $6x^2 + 23xy + 20y^2$ **43.** $x^2 + 10x + 25$ **45.** $y^2 - 18y + 81$

47. $36m^2 + 12mn + n^2$ **49.** $a^2 - 25$ **51.** $x^2 - 4y^2$ **53.** $25s^2 - 9t^2$

55. $x^2 + 10x + 25$ **57.** $4a^2 - 4a + 1$ **59.** $36m^2 + 12m + 1$

61. $9x^2 - 6xy + y^2$ **63.** $4r^2 + 20rs + 25s^2$ **65.** $x^2 + x + \dfrac{1}{4}$

67. $x^2 - 36$ **69.** $m^2 - 144$ **71.** $x^2 - \dfrac{1}{4}$ **73.** $p^2 - 0.16$

75. $a^2 - 9b^2$ **77.** $16r^2 - s^2$ **79.** False   **81.** True

83. $(6x^2 - 11x - 35)\,cm^2$ **85.** $6x^3 + 7x^2 - 18x + 5$

87. $3x^4 + x^3 + 20x^2 - 23x - 45$ **89.** $2{,}499$ **91.** 884 **93.** $3{,}575$

95. $25x^2 - 40x + 16$ **97.** Above and Beyond   **99.** Above and Beyond

101. $2x^3 + 2x^2 - 16x - 24$

3.5

Dividing Polynomials

< 3.5 Objectives >

1 > Find the quotient when a polynomial is divided by a monomial

2 > Find the quotient when a polynomial is divided by a binomial

In Section 3.1, we used the quotient property of exponents to divide one monomial by another monomial.

Step by Step		
To Divide a Monomial by a Monomial	Step 1	Divide the coefficients.
	Step 2	Use the quotient property of exponents to combine the variables.

 Example 1 | Dividing by a Monomial

< Objective 1 >

RECALL

The quotient property says: If x is not zero, then

$$\frac{x^m}{x^n} = x^{m-n}$$

Divide: $\frac{8}{2} = 4$

(a) $\dfrac{8x^4}{2x^2} = 4x^{4-2}$

Subtract the exponents.

$$= 4x^2$$

(b) $\dfrac{45a^5b^3}{9a^2b} = 5a^3b^2$

 Check Yourself 1

Divide.

(a) $\dfrac{16a^5}{8a^3}$ **(b)** $\dfrac{28m^4n^3}{7m^3n}$

NOTE

This step depends on the distributive property and the definition of division.

Now look at how this can be extended to divide any polynomial by a monomial. For example, to divide $12a^3 + 8a^2$ by $4a$, proceed as follows:

$$\frac{12a^3 + 8a^2}{4a} = \frac{12a^3}{4a} + \frac{8a^2}{4a}$$

Divide each term in the numerator by the denominator, $4a$.

Now do each division.

$$= 3a^2 + 2a$$

This work leads us to the following rule.

Step by Step

To Divide a Polynomial by a Monomial	Step 1	Divide each term of the polynomial by the monomial.
	Step 2	Simplify the results.

Example 2   **Dividing by a Monomial**

Divide each term by 2.

(a) $\dfrac{4a^2 + 8}{2} = \dfrac{4a^2}{2} + \dfrac{8}{2}$

$\qquad\qquad = 2a^2 + 4$

Divide each term by 6y.

(b) $\dfrac{24y^3 + (-18y^2)}{6y} = \dfrac{24y^3}{6y} + \dfrac{-18y^2}{6y}$

$\qquad\qquad\qquad = 4y^2 - 3y$

Remember the rules for signs in division.

(c) $\dfrac{15x^2 + 10x}{-5x} = \dfrac{15x^2}{-5x} + \dfrac{10x}{-5x}$

$\qquad\qquad\qquad = -3x - 2$

NOTE

With practice you can just write the quotient.

(d) $\dfrac{14x^4 + 28x^3 - 21x^2}{7x^2} = \dfrac{14x^4}{7x^2} + \dfrac{28x^3}{7x^2} - \dfrac{21x^2}{7x^2}$

$\qquad\qquad\qquad\qquad = 2x^2 + 4x - 3$

(e) $\dfrac{9a^3b^4 - 6a^2b^3 + 12ab^4}{3ab} = \dfrac{9a^3b^4}{3ab} - \dfrac{6a^2b^3}{3ab} + \dfrac{12ab^4}{3ab}$

$\qquad\qquad\qquad\qquad\qquad = 3a^2b^3 - 2ab^2 + 4b^3$

Check Yourself 2

Divide.

(a) $\dfrac{20y^3 - 15y^2}{5y}$ (b) $\dfrac{8a^3 - 12a^2 + 4a}{-4a}$

(c) $\dfrac{16m^4n^3 - 12m^3n^2 + 8mn}{4mn}$

We are now ready to look at dividing one polynomial by another polynomial (with more than one term). The process is very much like long division in arithmetic, as Example 3 illustrates.

Example 3	**Dividing by a Binomial**

< Objective 2 >

Compare the steps in these two division examples.

Divide $x^2 + 7x + 10$ by $x + 2$.   *Divide 2,176 by 32.*

Step 1
$$x + 2 \overline{\smash{\big)}\, x^2 + 7x + 10} \quad\quad \overset{x}{}$$

Divide x^2 by x to get x.

$$32\overline{\smash{\big)}\,2176} \quad \overset{6}{}$$

Step 2
$$x + 2 \overline{\smash{\big)}\, x^2 + 7x + 10} \quad \overset{x}{}$$
$$\underline{x^2 + 2x}$$

Multiply the divisor, $x + 2$, by x.

$$32\overline{\smash{\big)}\,2176} \quad \overset{6}{}$$
$$\underline{192}$$

Step 3
$$x + 2 \overline{\smash{\big)}\, x^2 + 7x + 10} \quad \overset{x}{}$$
$$\underline{x^2 + 2x}$$
$$5x + 10$$

Subtract and bring down 10.

$$32\overline{\smash{\big)}\,2176} \quad \overset{6}{}$$
$$\underline{192}$$
$$256$$

Step 4
$$x + 2 \overline{\smash{\big)}\, x^2 + 7x + 10} \quad \overset{x + \ 5}{}$$
$$\underline{x^2 + 2x}$$
$$5x + 10$$

Divide $5x$ by x to get 5.

$$32\overline{\smash{\big)}\,2176} \quad \overset{68}{}$$
$$\underline{192}$$
$$256$$

Step 5
$$x + 2 \overline{\smash{\big)}\, x^2 + 7x + 10} \quad \overset{x + \ 5}{}$$
$$\underline{x^2 + 2x}$$
$$5x + 10$$
$$\underline{5x + 10}$$
$$0$$

$$32\overline{\smash{\big)}\,2176} \quad \overset{68}{}$$
$$\underline{192}$$
$$256$$
$$\underline{256}$$
$$0$$

Multiply $x + 2$ by 5 and then subtract.

The quotient is $x + 5$.

 Check Yourself 3

Divide $x^2 + 9x + 20$ by $x + 4$.

In Example 3, we showed all the steps separately to help you see the process. In practice, the work can be shortened.

| **Example 4** | **Dividing by a Binomial**

Divide $x^2 + x - 12$ by $x - 3$.

NOTE

You might want to write out a problem like $408 \div 17$ to compare the steps.

$$
\begin{array}{r}
x + 4 \\
x - 3 \overline{\smash{)}x^2 + x - 12} \\
\underline{x^2 - 3x} \\
4x - 12 \\
\underline{4x - 12} \\
0
\end{array}
$$

Step 1 Divide x^2 by x to get x, the first term of the quotient.
Step 2 Multiply $x - 3$ by x.
Step 3 Subtract and bring down -12. Remember to mentally change the signs to $-x^2 + 3x$ and add.
Step 4 Divide $4x$ by x to get 4, the second term of the quotient.
Step 5 Multiply $x - 3$ by 4 and subtract.

The quotient is $x + 4$.

 Check Yourself 4

Divide.

$(x^2 + 2x - 24) \div (x - 4)$

You may have a remainder in algebraic long division just as in arithmetic. Consider Example 5.

 | **Example 5** | **Dividing by a Binomial**

Divide $4x^2 - 8x + 11$ by $2x - 3$.

$$
\begin{array}{r}
\overset{\text{Quotient}}{2x - 1} \\
2x - 3 \overline{\smash{)}4x^2 - 8x + 11} \\
\underline{4x^2 - 6x} \\
-2x + 11 \\
\underline{-2x + 3} \\
8
\end{array}
$$

Divisor ↗

↑ Remainder

We write this result as

$$
\frac{4x^2 - 8x + 11}{2x - 3} = 2x - 1 + \frac{8}{2x - 3}
$$

Remainder

Divisor

Quotient

 Check Yourself 5

Divide.

$(6x^2 - 7x + 15) \div (3x - 5)$

The division process shown in our previous examples can be extended to dividends of a higher degree. The steps involved in the division process are exactly the same, as Example 6 illustrates.

| Example 6 | Dividing by a Binomial |

Divide $6x^3 + x^2 - 4x - 5$ by $3x - 1$.

$$
\begin{array}{r}
2x^2 + x - 1 \\
3x - 1\overline{)6x^3 + x^2 - 4x - 5} \\
\underline{6x^3 - 2x^2} \\
3x^2 - 4x \\
\underline{3x^2 - x} \\
-3x - 5 \\
\underline{-3x + 1} \\
-6
\end{array}
$$

We write the result as

$$
\frac{6x^3 + x^2 - 4x - 5}{3x - 1} = 2x^2 + x - 1 + \frac{-6}{3x - 1}
$$

 Check Yourself 6

Divide $4x^3 - 2x^2 + 2x + 15$ by $2x + 3$.

Suppose that the dividend is "missing" a term in some power of the variable. You can use 0 as the coefficient for the missing term. Consider Example 7.

| Example 7 | Dividing by a Binomial |

NOTE

Think of $0x$ as a placeholder. Writing it in helps align like terms.

Divide $x^3 - 2x^2 + 5$ by $x + 3$.

$$
\begin{array}{r}
x^2 - 5x + 15 \\
x + 3\overline{)x^3 - 2x^2 + 0x + 5} \\
\underline{x^3 + 3x^2} \\
-5x^2 + 0x \\
\underline{-5x^2 - 15x} \\
15x + 5 \\
\underline{15x + 45} \\
-40
\end{array}
$$

Write $0x$ for the "missing" term in x.

This result can be written as

$$
\frac{x^3 - 2x^2 + 5}{x + 3} = x^2 - 5x + 15 + \frac{-40}{x + 3}
$$

 Check Yourself 7

Divide.

$(4x^3 + x + 10) \div (2x - 1)$

You should always arrange the terms of the divisor and dividend in descending order before starting the long-division process, as shown in Example 8.

Example 8   Dividing by a Binomial

Divide $5x^2 - x + x^3 - 5$ by $-1 + x^2$.

Write the divisor as $x^2 - 1$ and the dividend as $x^3 + 5x^2 - x - 5$.

$$
\begin{array}{r}
x + 5 \\
x^2 - 1\overline{)x^3 + 5x^2 - x - 5} \\
\underline{x^3 \qquad - x} \\
5x^2 \qquad - 5 \\
\underline{5x^2 \qquad - 5} \\
0
\end{array}
$$

Write $x^3 - x$, the product of x and $x^2 - 1$, so that like terms fall in the same columns.

The quotient is $x + 5$.

Check Yourself 8

Divide.

$(5x^2 + 10 + 2x^3 + 4x) \div (2 + x^2)$

Check Yourself ANSWERS

1. (a) $2a^2$; (b) $4mn^2$ 2. (a) $4y^2 - 3y$; (b) $-2a^2 + 3a - 1$;

(c) $4m^3n^2 - 3m^2n + 2$ 3. $x + 5$ 4. $x + 6$ 5. $2x + 1 + \dfrac{20}{3x - 5}$

6. $2x^2 - 4x + 7 + \dfrac{-6}{2x + 3}$ 7. $2x^2 + x + 1 + \dfrac{11}{2x - 1}$ 8. $2x + 5$

Reading Your Text

The following fill-in-the-blank exercises are designed to ensure that you understand some of the key vocabulary used in this section.

SECTION 3.5

(a) When dividing two monomials, we use the quotient property of exponents to combine the _____.

(b) When dividing a polynomial by a monomial, divide each _____ of the polynomial by the monomial.

(c) When dividing polynomials, we continue until the _____ of the remainder is less than that of the divisor.

(d) When the dividend is missing a term in some power of the variable, we use _____ as a coefficient for that missing term.

3.5 exercises

Name _____

Section _____ Date _____

Answers

1. _____   2. _____

3. _____   4. _____

5. _____   6. _____

7. _____   8. _____

9. _____   10. _____

11. _____   12. _____

13. _____

14. _____

15. _____

16. _____

17. _____

18. _____

19. _____

20. _____

21. _____

22. _____

23. _____

24. _____

< Objectives 1–2 >

Divide.

1. $\dfrac{18x^6}{9x^2}$

2. $\dfrac{20a^7}{5a^5}$

3. $\dfrac{35m^3n^2}{7mn^2}$

4. $\dfrac{42x^5y^2}{6x^3y}$

5. $\dfrac{3a + 6}{3}$

6. $\dfrac{4x - 8}{4}$

7. $\dfrac{9b^2 - 12}{3}$

8. $\dfrac{10m^2 + 5m}{5}$

9. $\dfrac{16a^3 - 24a^2}{4a}$

10. $\dfrac{9x^3 + 12x^2}{3x}$

11. $\dfrac{12m^2 + 6m}{-3m}$

12. $\dfrac{20b^3 - 25b^2}{-5b}$

13. $\dfrac{18a^4 + 12a^3 - 6a^2}{6a}$

14. $\dfrac{21x^5 - 28x^4 + 14x^3}{7x}$

15. $\dfrac{20x^4y^2 - 15x^2y^3 + 10x^3y}{5x^2y}$

16. $\dfrac{16m^3n^3 + 24m^2n^2 - 40mn^3}{8mn^2}$ > Videos

17. $\dfrac{27a^5b^5 + 9a^4b^4 - 3a^2b^3}{3a^2b^3}$

18. $\dfrac{7x^5y^5 - 21x^4y^4 + 14x^3y^3}{7x^3y^3}$

19. $\dfrac{3a^6b^4c^2 - 2a^4b^2c + 6a^3b^2c}{a^3b^2c}$

20. $\dfrac{2x^4y^4z^4 + 3x^3y^3z^3 - xy^2z^3}{xy^2z^3}$

21. $\dfrac{x^2 + 5x + 6}{x + 2}$

22. $\dfrac{x^2 + 8x + 15}{x + 3}$

23. $\dfrac{x^2 - x - 20}{x + 4}$ > Videos

24. $\dfrac{x^2 - 2x - 35}{x + 5}$

25. $\dfrac{2x^2 - 3x - 5}{x - 3}$

26. $\dfrac{3x^2 + 17x - 12}{x + 6}$

27. $\dfrac{6x^2 - x - 10}{3x - 5}$

28. $\dfrac{4x^2 + 6x - 25}{2x + 7}$

29. $\dfrac{x^3 + x^2 - 4x - 4}{x + 2}$

30. $\dfrac{x^3 - 2x^2 + 4x - 21}{x - 3}$

31. $\dfrac{4x^3 + 7x^2 + 10x + 5}{4x - 1}$

> Videos

32. $\dfrac{2x^3 - 3x^2 + 4x + 4}{2x + 1}$

33. $\dfrac{x^3 - x^2 + 5}{x - 2}$

34. $\dfrac{x^3 + 4x - 3}{x + 3}$

35. $\dfrac{25x^3 + x}{5x - 2}$

36. $\dfrac{8x^3 - 6x^2 + 2x}{4x + 1}$

37. $\dfrac{2x^2 - 8 - 3x + x^3}{x - 2}$

38. $\dfrac{x^2 - 18x + 2x^3 + 32}{x + 4}$

39. $\dfrac{x^4 - 1}{x - 1}$ > Videos

40. $\dfrac{x^4 + x^2 - 16}{x + 2}$

Basic Skills | **Challenge Yourself** | Calculator/Computer | Career Applications | Above and Beyond

---

41. $\dfrac{x^3 - 3x^2 - x + 3}{x^2 - 1}$

42. $\dfrac{x^3 + 2x^2 + 3x + 6}{x^2 + 3}$

43. $\dfrac{x^4 + 2x^2 - 2}{x^2 + 3}$ > Videos

44. $\dfrac{x^4 + x^2 - 5}{x^2 - 2}$

Answers

25. _____

26. _____

27. _____

28. _____

29. _____

30. _____

31. _____

32. _____

33. _____

34. _____

35. _____

36. _____

37. _____

38. _____

39. _____

40. _____

41. _____

42. _____

43. _____

44. _____

45. _____

46. _____

47. _____

48. _____

49. _____

50. _____

51. _____

52. _____

53. _____

45. $\dfrac{y^3 + 1}{y + 1}$

46. $\dfrac{y^3 - 8}{y - 2}$

47. $\dfrac{x^4 - 1}{x^2 - 1}$

48. $\dfrac{x^6 - 1}{x^3 - 1}$

| Basic Skills | Challenge Yourself | Calculator/Computer | Career Applications | **Above and Beyond** |

▲

49. Find the value of c so that $\dfrac{y^2 - y + c}{y + 1} = y - 2$.

50. Find the value of c so that $\dfrac{x^3 + x^2 + x + c}{x^2 + 1} = x + 1$.

51. Write a summary of your work with polynomials. Explain how a polynomial is recognized and explain the rules for the arithmetic of polynomials—how to add, subtract, multiply, and divide. What parts of this chapter do you feel you understand very well, and what parts do you still have questions about or feel unsure of? Exchange papers with another student and compare your questions.

52. A funny (and useful) thing about division of polynomials: To find out about it, do this division. Compare your answer with another student's.

$(x - 2)\overline{)2x^2 + 3x - 5}$ Is there a remainder?

Now, evaluate the polynomial $2x^2 + 3x - 5$ when $x = 2$. Is this value the same as the remainder?

Try $(x + 3)\overline{)5x^2 - 2x + 1}$ Is there a remainder?

Evaluate the polynomial $5x^2 - 2x + 1$ when $x = -3$. Is this value the same as the remainder?
What happens when there is no remainder?

Try $(x - 6)\overline{)3x^3 + 14x^2 - 23x + 6}$ Is the remainder zero?

Evaluate the polynomial $3x^3 + 14x^2 - 23x + 6$ when $x = 6$. Is this value zero? Write a description of the patterns you see. When does the pattern hold? Make up several more examples and test your conjecture.

53. **(a)** Divide $\dfrac{x^2 - 1}{x - 1}$.   **(b)** Divide $\dfrac{x^3 - 1}{x - 1}$.   **(c)** Divide $\dfrac{x^4 - 1}{x - 1}$.

(d) Based on your results on parts (a), (b), and (c), predict $\dfrac{x^{50} - 1}{x - 1}$.

54. (a) Divide $\dfrac{x^2 + x + 1}{x - 1}$.   **(b)** Divide $\dfrac{x^3 + x^2 + x + 1}{x - 1}$.

(c) Divide $\dfrac{x^4 + x^3 + x^2 + x + 1}{x - 1}$.

(d) Based on your results to (a), (b), and (c), predict

$$\dfrac{x^{10} + x^9 + x^8 + \cdots + x + 1}{x - 1}.$$

Answers

1. $2x^4$ **3.** $5m^2$ **5.** $a + 2$ **7.** $3b^2 - 4$ **9.** $4a^2 - 6a$

11. $-4m - 2$ **13.** $3a^3 + 2a^2 - a$ **15.** $4x^2y - 3y^2 + 2x$

17. $9a^3b^2 + 3a^2b - 1$ **19.** $3a^3b^2c - 2a + 6$ **21.** $x + 3$

23. $x - 5$ **25.** $2x + 3 + \dfrac{4}{x - 3}$ **27.** $2x + 3 + \dfrac{5}{3x - 5}$

29. $x^2 - x - 2$ **31.** $x^2 + 2x + 3 + \dfrac{8}{4x - 1}$

33. $x^2 + x + 2 + \dfrac{9}{x - 2}$ **35.** $5x^2 + 2x + 1 + \dfrac{2}{5x - 2}$

37. $x^2 + 4x + 5 + \dfrac{2}{x - 2}$ **39.** $x^3 + x^2 + x + 1$ **41.** $x - 3$

43. $x^2 - 1 + \dfrac{1}{x^2 + 3}$ **45.** $y^2 - y + 1$ **47.** $x^2 + 1$ **49.** $c = -2$

51. Above and Beyond   **53. (a)** $x + 1$; **(b)** $x^2 + x + 1$;
(c) $x^3 + x^2 + x + 1$; **(d)** $x^{49} + x^{48} + \cdots + x + 1$

54. _____

Definition/Procedure	Example	Reference

Exponents and Polynomials ... Section 3.1

Properties of Exponents

$a^m \cdot a^n = a^{m+n}$	Product property	$3^3 \cdot 3^4 = 3^7$	p. 184
$\dfrac{a^m}{a^n} = a^{m-n}$	Quotient property	$\dfrac{x^6}{x^2} = x^4$	p. 185
$(a^m)^n = a^{mn}$	Power to a power property	$(x^3)^5 = x^{15}$	p. 186
$(ab)^m = a^m b^m$	Product to a power property	$(3x)^2 = 9x^2$	p. 187
$\left(\dfrac{a}{b}\right)^m = \dfrac{a^m}{b^m}$	Quotient to a power property	$\left(\dfrac{2}{3}\right)^3 = \dfrac{8}{27}$	p. 188

Term

An expression that can be written as a number or the product of a number and variables.

$4x^3 - 3x^2 + 5x$ is a polynomial. The terms of $4x^3 - 3x^2 + 5x$ are $4x^3$, $-3x^2$, and $5x$.

p. 189

Polynomial

An algebraic expression made up of terms in which the exponents of the variables are whole numbers. These terms are connected by plus or minus signs. Each sign ($+$ or $-$) is attached to the term following that sign.

p. 189

Coefficient

In each term of a polynomial, the number factor is called the *numerical coefficient* or, more simply, the *coefficient,* of that term.

The coefficients of $4x^3 - 3x^2$ are 4 and -3.

p. 189

Types of Polynomials

A polynomial can be classified according to the number of terms it has.

p. 190

A *mono*mial has one term.
A *bi*nomial has two terms.
A *tri*nomial has three terms.

$2x^3$ is a monomial.
$3x^2 - 7x$ is a binomial.
$5x^5 - 5x^3 + 2$ is a trinomial.

Degree

The highest power of the variable appearing in any one term.

The degree of $4x^5 - 5x^3 + 3x$ is 5.

p. 190

Descending Order

The form of a polynomial when it is written with the highest-degree term first, the next highest-degree term second, and so on.

$4x^5 - 5x^3 + 3x$ is written in descending order.

p. 190

Definition/Procedure	Example	Reference
Negative Exponents and Scientific Notation		**Section 3.2**
The Zero Power Any nonzero expression raised to the 0 power equals 1.	$3^0 = 1$ $(5x)^0 = 1$	*p.* 198
Negative Powers An expression raised to a negative power equals its reciprocal taken to the absolute value of its power.	$\left(\dfrac{x}{3}\right)^{-4} = \left(\dfrac{3}{x}\right)^4 = \dfrac{3^4}{x^4}$	*p.* 199
Scientific Notation Any number written in the form $a \times 10^n$ in which $1 \le a < 10$ and n is an integer, is written in scientific notation.	6.2×10^{23}	*p.* 202
Adding and Subtracting Polynomials		**Section 3.3**
Removing Signs of Grouping **1.** If a plus sign ($+$) or no sign at all appears in front of parentheses, just remove the parentheses. No other changes are necessary.	$3x + (2x - 3) = 3x + 2x - 3$ $\qquad\qquad\qquad = 5x - 3$	*p.* 210
2. If a minus sign ($-$) appears in front of parentheses, the parentheses can be removed by changing the sign of each term inside the parentheses.	$2x - (x - 4) = 2x - x + 4$ $\qquad\qquad\quad = x + 4$	*p.* 212
Adding Polynomials Remove the signs of grouping. Then collect and combine any like terms.	$(2x + 3) + (3x - 5)$ $\quad = 2x + 3 + 3x - 5$ $\quad = 5x - 2$	*p.* 210
Subtracting Polynomials Remove the signs of grouping by changing the sign of each term in the polynomial being subtracted. Then combine any like terms.	$(3x^2 + 2x) - (2x^2 + 3x - 1)$ $\quad = 3x^2 + 2x - 2x^2 - 3x + 1$ <center>Sign changes</center> $\quad = 3x^2 - 2x^2 + 2x - 3x + 1$ $\quad = x^2 - x + 1$	*p.* 213
Multiplying Polynomials		**Section 3.4**
To Multiply a Polynomial by a Monomial Multiply each term of the polynomial by the monomial and simplify the results.	$3x(2x + 3) = 3x \cdot 2x + 3x \cdot 3$ $\qquad\qquad\ = 6x^2 + 9x$	*p.* 220

Continued

Definition/Procedure	Example	Reference
To Multiply a Binomial by a Binomial Use the FOIL method: $$\begin{array}{cccc} \text{F} & \text{O} & \text{I} & \text{L} \end{array}$$ $(a + b)(c + d) = a \cdot c + a \cdot d + b \cdot c + b \cdot d$	$(2x - 3)(3x + 5)$ $= 6x^2 + 10x - 9x - 15$ $ \begin{array}{cccc} \text{F} & \text{O} & \text{I} & \text{L} \end{array}$ $= 6x^2 + x - 15$	p. 222
To Multiply a Polynomial by a Polynomial Arrange the polynomials vertically. Multiply each term of the upper polynomial by each term of the lower polynomial and add the results.	$\begin{array}{r} x^2 - 3x + 5 \\ 2x - 3 \\ \hline -3x^2 + 9x - 15 \\ 2x^3 - 6x^2 + 10x \\ \hline 2x^3 - 9x^2 + 19x - 15 \end{array}$	p. 225
The Square of a Binomial $(a + b)^2 = a^2 + 2ab + b^2$	$(2x - 5)^2$ $= 4x^2 + 2 \cdot 2x \cdot (-5) + 25$ $= 4x^2 - 20x + 25$	p. 225
The Product of Binomials That Differ Only in Sign Subtract the square of the second term from the square of the first term. $(a + b)(a - b) = a^2 - b^2$	$(2x - 5y)(2x + 5y)$ $= (2x)^2 - (5y)^2$ $= 4x^2 - 25y^2$	p. 227
Dividing Polynomials		Section 3.5
To Divide a Polynomial by a Monomial **1.** Divide each term of the polynomial by the monomial. **2.** Simplify the result.	$\dfrac{27x^2y^2 + 9x^3y^4}{3xy^2}$ $= \dfrac{27x^2y^2}{3xy^2} + \dfrac{9x^3y^4}{3xy^2}$ $= 9x + 3x^2y^2$	p. 237

This summary exercise set is provided to give you practice with each of the objectives of this chapter. Each exercise is keyed to the appropriate chapter section. When you are finished, you can check your answer to the odd-numbered exercises against those presented in the back of the text. If you have difficulty with any of these questions, go back and reread the examples from that section. Your instructor will give you guidelines on how best to use these exercises in your instructional setting.

3.1 *Simplify each expression.*

1. $\dfrac{x^{10}}{x^3}$
  2. $\dfrac{a^5}{a^4}$
  3. $\dfrac{x^2 \cdot x^3}{x^4}$
  4. $\dfrac{m^2 \cdot m^3 \cdot m^4}{m^5}$

5. $\dfrac{18p^7}{9p^5}$
  6. $\dfrac{24x^{17}}{8x^{13}}$
  7. $\dfrac{30m^7n^5}{6m^2n^3}$
  8. $\dfrac{108x^9y^4}{9xy^4}$

9. $\dfrac{48p^5q^3}{6p^3q}$
  10. $\dfrac{52a^5b^3c^5}{13a^4c}$
  11. $(2ab)^2$
  12. $(p^2q^3)^3$

13. $(2x^2y^2)^3(3x^3y)^2$
  14. $\left(\dfrac{p^2q^3}{t^4}\right)^2$
  15. $\dfrac{(x^5)^2}{(x^3)^3}$
  16. $(4w^2t)^2\,(3wt^2)^3$

17. $(y^3)^2(3y^2)^3$
  18. $\left(\dfrac{4x^4}{3y}\right)^2$

Find the value of each polynomial for the given value of the variable.

19. $5x + 1$; $x = -1$
  20. $2x^2 + 7x - 5$; $x = 2$

21. $-x^2 + 3x - 1$; $x = 6$
  22. $4x^2 + 5x + 7$; $x = -4$

Classify each polynomial as a monomial, binomial, or trinomial, where possible.

23. $5x^3 - 2x^2$
  24. $7x^5$
  25. $4x^5 - 8x^3 + 5$

26. $x^3 + 2x^2 - 5x + 3$
  27. $9a^3 - 18a^2$

Arrange in descending order, if necessary, and give the degree of each polynomial.

28. $5x^5 + 3x^2$
  29. $9x$
  30. $6x^2 + 4x^4 + 6$

31. $5 + x$
  32. -8
  33. $9x^4 - 3x + 7x^6$

3.2 *Evaluate each expression.*

34. 4^0
  35. $(3a)^0$
  36. $6x^0$
  37. $(3a^4b)^0$

Write using positive exponents. Simplify when possible.

38. x^{-5}
  39. 3^{-3}
  40. 10^{-4}
  41. $4x^{-4}$

42. $\dfrac{x^6}{x^8}$
  43. m^7m^{-9}
  44. $\dfrac{a^{-4}}{a^{-9}}$
  45. $\dfrac{x^2y^{-3}}{x^{-3}y^2}$

46. $(3m^{-3})^2$
  47. $\dfrac{(a^4)^{-3}}{(a^{-2})^{-3}}$

Express each number in scientific notation.

48. The average distance from Earth to the Sun is 150,000,000,000 m.

49. A bat emits a sound with a frequency of 51,000 cycles per second.

50. The diameter of a grain of salt is 0.000062 m.

Compute the expression using scientific notation and express your answers in that form.

51. $(2.3 \times 10^{-3})(1.4 \times 10^{12})$

52. $(4.8 \times 10^{-10})(6.5 \times 10^{34})$

53. $\dfrac{(8 \times 10^{23})}{(4 \times 10^{6})}$

54. $\dfrac{(5.4 \times 10^{-12})}{(4.5 \times 10^{16})}$

3.3 *Add.*

55. $9a^2 - 5a$ and $12a^2 + 3a$

56. $5x^2 + 3x - 5$ and $4x^2 - 6x - 2$

57. $5y^3 - 3y^2$ and $4y + 3y^2$

Subtract.

58. $4x^2 - 3x$ from $8x^2 + 5x$

59. $2x^2 - 5x - 7$ from $7x^2 - 2x + 3$

60. $5x^2 + 3$ from $9x^2 - 4x$

Perform the indicated operations.

61. Subtract $5x - 3$ from the sum of $9x + 2$ and $-3x - 7$.

62. Subtract $5a^2 - 3a$ from the sum of $5a^2 + 2$ and $7a - 7$.

63. Subtract the sum of $16w^2 - 3w$ and $8w + 2$ from $7w^2 - 5w + 2$.

Add using the vertical method.

64. $x^2 + 5x - 3$ and $2x^2 + 4x - 3$

65. $9b^2 - 7$ and $8b + 5$

66. $x^2 + 7$, $3x - 2$, and $4x^2 - 8x$

Subtract using the vertical method.

67. $5x^2 - 3x + 2$ from $7x^2 - 5x - 7$

68. $8m - 7$ from $9m^2 - 7$

3.4 *Multiply.*

69. $(5a^3)(a^2)$

70. $(2x^2)(3x^5)$

71. $(-9p^3)(-6p^2)$

72. $(3a^2b^3)(-7a^3b^4)$

73. $5(3x - 8)$

74. $4a(3a + 7)$

75. $(-5rs)(2r^2s - 5rs)$

76. $7mn(3m^2n - 2mn^2 + 5mn)$

77. $(x + 5)(x + 4)$

78. $(w - 9)(w - 10)$

79. $(a - 7b)(a + 7b)$

80. $(p - 3q)^2$

81. $(a + 4b)(a + 3b)$

82. $(b - 8)(2b + 3)$

83. $(3x - 5y)(2x - 3y)$

84. $(5r + 7s)(3r - 9s)$

85. $(y + 2)(y^2 - 2y + 3)$

86. $(b + 3)(b^2 - 5b - 7)$

87. $(x - 2)(x^2 + 2x + 4)$

88. $(m^2 - 3)(m^2 + 7)$

89. $2x(x + 5)(x - 6)$

90. $a(2a - 5b)(2a - 7b)$

91. $(x + 7)^2$

92. $(a - 8)^2$

93. $(2w - 5)^2$

94. $(3p + 4)^2$

95. $(a + 7b)^2$

96. $(8x - 3y)^2$

97. $(x - 5)(x + 5)$

98. $(y + 9)(y - 9)$

99. $(2m + 3)(2m - 3)$

100. $(3r - 7)(3r + 7)$

101. $(5r - 2s)(5r + 2s)$

102. $(7a + 3b)(7a - 3b)$

103. $2x(x - 5)^2$

104. $3c(c + 5d)(c - 5d)$

3.5 *Divide.*

105. $\dfrac{9a^5}{3a^2}$

106. $\dfrac{24m^4n^2}{6m^2n}$

107. $\dfrac{15a - 10}{5}$

108. $\dfrac{32a^3 + 24a}{8a}$

109. $\dfrac{9r^2s^3 - 18r^3s^2}{-3rs^2}$

110. $\dfrac{35x^3y^2 - 21x^2y^3 + 14x^3y}{7x^2y}$

111. $\dfrac{x^2 - 2x - 15}{x + 3}$

112. $\dfrac{2x^2 + 9x - 35}{2x - 5}$

113. $\dfrac{x^2 - 8x + 17}{x - 5}$

114. $\dfrac{6x^2 - x - 10}{3x + 4}$

115. $\dfrac{6x^3 + 14x^2 - 2x - 6}{6x + 2}$

116. $\dfrac{4x^3 + x + 3}{2x - 1}$

117. $\dfrac{3x^2 + x^3 + 5 + 4x}{x + 2}$

118. $\dfrac{2x^4 - 2x^2 - 10}{x^2 - 3}$

Name _____

Section _____ **Date** _____

The purpose of this self-test is to help you assess your progress so that you can find concepts that you need to review before the next exam. Allow yourself about an hour to take this test. At the end of that hour, check your answers against those given in the back of this text. If you miss any, go back to the appropriate section to reread the examples until you have mastered that particular concept.

Answers

1. _____

2. _____

3. _____

4. _____

5. _____

6. _____

7. _____

8. _____

9. _____

10. _____

11. _____

12. _____

13. _____

14. _____

15. _____

16. _____

17. _____

18. _____

19. _____

20. _____

Use the properties of exponents to simplify each expression.

1. $a^5 \cdot a^9$

2. $3x^2y^3 \cdot 5xy^4$

3. $\dfrac{4x^5}{2x^2}$

4. $\dfrac{20a^3b^5}{5a^2b^2}$

5. $(3x^2y)^3$

6. $\left(\dfrac{2w^2}{3t^3}\right)^2$

7. $(2x^3y^2)^4(x^2y^3)^3$

8. $\dfrac{(5m^3n^2)^2}{2m^4n^5}$

Perform the indicated operations. Report your results in descending order.

9. $(3x^2 - 7x + 2) + (7x^2 - 5x - 9)$

10. $(7a^2 - 3a) + (7a^3 + 4a^2)$

11. $(8x^2 + 9x - 7) - (5x^2 - 2x + 5)$

12. $(3b^2 - 7b) - (2b^2 + 5)$

13. $(3a^2 - 5a) + (9a^2 - 4a) - (5a^2 + a)$

14. $(x^2 + 3) + (5x - 7) + (3x^2 - 2)$

15. $(5x^2 - 7x) - (3x^2 - 5)$

16. $5ab(3a^2b - 2ab + 4ab^2)$

17. $(x - 2)(3x + 7)$

18. $(2x + y)(x^2 + 3xy - 2y^2)$

19. $(4x + 3y)(2x - 5y)$

20. $x(3x - y)(4x + 5y)$

21. $(3m + 2n)^2$

22. $(a - 7b)(a + 7b)$

23. $\dfrac{14x^3y - 21xy^2}{7xy}$

24. $\dfrac{20c^3d - 30cd + 45c^2d^2}{5cd}$

25. $(x^2 - 2x - 24) \div (x + 4)$

26. $(2x^2 + x + 4) \div (2x - 3)$

27. $\dfrac{6x^3 - 7x^2 + 3x + 9}{3x + 1}$

28. $\dfrac{x^3 - 5x^2 + 9x - 9}{x - 1}$

Classify each polynomial as a **monomial, binomial,** *or* **trinomial.**

29. $6x^2 + 7x$

30. $5x^2 + 8x - 8$

31. Evaluate $-3x^2 - 5x + 8$ if $x = -2$.

32. Rewrite $-3x^2 + 8x^4 - 7$ in descending order, and then give the coefficients and degree of the polynomial.

Simplify, if possible, and rewrite each expression using only positive exponents.

33. y^{-5}

34. $3b^{-7}$

35. y^4y^{-8}

36. $\dfrac{p^{-5}}{p^5}$

Evaluate (assume any variables are nonzero).

37. 8^0

38. $6x^0$

Compute. Report your results in scientific notation.

39. $(2.1 \times 10^7)(8 \times 10^{12})$

40. $(6 \times 10^{-23})(5.2 \times 10^{12})$

41. $\dfrac{2.3 \times 10^{-6}}{9.2 \times 10^5}$

42. $\dfrac{7.28 \times 10^3}{1.4 \times 10^{-16}}$

Answers

21. _____

22. _____

23. _____

24. _____

25. _____

26. _____

27. _____

28. _____

29. _____

30. _____

31. _____

32. _____

33. _____

34. _____

35. _____

36. _____

37. _____

38. _____

39. _____

40. _____

41. _____

42. _____

Activity 3 ::
The Power of Compound Interest

Suppose that a wealthy uncle puts $500 in the bank for you. He never deposits money again, but the bank pays 5% interest on the money every year on your birthday. How much money is in the bank after 1 year? After 2 years?

After 1 year the amount is $500 + 500(0.05)$, which can be written as $500(1 + 0.05)$ because of the distributive property. $1 + 0.05 = 1.05$, so after 1 year the amount in the bank was 500(1.05). After 2 years, this amount was again multiplied by 1.05.

How much is in the bank after 8 years? Complete the following chart.

Birthday	Computation	Amount
0 (Day of birth)		$500
1	$500(1.05)$	
2	$500(1.05)(1.05)$	
3	$500(1.05)(1.05)(1.05)$	
4	$500(1.05)^4$	
5	$500(1.05)^5$	
6		
7		
8		

(a) Write a formula for the amount in the bank on your nth birthday. About how many years does it take for the money to double? How many years for it to double again? Can you see any connection between this and the rules for exponents? Explain why you think there may or may not be a connection.

(b) If the account earned 6% each year, how much more would it accumulate at the end of year 8? Year 21?

(c) Imagine that you start an Individual Retirement Account (IRA) at age 20, contributing $2,500 each year for 5 years (total $12,500) to an account that produces a return of 8% every year. You stop contributing and let the account grow. Using the information from the previous example, calculate the value of the account at age 65.

(d) Imagine that you don't start the IRA until you are 30. In an attempt to catch up, you invest $2,500 into the same account, 8% annual return, each year for 10 years. You then stop contributing and let the account grow. What will its value be at age 65?

(e) What have you discovered as a result of these computations?

The following questions are presented to help you review concepts from earlier chapters. This is meant as a review and not as a comprehensive exam. The answers are presented in the back of the text. Section references accompany the answers. If you have difficulty with any of these questions, be certain to at least read through the summary related to those sections.

Perform the indicated operations.

1. $8 + (+9)$ **2.** $-26 + 32$ **3.** $(-25)(-6)$ **4.** $(-48) \div (-12)$

Evaluate each expression if $x = -2$, $y = 5$, and $z = -2$.

5. $-5(-3y - 2z)$ **6.** $\dfrac{3x - 4y}{2z + 5y}$

Use the properties of exponents to simplify each expression.

7. $(3x^2)^2 (x^3)^4$ **8.** $\left(\dfrac{x^5}{y^3}\right)^2$ **9.** $(2x^3y)^3$

10. $7y^0$ **11.** $(3x^4y^5)^0$

Simplify each expression. Report your results using positive exponents only.

12. x^{-4} **13.** $3x^{-2}$ **14.** x^5x^{-9} **15.** $\dfrac{x^{-3}}{y^3}$

Simplify each expression.

16. $21x^5y - 17x^5y$ **17.** $(3x^2 + 4x - 5) - (2x^2 - 3x - 5)$

18. $3x + 2y - x - 4y$ **19.** $(x + 3)(x - 5)$

20. $(x + y)^2$ **21.** $(3x - 4y)^2$

22. $\dfrac{x^2 + 2x - 8}{x - 2}$ **23.** $x(x + y)(x - y)$

Name _____

Section _____ **Date** _____

Answers

1. _____   2. _____

3. _____   4. _____

5. _____

6. _____

7. _____

8. _____

9. _____

10. _____

11. _____

12. _____

13. _____

14. _____

15. _____

16. _____

17. _____

18. _____

19. _____

20. _____

21. _____

22. _____

23. _____

Answers

24. _____

25. _____

26. _____

27. _____

28. _____

29. _____

30. _____

31. _____

32. _____

33. _____

34. _____

Solve each equation.

24. $7x - 4 = 3x - 12$

25. $3x + 2 = 4x + 4$

26. $\dfrac{3}{4}x - 2 = 5 + \dfrac{2}{3}x$

27. $6(x - 1) - 3(1 - x) = 0$

28. Solve the equation $A = \dfrac{1}{2}(b + B)$ for B.

Solve each inequality.

29. $-5x - 7 \le 3x + 9$

30. $-3(x + 5) > -2x + 7$

Solve each application.

31. BUSINESS AND FINANCE Sam made $10 more than twice what Larry earned in one month. If together they earned $760, how much did each earn that month?

32. NUMBER PROBLEM The sum of two consecutive odd integers is 76. Find the two integers.

33. BUSINESS AND FINANCE Two-fifths of a woman's income each month goes to taxes. If she pays $848 in taxes each month, what is her monthly income?

34. BUSINESS AND FINANCE The retail selling price of a sofa is $806.25. What is the cost to the dealer if she sells at 25% markup on the cost?

CHAPTER

4

chapter 4 > Make the Connection

INTRODUCTION

Developing secret codes is big business because of the widespread use of computers and the Internet. Corporations all over the world sell encryption systems that are supposed to keep data secure and safe.

In 1977, three professors from the Massachusetts Institute of Technology developed an encryption system they called RSA, a name derived from the first letters of their last names. Their security code was based on a number that has *129 digits*. They called the code RSA-129. To break the code, the 129-digit number had to be factored into two prime numbers.

A data security company says that people who are using their system are safe because as yet no truly efficient algorithm for finding prime factors of massive numbers has been found, although one may someday exist. This company, hoping to test its encrypting system, now sponsors contests challenging people to factor very large numbers into two prime numbers. RSA-576 up to RSA-2048 are being worked on now.

The U.S. government does not allow any codes to be used unless it has the key. The software firms claim that this prohibition is costing them about $60 billion in lost sales because many companies will not buy an encryption system knowing they can be monitored by the U.S. government.

Factoring

CHAPTER 4 OUTLINE

Name _____

Section _____ Date _____

Answers

1. _____

2. _____

3. _____

4. _____

5. _____

6. _____

7. _____

8. _____

9. _____

10. _____

This prerequisite test provides some exercises requiring skills that you will need to be successful in the coming chapter. The answers for these exercises can be found in the back of this text. This prerequisite test can help you identify topics that you will need to review before beginning the chapter.

Find the prime factorization of each number.

1. 132

2. $1,240$

Perform the indicated operation.

3. $4(x - 8)$

4. $-2(3x^2 - 3x + 1)$

5. $2x(3x - 6)$

6. $7x^2(3x^2 + 4x - 9)$

7. $(x + 3)(2x - 1)$

8. $(3x - 5)(5x + 4)$

9. $\dfrac{6x^3 + 8x^2 - 2x}{2x}$

10. $\dfrac{2x^2 + 7x + 3}{x + 3}$

The Streeter/Hutchison Series in Mathematics   Beginning Algebra

4.1

An Introduction to Factoring

< 4.1 Objectives >

1 > Factor out the greatest common factor (GCF)

2 > Factor out a binomial GCF

3 > Factor a polynomial by grouping terms

> **Tips for Student Success**

Working Together

How many of your classmates do you know? Whether you are by nature outgoing or shy, you have much to gain by getting to know your classmates.

1. It is important to have someone to call when you miss class or are unclear on an assignment.

2. Working with another person is almost always beneficial to both people. If you don't understand something, it helps to have someone to ask about it. If you do understand something, nothing cements that understanding quite like explaining the idea to another person.

3. Sometimes we need to sympathize with others. If an assignment is particularly frustrating, it is reassuring to find that it is also frustrating for other students.

4. Have you ever thought you had the right answer, but it doesn't match the answer in the text? Frequently the answers are equivalent, but that's not always easy to see. A different perspective can help you see that. Occasionally there is an error in a textbook (here we are talking about *other* textbooks). In such cases it is wonderfully reassuring to find that someone else has the same answer you do.

In Chapter 3 you were given factors and asked to find a product. We are now going to reverse the process. You will be given a polynomial and asked to find its factors. This is called **factoring.**

We start with an example from arithmetic. To *multiply* $5 \cdot 7$, you write

$5 \cdot 7 = 35$

To *factor* 35, you write

$35 = 5 \cdot 7$

Factoring is the *reverse* of multiplication.

Now we look at factoring in algebra. We use the distributive property as

$a(b + c) = ab + ac$

For instance,

$3(x + 5) = 3x + 15$

NOTE

3 and $x + 5$ are the factors of $3x + 15$.

259

To use the distributive property in factoring, we reverse that property as

$$ab + ac = a(b + c)$$

This property lets us factor out the common factor a from the terms of $ab + ac$. To use this in factoring, the first step is to see whether each term of the polynomial has a common monomial factor. In our earlier example,

$$3x + 15 = 3 \cdot x + 3 \cdot 5$$

Common factor

So, by the distributive property,

$$3x + 15 = 3(x + 5)$$ The original terms are each divided by the greatest common factor to determine the terms in parentheses.

To check this, multiply $3(x + 5)$.

Multiplying

$$3(x + 5) = 3x + 15$$

Factoring

The first step in factoring polynomials is to identify the *greatest common factor* (GCF) of a set of terms. This factor is the product of the largest common numerical coefficient and the largest common factor of each variable.

Definition	
Greatest Common Factor	The **greatest common factor (GCF)** of a polynomial is the factor that is the product of the largest common numerical coefficient factor of the polynomial and each variable with the largest exponent that appears in all of the terms.

Example 1   Finding the GCF

< Objective 1 >

Find the GCF for each set of terms.

(a) 9 and 12   The largest number that is a factor of both is 3.

(b) 10, 25, 150   The GCF is 5.

(c) x^4 and x^7

$$x^4 = \boxed{x} \cdot \boxed{x} \cdot \boxed{x} \cdot \boxed{x}$$

$$x^7 = \boxed{x} \cdot \boxed{x} \cdot \boxed{x} \cdot \boxed{x} \cdot x \cdot x \cdot x$$

The largest power that divides both terms is x^4.

(d) $12a^3$ and $18a^2$

$$12a^3 = 2 \cdot \boxed{2} \cdot \boxed{3} \cdot \boxed{a} \cdot \boxed{a} \cdot a$$

$$18a^2 = \boxed{2} \cdot \boxed{3} \cdot 3 \cdot \boxed{a} \cdot \boxed{a}$$

The GCF is $6a^2$.

Check Yourself 1

Find the GCF for each set of terms.

(a) 14, 24   (b) 9, 27, 81

(c) a^9, a^5 (d) $10x^5$, $35x^4$

Step by Step

To Factor a Monomial from a Polynomial		
	Step 1	Find the GCF for all the terms.
	Step 2	Use the GCF to factor each term and then apply the distributive property.
	Step 3	Mentally check your factoring by multiplication. Checking your answer is always important and perhaps is never easier than after you have factored.

Example 2   **Finding the GCF of a Binomial**

(a) Factor $8x^2 + 12x$.

The largest common numerical factor of 8 and 12 is 4, and x is the common variable factor with the largest power. So $4x$ is the GCF. Write

$$8x^2 + 12x = 4x \cdot 2x + 4x \cdot 3$$

$$\text{GCF}$$

Now, by the distributive property, we have

$$8x^2 + 12x = 4x(2x + 3)$$

It is always a good idea to check your answer by multiplying to make sure that you get the original polynomial. Try it here. Multiply $4x$ by $2x + 3$.

(b) Factor $6a^4 - 18a^2$.

The GCF in this case is $6a^2$. Write

$$6a^4 - 18a^2 = 6a^2 \cdot a^2 - 6a^2 \cdot (3)$$

$$\text{GCF}$$

Again, using the distributive property yields

$$6a^4 - 18a^2 = 6a^2(a^2 - 3)$$

You should check this by multiplying.

NOTE

It is also true that

$6a^4 - 18a^2$
$= 3a(2a^3 - 6a)$.

However, this is *not completely factored.* Do you see why? You want to find the common monomial factor with the *largest possible* coefficient and the *largest exponent*, in this case $6a^2$.

Check Yourself 2

Factor each polynomial.

(a) $5x + 20$ (b) $6x^2 - 24x$ (c) $10a^3 - 15a^2$

The process is exactly the same for polynomials with more than two terms. Consider Example 3.

Example 3   Finding the GCF of a Polynomial

(a) Factor $5x^2 - 10x + 15$.

$$5x^2 - 10x + 15 = 5 \cdot x^2 - 5 \cdot 2x + 5 \cdot 3$$

$$\text{GCF}$$

$$= 5(x^2 - 2x + 3)$$

(b) Factor $6ab + 9ab^2 - 15a^2$.

$$6ab + 9ab^2 - 15a^2 = 3a \cdot 2b + 3a \cdot 3b^2 - 3a \cdot 5a$$

$$\text{GCF}$$

$$= 3a(2b + 3b^2 - 5a)$$

(c) Factor $4a^4 + 12a^3 - 20a^2$.

$$4a^4 + 12a^3 - 20a^2 = 4a^2 \cdot a^2 + 4a^2 \cdot 3a - 4a^2 \cdot 5$$

$$\text{GCF}$$

$$= 4a^2(a^2 + 3a - 5)$$

(d) Factor $\underbrace{6a^2b + 9ab^2 + 3ab}$.

Mentally note that 3, a, and b are factors of each term, so

$$6a^2b + 9ab^2 + 3ab = 3ab(2a + 3b + 1)$$

Check Yourself 3

Factor each polynomial.
(a) $8b^2 + 16b - 32$ **(b)** $4xy - 8x^2y + 12x^3$
(c) $7x^4 - 14x^3 + 21x^2$ **(d)** $5x^2y^2 - 10xy^2 + 15x^2y$

If the leading coefficient of a polynomial is negative, we usually choose to factor out a GCF with a negative coefficient. When factoring out a GCF with a negative coefficient, take care with the signs of the terms.

Example 4   Factoring Out a Negative Coefficient

Factor out the GCF with a negative coefficient.

(a) $-x^2 - 5x + 7$

Here, we factor out -1.

$$-x^2 - 5x + 7 = (-1)(x^2) + (-1)(5x) - (-1)(7)$$
$$= -1(x^2 + 5x - 7)$$

(b) $-10x^2y + 5xy^2 - 20xy$

$5xy$ is a factor of each term. Because the leading coefficient is negative, we factor out $-5xy$.

$$-10x^2y + 5xy^2 - 20xy = (-5xy)(2x) - (-5xy)(y) + (-5xy)(4)$$
$$= -5xy(2x - y + 4)$$

Check Yourself 4

Factor out the GCF with a negative coefficient.

(a) $-a^2 + 3a - 9$ (b) $-6m^3n^2 - 3m^2n + 12mn$

We can have two or more terms that have a binomial factor in common, as is the case in Example 5.

| Example 5 | Finding a Common Factor |

< Objective 2 >

(a) Factor $3x(x + y) + 2(x + y)$.

We see that *the binomial $x + y$ is a common factor* and can be removed.

$3x(x + y) + 2(x + y)$
$= (x + y) \cdot 3x + (x + y) \cdot 2$
$= (x + y)(3x + 2)$

> **NOTE**
>
> Because of the commutative property, the factors can be written in either order.

(b) Factor $3x^2(x - y) + 6x(x - y) + 9(x - y)$.

We note that here the GCF is $3(x - y)$. Factoring as before, we have

$3(x - y)(x^2 + 2x + 3)$.

Check Yourself 5

Completely factor each polynomial.

(a) $7a(a - 2b) + 3(a - 2b)$ (b) $4x^2(x + y) - 8x(x + y) - 16(x + y)$

Some polynomials can be factored by grouping the terms and finding common factors within each group. We explore this process, called **factoring by grouping.**

In Example 4, we looked at the expression

$3x(x + y) + 2(x + y)$

and found that we could factor out the common binomial, $(x + y)$, giving us

$(x + y)(3x + 2)$

That technique will be used in Example 6.

| Example 6 | Factoring by Grouping Terms |

< Objective 3 >

Suppose we want to factor the polynomial

$ax - ay + bx - by$

As you can see, the polynomial has no common factors. However, look at what happens if we separate the polynomial into *two groups of two terms.*

$ax - ay + bx - by$
$= \underline{ax - ay} + \underline{bx - by}$

> **NOTE**
>
> Our example has *four* terms. That is a clue for trying the factoring by grouping method.

Now *each* group has a common factor, and we can write the polynomial as

$a(x - y) + b(x - y)$

In this form, we can see that $x - y$ is the GCF. Factoring out $x - y$, we get

$a(x - y) + b(x - y) = (x - y)(a + b)$

Check Yourself 6

Use the factoring by grouping method.

$x^2 - 2xy + 3x - 6y$

Be particularly careful of your treatment of algebraic signs when applying the factoring by grouping method. Consider Example 7.

> **Example 7**   **Factoring by Grouping Terms**

NOTE

$9 = (-3)(-3)$

Factor $2x^3 - 3x^2 - 6x + 9$.

We group the polynomial as follows.

$$\underline{2x^3 - 3x^2} \ \underline{-6x + 9}$$

$$= x^2(2x - 3) - 3(2x - 3) \qquad \text{Factor out the common factor of } -3 \text{ from the second two terms.}$$

$$= (2x - 3)(x^2 - 3)$$

Check Yourself 7

Factor by grouping.

$3y^3 + 2y^2 - 6y - 4$

It may also be necessary to change the order of the terms as they are grouped. Look at Example 8.

> **Example 8**   **Factoring by Grouping Terms**

Factor $x^2 - 6yz + 2xy - 3xz$.

Grouping the terms as before, we have

$$\underline{x^2 - 6yz} \ \underline{+ 2xy - 3xz}$$

Do you see that we have accomplished nothing because there are no common factors in the first group?

We can, however, rearrange the terms to write the original polynomial as

$$\underline{x^2 + 2xy} \ \underline{- 3xz - 6yz}$$

$$= x(x + 2y) - 3z(x + 2y) \qquad \text{We can now factor out the common factor of } x + 2y \text{ from each group.}$$

$$= (x + 2y)(x - 3z)$$

Note: It is often true that the grouping can be done in more than one way. The factored form comes out the same.

Check Yourself 8

We can write the polynomial of Example 8 as

$x^2 - 3xz + 2xy - 6yz$

Factor and verify that the factored form is the same in either case.

Check Yourself ANSWERS

1. (a) 2; **(b)** 9; **(c)** a^5; **(d)** $5x^4$ **2. (a)** $5(x + 4)$; **(b)** $6x(x - 4)$;
(c) $5a^2(2a - 3)$ **3. (a)** $8(b^2 + 2b - 4)$; **(b)** $4x(y - 2xy + 3x^2)$;
(c) $7x^2(x^2 - 2x + 3)$; **(d)** $5xy(xy - 2y + 3x)$ **4. (a)** $-1(a^2 - 3a + 9)$;
(b) $-3mn(2m^2n + m - 4)$ **5. (a)** $(a - 2b)(7a + 3)$;
(b) $4(x + y)(x^2 - 2x - 4)$ **6.** $(x - 2y)(x + 3)$ **7.** $(3y + 2)(y^2 - 2)$
8. $(x - 3z)(x + 2y)$

Reading Your Text

The following fill-in-the-blank exercises are designed to ensure that you understand some of the key vocabulary used in this section.

SECTION 4.1

(a) We use the _____ property to remove the common factor a from the expression $ab + ac$.

(b) The first step in factoring a polynomial is to find the _____ of all of the terms.

(c) After factoring, you should check your result by _____ the factors.

(d) If a polynomial has four terms, you should try to factor by _____.

4.1 exercises

Name _____

Section _____ Date _____

Answers

< Objective 1 >

Find the greatest common factor for each set of terms.

1. 10, 12

2. 15, 35

3. 16, 32, 88

4. 55, 33, 132

5. x^2, x^5

6. y^7, y^9

7. a^3, a^6, a^9

8. b^4, b^6, b^8

9. $5x^4, 10x^5$

10. $8y^9, 24y^3$

11. $8a^4, 6a^6, 10a^{10}$

12. $9b^3, 6b^5, 12b^4$

13. $9x^2y, 12xy^2, 15x^2y^2$

14. $12a^3b^2, 18a^2b^3, 6a^4b^4$

15. $15ab^3, 10a^2bc, 25b^2c^3$

16. $9x^2, 3xy^3, 6y^3$

17. $15a^2bc^2, 9ab^2c^2, 6a^2b^2c^2$

> Videos

18. $18x^3y^2z^3, 27x^4y^2z^3, 81xy^2z$

19. $(x + y)^2, (x + y)^3$

20. $12(a + b)^4, 4(a + b)^3$

Factor each polynomial.

21. $8a + 4$

22. $5x - 15$

23. $24m - 32n$

24. $7p - 21q$

25. $12m + 8$

26. $24n - 32$

27. $10s^2 + 5s$

28. $12y^2 - 6y$

29. $12x^2 + 12x$

30. $14b^2 + 14b$

31. $15a^3 - 25a^2$

32. $36b^4 + 24b^2$

33. $6pq + 18p^2q$

34. $8ab - 24ab^2$

35. $6x^2 - 18x + 30$

36. $7a^2 + 21a - 42$

37. $3a^3 + 6a^2 - 12a$

38. $5x^3 - 15x^2 + 25x$

39. $6m + 9mn - 12mn^2$

40. $4s + 6st - 14st^2$

41. $10r^3s^2 + 25r^2s^2 - 15r^2s^3$

> Videos

42. $28x^2y^3 - 35x^2y^2 + 42x^3y$

43. $9a^5 - 15a^4 + 21a^3 - 27a$

44. $8p^6 - 40p^4 + 24p^3 - 16p^2$

Factor out the GCF with a negative coefficient.

45. $-x^2 + 6x + 10$

46. $-u^2 - 4u + 9$

47. $-4m^2n^3 - 6mn^3 - 10n^2$

48. $-8x^4y^2 + 4x^2y^3 + 12xy^3$

< Objective 2 >
Factor out the binomial in each expression.

49. $a(a + 2) - 3(a + 2)$

50. $b(b - 5) + 2(b - 5)$

51. $x(x - 2) + 3(x - 2)$ > Videos

52. $y(y + 5) - 3(y + 5)$

Answers

29. _____

30. _____

31. _____

32. _____

33. _____

34. _____

35. _____

36. _____

37. _____

38. _____

39. _____

40. _____

41. _____

42. _____

43. _____

44. _____

45. _____

46. _____

47. _____

48. _____

49. _____

50. _____

51. _____

52. _____

Answers

53. _____

54. _____

55. _____

56. _____

57. _____

58. _____

59. _____

60. _____

61. _____

62. _____

63. _____

64. _____

65. _____

66. _____

67. _____

68. _____

69. _____

70. _____

71. _____

72. _____

73. _____

74. _____

< Objective 3 >

Factor each polynomial by grouping the first two terms and the last two terms.

53. $x^3 - 4x^2 + 3x - 12$
54. $x^3 - 6x^2 + 2x - 12$

55. $a^3 - 3a^2 + 5a - 15$
56. $6x^3 - 2x^2 + 9x - 3$

57. $10x^3 + 5x^2 - 2x - 1$
58. $x^5 + x^3 - 2x^2 - 2$

59. $x^4 - 2x^3 + 3x - 6$
60. $x^3 - 4x^2 + 2x - 8$

Factor each polynomial completely by factoring out any common factors and then factor by grouping. Do not combine like terms.

61. $3x - 6 + xy - 2y$
62. $2x - 10 + xy - 5y$

63. $ab - ac + b^2 - bc$
64. $ax + 2a + bx + 2b$

65. $3x^2 - 2xy + 3x - 2y$
66. $xy - 5y^2 - x + 5y$

67. $5s^2 + 15st - 2st - 6t^2$
68. $3a^3 + 3ab^2 + 2a^2b + 2b^3$

69. $3x^3 + 6x^2y - x^2y - 2xy^2$
70. $2p^4 + 3p^3q - 2p^3q - 3p^2q^2$

| Basic Skills | **Challenge Yourself** | Calculator/Computer | Career Applications | Above and Beyond |

Complete each statement with **never, sometimes,** *or* **always.**

71. The GCF for two numbers is _____ a prime number.

72. The GCF of a polynomial _____ includes variables.

73. Multiplying the result of factoring will _____ result in the original polynomial.

74. Factoring a negative number from a negative term will _____ result in a negative term.

The Streeter/Hutchison Series in Mathematics   Beginning Algebra

Basic Skills | Challenge Yourself | Calculator/Computer | **Career Applications** | Above and Beyond
▲

75. **ALLIED HEALTH** A patient's protein secretion amount, in milligrams per day, is recorded over several days. Based on these observations, lab technicians determine that the polynomial $-t^3 - 6t^2 + 11t + 66$ provides a good approximation of the patient's protein secretion amounts t days after testing begins. Factor this polynomial.

76. **ALLIED HEALTH** The concentration, in micrograms per milliliter $\frac{\mu g}{mL}$, of the antibiotic chloramphenicol is given by $8t^2 - 2t^3$, where t is the number of hours after the drug is taken. Factor this polynomial.

77. **MANUFACTURING TECHNOLOGY** Polymer pellets need to be as perfectly round as possible. In order to avoid flat spots from forming during the hardening process, the pellets are kept off a surface by blasts of air. The height of a pellet above the surface t seconds after a blast is given by $v_0 t - 4.9t^2$. Factor this expression.

78. **INFORMATION TECHNOLOGY** The total time to transmit a packet is given by the expression $d + 2p$, in which d is the quotient of the distance and the propagation velocity and p is the quotient of the size of the packet and the information transfer rate. How long will it take to transmit a 1,500-byte packet 10 meters on an Ethernet if the information transfer rate is 100 MB per second and the propagation velocity is 2×10^8 m/s?

Basic Skills | Challenge Yourself | Calculator/Computer | Career Applications | **Above and Beyond**
▲

79. The GCF of $2x - 6$ is 2. The GCF of $5x + 10$ is 5. Find the GCF of the product $(2x - 6)(5x + 10)$.

80. The GCF of $3z + 12$ is 3. The GCF of $4z + 8$ is 4. Find the GCF of the product $(3z + 12)(4z + 8)$.

81. The GCF of $2x^3 - 4x$ is $2x$. The GCF of $3x + 6$ is 3. Find the GCF of the product $(2x^3 - 4x)(3x + 6)$.

82. State, in a sentence, the rule illustrated by exercises 79 to 81.

Find the GCF of each product.

83. $(2a + 8)(3a - 6)$

84. $(5b - 10)(2b + 4)$

85. $(2x^2 + 5x)(7x - 14)$

86. $(6y^2 - 3y)(y + 7)$

87. **GEOMETRY** The area of a rectangle with width t is given by $33t - t^2$. Factor the expression and determine the length of the rectangle in terms of t.

88. **GEOMETRY** The area of a rectangle of length x is given by $3x^2 + 5x$. Find the width of the rectangle.

Answers

75.

76.

77.

78.

79.

80.

81.

82.

83.

84.

85.

86.

87.

88.

Answers

89. NUMBER PROBLEM For centuries, mathematicians have found factoring numbers into prime factors a fascinating subject. A prime number is a number that cannot be written as a product of any numbers but 1 and itself. The list of primes begins with 2 because 1 is not considered a prime number and then goes on: 3, 5, 7, 11, ... What are the first 10 primes? What are the primes less than 100? If you list the numbers from 1 to 100 and then cross out all numbers that are multiples of 2, 3, 5, and 7, what is left? Are all the numbers not crossed out prime? Write a paragraph to explain why this might be so. You might want to investigate the Sieve of Eratosthenes, a system from 230 B.C.E. for finding prime numbers.

90. NUMBER PROBLEM If we could make a list of all the prime numbers, what number would be at the end of the list? Because there are an infinite number of prime numbers, there is no "largest prime number." But is there some formula that will give us all the primes? Here are some formulas proposed over the centuries:

$$n^2 + n + 17 \qquad 2n^2 + 29 \qquad n^2 - n + 11$$

In all these expressions, $n = +1, 2, 3, 4, \ldots$, that is, a positive integer beginning with 1. Investigate these expressions with a partner. Do the expressions give prime numbers when they are evaluated for these values of n? Do the expressions give *every* prime in the range of resulting numbers? Can you put in *any* positive number for n?

91. NUMBER PROBLEM How are primes used in coding messages and for security? Work together to decode the messages. The messages are coded using this code: After the numbers are factored into prime factors, the power of 2 gives the number of the letter in the alphabet. This code would be easy for a code breaker to figure out. Can you make up code that would be more difficult to break?

> chapter 4 > Make the Connection

(a) 1310720, 229376, 1572864, 1760, 460, 2097152, 336

(b) 786432, 286, 4608, 278528, 1344, 98304, 1835008, 352, 4718592, 5242880

(c) Code a message using this rule. Exchange your message with a partner to decode it.

Answers

1. 2 **3.** 8 **5.** x^2 **7.** a^3 **9.** $5x^4$ **11.** $2a^4$
13. $3xy$ **15.** $5b$ **17.** $3abc^2$ **19.** $(x + y)^2$ **21.** $4(2a + 1)$
23. $8(3m - 4n)$ **25.** $4(3m + 2)$ **27.** $5s(2s + 1)$ **29.** $12x(x + 1)$
31. $5a^2(3a - 5)$ **33.** $6pq(1 + 3p)$ **35.** $6(x^2 - 3x + 5)$
37. $3a(a^2 + 2a - 4)$ **39.** $3m(2 + 3n - 4n^2)$ **41.** $5r^2s^2(2r + 5 - 3s)$
43. $3a(3a^4 - 5a^3 + 7a^2 - 9)$ **45.** $-1(x^2 - 6x - 10)$
47. $-2n^2(2m^2n + 3mn + 5)$ **49.** $(a - 3)(a + 2)$ **51.** $(x + 3)(x - 2)$
53. $(x - 4)(x^2 + 3)$ **55.** $(a - 3)(a^2 + 5)$ **57.** $(2x + 1)(5x^2 - 1)$
59. $(x - 2)(x^3 + 3)$ **61.** $(x - 2)(3 + y)$ **63.** $(b - c)(a + b)$
65. $(x + 1)(3x - 2y)$ **67.** $(s + 3t)(5s - 2t)$ **69.** $x(x + 2y)(3x - y)$
71. sometimes **73.** always **75.** $(t + 6)(-t^2 + 11)$ **77.** $t(v_0 - 4.9t)$
79. 10 **81.** $6x$ **83.** 6 **85.** $7x$ **87.** $t(33 - t); 33 - t$
89. Above and Beyond **91.** Above and Beyond

4.2

Factoring Trinomials of the Form $x^2 + bx + c$

< 4.2 Objectives >

1 > Factor a trinomial of the form $x^2 + bx + c$

2 > Factor a trinomial containing a common factor

NOTE

The process used to factor here is frequently called the *trial-and-error method.* You should see the reason for the name as you work through this section.

You learned how to find the product of any two binomials by using the FOIL method in Section 3.4. Because factoring is the reverse of multiplication, we now want to use that pattern to find the factors of certain trinomials.

Recall that when we multiply the binomials $x + 2$ and $x + 3$, our result is

$$(x + 2)(x + 3) = x^2 + 5x + 6$$

The product of the first terms $(x \cdot x)$.

The sum of the products of the outer and inner terms ($3x$ and $2x$).

The product of the last terms ($2 \cdot 3$).

> CAUTION

Not every trinomial can be written as the product of two binomials.

Suppose now that you are given $x^2 + 5x + 6$ and want to find its factors. First, you know that the factors of a trinomial may be two binomials. So write

$$x^2 + 5x + 6 = (\qquad)(\qquad)$$

Because the first term of the trinomial is x^2, the first terms of the binomial factors must be x and x. We now have

$$x^2 + 5x + 6 = (x \qquad)(x \qquad)$$

NOTE

We are only interested in factoring polynomials *over the integers* (that is, with integer coefficients).

The product of the last terms must be 6. Because 6 is positive, the factors must have *like* signs. Here are the possibilities:

$$6 = 1 \cdot 6$$
$$= 2 \cdot 3$$
$$= (-1)(-6)$$
$$= (-2)(-3)$$

This means, if we can factor the polynomial, the possible factors of the trinomial are

$$(x + 1)(x + 6)$$
$$(x + 2)(x + 3) \quad 3x \quad 2x$$
$$(x - 1)(x - 6)$$
$$(x - 2)(x - 3)$$

How do we tell which is the correct pair? From the FOIL pattern we know that the sum of the outer and inner products must equal the middle term of the trinomial, in this case $5x$. This is the crucial step!

Possible Factorizations	Middle Terms
$(x + 1)(x + 6)$	$7x$
$(x + 2)(x + 3)$	$5x$
$(x - 1)(x - 6)$	$-7x$
$(x - 2)(x - 3)$	$-5x$

The correct middle term!

So we know that the correct factorization is

$$x^2 + 5x + 6 = (x + 2)(x + 3)$$

Are there any clues so far that will make this process quicker? Yes, there is an important one that you may have spotted. We started with a trinomial that had a positive middle term and a positive last term. The negative pairs of factors for 6 led to negative middle terms. So we do not need to bother with the negative factors if the middle term and the last term of the trinomial are both positive.

▶ **Example 1**	**Factoring a Trinomial**

< Objective 1 >

(a) Factor $x^2 + 9x + 8$.

Because the middle term and the last term of the trinomial are both positive, consider only the positive factors of 8, that is, $8 = 1 \cdot 8$ or $8 = 2 \cdot 4$.

Possible Factorizations	Middle Terms
$(x + 1)(x + 8)$	$9x$
$(x + 2)(x + 4)$	$6x$

NOTE

If you are wondering why we do not list $(x + 8)(x + 1)$ as a possibility, remember that multiplication is commutative. The order doesn't matter!

Because the first pair gives the correct middle term,

$$x^2 + 9x + 8 = (x + 1)(x + 8)$$

(b) Factor $x^2 + 12x + 20$.

Possible Factorizations	Middle Terms
$(x + 1)(x + 20)$	$21x$
$(x + 2)(x + 10)$	$12x$
$(x + 4)(x + 5)$	$9x$

NOTE

The factor-pairs of 20 are

$20 = 1 \cdot 20$

$ = 2 \cdot 10$

$ = 4 \cdot 5$

So

$$x^2 + 12x + 20 = (x + 2)(x + 10)$$

 Check Yourself 1

Factor.

(a) $x^2 + 6x + 5$ **(b)** $x^2 + 10x + 16$

What if the middle term of the trinomial is negative but the first and last terms are still positive? Consider

Positive   Positive

$$x^2 - 11x + 18$$

Negative

Because we want a negative middle term $(-11x)$ and a positive last term, we use *two negative factors* for 18. Recall that the product of two negative numbers is positive, and the sum of two negative numbers is negative.

| Example 2 | Factoring a Trinomial |

(a) Factor $x^2 - 11x + 18$.

NOTE

The negative factor pairs of 18 are

$18 = (-1)(-18)$

$\quad = (-2)(-9)$

$\quad = (-3)(-6)$

Possible Factorizations	Middle Terms
$(x - 1)(x - 18)$	$-19x$
$(x - 2)(x - 9)$	$-11x$
$(x - 3)(x - 6)$	$-9x$

So

$$x^2 - 11x + 18 = (x - 2)(x - 9)$$

(b) Factor $x^2 - 13x + 12$.

NOTE

The negative factors of 12 are

$12 = (-1)(-12)$

$\quad = (-2)(-6)$

$\quad = (-3)(-4)$

Possible Factorizations	Middle Terms
$(x - 1)(x - 12)$	$-13x$
$(x - 2)(x - 6)$	$-8x$
$(x - 3)(x - 4)$	$-7x$

So

$$x^2 - 13x + 12 = (x - 1)(x - 12)$$

A few more clues: We have listed all the possible factors in the above examples. In fact, you can just work until you find the right pair. Also, with practice much of this work can be done mentally.

Check Yourself 2

Factor.

(a) $x^2 - 10x + 9$ **(b)** $x^2 - 10x + 21$

Now we look at the process of factoring a trinomial whose last term is negative. For instance, to factor $x^2 + 2x - 15$, we can start as before:

$$x^2 + 2x - 15 = (x \quad ?)(x \quad ?)$$

Note that the product of the last terms must be negative (-15 here). So we must choose factors that have different signs.

What are our choices for the factors of -15?

$$-15 = (1)(-15)$$
$$= (-1)(15)$$
$$= (3)(-5)$$
$$= (-3)(5)$$

NOTE

Another clue: Some students prefer to look at the list of numerical factors rather than looking at the actual algebraic factors. Here you want the pair whose sum is 2, the coefficient of the middle term of the trinomial. That pair is -3 and 5, which leads us to the correct factors.

This means that the possible factors and the resulting middle terms are

Possible Factorizations	Middle Terms
$(x + 1)(x - 15)$	$-14x$
$(x - 1)(x + 15)$	$14x$
$(x + 3)(x - 5)$	$-2x$
$(x - 3)(x + 5)$	$2x$

So $x^2 + 2x - 15 = (x - 3)(x + 5)$.

In the next example, we practice factoring when the constant term is negative.

Example 3   **Factoring a Trinomial**

(a) Factor $x^2 - 5x - 6$.

First, list the factors of -6. Of course, one factor will be positive, and one will be negative.

$$-6 = (1)(-6)$$
$$= (-1)(6)$$
$$= (2)(-3)$$
$$= (-2)(3)$$

NOTE

You may be able to pick the factors directly from this list. You want the pair whose sum is -5 (the coefficient of the middle term).

For the trinomial, then, we have

Possible Factorizations	Middle Terms
$(x + 1)(x - 6)$	$-5x$
$(x - 1)(x + 6)$	$5x$
$(x + 2)(x - 3)$	$-x$
$(x - 2)(x + 3)$	x

So $x^2 - 5x - 6 = (x + 1)(x - 6)$.

(b) Factor $x^2 + 8xy - 9y^2$.

The process is similar if two variables are involved in the trinomial. Start with

$$x^2 + 8xy - 9y^2 = (x \qquad ?)(x \qquad ?).$$

The product of the last terms must be $-9y^2$.

$$-9y^2 = (-y)(9y)$$
$$= (y)(-9y)$$
$$= (3y)(-3y)$$

Possible Factorizations	Middle Terms
$(x - y)(x + 9y)$	$8xy$
$(x + y)(x - 9y)$	$-8xy$
$(x + 3y)(x - 3y)$	0

So $x^2 + 8xy - 9y^2 = (x - y)(x + 9y)$.

 Check Yourself 3

Factor.

(a) $x^2 + 7x - 30$ **(b)** $x^2 - 3xy - 10y^2$

As we pointed out in Section 4.1, any time that there is a common factor, that factor should be factored out *before* we try any other factoring technique. Consider Example 4.

Example 4   **Factoring a Trinomial**

< Objective 2 >

(a) Factor $3x^2 - 21x + 18$.

$3x^2 - 21x + 18 = 3(x^2 - 7x + 6)$ Factor out the common factor of 3.

We now factor the remaining trinomial. For $x^2 - 7x + 6$:

 > CAUTION

A common mistake is to forget to write the 3 that was factored out as the first step.

Possible Factorizations	Middle Terms	
$(x - 1)(x - 6)$	$-7x$	The correct middle term
$(x - 2)(x - 3)$	$-5x$	

So $3x^2 - 21x + 18 = 3(x - 1)(x - 6)$.

(b) Factor $2x^3 + 16x^2 - 40x$.

$2x^3 + 16x^2 - 40x = 2x(x^2 + 8x - 20)$ Factor out the common factor of $2x$.

To factor the remaining trinomial, which is $x^2 + 8x - 20$, we have

NOTE

Once we have found the desired middle term, there is no need to continue.

Possible Factorizations	Middle Terms	
$(x + 2)(x - 10)$	$-8x$	
$(x - 2)(x + 10)$	$8x$	The correct middle term

So $2x^3 + 16x^2 - 40x = 2x(x - 2)(x + 10)$.

 Check Yourself 4

Factor.

(a) $3x^2 - 3x - 36$ **(b)** $4x^3 + 24x^2 + 32x$

One further comment: Have you wondered whether all trinomials are factorable? Look at the trinomial

$$x^2 + 2x + 6$$

The only possible factors are $(x + 1)(x + 6)$ and $(x + 2)(x + 3)$. Neither pair is correct (you should check the middle terms), and so this trinomial does not have factors with integer coefficients. Of course, there are many other trinomials that cannot be factored. Can you find one?

 Check Yourself ANSWERS

1. **(a)** $(x + 1)(x + 5)$; **(b)** $(x + 2)(x + 8)$ 2. **(a)** $(x - 9)(x - 1)$; **(b)** $(x - 3)(x - 7)$ 3. **(a)** $(x + 10)(x - 3)$; **(b)** $(x + 2y)(x - 5y)$
4. **(a)** $3(x - 4)(x + 3)$; **(b)** $4x(x + 2)(x + 4)$

Reading Your Text

The following fill-in-the-blank exercises are designed to ensure that you understand some of the key vocabulary used in this section.

SECTION 4.2

(a) Factoring is the reverse of _____.

(b) From the FOIL pattern, we know that the sum of the inner and outer products must equal the _____ term of the trinomial.

(c) The product of two negative factors is always _____.

(d) Some trinomials do not have _____ with integer coefficients.

4.2 exercises

< Objective 1 >

Complete each statement.

1. $x^2 - 8x + 15 = (x - 3)(\quad)$

2. $y^2 - 3y - 18 = (y - 6)(\quad)$

3. $m^2 + 8m + 12 = (m + 2)(\quad)$

4. $x^2 - 10x + 24 = (x - 6)(\quad)$

5. $p^2 - 8p - 20 = (p + 2)(\quad)$

6. $a^2 + 9a - 36 = (a + 12)(\quad)$

7. $x^2 - 16x + 64 = (x - 8)(\quad)$

8. $w^2 - 12w - 45 = (w + 3)(\quad)$

9. $x^2 - 7xy + 10y^2 = (x - 2y)(\quad)$ > Videos

10. $a^2 + 18ab + 81b^2 = (a + 9b)(\quad)$

Factor each trinomial completely.

11. $x^2 + 8x + 15$

12. $x^2 - 11x + 24$

13. $x^2 - 11x + 28$

14. $y^2 - y - 20$

15. $s^2 + 13s + 30$

16. $b^2 + 14b + 33$

17. $a^2 - 2a - 48$

18. $x^2 - 17x + 60$

19. $x^2 - 8x + 7$

20. $x^2 + 7x - 18$

21. $m^2 + 3m - 28$ > Videos

22. $a^2 + 10a + 25$

23. $x^2 - 6x - 40$

24. $x^2 - 11x + 10$

25. $x^2 - 14x + 49$

26. $s^2 - 4s - 32$

Name _____

Section _____ Date _____

Answers

1. _____ 2. _____
3. _____ 4. _____
5. _____ 6. _____
7. _____ 8. _____
9. _____ 10. _____
11. _____
12. _____
13. _____
14. _____
15. _____
16. _____
17. _____
18. _____
19. _____
20. _____
21. _____
22. _____
23. _____
24. _____
25. _____
26. _____

Answers

27. $p^2 - 10p - 24$

28. $x^2 - 11x - 60$

29. $x^2 + 5x - 66$

30. $a^2 + 2a - 80$

31. $c^2 + 19c + 60$

32. $t^2 - 4t - 60$

33. $x^2 + 7xy + 10y^2$

34. $x^2 - 8xy + 12y^2$

35. $a^2 - ab - 42b^2$ ▸ Videos

36. $m^2 - 8mn + 16n^2$

37. $x^2 + x + 7$

38. $x^2 - 3x + 9$

39. $x^2 - 13xy + 40y^2$

40. $r^2 - 9rs - 36s^2$

41. $x^2 - 2xy - 8y^2$

42. $u^2 + 6uv - 55v^2$

43. $s^2 - 2st - 2t^2$

44. $x^2 + 5xy + y^2$

45. $25m^2 + 10mn + n^2$

46. $64m^2 - 16mn + n^2$

< Objective 2 >

47. $3a^2 - 3a - 126$

48. $2c^2 + 2c - 60$

49. $r^3 + 7r^2 - 18r$

50. $m^3 + 5m^2 - 14m$

51. $2x^3 - 20x^2 - 48x$

52. $3p^3 + 48p^2 - 108p$

53. $x^2y - 9xy^2 - 36y^3$ ▸ Videos

54. $4s^4 - 20s^3t - 96s^2t^2$

55. $m^3 - 29m^2n + 120mn^2$

56. $2a^3 - 52a^2b + 96ab^2$

Answers

27.
28.
29.
30.
31.
32.
33.
34.
35.
36.
37.
38.
39.
40.
41.
42.
43.
44.
45.
46.
47.
48.
49.
50.
51.
52.
53.
54.
55.
56.

Determine whether each statement is **true** *or* **false.**

57. Factoring is the reverse of division.

58. From the FOIL pattern, we know that the sum of the inner and outer products must equal the middle term of the trinomial.

59. The sum of two negative factors is always negative.

60. Every trinomial has integer coefficients.

| Basic Skills | Challenge Yourself | Calculator/Computer | **Career Applications** | Above and Beyond |

61. **MANUFACTURING TECHNOLOGY** The shape of a beam loaded with a single concentrated load is described by the expression $\dfrac{x^2 - 64}{200}$. Factor the numerator, $(x^2 - 64)$.

62. **ALLIED HEALTH** The concentration, in micrograms per milliliter (mcg/mL), of Vancocin, an antibiotic used to treat peritonitis, is given by the negative of the polynomial $t^2 - 8t - 20$, where t is the number of hours since the drug was administered via intravenous injection. Write this given polynomial in factored form.

| Basic Skills | Challenge Yourself | Calculator/Computer | Career Applications | **Above and Beyond** |

Find all positive values for k so that each expression can be factored.

63. $x^2 - kx + 16$ **64.** $x^2 - kx + 17$ **65.** $x^2 - kx - 5$

> Videos

66. $x^2 - kx - 7$ **67.** $x^2 + 3x + k$ **68.** $x^2 + 5x + k$

69. $x^2 + 2x - k$ **70.** $x^2 + x - k$

Answers

1. $x - 5$ **3.** $m + 6$ **5.** $p - 10$ **7.** $x - 8$ **9.** $x - 5y$
11. $(x + 3)(x + 5)$ **13.** $(x - 4)(x - 7)$ **15.** $(s + 3)(s + 10)$
17. $(a - 8)(a + 6)$ **19.** $(x - 1)(x - 7)$ **21.** $(m + 7)(m - 4)$
23. $(x + 4)(x - 10)$ **25.** $(x - 7)(x - 7)$ **27.** $(p - 12)(p + 2)$
29. $(x + 11)(x - 6)$ **31.** $(c + 4)(c + 15)$ **33.** $(x + 2y)(x + 5y)$
35. $(a + 6b)(a - 7b)$ **37.** Not factorable   **39.** $(x - 5y)(x - 8y)$
41. $(x + 2y)(x - 4y)$ **43.** Not factorable   **45.** $(5m + n)(5m + n)$
47. $3(a + 6)(a - 7)$ **49.** $r(r - 2)(r + 9)$ **51.** $2x(x - 12)(x + 2)$
53. $y(x + 3y)(x - 12y)$ **55.** $m(m - 5n)(m - 24n)$ **57.** False
59. True   **61.** $(x + 8)(x - 8)$ **63.** 8, 10, or 17   **65.** 4
67. 2   **69.** 3, 8, 15, 24, . . .

Answers

57. _____

58. _____

59. _____

60. _____

61. _____

62. _____

63. _____

64. _____

65. _____

66. _____

67. _____

68. _____

69. _____

70. _____

4.3

Factoring Trinomials of the Form $ax^2 + bx + c$

< 4.3 Objectives >

1 > Factor a trinomial of the form $ax^2 + bx + c$

2 > Completely factor a trinomial

3 > Use the *ac* test to determine factorability

4 > Use the results of the *ac* test to factor a trinomial

Factoring trinomials takes a little more work when the coefficient of the first term is not 1. Look at the following multiplication.

$$(5x + 2)(2x + 3) = 10x^2 + 19x + 6$$

Factors   Factors
of $10x^2$ of 6

Do you see the additional problem? We must consider all possible factors of the first coefficient (10 in our example) as well as those of the third term (6 in our example).

There is no easy way out! You need to form all possible combinations of factors and then check the middle term until the proper pair is found. If this seems a bit like guesswork, it is. In fact, some call this process factoring by *trial and error.*

We can simplify the work a bit by reviewing the sign patterns found in Section 4.2.

Property

Sign Patterns for Factoring Trinomials

1. If all terms of a trinomial are positive, the signs between the terms in the binomial factors are both plus signs.
2. If the third term of the trinomial is positive and the middle term is negative, the signs between the terms in the binomial factors are both minus signs.
3. If the third term of the trinomial is negative, the signs between the terms in the binomial factors are opposite (one is $+$ and one is $-$).

Example 1   Factoring a Trinomial

< Objective 1 >

Factor $3x^2 + 14x + 15$.

First, list the possible factors of 3, the coefficient of the first term.

$$3 = 1 \cdot 3$$

Now list the factors of 15, the last term.

$$15 = 1 \cdot 15$$
$$= 3 \cdot 5$$

Because the signs of the trinomial are all positive, we know any factors will have the form

The product of the numbers in the last blanks must be 15.

$$(\_x + \_)(\_x + \_)$$

The product of the numbers in the first blanks must be 3.

So the following are the possible factors and the corresponding middle terms:

NOTE

Take the time to multiply the binomial factors. This ensures that you have an expression equivalent to the original problem.

Possible Factorizations	Middle Terms	
$(x + 1)(3x + 15)$	$18x$	
$(x + 15)(3x + 1)$	$46x$	
$(3x + 3)(x + 5)$	$18x$	
$(3x + 5)(x + 3)$	$14x$	The correct middle term

So

$$3x^2 + 14x + 15 = (3x + 5)(x + 3)$$

 Check Yourself 1

Factor.

(a) $5x^2 + 14x + 8$ (b) $3x^2 + 20x + 12$

Example 2   **Factoring a Trinomial**

Factor $4x^2 - 11x + 6$.

Because only the middle term is negative, we know the factors have the form

$$(\ x - \_)(\_x - \_)$$

Both signs are negative.

Now look at the factors of the first coefficient and the last term.

$$4 = 1 \cdot 4 \qquad 6 = 1 \cdot 6$$
$$ = 2 \cdot 2 \qquad = 2 \cdot 3$$

This gives us the possible factors:

Possible Factorizations	Middle Terms	
$(x - 1)(4x - 6)$	$-10x$	
$(x - 6)(4x - 1)$	$-25x$	
$(x - 2)(4x - 3)$	$-11x$	The correct middle term

RECALL

Again, at least mentally, check your work by multiplying the factors.

Note that, in this example, we *stopped* as soon as the correct pair of factors was found. So

$$4x^2 - 11x + 6 = (x - 2)(4x - 3)$$

 Check Yourself 2

Factor.

(a) $2x^2 - 9x + 9$ (b) $6x^2 - 17x + 10$

Next, we factor a trinomial whose last term is negative.

Example 3   Factoring a Trinomial

Factor $5x^2 + 6x - 8$.

Because the last term is negative, the factors have the form

$(\_x + \_)(\_x - \_)$

Consider the factors of the first coefficient and the last term.

$5 = 1 \cdot 5 \qquad 8 = 1 \cdot 8$
$ = 2 \cdot 4$

The possible factors are then

Possible Factorizations	Middle Terms
$(x + 1)(5x - 8)$	$-3x$
$(x + 8)(5x - 1)$	$39x$
$(5x + 1)(x - 8)$	$-39x$
$(5x + 8)(x - 1)$	$3x$
$(x + 2)(5x - 4)$	$6x$

Again, we stop as soon as the correct pair of factors is found.

$5x^2 + 6x - 8 = (x + 2)(5x - 4)$

Check Yourself 3

Factor $4x^2 + 5x - 6$.

The same process is used to factor a trinomial with more than one variable.

Example 4   Factoring a Trinomial

Factor $6x^2 + 7xy - 10y^2$.

The form of the factors must be

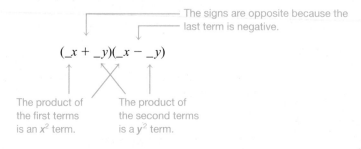

Again, look at the factors of the first and last coefficients.

$6 = 1 \cdot 6 \qquad 10 = 1 \cdot 10$
$ = 2 \cdot 3 \qquad = 2 \cdot 5$

Possible Factorizations	Middle Terms
$(x + y)(6x - 10y)$	$-4xy$
$(x + 10y)(6x - y)$	$59xy$
$(6x + y)(x - 10y)$	$-59xy$
$(6x + 10y)(x - y)$	$4xy$
$(x + 2y)(6x - 5y)$	$7xy$

We stop as soon as the correct factors are found.

$$6x^2 + 7xy - 10y^2 = (x + 2y)(6x - 5y)$$

 Check Yourself 4

Factor $15x^2 - 4xy - 4y^2$.

Example 5 illustrates a special kind of trinomial called a *perfect square trinomial*.

▶ **Example 5**   Factoring a Trinomial

Factor $9x^2 + 12xy + 4y^2$.

Because all terms are positive, the form of the factors must be

$$(\_x + \_y)(\_x + \_y)$$

Consider the factors of the first and last coefficients.

$$9 = 9 \cdot 1 \qquad 4 = 4 \cdot 1$$
$$ = 3 \cdot 3 \qquad = 2 \cdot 2$$

Possible Factorizations	Middle Terms
$(x + y)(9x + 4y)$	$13xy$
$(x + 4y)(9x + y)$	$37xy$
$(3x + 2y)(3x + 2y)$	$12xy$

So

$$9x^2 + 12xy + 4y^2 = (3x + 2y)(3x + 2y)$$
$$= (3x + 2y)^2$$

Square of $3x$ $2(3x)(2y)$ Square of $2y$

This trinomial is the result of squaring a binomial, thus the special name of perfect square trinomial.

 Check Yourself 5

Factor.

(a) $4x^2 + 28x + 49$ (b) $16x^2 - 40xy + 25y^2$

Before looking at Example 6, review one important point from Section 4.2. Recall that when you factor trinomials, you should not forget to look for a common factor as the first step. If there is a common factor, factor it out and then factor the remaining trinomial as before.

▶ **Example 6**	**Factoring a Trinomial**

< Objective 2 >

Factor $18x^2 - 18x + 4$.

First look for a common factor in all three terms. Here that factor is 2, so write

$$18x^2 - 18x + 4 = 2(9x^2 - 9x + 2)$$

By our earlier methods, we can factor the remaining trinomial as

$$9x^2 - 9x + 2 = (3x - 1)(3x - 2)$$

So

$$18x^2 - 18x + 4 = 2(3x - 1)(3x - 2)$$

Don't forget the 2 that was factored out!

NOTE

If you do not see why this is true, use your pencil to work it out before moving on!

 Check Yourself 6

Factor $16x^2 + 44x - 12$.

Now look at an example in which the common factor includes a variable.

▶ **Example 7**	**Factoring a Trinomial**

Factor

$$6x^3 + 10x^2 - 4x$$

The common factor is $2x$.

So

$$6x^3 + 10x^2 - 4x = 2x(3x^2 + 5x - 2)$$

Because

$$3x^2 + 5x - 2 = (3x - 1)(x + 2)$$

we have

$$6x^3 + 10x^2 - 4x = 2x(3x - 1)(x + 2)$$

RECALL

Be certain to include the monomial factor.

 Check Yourself 7

Factor $6x^3 - 27x^2 + 30x$.

You have now had a chance to work with a variety of factoring techniques. Your success in factoring polynomials depends on your ability to recognize when to use which technique. Here are some guidelines to help you apply the factoring methods you have studied in this chapter.

Step by Step

Factoring Polynomials	**Step 1** Look for a greatest common factor other than 1. If such a factor exists, factor out the GCF.
	Step 2 If the polynomial that remains is a *trinomial,* try to factor the trinomial by the trial-and-error methods of Sections 4.2 and 4.3.

Example 8 illustrates this strategy.

    Example 8   **Factoring a Trinomial**

(a) Factor $5m^2n + 20n$.

First, we see that the GCF is $5n$. Factoring it out gives

$$5m^2n + 20n = 5n(m^2 + 4)$$

NOTE

$m^2 + 4$ cannot be factored any further.

(b) Factor $3x^3 - 24x^2 + 48x$.

First, we see that the GCF is $3x$. Factoring out $3x$ yields

$$3x^3 - 24x^2 + 48x = 3x(x^2 - 8x + 16)$$
$$= 3x(x - 4)(x - 4) \quad \text{or} \quad 3x(x - 4)^2$$

(c) Factor $8r^2s + 20rs^2 - 12s^3$.

First, the GCF is $4s$, and we can write the original polynomial as

$$8r^2s + 20rs^2 - 12s^3 = 4s(2r^2 + 5rs - 3s^2)$$

Because the remaining polynomial is a trinomial, we can use the trial-and-error method to complete the factoring.

$$8r^2s + 20rs^2 - 12s^3 = 4s(2r - s)(r + 3s)$$

    Check Yourself 8

Factor each polynomial.

(a) $8a^3 + 32a^2b + 32ab^2$ **(b)** $7x^3 + 7x^2y - 42xy^2$
(c) $5m^4 + 15m^3 + 5m^2$

To this point we have used the trial-and-error method to factor trinomials. We have also learned that not all trinomials can be factored. In the remainder of this section we look at the same kinds of trinomials, but in a slightly different context. We first determine whether a trinomial is factorable, and then use the results of that analysis to factor the trinomial.

Some students prefer the trial-and-error method for factoring because it is generally faster and more intuitive. Other students prefer the method used in the remainder of this section (called the *ac* method) because it yields the answer in a systematic way. We let you determine which method you prefer.

We begin by looking at some factored trinomials.

Example 9   Matching Trinomials and Their Factors

Determine which statements are true.

(a) $x^2 - 2x - 8 = (x - 4)(x + 2)$

This is a true statement. Using the FOIL method, we see that

$$(x - 4)(x + 2) = x^2 + 2x - 4x - 8$$
$$= x^2 - 2x - 8$$

(b) $x^2 + 6x + 5 = (x + 2)(x + 3)$

This is not a true statement.

$$(x + 2)(x + 3) = x^2 + 3x + 2x + 6 = x^2 + 5x + 6$$

(c) $x^2 + 5x - 14 = (x - 2)(x + 7)$

This is true: $(x - 2)(x + 7) = x^2 + 7x - 2x - 14 = x^2 + 5x - 14$

(d) $x^2 - 8x - 15 = (x - 5)(x - 3)$

This is false: $(x - 5)(x - 3) = x^2 - 3x - 5x + 15 = x^2 - 8x + 15$

Check Yourself 9

Determine which statements are true.

(a) $2x^2 - 2x - 3 = (2x - 3)(x + 1)$
(b) $3x^2 + 11x - 4 = (3x - 1)(x + 4)$
(c) $2x^2 - 7x + 3 = (x - 3)(2x - 1)$

The first step in learning to factor a trinomial is to identify its coefficients. So that we are consistent, we first write the trinomial in standard form, $ax^2 + bx + c$, and then label the three coefficients as a, b, and c.

Example 10   Identifying the Coefficients of $ax^2 + bx + c$

RECALL

The negative sign is attached to the coefficient.

First, when necessary, rewrite the trinomial in $ax^2 + bx + c$ form. Then give the values for a, b, and c, in which a is the coefficient of the x^2 term, b is the coefficient of the x term, and c is the constant.

(a) $x^2 - 3x - 18$

$a = 1 \qquad b = -3 \qquad c = -18$

(b) $x^2 - 24x + 23$

$a = 1 \qquad b = -24 \qquad c = 23$

(c) $x^2 + 8 - 11x$

First rewrite the trinomial in descending order.

$x^2 - 11x + 8$

$a = 1 \qquad b = -11 \qquad c = 8$

Check Yourself 10

First, when necessary, rewrite the trinomials in $ax^2 + bx + c$ form. Then label a, b, and c, in which a is the coefficient of the x^2 term, b is the coefficient of the x term, and c is the constant.

(a) $x^2 + 5x - 14$ (b) $x^2 - 18x + 17$ (c) $x - 6 + 2x^2$

Not all trinomials can be factored. To discover whether a trinomial is factorable, we try the *ac* **test.**

Definition	
The *ac* Test	A trinomial of the form $ax^2 + bx + c$ is factorable if (and only if) there are two integers, m and n, such that $$ac = mn \quad \text{and} \quad b = m + n$$

In Example 11 we will look for m and n to determine whether each trinomial is factorable.

▶ Example 11   **Using the *ac* Test**

< Objective 3 >

Use the *ac* test to determine which trinomials can be factored. Find the values of m and n for each trinomial that can be factored.

(a) $x^2 - 3x - 18$

First, we find the values of a, b, and c, so that we can find ac.

$$a = 1 \quad b = -3 \quad c = -18$$

$$ac = 1(-18) = -18 \quad \text{and} \quad b = -3$$

Then, we look for two numbers, m and n, such that their product is ac and their sum is b. In this case, that means

$$mn = -18 \quad \text{and} \quad m + n = -3$$

We now look at all pairs of integers with a product of -18. We then look at the sum of each pair of integers, looking for a sum of -3.

mn	$m + n$	
$1(-18) = -18$	$1 + (-18) = -17$	
$2(-9) = -18$	$2 + (-9) = -7$	We need to look no
$3(-6) = -18$	$3 + (-6) = -3$	further than 3 and -6.
$6(-3) = -18$		
$9(-2) = -18$		
$18(-1) = -18$		

3 and -6 are the two integers with a product of ac and a sum of b. We can say that

$$m = 3 \quad \text{and} \quad n = -6$$

Because we found values for m and n, we know that $x^2 - 3x - 18$ is factorable.

(b) $x^2 - 24x + 23$

We find that

$$a = 1 \quad b = -24 \quad c = 23$$

$$ac = 1(23) = 23 \quad \text{and} \quad b = -24$$

NOTE

We could have chosen $m = -6$ and $n = 3$ as well.

So

$mn = 23$ and $m + n = -24$

We now calculate integer pairs, looking for two numbers with a product of 23 and a sum of -24.

mn	$m + n$
$1(23) = 23$	$1 + 23 = 24$
$-1(-23) = 23$	$-1 + (-23) = -24$

$m = -1$ and $n = -23$

So, $x^2 - 24x + 23$ is factorable.

(c) $x^2 - 11x + 8$

We find that $a = 1$, $b = -11$, and $c = 8$. Therefore, $ac = 8$ and $b = -11$. Thus $mn = 8$ and $m + n = -11$. We calculate integer pairs:

mn	$m + n$
$1(8) = 8$	$1 + 8 = 9$
$2(4) = 8$	$2 + 4 = 6$
$-1(-8) = 8$	$-1 + (-8) = -9$
$-2(-4) = 8$	$-2 + (-4) = -6$

There are no other pairs of integers with a product of 8, and none of these pairs has a sum of -11. The trinomial $x^2 - 11x + 8$ is not factorable.

(d) $2x^2 + 7x - 15$

We find that $a = 2$, $b = 7$, and $c = -15$. Therefore, $ac = 2(-15) = -30$ and $b = 7$. Thus $mn = -30$ and $m + n = 7$. We calculate integer pairs:

mn	$m + n$
$1(-30) = -30$	$1 + (-30) = -29$
$2(-15) = -30$	$2 + (-15) = -13$
$3(-10) = -30$	$3 + (-10) = -7$
$5(-6) = -30$	$5 + (-6) = -1$
$6(-5) = -30$	$6 + (-5) = 1$
$10(-3) = -30$	$10 + (-3) = 7$

There is no need to go any further. We see that 10 and -3 have a product of -30 and a sum of 7, so

$m = 10$ and $n = -3$

Therefore, $2x^2 + 7x - 15$ is factorable.

Check Yourself 11

Use the *ac* test to determine which trinomials can be factored. Find the values of *m* and *n* for each trinomial that can be factored.

(a) $x^2 - 7x + 12$ **(b)** $x^2 + 5x - 14$
(c) $3x^2 - 6x + 7$ **(d)** $2x^2 + x - 6$

So far we have used the results of the *ac* test to determine whether a trinomial is factorable. The results can also be used to help factor the trinomial.

Example 12	**Using the Results of the *ac* Test to Factor**

< Objective 4 >

Rewrite the middle term as the sum of two terms and then factor by grouping.

(a) $x^2 - 3x - 18$

We find that $a = 1$, $b = -3$, and $c = -18$, so $ac = -18$ and $b = -3$. We are looking for two numbers, *m* and *n*, where $mn = -18$ and $m + n = -3$. In Example 11, part (a), we looked at every pair of integers whose product (*mn*) was -18, to find a pair that had a sum ($m + n$) of -3. We found the two integers to be 3 and -6, because $3(-6) = -18$ and $3 + (-6) = -3$, so $m = 3$ and $n = -6$. We now use that result to rewrite the middle term as the sum of $3x$ and $-6x$.

$$x^2 + 3x - 6x - 18$$

We then factor by grouping:

$$= (x^2 + 3x) - (6x + 18)$$
$$x^2 + 3x - 6x - 18 = x(x + 3) - 6(x + 3)$$
$$= (x + 3)(x - 6)$$

(b) $x^2 - 24x + 23$

We use the results from Example 11, part (b), in which we found $m = -1$ and $n = -23$, to rewrite the middle term of the equation.

$$x^2 - 24x + 23 = x^2 - x - 23x + 23$$

Then we factor by grouping:

$$x^2 - x - 23x + 23 = (x^2 - x) - (23x - 23)$$
$$= x(x - 1) - 23(x - 1)$$
$$= (x - 1)(x - 23)$$

(c) $2x^2 + 7x - 15$

From Example 11, part (d), we know that this trinomial is factorable, and $m = 10$ and $n = -3$. We use that result to rewrite the middle term of the trinomial.

$$2x^2 + 7x - 15 = 2x^2 + 10x - 3x - 15$$
$$= (2x^2 + 10x) - (3x + 15)$$
$$= 2x(x + 5) - 3(x + 5)$$
$$= (x + 5)(2x - 3)$$

Note that we did not factor the trinomial in Example 11, part (c), $x^2 - 11x + 8$. Recall that, by the *ac* method, we determined that this trinomial is not factorable.

Check Yourself 12

Use the results of Check Yourself 11 to rewrite the middle term as the sum of two terms and then factor by grouping.

(a) $x^2 - 7x + 12$ **(b)** $x^2 + 5x - 14$ **(c)** $2x^2 + x - 6$

Next, we look at some examples that require us to first find m and n and then factor the trinomial.

| **Example 13** | Rewriting Middle Terms to Factor |

Rewrite the middle term as the sum of two terms and then factor by grouping.

(a) $2x^2 - 13x - 7$

We find $a = 2$, $b = -13$, and $c = -7$, so $mn = ac = -14$ and $m + n = b = -13$. Therefore,

mn	$m + n$
$1(-14) = -14$	$1 + (-14) = -13$

So, $m = 1$ and $n = -14$. We rewrite the middle term of the trinomial as

$$2x^2 - 13x - 7 = 2x^2 + x - 14x - 7$$
$$= (2x^2 + x) - (14x + 7)$$
$$= x(2x + 1) - 7(2x + 1)$$
$$= (2x + 1)(x - 7)$$

(b) $6x^2 - 5x - 6$

We find that $a = 6$, $b = -5$, and $c = -6$, so $mn = ac = -36$ and $m + n = b = -5$.

mn	$m + n$
$1(-36) = -36$	$1 + (-36) = -35$
$2(-18) = -36$	$2 + (-18) = -16$
$3(-12) = -36$	$3 + (-12) = -9$
$4(-9) = -36$	$4 + (-9) = -5$

So, $m = 4$ and $n = -9$. We rewrite the middle term of the trinomial as

$$6x^2 - 5x - 6 = 6x^2 + 4x - 9x - 6$$
$$= (6x^2 + 4x) - (9x + 6)$$
$$= 2x(3x + 2) - 3(3x + 2)$$
$$= (3x + 2)(2x - 3)$$

$2 \cdot -15 = -30$

Check Yourself 13

Rewrite the middle term as the sum of two terms and then factor by grouping.

(a) $2x^2 - 7x - 15$ **(b)** $6x^2 - 5x - 4$

Be certain to check trinomials and binomial factors for any common monomial factor. (There is no common factor in the binomial unless it is also a common factor in the original trinomial.) Example 14 shows the factoring out of monomial factors.

| Example 14 | Factoring Out Common Factors |

Completely factor the trinomial.

$3x^2 + 12x - 15$

We first factor out the common factor of 3.

$3x^2 + 12x - 15 = 3(x^2 + 4x - 5)$

Finding m and n for the trinomial $x^2 + 4x - 5$ yields $mn = -5$ and $m + n = 4$.

mn	$m + n$
$1(-5) = -5$	$1 + (-5) = -4$
$5(-1) = -5$	$-1 + (5) = 4$

So, $m = 5$ and $n = -1$. This gives us

$$3x^2 + 12x - 15 = 3(x^2 + 4x - 5)$$
$$= 3(x^2 + 5x - x - 5)$$
$$= 3[(x^2 + 5x) - (x + 5)]$$
$$= 3[x(x + 5) - (x + 5)]$$
$$= 3[(x + 5)(x - 1)]$$
$$= 3(x + 5)(x - 1)$$

Check Yourself 14

Completely factor the trinomial.

$6x^3 + 3x^2 - 18x$

You do not need to try all possible product pairs to find m and n. A look at the sign pattern of the trinomial eliminates many of the possibilities. Assuming the leading coefficient is positive, there are four possible sign patterns.

Pattern	Example	Conclusion
1. b and c are both positive.	$2x^2 + 13x + 15$	m and n must both be positive.
2. b is negative and c is positive.	$x^2 - 7x + 12$	m and n must both be negative.
3. b is positive and c is negative.	$x^2 + 3x - 10$	m and n are of opposite signs. (The value with the larger absolute value is positive.)
4. b and c are both negative.	$x^2 - 3x - 10$	m and n are of opposite signs. (The value with the larger absolute value is negative.)

Check Yourself ANSWERS

1. (a) $(5x + 4)(x + 2)$; (b) $(3x + 2)(x + 6)$ 2. (a) $(2x - 3)(x - 3)$;
(b) $(6x - 5)(x - 2)$ 3. $(4x - 3)(x + 2)$ 4. $(3x - 2y)(5x + 2y)$
5. (a) $(2x + 7)^2$; (b) $(4x - 5y)^2$ 6. $4(4x - 1)(x + 3)$
7. $3x(2x - 5)(x - 2)$ 8. (a) $8a(a + 2b)(a + 2b)$; (b) $7x(x + 3y)(x - 2y)$;
(c) $5m^2(m^2 + 3m + 1)$ 9. (a) False; (b) true; (c) true
10. (a) $a = 1, b = 5, c = -14$; (b) $a = 1, b = -18, c = 17$;
(c) $a = 2, b = 1, c = -6$ 11. (a) Factorable, $m = -3, n = -4$;
(b) factorable, $m = 7, n = -2$; (c) not factorable;
(d) factorable, $m = 4, n = -3$ 12. (a) $x^2 - 3x - 4x + 12 = (x - 3)(x - 4)$;
(b) $x^2 + 7x - 2x - 14 = (x + 7)(x - 2)$;
(c) $2x^2 + 4x - 3x - 6 = (x + 2)(2x - 3)$
13. (a) $2x^2 - 10x + 3x - 15 = (x - 5)(2x + 3)$;
(b) $6x^2 - 8x + 3x - 4 = (3x - 4)(2x + 1)$ 14. $3x(2x - 3)(x + 2)$

Reading Your Text

The following fill-in-the-blank exercises are designed to ensure that you understand some of the key vocabulary used in this section.

SECTION 4.3

(a) If all the terms of a trinomial are positive, the signs between the terms in the binomial factors are both _____ signs.

(b) If the third term of a trinomial is negative, the signs between the terms in the binomial factors are _____ .

(c) The first step in factoring a polynomial is to factor out the _____ .

(d) We use the _____ to determine whether a trinomial is factorable.

< Objective 1 >

Complete each statement.

1. $4x^2 - 4x - 3 = (2x + 1)(\quad)$

2. $3w^2 + 11w - 4 = (w + 4)(\quad)$

3. $6a^2 + 13a + 6 = (2a + 3)(\quad)$

4. $25y^2 - 10y + 1 = (5y - 1)(\quad)$

5. $15x^2 - 16x + 4 = (3x - 2)(\quad)$

6. $6m^2 + 5m - 4 = (3m + 4)(\quad)$

7. $16a^2 + 8ab + b^2 = (4a + b)(\quad)$

8. $6x^2 + 5xy - 4y^2 = (3x + 4y)(\quad)$ > Videos

9. $4m^2 + 5mn - 6n^2 = (m + 2n)(\quad)$

10. $10p^2 - pq - 3q^2 = (5p - 3q)(\quad)$

*Determine whether each equation is **true** or **false.***

11. $x^2 + 2x - 3 = (x + 3)(x - 1)$

12. $y^2 - 3y - 18 = (y - 6)(y + 3)$

13. $x^2 - 10x - 24 = (x - 6)(x + 4)$

14. $a^2 + 9a - 36 = (a - 12)(a + 4)$

15. $x^2 - 16x + 64 = (x - 8)(x - 8)$

16. $w^2 - 12w - 45 = (w - 9)(w - 5)$

17. $25y^2 - 10y + 1 = (5y - 1)(5y + 1)$ > Videos

Boost *your* **GRADE** at
ALEKS.com!

ALEKS®

- Practice Problems   • e-Professors
- Self-Tests   • Videos
- NetTutor

Name _____

Section _____ Date _____

Answers

1. _____

2. _____

3. _____

4. _____

5. _____

6. _____

7. _____

8. _____

9. _____

10. _____

11. _____

12. _____

13. _____

14. _____

15. _____

16. _____

17. _____

Answers

18. _____

19. _____

20. _____

21. _____

22. _____

23. _____

24. _____

25. _____

26. _____

27. _____

28. _____

29. _____

30. _____

31. _____

32. _____

33. _____

34. _____

35. _____

36. _____

37. _____

38. _____

39. _____

40. _____

41. _____

42. _____

43. _____

44. _____

18. $6x^2 + 5xy - 4y^2 = (6x - 2y)(x + 2y)$

19. $10p^2 - pq - 3q^2 = (5p - 3q)(2p + q)$

20. $6a^2 + 13a + 6 = (2a + 3)(3a + 2)$

For each trinomial, label a, b, and c.

21. $x^2 + 4x - 9$ **22.** $x^2 + 5x + 11$

23. $x^2 - 3x + 8$ **24.** $x^2 + 7x - 15$

25. $3x^2 + 5x - 8$ **26.** $2x^2 + 7x - 9$

27. $4x^2 + 11 + 8x$ **28.** $5x^2 - 9 + 7x$

29. $5x - 3x^2 - 10$ **30.** $9x - 7x^2 - 18$

< Objective 3 >

Use the ac test to determine which trinomials can be factored. Find the values of m and n for each trinomial that can be factored.

31. $x^2 + x - 6$ **32.** $x^2 + 2x - 15$

33. $x^2 + x + 2$ **34.** $x^2 - 3x + 7$

35. $x^2 - 5x + 6$ **36.** $x^2 - x + 2$

37. $2x^2 + 5x - 3$ **38.** $3x^2 - 14x - 5$

39. $6x^2 - 19x + 10$ > Videos   **40.** $4x^2 + 5x + 6$

< Objectives 2–4 >

Factor each polynomial completely.

41. $x^2 + 8x + 15$ **42.** $x^2 - 11x + 24$

43. $s^2 + 13s + 30$ **44.** $b^2 + 14b + 33$

45. $x^2 + 3x + 11$

46. $x^2 - 8x + 8$

47. $x^2 - 6x - 40$

48. $x^2 - 11x + 10$

49. $p^2 - 10p - 24$

50. $x^2 - 11x - 60$

51. $x^2 + 5x - 66$

52. $a^2 + 2a - 80$

53. $c^2 + 19c + 60$

54. $t^2 - 4t - 60$

55. $n^2 + 5n - 50$

56. $x^2 - 16x + 63$

57. $m^2 - 6m + 1$

58. $w^2 + w - 5$

59. $x^2 + 7xy + 10y^2$

60. $x^2 - 8xy + 12y^2$

61. $a^2 - ab - 42b^2$

62. $m^2 - 8mn + 16n^2$

63. $x^2 - 13xy + 40y^2$

64. $r^2 - 9rs - 36s^2$

65. $6x^2 + 19x + 10$

66. $6x^2 - 7x - 3$

67. $15x^2 + x - 6$

68. $12w^2 + 19w + 4$

69. $6m^2 + 25m - 25$

70. $8x^2 - 6x - 9$

71. $9x^2 - 12x + 4$

72. $20x^2 - 23x + 6$

Answers

45. _____
46. _____
47. _____
48. _____
49. _____
50. _____
51. _____
52. _____
53. _____
54. _____
55. _____
56. _____
57. _____
58. _____
59. _____
60. _____
61. _____
62. _____
63. _____
64. _____
65. _____
66. _____
67. _____
68. _____
69. _____
70. _____
71. _____
72. _____

Answers

73. _____

74. _____

75. _____

76. _____

77. _____

78. _____

79. _____

80. _____

81. _____

82. _____

83. _____

84. _____

85. _____

86. _____

87. _____

88. _____

89. _____

90. _____

91. _____

92. _____

93. _____

94. _____

95. _____

96. _____

97. _____

98. _____

99. _____

100. _____

73. $12x^2 - 8x - 15$

74. $16a^2 + 40a + 25$

75. $3y^2 + 7y - 6$

76. $12x^2 + 11x - 15$

77. $8x^2 - 27x - 20$ ▸ Videos

78. $24v^2 + 5v - 36$

79. $4x^2 + 3x + 11$

80. $6x^2 - x + 1$

81. $2x^2 + 3xy + y^2$

82. $3x^2 - 5xy + 2y^2$

83. $5a^2 - 8ab - 4b^2$

84. $5x^2 + 7xy - 6y^2$

85. $9x^2 + 4xy - 5y^2$

86. $16x^2 + 32xy + 15y^2$

87. $6m^2 - 17mn + 12n^2$

88. $15x^2 - xy - 6y^2$

89. $36a^2 - 3ab - 5b^2$

90. $3q^2 - 17qr - 6r^2$

91. $x^2 + 4xy + 4y^2$

92. $25b^2 - 80bc + 64c^2$

93. $2x^2 + 18x - 1$

94. $5x^2 - 12x - 6$

95. $20x^2 - 20x - 15$

96. $24x^2 - 18x - 6$

97. $8m^2 + 12m + 4$

98. $14x^2 - 20x + 6$

99. $15r^2 - 21rs + 6s^2$

100. $10x^2 + 5xy - 30y^2$

101. $2x^3 - 2x^2 - 4x$

102. $2y^3 + y^2 - 3y$

103. $2y^4 + 5y^3 + 3y^2$ > Videos

104. $4z^3 - 18z^2 - 10z$

Basic Skills | **Challenge Yourself** | Calculator/Computer | Career Applications | Above and Beyond

---

Complete each statement with **never, sometimes,** *or* **always.**

105. A trinomial with integer coefficients is _____ factorable.

106. If a trinomial with all positive terms is factored, the signs between the terms in the binomial factors will _____ be positive.

107. The product of two binomials _____ results in a trinomial.

108. If the GCF for the terms in a polynomial is not 1, it should _____ be factored out first.

Basic Skills | Challenge Yourself | Calculator/Computer | **Career Applications** | Above and Beyond

---

109. **AGRICULTURAL TECHNOLOGY** The yield of a crop is given by the equation

$Y = -0.05x^2 + 1.5x + 140$

Rewrite this equation by factoring the right-hand side.
Hint: Begin by factoring out –0.05.

110. **ALLIED HEALTH** The number of people who are sick t days after the outbreak of a flu epidemic is given by the polynomial

$50 + 25t - 3t^2$

Write this polynomial in factored form.

111. **MECHANICAL ENGINEERING** The bending moment in an overhanging beam is described by the expression

$218(x^2 - 20x + 36)$

Factor the $x^2 - 20x + 36$ portion of the expression.

112. **MANUFACTURING TECHNOLOGY** The flow rate through a hydraulic hose can be found from the equation

$2Q^2 + Q - 21 = 0$

Factor the left side of this equation.

Basic Skills | Challenge Yourself | Calculator/Computer | Career Applications | **Above and Beyond**

Find a positive value for k so that each polynomial can be factored.

113. $x^2 + kx + 8$

114. $x^2 + kx + 9$

115. $x^2 - kx + 16$

116. $x^2 - kx + 17$

Answers

101. _____
102. _____
103. _____
104. _____
105. _____
106. _____
107. _____
108. _____
109. _____
110. _____
111. _____
112. _____
113. _____
114. _____
115. _____
116. _____

Factor each polynomial completely.

117. $10(x + y)^2 - 11(x + y) - 6$ > Videos

118. $8(a - b)^2 + 14(a - b) - 15$

119. $5(x - 1)^2 - 15(x - 1) - 350$ **120.** $3(x + 1)^2 - 6(x + 1) - 45$

121. $15 + 29x - 48x^2$ **122.** $12 + 4a - 21a^2$

123. $-6x^2 + 19x - 15$ **124.** $-3s^2 - 10s + 8$

Answers

1. $2x - 3$ **3.** $3a + 2$ **5.** $5x - 2$ **7.** $4a + b$ **9.** $4m - 3n$
11. True **13.** False **15.** True **17.** False **19.** True
21. $a = 1, b = 4, c = -9$ **23.** $a = 1, b = -3, c = 8$
25. $a = 3, b = 5, c = -8$ **27.** $a = 4, b = 8, c = 11$
29. $a = -3, b = 5, c = -10$ **31.** Factorable; 3, −2
33. Not factorable **35.** Factorable; −3, −2 **37.** Factorable; 6, −1
39. Factorable; −15, −4 **41.** $(x + 3)(x + 5)$
43. $(s + 10)(s + 3)$ **45.** Not factorable **47.** $(x - 10)(x + 4)$
49. $(p - 12)(p + 2)$ **51.** $(x + 11)(x - 6)$ **53.** $(c + 4)(c + 15)$
55. $(n + 10)(n - 5)$ **57.** Not factorable **59.** $(x + 2y)(x + 5y)$
61. $(a - 7b)(a + 6b)$ **63.** $(x - 5y)(x - 8y)$
65. $(3x + 2)(2x + 5)$ **67.** $(5x - 3)(3x + 2)$ **69.** $(6m - 5)(m + 5)$
71. $(3x - 2)(3x - 2)$ **73.** $(6x + 5)(2x - 3)$ **75.** $(3y - 2)(y + 3)$
77. $(8x + 5)(x - 4)$ **79.** Not factorable **81.** $(2x + y)(x + y)$
83. $(5a + 2b)(a - 2b)$ **85.** $(9x - 5y)(x + y)$ **87.** $(3m - 4n)(2m - 3n)$
89. $(12a - 5b)(3a + b)$ **91.** $(x + 2y)^2$ **93.** Not factorable
95. $5(2x - 3)(2x + 1)$ **97.** $4(2m + 1)(m + 1)$ **99.** $3(5r - 2s)(r - s)$
101. $2x(x - 2)(x + 1)$ **103.** $y^2(2y + 3)(y + 1)$ **105.** sometimes
107. sometimes **109.** $Y = -0.05(x + 40)(x - 70)$
111. $(x - 18)(x - 2)$ **113.** 6 or 9 **115.** 8 or 10 or 17
117. $(5x + 5y + 2)(2x + 2y - 3)$ **119.** $5(x - 11)(x + 6)$
121. $(1 + 3x)(15 - 16x)$ **123.** $-(2x - 3)(3x - 5)$

4.4

Difference of Squares and Perfect Square Trinomials

< 4.4 Objectives >

1 > Factor a binomial that is the difference of squares

2 > Factor a perfect square trinomial

In Section 3.4, we introduced some special products. Recall the following formula for the product of a sum and difference of two terms:

$$(a + b)(a - b) = a^2 - b^2$$

This also means that a binomial of the form $a^2 - b^2$, called a **difference of squares,** has as its factors $a + b$ and $a - b$.

To use this idea for factoring, we can write

$$a^2 - b^2 = (a + b)(a - b)$$

A **perfect square** term has a coefficient that is a square (1, 4, 9, 16, 25, 36, and so on), and any variables have exponents that are multiples of 2 (x^2, y^4, z^6, and so on).

| Example 1 | Identifying Perfect Square Terms |

< Objective 1 >

Decide whether each is a perfect square term. If it is, rewrite the expression as an expression squared.

(a) $36x$ **(b)** $24x^6$ **(c)** $9x^4$ **(d)** $64x^6$ **(e)** $16x^9$

Only parts **(c)** and **(d)** are perfect square terms.

$9x^4 = (3x^2)^2$
$64x^6 = (8x^3)^2$

> ### Check Yourself 1
>
> Decide whether each is a perfect square term. If it is, rewrite the expression as an expression squared.
>
> **(a)** $36x^{12}$ **(b)** $4x^6$ **(c)** $9x^7$ **(d)** $25x^8$ **(e)** $16x^{25}$

In Example 2, we factor the difference between perfect square terms.

| Example 2 | Factoring the Difference of Squares |

Factor $x^2 - 16$.

Think $x^2 - 4^2$.

Because $x^2 - 16$ is a difference of squares, we have
$$x^2 - 16 = (x + 4)(x - 4)$$

NOTE

You could also write
$(x - 4)(x + 4)$. The order
doesn't matter because
multiplication is commutative.

> ### Check Yourself 2
>
> Factor $m^2 - 49$.

Any time an expression is a difference of squares, it can be factored.

Example 3	Factoring the Difference of Squares

Factor $4a^2 - 9$.

Think $(2a)^2 - 3^2$.

So

$$4a^2 - 9 = (2a)^2 - (3)^2 = (2a + 3)(2a - 3)$$

 Check Yourself 3

Factor $9b^2 - 25$. $(3b)^2 - (5)^2 = (3b+5)(3b-5)$

The process for factoring a difference of squares does not change when more than one variable is involved.

Example 4	Factoring the Difference of Squares

NOTE

Think $(5a)^2 - (4b^2)^2$.

Factor $25a^2 - 16b^4$.

$$25a^2 - 16b^4 = (5a + 4b^2)(5a - 4b^2)$$

 Check Yourself 4

Factor $49c^4 - 9d^2$. $(7c^2 + 3d)(7c^2 - 3d)$

Now consider an example that combines common-term factoring with difference-of-squares factoring. Note that the common factor is always factored out as the *first step*.

Example 5	Removing the GCF

NOTE

Step 1
Factor out the GCF.
Step 2
Factor the remaining binomial.

Factor $32x^2y - 18y^3$.

Note that $2y$ is a common factor, so

$$32x^2y - 18y^3 = 2y(\underbrace{16x^2 - 9y^2})$$

Difference of squares

$$= 2y(4x + 3y)(4x - 3y)$$

 Check Yourself 5

Factor $50a^3 - 8ab^2$. $2a(25a^2 - 4b^2) = 2a(5a+2b)(5a-2b)$

 > CAUTION

Note that this is different from the sum of squares (such as $x^2 + y^2$), which never has real factors.

Recall the multiplication pattern

$$(a + b)^2 = a^2 + 2ab + b^2$$

For example,

$$(x + 2)^2 = x^2 + 4x + 4$$
$$(x + 5)^2 = x^2 + 10x + 25$$
$$(2x + 1)^2 = 4x^2 + 4x + 1$$

Recognizing this pattern can simplify the process of factoring perfect square trinomials.

Example 6   **Factoring a Perfect Square Trinomial**

< Objective 2 >

Factor the trinomial $4x^2 + 12xy + 9y^2$.
  Note that this is a perfect square trinomial in which

$a = 2x$ and $b = 3y$.

The factored form is

$4x^2 + 12xy + 9y^2 = (2x + 3y)^2$

 Check Yourself 6

Factor the trinomial $16u^2 + 24uv + 9v^2$.

Handwritten: $4u+3v$ $4u+3v$
$4(u^2 + 12uv + 12uv + 9v^2$
$(4u+3v)^2$ $4u^2 + 24uv + 9v^2$

Recognizing the same pattern can simplify the process of factoring perfect square trinomials in which the second term is negative.

 Example 7   **Factoring a Perfect Square Trinomial**

Factor the trinomial $25x^2 - 10xy + y^2$.
  This is also a perfect square trinomial, in which

$a = 5x$ and $b = -y$.

The factored form is

$25x^2 - 10xy + y^2 = [5x + (-y)]^2 = (5x - y)^2$

 Check Yourself 7   *Handwritten:* $(2u - 3v)^2$

Factor the trinomial $4u^2 - 12uv + 9v^2$.

Check Yourself ANSWERS

1. **(a)** $(6x^6)^2$; **(b)** $(2x^3)^2$; **(d)** $(5x^4)^2$ **2.** $(m + 7)(m - 7)$
3. $(3b + 5)(3b - 5)$ **4.** $(7c^2 + 3d)(7c^2 - 3d)$
5. $2a(5a + 2b)(5a - 2b)$ **6.** $(4u + 3v)^2$ **7.** $(2u - 3v)^2$

Reading Your Text

The following fill-in-the-blank exercises are designed to ensure that you understand some of the key vocabulary used in this section.

SECTION 4.4

(a) A perfect square term has a coefficient that is a perfect square and any variables have exponents that are _____ of 2.

(b) Any time an expression is the difference of squares, it can be _____.

(c) When factoring, the first step is to factor out the _____.

(d) Although the difference of squares can be factored, the _____ of squares cannot.

4.4 exercises

Name _____

Section _____ Date _____

Answers

1. _____   2. _____

3. _____   4. _____

5. _____   6. _____

7. _____   8. _____

9. _____   10. _____

11. _____

12. _____

13. _____

14. _____

15. _____

16. _____

17. _____

18. _____

19. _____

20. _____

21. _____

22. _____

23. _____

24. _____

25. _____

26. _____

< Objective 1 >

For each binomial, is the binomial a difference of squares?

1. $3x^2 + 2y^2$

2. $5x^2 - 7y^2$

3. $16a^2 - 25b^2$

4. $9n^2 - 16m^2$

5. $16r^2 + 4$

6. $p^2 - 45$

7. $16a^2 - 12b^3$

8. $9a^2b^2 - 16c^2d^2$

9. $a^2b^2 - 25$ > Videos

10. $4a^3 - b^3$

Factor each binomial.

11. $m^2 - n^2$

12. $r^2 - 9$

13. $x^2 - 49$

14. $c^2 - d^2$

15. $49 - y^2$

16. $81 - b^2$

17. $9b^2 - 16$

18. $36 - x^2$

19. $16w^2 - 49$

20. $4x^2 - 25$

21. $4s^2 - 9r^2$

22. $64y^2 - x^2$

23. $9w^2 - 49z^2$ > Videos

24. $25x^2 - 81y^2$

25. $16a^2 - 49b^2$

26. $64m^2 - 9n^2$

27. $x^2 + 4$

28. $y^2 + 16$

29. $x^4 - 36$

30. $y^6 - 49$

31. $x^2y^2 - 16$

32. $m^2n^2 - 64$

33. $25 - a^2b^2$

34. $49 - w^2z^2$

35. $16x^2 + 49$

36. $9x^2 + 25$

37. $81a^2 - 100b^6$

38. $64x^4 - 25y^4$

39. $18x^3 - 2xy^2$ > Videos

40. $50a^2b - 2b^3$

41. $12m^3n - 75mn^3$

42. $63p^4 - 7p^2q^2$

< Objective 2 >

Determine whether each trinomial is a perfect square. If it is, factor the trinomial.

43. $x^2 - 14x + 49$

44. $x^2 + 9x + 16$

45. $x^2 - 18x - 81$

46. $x^2 + 10x + 25$

47. $x^2 - 18x + 81$

48. $x^2 - 24x + 48$

Factor each trinomial.

49. $x^2 + 4x + 4$

50. $x^2 + 6x + 9$

51. $x^2 - 10x + 25$

52. $x^2 - 8x + 16$

Answers

27. _____

28. _____

29. _____

30. _____

31. _____

32. _____

33. _____

34. _____

35. _____

36. _____

37. _____

38. _____

39. _____

40. _____

41. _____

42. _____

43. _____

44. _____

45. _____

46. _____

47. _____

48. _____

49. _____

50. _____

51. _____

52. _____

Answers

53. _____

54. _____

55. _____

56. _____

57. _____

58. _____

59. _____

60. _____

61. _____

62. _____

63. _____

64. _____

65. _____

66. _____

Basic Skills | **Challenge Yourself** | Calculator/Computer | Career Applications | Above and Beyond
▲

*Determine whether each statement is **true** or **false**.*

53. A perfect square term has a coefficient that is a square and any variables have exponents that are factors of 2.

54. Any time an expression is the difference of squares, it can be factored.

55. Although the difference of squares can be factored, the sum of squares cannot.

56. When factoring, the middle factor is always factored out as the first step.

Factor each polynomial.

57. $4x^2 + 12xy + 9y^2$

58. $16x^2 + 40xy + 25y^2$

59. $9x^2 - 24xy + 16y^2$ > Videos

60. $9w^2 - 30wv + 25v^2$

61. $y^3 - 10y^2 + 25y$

62. $12b^3 - 12b^2 + 3b$

Basic Skills | Challenge Yourself | Calculator/Computer | **Career Applications** | Above and Beyond
▲

63. **MANUFACTURING TECHNOLOGY** The difference d in the calculated maximum deflection between two similar cantilevered beams is given by the formula

$$d = \left(\frac{w}{8EI}\right)(l_1^2 - l_2^2)(l_2^2 + l_2^2)$$

Rewrite the formula in its completely factored form.

64. **MANUFACTURING TECHNOLOGY** The work done W by a steam turbine is given by the formula

$$W = \frac{1}{2} m(v_1^2 - v_2^2)$$

Factor the right-hand side of this equation.

65. **ALLIED HEALTH** A toxic chemical is introduced into a protozoan culture. The number of deaths per hour is given by the polynomial $338 - 2t^2$, in which t is the number of hours after the chemical is introduced. Factor this expression.

66. **ALLIED HEALTH** Radiation therapy is one technique used to control cancer. After treatment, the total number of cancerous cells, in thousands, can be estimated by $144 - 4t^2$, in which t is the number of days of treatment. Factor this expression.

Basic Skills | Challenge Yourself | Calculator/Computer | Career Applications | **Above and Beyond**

▲

Factor each expression.

67. $x^2(x + y) - y^2(x + y)$ > Videos   **68.** $a^2(b - c) - 16b^2(b - c)$

69. $2m^2(m - 2n) - 18n^2(m - 2n)$ **70.** $3a^3(2a + b) - 27ab^2(2a + b)$

71. Find a value for k so that $kx^2 - 25$ has the factors $2x + 5$ and $2x - 5$.

72. Find a value for k so that $9m^2 - kn^2$ has the factors $3m + 7n$ and $3m - 7n$.

73. Find a value for k so that $2x^3 - kxy^2$ has the factors $2x$, $x - 3y$, and $x + 3y$.

74. Find a value for k so that $20a^3b - kab^3$ has the factors $5ab$, $2a - 3b$, and $2a + 3b$.

75. Complete the statement "To factor a number, you. . . ."

76. Complete the statement "To factor an algebraic expression into prime factors means. . . ."

Answers

67. _____

68. _____

69. _____

70. _____

71. _____

72. _____

73. _____

74. _____

75. _____

76. _____

Answers

1. No   **3.** Yes   **5.** No   **7.** No   **9.** Yes   **11.** $(m + n)(m - n)$
13. $(x + 7)(x - 7)$ **15.** $(7 + y)(7 - y)$ **17.** $(3b + 4)(3b - 4)$
19. $(4w + 7)(4w - 7)$ **21.** $(2s + 3r)(2s - 3r)$ **23.** $(3w + 7z)(3w - 7z)$
25. $(4a + 7b)(4a - 7b)$ **27.** Not factorable   **29.** $(x^2 + 6)(x^2 - 6)$
31. $(xy + 4)(xy - 4)$ **33.** $(5 + ab)(5 - ab)$ **35.** Not factorable
37. $(9a + 10b^3)(9a - 10b^3)$ **39.** $2x(3x + y)(3x - y)$
41. $3mn(2m + 5n)(2m - 5n)$ **43.** Yes; $(x - 7)^2$ **45.** No
47. Yes; $(x - 9)^2$ **49.** $(x + 2)^2$ **51.** $(x - 5)^2$
53. False   **55.** True   **57.** $(2x + 3y)^2$ **59.** $(3x - 4y)^2$

61. $y(y - 5)^2$ **63.** $d = \left(\dfrac{w}{8EI}\right)(l_1 + l_2)(l_1 - l_2)(l_1^2 + l_2^2)$

65. $2(13 - t)(13 + t)$ **67.** $(x + y)^2(x - y)$
69. $2(m - 2n)(m + 3n)(m - 3n)$ **71.** 4   **73.** 18
75. Above and Beyond

4.5

Strategies in Factoring

< 4.5 Objectives >

1 > Recognize factoring patterns

2 > Apply appropriate factoring strategies

In Sections 4.1 to 4.4 you have seen a variety of techniques for factoring polynomials. This section reviews those techniques and presents some guidelines for choosing an appropriate strategy or combination of strategies.

1. Always look for a greatest common factor. If you find a GCF (other than 1), factor out the GCF as your first step. If the leading coefficient is negative, factor out -1 along with the GCF.

   To factor $5x^2y - 10xy + 25xy^2$, the GCF is $5xy$, so

$$5x^2y - 10xy + 25xy^2 = 5xy(x - 2 + 5y)$$

2. Now look at the number of terms in the polynomial you are trying to factor.

   (a) If the polynomial is a *binomial,* consider the formula for the difference of two squares. Recall that a sum of squares does not factor over the real numbers.

   (i) To factor $x^2 - 49y^2$, recognize the difference of squares, so

$$x^2 - 49y^2 = (x + 7y)(x - 7y)$$

   (ii) The binomial

$$x^2 + 64$$

   cannot be further factored.

NOTE

You may prefer to use the *ac* method shown in Section 4.3.

   (b) If the polynomial is a *trinomial,* try to factor the trinomial as a product of binomials, using trial and error.

   To factor $2x^2 - x - 6$, a consideration of possible factors of the first and last terms of the trinomial will lead to

$$2x^2 - x - 6 = (2x + 3)(x - 2)$$

   (c) If the polynomial has *more than three terms,* try factoring by grouping.

   To factor $2x^2 - 3xy + 10x - 15y$, group the first two terms, and then the last two, and factor out common factors.

$$2x^2 - 3xy + 10x - 15y = x(2x - 3y) + 5(2x - 3y)$$

   Now factor out the common factor $(2x - 3y)$.

$$2x^2 - 3xy + 10x - 15y = (2x - 3y)(x + 5)$$

3. You should always factor the given polynomial completely. So after you apply one of the techniques given in part 2, another one may be necessary.

   (a) To factor

$$6x^3 + 22x^2 - 40x$$

   first factor out the common factor of $2x$. So

$$6x^3 + 22x^2 - 40x = 2x(3x^2 + 11x - 20)$$

   Now continue to factor the trinomial as before and

$$6x^3 + 22x^2 - 40x = 2x(3x - 4)(x + 5)$$

(b) To factor

$$x^3 - x^2y - 4x + 4y$$

first we proceed by grouping:

$$x^3 - x^2y - 4x + 4y = x^2(x - y) - 4(x - y)$$
$$= (x - y)(x^2 - 4)$$

Because $x^2 - 4$ is a difference of squares, we continue to factor and obtain

$$x^3 - x^2y - 4x + 4y = (x - y)(x + 2)(x - 2)$$

Example 1 | **Recognizing Factoring Patterns**

< Objective 1 >

State the appropriate first step for factoring each polynomial.

(a) $9x^2 - 18x - 72$

Find the GCF.

(b) $x^2 - 3x + 2xy - 6y$

Group the terms.

(c) $x^4 - 81y^4$

Factor the difference of squares.

(d) $3x^2 + 7x + 2$

Use the *ac* method (or trial and error).

Check Yourself 1

State the appropriate first step for factoring each polynomial.

(a) $5x^2 + 2x - 3$ **(b)** $a^4b^4 - 16$
(c) $3x^2 + 3x - 60$ **(d)** $2a^2 - 5a + 4ab - 10b$

Example 2 | **Factoring Polynomials**

< Objective 2 >

Completely factor each polynomial.

(a) $9x^2 - 18x - 72$

The GCF is 9.

$$9x^2 - 18x - 72 = 9(x^2 - 2x - 8)$$
$$= 9(x - 4)(x + 2)$$

(b) $x^2 - 3x + 2xy - 6y$

Grouping the terms, we have

$$x^2 - 3x + 2xy - 6y = (x^2 - 3x) + (2xy - 6y)$$
$$= x(x - 3) + 2y(x - 3)$$
$$= (x - 3)(x + 2y)$$

(c) $x^4 - 81y^4$

Factoring the difference of squares, we find

$$x^4 - 81y^4 = (x^2 + 9y^2)(x^2 - 9y^2)$$
$$= (x^2 + 9y^2)(x - 3y)(x + 3y)$$

(d) $3x^2 + 7x + 2$

Using the *ac* method, we find $m = 1$ and $n = 6$.

$$3x^2 + 7x + 2 = 3x^2 + x + 6x + 2$$
$$= (3x^2 + x) + (6x + 2)$$
$$= x(3x + 1) + 2(3x + 1)$$
$$= (3x + 1)(x + 2)$$

 Check Yourself 2

Completely factor each polynomial.

(a) $5x^2 + 2x - 3$ **(b)** $a^4b^4 - 16$

(c) $3x^2 + 3x - 60$ **(d)** $2a^2 - 5a + 4ab - 10b$

Start with step 1: Factor out the GCF. If the leading coefficient is negative, remember to factor out −1 along with the GCF.

▶ **Example** 3	**Factoring Out a Negative Coefficient**

RECALL

Include the GCF when writing the final factored form.

Factor $-6x^2y + 18xy + 60y$.

The GCF is $6y$. Because the leading coefficient is negative, we factor out $-6y$.

$$-6x^2y + 18xy + 60y = -6y(x^2 - 3x - 10) \quad \text{Factor out the negative GCF.}$$
$$= -6y(x - 5)(x + 2) \quad \text{Use either trial and error or the } ac \text{ method.}$$

 Check Yourself 3

Factor $-5xy^2 - 15xy + 90x$.

There are other patterns that sometimes occur when factoring. Several of these relate to the factoring of expressions that contain terms that are perfect cubes. The most common are the sum or difference of cubes, shown here.

Factoring the sum of perfect cubes

$$x^3 + y^3 = (x + y)(x^2 - xy + y^2)$$

Factoring the difference of perfect cubes

$$x^3 - y^3 = (x - y)(x^2 + xy + y^2)$$

▶ **Example** 4	**Factoring Expressions Involving Perfect Cube Terms**

Factor each expression.

(a) $8x^3 + 27y^3$

$$8x^3 + 27y^3 = (2x)^3 + (3y)^3 \quad \text{Substitute these values into the given patterns.}$$
$$= [(2x) + (3y)][(2x)^2 - (2x)(3y) + (3y)^2] \quad \text{Simplify.}$$
$$= (2x + 3y)(4x^2 - 6xy + 9y^2)$$

(b) $a^3b^3 - 64c^3$

$$a^3b^3 - 64c^3 = (ab)^3 - (4c)^3$$
$$= [(ab) - (4c)][(ab)^2 + (ab)(4c) + (4c)^2]$$
$$= (ab - 4c)(a^2b^2 + 4abc + 16c^2)$$

Check Yourself 4

Factor each expression.

(a) $a^3 + 64b^3c^3$ **(b)** $27x^3 - 8y^3$

Do not become frustrated if factoring attempts do not seem to produce results. You may have a polynomial that does not factor. A polynomial that does not factor over the integers is called a **prime polynomial.**

| Example 5 | Factoring Polynomials |

Factor $9m^2 - 8$.

We cannot find a GCF greater than 1, so we proceed to step 2. We have a binomial, but it does not fit any special pattern. $9m^2 = (3m)^2$ is a perfect square, but 8 is not, so this is not a difference of squares.

8 is a perfect cube, but $9m^2$ is not.

We conclude that the given binomial is a prime polynomial.

Check Yourself 5

Factor $9x^2 + 100$.

Check Yourself ANSWERS

1. **(a)** *ac* method (or trial and error); **(b)** factor the difference of squares;
(c) find the GCF; **(d)** group the terms
2. **(a)** $(5x - 3)(x + 1)$; **(b)** $(a^2b^2 + 4)(ab - 2)(ab + 2)$; **(c)** $3(x + 5)(x - 4)$;
(d) $(2a - 5)(a + 2b)$ 3. $-5x(y + 6)(y - 3)$
4. **(a)** $(a + 4bc)(a^2 - 4abc + 16b^2c^2)$; **(b)** $(3x - 2y)(9x^2 - 6xy + 4y^2)$
5. Not factorable

Reading Your Text

The following fill-in-the-blank exercises are designed to ensure that you understand some of the key vocabulary used in this section.

SECTION 4.5

(a) The first step in factoring requires that we find the _____ of all the terms.

(b) The sum of two perfect squares is _____ factorable.

(c) A binomial that is the sum of two perfect _____ is factorable.

(d) When we multiply two binomial factors, we get the original _____.

Name _____

Section _____ Date _____

Answers

1. _____
2. _____
3. _____
4. _____
5. _____
6. _____
7. _____
8. _____
9. _____
10. _____
11. _____
12. _____
13. _____
14. _____
15. _____
16. _____
17. _____
18. _____

< Objectives 1–2 >

Factor each polynomial completely. To begin, state which method should be applied as the first step, given the guidelines of this section. Then factor each polynomial completely.

1. $x^2 - 3x$

2. $4y^2 - 9$

3. $x^2 - 5x - 24$

4. $8x^3 + 10x$

5. $x(x - y) + 2(x - y)$

6. $5a^2 - 10a + 25$

7. $2x^2y - 6xy + 8y^2$ > Videos

8. $2p - 6q + pq - 3q^2$

9. $y^2 - 13y + 40$

10. $m^3 + 27m^2n$

11. $3b^2 + 17b - 28$ > Videos

12. $3x^2 + 6x - 5xy - 10y$ > Videos

13. $3x^2 - 14xy - 24y^2$

14. $16c^2 - 49d^2$

15. $2a^2 + 11a + 12$

16. $m^3n^3 - mn$

17. $125r^3 + r^2$

18. $(x - y)^2 - 16$

19. $3x^2 - 30x + 63$

20. $3a^2 - 108$

21. $40a^2 + 5$

22. $4p^2 - 8p - 60$

Basic Skills | **Challenge Yourself** ▲ | Calculator/Computer | Career Applications | Above and Beyond

---

23. $2w^2 - 14w - 36$

24. $xy^3 - 9xy$

25. $3a^2b - 48b^3$ ⊙ > Videos

26. $12b^3 - 86b^2 + 14b$

27. $x^4 - 3x^2 - 10$

28. $m^4 - 9n^4$

29. $8p^3 - q^3r^3$

30. $27x^3 + 125y^3$

Basic Skills | Challenge Yourself | Calculator/Computer | Career Applications | **Above and Beyond** ▲

---

31. $(x - 5)^2 - 169$ ⊙ > Videos

32. $(x - 7)^2 - 81$

33. $x^2 + 4xy + 4y^2 - 16$

34. $9x^2 + 12xy + 4y^2 - 25$

35. $6(x - 2)^2 + 7(x - 2) - 5$

36. $12(x + 1)^2 - 17(x + 1) + 6$

Answers

1. GCF, $x(x - 3)$ **3.** Trial and error, $(x - 8)(x + 3)$
5. GCF, $(x + 2)(x - y)$ **7.** GCF, $2y(x^2 - 3x + 4y)$
9. Trial and error, $(y - 5)(y - 8)$ **11.** Trial and error, $(b + 7)(3b - 4)$
13. Trial and error, $(3x + 4y)(x - 6y)$ **15.** Trial and error, $(2a + 3)(a + 4)$
17. GCF, $r^2(125r + 1)$ **19.** GCF, then trial and error, $3(x - 3)(x - 7)$
21. GCF, $5(8a^2 + 1)$ **23.** GCF, then trial and error, $2(w - 9)(w + 2)$
25. GCF, then difference of squares, $3b(a + 4b)(a - 4b)$
27. Trial and error, $(x^2 - 5)(x^2 + 2)$ **29.** $(2p - qr)(4p^2 - 2pqr + q^2r^2)$
31. $(x + 8)(x - 18)$ **33.** $(x + 2y + 4)(x + 2y - 4)$
35. $(2x - 5)(3x - 1)$

Answers

19. _____

20. _____

21. _____

22. _____

23. _____

24. _____

25. _____

26. _____

27. _____

28. _____

29. _____

30. _____

31. _____

32. _____

33. _____

34. _____

35. _____

36. _____

4.6 Solving Quadratic Equations by Factoring

< 4.6 Objectives >

1 > Solve quadratic equations by factoring

2 > Solve applications involving quadratic equations

The factoring techniques you have learned provide us with tools for solving equations that can be written in the form

$$ax^2 + bx + c = 0 \qquad a \neq 0$$

This is a quadratic equation in one variable, here x. You can recognize such a quadratic equation by the fact that the highest power of the variable x is the second power.

in which a, b, and c are constants.

An equation written in the form $ax^2 + bx + c = 0$ is called a **quadratic equation in standard form.** Using factoring to solve quadratic equations requires the **zero-product principle,** which says that if the product of two factors is 0, then one or both of the factors must be equal to 0. In symbols:

Definition

Zero-Product Principle

If $a \cdot b = 0$, then $a = 0$ or $b = 0$ or $a = b = 0$.

We can now apply this principle to solve quadratic equations.

Example 1   Solving Equations by Factoring

< Objective 1 >

NOTE

To use the zero-product principle, 0 must be on one side of the equation.

Solve.

$$x^2 - 3x - 18 = 0$$

Factoring on the left, we have

$$(x - 6)(x + 3) = 0$$

By the zero-product principle, we know that one or both of the factors must be zero. We can then write

$$x - 6 = 0 \qquad \text{or} \qquad x + 3 = 0$$

Solving each equation gives

$$x = 6 \qquad \text{or} \qquad x = -3$$

The two solutions are 6 and -3.

Quadratic equations can be checked in the same way as linear equations were checked: by substitution. For instance, if $x = 6$, we have

$$6^2 - 3 \cdot 6 - 18 \overset{?}{=} 0$$
$$36 - 18 - 18 \overset{?}{=} 0$$
$$0 = 0$$

which is a true statement. We leave it to you to check the solution -3.

Check Yourself 1

Solve $x^2 - 9x + 20 = 0$.

Other factoring techniques are also used in solving quadratic equations. Example 2 illustrates this.

 Example 2 | Solving Equations by Factoring

 > CAUTION

A *common mistake* is to forget the statement $x = 0$ when you are solving equations of this type. Be sure to include both answers.

(a) Solve $x^2 - 5x = 0$.

Again, factor the left side of the equation and apply the zero-product principle.

$$x(x - 5) = 0$$

Now

$$x = 0 \qquad \text{or} \qquad x - 5 = 0$$
$$x = 5$$

The two solutions are 0 and 5.

(b) Solve $x^2 - 9 = 0$.

Factoring yields

$$(x + 3)(x - 3) = 0$$
$$x + 3 = 0 \qquad \text{or} \qquad x - 3 = 0$$
$$x = -3 \qquad\qquad x = 3$$

The solutions may be written as $x = \pm 3$.

NOTE

The symbol $\pm$ is read "plus or minus."

> ✓ **Check Yourself** 2
>
> Solve by factoring.
>
> **(a)** $x^2 + 8x = 0$ **(b)** $x^2 - 16 = 0$

Example 3 illustrates a crucial point. Our solution technique depends on the zero-product principle, which means that the product of factors *must be equal to* 0. The importance of this is shown now.

 Example 3 | Solving Equations by Factoring

> CAUTION

Consider the equation

$x(2x - 1) = 3$

Students are sometimes tempted to write

$x = 3$ or $2x - 1 = 3$

This is *not correct*. Instead, subtract 3 from both sides of the equation *as the first step* to write

$x(2x - 1) - 3 = 0$

Then proceed to write the equation in standard form. Only then can you factor and proceed as before.

Solve $2x^2 - x = 3$.

The first step in the solution is to write the equation in standard form (that is, write it so that one side of the equation is 0). So start by adding -3 to both sides of the equation.

Then,

$$2x^2 - x - 3 = 0 \qquad \text{Make sure all terms are on one side of the equation. The other side will be 0.}$$

You can now factor and solve by using the zero-product principle.

$$(2x - 3)(x + 1) = 0$$
$$2x - 3 = 0 \qquad \text{or} \qquad x + 1 = 0$$
$$2x = 3 \qquad\qquad\qquad x = -1$$
$$x = \frac{3}{2}$$

The solutions are $\dfrac{3}{2}$ and -1.

Check Yourself 3

Solve $3x^2 = 5x + 2$.

In all the previous examples, the quadratic equations had two distinct real-number solutions. That may not always be the case, as we shall see.

Example 4   **Solving Equations by Factoring**

Solve $x^2 - 6x + 9 = 0$.

Factoring, we have

$(x - 3)(x - 3) = 0$

and

$x - 3 = 0$ or $x - 3 = 0$

$x = 3$ $x = 3$

The solution is 3.

A quadratic (or second-degree) equation always has *two* solutions. When an equation such as this one has two solutions that are the same number, we call 3 the **repeated** (or **double**) **solution** of the equation.

Although a quadratic equation always has two solutions, they may not always be real numbers. You will learn more about this in a later course.

Check Yourself 4

Solve $x^2 + 6x + 9 = 0$.

Always examine the quadratic member of an equation for common factors. It will make your work much easier, as Example 5 illustrates.

Example 5   **Solving Equations by Factoring**

Solve $3x^2 - 3x - 60 = 0$.

Note the common factor 3 in the quadratic expression. Factoring out the 3 gives

$3(x^2 - x - 20) = 0$

Now, because the common factor has no variables, we can divide both sides of the equation by 3.

$$\frac{3(x^2 - x - 20)}{3} = \frac{0}{3}$$

or

$x^2 - x - 20 = 0$

We can now factor and solve as before.

$(x - 5)(x + 4) = 0$

$x - 5 = 0$ or $x + 4 = 0$

$x = 5$ $x = -4$

NOTE

The advantage of dividing both sides of the equation by 3 is that the coefficients in the quadratic expression become smaller and are easier to factor.

Check Yourself 5

Solve $2x^2 - 10x - 48 = 0$.

Many applications can be solved with quadratic equations.

 Example 6   **Solving an Application**

< Objective 2 >

The Microhard Corporation has found that the equation

$$P = x^2 - 7x - 94$$

describes the profit P, in thousands of dollars, for every x hundred computers sold. How many computers were sold if the profit was \$50,000?

    If the profit was \$50,000, then $P = 50$. We now set up and solve the equation.

NOTE

P is expressed in thousands so the value 50 is substituted for P, not 50,000.

$$50 = x^2 - 7x - 94$$
$$0 = x^2 - 7x - 144$$
$$0 = (x + 9)(x - 16)$$
$$x = -9 \quad \text{or} \quad x = 16$$

They cannot sell a negative number of computers, so $x = 16$. They sold 1,600 computers.

 Check Yourself 6

The Pureed Babyfood Corporation has found that the equation

$$P = x^2 - 6x - 7$$

describes the profit P, in hundreds of dollars, for every x thousand jars sold. How many jars were sold if the profit was \$2,000?

 Check Yourself ANSWERS

1. $4, 5$ **2.** (a) $0, -8$; (b) $4, -4$ **3.** $-\dfrac{1}{3}, 2$ **4.** -3 **5.** $-3, 8$
6. $9,000$ jars

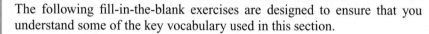 **Reading Your Text**

The following fill-in-the-blank exercises are designed to ensure that you understand some of the key vocabulary used in this section.

SECTION 4.6

(a) An equation written in the form $ax^2 + bx + c = 0$ is called a _____ equation in standard form.

(b) Using factoring to solve quadratic equations requires the _____ principle.

(c) To use the zero-product principle, it is important that the product of factors be equal to _____.

(d) When an equation has two solutions that are the same number, we call it a _____ solution.

4.6 exercises

Name _____

Section _____ Date _____

Answers

1. _____	2. _____
3. _____	4. _____
5. _____	6. _____
7. _____	8. _____
9. _____	10. _____
11. _____	12. _____
13. _____	14. _____
15. _____	16. _____
17. _____	18. _____
19. _____	20. _____
21. _____	22. _____
23. _____	24. _____
25. _____	26. _____
27. _____	28. _____
29. _____	30. _____
31. _____	32. _____

Basic Skills | Challenge Yourself | Calculator/Computer | Career Applications | Above and Beyond

< Objective 1 >

Solve each quadratic equation.

1. $(x - 3)(x - 4) = 0$

2. $(x - 7)(x + 1) = 0$

3. $(3x + 1)(x - 6) = 0$

4. $(5x - 4)(x - 6) = 0$

5. $x^2 - 2x - 3 = 0$

6. $x^2 + 5x + 4 = 0$

7. $x^2 - 7x + 6 = 0$

8. $x^2 + 3x - 10 = 0$

9. $x^2 + 8x + 15 = 0$

10. $x^2 - 3x - 18 = 0$

11. $x^2 + 4x - 21 = 0$

12. $x^2 - 12x + 32 = 0$

13. $x^2 - 4x = 12$ > Videos

14. $x^2 + 8x = -15$

15. $x^2 + 5x = 14$

16. $x^2 = 11x - 24$

17. $2x^2 + 5x - 3 = 0$

18. $3x^2 + 7x + 2 = 0$

19. $4x^2 - 24x + 35 = 0$

20. $6x^2 + 11x - 10 = 0$

21. $4x^2 + 11x = -6$

22. $5x^2 + 2x = 3$

23. $5x^2 + 13x = 6$

24. $4x^2 = 13x + 12$ > Videos

25. $x^2 - 2x = 0$

26. $x^2 + 5x = 0$

27. $x^2 = -8x$

28. $x^2 = 7x$

29. $5x^2 - 15x = 0$ > Videos

30. $4x^2 + 20x = 0$

31. $x^2 - 25 = 0$

32. $x^2 = 49$

33. $x^2 = 81$　　　　　　**34.** $x^2 = 64$

35. $2x^2 - 18 = 0$　　　　　**36.** $3x^2 - 75 = 0$

37. $3x^2 + 24x + 45 = 0$　　　**38.** $4x^2 - 4x = 24$

39. $2x(3x + 14) = 10$ > Videos　　**40.** $3x(5x + 9) = 6$

41. $(x + 3)(x - 2) = 14$　　　**42.** $(x - 5)(x + 2) = 18$

< Objective 2 >

Solve each problem.

43. NUMBER PROBLEM The product of two consecutive integers is 132. Find the two integers.

44. NUMBER PROBLEM The product of two consecutive positive even integers is 120. Find the two integers.  > Videos

45. NUMBER PROBLEM The sum of an integer and its square is 72. What is the integer?

46. NUMBER PROBLEM The square of an integer is 56 more than the integer. Find the integer.

47. GEOMETRY If the sides of a square are increased by 3 in., the area is increased by 39 in.². What were the dimensions of the original square?

48. GEOMETRY If the sides of a square are decreased by 2 cm, the area is decreased by 36 cm². What were the dimensions of the original square?

49. BUSINESS AND FINANCE The profit on a small appliance is given by $P = x^2 - 3x - 60$, in which x is the number of appliances sold per day. How many appliances were sold on a day when there was a $20 loss?

50. BUSINESS AND FINANCE The relationship between the number of calculators x that a company can sell per month and the price of each calculator p is given by $x = 1{,}700 - 100p$. Find the price at which a calculator should be sold to produce a monthly revenue of $7,000. (*Hint:* Revenue $= xp$.)

Basic Skills | Challenge Yourself | Calculator/Computer | **Career Applications** | Above and Beyond
▲

51. ALLIED HEALTH The concentration, C, in micrograms per milliliter (mcg/mL), of Tobrex, an antibiotic prescribed for burn patients, is given by the equation $C = 12 + t - t^2$, where t is the number of hours since the drug was administered via intravenous injection. Find the value of t when the concentration is $C = 0$.

Answers

33. _____

34. _____

35. _____

36. _____

37. _____

38. _____

39. _____

40. _____

41. _____

42. _____

43. _____

44. _____

45. _____

46. _____

47. _____

48. _____

49. _____

50. _____

51. _____

Answers

52. _____

53. _____

54. _____

55. _____

56. _____

52. ALLIED HEALTH The number of people who are sick t days after the outbreak of a flu epidemic is given by the equation $P = 50 + 25t - 3t^2$. Write the polynomial in factored form. Find the value of t when the number of people is $P = 0$.

53. MANUFACTURING TECHNOLOGY The maximum stress for a given allowable strain (deformation) for a certain material is given by the polynomial

$$S = 85.8x - 0.6x^2 - 1{,}537.2$$

in which x is the allowable strain in micrometers. Find the allowable strain in micrometers when the stress is $S = 0$.

   Hint: Rearrange the polynomial and factor out a common factor of -0.6 first.

54. AGRICULTURAL TECHNOLOGY The height (in feet) of a drop of water above an irrigation nozzle in terms of the time (in seconds) since the drop left the nozzle is given by the formula

$$h = v_0 t - 16t^2$$

in which v_0 is the initial velocity of the water when it comes out of the nozzle.

   If the initial velocity of a drop of water is 80 ft/s, how many seconds need to pass before the drop reaches a height of 75 ft?

Basic Skills | Challenge Yourself | Calculator/Computer | Career Applications | **Above and Beyond**
▲

---

55. Write a short comparison that explains the difference between $ax^2 + bx + c$ and $ax^2 + bx + c = 0$.

56. When solving quadratic equations, some people try to solve an equation in the manner shown below, but this does not work! Write a paragraph to explain what is wrong with this approach.

$$2x^2 + 7x + 3 = 52$$
$$(2x + 1)(x + 3) = 52$$
$$2x + 1 = 52 \quad \text{or} \quad x + 3 = 52$$
$$x = \frac{51}{2} \quad \text{or} \quad x = 49$$

Answers

1. $3, 4$ **3.** $-\dfrac{1}{3}, 6$ **5.** $-1, 3$ **7.** $1, 6$ **9.** $-3, -5$ **11.** $-7, 3$

13. $-2, 6$ **15.** $-7, 2$ **17.** $-3, \dfrac{1}{2}$ **19.** $\dfrac{5}{2}, \dfrac{7}{2}$ **21.** $-\dfrac{3}{4}, -2$

23. $-3, \dfrac{2}{5}$ **25.** $0, 2$ **27.** $0, -8$ **29.** $0, 3$ **31.** $-5, 5$

33. $-9, 9$ **35.** $-3, 3$ **37.** $-5, -3$ **39.** $-5, \dfrac{1}{3}$ **41.** $4, -5$

43. $11, 12$ or $-12, -11$ **45.** -9 or 8 **47.** 5 in. by 5 in.   **49.** 8

51. $t = 4$ hours   **53.** $x = 21$ or $x = 122$ micrometers

55. Above and Beyond

Definition/Procedure	Example	Reference
An Introduction to Factoring		Section 4.1
Common Monomial Factor		
A single term that is a factor of every term of the polynomial. The greatest common factor (GCF) of a polynomial is the factor that is a product of (a) the largest common numerical factor and (b) each variable with the smallest exponent in any term.	$4x^2$ is the greatest common monomial factor of $8x^4 - 12x^3 + 16x^2$.	*p. 260*
Factoring a Monomial from a Polynomial		
1. Determine the GCF for all terms. 2. Use the GCF to factor each term and then apply the distributive property in the form $$ab + ac = a(b + c)$$ $\uparrow$ The greatest common factor 3. Mentally check by multiplication.	$8x^4 - 12x^3 + 16x^2$ $= 4x^2(2x^2 - 3x + 4)$	*p. 261*
Factoring by Grouping		
When there are four terms of a polynomial, factor the first pair and factor the last pair. If these two pairs have a common binomial factor, factor that out. The result will be the product of two binomials.	$4x^2 - 6x + 10x - 15$ $= 2x(2x - 3) + 5(2x - 3)$ $= (2x - 3)(2x + 5)$	*p. 263*
Factoring Trinomials		Sections 4.2– 4.3
Trial and Error		
To factor a trinomial, find the appropriate sign pattern and then find integer values that yield the appropriate coefficients for the trinomial.	$x^2 - 5x - 24$ $= (x -\ )(x +\ )$ $= (x - 8)(x + 3)$	*p. 271*
Using the ac Method to Factor		
To factor a trinomial, first use the *ac* test to determine factorability. If the trinomial is factorable, the *ac* test will yield two terms (which have as their sum the middle term) that allow the factoring to be completed by using the grouping method.	$x^2 + 3x - 28$ $ac = -28; b = 3$ $mn = -28; m + n = 3$ $m = 7, n = -4$ $x^2 + 7x - 4x - 28$ $= x(x + 7) - 4(x + 7)$ $= (x - 4)(x + 7)$	*p. 287*

Continued

Definition/Procedure	Example	Reference
Difference of Squares and Perfect Square Trinomials		Section 4.4
Factoring a Difference of Squares		
Use the formula $$a^2 - b^2 = (a + b)(a - b)$$	To factor: $16x^2 - 25y^2$: Think: $(4x)^2 - (5y)^2$ so $16x^2 - 25y^2$ $= (4x + 5y)(4x - 5y)$	p. 299
Factoring a Perfect Square Trinomial		
Use the formula $$a^2 + 2ab + b^2 = (a + b)^2$$	$4x^2 + 12xy + 9y^2$ $= (2x)^2 + 2(2x)(3y) + (3y)^2$ $= (2x + 3y)^2$	p. 301
Strategies in Factoring		Section 4.5

When factoring a polynomial,
 	
 1. Factor out the GCF. If the leading coefficient is negative, factor out -1 along with the GCF.
 2. Consider the number of terms.
    a. If it is a binomial, look for a difference of squares.
    b. If it is trinomial, use the *ac* method or trial and error.
    c. If there are four or more terms, try grouping terms.
 3. Be certain that the polynomial is completely factored.
	Given $12x^3 - 86x^2 + 14x$,
	 factor out $2x$.
	 $2x(6x^2 - 43x + 7)$
	 $= 2x(6x - 1)(x - 7)$
		p. 306

Solving Quadratic Equations by Factoring		Section 4.6
1. Add or subtract the necessary terms on both sides of the equation so that the equation is in standard form (set equal to 0). 2. Factor the quadratic expression. 3. Set each factor equal to 0. 4. Solve the resulting equations to find the solutions. 5. Check each solution by substituting in the original equation.	To solve $x^2 + 7x = 30$ $x^2 + 7x - 30 = 0$ $(x + 10)(x - 3) = 0$ $x + 10 = 0$ or $x - 3 = 0$ $x = -10$ and $x = 3$ are solutions.	p. 312

This summary exercise set is provided to give you practice with each of the objectives of this chapter. Each exercise is keyed to the appropriate chapter section. When you are finished, you can check your answers to the odd-numbered exercises against those presented in the back of the text. If you have difficulty with any of these questions, go back and reread the examples from that section. Your instructor will give you guidelines on how best to use these exercises in your instructional setting.

4.1 Factor each polynomial.

1. $18a + 24$

2. $9m^2 - 21m$

3. $24s^2t - 16s^2$

4. $18a^2b + 36ab^2$

5. $35s^3 - 28s^2$

6. $3x^3 - 6x^2 + 15x$

7. $18m^2n^2 - 27m^2n + 18m^2n^3$

8. $121x^8y^3 + 77x^6y^3$

9. $8a^2b + 24ab - 16ab^2$

10. $3x^2y - 6xy^3 + 9x^3y - 12xy^2$

11. $x(2x - y) + y(2x - y)$

12. $5(w - 3z) - w(w - 3z)$

4.2 Factor each trinomial completely.

13. $x^2 + 9x + 20$

14. $x^2 - 10x + 24$

15. $a^2 - a - 12$

16. $w^2 - 13w + 40$

17. $x^2 + 12x + 36$

18. $r^2 - 9r - 36$

19. $b^2 - 4bc - 21c^2$

20. $m^2n + 4mn - 32n$

21. $m^3 + 2m^2 - 35m$

22. $2x^2 - 2x - 40$

23. $3y^3 - 48y^2 + 189y$

24. $3b^3 - 15b^2 - 42b$

4.3 Factor each trinomial completely.

25. $3x^2 + 8x + 5$

26. $5w^2 + 13w - 6$

27. $2b^2 - 9b + 9$

28. $8x^2 + 2x - 3$

29. $10x^2 - 11x + 3$

30. $4a^2 + 7a - 15$

31. $9y^2 - 3yz - 20z^2$

32. $8x^2 + 14xy - 15y^2$

33. $8x^3 - 36x^2 - 20x$

34. $9x^2 - 15x - 6$

35. $6x^3 - 3x^2 - 9x$

36. $5w^2 - 25wz + 30z^2$

4.4 *Factor each polynomial completely.*

37. $p^2 - 49$

38. $25a^2 - 16$

39. $m^2 - 9n^2$

40. $16r^2 - 49s^2$

41. $25 - z^2$

42. $a^4 - 16b^2$

43. $25a^2 - 36b^2$

44. $x^6 - 4y^2$

45. $3w^3 - 12wz^2$

46. $9a^4 - 49b^2$

47. $2m^2 - 72n^4$

48. $3w^3z - 12wz^3$

49. $x^2 + 8x + 16$

50. $x^2 - 18x + 81$

51. $4x^2 + 12x + 9$

52. $9x^2 - 12x + 4$

53. $16x^3 + 40x^2 + 25x$

54. $4x^3 - 4x^2 + x$

4.5

55. $x^2 - 4x + 5x - 20$

56. $x^2 + 7x - 2x - 14$

57. $6x^2 + 4x - 15x - 10$

58. $12x^2 - 9x - 28x + 21$

59. $6x^3 + 9x^2 - 4x^2 - 6x$

60. $3x^4 + 6x^3 + 5x^3 + 10x^2$

4.6 *Solve each quadratic equation.*

61. $(x - 1)(2x + 3) = 0$

62. $x^2 - 5x + 6 = 0$

63. $x^2 - 10x = 0$

64. $x^2 = 144$

65. $x^2 - 2x = 15$

66. $3x^2 - 5x - 2 = 0$

67. $4x^2 - 13x + 10 = 0$

68. $2x^2 - 3x = 5$

69. $3x^2 - 9x = 0$

70. $x^2 - 25 = 0$

71. $2x^2 - 32 = 0$

72. $2x^2 - x - 3 = 0$

The purpose of this self-test is to help you assess your progress so that you can find concepts that you need to review before the next exam. Allow yourself about an hour to take this test. At the end of that hour, check your answers against those given in the back of this text. If you miss any, go back to the appropriate section to reread the examples until you have mastered that particular concept.

Name _____

Section _____ Date _____

Answers

Factor each polynomial.

1. $12b + 18$

2. $9p^3 - 12p^2$

3. $5x^2 - 10x + 20$

4. $6a^2b - 18ab + 12ab^2$

5. $a^2 - 10a + 25$

6. $64m^2 - n^2$

7. $49x^2 - 16y^2$

8. $32a^2b - 50b^3$

9. $a^2 - 5a - 14$

10. $b^2 + 8b + 15$

11. $x^2 - 11x + 28$

12. $y^2 + 12yz + 20z^2$

13. $x^2 + 2x - 5x - 10$

14. $6x^2 + 2x - 9x - 3$

15. $2x^2 + 15x - 8$

16. $3w^2 + 10w + 7$

17. $8x^2 - 2xy - 3y^2$

18. $6x^3 + 3x^2 - 30x$

Solve each equation.

19. $x^2 - 8x + 15 = 0$

20. $x^2 - 3x = 4$

21. $3x^2 + x - 2 = 0$

22. $4x^2 - 12x = 0$

23. $x(x - 4) = 0$

24. $(x - 3)(x - 2) = 30$

25. $x^2 - 14x = -49$

26. $4x^2 + 25 = 20x$

1. _____
2. _____
3. _____
4. _____
5. _____
6. _____
7. _____
8. _____
9. _____
10. _____
11. _____
12. _____
13. _____
14. _____
15. _____
16. _____
17. _____
18. _____
19. _____ 20. _____
21. _____ 22. _____
23. _____ 24. _____
25. _____ 26. _____

Answers

27. _____

28. _____

27. **GEOMETRY** The length of a rectangle is 4 cm less than twice its width. If the area of the rectangle is 240 cm^2, what is the length of the rectangle?

28. **SCIENCE AND MEDICINE** If a ball is thrown upward from the roof of an 18-meter tall building with an initial velocity of 20 m/s, its height after t seconds is given by

$$h = -5t^2 + 20t + 18$$

How long does it take for the ball to reach a height of 38 m?

The Streeter/Hutchison Series in Mathematics Beginning Algebra

chapter 4 > Make the Connection

Activity 4 ::
ISBNs and the Check Digit

Each activity in this text is designed to either enhance your understanding of the topics of the preceding chapter, or to provide you with a mathematical extension of those topics, or both. The activities can be undertaken by one student, but they are better suited for a small-group project. Occasionally it is only through discussion that different facets of the activity become apparent.

If you look at the back of your textbook, you should see a long number and a bar code. The number is called the International Standard Book Number, or ISBN.

The ISBN system was first developed in 1966 by Gordon Foster at Trinity College in Dublin, Ireland. When first developed, ISBNs were 9 digits long, but by 1970, an international agreement extended them to 10 digits.

In 2007, 13 digits became the standard for ISBN numbers. This is the number on the back of your text. Each ISBN has five blocks of numbers. A common form is XXX-X-XX-XXXXXX-X, though it can vary.

- The first block or set of digits is either 978 or 979. This set was added in 2007 to increase the number of ISBNs available for new books.
- The second set of digits represents the language of the book. Zero represents English.
- The third set represents the publisher. This block is usually two or three digits long.
- The fourth set is the book code and is assigned by the publisher. This block is usually five or six digits long.
- The fifth and final block is a one-digit *check digit*.

Consider the ISBN assigned to this text: 978-0-07-338418-4. The check digit in this ISBN is the final digit, 4. It ensures that the book has a valid ISBN.

To use the check digit, we use the algorithm that follows.

Step by Step: Validating an ISBN

Step 1 Identify the first 12 digits of the ISBN (omit the check digit).
Step 2 Multiply the first digit by 1, the second by 3, the third by 1, the fourth by 3, and continue alternating until each of the first 12 digits has been multiplied.
Step 3 Add all 12 of these products together.
Step 4 Take only the units digit of this sum and subtract it from 10.
Step 5 If the difference found in step 4 is the same as the check digit, then the ISBN is valid.

We can use the ISBN from this text, 978-0-07-338418, to see how this works.

To do so, we multiply the first digit by 1, the second by 3, the third by 1, the fourth by 3, again, and so on. Then we add these products together. We call this a *weighted sum*.

$$9 \cdot 1 + 7 \cdot 3 + 8 \cdot 1 + 0 \cdot 3 + 0 \cdot 1 + 7 \cdot 3 + 3 \cdot 1 + 3 \cdot 3 + 8 \cdot 1 + 4 \cdot 3 + 1 \cdot 1 + 8 \cdot 3$$
$$= 9 + 21 + 8 + 0 + 0 + 21 + 3 + 9 + 8 + 12 + 1 + 24$$
$$= 116$$

The units digit is 6. We subtract this from 10.

$$10 - 6 = 4$$

325

The last digit in the ISBN 978-0-07-338418-4 is 4. This matches the difference above and so this text has a valid ISBN number.

Determine whether each set of numbers represents a valid ISBN.

1. 978-0-07-038023-6

2. 978-0-07-327374-7

3. 978-0-553-34948-1

4. 978-0-07-000317-3

5. 978-0-14-200066-3

For each valid ISBN, go online and find the book associated with that ISBN.

The following exercises are presented to help you review concepts from earlier chapters. This is meant as a review and not as a comprehensive exam. The answers are presented in the back of the text. Section references accompany the answers. If you have difficulty with any of these exercises, be certain to at least read through the summary related to those sections.

Name _____

Section _____ Date _____

Answers

Perform the indicated operations.

1. $7 - (-10)$

2. $(-34) \div (17)$

Perform each of the indicated operations.

3. $(7x^2 + 5x - 4) + (2x^2 - 6x - 1)$

4. $(3a^2 - 2a) - (7a^2 + 5)$

5. Subtract $4b^2 - 3b$ from the sum of $6b^2 + 5b$ and $4b^2 - 3$.

6. $3rs(5r^2s - 4rs + 6rs^2)$

7. $(2a - b)(3a^2 - ab + b^2)$

8. $\dfrac{7xy^3 - 21x^2y^2 + 14x^3y}{-7xy}$

9. $\dfrac{3a^2 - 10a - 8}{a - 4}$

10. $\dfrac{2x^3 - 8x + 5}{2x + 4}$

Solve the equation for x.

11. $2 - 4(3x + 1) = 8 - 7x$

Solve the inequality.

12. $4(x - 7) \le -(x - 5)$

Solve the equation for the indicated variable.

13. $S = \dfrac{n}{2}(a + t)$ for t

1. _____

2. _____

3. _____

4. _____

5. _____

6. _____

7. _____

8. _____

9. _____

10. _____

11. _____

12. _____

13. _____

Answers

14.
_____

15.
_____

16.
_____

17.
_____

18.
_____

19.
_____

20.
_____

21.
_____

22.
_____

23.
_____

24.
_____

25.
_____

26.
_____

27.
_____

28.
_____

29.
_____

30.
_____

Simplify each expression.

14. $x^6 x^{11}$

15. $(3x^2 y^3)(2x^3 y^4)$

16. $(3x^2 y^3)^2(-4x^3 y^2)^0$

17. $\dfrac{16x^2 y^5}{4xy^3}$

18. $(3x^2)^3(2x)^2$

Factor each polynomial completely.

19. $36w^5 - 48w^4$

20. $5x^2 y - 15xy + 10xy^2$

21. $25x^2 + 30xy + 9y^2$

22. $4p^3 - 144pq^2$

23. $a^2 + 4a + 3$

24. $2w^3 - 4w^2 - 24w$

25. $3x^2 + 11xy + 6y^2$

Solve each equation.

26. $a^2 - 7a + 12 = 0$

27. $3w^2 - 48 = 0$

28. $15x^2 + 5x = 10$

Solve each problem.

29. NUMBER PROBLEM Twice the square of a positive integer is 12 more than 10 times that integer. What is the integer?

30. GEOMETRY The length of a rectangle is 1 in. more than 4 times its width. If the area of the rectangle is 105 in.^{2}, find the dimensions of the rectangle.

CHAPTER

5

chapter 5 > Make the Connection

INTRODUCTION

The House of Representatives is made up of officials elected from congressional districts in each state. The number of representatives a state sends to the House depends on the state's population. The total number of representatives grew from 106 in 1790 to 435, the maximum number established in 1930. (At the time of this writing, Congress is discussing adding two more representatives, one of whom will represent Washington, D.C., residents.) These 435 representatives are apportioned to the 50 states on the basis of population. This apportionment is revised after every decennial (10-year) census.

If a particular state has population A and its number of representatives is equal to a, then $\dfrac{A}{a}$ represents the ratio of people in the state to their total number of representatives in the U.S. House.

A recent comparison of these ratios for states finds Pennsylvania with 652,959 people per representative and Arizona with 717,979—the national average was 687,080 people per representative. The difference is a result of ratios that do not divide evenly. Should the numbers be rounded up or down? If they are all rounded down, the total is too small, if rounded up, the total number of representatives would be more than the 435 seats in the House. Because all the states cannot be treated equally, the question of what is fair and how to decide who gets an additional representative has been debated in Congress since its inception.

Rational Expressions

The Streeter/Hutchison Series in Mathematics   Beginning Algebra

329

5 prerequisite test

This prerequisite test provides some exercises requiring skills that you will need to be successful in the coming chapter. The answers for these exercises can be found in the back of this text. This prerequisite test can help you identify topics that you will need to review before beginning the chapter.

Name _____

Section _____ Date _____

Answers

1. _____   2. _____

3. _____   4. _____

5. _____   6. _____

7. _____   8. _____

9. _____   10. _____

11. _____   12. _____

13. _____   14. _____

15. _____

16. _____

17. _____

18. _____

19. _____

20. _____

Simplify each fraction.

1. $\dfrac{14}{21}$

2. $\dfrac{156}{72}$

3. $\dfrac{3+5}{15-3}$

4. $-\dfrac{24}{56}$

Write each mixed number as an improper fraction.

5. $4\dfrac{3}{8}$

6. $1\dfrac{17}{32}$

Perform the indicated operation.

7. $\dfrac{3}{4} \cdot \dfrac{7}{10}$

8. $\dfrac{10}{21} \cdot \dfrac{6}{5}$

9. $\dfrac{3}{4} \div \dfrac{7}{10}$

10. $\dfrac{10}{21} \div \dfrac{6}{5}$

11. $\dfrac{5}{8} + \dfrac{5}{12}$

12. $3\dfrac{1}{2} + 7\dfrac{1}{3}$

13. $\dfrac{2}{3} - \dfrac{4}{5}$

14. $\dfrac{5}{6} - \dfrac{3}{10}$

Simplify each expression by removing the parentheses.

15. $8(3x + 4)$

16. $-(4x + 1)$

17. $6x - 3x(x - 5)$

18. $-(x - 1)$

Solve each application.

19. **CONSTRUCTION** A $6\dfrac{1}{2}$-in. bolt is placed through a $5\dfrac{7}{8}$-in.-thick wall. How far does the bolt extend beyond the wall?

20. **CONSTRUCTION** An 18-acre piece of land is to be divided into $\dfrac{3}{8}$-acre home lots. How many lots will be formed?

5.1

Simplifying Rational Expressions

< 5.1 Objectives >

1 > Find the GCF for two monomials and simplify a rational expression

2 > Find the GCF for two polynomials and simplify a rational expression

Much of our work with rational expressions (also called algebraic fractions) is similar to your work in arithmetic. For instance, in algebra, as in arithmetic, many fractions name the same number. Recall

$$\frac{1}{4} = \frac{1 \cdot 2}{4 \cdot 2} = \frac{2}{8}$$

and

$$\frac{1}{4} = \frac{1 \cdot 3}{4 \cdot 3} = \frac{3}{12}$$

So $\frac{1}{4}, \frac{2}{8}$, and $\frac{3}{12}$ all name the same number; they are called **equivalent fractions.** These examples illustrate what is called the **Fundamental Principle of Fractions.** In algebra it becomes the **Fundamental Principle of Rational Expressions.**

NOTE

A rational expression is sometimes called an algebraic fraction, or simply a fraction.

Property

Fundamental Principle of Rational Expressions

For polynomials P, Q, and R,

$$\frac{P}{Q} = \frac{PR}{QR} \qquad \text{when } Q \neq 0 \text{ and } R \neq 0$$

This principle allows us to multiply or divide the numerator and denominator of a fraction by the same nonzero polynomial. The result will be an expression that is equivalent to the original one.

Our objective in this section is to simplify rational expressions by using the fundamental principle. In algebra, as in arithmetic, to write a fraction in simplest form, you divide the numerator and denominator of the fraction by their greatest common factor (GCF). The numerator and denominator of the resulting fraction will have no common factors other than 1, and the fraction is then in **simplest form.** The following rule summarizes this procedure.

Step by Step

To Write Rational Expressions in Simplest Form

Step 1 Factor the numerator and denominator.
Step 2 Divide the numerator and denominator by the GCF. The resulting fraction will be in lowest terms.

NOTE

Step 2 uses the Fundamental Principle of Fractions. The GCF is R in the Fundamental Principle of Rational Expressions rule.

In Example 1, we simplify both numeric and algebraic fractions using the steps provided above.

Example 1   **Writing Fractions in Simplest Form**

< Objective 1 >

(a) Write $\dfrac{18}{30}$ in simplest form.

$$\frac{18}{30} = \frac{2 \cdot 3 \cdot 3}{2 \cdot 3 \cdot 5} = \frac{\cancel{2} \cdot \cancel{3} \cdot 3}{\cancel{2} \cdot \cancel{3} \cdot 5} = \frac{3}{5}$$

Divide by the GCF. The slash lines indicate that we have divided the numerator and denominator by 2 and by 3.

(b) Write $\dfrac{4x^3}{6x}$ in simplest form.

$$\frac{4x^3}{6x} = \frac{\cancel{2} \cdot 2 \cdot \cancel{x} \cdot x \cdot x}{\cancel{2} \cdot 3 \cdot \cancel{x}} = \frac{2x^2}{3}$$

(c) Write $\dfrac{15x^3y^2}{20xy^4}$ in simplest form.

$$\frac{15x^3y^2}{20xy^4} = \frac{3 \cdot \cancel{5} \cdot \cancel{x} \cdot x \cdot x \cdot \cancel{y} \cdot \cancel{y}}{2 \cdot 2 \cdot \cancel{5} \cdot \cancel{x} \cdot \cancel{y} \cdot \cancel{y} \cdot y \cdot y} = \frac{3x^2}{4y^2}$$

We can also simplify directly by finding the GCF. In this case, we have

$$\frac{15x^3y^2}{20xy^4} = \frac{(5xy^2)(3x^2)}{(5xy^2)(4y^2)} = \frac{3x^2}{4y^2}$$

(d) Write $\dfrac{3a^2b}{9a^3b^2}$ in simplest form.

$$\frac{3a^2b}{9a^3b^2} = \frac{(3a^2b)}{(3a^2b)(3ab)} = \frac{1}{3ab}$$

(e) Write $\dfrac{10a^5b^4}{2a^2b^3}$ in simplest form.

$$\frac{10a^5b^4}{2a^2b^3} = \frac{(2a^2b^3)(5a^3b)}{(2a^2b^3)} = \frac{(5a^3b)}{1} = 5a^3b$$

 Check Yourself 1

Write each fraction in simplest form.

(a) $\dfrac{30}{66}$ **(b)** $\dfrac{5x^4}{15x}$ **(c)** $\dfrac{12xy^4}{18x^3y^2}$

(d) $\dfrac{5m^2n}{10m^3n^3}$ **(e)** $\dfrac{12a^4b^6}{2a^3b^4}$

In simplifying arithmetic fractions, common factors are generally easy to recognize. With rational expressions, the factoring techniques you studied in Chapter 4 are often the *first step* in determining those factors.

|  **Example 2** | **Writing Fractions in Simplest Form** |

< Objective 2 >

Write each fraction in simplest form.

(a) $\dfrac{2x - 4}{x^2 - 4} = \dfrac{2(x - 2)}{(x + 2)(x - 2)}$ Factor the numerator and denominator.

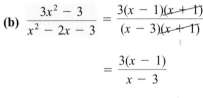

$$= \dfrac{2\cancel{(x - 2)}}{(x + 2)\cancel{(x - 2)}}$$ Divide by the GCF $x - 2$. The slash lines indicate that we have divided by that common factor.

$$= \dfrac{2}{x + 2}$$

NOTE

$3x^2 - 3$
$= 3(x^2 - 1)$
$= 3(x - 1)(x + 1)$

(b) $\dfrac{3x^2 - 3}{x^2 - 2x - 3} = \dfrac{3(x - 1)\cancel{(x + 1)}}{(x - 3)\cancel{(x + 1)}}$

$$= \dfrac{3(x - 1)}{x - 3}$$

(c) $\dfrac{2x^2 + x - 6}{2x^2 - x - 3} = \dfrac{(x + 2)\cancel{(2x - 3)}}{(x + 1)\cancel{(2x - 3)}}$

$$= \dfrac{x + 2}{x + 1}$$

> **CAUTION**

Pick any value, other than 0, for x and substitute. You will quickly see that

$$\dfrac{x + 2}{x + 1} \neq \dfrac{2}{1}$$

For example, if $x = 4$,

$$\dfrac{4 + 2}{4 + 1} = \dfrac{6}{5}$$

Be careful! The expression $\dfrac{x + 2}{x + 1}$ is already in simplest form. Students are often tempted to divide as follows:

$$\dfrac{\cancel{x} + 2}{\cancel{x} + 1} \quad \text{is } not \text{ equal to} \quad \dfrac{2}{1}$$

The x's are *terms* in the numerator and denominator. They *cannot* be divided out. Only *factors* can be divided. The fraction

$$\dfrac{x + 2}{x + 1}$$

is simplified.

 Check Yourself 2

Write each fraction in simplest form.

(a) $\dfrac{5x - 15}{x^2 - 9}$ (b) $\dfrac{a^2 - 5a + 6}{3a^2 - 6a}$

(c) $\dfrac{3x^2 + 14x - 5}{3x^2 + 2x - 1}$ (d) $\dfrac{5p - 15}{p^2 - 4}$

Remember the rules for signs in division. The quotient of a positive number and a negative number is always negative. Thus there are three equivalent ways to write such a quotient. For instance,

$$\dfrac{-2}{3} = \dfrac{2}{-3} = -\dfrac{2}{3}$$

The quotient of two positive numbers or two negative numbers is always positive. For example,

$$\dfrac{-2}{-3} = \dfrac{2}{3}$$

Example 3   **Writing Fractions in Simplest Form**

Write each fraction in simplest form.

(a) $\dfrac{6x^2}{-3xy} = \dfrac{2 \cdot \cancel{3} \cdot \cancel{x} \cdot x}{(-1) \cdot \cancel{3} \cdot \cancel{x} \cdot y} = \dfrac{2x}{-y} = -\dfrac{2x}{y}$

(b) $\dfrac{-5a^2b}{-10b^2} = \dfrac{(\cancel{-1}) \cdot \cancel{5} \cdot a \cdot a \cdot \cancel{b}}{(\cancel{-1}) \cdot 2 \cdot \cancel{5} \cdot \cancel{b} \cdot b} = \dfrac{a^2}{2b}$

> ### Check Yourself 3
>
> Write each fraction in simplest form.
>
> **(a)** $\dfrac{8x^3y}{-4xy^2}$ **(b)** $\dfrac{-16a^4b^2}{-12a^2b^5}$

It is sometimes necessary to factor out a monomial before simplifying the fraction.

Example 4   **Writing Fractions in Simplest Form**

Write each fraction in simplest form.

(a) $\dfrac{6x^2 + 2x}{2x^2 + 12x} = \dfrac{2x(3x + 1)}{2x(x + 6)} = \dfrac{3x + 1}{x + 6}$

(b) $\dfrac{x^2 - 4}{x^2 + 6x + 8} = \dfrac{(x + 2)(x - 2)}{(x + 2)(x + 4)} = \dfrac{x - 2}{x + 4}$

(c) $\dfrac{x + 3}{x^2 + 7x + 12} = \dfrac{x + 3}{(x + 3)(x + 4)} = \dfrac{1}{x + 4}$

> ### Check Yourself 4
>
> Simplify each fraction.
>
> **(a)** $\dfrac{3x^3 - 6x^2}{9x^4 - 3x^2}$ **(b)** $\dfrac{x^2 - 9}{x^2 - 12x + 27}$

Simplifying certain rational expressions is easier with the following result. First, verify for yourself that

$$5 - 8 = -(8 - 5)$$

More generally,

$$a - b = -(b - a)$$

If we take this equation and divide both sides by $b - a$, we get

$$\frac{a - b}{b - a} = \frac{-(b - a)}{b - a} = \frac{-1}{1} = -1$$

Therefore, we have the result

$$\frac{a - b}{b - a} = -1$$

| Example 5 | Writing Rational Expressions in Simplest Form |

Write each fraction in simplest form.

(a) $\dfrac{2x - 4}{4 - x^2} = \dfrac{2(x - 2)}{(2 + x)(2 - x)}$ This is equal to -1.

$= \dfrac{2(-1)}{2 + x}$

$= \dfrac{-2}{2 + x}$

(b) $\dfrac{9 - x^2}{x^2 + 2x - 15} = \dfrac{(3 + x)(3 - x)}{(x + 5)(x - 3)}$ This is equal to -1.

$= \dfrac{(3 + x)(-1)}{x + 5}$

$= \dfrac{-x - 3}{x + 5}$

Check Yourself 5

Write each fraction in simplest form.

(a) $\dfrac{3x - 9}{9 - x^2}$ (b) $\dfrac{x^2 - 6x - 27}{81 - x^2}$

Check Yourself ANSWERS

1. (a) $\dfrac{5}{11}$; (b) $\dfrac{x^3}{3}$; (c) $\dfrac{2y^2}{3x^2}$; (d) $\dfrac{1}{2mn^2}$; (e) $6ab^2$ 2. (a) $\dfrac{5}{x + 3}$; (b) $\dfrac{a - 3}{3a}$;

(c) $\dfrac{x + 5}{x + 1}$; (d) $\dfrac{5(p - 3)}{(p + 2)(p - 2)}$ 3. (a) $-\dfrac{2x^2}{y}$; (b) $\dfrac{4a^2}{3b^3}$

4. (a) $\dfrac{x - 2}{3x^2 - 1}$; (b) $\dfrac{x + 3}{x - 9}$ 5. (a) $-\dfrac{3}{x + 3}$; (b) $\dfrac{-x - 3}{x + 9}$

Reading Your Text

The following fill-in-the-blank exercises are designed to ensure that you understand some of the key vocabulary used in this section.

SECTION 5.1

(a) Fractions that name the same number are called _____ fractions.

(b) When simplifying a rational expression, we divide the numerator and denominator by any common _____.

(c) When the numerator and denominator of a fraction have no common factors other than 1, it is said to be in _____ form.

(d) The quotient of a positive number and a negative number is always _____.

5.1 exercises

Name _____

Section _____ Date _____

Answers

1. _____ 2. _____

3. _____ 4. _____

5. _____ 6. _____

7. _____ 8. _____

9. _____ 10. _____

11. _____ 12. _____

13. _____ 14. _____

15. _____ 16. _____

17. _____

18. _____

19. _____

20. _____

21. _____

22. _____

< Objective 1 >

Write each fraction in simplest form.

1. $\dfrac{16}{24}$

2. $\dfrac{56}{64}$

3. $\dfrac{80}{180}$

4. $\dfrac{18}{30}$

5. $\dfrac{4x^5}{6x^2}$

6. $\dfrac{10x^2}{15x^4}$

7. $\dfrac{9x^3}{27x^6}$

8. $\dfrac{25w^6}{20w^2}$

9. $\dfrac{10a^2b^5}{25ab^2}$

10. $\dfrac{18x^4y^3}{24x^2y^3}$

11. $\dfrac{42x^3y}{14xy^3}$

12. $\dfrac{18pq}{45p^2q^2}$

13. $\dfrac{2xyw^2}{6x^2y^3w^3}$ > Videos

14. $\dfrac{3c^2d^2}{6bc^3d^3}$

15. $\dfrac{10x^5y^5}{2x^3y^4}$

16. $\dfrac{3bc^6d^3}{bc^3d}$

17. $\dfrac{-4m^3n}{6mn^2}$ > Videos

18. $\dfrac{-15x^3y^3}{-20xy^4}$

19. $\dfrac{-8ab^3}{-16a^3b}$

20. $\dfrac{14x^2y}{-21xy^4}$

21. $\dfrac{8r^2s^3t}{-16rs^4t^3}$

22. $\dfrac{-10a^3b^2c^3}{15ab^4c}$

< Objective 2 >

Write each expression in simplest form.

23. $\dfrac{3x + 18}{5x + 30}$

24. $\dfrac{4x - 28}{5x - 35}$

25. $\dfrac{6a - 24}{a^2 - 16}$

26. $\dfrac{5x - 5}{x^2 - 4}$

27. $\dfrac{x^2 + 3x + 2}{5x + 10}$ > Videos

28. $\dfrac{4w^2 - 20w}{w^2 - 2w - 15}$

29. $\dfrac{2m^2 + 3m - 5}{2m^2 + 11m + 15}$

30. $\dfrac{6x^2 - x - 2}{3x^2 - 5x + 2}$

31. $\dfrac{p^2 + 2pq - 15q^2}{p^2 - 25q^2}$

32. $\dfrac{4r^2 - 25s^2}{2r^2 + 3rs - 20s^2}$

33. $\dfrac{y - 7}{7 - y}$ > Videos

34. $\dfrac{5 - y}{y - 5}$

35. $\dfrac{2x - 10}{25 - x^2}$

36. $\dfrac{3a - 12}{16 - a^2}$

37. $\dfrac{25 - a^2}{a^2 + a - 30}$

38. $\dfrac{2x^2 - 7x + 3}{9 - x^2}$

39. $\dfrac{x^2 + xy - 6y^2}{4y^2 - x^2}$

40. $\dfrac{16z^2 - w^2}{2w^2 - 5wz - 12z^2}$

41. $\dfrac{x^2 + 4x + 4}{x + 2}$

42. $\dfrac{4x^2 + 12x + 9}{2x + 3}$

| Basic Skills | **Challenge Yourself** | Calculator/Computer | Career Applications | Above and Beyond |

Complete each statement with **never, sometimes,** *or* **always.**

43. The quotient of two negative values is _____ negative.

44. The expression $\dfrac{x + 2}{x + 1}$ is _____ equal to zero.

45. The expression $\dfrac{a - b}{b - a}$ is _____ equal to -1 when $a \neq b$.

46. The quotient of a positive value and a negative value is _____ negative.

Answers

23. _____ 24. _____

25. _____

26. _____

27. _____

28. _____

29. _____

30. _____

31. _____

32. _____

33. _____

34. _____

35. _____

36. _____

37. _____

38. _____

39. _____

40. _____

41. _____

42. _____

43. _____ 44. _____

45. _____ 46. _____

Answers

47. _____

48. _____

49. _____

50. _____

51. _____

52. _____

53. _____

Simplify each expression.

47. $\dfrac{xy - 2y + 4x - 8}{2y + 6 - xy - 3x}$ > Videos

48. $\dfrac{ab - 3a + 5b - 15}{15 + 3a^2 - 5b - a^2 b}$

49. GEOMETRY The area of the rectangle is represented by $6x^2 + 19x + 10$. What is the length?

$3x + 2$

50. GEOMETRY The volume of the box is represented by $(x^2 + 5x + 6)(x + 5)$. Find the polynomial that represents the area of the bottom of the box.

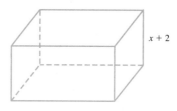

$x + 2$

| Basic Skills | Challenge Yourself | Calculator/Computer | **Career Applications** | Above and Beyond |

▲

51. BUSINESS AND FINANCE A company has a fixed setup cost of $3,500 for a new product. The marginal cost (or cost to produce a single unit) is $8.75.

  (a) Write an expression that gives the average cost per unit when x units are produced.

  (b) Find the average cost when 50 units are produced.

52. BUSINESS AND FINANCE The total revenue, in hundreds of dollars, from the sale of a popular video is approximated by the expression

$$\frac{300t^2}{t^2 + 9}$$

in which t is the number of months since the video was released.

  (a) Find the revenue generated by the end of the first month.

  (b) Find the total revenue generated by the end of the second month.

  (c) Find the total revenue generated by the end of the third month.

  (d) Find the revenue generated in the second month only.

53. MANUFACTURING TECHNOLOGY The safe load of a drop-hammer-style pile driver is given by the expression

$$\frac{6wsh + 6wh}{3s^2 + 6s + 3}$$

Simplify this expression.

54. **MECHANICAL ENGINEERING** The shape of a beam loaded with a single concentrated load is described by the expression

$$\frac{x^2 - 64}{200}$$

Rewrite this expression by factoring the numerator.

| Basic Skills | Challenge Yourself | Calculator/Computer | Career Applications | **Above and Beyond** |

▲

55. To work with rational expressions correctly, it is important to understand the difference between a *factor* and a *term* of an expression. In your own words, write definitions for both, explaining the difference between the two.

56. Give some examples of terms and factors in rational expressions and explain how both are affected when a fraction is simplified.

57. Show how the following rational expression can be simplified:

$$\frac{x^2 - 9}{4x + 12}$$

Note that your simplified fraction is equivalent to the given fraction. Are there other rational expressions equivalent to this one? Write another rational expression that you think is equivalent to this one. Exchange papers with another student. Do you agree that the other student's fraction is equivalent to yours? Why or why not?

58. Explain the reasoning involved in each step when simplifying the fraction $\frac{42}{56}$.

59. Describe why $\frac{3}{5}$ and $\frac{27}{45}$ are *equivalent fractions*.

Answers

54. _____

55. _____

56. _____

57. _____

58. _____

59. _____

Answers

1. $\frac{2}{3}$ 3. $\frac{4}{9}$ 5. $\frac{2x^3}{3}$ 7. $\frac{1}{3x^3}$ 9. $\frac{2ab^3}{5}$ 11. $\frac{3x^2}{y^2}$

13. $\frac{1}{3xy^2w}$ 15. $5x^2y$ 17. $-\frac{2m^2}{3n}$ 19. $\frac{b^2}{2a^2}$ 21. $-\frac{r}{2st^2}$

23. $\frac{3}{5}$ 25. $\frac{6}{a + 4}$ 27. $\frac{x + 1}{5}$ 29. $\frac{m - 1}{m + 3}$ 31. $\frac{p - 3q}{p - 5q}$

33. -1 35. $-\frac{2}{x + 5}$ 37. $\frac{-a - 5}{a + 6} = -\frac{a + 5}{a + 6}$

39. $\frac{-x - 3y}{2y + x} = -\frac{x + 3y}{2y + x}$ 41. $x + 2$ 43. never 45. always

47. $-\frac{(y + 4)}{y + 3}$ 49. $2x + 5$ 51. (a) $\frac{8.75x + 3,500}{x}$; (b) $78.75

53. $\frac{2wh}{s + 1}$ 55. Above and Beyond 57. Above and Beyond

59. Above and Beyond

5.2

Multiplying and Dividing Rational Expressions

< 5.2 Objectives >

1 > Write the product of two rational expressions in simplest form

2 > Write the quotient of two rational expressions in simplest form

In arithmetic, you found the product of two fractions by multiplying the numerators and the denominators. For example,

$$\frac{2}{5} \cdot \frac{3}{7} = \frac{2 \cdot 3}{5 \cdot 7} = \frac{6}{35}$$

In symbols, we have

Property

Multiplying Rational Expressions

$$\frac{P}{Q} \cdot \frac{R}{S} = \frac{PR}{QS}$$

when $Q \neq 0$ and $S \neq 0$ *P, Q, R*, and *S* represent polynomials.

NOTE

Divide by the common factors of 3 and 4. The alternative is to multiply *first:*

$$\frac{3}{8} \cdot \frac{4}{9} = \frac{12}{72}$$

and then use the GCF to reduce to lowest terms

$$\frac{12}{72} = \frac{1}{6}$$

It is easier to divide the numerator and denominator by any common factors *before* multiplying. Consider the following.

$$\frac{3}{8} \cdot \frac{4}{9} = \frac{\overset{1}{\cancel{3}} \cdot \overset{1}{\cancel{4}}}{\underset{2}{\cancel{8}} \cdot \underset{3}{\cancel{9}}} = \frac{1}{6}$$

In algebra, we multiply fractions in exactly the same way.

Step by Step

To Multiply Rational Expressions

Step 1 Factor the numerators and denominators.

Step 2 Write the product of the factors of the numerators over the product of the factors of the denominators.

Step 3 Divide the numerator and denominator by any common factors.

We illustrate this method in Example 1.

▶ Example 1   Multiplying Rational Expressions

< Objective 1 >

Multiply.

(a) $\dfrac{2x^3}{5y^2} \cdot \dfrac{10y}{3x^2} = \dfrac{2x^3 \cdot 10y}{5y^2 \cdot 3x^2} = \dfrac{20x^3y}{15x^2y^2} = \dfrac{4x}{3y}$

NOTES

In (a), divide by the common factors of 5, x^2, and y.

In (b), divide by the common factors of 3, x, and $x - 3$.

RECALL

$$\frac{2 - x}{x - 2} = \frac{-(x - 2)}{x - 2} = -1$$

NOTE

In (d), divide by the common factors of $x - 4$, x, and 3.

(b) $\dfrac{x}{x^2 - 3x} \cdot \dfrac{6x - 18}{9x} = \dfrac{x}{x(x - 3)} \cdot \dfrac{6(x - 3)}{9x}$

Factor

$$= \frac{\overset{1}{\cancel{x}} \cdot \overset{2}{\cancel{6}}\overset{1}{(\cancel{x - 3})}}{\underset{1}{\cancel{x}}\underset{1}{(\cancel{x - 3})} \cdot \underset{3}{\cancel{9}}x} = \frac{2}{3x}$$

(c) $\dfrac{4}{x^2 - 2x} \cdot \dfrac{10 - 5x}{8} = \dfrac{4}{x(x - 2)} \cdot \dfrac{5(2 - x)}{8}$

$$= \frac{\overset{1}{\cancel{4}} \cdot 5\overset{-1}{\cancel{(2 - x)}}}{x\cancel{(x - 2)} \cdot \underset{2}{\cancel{8}}} = \frac{-5}{2x}$$

(d) $\dfrac{x^2 - 2x - 8}{3x^2} \cdot \dfrac{6x}{3x - 12} = \dfrac{(x - 4)(x + 2) \cdot 6x}{3x^2 \cdot 3(x - 4)}$

$$= \frac{\overset{1}{\cancel{(x - 4)}}(x + 2) \cdot \overset{2}{\cancel{6x}}}{\underset{x}{\cancel{3x^2}} \cdot \underset{1}{3\cancel{(x - 4)}}}$$

$$= \frac{2(x + 2)}{3x}$$

(e) $\dfrac{x^2 - y^2}{5x - 5y} \cdot \dfrac{10xy}{x^2 + 2xy + y^2} = \dfrac{(x - y)(x + y) \cdot 10xy}{5(x - y) \cdot (x + y)(x + y)}$

$$= \frac{\overset{1}{\cancel{(x - y)}}\overset{1}{\cancel{(x + y)}} \cdot \overset{2}{\cancel{10}}xy}{\underset{1}{5\cancel{(x - y)}} \cdot \underset{1}{\cancel{(x + y)}}(x + y)}$$

$$= \frac{2xy}{x + y}$$

Check Yourself 1

Multiply.

(a) $\dfrac{3x^2}{5y^2} \cdot \dfrac{10y^5}{15x^3}$ **(b)** $\dfrac{5x + 15}{x} \cdot \dfrac{2x^2}{x^2 + 3x}$ **(c)** $\dfrac{x}{2x - 6} \cdot \dfrac{3x - x^2}{2}$

(d) $\dfrac{3x - 15}{6x^2} \cdot \dfrac{2x}{x^2 - 25}$ **(e)** $\dfrac{x^2 - 5x - 14}{4x^2} \cdot \dfrac{8x}{x^2 - 49}$

You can also use your experience from arithmetic in dividing fractions. Recall that, to divide fractions, we *invert the divisor* (the *second* fraction) and multiply. For example,

$$\frac{2}{3} \div \frac{5}{6} = \frac{2}{3} \cdot \frac{6}{5} = \frac{2 \cdot 6}{3 \cdot 5} = \frac{12}{15} = \frac{4}{5}$$

In symbols, we have

RECALL

$\dfrac{6}{5}$ is the reciprocal of $\dfrac{5}{6}$.

Property

Dividing Rational Expressions

$$\frac{P}{Q} \div \frac{R}{S} = \frac{P}{Q} \cdot \frac{S}{R} = \frac{PS}{QR}$$

P, Q, R, and S are polynomials.

when $Q \neq 0$, $R \neq 0$, and $S \neq 0$.

We divide rational expressions in exactly the same way.

To Divide Rational Expressions

Step 1   Invert the divisor and change the operation to multiplication.
Step 2   Proceed, using the steps for multiplying rational expressions.

Example 2 illustrates this approach.

Example 2   **Dividing Rational Expressions**

< Objective 2 >

Divide.

(a) $\dfrac{6}{x^2} \div \dfrac{9}{x^3} = \dfrac{6}{x^2} \cdot \dfrac{x^3}{9}$ Invert the divisor and multiply.

$= \dfrac{\overset{2}{\cancel{6}}\,\overset{x}{\cancel{x^3}}}{\underset{3}{\cancel{9}}\,\underset{1}{\cancel{x^2}}}$ No simplification can be done until the divisor is inverted. Then divide by the common factors of 3 and x^2.

$= \dfrac{2x}{3}$

(b) $\dfrac{3x^2y}{8xy^3} \div \dfrac{9x^3}{4y^4} = \dfrac{3x^2y}{8xy^3} \cdot \dfrac{4y^4}{9x^3}$

$= \dfrac{y^2}{6x^2}$

NOTE

Factor all numerators and denominators *before* dividing out any common factors.

(c) $\dfrac{2x + 4y}{9x - 18y} \div \dfrac{4x + 8y}{3x - 6y} = \dfrac{2x + 4y}{9x - 18y} \cdot \dfrac{3x - 6y}{4x + 8y}$

$= \dfrac{\overset{1}{\cancel{2}}\overset{1}{\cancel{(x + 2y)}} \cdot \overset{1}{\cancel{3}}\overset{1}{\cancel{(x - 2y)}}}{\underset{3}{\cancel{9}}\underset{1}{\cancel{(x - 2y)}} \cdot \underset{2}{\cancel{4}}\underset{1}{\cancel{(x + 2y)}}}$

$= \dfrac{1}{6}$

(d) $\dfrac{x^2 - x - 6}{2x - 6} \div \dfrac{x^2 - 4}{4x^2} = \dfrac{x^2 - x - 6}{2x - 6} \cdot \dfrac{4x^2}{x^2 - 4}$

$= \dfrac{\overset{1}{\cancel{(x - 3)}}(x + 2) \cdot \overset{2}{\cancel{4}}x^2}{\underset{1}{\cancel{2}}\underset{1}{\cancel{(x - 3)}} \cdot (x + 2)(x - 2)}$

$= \dfrac{2x^2}{x - 2}$

Check Yourself 2

Divide.

(a) $\dfrac{4}{x^5} \div \dfrac{12}{x^3}$ (b) $\dfrac{5xy^2}{7x^3y} \div \dfrac{10y^2}{14x^3}$

(c) $\dfrac{3x - 9y}{2x + 10y} \div \dfrac{x^2 - 3xy}{4x + 20y}$ (d) $\dfrac{x^2 - 9}{4x} \div \dfrac{x^2 - 2x - 15}{2x - 10}$

Check Yourself ANSWERS

1. (a) $\dfrac{2y^3}{5x}$; (b) 10; (c) $\dfrac{-x^2}{4}$; (d) $\dfrac{1}{x(x+5)}$; (e) $\dfrac{2(x+2)}{x(x+7)}$

2. (a) $\dfrac{1}{3x^2}$; (b) $\dfrac{x}{y}$; (c) $\dfrac{6}{x}$; (d) $\dfrac{x-3}{2x}$

Reading Your Text

The following fill-in-the-blank exercises are designed to ensure that you understand some of the key vocabulary used in this section.

SECTION 5.2

(a) In arithmetic, we find the product of two fractions by _____ the numerators and the denominators.

(b) The first step when multiplying rational expressions is to _____ the numerators and the denominators.

(c) When dividing two rational expressions, _____ the divisor and multiply.

(d) When dividing rational expressions, the divisor cannot equal _____.

5.2 exercises

Name _____

Section _____ Date _____

Answers

< Objective 1 >

Multiply.

1. $\dfrac{3}{7} \cdot \dfrac{14}{27}$

2. $\dfrac{9}{20} \cdot \dfrac{5}{36}$

3. $\dfrac{x}{2} \cdot \dfrac{y}{6}$

4. $\dfrac{w}{2} \cdot \dfrac{5}{14}$

5. $\dfrac{3a}{2} \cdot \dfrac{4}{a^2}$

6. $\dfrac{5x^3}{3x} \cdot \dfrac{9}{20x}$

7. $\dfrac{3x^3y}{10xy^3} \cdot \dfrac{5xy^2}{9xy^2}$

8. $\dfrac{8xy^5}{5x^3y^2} \cdot \dfrac{15y^2}{16xy^3}$

9. $\dfrac{3a^2b^3}{2ab} \cdot \dfrac{8a^3b}{6ab^3}$

10. $\dfrac{4x^4y^3}{8xy^3} \cdot \dfrac{12xy}{6x^3}$

11. $\dfrac{x^2y^3}{5a^3b} \cdot \dfrac{10ab^3}{3x^3}$

12. $\dfrac{9a^4b^{10}}{xy^3} \cdot \dfrac{2xy}{6b^7}$

13. $\dfrac{-4ab^2}{15a^3} \cdot \dfrac{25ab}{-16b^3}$

14. $\dfrac{-7xy^2}{12x^2y} \cdot \dfrac{24x^3y^5}{-21x^2y^7}$

15. $\dfrac{-3m^3n}{10mn^3} \cdot \dfrac{5mn^2}{-9mn^3}$ > Videos

16. $\dfrac{3x}{2x-6} \cdot \dfrac{x^2-3x}{6}$

17. $\dfrac{x^2+5x}{3x^2} \cdot \dfrac{10x}{5x+25}$

18. $\dfrac{x^2-3x-10}{5x} \cdot \dfrac{15x^2}{3x-15}$

19. $\dfrac{m^2-4m-21}{3m^2} \cdot \dfrac{m^2+7m}{m^2-49}$ > Videos

20. $\dfrac{2x^2-x-3}{3x^2+7x+4} \cdot \dfrac{3x^2-11x-20}{4x^2-9}$

21. $\dfrac{4r^2-1}{2r^2-9r-5} \cdot \dfrac{3r^2-13r-10}{9r^2-4}$

22. $\dfrac{a^2+ab}{2a^2-ab-3b^2} \cdot \dfrac{4a^2-9b^2}{5a^2-4ab}$

23. $\dfrac{x^2-4y^2}{x^2-xy-6y^2} \cdot \dfrac{7x^2-21xy}{5x-10y}$

24. $\dfrac{a^2-9b^2}{a^2+ab-6b^2} \cdot \dfrac{6a^2-12ab}{7a-21b}$

25. $\dfrac{2x-6}{x^2+2x} \cdot \dfrac{3x}{3-x}$

26. $\dfrac{3x-15}{x^2+3x} \cdot \dfrac{4x}{5-x}$

< Objective 2 >

Divide.

27. $\dfrac{5}{8} \div \dfrac{15}{16}$

28. $\dfrac{4}{9} \div \dfrac{12}{18}$

29. $\dfrac{5}{x^2} \div \dfrac{10}{x}$

30. $\dfrac{w^2}{3} \div \dfrac{w}{9}$

31. $\dfrac{4x^2y^2}{9x^3} \div \dfrac{8y^2}{27xy}$

32. $\dfrac{8x^3y}{27xy^3} \div \dfrac{16x^3y}{45y}$

33. $\dfrac{3x + 6}{8} \div \dfrac{5x + 10}{6}$

34. $\dfrac{x^2 - 2x}{4x} \div \dfrac{6x - 12}{8}$

35. $\dfrac{4a - 12}{5a + 15} \div \dfrac{8a^2}{a^2 + 3a}$

36. $\dfrac{6p - 18}{9p} \div \dfrac{3p - 9}{p^2 + 2p}$

| Basic Skills | **Challenge Yourself** | Calculator/Computer | Career Applications | Above and Beyond |

Determine whether each statement is **true** *or* **false.**

37. The product of three negative values is negative.

38. Order of operations states that we multiply and divide before applying powers.

39. Division by zero results in a quotient of zero.

40. A fraction can always be simplified if the expression in the numerator contains the denominator.

Divide.

41. $\dfrac{x^2 + 2x - 8}{9x^2} \div \dfrac{x^2 - 16}{3x - 12}$ > Videos

42. $\dfrac{16x}{4x^2 - 16} \div \dfrac{4x - 24}{x^2 - 4x - 12}$

43. $\dfrac{x^2 - 9}{2x^2 - 6x} \div \dfrac{2x^2 + 5x - 3}{4x^2 - 1}$

44. $\dfrac{2m^2 - 5m - 7}{4m^2 - 9} \div \dfrac{5m^2 + 5m}{2m^2 + 3m}$

45. $\dfrac{a^2 - 9b^2}{4a^2 + 12ab} \div \dfrac{a^2 - ab - 6b^2}{12ab}$

46. $\dfrac{r^2 + 2rs - 15s^2}{r^3 + 5r^2s} \div \dfrac{r^2 - 9s^2}{5r^3}$

Answers

27. _____

28. _____

29. _____

30. _____

31. _____

32. _____

33. _____

34. _____

35. _____

36. _____

37. _____

38. _____

39. _____

40. _____

41. _____

42. _____

43. _____

44. _____

45. _____

46. _____

Beginning Algebra   The Streeter/Hutchison Series in Mathematics

The Streeter/Hutchison Series in Mathematics   Beginning Algebra

© The McGraw-Hill Companies. All Rights Reserved.

Answers

47. _____

48. _____

49. _____

50. _____

51. _____

52. _____

53. _____

54. _____

55. _____

56. _____

57. _____

47. $\dfrac{x^2 - 16y^2}{3x^2 - 12xy} \div (x^2 + 4xy)$

48. $\dfrac{p^2 - 4pq - 21q^2}{4p - 28q} \div (2p^2 + 6pq)$

49. $\dfrac{x - 7}{2x + 6} \div \dfrac{21 - 3x}{x^2 + 3x}$

50. $\dfrac{x - 4}{x^2 + 2x} \div \dfrac{16 - 4x}{3x + 6}$

| Basic Skills | Challenge Yourself | Calculator/Computer | Career Applications | **Above and Beyond** |

Perform the indicated operations.

51. $\dfrac{x^2 + 5x}{3x - 6} \cdot \dfrac{x^2 - 4}{3x^2 + 15x} \cdot \dfrac{6x}{x^2 + 6x + 8}$

52. $\dfrac{m^2 - n^2}{m^2 - mn} \cdot \dfrac{6m}{2m^2 + mn - n^2} \cdot \dfrac{8m - 4n}{12m^2 + 12mn}$

53. $\dfrac{x^2 - 2x - 8}{2x - 8} \cdot \dfrac{x^2 + 5x}{x^2 + 5x + 6} \div \dfrac{x^2 + 2x - 15}{x^2 - 9}$ > Videos

54. $\dfrac{14x - 7}{x^2 + 3x - 4} \cdot \dfrac{x^2 + 6x + 8}{2x^2 + 5x - 3} \div \dfrac{x^2 + 2x}{x^2 + 2x - 3}$

Solve each application.

55. **SCIENCE AND MEDICINE** Herbicides constitute $\dfrac{2}{3}$ of all pesticides used in the United States. Insecticides are $\dfrac{1}{4}$ of all pesticides used in the United States. The ratio of herbicides to insecticides used in the United States can be written $\dfrac{2}{3} \div \dfrac{1}{4}$. Write this ratio in simplest form.

56. **SCIENCE AND MEDICINE** Fungicides account for $\dfrac{1}{10}$ of the pesticides used in the United States. Insecticides account for $\dfrac{1}{4}$ of all the pesticides used in the United States. The ratio of fungicides to insecticides used in the United States can be written $\dfrac{1}{10} \div \dfrac{1}{4}$. Write this ratio in simplest form.

57. **SCIENCE AND MEDICINE** The ratio of insecticides to herbicides applied to wheat, soybeans, corn, and cotton can be expressed as $\dfrac{7}{10} \div \dfrac{4}{5}$. Simplify this ratio.

58. **GEOMETRY** Find the area of the rectangle shown.

$$\frac{2x - 4}{x - 1}$$

$$\frac{3x - 2}{x - 2}$$

Answers

1. $\dfrac{2}{9}$ **3.** $\dfrac{xy}{12}$ **5.** $\dfrac{6}{a}$ **7.** $\dfrac{x^2}{6y^2}$ **9.** $2a^3$ **11.** $\dfrac{2y^3 b^2}{3xa^2}$ **13.** $\dfrac{5}{12a}$

15. $\dfrac{m^2}{6n^3}$ **17.** $\dfrac{2}{3}$ **19.** $\dfrac{m + 3}{3m}$ **21.** $\dfrac{2r - 1}{3r - 2}$ **23.** $\dfrac{7x}{5}$ **25.** $\dfrac{-6}{x + 2}$

27. $\dfrac{2}{3}$ **29.** $\dfrac{1}{2x}$ **31.** $\dfrac{3y}{2}$ **33.** $\dfrac{9}{20}$ **35.** $\dfrac{a - 3}{10a}$ **37.** True

39. False **41.** $\dfrac{x - 2}{3x^2}$ **43.** $\dfrac{2x + 1}{2x}$ **45.** $\dfrac{3b}{a + 2b}$ **47.** $\dfrac{1}{3x^2}$

49. $\dfrac{-x}{6}$ **51.** $\dfrac{2x}{3(x + 4)}$ **53.** $\dfrac{x}{2}$ **55.** $\dfrac{8}{3}$ **57.** $\dfrac{7}{8}$

5.3

Adding and Subtracting Like Rational Expressions

< 5.3 Objectives >

1 > Write the sum or difference of two rational expressions whose numerator and denominator are monomials

2 > Write the sum or difference of two rational expressions whose numerator and denominator are polynomials

You probably remember from arithmetic that **like fractions** are fractions that have the same denominator. The same is true in algebra.

$\dfrac{2}{5}, \dfrac{12}{5},$ and $\dfrac{4}{5}$ are like fractions.

$\dfrac{x}{3(x + y)}, \dfrac{y}{3(x + y)},$ and $\dfrac{z - 5}{3(x + y)}$ are like fractions.

$\dfrac{3x}{2}, \dfrac{x}{4},$ and $\dfrac{3x}{8}$ are unlike fractions.

$\dfrac{3}{x}, \dfrac{2}{x^2},$ and $\dfrac{x + 1}{x^3}$ are unlike fractions.

> **NOTE**
>
> The fractions have different denominators.

In arithmetic, the sum or difference of like fractions is found by adding or subtracting the numerators and writing the result over the common denominator. For example,

$$\frac{3}{11} + \frac{5}{11} = \frac{3 + 5}{11} = \frac{8}{11}$$

In symbols, we have

Property

To Add or Subtract Like Rational Expressions	$\dfrac{P}{R} + \dfrac{Q}{R} = \dfrac{P + Q}{R} \qquad R \neq 0$
	$\dfrac{P}{R} - \dfrac{Q}{R} = \dfrac{P - Q}{R} \qquad R \neq 0$

Adding or subtracting like rational expressions is just as straightforward. You can use the following steps.

Step by Step

To Add or Subtract Like Rational Expressions	Step 1	Add or subtract the numerators.
	Step 2	Write the sum or difference over the common denominator.
	Step 3	Write the resulting fraction in simplest form.

The Streeter/Hutchison Series in Mathematics   Beginning Algebra

| | Example 1 | Adding and Subtracting Rational Expressions |

< Objective 1 >

Add or subtract as indicated. Express your results in simplest form.

(a) $\dfrac{2x}{15} + \dfrac{x}{15} = \dfrac{\overline{2x + x}}{15}$ Add the numerators.

$= \dfrac{3x}{15} = \dfrac{x}{5}$

↖ Simplify

(b) $\dfrac{5y}{6} - \dfrac{y}{6} = \dfrac{\overline{5y - y}}{6}$ Subtract the numerators.

$= \dfrac{4y}{6} = \dfrac{2y}{3}$

↖ Simplify

(c) $\dfrac{3}{x} + \dfrac{5}{x} = \dfrac{3 + 5}{x} = \dfrac{8}{x}$

(d) $\dfrac{9b}{a^2} - \dfrac{7b}{a^2} = \dfrac{9b - 7b}{a^2} = \dfrac{2b}{a^2}$

(e) $\dfrac{7}{2ab} - \dfrac{5}{2ab} = \dfrac{7 - 5}{2ab}$

$= \dfrac{2}{2ab}$

$= \dfrac{1}{ab}$

Check Yourself 1

Add or subtract as indicated.

(a) $\dfrac{3a}{10} + \dfrac{2a}{10}$ **(b)** $\dfrac{7b}{8} - \dfrac{3b}{8}$ **(c)** $\dfrac{4}{x} + \dfrac{3}{x}$ **(d)** $\dfrac{5}{3xy} - \dfrac{2}{3xy}$

If polynomials are involved in the numerators or denominators, the process is exactly the same.

| | Example 2 | Adding and Subtracting Rational Expressions |

< Objective 2 >

Add or subtract as indicated. Express your results in simplest form.

(a) $\dfrac{5}{x + 3} + \dfrac{2}{x + 3} = \dfrac{5 + 2}{x + 3} = \dfrac{7}{x + 3}$

(b) $\dfrac{4x}{x-4} - \dfrac{16}{x-4} = \dfrac{4x-16}{x-4}$

Factor and simplify.

$$= \dfrac{4(x-4)}{x-4} = 4$$

RECALL

Always report the final result in simplest form.

(c) $\dfrac{a-b}{3} + \dfrac{2a+b}{3} = \dfrac{(a-b)+(2a+b)}{3}$

$$= \dfrac{a-b+2a+b}{3}$$

$$= \dfrac{3a}{3} = a$$

> C A U T I O N

Notice what happens if parentheses are not used for the second numerator.

$(3x+y) - x - 3y$
$= 3x + y - x - 3y$
$= 2x - 2y$

We get a different (and wrong) result!

Be sure to enclose the second numerator in parentheses!

(d) $\dfrac{3x+y}{2x} - \dfrac{x-3y}{2x} = \dfrac{(3x+y)-(x-3y)}{2x}$

Change both signs.

$$= \dfrac{3x+y-x+3y}{2x}$$

$$= \dfrac{2x+4y}{2x}$$

$$= \dfrac{2(x+2y)}{2x}$$

Factor and divide by the common factor of 2.

$$= \dfrac{x+2y}{x}$$

(e) $\dfrac{3x-5}{x^2+x-2} - \dfrac{2x-4}{x^2+x-2} = \dfrac{(3x-5)-(2x-4)}{x^2+x-2}$

Put the second numerator in parentheses.

Change both signs.

$$= \dfrac{3x-5-2x+4}{x^2+x-2}$$

$$= \dfrac{x-1}{x^2+x-2}$$

$$= \dfrac{(x-1)}{(x+2)(x-1)}$$

Factor and divide by the common factor of $x-1$.

$$= \dfrac{1}{x+2}$$

(f) $\dfrac{2x + 7y}{x + 3y} - \dfrac{x + 4y}{x + 3y} = \dfrac{(2x + 7y) - (x + 4y)}{x + 3y}$

Change both signs.

$$= \dfrac{2x + 7y - x - 4y}{x + 3y}$$

$$= \dfrac{x + 3y}{x + 3y} = 1$$

Check Yourself 2

Add or subtract as indicated.

(a) $\dfrac{4}{x - 5} - \dfrac{2}{x - 5}$ **(b)** $\dfrac{3x}{x + 3} + \dfrac{9}{x + 3}$

(c) $\dfrac{5x - y}{3y} - \dfrac{2x - 4y}{3y}$ **(d)** $\dfrac{5x + 8}{x^2 - 2x - 15} - \dfrac{4x + 5}{x^2 - 2x - 15}$

Check Yourself ANSWERS

1. (a) $\dfrac{a}{2}$; **(b)** $\dfrac{b}{2}$; **(c)** $\dfrac{7}{x}$; **(d)** $\dfrac{1}{xy}$ **2. (a)** $\dfrac{2}{x - 5}$; **(b)** 3; **(c)** $\dfrac{x + y}{y}$; **(d)** $\dfrac{1}{x - 5}$

Reading Your Text

The following fill-in-the-blank exercises are designed to ensure that you understand some of the key vocabulary used in this section.

SECTION 5.3

(a) Fractions with the same denominator are called _____ fractions.

(b) When adding rational expressions, the final step is to write the result in _____ form.

(c) When subtracting fractions, the second numerator is enclosed in _____ before subtracting.

(d) Rational expressions can be simplified if the numerator and denominator have a common _____.

Name _____

Section _____ Date _____

Answers

Basic Skills | Challenge Yourself | Calculator/Computer | Career Applications | Above and Beyond

< Objectives 1–2 >

Add or subtract as indicated. Express your results in simplest form.

1. $\dfrac{7}{18} + \dfrac{5}{18}$

2. $\dfrac{5}{18} - \dfrac{2}{18}$

3. $\dfrac{13}{16} - \dfrac{9}{16}$

4. $\dfrac{5}{12} + \dfrac{11}{12}$

5. $\dfrac{x}{8} + \dfrac{3x}{8}$

6. $\dfrac{5y}{16} + \dfrac{7y}{16}$

7. $\dfrac{7a}{10} - \dfrac{3a}{10}$

8. $\dfrac{5x}{12} - \dfrac{x}{12}$

9. $\dfrac{5}{x} + \dfrac{3}{x}$

10. $\dfrac{9}{y} - \dfrac{3}{y}$

11. $\dfrac{8}{w} - \dfrac{2}{w}$

12. $\dfrac{7}{z} + \dfrac{9}{z}$

13. $\dfrac{2}{xy} + \dfrac{3}{xy}$

14. $\dfrac{8}{ab} + \dfrac{4}{ab}$

15. $\dfrac{2}{3cd} + \dfrac{4}{3cd}$

16. $\dfrac{5}{4cd} + \dfrac{11}{4cd}$

17. $\dfrac{7}{x-5} + \dfrac{9}{x-5}$

18. $\dfrac{11}{x+7} - \dfrac{4}{x+7}$

19. $\dfrac{2x}{x-2} - \dfrac{4}{x-2}$ > Videos

20. $\dfrac{7w}{w+3} + \dfrac{21}{w+3}$

21. $\dfrac{8p}{p+4} + \dfrac{32}{p+4}$

22. $\dfrac{5a}{a-3} - \dfrac{15}{a-3}$

23. $\dfrac{x^2}{x+4} + \dfrac{3x-4}{x+4}$ > Videos

24. $\dfrac{x^2}{x-3} - \dfrac{9}{x-3}$

25. $\dfrac{m^2}{m-5} - \dfrac{25}{m-5}$

26. $\dfrac{s^2}{s+3} + \dfrac{2s-3}{s+3}$

Basic Skills | **Challenge Yourself** | Calculator/Computer | Career Applications | Above and Beyond
▲

Complete each statement with **never, sometimes,** *or* **always.**

27. The sum of two negative values is _____ negative.

28. The sum of a negative value and a positive value is _____ negative.

29. The difference of two negative values is _____ negative.

30. The difference of two positive values is _____ negative.

Add or subtract as indicated.

31. $\dfrac{4m + 7}{6m} - \dfrac{2m + 5}{6m}$ > Videos

32. $\dfrac{6x - y}{4y} - \dfrac{2x + 3y}{4y}$

33. $\dfrac{4w - 7}{w - 5} - \dfrac{2w + 3}{w - 5}$

34. $\dfrac{3b - 8}{b - 6} + \dfrac{b - 16}{b - 6}$

35. $\dfrac{x - 7}{x^2 - x - 6} + \dfrac{2x - 2}{x^2 - x - 6}$

36. $\dfrac{5a - 12}{a^2 - 8a + 15} - \dfrac{3a - 2}{a^2 - 8a + 15}$

37. $\dfrac{y^2}{2y + 8} + \dfrac{3y - 4}{2y + 8}$ > Videos

38. $\dfrac{x^2}{4x - 12} - \dfrac{9}{4x - 12}$

39. $\dfrac{7w}{w + 3} + \dfrac{21}{w + 3}$

40. $\dfrac{2x}{x - 3} - \dfrac{6}{x - 3}$

41. $\dfrac{x^2}{x^2 + x - 6} - \dfrac{6}{(x + 3)(x - 2)} + \dfrac{x}{(x^2 + x - 6)}$ > Videos

42. $\dfrac{-12}{x^2 + x - 12} + \dfrac{x^2}{(x + 4)(x - 3)} + \dfrac{x}{x^2 + x - 12}$

Answers

27. _____

28. _____

29. _____

30. _____

31. _____

32. _____

33. _____

34. _____

35. _____

36. _____

37. _____

38. _____

39. _____

40. _____

41. _____

42. _____

The Streeter/Hutchison Series in Mathematics   Beginning Algebra

Answers

43. _____

44. _____

43. GEOMETRY Find the perimeter of the given figure.

$$\frac{2x}{x+3}$$

$$\frac{6}{x+3}$$

44. GEOMETRY Find the perimeter of the given figure.

$$\frac{x}{2x-5}$$

$$\frac{8}{2x-5}$$

Answers

1. $\dfrac{2}{3}$ **3.** $\dfrac{1}{4}$ **5.** $\dfrac{x}{2}$ **7.** $\dfrac{2a}{5}$ **9.** $\dfrac{8}{x}$ **11.** $\dfrac{6}{w}$ **13.** $\dfrac{5}{xy}$

15. $\dfrac{2}{cd}$ **17.** $\dfrac{16}{x-5}$ **19.** 2   **21.** 8   **23.** $x-1$ **25.** $m+5$

27. always   **29.** sometimes   **31.** $\dfrac{m+1}{3m}$ **33.** 2   **35.** $\dfrac{3}{x+2}$

37. $\dfrac{y-1}{2}$ **39.** 7   **41.** 1   **43.** 4

The Streeter/Hutchison Series in Mathematics   Beginning Algebra

5.4 Adding and Subtracting Unlike Rational Expressions

< 5.4 Objectives >

1 > Write the sum of two unlike rational expressions in simplest form

2 > Write the difference of two unlike rational expressions in simplest form

Adding or subtracting **unlike rational expressions** (fractions that do not have the same denominator) requires a bit more work than adding or subtracting the like rational expressions of Section 5.3. When the denominators are not the same, we must use the idea of the *least common denominator* (LCD). Each fraction is "built up" to an equivalent fraction having the LCD as a denominator. You can then add or subtract as before.

Example 1   Finding the LCD and Adding Fractions

< Objective 1 >

Add $\dfrac{5}{9} + \dfrac{1}{6}$.

Step 1   To find the LCD, factor each denominator.

$9 = 3 \cdot 3$ ⟵ 3 appears twice.

$6 = 2 \cdot 3$

To form the LCD, include each factor the greatest number of times it appears in any single denominator. In this example, use one 2, because 2 appears only once in the factorization of 6. Use two 3's, because 3 appears twice in the factorization of 9. Thus the LCD for the fractions is $2 \cdot 3 \cdot 3 = 18$.

Step 2   "Build up" each fraction to an equivalent fraction with the LCD as the denominator. Do this by multiplying the numerator and denominator of the given fractions by the same number.

$$\frac{5}{9} = \frac{5 \cdot 2}{9 \cdot 2} = \frac{10}{18}$$

$$\frac{1}{6} = \frac{1 \cdot 3}{6 \cdot 3} = \frac{3}{18}$$

NOTE

Do you see that this uses the fundamental principle?

$$\frac{P}{Q} = \frac{PR}{QR}$$

Step 3   Add the fractions.

$$\frac{5}{9} + \frac{1}{6} = \frac{10}{18} + \frac{3}{18} = \frac{13}{18}$$

$\dfrac{13}{18}$ is in simplest form and so we are done!

✓ **Check Yourself 1**

Add the fractions.

(a) $\dfrac{1}{6} + \dfrac{3}{8}$ (b) $\dfrac{3}{10} + \dfrac{4}{15}$

The process of finding the sum or difference is exactly the same in algebra as it is in arithmetic. We can summarize the steps with the following rule.

Step by Step

To Add or Subtract Unlike Rational Expressions		
	Step 1	Find the least common denominator of all the fractions.
	Step 2	Convert each fraction to an equivalent fraction with the LCD as a denominator.
	Step 3	Add or subtract the like fractions formed in step 2.
	Step 4	Write the sum or difference in simplest form.

▶   **Example 2**   **Adding and Subtracting Unlike Rational Expressions**

< Objectives 1–2 >

(a) Add $\dfrac{3}{2x} + \dfrac{4}{x^2}$.

Step 1 Factor the denominators.

$2x = 2 \cdot x$

$x^2 = x \cdot x$

NOTE

Although the product of the denominators is a common denominator, it is not necessarily the *least* common denominator (LCD).

The LCD must contain the factors 2 and x. The factor x must appear *twice* because it appears twice as a factor in the second denominator.

The LCD is $2 \cdot x \cdot x$, or $2x^2$.

Step 2

$$\frac{3}{2x} = \frac{3 \cdot x}{2x \cdot x} = \frac{3x}{2x^2}$$

$$\frac{4}{x^2} = \frac{4 \cdot 2}{x^2 \cdot 2} = \frac{8}{2x^2}$$

Step 3

$$\frac{3}{2x} + \frac{4}{x^2} = \frac{3x}{2x^2} + \frac{8}{2x^2}$$

$$= \frac{3x + 8}{2x^2}$$

The sum is in simplest form.

(b) Subtract $\dfrac{4}{3x^2} - \dfrac{3}{2x^3}$.

Step 1   Factor the denominators.

$3x^2 = 3 \cdot x \cdot x$

$2x^3 = 2 \cdot x \cdot x \cdot x$

The LCD must contain the factors 2, 3, and x. The LCD is

$2 \cdot 3 \cdot x \cdot x \cdot x$ or $6x^3$ The factor x must appear 3 times. Do you see why?

Step 2

$\dfrac{4}{3x^2} = \dfrac{4 \cdot 2x}{3x^2 \cdot 2x} = \dfrac{8x}{6x^3}$

$\dfrac{3}{2x^3} = \dfrac{3 \cdot 3}{2x^3 \cdot 3} = \dfrac{9}{6x^3}$

Step 3

$\dfrac{4}{3x^2} - \dfrac{3}{2x^3} = \dfrac{8x}{6x^3} - \dfrac{9}{6x^3}$

$= \dfrac{8x - 9}{6x^3}$

The difference is in simplest form.

RECALL

Both the numerator and the denominator must be multiplied by the same quantity.

Check Yourself 2

Add or subtract as indicated.

(a) $\dfrac{5}{x^2} + \dfrac{3}{x^3}$ **(b)** $\dfrac{3}{5x} - \dfrac{1}{4x^2}$

We can also add fractions with more than one variable in the denominator. Example 3 illustrates this type of sum.

Example 3   Adding Unlike Rational Expressions

Add $\dfrac{2}{3x^2y} + \dfrac{3}{4x^3}$.

Step 1   Factor the denominators.

$3x^2y = 3 \cdot x \cdot x \cdot y$

$4x^3 = 2 \cdot 2 \cdot x \cdot x \cdot x$

The LCD is $12x^3y$. Do you see why?

Step 2

$\dfrac{2}{3x^2y} = \dfrac{2 \cdot 4x}{3x^2y \cdot 4x} = \dfrac{8x}{12x^3y}$

$\dfrac{3}{4x^3} = \dfrac{3 \cdot 3y}{4x^3 \cdot 3y} = \dfrac{9y}{12x^3y}$

Step 3

$$\frac{2}{3x^2y} + \frac{3}{4x^3} = \frac{8x}{12x^3y} + \frac{9y}{12x^3y}$$

$$= \frac{8x + 9y}{12x^3y}$$

Beginning Algebra   The Streeter/Hutchison Series in Mathematics   © The McGraw-Hill Companies. All Rights Reserved.

NOTE

The y in the numerator and that in the denominator cannot be divided out because y is not a factor of the numerator.

Check Yourself 3

Add.

$$\frac{2}{3x^2y} + \frac{1}{6xy^2}$$

Rational expressions with binomials in the denominator can also be added by taking the approach shown in Example 3. Example 4 illustrates this approach with binomials.

Example 4   Adding and Subtracting Unlike Rational Expressions

(a) Add $\dfrac{5}{x} + \dfrac{2}{x - 1}$.

Step 1   The LCD must have factors of x and $x - 1$. The LCD is $x(x - 1)$.

Step 2

$$\frac{5}{x} = \frac{5(x - 1)}{x(x - 1)}$$

$$\frac{2}{x - 1} = \frac{2x}{(x - 1)x} = \frac{2x}{x(x - 1)}$$

Step 3

$$\frac{5}{x} + \frac{2}{x - 1} = \frac{5(x - 1)}{x(x - 1)} + \frac{2x}{x(x - 1)}$$

$$= \frac{5x - 5 + 2x}{x(x - 1)}$$

$$= \frac{7x - 5}{x(x - 1)}$$

(b) Subtract $\dfrac{3}{x - 2} - \dfrac{4}{x + 2}$.

Step 1   The LCD must have factors of $x - 2$ and $x + 2$. The LCD is $(x - 2)(x + 2)$.

Step 2

$$\frac{3}{x - 2} = \frac{3(x + 2)}{(x - 2)(x + 2)}$$ Multiply the numerator and denominator by $x + 2$.

$$\frac{4}{x + 2} = \frac{4(x - 2)}{(x + 2)(x - 2)}$$ Multiply the numerator and denominator by $x - 2$.

Step 3

$$\frac{3}{x-2} - \frac{4}{x+2} = \frac{3(x+2) - 4(x-2)}{(x+2)(x-2)}$$

Note that the *x*-term becomes negative and the constant term becomes positive.

$$= \frac{3x+6-4x+8}{(x+2)(x-2)}$$

$$= \frac{-x+14}{(x+2)(x-2)}$$

Check Yourself 4

Add or subtract as indicated.

(a) $\dfrac{3}{x+2} + \dfrac{5}{x}$ (b) $\dfrac{4}{x+3} - \dfrac{2}{x-3}$

Example 5 shows how factoring must sometimes be used in forming the LCD.

Example 5   Adding and Subtracting Unlike Rational Expressions

(a) Add $\dfrac{3}{2x-2} + \dfrac{5}{3x-3}$.

Step 1   Factor the denominators.

$$2x - 2 = 2(x-1)$$
$$3x - 3 = 3(x-1)$$

> CAUTION

$x-1$ is not used twice in forming the LCD.

The LCD must have factors of 2, 3, and $x-1$. The LCD is $2 \cdot 3(x-1)$, or $6(x-1)$.

Step 2

$$\frac{3}{2x-2} = \frac{3}{2(x-1)} = \frac{3 \cdot 3}{2(x-1) \cdot 3} = \frac{9}{6(x-1)}$$

$$\frac{5}{3x-3} = \frac{5}{3(x-1)} = \frac{5 \cdot 2}{3(x-1) \cdot 2} = \frac{10}{6(x-1)}$$

Step 3

$$\frac{3}{2x-2} + \frac{5}{3x-3} = \frac{9}{6(x-1)} + \frac{10}{6(x-1)}$$

$$= \frac{9+10}{6(x-1)}$$

$$= \frac{19}{6(x-1)}$$

(b) Subtract $\dfrac{3}{2x-4} - \dfrac{6}{x^2-4}$.

Step 1   Factor the denominators.

$2x - 4 = 2(x - 2)$

$x^2 - 4 = (x + 2)(x - 2)$

The LCD must have factors of 2, $x - 2$, and $x + 2$. The LCD is $2(x - 2)(x + 2)$.

NOTES

Multiply numerator and denominator by $x + 2$.

Multiply numerator and denominator by 2.

Step 2

$$\frac{3}{2x-4} = \frac{3}{2(x-2)} = \frac{3(x+2)}{2(x-2)(x+2)}$$

$$\frac{6}{x^2-4} = \frac{6}{(x+2)(x-2)} = \frac{6 \cdot 2}{2(x+2)(x-2)}$$

$$= \frac{12}{2(x+2)(x-2)}$$

Step 3

NOTE

Remove the parentheses and combine like terms in the numerator.

$$\frac{3}{2x-4} - \frac{6}{x^2-4} = \frac{3(x+2) - 12}{2(x-2)(x+2)}$$

$$= \frac{3x + 6 - 12}{2(x-2)(x+2)}$$

$$= \frac{3x - 6}{2(x-2)(x+2)}$$

Step 4   Simplify the difference.

NOTE

Factor the numerator and divide by the common factor, $x - 2$.

$$\frac{3x - 6}{2(x-2)(x+2)} = \frac{3(x-2)}{2(x-2)(x+2)}$$

$$= \frac{3}{2(x+2)}$$

(c) Subtract $\dfrac{5}{x^2-1} - \dfrac{2}{x^2+2x+1}$.

Step 1   Factor the denominators.

$x^2 - 1 = (x - 1)(x + 1)$

$x^2 + 2x + 1 = (x + 1)(x + 1)$

The LCD is $(x - 1)\underbrace{(x + 1)(x + 1)}$.

This factor is needed twice.

Step 2

$$\frac{5}{(x-1)(x+1)} = \frac{5(x+1)}{(x-1)(x+1)(x+1)}$$

$$\frac{2}{(x+1)(x+1)} = \frac{2(x-1)}{(x+1)(x+1)(x-1)}$$

Step 3

$$\frac{5}{x^2 - 1} - \frac{2}{x^2 + 2x + 1} = \frac{5(x + 1) - 2(x - 1)}{(x - 1)(x + 1)(x + 1)}$$

$$= \frac{5x + 5 - 2x + 2}{(x - 1)(x + 1)(x + 1)}$$

$$= \frac{3x + 7}{(x - 1)(x + 1)(x + 1)}$$

NOTE

Remove the parentheses and simplify in the numerator.

Check Yourself 5

Add or subtract as indicated.

(a) $\dfrac{5}{2x + 2} + \dfrac{1}{5x + 5}$ (b) $\dfrac{3}{x^2 - 9} - \dfrac{1}{2x - 6}$

(c) $\dfrac{4}{x^2 - x - 2} - \dfrac{3}{x^2 + 4x + 3}$

Recall from Section 5.1 that

$$a - b = -(b - a)$$

We can use this when adding or subtracting rational expressions.

 Example 6   **Adding Unlike Rational Expressions**

Add $\dfrac{4}{x - 5} + \dfrac{2}{5 - x}$.

Rather than try a denominator of $(x - 5)(5 - x)$, we rewrite one of the denominators.

NOTE

Replace $5 - x$ with $-(x - 5)$. We now use the fact that

$$\frac{a}{-b} = \frac{-a}{b}$$

$$\frac{4}{x - 5} + \frac{2}{5 - x} = \frac{4}{x - 5} + \frac{2}{-(x - 5)}$$

$$= \frac{4}{x - 5} + \frac{-2}{x - 5}$$

The LCD is now $x - 5$, and we can combine the rational expressions as

$$\frac{4 - 2}{x - 5} = \frac{2}{x - 5}$$

Check Yourself 6

Subtract the fractions.

$$\frac{3}{x - 3} - \frac{1}{3 - x}$$

✓ **Check Yourself ANSWERS**

1. (a) $\dfrac{13}{24}$; (b) $\dfrac{17}{30}$
2. (a) $\dfrac{5x + 3}{x^3}$; (b) $\dfrac{12x - 5}{20x^2}$
3. $\dfrac{4y + x}{6x^2y^2}$

4. (a) $\dfrac{8x + 10}{x(x + 2)}$; (b) $\dfrac{2x - 18}{(x + 3)(x - 3)}$
5. (a) $\dfrac{27}{10(x + 1)}$; (b) $\dfrac{-1}{2(x + 3)}$;

(c) $\dfrac{x + 18}{(x + 1)(x - 2)(x + 3)}$
6. $\dfrac{4}{x - 3}$

Reading Your Text

The following fill-in-the-blank exercises are designed to ensure that you understand some of the key vocabulary used in this section.

SECTION 5.4

(a) Algebraic fractions that do not have the same denominator are called unlike _____ expressions.

(b) When adding unlike fractions, it is necessary to find a _____ denominator.

(c) The final step in subtracting fractions is to write the difference in _____ form.

(d) The expression $a - b$ is the _____ of the expression $b - a$.

5.4 exercises

< Objectives 1–2 >

Add or subtract as indicated. Express your result in simplest form.

1. $\dfrac{3}{7} + \dfrac{5}{6}$

2. $\dfrac{7}{12} - \dfrac{4}{9}$

3. $\dfrac{13}{25} - \dfrac{7}{20}$

4. $\dfrac{3}{5} + \dfrac{7}{9}$

5. $\dfrac{y}{4} + \dfrac{3y}{5}$

6. $\dfrac{5x}{6} - \dfrac{2x}{3}$

7. $\dfrac{7a}{3} - \dfrac{a}{7}$

8. $\dfrac{3m}{4} + \dfrac{m}{9}$

9. $\dfrac{3}{x} - \dfrac{4}{5}$

10. $\dfrac{5}{x} + \dfrac{2}{3}$

11. $\dfrac{5}{a} + \dfrac{a}{5}$

12. $\dfrac{y}{3} - \dfrac{3}{y}$

13. $\dfrac{5}{m} + \dfrac{3}{m^2}$

14. $\dfrac{4}{x^2} - \dfrac{3}{x}$

15. $\dfrac{2}{x^2} - \dfrac{5}{7x}$ > Videos

16. $\dfrac{7}{3w} + \dfrac{5}{w^3}$

17. $\dfrac{7}{9s} + \dfrac{5}{s^2}$

18. $\dfrac{11}{x^2} - \dfrac{5}{7x}$

19. $\dfrac{3}{4b^2} + \dfrac{5}{3b^3}$

20. $\dfrac{4}{5x^3} - \dfrac{3}{2x^2}$

21. $\dfrac{x}{x + 2} + \dfrac{2}{5}$

22. $\dfrac{3}{4} - \dfrac{a}{a - 1}$

23. $\dfrac{y}{y - 4} - \dfrac{3}{4}$

24. $\dfrac{m}{m + 3} + \dfrac{2}{3}$

25. $\dfrac{4}{x} + \dfrac{3}{x + 1}$

26. $\dfrac{2}{x} - \dfrac{1}{x - 2}$

Name _____

Section _____ Date _____

Answers

1. _____   2. _____

3. _____   4. _____

5. _____   6. _____

7. _____   8. _____

9. _____   10. _____

11. _____   12. _____

13. _____   14. _____

15. _____   16. _____

17. _____   18. _____

19. _____   20. _____

21. _____   22. _____

23. _____   24. _____

25. _____   26. _____

27. $\dfrac{5}{a-1} - \dfrac{2}{a}$

28. $\dfrac{4}{x+2} + \dfrac{3}{x}$

29. $\dfrac{4}{2x-3} + \dfrac{2}{3x}$

30. $\dfrac{7}{2y-1} - \dfrac{3}{2y}$

31. $\dfrac{2}{x+1} + \dfrac{3}{x+3}$

32. $\dfrac{5}{x-1} + \dfrac{2}{x+2}$

33. $\dfrac{4}{y-2} - \dfrac{1}{y+1}$ > Videos

34. $\dfrac{5}{x+4} - \dfrac{3}{x-1}$

| Basic Skills | **Challenge Yourself** ▲ | Calculator/Computer | Career Applications | Above and Beyond |

Determine whether each statement is **true** *or* **false**.

35. The expression $a - b$ is the opposite of the expression $b - a$.

36. The expression $a - b$ is the opposite of the expression $a + b$.

37. You must find the greatest common factor in order to add unlike fractions.

38. To add two like fractions, add the denominators and place the sum under the common numerator.

Evaluate and simplify.

39. $\dfrac{x}{x+4} - \dfrac{2}{3x+12}$

40. $\dfrac{x}{x-3} + \dfrac{5}{2x-6}$

41. $\dfrac{4}{5x-10} - \dfrac{1}{3x-6}$

42. $\dfrac{2}{3w+3} + \dfrac{5}{2w+2}$

43. $\dfrac{7}{3c+6} - \dfrac{2c}{7c+14}$

44. $\dfrac{5}{3c-12} + \dfrac{4c}{5c-20}$

45. $\dfrac{y-1}{y+1} - \dfrac{y}{3y+3}$

46. $\dfrac{x+2}{x-2} - \dfrac{x}{3x-6}$

47. $\dfrac{3}{x^2-4} + \dfrac{2}{x+2}$

48. $\dfrac{4}{x-2} + \dfrac{3}{x^2-x-2}$

49. $\dfrac{3x}{x^2-3x+2} - \dfrac{1}{x-2}$

50. $\dfrac{a}{a^2-1} - \dfrac{4}{a+1}$

Answers

27.

28.

29.

30.

31.

32.

33.

34.

35.   36.

37.   38.

39.   40.

41.   42.

43.   44.

45.   46.

47.

48.

49.

50.

The Streeter/Hutchison Series in Mathematics   Beginning Algebra

51. $\dfrac{2x}{x^2 - 5x + 6} + \dfrac{4}{x - 2}$ > Videos

52. $\dfrac{7a}{a^2 + a - 12} - \dfrac{4}{a + 4}$

53. $\dfrac{2}{3x - 3} - \dfrac{1}{4x + 4}$

54. $\dfrac{2}{5w + 10} - \dfrac{3}{2w - 4}$

55. $\dfrac{4}{3a - 9} - \dfrac{3}{2a + 4}$

56. $\dfrac{2}{3b - 6} + \dfrac{3}{4b + 8}$

57. $\dfrac{5}{x^2 - 16} - \dfrac{3}{x^2 - x - 12}$

58. $\dfrac{3}{x^2 + 4x + 3} - \dfrac{1}{x^2 - 9}$

59. $\dfrac{2}{y^2 + y - 6} + \dfrac{3y}{y^2 - 2y - 15}$

 > Videos

60. $\dfrac{2a}{a^2 - a - 12} - \dfrac{3}{a^2 - 2a - 8}$

61. $\dfrac{6x}{x^2 - 9} - \dfrac{5x}{x^2 + x - 6}$

62. $\dfrac{4y}{y^2 + 6y + 5} + \dfrac{2y}{y^2 - 1}$

63. $\dfrac{3}{a - 7} + \dfrac{2}{7 - a}$

64. $\dfrac{5}{x - 5} - \dfrac{3}{5 - x}$

65. $\dfrac{2x}{2x - 3} - \dfrac{1}{3 - 2x}$

 > Videos

66. $\dfrac{9m}{3m - 1} + \dfrac{3}{1 - 3m}$

| Basic Skills | Challenge Yourself | Calculator/Computer | Career Applications | **Above and Beyond** |

▲

67. $\dfrac{1}{a - 3} - \dfrac{1}{a + 3} + \dfrac{2a}{a^2 - 9}$

68. $\dfrac{1}{p + 1} + \dfrac{1}{p - 3} - \dfrac{4}{p^2 - 2p - 3}$

69. $\dfrac{2x^2 + 3x}{x^2 - 2x - 63} + \dfrac{7 - x}{x^2 - 2x - 63} - \dfrac{x^2 - 3x + 21}{x^2 - 2x - 63}$

70. $-\dfrac{3 - 2x^2}{x^2 - 9x + 20} - \dfrac{4x^2 + 2x + 1}{x^2 - 9x + 20} + \dfrac{2x^2 + 3x}{x^2 - 9x + 20}$

Answers

51. _____
52. _____
53. _____
54. _____
55. _____
56. _____
57. _____
58. _____
59. _____
60. _____
61. _____
62. _____
63. _____
64. _____
65. _____
66. _____
67. _____
68. _____
69. _____
70. _____

Answers

71. _____

72. _____

73. _____

74. _____

Solve each application.

71. NUMBER PROBLEM Use a rational expression to represent the sum of the reciprocals of two consecutive even integers.

72. NUMBER PROBLEM One number is two less than another. Use a rational expression to represent the sum of the reciprocals of the two numbers.

73. GEOMETRY Refer to the rectangle in the figure. Find an expression that represents its perimeter.

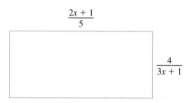

74. GEOMETRY Refer to the triangle in the figure. Find an expression that represents its perimeter.

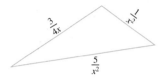

Answers

1. $\dfrac{53}{42}$ 3. $\dfrac{17}{100}$ 5. $\dfrac{17y}{20}$ 7. $\dfrac{46a}{21}$ 9. $\dfrac{15 - 4x}{5x}$ 11. $\dfrac{25 + a^2}{5a}$

13. $\dfrac{5m + 3}{m^2}$ 15. $\dfrac{14 - 5x}{7x^2}$ 17. $\dfrac{7s + 45}{9s^2}$ 19. $\dfrac{9b + 20}{12b^3}$

21. $\dfrac{7x + 4}{5(x + 2)}$ 23. $\dfrac{y + 12}{4(y - 4)}$ 25. $\dfrac{7x + 4}{x(x + 1)}$ 27. $\dfrac{3a + 2}{a(a - 1)}$

29. $\dfrac{2(8x - 3)}{3x(2x - 3)}$ 31. $\dfrac{5x + 9}{(x + 1)(x + 3)}$ 33. $\dfrac{3(y + 2)}{(y - 2)(y + 1)}$

35. True   37. False   39. $\dfrac{3x - 2}{3(x + 4)}$ 41. $\dfrac{7}{15(x - 2)}$

43. $\dfrac{49 - 6c}{21(c + 2)}$ 45. $\dfrac{2y - 3}{3(y + 1)}$ 47. $\dfrac{2x - 1}{(x - 2)(x + 2)}$

49. $\dfrac{2x + 1}{(x - 1)(x - 2)}$ 51. $\dfrac{6}{x - 3}$ 53. $\dfrac{5x + 11}{12(x - 1)(x + 1)}$

55. $\dfrac{-a + 43}{6(a - 3)(a + 2)}$ 57. $\dfrac{2x + 3}{(x + 4)(x - 4)(x + 3)}$

59. $\dfrac{3y^2 - 4y - 10}{(y + 3)(y - 2)(y - 5)}$ 61. $\dfrac{x}{(x - 3)(x - 2)}$ 63. $\dfrac{1}{a - 7}$

65. $\dfrac{2x + 1}{2x - 3}$ 67. $\dfrac{2}{a - 3}$ 69. $\dfrac{x - 2}{x - 9}$ 71. $\dfrac{2x + 2}{x(x + 2)}$

73. $\dfrac{2(6x^2 + 5x + 21)}{5(3x + 1)}$

5.5

Complex Rational Expressions

< 5.5 Objectives >

1 > Simplify a complex arithmetic fraction

2 > Simplify a complex rational expression

Recall the way you were taught to divide fractions. The rule was referred to as invert-and-multiply. We will see why this rule works.

$$\frac{3}{5} \div \frac{2}{3}$$

We can write

$$\frac{3}{5} \div \frac{2}{3} = \frac{\dfrac{3}{5}}{\dfrac{2}{3}} = \frac{\dfrac{3}{5} \cdot \dfrac{3}{2}}{\dfrac{2}{3} \cdot \dfrac{3}{2}}$$ We are multiplying by 1.

Interpret the division as a fraction.

$$= \frac{\dfrac{3}{5} \cdot \dfrac{3}{2}}{1}$$

$$\frac{2}{3} \cdot \frac{3}{2} = 1$$

$$= \frac{3}{5} \cdot \frac{3}{2}$$

We then have

$$\frac{3}{5} \div \frac{2}{3} = \frac{3}{5} \cdot \frac{3}{2} = \frac{9}{10}$$

By comparing these expressions, you should see the rule for dividing fractions. Invert the fraction that follows the division symbol and multiply.

A fraction that has a fraction in its numerator, in its denominator, or in both is called a **complex fraction.** For example, the following are complex fractions.

$$\frac{\dfrac{5}{6}}{\dfrac{3}{4}}, \qquad \frac{\dfrac{4}{x}}{\dfrac{3}{x^2}}, \qquad \text{and} \qquad \frac{\dfrac{a+2}{3}}{\dfrac{a-2}{5}}$$

Remember that we can always multiply the numerator and the denominator of a fraction by the same nonzero term.

RECALL

This is the Fundamental Principle of Fractions.

$$\frac{P}{Q} = \frac{P \cdot R}{Q \cdot R} \qquad \text{in which } Q \neq 0 \text{ and } R \neq 0$$

To simplify a complex fraction, multiply the numerator and denominator by the LCD of all fractions that appear within the complex fraction.

367

> **Example 1**   **Simplifying Complex Fractions**

< Objective 1 >

Simplify $\dfrac{\dfrac{3}{4}}{\dfrac{5}{8}}$.

The LCD of $\dfrac{3}{4}$ and $\dfrac{5}{8}$ is 8. So multiply the numerator and denominator by 8.

$$\frac{\dfrac{3}{4}}{\dfrac{5}{8}} = \frac{\dfrac{3}{4} \cdot 8}{\dfrac{5}{8} \cdot 8} = \frac{3 \cdot 2}{5 \cdot 1} = \frac{6}{5}$$

 Check Yourself 1

Simplify.

(a) $\dfrac{\dfrac{4}{7}}{\dfrac{3}{7}}$ (b) $\dfrac{\dfrac{3}{8}}{\dfrac{5}{6}}$

The same method can be used to simplify a complex fraction when variables are involved in the expression. Consider Example 2.

> **Example 2**   **Simplifying Complex Rational Expressions**

< Objective 2 >

Simplify $\dfrac{\dfrac{5}{x}}{\dfrac{10}{x^2}}$.

The LCD of $\dfrac{5}{x}$ and $\dfrac{10}{x^2}$ is x^2, so multiply the numerator and denominator by x^2.

NOTE

Be sure to write the result in simplest form.

$$\frac{\dfrac{5}{x}}{\dfrac{10}{x^2}} = \frac{\left(\dfrac{5}{x}\right)x^2}{\left(\dfrac{10}{x^2}\right)x^2} = \frac{5x}{10} = \frac{x}{2}$$

 Check Yourself 2

Simplify.

(a) $\dfrac{\dfrac{6}{x^3}}{\dfrac{9}{x^2}}$ (b) $\dfrac{\dfrac{m^4}{15}}{\dfrac{m^3}{20}}$

We may also have a sum or a difference in the numerator or denominator of a complex fraction. The simplification steps are exactly the same. Consider Example 3.

Example 3 — **Simplifying Complex Fractions**

Simplify $\dfrac{1 + \dfrac{x}{y}}{1 - \dfrac{x}{y}}$.

The LCD of $1, \dfrac{x}{y}, 1$, and $\dfrac{x}{y}$ is y, so multiply the numerator and denominator by y.

NOTE

We use the distributive property to multiply *each term* in the numerator and in the denominator by y.

$$\frac{1 + \dfrac{x}{y}}{1 - \dfrac{x}{y}} = \frac{\left(1 + \dfrac{x}{y}\right)y}{\left(1 - \dfrac{x}{y}\right)y} = \frac{1 \cdot y + \dfrac{x}{y} \cdot y}{1 \cdot y - \dfrac{x}{y} \cdot y}$$

$$= \frac{y + x}{y - x}$$

 Check Yourself 3

Simplify.

$$\frac{\dfrac{x}{y} - 2}{\dfrac{x}{y} + 2}$$

The following algorithm summarizes our work to this point with simplifying complex fractions.

Step by Step

To Simplify Complex Rational Expressions

Step 1   Multiply the numerator and denominator of the complex rational expression by the LCD of all the fractions that appear within the complex rational expression.

Step 2   Write the resulting fraction in simplest form.

RECALL

To divide by a fraction, we invert the divisor (it *follows* the division sign) and multiply.

A second method for simplifying complex fractions uses the fact that

$$\frac{\dfrac{P}{Q}}{\dfrac{R}{S}} = \frac{P}{Q} \div \frac{R}{S} = \frac{P}{Q} \cdot \frac{S}{R}$$

To use this method, we must write the numerator and denominator of the complex fraction as single fractions. We can then divide the numerator by the denominator as before.

In Example 4, we use this method to simplify the complex rational expression we saw in Example 3.

Example 4 — **Simplifying Complex Fractions**

Simplify $\dfrac{1 + \dfrac{x}{y}}{1 - \dfrac{x}{y}}$.

To use this method, we rewrite both the numerator and the denominator as single fractions.

$$\frac{1 + \dfrac{x}{y}}{1 - \dfrac{x}{y}} = \frac{\dfrac{y}{y} + \dfrac{x}{y}}{\dfrac{y}{y} - \dfrac{x}{y}} = \frac{\dfrac{y + x}{y}}{\dfrac{y - x}{y}}$$

Now we invert and multiply.

$$\frac{\dfrac{y + x}{y}}{\dfrac{y - x}{y}} = \frac{y + x}{y} \cdot \frac{y}{y - x} = \frac{y + x}{y - x}$$

Not surprisingly, we have the same result as we found in Example 3.

Check Yourself 4

Simplify using the second method $\dfrac{\dfrac{x}{y} - 2}{\dfrac{x}{y} + 2}$.

Check Yourself ANSWERS

1. (a) $\dfrac{4}{3}$; (b) $\dfrac{9}{20}$ 2. (a) $\dfrac{2}{3x}$; (b) $\dfrac{4m}{3}$ 3. $\dfrac{x - 2y}{x + 2y}$ 4. $\dfrac{x - 2y}{x + 2y}$

Reading Your Text

The following fill-in-the-blank exercises are designed to ensure that you understand some of the key vocabulary used in this section.

SECTION 5.5

(a) The rule for dividing fractions is referred to as _____ and multiply.

(b) We can always multiply the numerator and denominator of a fraction by the same _____ term.

(c) A fraction that has a _____ in its numerator, in its denominator, or in both is called a complex fraction.

(d) To simplify a complex fraction, multiply the numerator and denominator by the _____ of all fractions that appear within the complex fraction.

Basic Skills | Challenge Yourself | Calculator/Computer | Career Applications | Above and Beyond

< Objectives 1–2 >

Simplify each complex fraction.

1. $\dfrac{\frac{2}{3}}{\frac{6}{8}}$

2. $\dfrac{\frac{5}{6}}{\frac{10}{15}}$

3. $\dfrac{1 + \frac{1}{2}}{2 + \frac{1}{4}}$ > Videos

4. $\dfrac{1 + \frac{3}{4}}{2 - \frac{1}{8}}$

5. $\dfrac{\frac{x}{8}}{\frac{x^2}{4}}$

6. $\dfrac{\frac{m^2}{10}}{\frac{m^3}{15}}$

7. $\dfrac{\frac{3}{a}}{\frac{2}{a^2}}$

8. $\dfrac{\frac{6}{x^2}}{\frac{9}{x^3}}$

9. $\dfrac{\frac{y+1}{y}}{\frac{y-1}{2y}}$ > Videos

10. $\dfrac{\frac{w+3}{4w}}{\frac{w-3}{2w}}$

11. $\dfrac{2 - \frac{1}{x}}{2 + \frac{1}{x}}$ > Videos

12. $\dfrac{3 + \frac{1}{a}}{3 - \frac{1}{a}}$

13. $\dfrac{3 - \frac{x}{y}}{\frac{6}{y}}$

14. $\dfrac{2 + \frac{x}{y}}{\frac{4}{y}}$

Basic Skills | **Challenge Yourself** | Calculator/Computer | Career Applications | Above and Beyond

15. $\dfrac{\frac{x^2}{y^2} - 1}{\frac{x}{y} + 1}$

16. $\dfrac{\frac{a}{b} + 2}{\frac{a^2}{b^2} - 4}$

Boost *your* GRADE at ALEKS.com!

ALEKS

- Practice Problems
- Self-Tests
- NetTutor
- e-Professors
- Videos

Name _____

Section _____ Date _____

Answers

1. _____   2. _____

3. _____   4. _____

5. _____

6. _____

7. _____

8. _____

9. _____

10. _____

11. _____

12. _____

13. _____

14. _____

15. _____

16. _____

Answers

17. _____

18. _____

19. _____

20. _____

21. _____

22. _____

23. _____

24. _____

25. _____

26. _____

27. _____

28. _____

17. $\dfrac{1 + \dfrac{3}{x} - \dfrac{4}{x^2}}{1 + \dfrac{2}{x} - \dfrac{3}{x^2}}$

18. $\dfrac{1 - \dfrac{2}{r} - \dfrac{8}{r^2}}{1 - \dfrac{1}{r} - \dfrac{6}{r^2}}$

19. $\dfrac{\dfrac{2}{x} - \dfrac{1}{xy}}{\dfrac{1}{xy} + \dfrac{2}{y}}$

20. $\dfrac{\dfrac{1}{xy} + \dfrac{2}{x}}{\dfrac{3}{y} - \dfrac{1}{xy}}$

21. $\dfrac{\dfrac{2}{x - 1} + 1}{1 - \dfrac{3}{x - 1}}$ › Videos

22. $\dfrac{\dfrac{3}{a + 2} - 1}{1 + \dfrac{2}{a + 2}}$

23. $\dfrac{1 - \dfrac{1}{y - 1}}{y - \dfrac{8}{y + 2}}$

24. $\dfrac{1 + \dfrac{1}{x + 2}}{x - \dfrac{18}{x - 3}}$

25. $1 + \dfrac{1}{1 + \dfrac{1}{x}}$ › Videos

26. $1 + \dfrac{1}{1 - \dfrac{1}{y}}$

Solve each application.

27. **GEOMETRY** The area of the rectangle shown here is $\dfrac{2}{3}$. Find the width.

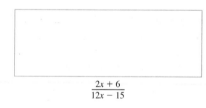

$\dfrac{2x + 6}{12x - 15}$

28. **GEOMETRY** The area of the rectangle shown here is $\dfrac{2(3x - 2)}{x - 1}$. Find the width.

$\dfrac{3x - 2}{x - 2}$

Basic Skills | Challenge Yourself | Calculator/Computer | Career Applications | **Above and Beyond**
▲

29. Complex fractions have some interesting patterns. Work with a partner to evaluate each complex fraction in the sequence below. This is an interesting sequence of fractions because the numerators and denominators are a famous sequence of whole numbers, and the fractions get closer and closer to a number called "the golden mean."

$$1, \quad 1 + \frac{1}{1}, \quad 1 + \frac{1}{1 + \frac{1}{1}}, \quad 1 + \frac{1}{1 + \frac{1}{1 + \frac{1}{1}}}, \quad 1 + \frac{1}{1 + \frac{1}{1 + \frac{1}{1 + \frac{1}{1}}}}, \ldots$$

After you have evaluated these first five, you no doubt will see a pattern in the resulting fractions that allows you to go on indefinitely without having to evaluate more complex fractions. Write each of these fractions as decimals. Write your observations about the sequence of fractions and about the sequence of decimal fractions.

30. This inequality is used when the U.S. House of Representatives seats are apportioned (see the chapter opener for more information).

$$\frac{\dfrac{E}{e} - \dfrac{A}{a + 1}}{\dfrac{A}{a + 1}} < \frac{\dfrac{A}{a} - \dfrac{E}{e + 1}}{\dfrac{E}{e + 1}}$$

chapter 5 > Make the Connection

Show that this inequality can be simplified to

$$\frac{A}{\sqrt{a(a + 1)}} > \frac{E}{\sqrt{e(e + 1)}}.$$

Here, A and E represent the populations of two states of the United States, and a and e are the number of representatives each of these two states have in the U.S. House of Representatives.

Mathematicians have shown that there are situations in which the method for apportionment described in the chapter's introduction does not work, and a state may not even get its basic quota of representatives. They give the table below of a hypothetical seven states and their populations as an example.

State	Population	Exact Quota	Number of Reps.
A	325	1.625	2
B	788	3.940	4
C	548	2.740	3
D	562	2.810	3
E	4,263	21.315	21
F	3,219	16.095	15
G	295	1.475	2
Total	10,000	50	50

In this case, the total population of all states is 10,000, and there are 50 representatives in all, so there should be no more than 10,000/50 or 200 people per representative. The quotas are found by dividing the population by 200. Whether a state, A, should get an additional representative before another state, E, should get one is decided in this method by using the simplified inequality below. If the ratio

$$\frac{A}{\sqrt{a(a+1)}} > \frac{E}{\sqrt{e(e+1)}}$$

is true, then A gets an extra representative before E does.

(a) If you go through the process of comparing the inequality above for each pair of states, state F loses a representative to state G. Do you see how this happens? Will state F complain?

(b) Alexander Hamilton, one of the signers of the Constitution, proposed that the extra representative positions be given one at a time to states with the largest remainder until all the "extra" positions were filled. How would this affect the table? Do you agree or disagree?

Answers

1. $\dfrac{8}{9}$ 3. $\dfrac{2}{3}$ 5. $\dfrac{1}{2x}$ 7. $\dfrac{3a}{2}$ 9. $\dfrac{2(y+1)}{y-1}$ 11. $\dfrac{2x-1}{2x+1}$

13. $\dfrac{3y-x}{6}$ 15. $\dfrac{x-y}{y}$ 17. $\dfrac{x+4}{x+3}$ 19. $\dfrac{2y-1}{1+2x}$ 21. $\dfrac{x+1}{x-4}$

23. $\dfrac{y+2}{(y-1)(y+4)}$ 25. $\dfrac{2x+1}{x+1}$ 27. $\dfrac{4x-5}{x+3}$

29. Above and Beyond

5.6 Equations Involving Rational Expressions

< 5.6 Objectives >

1 > Solve a rational equation with integer denominators

2 > Determine the excluded values for the variables of a rational expression

3 > Solve a rational equation

4 > Solve a proportion for an unknown

In Chapter 2, you learned how to solve a variety of equations. We now want to extend that work to solving **rational equations** or equations involving rational expressions.

To solve a rational equation, we multiply each term of the equation by the LCD of any fractions in the equation. The resulting equation should be equivalent to the original equation but cleared of all fractions.

| Example 1 | Solving Equations with Integer Denominators |

< Objective 1 >

Solve.

$$\frac{x}{2} - \frac{1}{3} = \frac{2x + 3}{6}$$

NOTE

This equation has three terms: $\frac{x}{2}$, $-\frac{1}{3}$, and $\frac{2x + 3}{6}$. The sign of the term is not used to find the LCD.

The LCD for $\frac{x}{2}, \frac{1}{3}$, and $\frac{2x + 3}{6}$ is 6. Multiply both sides of the equation by 6. Using the distributive property, we multiply *each* term by 6.

$$6 \cdot \frac{x}{2} - 6 \cdot \frac{1}{3} = 6\left(\frac{2x + 3}{6}\right)$$

or

$$3x - 2 = 2x + 3$$

Solving as before, we have

$$3x - 2x = 3 + 2 \qquad \text{or} \qquad x = 5$$

To check, substitute 5 for x in the *original* equation.

$$\frac{(5)}{2} - \frac{1}{3} \overset{?}{=} \frac{2(5) + 3}{6}$$

$$\frac{13}{6} \overset{\checkmark}{=} \frac{13}{6} \qquad \text{(True)}$$

 >CAUTION

Be careful! Many students have difficulty because they do not distinguish between adding and subtracting *expressions* (as we did in Sections 5.3 and 5.4) and solving equations. In the **expression**

$$\frac{x + 1}{2} + \frac{x}{3}$$

we want to add the two fractions to form a single fraction. In the **equation**

$$\frac{x + 1}{2} = \frac{x}{3} + 1$$

we want to solve for x.

Check Yourself 1

Solve and check.

$$\frac{x}{4} - \frac{1}{6} = \frac{4x - 5}{12}$$

In Example 1, all the denominators were integers. What happens when we allow variables in the denominator? Recall that, for any fraction, the denominator must not be equal to zero. When a fraction has a variable in the denominator, we must exclude any value for the variable that results in division by zero.

| **Example 2** | **Finding Excluded Values for x** |

< Objective 2 >

In each rational expression, what values for x must be excluded?

(a) $\dfrac{x}{5}$ Here x can have any value, so there are no excluded values.

(b) $\dfrac{3}{x}$ If $x = 0$, then $\dfrac{3}{x}$ is undefined; 0 is the excluded value.

(c) $\dfrac{5}{x - 2}$ If $x = 2$, then $\dfrac{5}{x - 2} = \dfrac{5}{(2) - 2} = \dfrac{5}{0}$, which is undefined, so 2 is the excluded value.

Check Yourself 2

What values for x, if any, must be excluded?

(a) $\dfrac{x}{7}$ (b) $\dfrac{5}{x}$ (c) $\dfrac{7}{x - 5}$

If the denominator of a rational expression contains a product of two or more variable factors, the zero-product principle must be used to determine the excluded values for the variable.

In some cases, you have to factor the denominator to see the restrictions on the values for the variable.

| **Example 3** | **Finding Excluded Values for x** |

What values for x must be excluded in each fraction?

(a) $\dfrac{3}{x^2 - 6x - 16}$

Factoring the denominator, we have

$$\frac{3}{x^2 - 6x - 16} = \frac{3}{(x - 8)(x + 2)}$$

Letting $x - 8 = 0$ or $x + 2 = 0$, we see that 8 and -2 make the denominator 0, so both 8 and -2 must be excluded.

(b) $\dfrac{3}{x^2 + 2x - 48}$

The denominator is zero when

$$x^2 + 2x - 48 = 0$$

Factoring, we find

$$(x - 6)(x + 8) = 0$$

The Streeter/Hutchison Series in Mathematics   Beginning Algebra

The denominator is zero when

$x = 6$ or $x = -8$

The excluded values are 6 and -8.

Check Yourself 3

What values for x must be excluded in each fraction?

(a) $\dfrac{5}{x^2 - 3x - 10}$ (b) $\dfrac{7}{x^2 + 5x - 14}$

Here are the steps for solving an equation involving fractions.

Step by Step

To Solve a Rational Equation

Step 1 Remove the fractions in the equation by multiplying each term by the LCD of all the fractions.

Step 2 Solve the equation resulting from step 1 by the methods of Sections 2.3 and 4.6.

Step 3 Check the solution in the *original equation.*

We can also solve rational equations with variables in the denominator by using the above algorithm. Example 4 illustrates this approach.

  Example 4   **Solving Rational Equations**

< Objective 3 >

Solve.

$$\frac{7}{4x} - \frac{3}{x^2} = \frac{1}{2x^2}$$

The LCD of the three terms in the equation is $4x^2$, so we multiply both sides of the equation by $4x^2$.

NOTE

The factor x appears twice in the LCD.

$$4x^2 \cdot \frac{7}{4x} - 4x^2 \cdot \frac{3}{x^2} = 4x^2 \cdot \frac{1}{2x^2}$$

Simplifying, we have

$$7x - 12 = 2$$
$$7x = 14$$
$$x = 2$$

We leave the check to you. Be sure to return to the original equation.

Check Yourself 4

Solve and check.

$$\frac{5}{2x} - \frac{4}{x^2} = \frac{7}{2x^2}$$

The process of solving rational equations is exactly the same when there are binomials in the denominators.

Example 5   Solving Rational Equations

(a) Solve.

$$\frac{x}{x-3} - 2 = \frac{1}{x-3}$$

The LCD is $x - 3$, so we multiply each side (every term) by $x - 3$.

$$(x-3) \cdot \left(\frac{x}{x-3}\right) - 2(x-3) = (x-3) \cdot \left(\frac{1}{x-3}\right)$$

Simplifying, we have

$$x - 2(x - 3) = 1$$
$$x - 2x + 6 = 1$$
$$-x = -5$$
$$x = 5$$

To check, substitute 5 for x in the original equation.

$$\frac{(5)}{(5)-3} - 2 \stackrel{?}{=} \frac{1}{(5)-3}$$

$$\frac{5}{2} - 2 \stackrel{?}{=} \frac{1}{2}$$

$$\frac{1}{2} \stackrel{\checkmark}{=} \frac{1}{2}$$

(b) Solve.

$$\frac{3}{x-3} - \frac{7}{x+3} = \frac{2}{x^2-9}$$

In factored form, the three denominators are $x - 3$, $x + 3$, and $(x + 3)(x - 3)$. This means that the LCD is $(x + 3)(x - 3)$, and so we multiply:

$$(x-3)(x+3)\left(\frac{3}{x-3}\right) - (x+3)(x-3)\left(\frac{7}{x+3}\right) = (x+3)(x-3)\left(\frac{2}{x^2-9}\right)$$

Simplifying, we have

$$3(x + 3) - 7(x - 3) = 2$$
$$3x + 9 - 7x + 21 = 2$$
$$-4x + 30 = 2$$
$$-4x = -28$$
$$x = 7$$

 Check Yourself 5

Solve and check.

(a) $\dfrac{x}{x-5} - 2 = \dfrac{2}{x-5}$ **(b)** $\dfrac{4}{x-4} - \dfrac{3}{x+1} = \dfrac{5}{x^2-3x-4}$

You should be aware that some rational equations have no solutions. Example 6 shows that possibility.

| Example 6 | Solving Rational Equations |

Solve.

$$\frac{x}{x-2} - 7 = \frac{2}{x-2}$$

The LCD is $x - 2$, and so we multiply each side (every term) by $x - 2$.

$$(x-2)\left(\frac{x}{x-2}\right) - 7(x-2) = (x-2)\left(\frac{2}{x-2}\right)$$

Simplifying, we have

$$x - 7x + 14 = 2$$
$$-6x = -12$$
$$x = 2$$

Now, when we try to check our result, we have

NOTE

2 is substituted for x in the original equation.

$$\frac{(2)}{(2)-2} - 7 \stackrel{?}{=} \frac{2}{(2)-2} \qquad \text{or} \qquad \frac{2}{0} - 7 \stackrel{?}{=} \frac{2}{0}$$

These terms are undefined.

What went wrong? Remember that two of the terms in our original equation were $\frac{x}{x-2}$ and $\frac{2}{x-2}$. The variable x cannot have the value 2 because 2 is an excluded value (it makes the denominator 0). So our original equation has *no solution*.

Check Yourself 6

Solve, if possible.

$$\frac{x}{x+3} - 6 = \frac{-3}{x+3}$$

Equations involving fractions may also lead to quadratic equations, as Example 7 illustrates.

| Example 7 | Solving Rational Equations |

Solve.

$$\frac{x}{x-4} = \frac{15}{x-3} - \frac{2x}{x^2 - 7x + 12}$$

The LCD is $(x-4)(x-3)$. Multiply each side (every term) by $(x-4)(x-3)$.

$$\frac{x}{(x-4)}(x-4)(x-3) = \frac{15}{(x-3)}(x-4)(x-3) - \frac{2x}{(x-4)(x-3)}(x-4)(x-3)$$

Simplifying, we have

$$x(x-3) = 15(x-4) - 2x$$

Multiply to remove the parentheses:

$$x^2 - 3x = 15x - 60 - 2x$$

Beginning Algebra   The Streeter/Hutchison Series in Mathematics   © The McGraw-Hill Companies. All Rights Reserved.

NOTE

This equation is *quadratic*. It can be solved by the methods of Section 4.4.

In standard form, the equation is

$$x^2 - 16x + 60 = 0$$

or

$$(x - 6)(x - 10) = 0$$

Setting the factors to 0, we have

$$x - 6 = 0 \quad \text{or} \quad x - 10 = 0$$
$$x = 6 \qquad\qquad x = 10$$

So $x = 6$ and $x = 10$ are possible solutions. We leave the check of *each* solution to you.

Check Yourself 7

Solve and check.

$$\frac{3x}{x + 2} - \frac{2}{x + 3} = \frac{36}{x^2 + 5x + 6}$$

The following equation is a special kind of equation involving fractions:

$$\frac{135}{t} = \frac{180}{t + 1}$$

An equation of the form $\dfrac{a}{b} = \dfrac{c}{d}$ is said to be in **proportion form,** or, more simply, it is called a **proportion.** This type of equation occurs often enough in algebra that it is worth developing some special methods for its solution. First, we need some definitions.

A **ratio** is a means of comparing two quantities. A ratio can be written as a fraction. For instance, the ratio of 2 to 3 can be written as $\dfrac{2}{3}$. A statement that two ratios are equal is called a *proportion.* A proportion has the form

$$\frac{a}{b} = \frac{c}{d}$$ In this proportion, *a* and *d* are called the **extremes** of the proportion, and *b* and *c* are called the **means.**

NOTE

bd is the LCD of the denominators.

A useful property of proportions is easily developed. If

$$\frac{a}{b} = \frac{c}{d}$$ We multiply both sides by $b \cdot d$.

$$\left(\frac{a}{b}\right)bd = \left(\frac{c}{d}\right)bd \quad \text{or} \quad ad = bc$$

Property

Proportions   If $\dfrac{a}{b} = \dfrac{c}{d}$, then $ad = bc$.

In words:

In any proportion, the product of the extremes (*ad*) is equal to the product of the means (*bc*).

Because a proportion is a special kind of rational equation, this rule gives us an alternative approach to solving equations that are in the proportion form.

Example 8	**Solving a Proportion for an Unknown**

< Objective 4 >

Solve each equation.

(a) $\dfrac{x}{5} = \dfrac{12}{15}$

NOTE

The extremes are x and 15.
The means are 5 and 12.

Set the product of the extremes equal to the product of the means.

$15x = 5 \cdot 12$
$15x = 60$
$x = 4$

Our solution is 4. You can check as before, by substituting in the original proportion.

(b) $\dfrac{x + 3}{10} = \dfrac{x}{7}$

Set the product of the extremes equal to the product of the means. Be certain to use parentheses with a numerator with more than one term.

$7(x + 3) = 10x$
$7x + 21 = 10x$
$21 = 3x$
$7 = x$

We leave it to you to check this result.

Check Yourself 8

Solve each equation.

(a) $\dfrac{x}{8} = \dfrac{3}{4}$ **(b)** $\dfrac{x - 1}{9} = \dfrac{x + 1}{12}$

As the examples of this section illustrated, *whenever* an equation involves rational expressions, the *first step* of the solution is to clear the equation of fractions by multiplication.

The following algorithm summarizes our work in solving equations that involve rational expressions.

Step by Step

To Solve an Equation Involving Fractions

Step 1 Remove the fractions appearing in the equation by multiplying each side (every term) by the LCD of all the fractions.

Step 2 Solve the equation resulting from step 1. If the equation is linear, use the methods of Section 2.3 for the solution. If the equation is quadratic, use the methods of Section 4.6.

Step 3 Check all solutions by substitution in the *original equation*. Be sure to discard any *extraneous* solutions, that is, solutions that result in a zero denominator in the original equation.

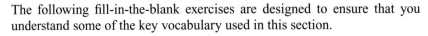

Check Yourself ANSWERS

1. $x = 3$ 2. **(a)** None; **(b)** 0; **(c)** 5   3. **(a)** $-2, 5$; **(b)** $-7, 2$ 4. $x = 3$

5. **(a)** $x = 8$; **(b)** $x = -11$ 6. No solution   7. $x = -5$ or $x = \dfrac{8}{3}$

8. **(a)** $x = 6$; **(b)** $x = 7$

Reading Your Text

The following fill-in-the-blank exercises are designed to ensure that you understand some of the key vocabulary used in this section.

SECTION 5.6

(a) Rational equations are equations that involve rational _____.

(b) To solve a rational equation, we multiply each term by the _____ of any fractions in the equation.

(c) If the denominator of a rational equation contains a product of two or more variable factors, the zero-product principle is used to determine the _____ values for the variable.

(d) The final step in solving a rational equation is to check the solution in the _____ equation.

< Objective 2 >

What values for x, if any, must be excluded in each algebraic fraction?

1. $\dfrac{x}{15}$

2. $\dfrac{8}{x}$

3. $\dfrac{17}{x}$

4. $\dfrac{x}{8}$

5. $\dfrac{3}{x-2}$

6. $\dfrac{x-1}{5}$

7. $\dfrac{-5}{x+4}$

8. $\dfrac{4}{x+3}$

9. $\dfrac{x-5}{2}$

10. $\dfrac{x-1}{x-5}$

11. $\dfrac{3x}{(x+1)(x-2)}$

12. $\dfrac{5x}{(x-3)(x+7)}$

13. $\dfrac{x-1}{(2x-1)(x+3)}$

14. $\dfrac{x+3}{(3x+1)(x-2)}$

15. $\dfrac{7}{x^2-9}$

16. $\dfrac{5x}{x^2+x-2}$

17. $\dfrac{x+3}{x^2-7x+12}$

18. $\dfrac{3x-4}{x^2-49}$

19. $\dfrac{2x-1}{3x^2+x-2}$  > Videos

20. $\dfrac{3x+1}{4x^2-11x+6}$

< Objectives 1 and 3 >

Solve and check each equation.

21. $\dfrac{x}{2}+3=6$

22. $\dfrac{x}{3}-2=1$

23. $\dfrac{x}{2}-\dfrac{x}{3}=2$

24. $\dfrac{x}{6}-\dfrac{x}{8}=1$

25. $\dfrac{x}{5}-\dfrac{1}{3}=\dfrac{x-7}{3}$

26. $\dfrac{x}{6}+\dfrac{3}{4}=\dfrac{x-1}{4}$

Name _____

Section _____ Date _____

Answers

1. _____   2. _____

3. _____   4. _____

5. _____   6. _____

7. _____   8. _____

9. _____   10. _____

11. _____   12. _____

13. _____   14. _____

15. _____   16. _____

17. _____   18. _____

19. _____   20. _____

21. _____

22. _____

23. _____

24. _____

25. _____

26. _____

Answers

27. _____

28. _____

29. _____

30. _____

31. _____

32. _____

33. _____

34. _____

35. _____

36. _____

37. _____

38. _____

39. _____

40. _____

41. _____

42. _____

43. _____

44. _____

45. _____

46. _____

47. _____

48. _____

27. $\dfrac{x}{4} - \dfrac{1}{5} = \dfrac{4x + 3}{20}$

28. $\dfrac{x}{12} - \dfrac{1}{6} = \dfrac{2x - 7}{12}$

29. $\dfrac{3}{x} + 2 = \dfrac{7}{x}$

30. $\dfrac{4}{x} - 3 = \dfrac{16}{x}$

31. $\dfrac{4}{x} + \dfrac{3}{4} = \dfrac{10}{x}$ > Videos

32. $\dfrac{3}{x} = \dfrac{5}{3} - \dfrac{7}{x}$

33. $\dfrac{5}{2x} - \dfrac{1}{x} = \dfrac{9}{2x^2}$

34. $\dfrac{4}{3x} + \dfrac{1}{x} = \dfrac{14}{3x^2}$

35. $\dfrac{2}{x - 3} + 1 = \dfrac{7}{x - 3}$

36. $\dfrac{x}{x + 1} + 2 = \dfrac{14}{x + 1}$

Basic Skills | **Challenge Yourself** | Calculator/Computer | Career Applications | Above and Beyond
▲

Complete each statement with **never, sometimes,** *or* **always.**

37. The value of the term $\dfrac{5}{x}$, $x \neq 0$, is _____ equal to 1.

38. The value of the term $\dfrac{0}{x}$, $x \neq 0$, is _____ equal to 1.

39. The value of the term $\dfrac{5}{x}$, $x \neq 0$, is _____ equal to 0.

40. The value of the term $\dfrac{0}{x}$, $x \neq 0$, is _____ equal to 0.

Solve and check each equation.

41. $\dfrac{2}{x + 3} + \dfrac{1}{2} = \dfrac{x + 6}{x + 3}$

42. $\dfrac{6}{x - 5} - \dfrac{2}{3} = \dfrac{x - 9}{x - 5}$

43. $\dfrac{x}{3x + 12} + \dfrac{x - 1}{x + 4} = \dfrac{5}{3}$ > Videos

44. $\dfrac{x}{4x - 12} - \dfrac{x - 4}{x - 3} = \dfrac{1}{8}$

45. $\dfrac{x}{x - 3} - 2 = \dfrac{3}{x - 3}$

46. $\dfrac{x}{x - 5} + 2 = \dfrac{5}{x - 5}$

47. $\dfrac{x - 1}{x + 3} - \dfrac{x - 3}{x} = \dfrac{3}{x^2 + 3x}$

48. $\dfrac{x}{x - 2} - \dfrac{x + 1}{x} = \dfrac{8}{x^2 - 2x}$

49. $\dfrac{1}{x-2} - \dfrac{2}{x+2} = \dfrac{2}{x^2-4}$

50. $\dfrac{1}{x+4} + \dfrac{1}{x-4} = \dfrac{12}{x^2-16}$

51. $\dfrac{5}{x-4} = \dfrac{1}{x+2} - \dfrac{2}{x^2-2x-8}$

52. $\dfrac{11}{x+2} = \dfrac{5}{x^2-x-6} + \dfrac{1}{x-3}$

53. $\dfrac{3}{x-1} - \dfrac{1}{x+9} = \dfrac{18}{x^2+8x-9}$

54. $\dfrac{2}{x+2} = \dfrac{3}{x+6} + \dfrac{9}{x^2+8x+12}$

55. $\dfrac{3}{x+3} + \dfrac{25}{x^2+x-6} = \dfrac{5}{x-2}$

56. $\dfrac{5}{x+6} + \dfrac{2}{x^2+7x+6} = \dfrac{3}{x+1}$

57. $\dfrac{7}{x-5} - \dfrac{3}{x+5} = \dfrac{40}{x^2-25}$

58. $\dfrac{3}{x-3} - \dfrac{18}{x^2-9} = \dfrac{5}{x+3}$

59. $\dfrac{2x}{x-3} + \dfrac{2}{x-5} = \dfrac{3x}{x^2-8x+15}$

> Videos

60. $\dfrac{x}{x-4} = \dfrac{5x}{x^2-x-12} - \dfrac{3}{x+3}$

61. $\dfrac{2x}{x+2} = \dfrac{5}{x^2-x-6} - \dfrac{1}{x-3}$

62. $\dfrac{3x}{x-1} = \dfrac{2}{x-2} - \dfrac{2}{x^2-3x+2}$

63. $\dfrac{7}{x-2} + \dfrac{16}{x+3} = 3$

64. $\dfrac{5}{x-2} + \dfrac{6}{x+2} = 2$

65. $\dfrac{11}{x-3} - 1 = \dfrac{10}{x+3}$

66. $\dfrac{17}{x-4} - 2 = \dfrac{10}{x+2}$

< Objective 4 >

67. $\dfrac{x}{11} = \dfrac{12}{33}$

68. $\dfrac{4}{x} = \dfrac{16}{20}$

69. $\dfrac{5}{8} = \dfrac{20}{x}$

70. $\dfrac{x}{10} = \dfrac{9}{30}$

71. $\dfrac{x+1}{5} = \dfrac{20}{25}$

72. $\dfrac{2}{5} = \dfrac{x-2}{20}$

Answers

49. _____

50. _____

51. _____

52. _____

53. _____

54. _____

55. _____

56. _____

57. _____

58. _____

59. _____

60. _____

61. _____

62. _____

63. _____

64. _____

65. _____

66. _____

67. _____

68. _____

69. _____

70. _____

71. _____

72. _____

Answers

73. _____

74. _____

75. _____

76. _____

77. _____

78. _____

79. _____

80. _____

81. _____

82. _____

73. $\dfrac{3}{5} = \dfrac{x-1}{20}$

74. $\dfrac{5}{x-3} = \dfrac{15}{21}$

75. $\dfrac{x}{6} = \dfrac{x+5}{16}$

76. $\dfrac{x-2}{x+2} = \dfrac{12}{20}$ > Videos

77. $\dfrac{x}{x+7} = \dfrac{10}{17}$

78. $\dfrac{x}{10} = \dfrac{x+6}{30}$

79. $\dfrac{2}{x-1} = \dfrac{6}{x+9}$

80. $\dfrac{3}{x-3} = \dfrac{4}{x-5}$

81. $\dfrac{1}{x+3} = \dfrac{7}{x^2-9}$

82. $\dfrac{1}{x+5} = \dfrac{4}{x^2+3x-10}$

Answers

1. None **3.** 0 **5.** 2 **7.** −4 **9.** None **11.** −1, 2

13. $-3, \dfrac{1}{2}$ **15.** −3, 3 **17.** 3, 4 **19.** $-1, \dfrac{2}{3}$ **21.** 6

23. 12 **25.** 15 **27.** 7 **29.** 2 **31.** 8 **33.** 3
35. 8 **37.** sometimes **39.** never **41.** −5 **43.** −23
45. No solution **47.** 6 **49.** 4 **51.** −4 **53.** −5

55. No solution **57.** $-\dfrac{5}{2}$ **59.** $-\dfrac{1}{2}, 6$ **61.** $-\dfrac{1}{2}$ **63.** $-\dfrac{1}{3}, 7$

65. −8, 9 **67.** 4 **69.** 32 **71.** 3 **73.** 13 **75.** 3 **77.** 10

79. 6 **81.** 10

5.7

Applications of Rational Expressions

< 5.7 Objectives >

1 > Solve a word problem that leads to a rational equation

2 > Use a proportion to solve a word problem

Many word problems lead to rational equations that can be solved using the methods of Section 5.6. The five steps in solving word problems are, of course, the same as you saw earlier.

| Example 1 | Solving a Numerical Application |

< Objective 1 >

If one-third of a number is added to three-fourths of that same number, the sum is 26. Find the number.

Step 1 Read the problem carefully. You want to find the unknown number.

Step 2 Choose a letter to represent the unknown. Let x be the unknown number.

Step 3 Form an equation.

$$\frac{1}{3}x + \frac{3}{4}x = 26$$

One-third of number   Three-fourths of number

Step 4 Solve the equation. Multiply each side (every term) of the equation by 12, the LCD.

$$12 \cdot \frac{1}{3}x + 12 \cdot \frac{3}{4}x = 12 \cdot 26$$

Simplifying yields

$$4x + 9x = 312$$
$$13x = 312$$
$$x - 24$$

Step 5 The number is 24.

Check your solution by returning to the *original problem.* If the number is 24, we have

$$\frac{1}{3}(24) + \frac{3}{4}(24) = 8 + 18 = 26$$

and the solution is verified.

NOTE

Be sure to answer the question raised in the problem.

 Check Yourself 1

 The sum of two-fifths of a number and one-half of that number is 18. Find the number.

Number problems that involve reciprocals can be solved by using rational equations. Example 2 illustrates this approach.

> **Example 2**   Solving a Numerical Application

One number is twice another number. If the sum of their reciprocals is $\dfrac{3}{10}$, what are the two numbers?

Step 1   You want to find the two numbers.

Step 2   Let x be one number. Then $2x$ is the other number.

Twice the first

RECALL

The reciprocal of a fraction is the fraction obtained by switching the numerator and denominator.

Step 3

$$\frac{1}{x} + \frac{1}{2x} = \frac{3}{10}$$

The reciprocal   The reciprocal
of the first   of the second
number, x number, $2x$

Step 4   The LCD of the fractions is $10x$, so we multiply by $10x$.

$$10x\left(\frac{1}{x}\right) + 10x\left(\frac{1}{2x}\right) = 10x\left(\frac{3}{10}\right)$$

Simplifying, we have

$$10 + 5 = 3x$$
$$15 = 3x$$
$$5 = x$$

NOTE

x is one number, and $2x$ is the other.

Step 5   The numbers are 5 and 10.

Again check the result by returning to the original problem. If the numbers are 5 and 10, we have

$$\frac{1}{(5)} + \frac{1}{2(5)} = \frac{2+1}{10} = \frac{3}{10}$$

The sum of the reciprocals is $\dfrac{3}{10}$.

   Check Yourself 2

One number is 3 times another. If the sum of their reciprocals is $\dfrac{2}{9}$, find the two numbers.

Motion problems often involve rational expressions. Recall that the key equation for solving all motion problems relates the distance traveled, the speed or rate, and the time:

Definition

Motion Problem Relationships

$d = r \cdot t$

Often we use this equation in different forms by solving for r or for t.

$$r = \frac{d}{t} \quad \text{and} \quad t = \frac{d}{r}$$

| Example 3 | Solving an Application Involving $r = d/t$ |

NOTE

It is often helpful to choose your variable to "suggest" the unknown quantity—here *t* for time.

RECALL

Rate is distance divided by time. The rate column is formed by using that relationship.

NOTE

The equation is in proportion form. So we could solve by setting the product of the means equal to the product of the extremes.

Vince took 2 h longer to drive 225 mi than he did on a trip of 135 mi. If his speed was the same both times, how long did each trip take?

225 miles
135 miles

Step 1   You want to find the times taken for the 225-mi trip and for the 135-mi trip.

Step 2   Let *t* be the time for the 135-mi trip (in hours).

2 h longer

Then $t + 2$ is the time for the 225-mi trip.

It is often helpful to arrange the information in tabular form such as that shown.

	Distance	Rate	Time
135-mi trip	135	$\dfrac{135}{t}$	t
225-mi trip	225	$\dfrac{225}{t + 2}$	$t + 2$

Step 3   In forming the equation, remember that the speed (or rate) for each trip was the same. That is the *key* idea. We can equate the rates for the two trips that were found in step 2. The two rates are shown in the third column of the table. Thus we can write

$$\frac{135}{t} = \frac{225}{t + 2}$$

Step 4   To solve the equation from step 3, multiply each side by $t(t + 2)$, the LCD of the fractions.

$$t(t + 2)\left(\frac{135}{t}\right) = t(t + 2)\left(\frac{225}{t + 2}\right)$$

Simplifying, we have

$$135(t + 2) = 225t$$
$$135t + 270 = 225t$$
$$270 = 90t$$
$$t = 3 \text{ h}$$

Step 5   The time for the 135-mi trip was 3 h, and the time for the 225-mi trip was 5 h. We leave it to you to check this result.

Check Yourself 3

Cynthia took 2 h longer to bicycle 75 mi than she did on a trip of 45 mi. If her speed was the same each time, find the time for each trip.

Example 4 uses the $t = \dfrac{d}{r}$ form of the $d = r \cdot t$ relationship to find the speed.

| | Example 4 | Solving an Application Involving Distance, Rate, and Time |

A train makes a trip of 300 mi in the same time that a bus can travel 250 mi. If the speed of the train is 10 mi/h faster than the speed of the bus, find the speed of each.

Step 1   You want to find the speeds of the train and of the bus.

Step 2   Let r be the speed (or rate) of the bus (in miles per hour).

Then $\underline{r + 10}$ is the rate of the train.

        10 mi/h faster

Again, form a chart of the information.

	Distance	Rate	Time
Train	300	$r + 10$	$\dfrac{300}{r + 10}$
Bus	250	r	$\dfrac{250}{r}$

Step 3   To form an equation, remember that the times for the train and bus are the same. We can equate the expressions for time found in step 2. Working from the rightmost column, we have

$$\frac{300}{r + 10} = \frac{250}{r}$$

Step 4   We multiply each side by $r(r + 10)$, the LCD of the fractions.

$$\cancel{r}(r + 10)\left(\frac{250}{\cancel{r}}\right) = r\cancel{(r + 10)}\left(\frac{300}{\cancel{r + 10}}\right)$$

Simplifying, we have

$$250(r + 10) = 300r$$
$$250r + 2500 = 300r$$
$$2500 = 50r$$
$$r = 50 \text{ mi/h}$$

Step 5   The rate of the bus is 50 mi/h, and the rate of the train is 60 mi/h. You can check this result.

 Check Yourself 4

A car makes a trip of 280 mi in the same time that a truck travels 245 mi. If the speed of the truck is 5 mi/h slower than that of the car, find the speed of each.

Example 5 involves fractions in decimal form. Mixture problems often use percentages, and those percentages can be written as decimals. Example 5 illustrates this method.

| Example 5 | Solving an Application Involving Solutions |

A solution of antifreeze is 20% alcohol. How much pure alcohol must be added to 12 quarts (qt) of the solution to make a 40% solution?

Step 1   You want to find the number of quarts of pure alcohol that must be added.

Step 2   Let x be the number of quarts of pure alcohol to be added.

Step 3   To form our equation, note that the amount of alcohol present before mixing *must be the same* as the amount in the combined solution.

A picture will help.

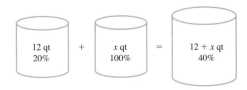

So

$$12(0.20) + x(1.00) = (12 + x)(0.40)$$

The amount of alcohol in the first solution (20% is 0.20)

The amount of pure alcohol ("pure" is 100%, or 1.00)

The amount of alcohol in the mixture

> **NOTE**
>
> Express the percentages as decimals in the equation.

Step 4   Most students prefer to clear the decimals at this stage. Multiplying by 100 moves the decimal point *two places to the right.* We then have

$$12(20) + x(100) = (12 + x)(40)$$
$$240 + 100x = 480 + 40x$$
$$60x = 240$$
$$x = 4 \text{ qt}$$

Step 5   We should add 4 qt of pure alcohol.

 Check Yourself 5

How much pure alcohol must be added to 500 cm³ of a 40% alcohol mixture to make a solution that is 80% alcohol?

There are many types of applications that lead to proportions in their solution. Typically these applications involve a common ratio, such as miles to gallons or miles to hours, and they can be solved with three basic steps.

Step by Step	
To Solve an Application Using a Proportion	**Step 1** Assign a variable to represent the unknown quantity.
	Step 2 Write a proportion, using the known and unknown quantities. Be sure each ratio involves the same units.
	Step 3 Solve the proportion written in step 2 for the unknown quantity.

Example 6 illustrates this approach.

Example 6   **Solving an Application Using a Proportion**

< Objective 2 >

A car uses 3 gal of gas to travel 105 mi. At that mileage rate, how many gallons will be used on a trip of 385 mi?

Step 1   Assign a variable to represent the unknown quantity. Let x be the number of gallons of gas that will be used on the 385-mi trip.

Step 2   Write a proportion. Note that the ratio of miles to gallons must stay the same.

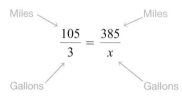

$$\frac{105}{3} = \frac{385}{x}$$

Step 3   Solve the proportion. The product of the extremes is equal to the product of the means.

$$105x = 3 \cdot 385$$
$$105x = 1{,}155$$
$$\frac{105x}{105} = \frac{1{,}155}{105}$$
$$x = 11 \text{ gal}$$

So 11 gal of gas will be used for the 385-mi trip.

NOTE

To verify your solution, return to the original problem and check that the two ratios are equivalent.

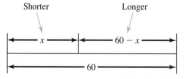 **Check Yourself 6**

A car uses 8 L of gasoline in traveling 100 km. At that rate, how many liters of gas will be used on a trip of 250 km?

Proportions can also be used to solve problems in which a quantity is divided using a specific ratio. Example 7 shows how.

  Example 7   **Solving an Application Using a Proportion**

A 60-in.-long piece of wire is to be cut into two pieces whose lengths have the ratio 5 to 7. Find the length of each piece.

Step 1   Let x represent the length of the shorter piece. Then $60 - x$ is the length of the longer piece.

Shorter   Longer

|← x →|← $60 - x$ →|
|← 60 →|

RECALL

A picture of the problem always helps.

Step 2   The two pieces have the ratio $\dfrac{5}{7}$, so

$$\frac{x}{60 - x} = \frac{5}{7}$$

Step 3   Solving as before, we get

$$7x = (60 - x)5$$
$$7x = 300 - 5x$$
$$12x = 300$$
$$x = 25 \quad \text{Shorter piece}$$
$$60 - x = 35 \quad \text{Longer piece}$$

The pieces have lengths of 25 in. and 35 in.

Check Yourself 7

A 21-ft-long board is to be cut into two pieces so that the ratio of their lengths is 3 to 4. Find the lengths of the two pieces.

Check Yourself ANSWERS

1. The number is 20.   2. The numbers are 6 and 18.
3. 75-mi trip: 5 h; 45-mi trip: 3 h   4. Car: 40 mi/h; truck: 35 mi/h
5. $1,000 \text{ cm}^3$ 6. 20 L   7. 9 ft; 12 ft

Reading Your Text

The following fill-in-the-blank exercises are designed to ensure that you understand some of the key vocabulary used in this section.

SECTION 5.7

(a) When solving a rational equation, the solution is checked by returning to the _____ problem.

(b) The key equation for solving motion problems relates the distance traveled, the speed, and the _____.

(c) Time is distance divided by _____.

(d) To solve an application using a proportion, first assign a _____ to represent the unknown quantity.

Name _____

Section _____ Date _____

Answers

1. _____

2. _____

3. _____

4. _____

5. _____

6. _____

7. _____

8. _____

9. _____

10. _____

11. _____

12. _____

13. _____

14. _____

| Basic Skills | Challenge Yourself | Calculator/Computer | Career Applications | Above and Beyond |

< Objectives 1–2 >

Solve each word problem.

1. **NUMBER PROBLEM** If two-thirds of a number is added to one-half of that number, the sum is 35. Find the number.

2. **NUMBER PROBLEM** If one-third of a number is subtracted from three-fourths of that number, the difference is 15. What is the number?

3. **NUMBER PROBLEM** If one-fourth of a number is subtracted from two-fifths of the number, the difference is 3. Find the number.

4. **NUMBER PROBLEM** If five-sixths of a number is added to one-fifth of the number, the sum is 31. What is the number?

5. **NUMBER PROBLEM** If one-third of an integer is added to one-half of the next consecutive integer, the sum is 13. What are the two integers?

6. **NUMBER PROBLEM** If one-half of one integer is subtracted from three-fifths of the next consecutive integer, the difference is 3. What are the two integers?

7. **NUMBER PROBLEM** One number is twice another number. If the sum of their reciprocals is $\frac{1}{4}$, find the two numbers.

8. **NUMBER PROBLEM** One number is 3 times another. If the sum of their reciprocals is $\frac{1}{6}$, find the two numbers.

9. **NUMBER PROBLEM** One number is 4 times another. If the sum of their reciprocals is $\frac{5}{12}$, find the two numbers.

10. **NUMBER PROBLEM** One number is 3 times another. If the sum of their reciprocals is $\frac{4}{15}$, what are the two numbers?

11. **NUMBER PROBLEM** One number is 5 times another number. If the sum of their reciprocals is $\frac{6}{35}$, what are the two numbers?

12. **NUMBER PROBLEM** One number is 4 times another. The sum of their reciprocals is $\frac{5}{24}$. What are the two numbers?

13. **NUMBER PROBLEM** If the reciprocal of 5 times a number is subtracted from the reciprocal of that number, the result is $\frac{4}{25}$. What is the number?

14. **NUMBER PROBLEM** If the reciprocal of a number is added to 4 times the reciprocal of that number, the result is $\frac{5}{9}$. Find the number.

15. **SCIENCE AND MEDICINE** Lee can ride his bicycle 50 mi in the same time it takes him to drive 125 mi. If his driving rate is 30 mi/h faster than his rate bicycling, find each rate.

16. **SCIENCE AND MEDICINE** Tina can run 12 mi in the same time it takes her to bicycle 72 mi. If her bicycling rate is 20 mi/h faster than her running rate, find each rate.

15. _____

17. **SCIENCE AND MEDICINE** An express bus can travel 275 mi in the same time that it takes a local bus to travel 225 mi. If the rate of the express bus is 10 mi/h faster than that of the local bus, find the rate for each bus.

16. _____

17. _____

18. **SCIENCE AND MEDICINE** A passenger train can travel 325 mi in the same time a freight train takes to travel 200 mi. If the speed of the passenger train is 25 mi/h faster than the speed of the freight train, find the speed of each.

18. _____

19. _____

19. **SCIENCE AND MEDICINE** A light plane took 1 h longer to travel 450 mi on the first portion of a trip than it took to fly 300 mi on the second. If the speed was the same for each portion, what was the flying time for each part of the trip?

20. _____

20. **SCIENCE AND MEDICINE** A small business jet took 1 h longer to fly 810 mi on the first part of a flight than to fly 540 mi on the second portion. If the jet's rate was the same for each leg of the flight, what was the flying time for each leg?

21. _____

22. _____

23. _____

21. **SCIENCE AND MEDICINE** Charles took 2 h longer to drive 240 mi on the first day of a vacation trip than to drive 144 mi on the second day. If his average driving rate was the same on both days, what was his driving time for each of the days?

24. _____

25. _____

22. **SCIENCE AND MEDICINE** Ariana took 2 h longer to drive 360 mi on the first day of a trip than she took to drive 270 mi on the second day. If her speed was the same on both days, what was the driving time each day?

26. _____

23. **SCIENCE AND MEDICINE** An airplane took 3 h longer to fly 1,200 mi than it took for a flight of 480 mi. If the plane's rate was the same on each trip, what was the time of each flight?

24. **SCIENCE AND MEDICINE** A train travels 80 mi in the same time that a light plane can travel 280 mi. If the speed of the plane is 100 mi/h faster than that of the train, find each of the rates.

25. **SCIENCE AND MEDICINE** Jan and Tariq took a canoeing trip, traveling 6 mi upstream against a 2-mi/h current. They then returned to the same point downstream. If their entire trip took 4 h, how fast can they paddle in still water? [*Hint:* If r is their rate (in miles per hour) in still water, their rate upstream is $r - 2$ and their rate downstream is $r + 2$.]

26. **SCIENCE AND MEDICINE** A plane flies 720 mi against a steady 30-mi/h headwind and then returns to the same point with the wind. If the entire trip takes 10 h, what is the plane's speed in still air?

Answers

27. _____

28. _____

29. _____

30. _____

31. _____

32. _____

33. _____

34. _____

35. _____

36. _____

37. _____

38. _____

27. SCIENCE AND MEDICINE How much pure alcohol must be added to 40 oz of a 25% solution to produce a mixture that is 40% alcohol?

28. SCIENCE AND MEDICINE How many centiliters (cL) of pure acid must be added to 200 cL of a 40% acid solution to produce a 50% solution?

29. SCIENCE AND MEDICINE A speed of 60 mi/h corresponds to 88 ft/s. If a light plane's speed is 150 mi/h, what is its speed in feet per second?

30. BUSINESS AND FINANCE If 342 cups of coffee can be made from 9 lb of coffee, how many cups can be made from 6 lb of coffee?

31. SOCIAL SCIENCE A car uses 5 gal of gasoline on a trip of 160 mi. At the same mileage rate, how much gasoline will a 384-mi trip require?

32. SOCIAL SCIENCE A car uses 12 L of gasoline in traveling 150 km. At that rate, how many liters of gasoline will be used in a trip of 400 km?

33. BUSINESS AND FINANCE Sveta earns $13,500 commission in 20 weeks in her new sales position. At that rate, how much will she earn in 1 year (52 weeks)?

34. BUSINESS AND FINANCE Kevin earned $165 interest for 1 year on an investment of $1,500. At the same rate, what amount of interest would be earned by an investment of $2,500?

35. SOCIAL SCIENCE A company is selling a natural insect control that mixes ladybug beetles and praying mantises in the ratio of 7 to 4. If there are a total of 110 insects per package, how many of each type of insect is in a package?

36. SOCIAL SCIENCE A woman casts a 4-ft shadow. At the same time, a 72-ft building casts a 48-ft shadow. How tall is the woman?

37. BUSINESS AND FINANCE A brother and sister are to divide an inheritance of $12,000 in the ratio of 2 to 3. What amount will each receive?

38. BUSINESS AND FINANCE In Bucks County, the property tax rate is $25.32 per $1,000 of assessed value. If a house and property have a value of $128,000, find the tax the owner will have to pay.

Assessed Value: $128,000

Tax Rate: $25.32 per $1,000

Answers

1. 30   **3.** 20   **5.** 15, 16   **7.** 6, 12   **9.** 3, 12   **11.** 7, 35
13. 5   **15.** 20 mi/h bicycling, 50 mi/h driving   **17.** Express 55 mi/h, local 45 mi/h   **19.** 3 h, 2 h   **21.** 5 h, 3 h   **23.** 5 h, 2 h
25. 4 mi/h   **27.** 10 oz   **29.** 220 ft/s   **31.** 12 gal   **33.** $35,100
35. 70 ladybugs, 40 praying mantises   **37.** Brother $4,800, sister $7,200

Definition/Procedure	Example	Reference

Simplifying Rational Expressions

Rational Expressions

These have the form

$$\frac{P}{Q}$$

Numerator ↗ (pointing to P)

Fraction bar → (pointing to fraction bar)

Denominator ↘ (pointing to Q)

in which P and Q are polynomials and Q cannot have the value 0.

$\dfrac{x^2 - 3x}{x - 2}$ is a rational expression. The variable x cannot have the value 2.

Section 5.1

p. 331

Writing in Simplest Form

A fraction is in simplest form if its numerator and denominator have no common factors other than 1.
To write in simplest form:

1. Factor the numerator and denominator.
2. Divide the numerator and denominator by all common factors. The resulting fraction will be in simplest form.

$\dfrac{x + 2}{x - 1}$ is in simplest form.

$$\frac{x^2 - 4}{x^2 - 2x - 8}$$

$$= \frac{(x - 2)(x + 2)}{(x - 4)(x + 2)}$$

$$= \frac{(x - 2)\cancel{(x + 2)}}{(x - 4)\cancel{(x + 2)}}$$

$$= \frac{x - 2}{x - 4}$$

p. 331

Multiplying and Dividing Rational Expressions

Multiplying Rational Expressions

$$\frac{P}{Q} \cdot \frac{R}{S} = \frac{PR}{QS}$$

in which $Q \neq 0$ and $S \neq 0$.

$$\frac{2}{3} \cdot \frac{4}{5} = \frac{2 \cdot 4}{3 \cdot 5} = \frac{8}{15}$$

Section 5.2

p. 340

Multiplying Rational Expressions

Step 1 Factor the numerators and denominators.

Step 2 Write the product of the factors of the numerators over the product of the factors of the denominators.

Step 3 Divide the numerator and denominator by any common factors.

$$\frac{2x - 4}{x^2 - 4} \cdot \frac{x^2 + 2x}{6x + 18}$$

$$= \frac{2(x - 2) \cdot x(x + 2)}{(x - 2)(x + 2) \cdot 6(x + 3)}$$

$$= \frac{2\cancel{(x - 2)} \cdot x\cancel{(x + 2)}}{\cancel{(x - 2)}\cancel{(x + 2)} \cdot 6(x + 3)}$$

$$= \frac{x}{3(x + 3)}$$

p. 340

Continued

Definition/Procedure	Example	Reference

Dividing Rational Expressions

$$\frac{P}{Q} \div \frac{R}{S} = \frac{P}{Q} \cdot \frac{S}{R}$$

in which $Q \neq 0$, $R \neq 0$, and $S \neq 0$. In words, invert the divisor (the second fraction) and multiply.

$$\frac{4}{9} \div \frac{8}{12}$$

$$= \frac{4}{9} \cdot \frac{12}{8} = \frac{2}{3}$$

$$\frac{3x}{2x - 6} \div \frac{9x^2}{x^2 - 9}$$

$$= \frac{3x}{2x - 6} \cdot \frac{x^2 - 9}{9x^2}$$

$$= \frac{3x \cdot (x + 3)(x - 3)}{2(x - 3) \cdot 9x^2}$$

$$= \frac{x + 3}{6x}$$

p. 341

Adding and Subtracting Like Rational Expressions

Like Rational Expressions

1. Add or subtract the numerators.
2. Write the sum or difference over the common denominator.
3. Write the resulting fraction in simplest form.

$$\frac{2x}{x^2 + 3x} + \frac{6}{x^2 + 3x}$$

$$= \frac{2x + 6}{x^2 + 3x}$$

$$= \frac{2(x + 3)}{x(x + 3)} = \frac{2}{x}$$

Section 5.3

p. 348

Adding and Subtracting Unlike Rational Expressions

The Least Common Denominator

Finding the LCD:
1. Factor each denominator.
2. Write each factor the greatest number of times it appears in any single denominator.
3. The LCD is the product of the factors found in step 2.

For $\dfrac{2}{x^2 + 2x + 1}$

and $\dfrac{3}{x^2 + x}$

Factor:

$$x^2 + 2x + 1 = (x + 1)(x + 1)$$
$$x^2 + x = x(x + 1)$$

The LCD is $x(x + 1)(x + 1)$.

Section 5.4

p. 355

Definition/Procedure	Example	Reference

Unlike Rational Expressions

To add or subtract unlike rational expressions:

1. Find the LCD.
2. Convert each rational expression to an equivalent rational expression with the LCD as a common denominator.
3. Add or subtract the like rational expressions formed.
4. Write the sum or difference in simplest form.

$$\frac{2}{x^2 + 2x + 1} - \frac{3}{x^2 + x}$$

$$= \frac{2x}{x(x + 1)(x + 1)}$$

$$- \frac{3(x + 1)}{x(x + 1)(x + 1)}$$

$$= \frac{2x - 3x - 3}{x(x + 1)(x + 1)}$$

$$= \frac{-x - 3}{x(x + 1)(x + 1)}$$

p. 356

Complex Rational Expressions

Section 5.5

Simplifying Complex Fractions

$$\frac{\dfrac{a}{b}}{\dfrac{c}{d}} = \frac{a}{b} \div \frac{c}{d} = \frac{a}{b} \cdot \frac{d}{c}$$

$$\frac{\dfrac{3}{8}}{\dfrac{5}{6}} = \frac{3}{8} \div \frac{5}{6}$$

$$= \frac{3}{\overset{}{\underset{4}{8}}} \cdot \frac{\overset{3}{\cancel{6}}}{5}$$

$$= \frac{9}{20}$$

p. 369

Equations Involving Rational Expressions

1. Remove the fractions in the equation by multiplying both sides by the LCD of all the fractions.
2. Solve the equation resulting from step 1.
3. Check the solution using the original equation. Discard any extraneous solutions.

$$2 - \frac{4}{x} = \frac{2}{3}$$

LCD: $3x$

$$2(3x) - \frac{4}{x}(3x) = \frac{2}{3}(3x)$$

$$6x - 12 = 2x$$

$$4x = 12$$

$$x = 3$$

Section 5.6

p. 377

This summary exercise set is provided to give you practice with each of the objectives of this chapter. Each exercise is keyed to the appropriate chapter section. When you are finished, you can check your answers to the odd-numbered exercises against those presented in the back of the text. If you have difficulty with any of these questions, go back and reread the examples from that section. Your instructor will give you guidelines on how best to use these exercises in your instructional setting.

5.1 Write each fraction in simplest form.

1. $\dfrac{6a^2}{9a^3}$

2. $\dfrac{-12x^4y^3}{18x^2y^2}$

3. $\dfrac{w^2 - 25}{2w - 8}$

4. $\dfrac{3x^2 + 11x - 4}{2x^2 + 11x + 12}$

5. $\dfrac{m^2 - 2m - 3}{9 - m^2}$

6. $\dfrac{3c^2 - 2cd - d^2}{6c^2 + 2cd}$

5.2 Multiply or divide as indicated.

7. $\dfrac{6x}{5} \cdot \dfrac{10}{18x^2}$

8. $\dfrac{-2a^2}{ab^3} \cdot \dfrac{3ab^2}{-4ab}$

9. $\dfrac{2x + 6}{x^2 - 9} \cdot \dfrac{x^2 - 3x}{4}$

10. $\dfrac{a^2 + 5a + 4}{2a^2 + 2a} \cdot \dfrac{a^2 - a - 12}{a^2 - 16}$

11. $\dfrac{3p}{5} \div \dfrac{9p^2}{10}$

12. $\dfrac{8m^3}{5mn} \div \dfrac{12m^2n^2}{15mn^3}$

13. $\dfrac{x^2 + 7x + 10}{x^2 + 5x} \div \dfrac{x^2 - 4}{2x^2 - 7x + 6}$

14. $\dfrac{2w^2 + 11w - 21}{w^2 - 49} \div (4w - 6)$

15. $\dfrac{a^2b + 2ab^2}{a^2 - 4b^2} \div \dfrac{4a^2b}{a^2 - ab - 2b^2}$

16. $\dfrac{2x^2 + 6x}{4x} \cdot \dfrac{6x + 12}{x^2 + 2x - 3} \div \dfrac{x^2 - 4}{x^2 - 3x + 2}$

5.3 Add or subtract as indicated.

17. $\dfrac{x}{9} + \dfrac{2x}{9}$

18. $\dfrac{7a}{15} - \dfrac{2a}{15}$

19. $\dfrac{8}{x + 2} + \dfrac{3}{x + 2}$

20. $\dfrac{y - 2}{5} - \dfrac{2y + 3}{5}$

21. $\dfrac{7r - 3s}{4r} + \dfrac{r - s}{4r}$

22. $\dfrac{x^2}{x - 4} - \dfrac{16}{x - 4}$

23. $\dfrac{5w - 6}{w - 4} - \dfrac{3w + 2}{w - 4}$

24. $\dfrac{x + 3}{x^2 - 2x - 8} + \dfrac{2x + 3}{x^2 - 2x - 8}$

5.4 *Add or subtract as indicated.*

25. $\dfrac{5x}{6} + \dfrac{x}{3}$

26. $\dfrac{3y}{10} - \dfrac{2y}{5}$

27. $\dfrac{5}{2m} - \dfrac{3}{m^2}$

28. $\dfrac{x}{x-3} - \dfrac{2}{3}$

29. $\dfrac{4}{x-3} - \dfrac{1}{x}$

30. $\dfrac{2}{s+5} + \dfrac{3}{s+1}$

31. $\dfrac{5}{w-5} - \dfrac{2}{w-3}$

32. $\dfrac{4x}{2x-1} + \dfrac{2}{1-2x}$

33. $\dfrac{2}{3x-3} - \dfrac{5}{2x-2}$

34. $\dfrac{4y}{y^2-8y+15} + \dfrac{6}{y-3}$

35. $\dfrac{3a}{a^2+5a+4} + \dfrac{2a}{a^2-1}$

36. $\dfrac{3x}{x^2+2x-8} - \dfrac{1}{x-2} + \dfrac{1}{x+4}$

5.5 *Simplify the complex fractions.*

37. $\dfrac{\dfrac{x^2}{12}}{\dfrac{x^3}{8}}$

38. $\dfrac{3 + \dfrac{1}{a}}{3 - \dfrac{1}{a}}$

39. $\dfrac{1 + \dfrac{x}{y}}{1 - \dfrac{x}{y}}$

40. $\dfrac{1 + \dfrac{1}{p}}{p^2 - 1}$

41. $\dfrac{\dfrac{1}{m} - \dfrac{1}{n}}{\dfrac{1}{m} + \dfrac{1}{n}}$

42. $\dfrac{2 - \dfrac{x}{y}}{4 - \dfrac{x^2}{y^2}}$

43. $\dfrac{\dfrac{2}{a+1} + 1}{1 - \dfrac{4}{a+1}}$

44. $\dfrac{\dfrac{a}{b} - 1 - \dfrac{2b}{a}}{\dfrac{1}{b^2} - \dfrac{1}{a^2}}$

5.6 *What values for x, if any, must be excluded in each rational expression?*

45. $\dfrac{x}{5}$

46. $\dfrac{3}{x-4}$

47. $\dfrac{2}{(x+1)(x-2)}$

48. $\dfrac{7}{x^2-16}$

49. $\dfrac{x-1}{x^2+3x+2}$

50. $\dfrac{2x+3}{3x^2+x-2}$

Solve each proportion.

51. $\dfrac{x-3}{8}=\dfrac{x-2}{10}$

52. $\dfrac{1}{x-3}=\dfrac{7}{x^2-x-6}$

Solve each equation.

53. $\dfrac{x}{4}-\dfrac{x}{5}=2$

54. $\dfrac{13}{4x}+\dfrac{3}{x^2}=\dfrac{5}{2x}$

55. $\dfrac{x}{x-2}+1=\dfrac{x+4}{x-2}$

56. $\dfrac{x}{x-4}-3=\dfrac{4}{x-4}$

57. $\dfrac{x}{2x-6}-\dfrac{x-4}{x-3}=\dfrac{1}{8}$

58. $\dfrac{7}{x}-\dfrac{1}{x-3}=\dfrac{9}{x^2-3x}$

59. $\dfrac{x}{x-5}=\dfrac{3x}{x^2-7x+10}+\dfrac{8}{x-2}$

60. $\dfrac{6}{x+5}+1=\dfrac{3}{x-5}$

61. $\dfrac{24}{x+2}-2=\dfrac{2}{x-3}$

5.7 *Solve each application.*

62. **NUMBER PROBLEM** If two-fifths of a number is added to one-half of that number, the sum is 27. Find the number.

63. **NUMBER PROBLEM** One number is 3 times another. If the sum of their reciprocals is $\dfrac{1}{3}$, what are the two numbers?

64. **NUMBER PROBLEM** If the reciprocal of 4 times a number is subtracted from the reciprocal of that number, the result is $\dfrac{1}{8}$. What is the number?

65. **SCIENCE AND MEDICINE** Robert made a trip of 240 mi. Returning by a different route, he found that the distance was only 200 mi, but traffic slowed his speed by 8 mi/h. If the trip took the same amount of time in both directions, what was Robert's rate each way?

66. **SCIENCE AND MEDICINE** On the first day of a vacation trip, Jovita drove 225 mi. On the second day it took her 1 h longer to drive 270 mi. If her average speed was the same on both days, how long did she drive each day?

67. **SCIENCE AND MEDICINE** A light plane flies 700 mi against a steady 20-mi/h headwind and then returns, with the wind, to the same point. If the entire trip took 12 h, what was the speed of the plane in still air?

68. **SCIENCE AND MEDICINE** How much pure alcohol should be added to 300 mL of a 30% solution to obtain a 40% solution?

69. **SCIENCE AND MEDICINE** A chemist has a 10% acid solution and a 40% solution. How much of the 40% solution should be added to 300 mL of the 10% solution to produce a mixture with a concentration of 20%?

70. **BUSINESS AND FINANCE** Melina wants to invest a total of $10,800 in two types of savings accounts. If she wants the ratio of the amounts deposited in the two accounts to be 4 to 5, what amount should she invest in each account?

The Streeter/Hutchison Series in Mathematics

Beginning Algebra

Name _____

Section _____ Date _____

Answers

The purpose of this self-test is to help you assess your progress so that you can find concepts that you need to review before the next exam. Allow yourself about an hour to take this test. At the end of that hour, check your answers against those given in the back of this text. If you miss any, go back to the appropriate section to reread the examples until you have mastered that particular concept.

1. _____

2. _____

3. _____

4. _____

5. _____

6. _____

7. _____

8. _____

9. _____

10. _____

11. _____

12. _____

13. _____

14. _____

15. _____

16. _____

17. _____

18. _____

Write each fraction in simplest form.

1. $\dfrac{-21x^5y^3}{28xy^5}$

2. $\dfrac{4a - 24}{a^2 - 6a}$

3. $\dfrac{3x^2 + x - 2}{3x^2 - 8x + 4}$

Perform the indicated operations and simplify your results.

4. $\dfrac{3a}{8} + \dfrac{5a}{8}$

5. $\dfrac{2x}{x + 3} + \dfrac{6}{x + 3}$

6. $\dfrac{7x - 3}{x - 2} - \dfrac{2x + 7}{x - 2}$

7. $\dfrac{x}{3} + \dfrac{4x}{5}$

8. $\dfrac{3}{s} - \dfrac{2}{s^2}$

9. $\dfrac{5}{x - 2} - \dfrac{1}{x + 3}$

10. $\dfrac{6}{w - 2} + \dfrac{9w}{w^2 - 7w + 10}$

11. $\dfrac{3pq^2}{5pq^3} \cdot \dfrac{20p^2q}{21q}$

12. $\dfrac{x^2 - 3x}{5x^2} \cdot \dfrac{10x}{x^2 - 4x + 3}$

13. $\dfrac{2x^2}{3xy} \div \dfrac{8x^2y}{9xy}$

14. $\dfrac{3m - 9}{m^2 - 2m} \div \dfrac{m^2 - m - 6}{m^2 - 4}$

15. $\dfrac{\dfrac{x^2}{18}}{\dfrac{x^3}{12}}$

16. $\dfrac{2 - \dfrac{m}{n}}{4 - \dfrac{m^2}{n^2}}$

What values for x, if any, must be excluded in each rational expression?

17. $\dfrac{8}{x - 4}$

18. $\dfrac{3}{x^2 - 9}$

Solve each equation.

19. $\dfrac{x}{3} - \dfrac{x}{4} = 3$

20. $\dfrac{5}{x} - \dfrac{x-3}{x+2} = \dfrac{22}{x^2 + 2x}$

21. $\dfrac{x-1}{5} = \dfrac{x+2}{8}$

22. $\dfrac{2x-1}{7} = \dfrac{x}{4}$

Solve each application.

23. **NUMBER PROBLEM** One number is 3 times another. If the sum of their reciprocals is $\dfrac{1}{3}$, find the two numbers.

24. **SCIENCE AND MEDICINE** Mark drove 250 mi to visit Sandra. Returning by a shorter route, he found that the trip was only 225 mi, but traffic slowed his speed by 5 mi/h. If the two trips took exactly the same time, what was his rate each way?

25. **CONSTRUCTION** A cable that is 55 ft long is to be cut into two pieces whose lengths have the ratio 4 to 7. Find the lengths of the two pieces.

Answers

19. _____

20. _____

21. _____

22. _____

23. _____

24. _____

25. _____

Activity 5 ::
Determining State Apportionment

The introduction to this chapter referred to the ratio of the people in a particular state to their total number of representatives in the U.S. House based on the 2000 census. It was noted that the ratio of the total population of the country to the 435 representatives in Congress should equal the state apportionment if it is fair. That is, $\dfrac{A}{a} = \dfrac{P}{r}$, where A is the population of the state, a is the number of representatives for that state, P is the total population of the U.S., and r is the total number of representatives in Congress (435).

Pick 5 states (your own included) and search the Internet to find the following.

1. Determine the year 2000 population of each state.

2. Note the number of representatives for each state and any increase or decrease.

3. Find the number of people per representative for each state.

4. Compare that with the national average of the number of people per representative.

5. Solve the rational equation $\dfrac{A}{a} = \dfrac{P}{r}$ for a. For each state substitute the number values for the variables, A, P, and r. Find a. Based on your findings which states have

    (a) a greater number of representatives than they should (that is, the number has been rounded up), and

    (b) which states have a smaller number of representative than they should (that is, the number has been rounded down)?

You can find out more about apportionment counts and how they are determined from the U.S. Census website.

The following questions are presented to help you review concepts from earlier chapters. This is meant as a review and not as a comprehensive exam. The answers are presented in the back of the text. Section references accompany the answers. If you have difficulty with any of these questions, be certain to at least read through the summary related to those sections.

Name _____

Section _____ Date _____

Answers

Perform the indicated operation.

1. $x^2y - 4xy - x^2y + 2xy$

2. $\dfrac{12a^3b}{9ab}$

3. $(5x^2 - 2x + 1) - (3x^2 + 3x - 5)$

4. $(5a^2 + 6a) - (2a^2 - 1)$

5. $4 + 3(7 - 4)^2$

6. $|3 - 5| - |-4 + 3|$

Multiply.

7. $(x - 2y)(2x + 3y)$

8. $(x + 7)(x + 4)$

Divide.

9. $(2x^2 + 3x - 1) \div (x + 2)$

10. $(x^2 - 5) \div (x - 1)$

Solve each equation and check your results.

11. $4x - 3 = 2x + 5$

12. $2 - 3(2x + 1) = 11$

Factor each polynomial completely.

13. $x^2 - 5x - 14$

14. $3m^2n - 6mn^2 + 9mn$

15. $a^2 - 9b^2$

16. $2x^3 - 28x^2 + 96x$

Solve each word problem. Show the equation used for each solution.

17. NUMBER PROBLEM 2 more than 4 times a number is 30. Find the number.

1. _____

2. _____

3. _____

4. _____

5. _____

6. _____

7. _____

8. _____

9. _____

10. _____

11. _____

12. _____

13. _____

14. _____

15. _____

16. _____

17. _____

Answers

18. _____

19. _____

20. _____

21. _____

22. _____

23. _____

24. _____

25. _____

26. _____

27. _____

28. _____

29. _____

30. _____

18. **Number Problem** If the reciprocal of 4 times a number is subtracted from the reciprocal of that number, the result is $\dfrac{3}{16}$. What is the number?

19. **Science and Medicine** A speed of 60 mi/h corresponds to 88 ft/s. If a race car is traveling at 180 mi/h, what is its speed in feet per second?

20. **Geometry** The length of a rectangle is 3 in. less than twice its width. If the area of the rectangle is 35 in.^{2}, find the dimensions of the rectangle.

Write each rational expression in simplest form.

21. $\dfrac{m^2 - 4m}{3m - 12}$

22. $\dfrac{a^2 - 49}{3a^2 + 22a + 7}$

Perform the indicated operations.

23. $\dfrac{4}{3r} + \dfrac{1}{2r^2}$

24. $\dfrac{2}{x - 3} - \dfrac{5}{3x + 9}$

25. $\dfrac{3x^2 + 9x}{x^2 - 9} \cdot \dfrac{2x^2 - 9x + 9}{2x^3 - 3x^2}$

26. $\dfrac{4w^2 - 25}{2w^2 - 5w} \div (6w + 15)$

Simplify each complex rational expression.

27. $\dfrac{1 - \dfrac{1}{x}}{2 + \dfrac{1}{x}}$

28. $\dfrac{3 - \dfrac{m}{n}}{9 - \dfrac{m^2}{n^2}}$

Solve each equation.

29. $\dfrac{5}{3x} + \dfrac{1}{x^2} = \dfrac{5}{2x}$

30. $\dfrac{10}{x - 3} - 2 = \dfrac{5}{x + 3}$

CHAPTER 6

An Introduction to Graphing

INTRODUCTION

Graphs are used to discern patterns that may be difficult to see when looking at a list of numbers or other kinds of data. The word *graph* has Latin and Greek roots and means "to draw a picture." A graph in mathematics is a picture of a relationship between variables. Graphs are used in every field that uses numbers.

In the field of pediatric medicine there has been controversy over the use of human growth hormone to help children whose growth has been impeded by health problems. The reason for the controversy is that some doctors are giving therapy to children who are simply shorter than average or shorter than their parents want them to be. The determination of which children are healthy but small in stature and which children have health defects that keep them from growing is an issue that has been vigorously argued in medical research.

Measures used to distinguish between the two groups include blood tests and age and height measurements. These measurements are graphed and monitored over several years to gauge the child's growth rate. If this rate of growth is below 4.5 centimeters per year, then there may be a problem. The graph can also indicate whether the child's size is within a range considered normal at each age of the child's life.

CHAPTER 6 OUTLINE

Name _____

Section _____ Date _____

This prerequisite test provides some exercises requiring skills that you will need to be successful in the coming chapter. The answers for these exercises can be found in the back of this text. This prerequisite test can help you identify topics that you will need to review before beginning the chapter.

Answers

1. _____

2. _____

3. _____

4. _____

5. _____

6. _____

7. _____

8. _____

9. _____

10. _____

Solve each equation.

1. $2 - 5x = 12$

2. $-3 + 5x = 1$

3. $2x - 2 = 6$

4. $7y + 10 = -11$

5. $6 - 3x = 8$

6. $-4y + 6 = 3$

Evaluate each expression.

7. $\dfrac{-9 - 5}{-4 - 3}$

8. $\dfrac{4 - (-2)}{6 - 2}$

9. $\dfrac{7 - 3}{8 - 4}$

10. $\dfrac{-4 - (-4)}{8 - 2}$

Beginning Algebra

The Streeter/Hutchison Series in Mathematics

6.1

Solutions of Equations in Two Variables

< 6.1 Objectives >

1 > Find solutions for an equation in two variables

2 > Use ordered-pair notation to write solutions for equations in two variables

We discussed finding solutions for equations in Chapter 2. Recall that a solution is a value for the variable that "satisfies" the equation or makes the equation a true statement. For instance, we know that 4 is a solution of the equation

$$2x + 5 = 13$$

We know this is true because, when we replace x with 4, we have

$$2(4) + 5 \overset{?}{=} 13$$
$$8 + 5 \overset{?}{=} 13$$
$$13 = 13 \qquad \text{A true statement}$$

RECALL

An equation consists of two expressions separated by an equal sign.

We now want to consider **equations in two variables.** In this chapter we study equations of the form $Ax + By = C$, in which A and B are not both 0. Such equations are called **linear equations,** and are said to be in **standard form.** An example is

$$x + y = 5$$

What will a solution look like? It is not going to be a single number, because there are two variables. Here a solution is a pair of numbers—one value for each of the variables, x and y. Suppose that x has the value 3. In the equation $x + y = 5$, you can substitute 3 for x.

$$(3) + y = 5$$

Solving for y gives

$$y = 2$$

NOTE

The solution of an equation in two variables "pairs" two numbers, one for x and one for y.

So the pair of values $x = 3$ and $y = 2$ satisfies the equation because

$$3 + 2 = 5$$

The pair of numbers that satisfies an equation is called a **solution** for the equation in two variables.

How many such pairs are there? Choose any value for x (or for y). You can always find the other *paired* or *corresponding* value in an equation of this form. We say that there are an *infinite* number of pairs that satisfy the equation. Each of these pairs is a solution. We find some other solutions for the equation $x + y = 5$ in Example 1.

Example 1   Solving for Corresponding Values

< Objective 1 >

For the equation $x + y = 5$, find (a) y if $x = 5$ and (b) x if $y = 4$.

(a) If $x = 5$,

$(5) + y = 5$ or $y = 0$

(b) If $y = 4$,

$x + (4) = 5$ or $x = 1$

So the pairs $x = 5$, $y = 0$ and $x = 1$, $y = 4$ are both solutions.

 Check Yourself 1

For the equation $2x + 3y = 26$,

(a) If $x = 4$, $y = ?$ (b) If $y = 0$, $x = ?$

To simplify writing the pairs that satisfy an equation, we use **ordered-pair notation.** The numbers are written in parentheses and are separated by a comma. For example, we know that the values $x = 3$ and $y = 2$ satisfy the equation $x + y = 5$. So we write the pair as

$$(3, 2)$$

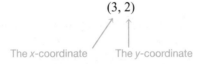

The x-coordinate   The y-coordinate

 > CAUTION

$(3, 2)$ means $x = 3$ and $y = 2$.
$(2, 3)$ means $x = 2$ and $y = 3$.
$(3, 2)$ and $(2, 3)$ are different, which
is why we call them *ordered pairs*.

The first number of the pair is *always* the value for x and is called the **x-coordinate.** The second number of the pair is *always* the value for y and is the **y-coordinate.**

Using this ordered-pair notation, we can say that $(3, 2)$, $(5, 0)$, and $(1, 4)$ are all *solutions* for the equation $x + y = 5$. Each pair gives values for x and y that satisfy the equation.

 Example 2   Identifying Solutions of Two-Variable Equations

< Objective 2 >

Which of the ordered pairs (a) $(2, 5)$, (b) $(5, -1)$, and (c) $(3, 4)$ are solutions for the equation $2x + y = 9$?

(a) To check whether $(2, 5)$ is a solution, let $x = 2$ and $y = 5$ and see whether the equation is satisfied.

$2x + y = 9$ The original equation

$2(2) + (5) \overset{?}{=} 9$ Substitute 2 for x and 5 for y.

$4 + 5 \overset{?}{=} 9$

$9 = 9$ A true statement

NOTE

$(2, 5)$ is a solution because a
true statement results.

$(2, 5)$ is a solution for the equation.

(b) For $(5, -1)$, let $x = 5$ and $y = -1$.

$$2(5) + (-1) \stackrel{?}{=} 9$$
$$10 - 1 \stackrel{?}{=} 9$$
$$9 = 9 \qquad \text{A true statement}$$

So $(5, -1)$ is a solution.

(c) For $(3, 4)$, let $x = 3$ and $y = 4$. Then

$$2(3) + (4) \stackrel{?}{=} 9$$
$$6 + 4 \stackrel{?}{=} 9$$
$$10 \stackrel{?}{=} 9 \qquad \textit{Not} \text{ a true statement}$$

So $(3, 4)$ is *not* a solution for the equation.

 Check Yourself 2

Which of the ordered pairs (3, 4), (4, 3), (1, −2), and (0, −5) are solutions for the equation

$3x - y = 5$

If the equation contains only one variable, then the missing variable can take on any value.

Example 3   Identifying Solutions of One-Variable Equations

NOTE

Think of this equation as

$1 \cdot x + 0 \cdot y = 2$

Which of the ordered pairs $(2, 0)$, $(0, 2)$, $(5, 2)$, $(2, 5)$, and $(2, -1)$ are solutions for the equation $x = 2$?

A solution is any ordered pair in which the x-coordinate is 2. That makes $(2, 0)$, $(2, 5)$, and $(2, -1)$ solutions for the given equation.

 Check Yourself 3

Which of the ordered pairs (3, 0), (0, 3), (3, 3), (−1, 3), and (3, −1) are solutions for the equation y = 3?

Remember that, when an ordered pair is presented, the first number is always the x-coordinate and the second number is always the y-coordinate.

Example 4   Completing Ordered-Pair Solutions

NOTE

The x-coordinate is also called the **abscissa** and the y-coordinate the **ordinate.**

Complete the ordered pairs (a) $(9, \ )$, (b) $(\ , -1)$, (c) $(0, \ )$, and (d) $(\ , 0)$ for the equation $x - 3y = 6$.

(a) The first number, 9, appearing in $(9, \ )$ represents the x-value. To complete the pair $(9, \ )$, substitute 9 for x and then solve for y.

$$(9) - 3y = 6$$
$$-3y = -3$$
$$y = 1$$
$(9, 1)$ is a solution.

(b) To complete the pair (, -1), let y be -1 and solve for x.

$$x - 3(-1) = 6$$
$$x + 3 = 6$$
$$x = 3$$

$(3, -1)$ is a solution.

(c) To complete the pair $(0,)$, let x be 0.

$$(0) - 3y = 6$$
$$-3y = 6$$
$$y = -2$$

$(0, -2)$ is a solution.

(d) To complete the pair (, 0), let y be 0.

$$x - 3(0) = 6$$
$$x - 0 = 6$$
$$x = 6$$

$(6, 0)$ is a solution.

Check Yourself 4

Complete the given ordered pairs so that each is a solution for the equation $2x + 5y = 10$.

$(10,)$, $(, 4)$, $(0,)$,   and $(, 0)$

Example 5   **Finding Some Solutions of a Two-Variable Equation**

Find four solutions for the equation

$$2x + y = 8$$

In this case the values used to form the solutions are *up to you*. You can assign any value for x (or for y). We demonstrate with some possible choices.

NOTE

Generally, you want to pick values for x (or for y) so that the resulting equation in one variable is easy to solve.

Solution with $x = 2$:

$$2(2) + y = 8$$
$$4 + y = 8$$
$$y = 4$$

$(2, 4)$ is a solution.

Solution with $y = 6$:

$$2x + (6) = 8$$
$$2x = 2$$
$$x = 1$$

$(1, 6)$ is a solution.

NOTE

The solutions $(0, 8)$ and $(4, 0)$ have special significance when graphing. They are also easy to find!

Solution with $x = 0$:

$$2(0) + y = 8$$
$$y = 8$$

$(0, 8)$ is a solution.

Solution with $y = 0$:

$$2x + (0) = 8$$
$$2x = 8$$
$$x = 4$$

$(4, 0)$ is a solution.

Check Yourself 5

Find four solutions for $x - 3y = 12$.

Applications involving two-variable equations are fairly common.

| Example 6 | Applications of Two-Variable Equations |

NOTE

We will look at variable and fixed costs in more detail in Section 7.1.

Suppose that it costs the manufacturer $1.25 for each stapler that is produced. In addition, fixed costs (related to staplers) are $110 per day.

(a) Write an equation relating the total daily costs C to the number x of staplers produced in a day.

Because each stapler costs $1.25 to produce, the cost of producing staplers is $1.25x$. Adding the fixed cost to this gives us an equation for the total daily costs.

$$C = 1.25x + 110$$

(b) What is the total cost of producing 500 staplers in a day?

We substitute 500 for x in the equation from part (a) and calculate the total cost.

$$C = 1.25(500) + 110$$
$$= 625 + 110$$
$$= 735$$

It costs the manufacturer a total of $735 to produce 500 staplers in one day.

(c) How many staplers can be produced for $1,110?

In this case, we substitute 1,110 for C in the equation from part (a) and solve for x.

RECALL

Divide both sides by 1.25.

$$\frac{1,000}{1.25} = \frac{1.25x}{1.25}$$
$$800 = x$$

$(1,110) = 1.25x + 110$ Subtract 110 from both sides.

$1,000 - 1.25x$ Divide both sides by 1.25.

$800 = x$

800 staplers can be produced at a cost of $1,110.

Check Yourself 6

Suppose that the stapler manufacturer earns a profit of $1.80 on each stapler shipped. However, it costs $120 to operate each day.

(a) Write an equation relating the daily profit P to the number x of staplers shipped in a day.

(b) What is the total profit of shipping 500 staplers in a day?

(c) How many staplers need to be shipped to produce a profit of $1,500?

We close this section with an application from the field of medicine.

Example 7	An Allied Health Application

For a particular patient, the weight w, in grams, of a tumor is related to the number of days d of chemotherapy treatment by the equation

$$w = -1.75d + 25$$

(a) What was the original weight of the tumor?

The original weight of the tumor is the value of w when $d = 0$. Substituting 0 for d in the equation gives

$$w = -1.75(0) + 25$$
$$= 25$$

The original weight of the tumor was 25 grams.

(b) How many days of chemotherapy are required to eliminate the tumor?
The tumor will be eliminated when the weight (w) is 0.

$$(0) = -1.75d + 25$$
$$-25 = -1.75d$$
$$d \approx 14.3$$

It will take about 14.3 days to eliminate the tumor.

Check Yourself 7

For a particular patient, the weight (w), in grams, of a tumor is related to the number of days (d) of chemotherapy treatment by the equation

$$w = -1.6d + 32$$

(a) Find the original weight of the tumor.
(b) Determine the number of days of chemotherapy required to eliminate the tumor.

Check Yourself ANSWERS

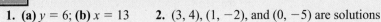

1. (a) $y = 6$; **(b)** $x = 13$ **2.** $(3, 4)$, $(1, -2)$, and $(0, -5)$ are solutions
3. $(0, 3)$, $(3, 3)$, and $(-1, 3)$ are solutions   **4.** $(10, -2)$, $(-5, 4)$, $(0, 2)$, and $(5, 0)$
5. $(6, -2)$, $(3, -3)$, $(0, -4)$, and $(12, 0)$ are four possibilities
6. (a) $P = 1.80x - 120$; **(b)** \$780; **(c)** 900   **7. (a)** 32 g; **(b)** 20 days

Reading Your Text

The following fill-in-the-blank exercises are designed to ensure that you understand some of the key vocabulary used in this section.

SECTION 6.1

(a) A _____ is a value for the variable that makes the equation a true statement.

(b) An equation of the form $Ax + By = C$, in which A and B are not both 0, is called a _____ equation.

(c) To simplify writing the pairs that satisfy an equation, we use _____ notation.

(d) When an ordered pair is presented, the _____ number is always the x-coordinate.

< Objectives 1–2 >

Determine which of the ordered pairs are solutions for the given equation.

1. $x + y = 6$ $(4, 2), (-2, 4), (0, 6), (-3, 9)$

2. $x - y = 12$ $(13, 1), (13, -1), (12, 0), (6, 6)$

3. $2x - y = 8$ $(5, 2), (4, 0), (0, 8), (6, 4)$

4. $x + 5y = 20$ $(10, -2), (10, 2), (20, 0), (25, -1)$

5. $3x - 2y = 12$ $(4, 0), \left(\frac{2}{3}, -5\right), (0, 6), \left(5, \frac{3}{2}\right)$ › Videos

6. $3x + 4y = 12$ $(-4, 0), \left(\frac{2}{3}, \frac{5}{2}\right), (0, 3), \left(\frac{2}{3}, 2\right)$

7. $y = 4x$ $(0, 0), (1, 3), (2, 8), (8, 2)$

8. $y = 2x - 1$ $(0, -2), (0, -1), \left(\frac{1}{2}, 0\right), (3, -5)$

9. $x = 3$ $(3, 5), (0, 3), (3, 0), (3, 7)$

10. $y = 5$ $(0, 5), (3, 5), (-2, -5), (5, 5)$

Complete the ordered pairs so that each is a solution for the given equation.

11. $x + y = 12$ $(4,\ ), (\ , 5), (0,\ ), (\ , 0)$

12. $x - y = 7$ $(\ , 4), (15,\ ), (0,\ ), (\ , 0)$

13. $3x - 2y = 12$ $(\ , 0), (\ , -6), (2,\ ), (\ , 3)$ › Videos

14. $2x + 5y = 20$ $(0,\ ), (5,\ ), (\ , 0), (\ , 6)$

15. $y = 3x + 9$ $(\ , 0), \left(\frac{2}{3},\ \right), (0,\ ), \left(-\frac{2}{3},\ \right)$

Boost *your* GRADE at ALEKS.com!

ALEKS

• Practice Problems   • e-Professors
• Self-Tests   • Videos
• NetTutor

Name _____

Section _____ Date _____

Answers

1. _____
2. _____
3. _____
4. _____
5. _____
6. _____
7. _____
8. _____
9. _____
10. _____
11. _____
12. _____
13. _____
14. _____
15. _____

Answers

16. _____

17. _____

18. _____

19. _____

20. _____

21. _____

22. _____

23. _____

24. _____

25. _____

26. _____

27. _____

28. _____

29. _____

30. _____

31. _____

32. _____

16. $3x + 4y = 12$ $(0, \ )$, $\left(\ , \dfrac{3}{4} \right)$, $(\ , 0)$, $\left(\dfrac{8}{3}, \ \right)$

17. $y = 3x - 4$ $(0, \ )$, $(\ , 5)$, $(\ , 0)$, $\left(\dfrac{5}{3}, \ \right)$

18. $y = -2x + 5$ $(0, \ )$, $(\ , 5)$, $\left(\dfrac{3}{2}, \ \right)$, $(\ , 1)$

Find four solutions for each equation. **Note:** *Your answers may vary from those shown in the answer section.*

19. $x - y = 7$ ▸ Videos

20. $x + y = 18$

21. $2x - y = 6$

22. $3x - y = 12$

23. $2x - 5y = 10$

24. $2x + 7y = 14$

25. $y = 2x + 3$

26. $y = 8x - 5$

27. $x = -5$ ▸ Videos

28. $y = 8$

29. **BUSINESS AND FINANCE** When an employee produces x units per hour, the hourly wage in dollars is given by $y = 0.75x + 8$. What are the hourly wages for each number of units: 2, 5, 10, 15, and 20?

30. **SCIENCE AND MEDICINE** Celsius temperature readings can be converted to Fahrenheit readings using the formula $F = \dfrac{9}{5}C + 32$. What is the Fahrenheit temperature that corresponds to each Celsius temperature: -10, 0, 15, 100?

31. **GEOMETRY** The perimeter of a square is given by $P = 4s$. What are the perimeters of the squares whose sides are 5, 10, 12, and 15 cm?

32. **BUSINESS AND FINANCE** When x units are sold, the price of each unit (in dollars) is given by $p = \dfrac{-x}{2} + 75$. Find the unit price when each quantity is sold: 2, 7, 9, 11.

33. **STATISTICS** The number of programs for the disabled in the United States for a 5-year period is approximated by the equation $y = 162x + 4{,}365$, where x represents particular years. Complete the table.

x	1	2	3	4	6
y					

34. **BUSINESS AND FINANCE** Your monthly pay as a car salesperson is determined using the equation $S = 200x + 1{,}500$ in which x is the number of cars you can sell each month.

    (a) Complete the table.

x	12	15	17	18
S				

    (b) You are offered a job at a salary of $56,400 per year. How many cars would you have to sell per month to equal this salary?

Basic Skills	**Challenge Yourself**	Calculator/Computer	Career Applications	Above and Beyond

Determine whether each statement is **true** *or* **false.**

35. When finding solutions for the equation $1 \cdot x + 0 \cdot y = 5$, you can choose any number for x.

36. When finding solutions for the equation $1 \cdot x + 1 \cdot y = 5$, you can choose any number for x.

Complete each statement with **never, sometimes,** *or* **always.**

37. If (a, b) is a solution to a particular two-variable equation, then (b, a) is _____ a solution to the same equation.

38. There are _____ an infinite number of solutions to a two-variable equation in standard form.

Find two solutions for each equation. **Note:** *Your answers may vary from those shown in the answer section.*

39. $\dfrac{1}{2}x + \dfrac{1}{3}y = 1$

40. $\dfrac{1}{3}x - \dfrac{1}{4}y = 1$

41. $0.3x + 0.5y = 2$

42. $0.6x - 0.2y = 5$

Answers

33. _____

34. _____

35. _____

36. _____

37. _____

38. _____

39. _____

40. _____

41. _____

42. _____

The Streeter/Hutchison Series in Mathematics   Beginning Algebra

Answers

43. _____

44. _____

45. _____

46. _____

47. _____

48. _____

49. _____

50. _____

51. _____

52. _____

53. _____

54. _____

55. _____

56. _____

57. _____

43. $\dfrac{3}{4}x - \dfrac{2}{5}y = 6$ > Videos

44. $\dfrac{4}{5}x + \dfrac{2}{3}y = 8$

45. $0.4x - 0.7y = 3$

46. $0.8x + 0.9y = 2$

An equation in three variables has an ordered triple as a solution. For example, (1, 2, 2) is a solution to the equation $x + 2y - z = 3$. Complete the ordered-triple solutions for each equation.

47. $x + y + z = 0$ $(2, -3, \ )$

48. $2x + y + z = 2$ $(\ , -1, 3)$

49. $x + y + z = 0$ $(1, \ , 5)$

50. $x + y - z = 1$ $(4, \ , 3)$

51. $2x + y + z = 2$ $(-2, \ , 1)$

52. $x + y - z = 1$ $(-2, 1, \ )$

| Basic Skills | Challenge Yourself | Calculator/Computer | **Career Applications** | Above and Beyond |

53. **ALLIED HEALTH** The recommended dosage (d), in milligrams (mg) of the antibiotic ampicillin sodium for children weighing less than 40 kilograms is given by the linear equation $d = 7.5w$, where w represents the child's weight in kilograms (kg).

 (a) What dose should be given to a child weighing 30 kg?

 (b) What size child requires a dose of 150 mg?

 Chapter 6 > Make the Connection

54. **ALLIED HEALTH** The recommended dosage (d), in micrograms (mcg) of Neupogen, a medication given to bone marrow transplant patients, is given by the linear equation $d = 8w$, where w represents the patient's weight in kilograms (kg).

 Chapter 6 > Make the Connection

 (a) What dose should be given to a male patient weighing 92 kg?

 (b) What size patient requires a dose of 250 mcg?

55. **MANUFACTURING TECHNOLOGY** The number of board feet b of lumber in a 2×6 of length L feet is given by the equation

$$b = \frac{8.25}{144}L$$

 Determine the number of board feet in 2×6 boards of length 12 ft, 16 ft, and 20 ft. Round to the nearest hundredth.

56. **MANUFACTURING TECHNOLOGY** The number of studs s (16 inches on center) required to build a wall that is L feet long is given by the formula

$$s = \frac{3}{4}L + 1$$

 Determine the number of studs required to build walls of length 12 ft, 20 ft, and 24 ft.

57. **MECHANICAL ENGINEERING** The force that a coil exerts on an object is related to the distance that the coil is pulled from its natural position. The formula to describe this is $F = kx$. If $k = 72$ pounds per foot for a certain coil, determine the force exerted if $x = 3$ ft or $x = 5$ ft.

58. MECHANICAL ENGINEERING If a machine is to be operated under water, it must be designed to handle the pressure (p) measured in pounds, which depends on the depth (d), measured in feet, of the water. The relationship is approximated by the formula $p = 59d + 13$. Determine the pressure at depths of 10 ft, 20 ft, and 30 ft.

Basic Skills | Challenge Yourself | Calculator/Computer | Career Applications | **Above and Beyond**
▲

59. You now have had practice solving equations with one variable and equations with two variables. Compare equations with one variable to equations with two variables. How are they alike? How are they different?

60. Each sentence describes pairs of numbers that are related. After completing the sentences in parts **(a)** to **(g)**, write two of your own sentences in parts **(h)** and **(i)**.

   (a) The *number of hours you work* determines the *amount you are* _____.

   (b) The *number of gallons of gasoline* you put in your car determines *the amount you* _____.

   (c) The *amount of the* _____ in a restaurant is related to *the amount of the tip.*

   (d) The *sale amount of a purchase in a store* determines _____.

   (e) The *age of an automobile* is related to _____.

   (f) The *amount of electricity you use in a month* determines _____.

   (g) The *cost of food for a family of four* is related to _____.

   Think of two more related pairs:

   (h) _____.

   (i) _____.

Answers

Answers

58. _____

59. _____

60. _____

1. $(4, 2), (0, 6), (-3, 9)$ **3.** $(5, 2), (4, 0), (6, 4)$

5. $(4, 0), \left(\frac{2}{3}, -5\right), \left(5, \frac{3}{2}\right)$ **7.** $(0, 0), (2, 8)$

9. $(3, 5), (3, 0), (3, 7)$ **11.** $(4, 8), (7, 5), (0, 12), (12, 0)$

13. $(4, 0), (0, -6), (2, -3), (6, 3)$ **15.** $(-3, 0), \left(\frac{2}{3}, 11\right), (0, 9), \left(-\frac{2}{3}, 7\right)$

17. $(0, -4), (3, 5), \left(\frac{4}{3}, 0\right), \left(\frac{5}{3}, 1\right)$ **19.** $(0, -7), (2, -5), (4, -3), (6, -1)$

21. $(0, -6), (3, 0), (6, 6), (9, 12)$ **23.** $(-5, -4), (0, -2), (5, 0), (10, 2)$
25. $(0, 3), (1, 5), (2, 7), (3, 9)$ **27.** $(-5, 0), (-5, 1), (-5, 2), (-5, 3)$
29. \$9.50, \$11.75, \$15.50, \$19.25, \$23   **31.** 20 cm, 40 cm, 48 cm, 60 cm
33. 4,527, 4,689, 4,851, 5,013, 5,337   **35.** False   **37.** sometimes

39. $(2, 0), (0, 3)$ **41.** $(0, 4), \left(\frac{20}{3}, 0\right)$ **43.** $(8, 0), (0, -15)$

45. $\left(\frac{15}{2}, 0\right), \left(0, -\frac{30}{7}\right)$ **47.** $(2, -3, 1)$ **49.** $(1, -6, 5)$ **51.** $(-2, 5, 1)$

53. (a) 225 mg; **(b)** 20 kg   **55.** 0.69 bd ft, 0.92 bd ft, 1.15 bd ft
57. 216 lb, 360 lb   **59.** Above and Beyond

6.2

The Rectangular Coordinate System

< 6.2 Objectives >

1 > Give the coordinates of a set of points in the plane

2 > Graph the points corresponding to a set of ordered pairs

In Section 6.1, we saw that ordered pairs could be used to write the solutions of equations in two variables. The next step is to graph those ordered pairs as points in a plane.

Because there are two numbers (one for *x* and one for *y*), we need two number lines. We draw one line horizontally, and the other is drawn vertically; their point of intersection (at their respective zero points) is called the *origin*. The horizontal line is called the ***x*-axis,** and the vertical line is called the ***y*-axis.** Together the lines form the **rectangular coordinate system.**

The axes divide the plane into four regions called **quadrants,** which are numbered (usually by Roman numerals) counterclockwise from the upper right.

NOTE

This system is also called the **Cartesian coordinate system,** named in honor of its inventor, René Descartes (1596–1650), a French mathematician and philosopher.

y-axis

Quadrant II   Quadrant I

Origin   *x*-axis

The origin is the point with coordinates (0, 0).

Quadrant III   Quadrant IV

We now want to establish correspondences between ordered pairs of numbers (x, y) and points in the plane. For any ordered pair

(x, y)

x-coordinate   *y*-coordinate

the following are true:

1. If the *x*-coordinate is

Positive, the point corresponding to that pair is located *x* units to the *right* of the *y*-axis.

Negative, the point is *x* units to the *left* of the *y*-axis.

Zero, the point is on the *y*-axis.

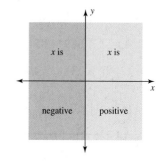

2. If the *y*-coordinate is

Positive, the point is *y* units *above* the *x*-axis.

Negative, the point is *y* units *below* the *x*-axis.

Zero, the point is on the *x*-axis.

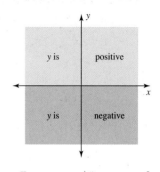

Putting this together we see the relationship

In Quadrant I, *x* is positive and *y* is positive.

In Quadrant II, *x* is negative and *y* is positive.

In Quadrant III, *x* is negative and *y* is negative.

In Quadrant IV, *x* is positive and *y* is negative.

Example 1 illustrates how to use these guidelines to give coordinates to points in the plane.

Example 1   **Identifying the Coordinates for a Given Point**

< Objective 1 >

RECALL

The *x*-coordinate gives the *horizontal* distance from the *y*-axis. The *y*-coordinate gives the *vertical* distance from the *x*-axis.

Give the coordinates for the given point.

(a)

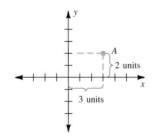

Point *A* is 3 units to the *right* of the *y*-axis and 2 units *above* the *x*-axis. Point *A* has coordinates (3, 2).

(b)

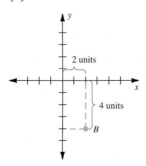

Point *B* is 2 units to the *right* of the *y*-axis and 4 units *below* the *x*-axis. Point *B* has coordinates (2, −4).

(c)

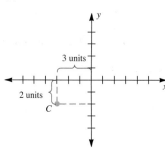

Point *C* is 3 units to the *left* of the *y*-axis and 2 units *below* the *x*-axis. *C* has coordinates $(-3, -2)$.

(d)

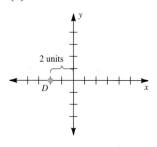

Point *D* is 2 units to the *left* of the *y*-axis and *on* the *x*-axis. Point *D* has coordinates $(-2, 0)$.

Check Yourself 1

Give the coordinates of points *P, Q, R,* and *S.*

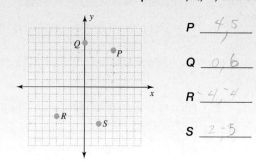

P _____ 4, 5 _____

Q _____ 0, 6 _____

R _____ 4, ¯4 _____

S _____ 2, ¯5 _____

Reversing the previous process allows us to graph (or plot) a point in the plane given the coordinates of the point. You can use these steps.

Step by Step

To Graph a Point in the Plane		
	Step 1	Start at the origin.
	Step 2	Move right or left according to the value of the *x*-coordinate.
	Step 3	Move up or down according to the value of the *y*-coordinate.

> **Example 2**   **Graphing Points**

< Objective 2 >

> **NOTE**
>
> The graphing of individual points is sometimes called **point plotting**.

(a) Graph the point corresponding to the ordered pair (4, 3).

Move 4 units to the right on the *x*-axis. Then move 3 units up from the point you stopped at on the *x*-axis. This locates the point corresponding to (4, 3).

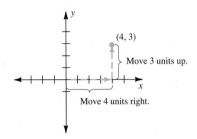

(b) Graph the point corresponding to the ordered pair (−5, 2).

In this case move 5 units *left* (because the *x*-coordinate is negative) and then 2 units *up*.

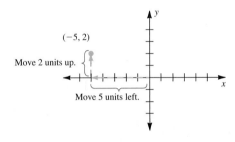

(c) Graph the point corresponding to (−4, −2).

Here move 4 units *left* and then 2 units *down* (the *y*-coordinate is negative).

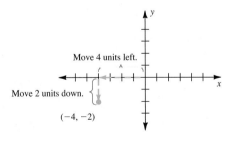

(d) Graph the point corresponding to (0, −3).

> **NOTE**
>
> Any point on an axis has 0 for one of its coordinates.

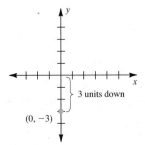

There is *no* horizontal movement because the *x*-coordinate is 0. Move 3 units *down*.

(e) Graph the point corresponding to (5, 0).

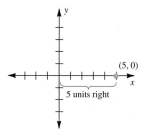

Move 5 units *right*. The desired point is on the *x*-axis because the *y*-coordinate is 0.

Check Yourself 2

Graph the points corresponding to *M*(4, 3), *N*(−2, 4), *P*(−5, −3), and *Q*(0, −3).

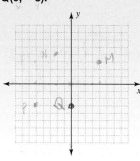

Example 3 gives an application from the field of manufacturing.

Example 3   **A Manufacturing Technology Application**

A computer-aided design (CAD) operator has located three corners of a rectangle. The corners are at (5, 9), (−2, 9), and (5, 2). Find the location of the fourth corner.

We plot the three indicated points on graph paper.

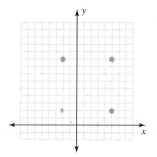

The fourth corner must lie directly underneath the point (−2, 9), so the *x*-coordinate must be −2. The corner must lie on the same horizontal as the point (5, 2), so the *y*-coordinate must be 2. Therefore the coordinates of the fourth corner are (−2, 2).

Check Yourself 3

A CAD operator has located three corners of a rectangle. The corners are at (−3, 4), (6, 4), and (−3, −7). Find the location of the fourth corner.

Check Yourself ANSWERS

1. $P(4, 5)$, $Q(0, 6)$, $R(-4, -4)$, and $S(2, -5)$

2.   **3.** $(6, -7)$

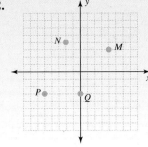

Reading Your Text

The following fill-in-the-blank exercises are designed to ensure that you understand some of the key vocabulary used in this section.

SECTION 6.2

(a) The point of intersection of the x-axis and y-axis is called the _____.

(b) The axes divide the plane into four regions called _____.

(c) If the y-coordinate is _____, the point is y units below the x-axis.

(d) Any point on an axis will have _____ for one of its coordinates.

6.2 exercises

Name _____

Section _____ Date _____

Answers

1. _____ 2. _____

3. _____ 4. _____

5. _____ 6. _____

7. _____ 8. _____

9. _____ 10. _____

11. _____

12. _____

13. _____

14. _____

15. _____

16. _____

17. _____

18. _____

19. _____

20. _____

Basic Skills | Challenge Yourself | Calculator/Computer | Career Applications | Above and Beyond

< Objective 1 >

Give the coordinates of the points graphed below and name the quadrant or axis where the point is located.

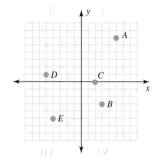

1. A 5, 6

2. B

3. C 2   > Videos

4. D

5. E > Videos
   -4,5 III

Give the coordinates of the points graphed below and name the quadrant or axis where the point is located.

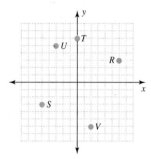

6. R

7. S -5, -3 III

8. T

9. U -3, 5 II

10. V

< Objective 2 >

Plot the points on the graph below.

11. $M(5, 3)$ 12. $N(0, -3)$

13. $P(-2, 6)$ 14. $Q(5, 0)$

15. $R(-4, -6)$ 16. $S(-3, -4)$

Plot the points on the given graph.

17. $F(-3, -1)$ 18. $G(4, 3)$

19. $H(5, -2)$ 20. $I(-3, 0)$

21. $J(-5, 3)$ 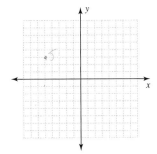 > Videos **22.** $K(0, 6)$

23. **SCIENCE AND MEDICINE** A local plastics company is sponsoring a plastics recycling contest for the local community. The focus of the contest is collecting plastic milk, juice, and water jugs. The company will award $200 plus the current market price of the jugs collected to the group that collects the most jugs in a single month. The number of jugs collected and the amount of money won can be represented as an ordered pair.

(a) In April, group A collected 1,500 lb of jugs to win first place. The prize for the month was $350. On the graph, x represents the pounds of jugs and y represents the amount of money that the group won. Graph the point that represents the winner for April.

(b) In May, group B collected 2,300 lb of jugs to win first place. The prize for the month was $430. Graph the point that represents the May winner on the same axes you used in part (a).

(c) In June, group C collected 1,200 lb of jugs to win the contest. The prize for the month was $320. Graph the point that represents the June winner on the same axes as used before.

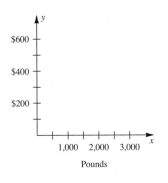

Answers

24.

25.

26.

24. STATISTICS The table gives the hours x that Damien studied for five different math exams and the resulting grades y. Plot the data given in the table.

x	4	5	5	2	6
y	83	89	93	75	95

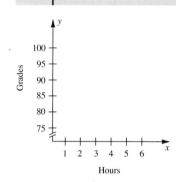

25. SCIENCE AND MEDICINE The table gives the average temperature y (in degrees Fahrenheit) for each of the first 6 months of the year, x. The months are numbered 1 through 6, with 1 corresponding to January. Plot the data given in the table.

> Videos

x	1	2	3	4	5	6
y	4	14	26	33	42	51

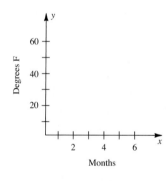

26. BUSINESS AND FINANCE The table gives the total salary of a salesperson, y, for each of the four quarters of the year, x. Plot the data given in the table.

x	1	2	3	4
y	$6,000	$5,000	$8,000	$9,000

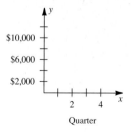

27. STATISTICS The table shows the number of runs scored by the Anaheim Angels in each game of the 2002 World Series.

Game	1	2	3	4	5	6	7
Runs	3	11	10	3	4	6	4

Source: Major League Baseball.

27. _____

28. _____

Plot the data given in the table.

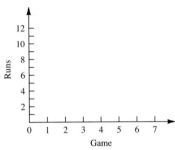

28. STATISTICS The table shows the number of wins and total points for the five teams in the Atlantic Division of the National Hockey League in the early part of a recent season.

Team	Wins	Points
New Jersey Devils	5	12
Philadelphia Flyers	4	10
New York Rangers	4	9
Pittsburgh Penguins	2	6
New York Islanders	2	5

Plot the data given in the table.

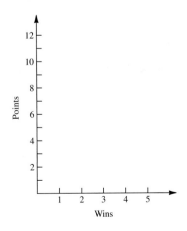

Answers

29. _____

30. _____

31. _____

32. _____

33. _____

34. _____

Determine whether each statement is **true** *or* **false.**

29. If the x- and y-coordinates of an ordered pair are both negative, then the plotted point must lie in Quadrant III.

30. If the y-coordinate of an ordered pair is 0, then the plotted point must lie on the y-axis.

Complete each statement with **never, sometimes,** *or* **always.**

31. The ordered pair (a, b) is _____ equal to the ordered pair (b, a).

32. If, in the ordered pair (a, b), a and b have different signs, then the point (a, b) is _____ in the second quadrant.

33. Plot points with coordinates $(2, 3)$, $(3, 4)$, and $(4, 5)$ on the given graph. What do you observe? Can you give the coordinates of another point with the same property? > Videos

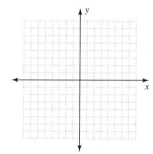

34. Plot points with coordinates $(-1, 4)$, $(0, 3)$, and $(1, 2)$ on the given graph. What do you observe? Can you give the coordinates of another point with the same property?

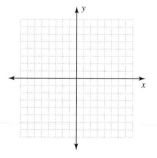

For exercises 35–38, do the following:

(a) Give the coordinates of the plotted points.

(b) Describe in words the relationship between the *y*-coordinate and the *x*-coordinate.

(c) Write an equation for the relationship described in part (b).

35.

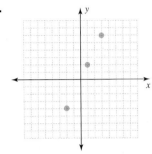

36.

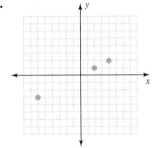

37.

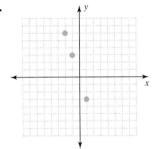

38.

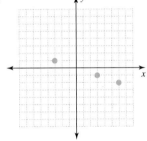

Basic Skills | Challenge Yourself | Calculator/Computer | **Career Applications** | Above and Beyond
▲

Plot the points given in the tables.

39. ALLIED HEALTH Medical lab technicians analyzed several concentrations, in milligrams per deciliter (mg/dL), of glucose solutions to determine the percent transmittance, which measures the percent of light that filters through the solution. The results are summarized in the table.

Glucose concentration (mg/dL)	0	80	160	240	320	400
Percent transmittance (% T)	100	62	40	25	15	10

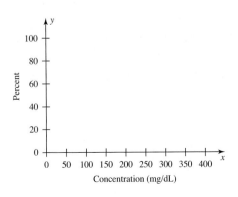

Answers

40. _____

41. _____

42. _____

40. **ALLIED HEALTH** Daniel's weight, in pounds, has been recorded at various well-baby checkups. The results are summarized in the table.

Age (months)	0	0.5	1	2	7	9
Weight (pounds)	7.8	7.14	9.25	12.5	20.25	21.25

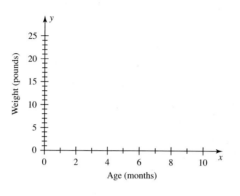

41. **ELECTRONICS** A certain project requires the use of a solenoid, a device that uses an applied electromagnetic force to cause mechanical force. Typically, a conductor such as wire is coiled and current is applied, creating an electromagnet. The magnetic field induced by the energized coil attracts a piece of ferrous material (iron), creating mechanical movement.

Plot the force (in newtons), y, versus applied voltage (in volts), x, of a solenoid using the values given in the table.

x	5	10	15	20
y	0.12	0.24	0.36	0.49

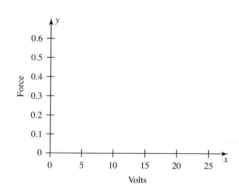

42. **MANUFACTURING TECHNOLOGY** The temperature and pressure relationship for a coolant is described by the table.

Temperature (°F)	−10	10	30	50	70	90
Pressure (psi)	4.6	14.9	28.3	47.1	71.1	99.2

(a) Graph the points.

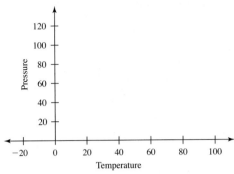

(b) Predict what the pressure will be when the temperature is 60°F.

(c) At what temperature would you expect the coolant to be when the pressure is 37 psi?

43. AUTOMOTIVE TECHNOLOGY The table lists the travel time and the distance for several business trips.

Travel time (hours)	6	2	7	9	4
Distance (miles)	320	90	410	465	235

Plot these points on a graph.

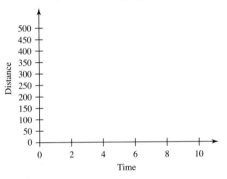

44. MANUFACTURING TECHNOLOGY The layout of a jobsite is shown here.

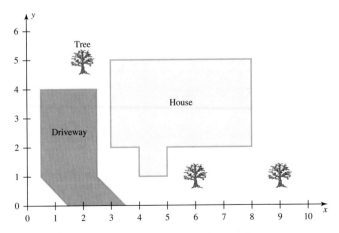

(a) What are the coordinates for each corner of the house?

(b) What is located at (6, 1)?

Answers

45. _____

46. _____

47. _____

45. The map shown here uses letters and numbers to label a grid that helps to locate a city. For instance, Salem is located at E4.

    (a) Find the coordinates for the following: White Swan, Newport, and Wheeler.

    (b) What cities correspond to the following coordinates: A2, F4, and A5?

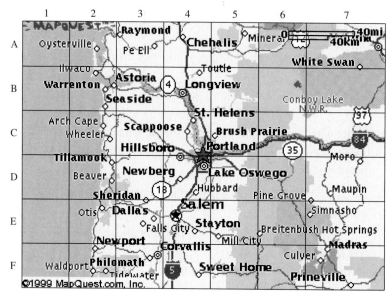

| Basic Skills | Challenge Yourself | Calculator/Computer | Career Applications | **Above and Beyond** |

46. How would you describe a rectangular coordinate system? Explain what information is needed to locate a point in a coordinate system.

47. Some newspapers have a special day that they devote to automobile ads. Use this special section or the Sunday classified ads from your local newspaper to find all the want ads for a particular automobile model. Make a list of the model year and asking price for 10 ads, being sure to get a variety of ages for this model. After collecting the information, make a graph of the age and the asking price for the car.

    Describe your graph, including an explanation of how you decided which variable to put on the vertical axis and which on the horizontal axis. What trends or other information are given by the graph?

Answers

1. $(5, 6)$; I   **3.** $(2, 0)$; x-axis   **5.** $(-4, -5)$; III
7. $(-5, -3)$; III   **9.** $(-3, 5)$; II
11–21.

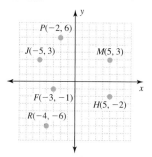

23. (a) $(1,500, 350)$; **(b)** $(2,300, 430)$; **(c)** $(1,200, 320)$

25.

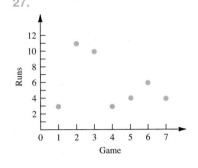

27.

29. True **31.** sometimes
33. The points lie on a line; e.g., $(1, 2)$

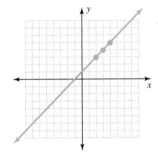

35. (a) $(-2, -4)$, $(1, 2)$, $(3, 6)$; **(b)** The y-value is twice the x-value; **(c)** $y = 2x$
37. (a) $(-2, 6)$, $(-1, 3)$, $(1, -3)$; **(b)** The y-value is -3 times the x-value;
(c) $y = -3x$
39.   **41.**

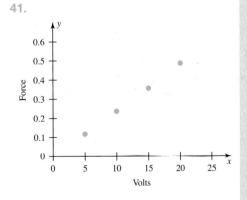

43.

45. (a) A7, F2, C2; **(b)** Oysterville, Sweet Home, Mineral
47. Above and Beyond

6.3

Graphing Linear Equations

< 6.3 Objectives >

1 > Graph a linear equation by plotting points

2 > Graph a linear equation that results in a vertical or horizontal line

3 > Graph a linear equation by the intercept method

4 > Graph a linear equation by solving the equation for y

We are now ready to combine our work in Sections 6.1 and 6.2. In Section 6.1 you learned to write solutions of equations in two variables as ordered pairs. Then, in Section 6.2, these ordered pairs were graphed in the plane. Putting these ideas together will help us to graph equations. Example 1 illustrates this approach.

  | **Example 1** | **Graphing a Linear Equation**

< Objective 1 >

Graph $x + 2y = 4$.

NOTE

We find *three* solutions for the equation. We'll point out why shortly.

Step 1   Find some solutions for $x + 2y = 4$. To find solutions, we choose any convenient values for x, say, $x = 0$, $x = 2$, and $x = 4$. Given these values for x, we can substitute and then solve for the corresponding value for y. So

If $x = 0$, then $y = 2$, so $(0, 2)$ is a solution.

If $x = 2$, then $y = 1$, so $(2, 1)$ is a solution.

If $x = 4$, then $y = 0$, so $(4, 0)$ is a solution.

A handy way to show this information is in a table such as

x	y
0	2
2	1
4	0

NOTE

The table is just a convenient way to display the information. It is the same as writing (0, 2), (2, 1), and (4, 0).

Step 2   We now graph the solutions found in step 1.

$x + 2y = 4$

x	y
0	2
2	1
4	0

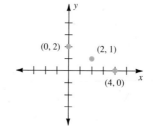

What pattern do you see? It appears that the three points lie on a straight line, which is in fact the case.

Step 3 Draw a straight line through the three points graphed in step 2.

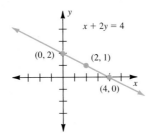

The line shown is the **graph** of the equation $x + 2y = 4$. It represents *all* of the ordered pairs that are solutions (an infinite number) for that equation.

Every ordered pair that is a solution lies on this line. Any point on the line will have coordinates that are a solution for the equation.

Note: Why did we suggest finding *three* solutions in step 1? Two points determine a line, so technically you need only two. The third point that we find is a check to catch any possible errors.

Check Yourself 1

Graph $2x - y = 6$, using the steps shown in Example 1.

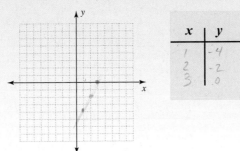

In Section 6.1, we mentioned that an equation that can be written in the form

$$Ax + By = C$$

in which A, B, and C are real numbers and A and B are not both 0 is called a **linear equation in two variables.** The graph of this equation is a *straight line.*

The steps for graphing follow.

Step by Step

To Graph a Linear Equation

Step 1 Find at least three solutions for the equation and put your results in tabular form.

Step 2 Graph the solutions found in step 1.

Step 3 Draw a straight line through the points determined in step 2 to form the graph of the equation.

Example 2   Graphing a Linear Equation

Graph $y = 3x$.

Step 1   Some solutions are

x	y
0	0
1	3
2	6

Step 2   Graph the points.

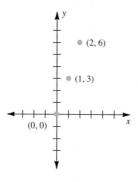

Step 3   Draw a line through the points.

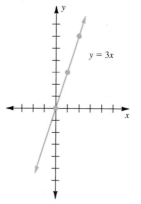

    Check Yourself 2

Graph the equation $y = -2x$ after completing the table of values.

x	y
0	0
1	-2
2	-4

Let's work through another example of graphing a line from its equation.

 Example 3 | Graphing a Linear Equation

Graph $y = 2x + 3$.

Step 1   Some solutions are

x	y
0	3
1	5
2	7

Step 2   Graph the points corresponding to these values.

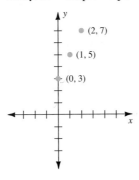

Step 3   Draw a line through the points.

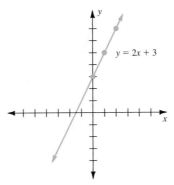

$y = 2x + 3$

Check Yourself 3

Graph the equation $y = 3x - 2$ after completing the table of values.

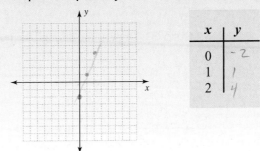

x	y
0	-2
1	1
2	4

When graphing equations, particularly if fractions are involved, a careful choice of values for x can simplify the process. Consider Example 4.

| **Example 4** | **Graphing a Linear Equation** |

Graph

$$y = \frac{3}{2}x - 2$$

As before, we want to find solutions for the given equation by picking convenient values for x. Note that in this case, choosing *multiples of* 2 will avoid fractional values for y and make the plotting of those solutions much easier. For instance, here we might choose values of -2, 0, and 2 for x.

Step 1   If $x = -2$:

$$y = \frac{3}{2}(-2) - 2 = -3 - 2 = -5$$

If $x = 0$:

$$y = \frac{3}{2}(0) - 2 \;\; = 0 - 2 = -2$$

If $x = 2$:

$$y = \frac{3}{2}(2) - 2 = 3 - 2 = 1$$

In tabular form, the solutions are

x	y
-2	-5
0	-2
2	1

Step 2   Graph the points determined in the table.

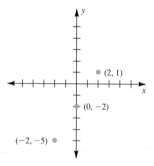

Step 3   Draw a line through the points.

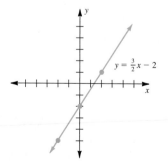

Check Yourself 4

Graph the equation $y = -\frac{1}{3}x + 3$ after completing the table of values.

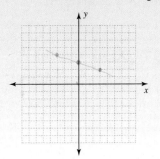

x	y
-3	4
0	3
3	2

Some special cases of linear equations are illustrated in Examples 5 and 6.

▶ **Example** 5   **Graphing an Equation That Results in a Vertical Line**

< Objective 2 >

Graph $x = 3$.   (3, 0)

The equation $x = 3$ is equivalent to $1 \cdot x + 0 \cdot y = 3$, or $x + 0(y) = 3$. Some solutions follow.

If $y = 1$:   If $y = 4$:   If $y = -2$:

$x + 0(1) = 3$ $x + 0(4) = 3$ $x + 0(-2) = 3$

$\qquad x = 3$ $\qquad x = 3$ $\qquad x = 3$

In tabular form,

x	y
3	1
3	4
3	-2

What do you observe? The variable x has the value 3, regardless of the value of y. Consider the graph.

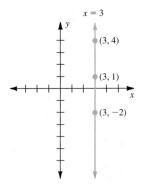

The graph of $x = 3$ is a vertical line crossing the x-axis at $(3, 0)$.

Note that graphing (or plotting) points in this case is not really necessary. Simply recognize that the graph of $x = 3$ *must* be a vertical line (parallel to the y-axis) that intercepts the x-axis at $(3, 0)$.

Check Yourself 5

Graph the equation $x = -2$.

Example 6 is a related example involving a horizontal line.

| **Example 6** | **Graphing an Equation That Results in a Horizontal Line** |

Graph $y = 4$.

Because $y = 4$ is equivalent to $0 \cdot x + 1 \cdot y = 4$, or $0(x) + y = 4$, any value for x paired with 4 for y will form a solution. A table of values might be

x	y
-2	4
0	4
2	4

Here is the graph.

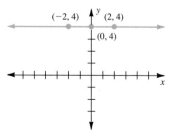

This time the graph is a horizontal line that crosses the y-axis at $(0, 4)$. Again, you do not need to graph points. The graph of $y = 4$ *must* be horizontal (parallel to the x-axis) and intercepts the y-axis at $(0, 4)$.

Check Yourself 6

Graph the equation $y = -3$.

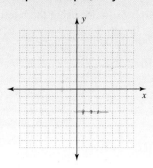

This box summarizes our work in Examples 5 and 6.

<table>
<tr><td>**Property**</td><td></td></tr>
</table>

Vertical and Horizontal Lines	**1.** The graph of $x = a$ is a *vertical line* crossing the x-axis at $(a, 0)$.
	2. The graph of $y = b$ is a *horizontal line* crossing the y-axis at $(0, b)$.

NOTE

With practice, this can be done mentally, which is the big advantage of this method.

To simplify the graphing of certain linear equations, some students prefer the **intercept method** of graphing. This method makes use of the fact that the solutions that are easiest to find are those with an x-coordinate or a y-coordinate of 0. For instance, to graph the equation

$$4x + 3y = 12$$

first, let $x = 0$ and then solve for y.

$$4(0) + 3y = 12$$
$$3y = 12$$
$$y = 4$$

So $(0, 4)$ is one solution. Now we let $y = 0$ and solve for x.

$$4x + 3(0) = 12$$
$$4x = 12$$
$$x = 3$$

A second solution is $(3, 0)$.

RECALL

Only two points are needed to graph a line. A third point is used only as a check.

The two points corresponding to these solutions can now be used to graph the equation.

$4x + 3y = 12$

NOTE

The intercepts are the points where the line crosses the x- and y-axes.

The ordered pair $(3, 0)$ is called the **x-intercept,** and the ordered pair $(0, 4)$ is the **y-intercept** of the graph. Using these points to draw the graph gives the name to this method. Here is a second example of graphing by the intercept method.

(▶) **Example 7**	**Using the Intercept Method to Graph a Line**

< Objective 3 >

Graph $3x - 5y = 15$, using the intercept method.

To find the x-intercept, let $y = 0$.

$3x - 5(0) = 15$

$\qquad x = 5$

The x-intercept is (5, 0).

To find the y-intercept, let $x = 0$.

$3(0) - 5y = 15$

$\qquad y = -3$

The y-intercept is (0, −3).

So (5, 0) and (0, −3) are solutions for the equation, and we can use the corresponding points to graph the equation.

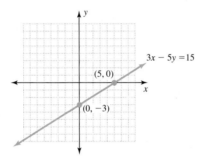

Check Yourself 7

Graph $4x + 5y = 20$, using the intercept method.

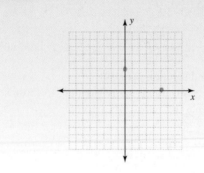

NOTE

Finding a third "checkpoint" is always a good idea.

This all looks quite easy, and for many equations it is. What are the drawbacks? For one, you don't have a third checkpoint, and it is possible for errors to occur. You can, of course, still find a third point (other than the two intercepts) to be sure your graph is correct. A second difficulty arises when the x- and y-intercepts are very close to one another (or are actually the same point—the origin). For instance, if we have the equation

$3x + 2y = 1$

the intercepts are $\left(\dfrac{1}{3}, 0\right)$ and $\left(0, \dfrac{1}{2}\right)$. It is difficult to draw a line accurately through these intercepts, so choose other solutions farther away from the origin for your points.

We summarize the steps of graphing by the intercept method for appropriate equations.

Step by Step		
Graphing a Line by the Intercept Method	Step 1	To find the x-intercept, let $y = 0$, then solve for x.
	Step 2	To find the y-intercept, let $x = 0$, then solve for y.
	Step 3	Graph the x- and y-intercepts.
	Step 4	Draw a straight line through the intercepts.

A third method of graphing linear equations involves **solving the equation for y.** The reason we use this extra step is that it often makes finding solutions for the equation much easier.

Example 8	Graphing a Linear Equation by Solving for y

< Objective 4 >

RECALL

Solving for y means that we want to have y isolated on the left.

Graph $2x + 3y = 6$.

Rather than finding solutions for the equation in this form, we solve for y.

$$2x + 3y = 6$$
$$3y = 6 - 2x \qquad \text{Subtract } 2x.$$
$$y = \frac{6 - 2x}{3} \qquad \text{Divide by 3.}$$
or $\quad y = 2 - \frac{2}{3}x$

Now find your solutions by picking convenient values for x.

NOTE

Again, to choose convenient values for x, we suggest you look at the equation carefully. Here, for instance, choosing multiples of 3 for x makes the work much easier.

If $x = -3$:

$$y = 2 - \frac{2}{3}(-3)$$
$$= 2 + 2 = 4$$

So $(-3, 4)$ is a solution.

If $x = 0$:

$$y = 2 - \frac{2}{3}(0)$$
$$= 2$$

So $(0, 2)$ is a solution.

If $x = 3$:

$$y = 2 - \frac{2}{3}(3)$$
$$= 2 - 2 = 0$$

So $(3, 0)$ is a solution.

We can now plot the points that correspond to these solutions and form the graph of the equation as before.

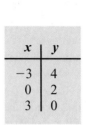

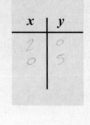

$2x + 3y = 6$

x	y
−3	4
0	2
3	0

Check Yourself 8

Graph the equation $5x + 2y = 10$. Solve for y to determine solutions.

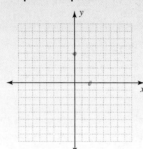

x	y
2	0
0	5

In applications, we often use graphs that do not fit nicely into a standard 8-by-8 grid. In these cases, we should show the portion of the graph that displays the interesting properties of the graph. We may need to scale the x- or y-axis to meet our needs or even set the axes so that part of an axis seems to be "cut out."

When we scale the axes, it is important to include numbers on the axes at convenient grid lines. If we set the axes so that part of an axis is removed, we include a mark to indicate this. Both of these situations are illustrated in Example 9.

    Example 9   Graphing in Nonstandard Windows

NOTE

In business, the constant, 2,500, is called the **fixed cost**. The coefficient, 45, is referred to as the **marginal cost**.

NOTE

Observe the mark on the y-axis that is used to show a break in the axis.

The cost, y, to produce x CD players is given by the equation $y = 45x + 2,500$. Graph the cost equation, with appropriately scaled and set axes.

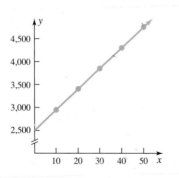

The *y*-intercept is (0, 2,500). We find more points to plot by creating a table.

x	10	20	30	40	50
y	2,950	3,400	3,850	4,300	4,750

Check Yourself 9

Graph the cost equation given by $y = 60x + 1{,}200$, with appropriately scaled and set axes.

Here is an example of an application from the field of medicine.

| Example 10 | An Allied Health Application |

The arterial oxygen tension (P_aO_2), in millimeters of mercury (mm Hg), of a patient can be estimated based on the patient's age (A), in years. If the patient is lying down, use the equation $P_aO_2 = 103.5 - 0.42A$. Draw the graph of this equation, using appropriately scaled and set axes.

We begin by creating a table. Using a calculator here is very helpful.

A	0	10	20	30	40	50	60	70	80
P_aO_2	103.5	99.3	95.1	90.9	86.7	82.5	78.3	74.1	69.9

Seeing these values allows us to decide upon the vertical axis scaling. We scale from 60 to 110 and include a mark to show a break in the axis. We estimate the locations of these coordinates, and draw the line.

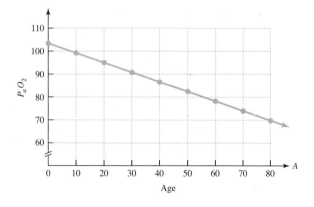

Check Yourself 10

The arterial oxygen tension (P_aO_2), in millimeters of mercury (mm Hg), of a patient can be estimated based on the patient's age (A), in years. If the patient is seated, use the equation $P_aO_2 = 104.2 - 0.27A$. Draw the graph of this equation, using appropriately scaled and set axes.

✓ **Check Yourself** ANSWERS

1.

x	y
1	-4
2	-2
3	0

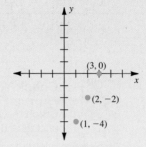

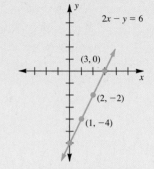

2.

x	y
0	0
1	-2
2	-4

3.

x	y
0	-2
1	1
2	4

4.

x	y
-3	4
0	3
3	2

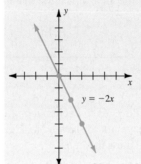

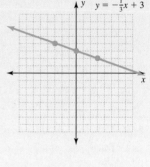

5.

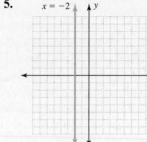

6.

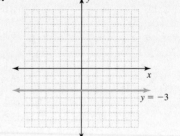

7.

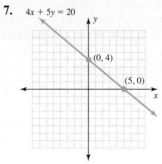

8.

x	y
0	5
2	0
4	−5

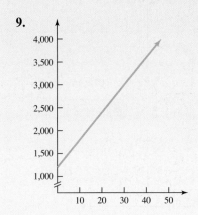

$y = -\frac{5}{2}x + 5$

9.

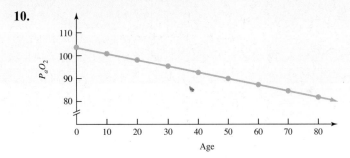

10.

Reading Your Text

The following fill-in-the-blank exercises are designed to ensure that you understand some of the key vocabulary used in this section.

SECTION 6.3

(a) An equation that can be written in the form $Ax + By = C$ is called a _____ equation in two variables.

(b) The graph of $x = a$ is a _____ line crossing the x-axis at $(a, 0)$.

(c) The graph of $y = b$ is a _____ line crossing the y-axis at $(0, b)$.

(d) The _____ method makes use of the fact that the solutions easiest to find are those with an x-coordinate or a y-coordinate of 0.

Name _____

Section _____ Date _____

Answers

1. _____

2. _____

3. _____

4. _____

5. _____

6. _____

7. _____

8. _____

< Objectives 1–2 >

Graph each equation.

1. $x + y = 6$

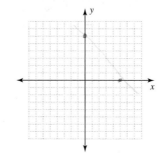

2. $x - y = 5$

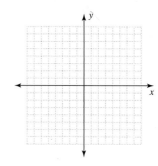

3. $x - y = -3$ > Videos

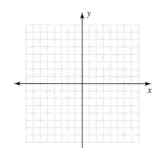

4. $x + y = -3$

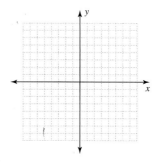

5. $2x + y = 2$

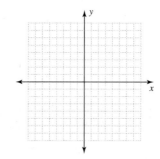

6. $x - 2y = 6$

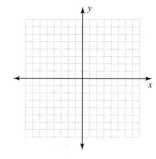

7. $3x + y = 0$

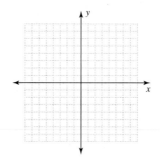

8. $3x - y = 6$

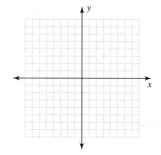

9. $x + 4y = 8$

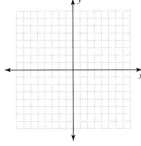

10. $2x - 3y = 6$

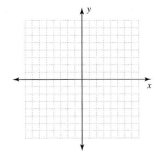

11. $y = 5x$

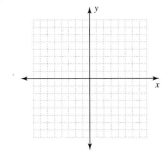

12. $y = -4x$

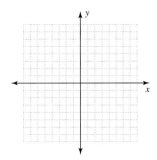

13. $y = 2x - 1$

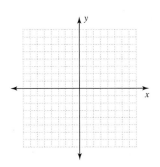

14. $y = 4x + 3$

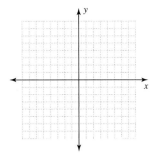

15. $y = -3x + 1$ > Videos

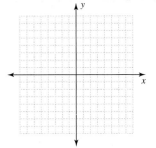

16. $y = -3x - 3$

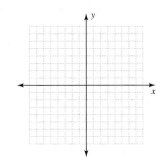

Answers

17. _____

18. _____

19. _____

20. _____

21. _____

22. _____

23. _____

24. _____

17. $y = \dfrac{1}{3}x$

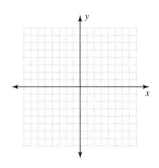

18. $y = -\dfrac{1}{4}x$

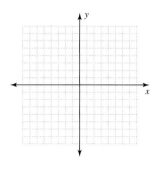

19. $y = \dfrac{2}{3}x - 3$

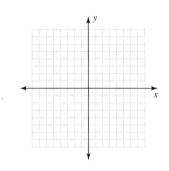

20. $y = \dfrac{3}{4}x + 2$

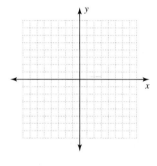

21. $x = 5$ > Videos

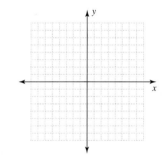

22. $y = -3$

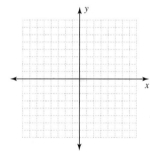

23. $y = 1$

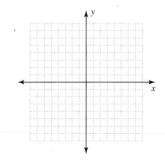

24. $x = -2$

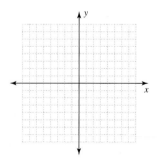

< Objective 3 >

Graph each equation using the intercept method.

25. $x - 2y = 4$

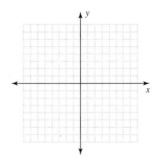

26. $6x + y = 6$

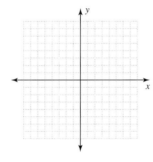

25. _____

26. _____

27. _____

28. _____

29. _____

30. _____

31. _____

32. _____

27. $5x + 2y = 10$

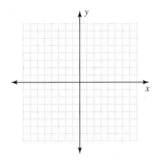

28. $2x + 3y = 6$

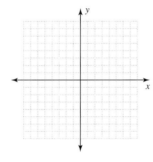

29. $3x + 5y = 15$ > Videos

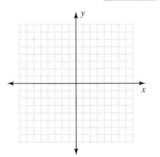

30. $4x + 3y = 12$

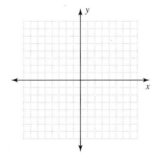

< Objective 4 >

Graph each equation by first solving for y.

31. $x + 3y = 6$ > Videos

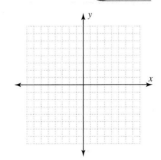

32. $x - 2y = 6$

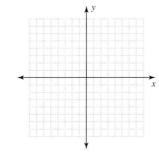

Answers

33. _____

34. _____

35. _____

36. _____

37. _____

38. _____

33. $3x + 4y = 12$

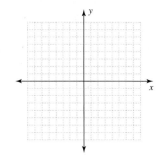

34. $2x - 3y = 12$

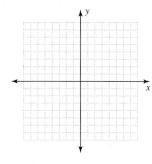

35. $5x - 4y = 20$

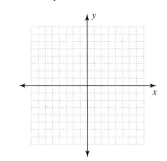

36. $7x + 3y = 21$

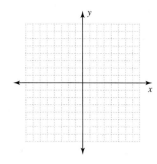

37. Science and Medicine The equation $y = 0.10x + 200$ describes the amount of winnings a group earns for collecting plastic jugs in the recycling contest described in exercise 23 at the end of Section 6.2. Sketch the graph of the line on the given coordinate system.

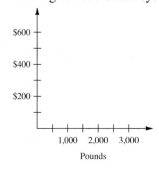

38. Business and Finance A car rental agency charges $40 per day for a certain type of car, plus 15¢ per mile driven. The equation $y = 0.15x + 40$ indicates the charge, y, for a day's rental where x miles are driven. Sketch the graph of this equation on the given axes.

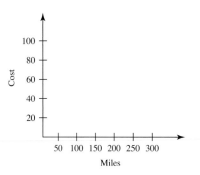

39. BUSINESS AND FINANCE A high school class wants to raise some money by recycling newspapers. The class decides to rent a truck for a weekend and to collect the newspapers from homes in the neighborhood. The market price for recycled newsprint is currently $11 per ton. The equation $y = 11x - 100$ describes the amount of money the class will make, in which y is the amount of money made in dollars, x is the number of tons of newsprint collected, and 100 is the cost in dollars to rent the truck.

(a) Using the given axes, draw a graph that represents the relationship between newsprint collected and money earned.

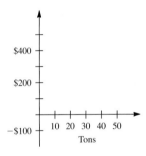

(b) The truck costs the class $100. How many tons of newspapers must the class collect to break even on this project?

(c) If the class members collect 16 tons of newsprint, how much money will they earn?

(d) Six months later the price of newsprint is $17 a ton, and the cost to rent the truck has risen to $125. Write the equation that describes the amount of money the class might make at that time.

40. BUSINESS AND FINANCE A car rental agency charges $30 per day for a certain type of car, plus 20¢ per mile driven. The equation $y = 0.20x + 30$ indicates the charge, y, for a day's rental where x miles are driven. Sketch the graph of this equation on the given axes.

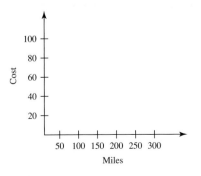

Answers

39. _____

40. _____

Answers

41. _____

42. _____

43. _____

44. _____

45. _____

46. _____

41. **BUSINESS AND FINANCE** The cost of producing x items is given by $C = mx + b$, in which b is the fixed cost and m is the marginal cost (the cost of producing one more item).

   (a) If the fixed cost is $200 and the marginal cost is $15, write the cost equation.

   (b) Graph the cost equation on the given axes.

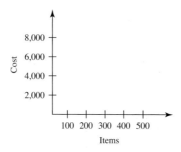

42. **BUSINESS AND FINANCE** The cost of producing x items is given by $C = mx + b$, in which b is the fixed cost and m is the marginal cost (the cost of producing one more item).

   (a) If the fixed cost is $150 and the marginal cost is $20, write the cost equation.

   (b) Graph the cost equation on the given axes.

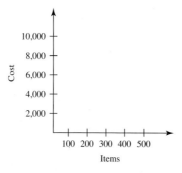

| Basic Skills | **Challenge Yourself** | Calculator/Computer | Career Applications | Above and Beyond |

Determine whether each statement is **true** *or* **false.**

43. If the ordered pair (x, y) is a solution to an equation, then the point (x, y) is always on the graph of the equation.

44. The graph of a linear equation must pass through the origin.

Complete each statement with **never, sometimes,** *or* **always.**

45. If the graph of a linear equation $Ax + By = C$ passes through the origin, then C _____ equals zero.

46. The graph of a linear equation _____ has an x-intercept.

Write an equation that describes each relationship between x and y. Then graph each relationship.

Answers

47. *y* is twice *x*.

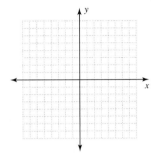

48. *y* is 2 less than *x*.

47. _____

48. _____

49. _____

50. _____

51. _____

52. _____

49. *y* is 3 less than 3 times *x*.

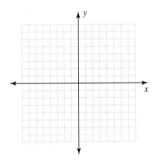

50. *y* is 4 more than twice *x*.

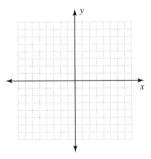

51. The difference of *x* and the product of 4 and *y* is 12.

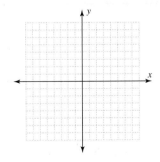

52. The difference of twice *x* and *y* is 6.

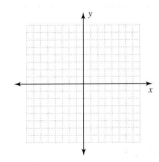

Answers

53. _____

54. _____

55. _____

56. _____

57. _____

Graph each pair of equations on the same axes. Give the coordinates of the point where the lines intersect.

53. $x + y = 4$
$x - y = 2$

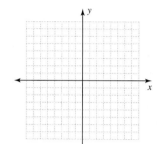

54. $x - y = 3$
$x + y = 5$

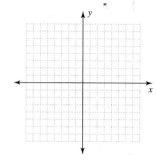

Graph each set of equations on the same coordinate system. Do the lines intersect? What are the y-intercepts?

55. $y = 3x$
$y = 3x + 4$
$y = 3x - 5$

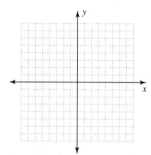

56. $y = -2x$
$y = -2x + 3$
$y = -2x - 5$

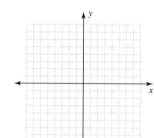

| Basic Skills | Challenge Yourself | Calculator/Computer | **Career Applications** | Above and Beyond |

57. **ALLIED HEALTH** The weight (w), in kilograms, of a tumor is related to the number of days (d) of chemotherapy treatment by the linear equation $w = -1.75d + 25$. Sketch a graph of this linear equation.

chapter 6 > Make the Connection

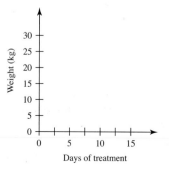

58. **ALLIED HEALTH** The weight (w), in kilograms, of a tumor is related to the number of days (d) of chemotherapy treatment by the linear equation $w = -1.6d + 32$. Sketch a graph of the linear equation.

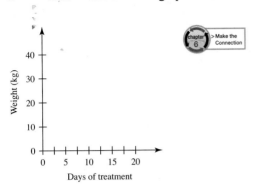

59. **MECHANICAL ENGINEERING** The force that a coil exerts on an object is related to the distance that the coil is pulled from its natural position. The formula to describe this is $F = kx$. Graph this relationship for a coil with $k = 72$ pounds per foot.

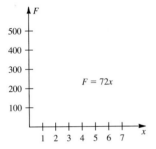

60. **MECHANICAL ENGINEERING** If a machine is to be operated under water, it must be designed to handle the pressure (p), which depends on the depth (d) of the water. The relationship is approximated by the formula $p = 59d + 13$. Graph the relationship between pressure and depth.

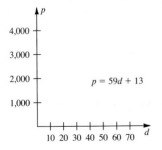

61. **MANUFACTURING TECHNOLOGY** The number of studs, s (16 inches on center), required to build a wall of length L (in feet) is given by the formula

$$s = \frac{3}{4}L + 1$$

Beginning Algebra   The Streeter/Hutchison Series in Mathematics   © The McGraw-Hill Companies. All Rights Reserved.

Answers

62. _____

63. _____

64. _____

Graph the number of studs as a function of the length of the wall.

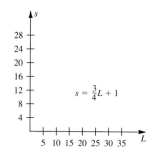

$$s = \frac{3}{4}L + 1$$

62. **MANUFACTURING TECHNOLOGY** The number of board feet of lumber, b, in a 2×6 of length L (in feet) is given by the equation

$$b = \frac{8.25}{144}L$$

Graph the number of board feet in terms of the length of the board.

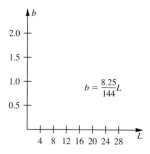

$$b = \frac{8.25}{144}L$$

Basic Skills | Challenge Yourself | Calculator/Computer | Career Applications | **Above and Beyond** ▲

In exercises 63 and 64:

    (a) Graph both given equations on the same coordinate system.

    (b) Describe any observations you can make concerning the two graphs and their equations. How do the two graphs relate to each other?

63. $y = -3x$

$y = \frac{1}{3}x$

64. $y = \frac{2}{5}x$

$y = -\frac{5}{2}x$

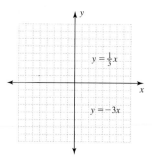

Answers

1. $x + y = 6$

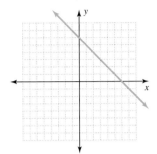

3. $x - y = -3$

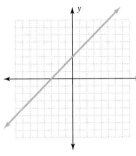

5. $2x + y = 2$

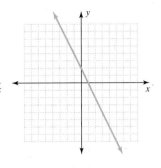

7. $3x + y = 0$

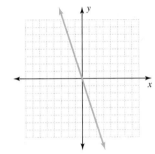

9. $x + 4y = 8$

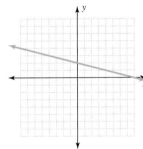

11. $y = 5x$

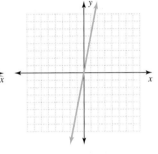

13. $y = 2x - 1$

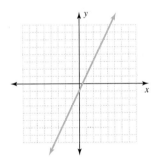

15. $y = -3x + 1$

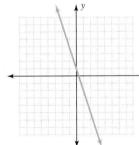

17. $y = \dfrac{1}{3}x$

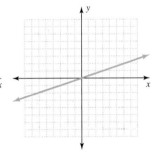

19. $y = \dfrac{2}{3}x - 3$

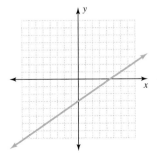

21. $x = 5$

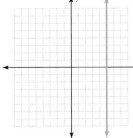

23. $y = 1$

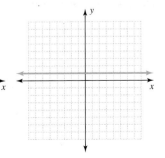

25. $x - 2y = 4$　　　　**27.** $5x + 2y = 10$　　　　**29.** $3x + 5y = 15$

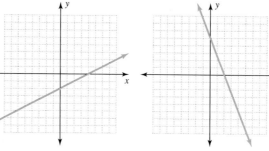

31. $y = 2 - \dfrac{x}{3}$　　　**33.** $y = 3 - \dfrac{3}{4}x$　　　**35.** $y = -5 + \dfrac{5}{4}x$

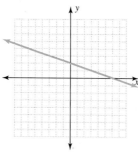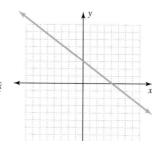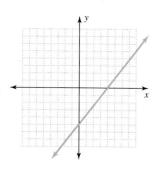

37.

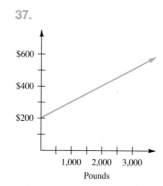

39. **(a)** See graph below;

(b) $\dfrac{100}{11} \approx 9$ tons; **(c)** \$76;

(d) $y = 17x - 125$

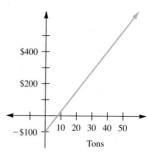

41. **(a)** $C = 15x + 200$;

(b)

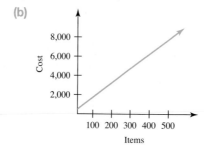

43. True

45. always

47. $y = 2x$

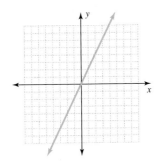

49. $y = 3x - 3$

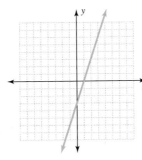

51. $x - 4y = 12$

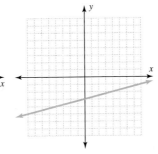

53. $(3, 1)$

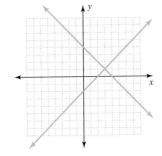

55. The lines do not intersect.
The y-intercepts are $(0, 0)$, $(0, 4)$, and $(0, -5)$.

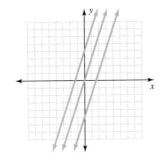

57.

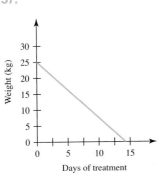

59.

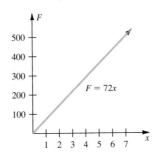

61.

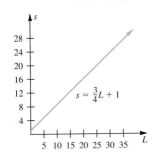

63.

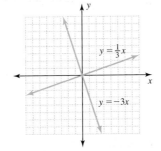

6.4

The Slope of a Line

< 6.4 Objectives >

1 > Find the slope of a line through two given points

2 > Find the slopes of horizontal and vertical lines

3 > Find the slope of a line from its graph

4 > Write and graph the equation for a direct-variation relationship

RECALL

An equation such as $y = 2x + 3$ is a *linear equation in two variables*. Its graph is always a straight line.

In Section 6.3 we saw that the graph of an equation such as

$$y = 2x + 3$$

is a straight line. In this section we want to develop an important idea related to the equation of a line and its graph, called the **slope** of a line. The slope of a line is a numerical measure of the "steepness," or inclination, of that line.

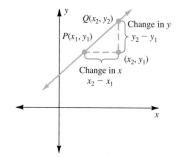

NOTES

x_1 is read "x sub 1," x_2 is read "x sub 2," and so on. The 1 in x_1 and the 2 in x_2 are called **subscripts**.

The difference $x_2 - x_1$ is sometimes called the **run** between points P and Q. The difference $y_2 - y_1$ is called the **rise**. So the slope may be thought of as "rise over run."

To find the slope of a line, we first let $P(x_1, y_1)$ and $Q(x_2, y_2)$ be any two distinct points on that line. The **horizontal change** (or the change in x) between the points is $x_2 - x_1$. The **vertical change** (or the change in y) between the points is $y_2 - y_1$.

We call the ratio of the vertical change, $y_2 - y_1$, to the horizontal change, $x_2 - x_1$, the *slope* of the line as we move along the line from P to Q. That ratio is usually denoted by the letter m which we use in a formula.

Definition

The Slope of a Line

If $P(x_1, y_1)$ and $Q(x_2, y_2)$ are any two points on a line, then m, the slope of the line, is given by

$$m = \frac{\text{vertical change}}{\text{horizontal change}} = \frac{y_2 - y_1}{x_2 - x_1} \qquad \text{when } x_2 \neq x_1$$

This definition provides exactly the numerical measure of "steepness" that we want. If a line "rises" as we move from left to right, the slope is positive—the steeper the line, the larger the numerical value of the slope. If the line "falls" from left to right, the slope is negative.

Let's proceed to some examples.

| Example 1 | Finding the Slope Given Two Points |

< Objective 1 >

Find the slope of the line containing points with coordinates $(1, 2)$ and $(5, 4)$.
   Let $P(x_1, y_1) = (1, 2)$ and $Q(x_2, y_2) = (5, 4)$. By the definition of slope, we have

$$m = \frac{y_2 - y_1}{x_2 - x_1} = \frac{(4) - (2)}{(5) - (1)} = \frac{2}{4} = \frac{1}{2}$$

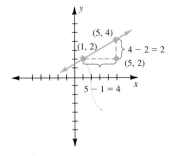

Note: We would have found the same slope if we had reversed P and Q and subtracted in the other order. In that case, $P(x_1, y_1) = (5, 4)$ and $Q(x_2, y_2) = (1, 2)$, so

$$m = \frac{(2) - (4)}{(1) - (5)} = \frac{-2}{-4} = \frac{1}{2}$$

It makes no difference which point is labeled (x_1, y_1) and which is (x_2, y_2). The resulting slope is the same. You must simply stay with your choice once it is made and *not* reverse the order of the subtraction in your calculations.

Check Yourself 1

Find the slope of the line containing points with coordinates (2, 3) and (5, 5).

By now you should be comfortable subtracting negative numbers. We apply that skill to finding a slope.

| Example 2 | Finding the Slope |

Find the slope of the line containing points with the coordinates $(-1, -2)$ and $(3, 6)$.
   Again, applying the definition, we have

$$m = \frac{(6) - (-2)}{(3) - (-1)} = \frac{6 + 2}{3 + 1} = \frac{8}{4} = 2$$

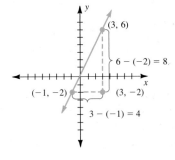

The next figure compares the slopes found in Example 1 and this example. Line l_1, from Example 1, has slope $\frac{1}{2}$. Line l_2, from this example, has slope 2. Do you see the idea of slope measuring steepness? The greater the slope, the more steeply the line is inclined upward.

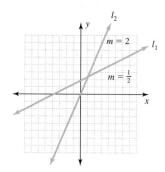

 Check Yourself 2

Find the slope of the line containing points with coordinates $(-1, 2)$ and $(2, 7)$. Draw a sketch of this line and the line of Check Yourself 1 on the same set of axes. Compare the lines and the two slopes.

Next, we look at lines with a negative slope.

► **Example 3** | **Finding a Negative Slope**

Find the slope of the line containing points with coordinates $(-2, 3)$ and $(1, -3)$.
By the definition,

$$m = \frac{(-3) - (3)}{(1) - (-2)} = \frac{-6}{3} = -2$$

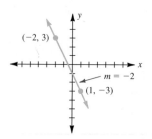

This line has a *negative* slope. The line *falls* as we move from left to right.

 Check Yourself 3

Find the slope of the line containing points with coordinates $(-1, 3)$ and $(1, -3)$.

We have seen that lines with positive slope rise from left to right and lines with negative slope fall from left to right. What about lines with a slope of zero? A line with a slope of 0 is especially important in mathematics.

| | Example 4 | Finding the Slope of a Horizontal Line |

< Objective 2 >

Find the slope of the line containing points with coordinates $(-5, 2)$ and $(3, 2)$.
By the definition,

$$m = \frac{(2) - (2)}{(3) - (-5)} = \frac{0}{8} = 0$$

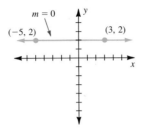

The slope of the line is 0. In fact, that is the case for any horizontal line. Because any two points on the line have the same y-coordinate, the vertical change $y_2 - y_1$ must always be 0, and so the resulting slope is 0.

 Check Yourself 4

Find the slope of the line containing points with coordinates $(-2, -4)$ and $(3, -4)$.

Because division by 0 is undefined, it is possible to have a line with an undefined slope.

| | Example 5 | Finding the Slope of a Vertical Line |

Find the slope of the line containing points with coordinates $(2, -5)$ and $(2, 5)$.
By the definition,

$$m = \frac{(5) - (-5)}{(2) - (2)} = \frac{10}{0}$$ Remember that division by 0 is undefined.

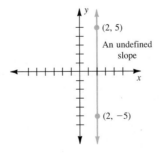

We say that the vertical line has an undefined slope. On a vertical line, any two points have the same x-coordinate. This means that the horizontal change $x_2 - x_1$ must always be 0, and because division by 0 is undefined, the slope of a vertical line will always be undefined.

Check Yourself 5

Find the slope of the line containing points with the coordinates (−3, −5) and (−3, 2).

Given the graph of a line, we can find the slope of that line. Example 6 illustrates this.

| **Example 6** | Finding the Slope from the Graph |

< Objective 3 >

Find the slope of the graphed line.

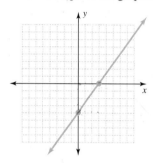

We can find the slope by identifying any two points. It is usually easiest to use the x- and y-intercepts. In this case, those intercepts are (3, 0) and (0, −4).

Using the definition of slope, we find

$$m = \frac{(0) - (-4)}{(3) - (0)} = \frac{4}{3}$$

The slope of the line is $\frac{4}{3}$.

Check Yourself 6

Find the slope of the graphed line.

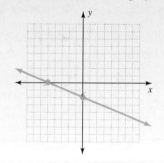

This sketch summarizes the results of our previous examples.

NOTE

As the slope gets closer to 0, the line gets "flatter."

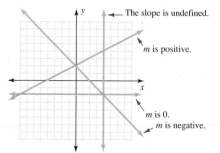

The slope is undefined.
m is positive.
m is 0.
m is negative.

Four lines are illustrated in the figure. Note that:

1. The slope of a line that rises from left to right is positive.

2. The slope of a line that falls from left to right is negative.

3. The slope of a horizontal line is 0.

4. A vertical line has an undefined slope.

In Section 6.3, we saw that a line can be drawn using two ordered pairs, and that it is helpful to plot a third point as a check. In the next few examples, we will focus on equations of the form $y = ax$. Note that, for any such equation, if $x = 0$ then $y = 0$. This means that all graphs of such equations pass through the origin. For these graphs, we know that $(0, 0)$ is a point, and we need find only two others in order to have three points to plot.

Example 7	**Graphing an Equation of the Form $y = ax$**

(a) Sketch the graph of the equation $y = -2x$.

From the table below, we know that the ordered pairs $(-1, 2)$, $(0, 0)$, and $(1, -2)$ are solutions to the equation.

x	y
-1	2
0	0
1	-2

The graph is displayed here.

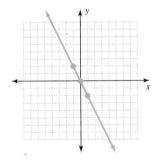

Note that the slope of the line that passes through the points $(0, 0)$ and $(1, -2)$ is

$$m = \frac{(0) - (-2)}{(0) - (1)} = \frac{2}{-1} = -2$$

(b) Sketch the graph of the equation $y = \frac{1}{3}x$.

From the table below, we know that the ordered pairs $(-3, -1)$, $(0, 0)$, and $(3, 1)$ are solutions to the equation.

x	y
-3	-1
0	0
3	1

The graph is displayed here.

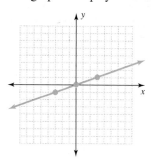

Note that the slope of the line that passes through the points (0, 0) and (3, 1) is

$$m = \frac{(0) - (1)}{(0) - (3)} = \frac{-1}{-3} = \frac{1}{3}$$

Check Yourself 7

Sketch the graph of the equation $y = -\frac{1}{2}x.$

In Example 7, we noted that the slope of the line for the equation $y = -2x$ is -2, and the slope of the line for the equation $y = \frac{1}{3}x$ is $\frac{1}{3}$. This leads us to an observation.

The slope of a line for an equation of the form $y = ax$ is always a. Because m is the slope, we generally write equations of this type as

$$y = mx$$

Again, note that (0, 0) is a solution for any equation of this form and the line for an equation of the form $y = mx$ always passes through the origin.

There are numerous applications involving lines that pass through the origin. Consider the following scenario. Pedro makes $25 an hour as an electrician. If he works for 1 h, he makes $25; if he works for 2 h, he makes $50; and so on. We say his total pay **varies directly** with the number of hours worked. This type of situation occurs so frequently that we use special terminology to describe it.

Definition

Direct Variation

If y is a constant multiple of x, we write

$y = kx$ in which k is a positive constant.

We say that y *varies directly* with x, or that y is *directly proportional* to x. The constant k is called the **constant of variation**.

Example 8   Writing an Equation for Direct Variation

< Objective 4 >

Marina earns $9 an hour as a tutor. Write the equation that describes the relationship between the number of hours she works and her pay.

Her pay (P) is equal to the rate of pay (r) times the number of hours worked (h), so

$$P = r \cdot h \qquad \text{or} \qquad P = 9h$$

RECALL

k is the constant of variation.

Check Yourself 8

Sorina is driving at a constant rate of 50 mi/h. Write the equation that shows the distance she travels (*d*) in *h* hours.

If two things vary directly and values are given for *x* and *y*, we can find *k*. This property is illustrated in Example 9.

Example 9   Finding the Constant of Variation

NOTE

The direct variation equation is *y* = *kx*. (6, 30) is one ordered pair that satisfies the equation.

If *y* varies directly with *x*, and *y* = 30 when *x* = 6, find *k*.
    Because *y* varies directly with *x*, we know from the definition that

$$y = kx$$

We need to find *k*. We do this by substituting 30 for *y* and 6 for *x*.

$$(30) = k(6) \qquad \text{or} \qquad k = 5$$

Check Yourself 9

If *y* varies directly with *x*, and *y* = 100 when *x* = 25, find the constant of variation.

The graph for a linear equation of direct variation always passes through the origin. Example 10 illustrates this.

Example 10   Graphing an Equation of Direct Variation

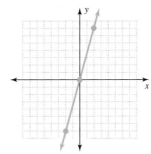

Let *y* vary directly with *x*, with a constant of variation *k* = 3.5. Graph the equation of variation.
    The equation of variation is *y* = 3.5*x*, so the graph has a slope of 3.5. Three points that satisfy the relationship are (−2, −7), (0, 0), and (2, 7).

Check Yourself 10

Let *y* vary directly with *x*, with a constant of variation $k = \dfrac{7}{3}$. Graph the equation of variation.

Here is an example from the field of electronics.

Example 11   An Electrical Engineering Application

The graph depicts the relationship between the position of a linear potentiometer (variable resistor) and the output voltage of some DC source. Consider the potentiometer to be a slider control, possibly to control volume of a speaker or the speed of a motor.

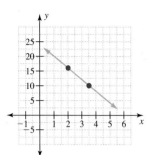

The linear position of the potentiometer is represented on the *x*-axis, and the resulting output voltage is represented on the *y*-axis. At the 2-cm position, the output voltage measured with a voltmeter is 16 VDC. At a position of 3.5 cm, the measured output is 10 VDC.

What is the slope of the resulting line?

We see that we have two ordered pairs: (2, 16) and (3.5, 10). Using our formula for slope, we have

$$m = \frac{16 - 10}{2 - 3.5} = \frac{6}{-1.5} = -4$$

The slope is -4.

 Check Yourself 11

The same potentiometer described in Example 11 is used in another circuit. This time, though, when at position 0 cm, the output voltage is 12 volts. At position 5 cm, the output voltage is 3 volts. Draw a graph using the new data and determine the slope.

 Check Yourself ANSWERS

1. $m = \frac{2}{3}$ **2.** $m = \frac{5}{3}$

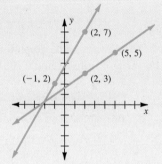

3. $m = -3$ **4.** $m = 0$ **5.** m is undefined **6.** $m = -\frac{2}{5}$

7.

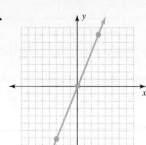

8. $d = 50h$ **9.** $k = 4$

10.

11. Slope $= -\frac{9}{5}$

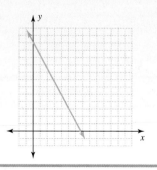

Reading Your Text

The following fill-in-the-blank exercises are designed to ensure that you understand some of the key vocabulary used in this section.

SECTION 6.4

(a) The _____ of a line gives us a numerical measure of the steepness of that line.

(b) The difference $x_2 - x_1$ is sometimes called the _____ between points P and Q.

(c) We say that a vertical line has an _____ slope.

(d) When y varies directly with x, we write $y = kx$, and k is called the constant of _____.

6.4 exercises

Name _____

Section _____ Date _____

Answers

< Objective 1 >

Find the slope of the line through each pair of points.

1. (5, 7) and (9, 11)

2. (4, 9) and (8, 17)

3. (−2, 3) and (3, 7)

4. (−3, −4) and (3, −2)

5. (3, 3) and (5, 0)   > Videos

6. (−2, 4) and (3, 1)

7. (5, −4) and (5, 2)

8. (−5, 4) and (2, 4)

9. (−4, −2) and (3, 3)

10. (−5, −3) and (−5, 2)

11. (−1, 7) and (2, 3)

12. (−4, −2) and (6, 4)

In exercises 13–18, two points are shown. Find the slope of the line through the given points.

13.

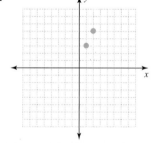

14.

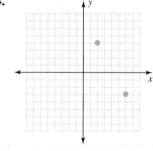

15.

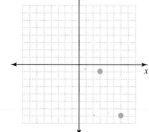

16.

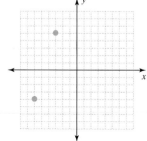

17.

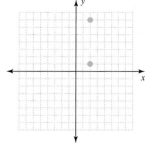

> Videos

18.

Answers

17. _____

18. _____

19. _____

20. _____

21. _____

22. _____

23. _____

24. _____

< Objectives 2–3 >

Find the slope of the lines graphed.

19.

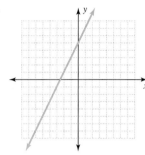

> Videos

20.

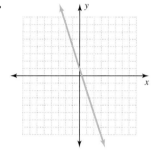

21.

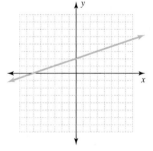

22.

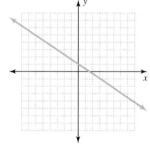

Sketch the graph of each equation.

23. $y = -4x$

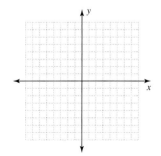

24. $y = 3x$

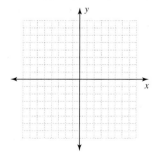

Answers

25. _____

26. _____

27. _____

28. _____

29. _____

30. _____

31. _____

32. _____

33. _____

34. _____

25. $y = \dfrac{2}{3}x$

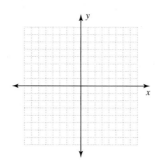

26. $y = -\dfrac{3}{4}x$

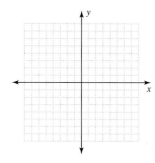

27. $y = \dfrac{5}{4}x$

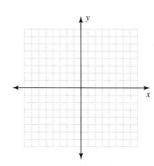

28. $y = -\dfrac{4}{5}x$

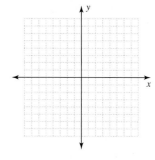

> Videos

Find the constant of variation k.

29. y varies directly with x; $y = 54$ when $x = 6$.

30. m varies directly with n; $m = 144$ when $n = 8$.

31. y varies directly with x; $y = 2{,}100$ when $x = 600$.

32. y varies directly with x; $y = 400$ when $x = 1{,}000$.

In exercises 33–36, y varies directly with x and the value of k is given. Graph the equation of variation.

33. $k = 2$

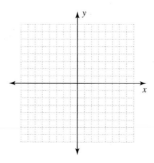

34. $k = 4$

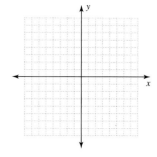

35. $k = 2.5$

36. $k = 2.2$

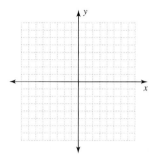

Answers

35. _____

36. _____

37. _____

38. _____

39. _____

40. _____

Complete each exercise.

37. BUSINESS AND FINANCE Robin earns $12 per hour. Write an equation that shows how much she makes (S) in h hours.

38. BUSINESS AND FINANCE Kwang earns $11.50 per hour. Write an equation that shows how much he earns (S) in h hours.

39. BUSINESS AND FINANCE At a factory that makes grinding wheels, Kalila makes $0.20 for each wheel completed. Sketch the equation of direct variation.

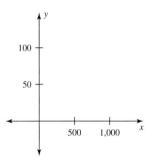

40. BUSINESS AND FINANCE Palmer makes $1.25 per page for each page that he types. Sketch the equation of direct variation.

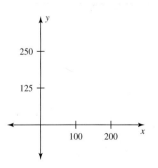

Answers

41. _____

42. _____

43. _____

44. _____

45. _____

46. _____

47. _____

48. _____

41. BUSINESS AND FINANCE Josephine works part-time in a local video store. Her salary varies directly as the number of hours worked. Last week she earned $75.20 for working 8 hours. This week she earned $206.80. How many hours did she work this week?

42. BUSINESS AND FINANCE The revenue for a sandwich shop is directly proportional to its advertising budget. When the owner spent $2,000 a month on advertising, the revenue was $120,000. If the revenue is now $180,000, how much is the owner spending on advertising?

| Basic Skills | **Challenge Yourself** | Calculator/Computer | Career Applications | Above and Beyond |

Determine whether each statement is **true** *or* **false.**

43. The slope of a horizontal line is undefined.

44. A negative slope means that the line falls from left to right.

Complete each statement with **never, sometimes,** *or* **always.**

45. The graph of $y = mx$ _____ passes through the origin.

46. If y varies directly with x, the constant of variation k is _____ negative.

47. Consider the equation $y = 2x - 5$.   > Videos

   (a) Complete the table.

x	y
3	
4	

   (b) Use the ordered pairs found in part (a) to calculate the slope of the line.

   (c) What do you observe concerning the slope found in part (b) and the given equation?

48. Repeat exercise 47 for $y = \dfrac{3}{2}x + 5$ and

x	y
2	
4	

49. Repeat exercise 47 for $y = -\dfrac{1}{3}x + 2$ and

x	y
3	
6	

50. Repeat exercise 47 for $y = -4x - 6$ and

x	y
-1	
-3	

49. _____

51. Consider the equation $y = 2x + 3$.

    (a) Complete the table of values.

Point	x	y
A	5	
B	6	
C	7	
D	8	
E	9	

50. _____

51. _____

    (b) As the x-coordinate changes by 1 (for example, as you move from point A to point B), by how much do the corresponding y-coordinates change?

    (c) Is your answer to part (b) the same if you move from B to C? from C to D? from D to E?

    (d) Describe the "growth rate" of the line using these observations. Complete the statement: When the x-value grows by 1 unit, the y-value _____.

52. _____

52. Repeat exercise 51 using $y = 2x + 5$.

53. Repeat exercise 51 using $y = -4x + 50$.

53. _____

54. Repeat exercise 51 using $y = -4x + 40$.

54. _____

In each exercise, (a) plot the given point; (b) using the given slope, move from the point plotted in (a) to plot a new point; (c) draw the line that passes through the points plotted in (a) and (b).

55. _____

56. _____

55. $(3, 1)$, $m = 2$ **56.** $(-1, 4)$, $m = -2$

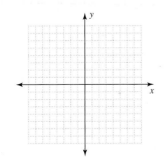   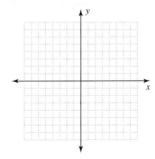

Answers

57. _____

58. _____

59. _____

60. _____

61. _____

57. $(-2, -1)$, $m = -4$

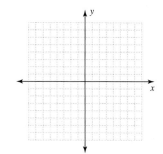

58. $(-3, 5)$, $m = 2$

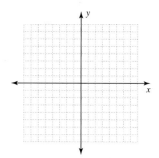

Basic Skills | Challenge Yourself | Calculator/Computer | **Career Applications** | Above and Beyond
▲

59. ALLIED HEALTH The recommended dosage, d, in milligrams (mg) of the antibiotic ampicillin sodium for children weighing less than 40 kilograms is given by the linear equation $d = 7.5w$, where w represents the child's weight in kilograms (kg). Sketch a graph of the linear equation.

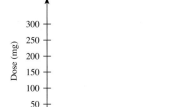

60. ALLIED HEALTH The recommended dosage, d, in micrograms (mcg) of Neupogen, a medication given to bone marrow transplant patients, is given by the linear equation $d = 8w$, where w represents the patient's weight in kilograms (kg). Sketch a graph of the linear equation.

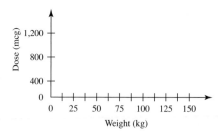

61. MANUFACTURING TECHNOLOGY This graph shows the bending moment in a wood beam.

Calculate the slope of the moment graph:

(a) Between 0 and 4 feet

(b) Between 4 and 11 feet

(c) Between 11 and 19 feet

(d) Between 19 and 24 feet

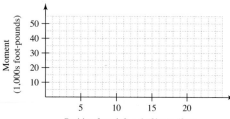

Basic Skills | Challenge Yourself | Calculator/Computer | Career Applications | **Above and Beyond**
▲

62. Summarize the results of exercises 51 to 54. In particular, how does the concept of "growth rate" connect to the concept of slope?

63. Complete the statement: "The difference between undefined slope and zero slope is . . . ."

64. Complete the statement: "The slope of a line tells you . . . ."

Answers

62. _____

63. _____

64. _____

Answers

1. 1   **3.** $\dfrac{4}{5}$ **5.** $-\dfrac{3}{2}$ **7.** Undefined   **9.** $\dfrac{5}{7}$ **11.** $-\dfrac{4}{3}$

13. 2   **15.** -2 **17.** 0   **19.** 2   **21.** $\dfrac{1}{3}$

23. $y = -4x$ **25.** $y = \dfrac{2}{3}x$

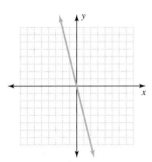   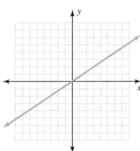

27. $y = \dfrac{5}{4}x$ **29.** 9   **31.** 3.5

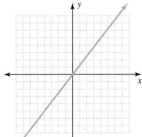

33.   **35.**

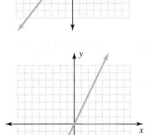   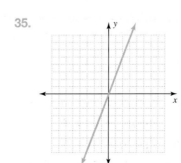

37. $S = 12h$ **39.**

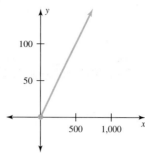

41. 22 h   **43.** False   **45.** always

47. **(a)** (3, 1), (4, 3); **(b)** 2; **(c)** slope equals coefficient of x

49. **(a)** (3, 1), (6, 0); **(b)** $-\dfrac{1}{3}$; **(c)** slope equals coefficient of x

51. **(a)** (5, 13), (6, 15), (7, 17), (8, 19), (9, 21); **(b)** 2; **(c)** yes; **(d)** increases by 2

53. **(a)** (5, 30), (6, 26), (7, 22), (8, 18), (9, 14); **(b)** 4; **(c)** yes; **(d)** decreases by 4

55.

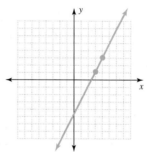

57.

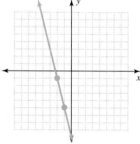

59.

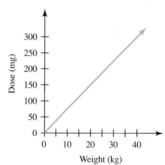

61. **(a)** 6,250 foot-pounds per foot;
(b) 2,857.14 foot-pounds per foot;
(c) −2,500 foot-pounds per foot;
(d) −5,000 foot-pounds per foot

63. Above and Beyond

6.5

< 6.5 Objectives >

Reading Graphs*

1 > Read and interpret a table

2 > Read and interpret a pie chart

3 > Read different types of bar graphs

4 > Read and interpret a line graph

5 > Create and make a prediction from a line graph

NOTE

Spreadsheets and databases are commonly used software tools for working with tables of data.

NOTE

Antarctica has been omitted from this table.

A **table** is a display of information in parallel rows or columns. Tables can be used anywhere that information is to be summarized.

Here is a table describing land area and world population. Each entry in a table is called a **cell.** This table will be used for Examples 1 and 2.

Continent or Region	Land Area (1,000 mi²)	Population (millions)		
		1900	1950	2000
North America	9,400	106	221	305
South America	6,900	38	111	515
Europe	3,800	400	392	510
Asia (including Russia)	17,400	932	1,591	4,178
Africa	11,700	118	229	889
Oceania (including Australia)	3,300	6	12	32
World totals	52,500	1,600	2,556	6,429

Source: Bureau of the Census: U.S. Dept. of Commerce.

Example 1 | Reading a Table

< Objective 1 >

Use the land area and world population table above to answer each question.

(a) What was the population of Africa in 1950?

Continent or Region	Land Area (1,000 mi²)	Population (millions)		
		1900	1950	2000
North America	9,400	106	221	305
South America	6,900	38	111	515
Europe	3,800	400	392	510
Asia (including Russia)	17,400	932	1,591	4,178
Africa	11,700	118	229	889
Oceania (including Australia)	3,300	6	12	32
World totals	52,500	1,600	2,556	6,429

Source: Bureau of the Census: U.S. Dept. of Commerce.

* This section is included for those instructors who have it in their curricula. It is optional in the sense that it is not required for the remainder of the text.

485

Looking at the cell in the row labeled Africa and the column labeled 1950, we find the number 229. Because we are told the population is given in millions, we conclude that the population of Africa in 1950 was 229,000,000.

(b) What is the land area of Asia, in square miles?

The cell in the row Asia and column Land Area reads 17,400. The land area is given in 1,000 mi^2 units, so the actual land area of Asia is 17,400,000 mi^2.

 Check Yourself 1

Use the land area and world population table to answer each question.

(a) What was the population of South America in 1900?
(b) What is the land area of Europe?

We frequently use a table to find information not explicitly given in the table. We will use the land area and world population table again to illustrate this point.

 Example 2   **Interpreting a Table**

Use the land area and world population table to answer each question.

(a) By what percentage did the population of Africa change between 1950 and 2000?

In Example 1, we found that the 1950 population of Africa was 229 million. The 2000 population of Africa was 889 million, or 660 million more people than in 1950. The percentage change is

$$\frac{660,000,000}{229,000,000} \approx 2.882 = 288.2\%$$

So, the population of Africa increased by about 288.2% between 1950 and 2000.

(b) What is the land area of Asia as a percentage of the earth's total land area (excluding Antarctica)?

In Example 1, we concluded that Asia is 17,400,000 mi^2. We find that there are 52,500,000 mi^2 of land area (excluding Antarctica) from the row labeled World totals. Therefore, the percentage of the earth's total land area made up by Asia is

$$\frac{17,400,000}{52,500,000} \approx 33.1\%$$

(c) What was the density of the population (people per square mile) in North America in 2000?

North America had a population of 305 million in 2000. With a land area of 9,400,000 mi^2, the population density of North America is given by

$$\frac{305,000,000}{9,400,000} \approx 32.4 \text{ people per square mile}$$

Check Yourself 2

Use the land area and world population table to answer each question.

(a) By what percentage did the population of South America change between 1900 and 1950?

(b) What is the land area of Europe as a percentage of the earth's total land area (excluding Antarctica)?

(c) Did the world population increase by a greater percentage between 1900 and 1950 or between 1950 and 2000?

NOTE

Pie charts are sometimes called *circle graphs*.

If we need to present the relationship between two sets of data in only a column or two of a table, it is often easier to use pictures. A **graph** is a diagram that represents the connection between two or more quantities.

The first graph we will look at is called a **pie chart.** We use a circle to represent some total that interests us. Wedges (or sectors) are drawn in the circle to show how much of the whole each part makes up.

Example 3	Reading a Pie Chart

< Objective 2 >

NOTE

The total of the percentages of the wedges is always 100%.

This pie chart represents the results of a survey that asked students how they get to school most often.

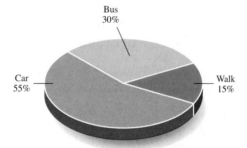

(a) What percentage of the students walk to school?

We see that 15% walk to school.

(b) What percentage of the students do not arrive by car?

Because 55% arrive by car, there are 100% − 55%, or 45%, who do not.

Check Yourself 3

This pie chart represents the results of a survey that asked students whether they bought lunch, brought it, or skipped lunch altogether.

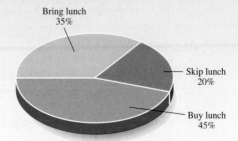

(a) What percentage of the students skipped lunch?

(b) What percentage of the students did not buy lunch?

If we know what the whole pie represents, we can also find out more about what each wedge represents, as illustrated by Example 4.

Example 4	Interpreting a Pie Chart

This pie chart shows how Sarah spent her $12,000 college scholarship.

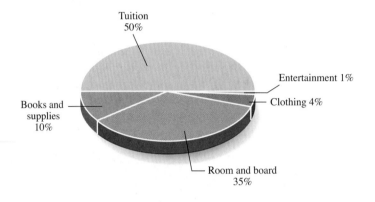

(a) How much did she spend on tuition?

She spent 50% of her $12,000 scholarship, or $6,000, on tuition.

(b) How much did she spend on clothing and entertainment?

Together, 5% of the money was spent on clothing and entertainment, and 0.05 × 12,000 = 600. Therefore, $600 was spent on clothing and entertainment.

Check Yourself 4

This pie chart shows how Rebecca spends an average 24-hour school day.

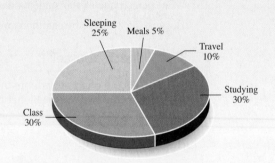

(a) How many hours does she spend sleeping each day?
(b) How many hours does she spend altogether studying and in class?

A **bar graph** provides yet another way to present information. It shows the relationship between two sets of data.

	Example 5	Reading a Bar Graph

< Objective 3 >

This bar graph represents the number of cars sold in the United States in each year listed (sales are in thousands).

U.S. Car Sales

(a) How many cars were sold in the United States in 1990?

We frequently have to estimate our answer when reading a bar graph. In this case, there were approximately 9,500,000 cars sold in 1990.

(b) In which year listed did the most car sales occur?

The tallest bar occurs in 1985; therefore, more cars were sold in 1985 than in any other year represented on the graph.

NOTE

Remember that sales numbers are in thousands.

Check Yourself 5

The bar graph represents the response to a recent Gallup poll that asked people to name their favorite spectator sport.

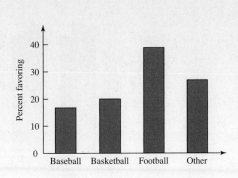

(a) Find the percentage of people for whom football is their favorite spectator sport.

(b) Of the three major sports listed, which was named as a favorite by the fewest people?

When we use bar graphs to display additional information, we often use different colors for different bars. With such graphs, it is necessary to include a legend. A **legend** is a key describing what each color or shade of bar represents.

> **Example 6**   Reading a Bar Graph

This bar graph compares the number of cars sold in the world and in the United States in each year listed (sales are in thousands).

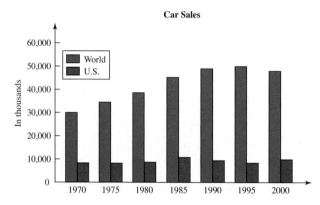

(a) How many cars were sold in the world in 1985? In the United States in 1985?

The legend tells us that the blue bar represents worldwide sales and the pink bar represents U.S. sales.

It would appear that there were about 45,000,000 cars sold in the world in 1985. Approximately 11,000,000 of these were sold in the United States.

(b) What percentage of global sales did the United States account for in 1985?

$$\frac{11}{45} \approx 0.244 = 24.4\%$$

 Check Yourself 6

The bar graph represents the average student age at Berndt Community College.

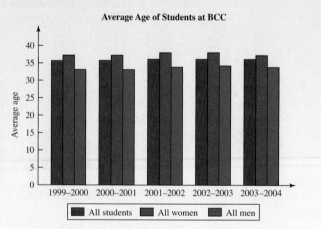

(a) What was the average age of female students in 2003–2004?
(b) Did the average age of female students increase or decrease between 2002–2003 and 2003–2004?
(c) Who tends to be older, male students or female students?

Another useful type of graph is called a **line graph.** In a line graph, one set of data is usually related to time. Line graphs give us a way of visualizing changes to something over time.

| ⦿ | **Example** 7 | Reading and Interpreting a Line Graph |

< Objective 4 >

This line graph represents the number of Social Security beneficiaries in each of the years listed.

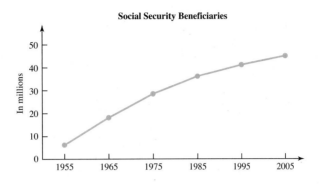

(a) How many Social Security beneficiaries were there in 1975?

We find the point on the line graph that lies above 1975. We then determine that this point lies at about 28 on the vertical axis.

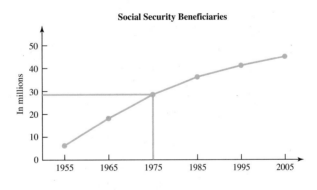

Therefore, we conclude that there were approximately 28 million Social Security beneficiaries in 1975.

(b) How many beneficiaries were there in 1990?

The line connecting 1985 and 1995 passes through 1990 at about 39 million.

(c) Approximate the mean number of annual beneficiaries between 1955 and 2005.

We estimate the number of beneficiaries in each of the years 1955, 1965, 1975, 1985, 1995, and 2005 to be 6 million, 18 million, 28 million, 36 million, 41 million, and 45 million, respectively.

Therefore, we calculate the mean number of beneficiaries as

$$\frac{6 + 18 + 28 + 36 + 41 + 45}{6} = \frac{174}{6} = 29$$

So, the mean number of annual beneficiaries is 29 million.

RECALL

The mean is computed by adding all of the values in a set and dividing by the number of values in the set.

Check Yourself 7

The graph indicates the high temperatures in Baltimore, Maryland, for a week in September.

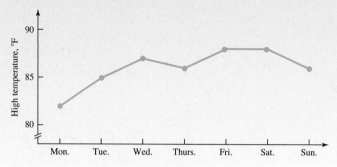

(a) What was the high temperature on Friday?
(b) Find the mean high temperature for that week.

We use many of the techniques learned in Section 6.2 to create a line graph. We start with the rectangular coordinate system. Generally, we make the *x*-axis represent the time quantity and scale the *y*-axis accordingly.

For each time given, we plot the point corresponding to whatever it is we are measuring. Finally, we "connect the dots." That is, we draw a line segment from each point to the next point immediately to the right.

 Example 8   Constructing a Line Graph

< Objective 5 >

Construct a line graph to summarize the data shown.

FBI: Larceny-Theft Cases
(in hundred thousands)

Year	Thefts
1985	69
1990	79
1995	80
2000	70
2005	64

Source: Bureau of Justice Statistics

We set, scale, and label our axes appropriately. Then, we plot the points given in the table, with *x* as the year and *y* as the number of thefts.

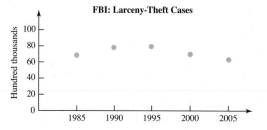

FBI: Larceny-Theft Cases

Finally, we sketch line segments from point to point (that is, we "connect the dots").

FBI: Larceny-Theft Cases

Check Yourself 8

Construct a line graph based on the table of data describing the cost of a first-class postage stamp, by year.

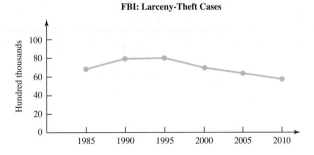

First-Class Postage Stamps (in cents)	
Year	**Cost**
1960	4
1965	5
1970	6
1975	10
1980	15
1985	22
1990	25
1995	32
2000	33
2005	37

Source: United States Postal Service.

An important feature of line graphs is that they allow us to (cautiously) predict results beyond the data given. To do this, we extend the sketched curve as needed.

Example 9	Making a Prediction from a Line Graph

Use the line graph in Example 8 to predict the number of larceny-theft cases that will be reported to the FBI in 2010.

We extend the line segment connecting the points at 2000 and 2005 to 2010.

FBI: Larceny-Theft Cases

We predict that there will be 58 hundred thousand, or 5,800,000, larceny-theft cases reported to the FBI in 2010.

Check Yourself 9

Use the postage stamp line graph constructed in Check Yourself 8
to predict the cost of a first-class postage stamp in the year 2010.

You should not use a line graph to predict occurrences too far out of range of the original data (at least not if you are looking for an accurate prediction). To illustrate this, consider the following.

Between the ages of 3 and 10 months, most kittens gain about 1 lb per month. Therefore, if a kitten weighs 2 lb at 3 months, it would be reasonable to predict that the kitten will weigh 9 lb at 10 months. We could even guess that the kitten might weigh around 11 lb after 1 year.

But, if we extrapolate further, we would then predict that this same kitten will weigh 59 lb when it is 5 years old. This is, of course, ridiculous. The problem is that our initial data describe only a small segment (within the first year). When we try to move too far beyond the original data, we run into difficulties because the original data no longer apply to our prediction.

Here is an example from the field of manufacturing.

Example 10   **A Business and Finance Application**

This pie chart shows the breakdown of the costs of producing a part.

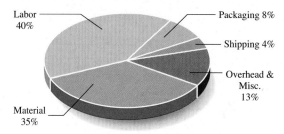

(a) What percentage of the cost of producing a product is the actual "material" for the product?

We see in the chart that 35% of the production cost goes to "material."

(b) For a part that costs $12.40 to produce, how much would you expect the labor cost to be?

Since 40% of production costs go to "labor," we find 40% of $12.40.

$$0.40 \times 12.40 = 4.96$$

So, we would expect the labor cost to be $4.96.

(c) What is the largest expense in producing a part?

From the chart, we see that the category with the greatest percentage is "labor."

(d) A part uses $4.30 in materials. What would you expect the total cost of producing the part to be?

From the chart, we can write that 35% of the total cost is $4.30. Writing this as an equation, we have

$$0.35c = 4.30$$

$$c \approx 12.29$$

We would expect the total cost of producing the part to be $12.29.

Check Yourself 10

Use the pie chart from Example 10.

(a) Find the shipping cost if the total production cost is $17.89.
(b) Find the total production cost if the "overhead and miscellaneous" cost is $3.15.

Check Yourself ANSWERS

1. (a) 38,000,000; **(b)** 3,800,000 mi²   **2. (a)** 192.1%; **(b)** 7.2%;
(c) 1950–2000 (152% vs. 60%)   **3. (a)** 20%; **(b)** 55%   **4. (a)** 6 h;
(b) 14.4 h   **5. (a)** 38%; **(b)** baseball   **6. (a)** 37 years; **(b)** it decreased;
(c) female students   **7. (a)** 88°F; **(b)** 86°F

8.

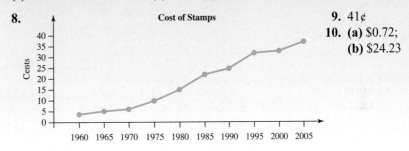

9. 41¢
10. (a) $0.72;
    (b) $24.23

Reading Your Text

The following fill-in-the-blank exercises are designed to ensure that you understand some of the key vocabulary used in this section.

SECTION 6.5

(a) A _____ is a display of information in parallel rows or columns.

(b) Pie charts are sometimes called _____ graphs.

(c) In a bar graph, a _____ is a key describing what each color or shade of bar represents.

(d) In a _____ graph, one set of data is usually related to time.

6.5 exercises

Name _____

Section _____ Date _____

Answers

1. _____

2. _____

3. _____

4. _____

5. _____

6. _____

7. _____

8. _____

Basic Skills | Challenge Yourself | Calculator/Computer | Career Applications | Above and Beyond
▲

< Objective 1 >

Use the table for exercises 1–8.

World Motor Vehicle Production, 1950–1997						
Production (in thousands)						
Year	United States	Canada	Europe	Japan	Other	World Total
1997	12,119	2,571	17,773	10,975	10,024	53,463
1996	11,799	2,397	17,550	10,346	9,241	51,332
1995	11,985	2,408	17,045	10,196	8,349	49,983
1994	12,263	2,321	16,195	10,554	8,167	49,500
1993	10,898	2,246	15,208	11,228	7,205	46,785
1992	9,729	1,961	17,628	12,499	6,269	48,088
1991	8,811	1,888	17,804	13,245	5,180	46,928
1990	9,783	1,928	18,866	13,487	4,496	48,554
1985	11,653	1,933	16,113	12,271	2,939	44,909
1980	8,010	1,324	15,496	11,043	2,692	38,565
1970	8,284	1,160	13,049	5,289	1,637	29,419
1960	7,905	398	6,837	482	866	16,488
1950	8,006	388	1,991	32	160	10,577

Note: As far as can be determined, production refers to vehicles locally manufactured.
Source: American Automobile Manufacturers Assn.

1. What was the motor vehicle production in Japan in 1950? 1997?

2. What was the motor vehicle production in countries outside the United States in 1950? 1997?

3. What was the percentage increase in motor vehicle production in the United States from 1950 to 1997? ⊙ > Videos

4. What was the percentage increase in motor vehicle production in countries outside the United States from 1950 to 1997?

5. What percentage of world motor vehicle production occurred in Japan in 1997?

6. What percentage of world motor vehicle production occurred in the United States in 1997?

7. What percentage of world motor vehicle production occurred outside the United States and Japan in 1997?

8. Between 1950 and 1997, did the production of motor vehicles increase by a greater percentage in Canada or Europe?

< Objective 2 >

This pie chart represents the way a new company ships its goods. Use the information presented to complete exercises 9–12.

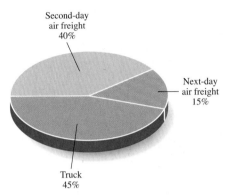

Second-day air freight 40%

Next-day air freight 15%

Truck 45%

9. What percentage is shipped by air freight? > Videos

10. What percentage is shipped by truck?

11. What percentage is shipped by truck or second-day air freight?

12. If the company shipped a total of 550 items last month, how many were shipped using second-day air freight?

< Objective 3 >

In exercises 13–18, use the given bar graph representing the number of bankruptcy filings during a recent 5-year period.

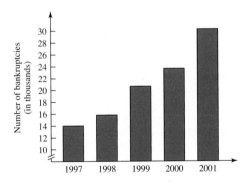

13. How many people filed for bankruptcy in 1998?

14. How many people filed for bankruptcy in 2001?

15. What was the increase in filings from 1999 to 2001? > Videos

16. What was the increase in filings from 1997 to 2001?

17. Which year had the greatest increase in filings?

18. In which year did the greatest percentage of increase in filings occur?

Answers

9. _____

10. _____

11. _____

12. _____

13. _____

14. _____

15. _____

16. _____

17. _____

18. _____

Answers

19. _____

20. _____

21. _____

22. _____

23. _____

24. _____

< Objective 4 >

Use the line graph showing ticket sales for the last 6 months of the year to complete exercises 19–20.

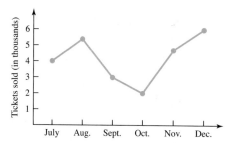

19. What month had the greatest number of ticket sales?

20. Between what two months did the greatest decrease in ticket sales occur?

< Objective 5 >

Use the table to complete exercises 21–22.

Year	Population of United States (in millions)
1950	151
1960	179
1970	203
1980	227
1990	249
2000	281

21. Create a line graph to present the data given in the table.

22. Use the line graph to predict the U.S. population in the year 2010.

Basic Skills	**Challenge Yourself**	Calculator/Computer	Career Applications	Above and Beyond
	▲			

Determine whether each statement is **true** *or* **false.**

23. A pie chart is the best tool for visualizing changes to something over time.

24. Caution must be used when making predictions beyond the range of the given data.

Complete each statement with **never, sometimes,** *or* **always.**

25. In a pie chart, the total of the percentages of the wedges is _____ 100%.

26. You should _____ use a line graph to predict occurrences that are far out of the range of the original data.

Basic Skills | Challenge Yourself | Calculator/Computer | **Career Applications** | Above and Beyond
▲

27. **SCIENCE AND MEDICINE** The table gives the number of live births, broken down by the race of the mother, in the United States from 2003 to 2007. Round your results to the nearest tenth of a percent in each exercise.

Year	Total Births	Non-Hispanic Whites	Non-Hispanic Blacks	American Indian or Alaskan Native	Asian or Pacific Islander	Hispanic
2007	4,317,119	2,312,473	627,230	49,284	254,734	1,061,970
2006	4,265,555	2,308,640	617,247	47,721	241,045	1,039,077
2005	4,140,419	2,284,505	583,907	44,767	231,244	982,862
2004	4,112,052	2,296,683	578,772	43,927	229,123	946,349
2003	4,089,950	2,321,904	576,033	43,052	221,203	912,329

Source: National Vital Statistics Reports.

(a) What percentage of the live births was to Hispanic mothers in 2007?

(b) What percentage of live births was to non-Hispanic White mothers in 2004?

(c) Determine the percentage increase or decrease in live births from 2003 to 2007 for non-Hispanic Black mothers.

28. **ELECTRONICS** The graph shows typical home energy uses in the United States.

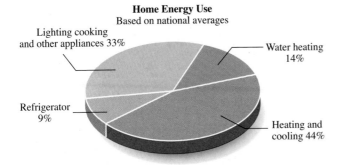

Home Energy Use
Based on national averages

Lighting cooking and other appliances 33%

Water heating 14%

Refrigerator 9%

Heating and cooling 44%

From "Small Electric Wind Systems: A U.S. Consumer's Guide," May 2001, Revised October 2002. U.S. Department of Energy, Office of Energy Efficiency and Renewable Energy, Wind and Hydropower Technologies Program.

Assume that a household uses 10,600 kWh (kilowatt-hours) of electricity annually. Calculate the energy (in kWh) used annually for each category.

Answers

29. **AUTOMOTIVE TECHNOLOGY** This table shows the effects of adjusting the spark advance on the horsepower and the exhaust temperature of an engine.

Spark Advance	Brake Horsepower	Exhaust Temperature
10°	59 hp	1,340°F
20°	76 hp	1,275°F
30°	84 hp	1,250°F
40°	87 hp	1,255°F
50°	82 hp	1,300°F

(a) Which degree of spark advance results in the greatest horsepower?

(b) Which degree of spark advance results in the lowest exhaust temperature?

30. **MECHANICAL ENGINEERING** The table shows properties of five different metals.

Metal	Density (g/cm^3)	Melting point (°C)
Iron	7.87	1,538
Aluminum	2.699	660.4
Copper	8.93	1,084.9
Tin	5.765	231.9
Titanium	4.507	1,668

(a) What is the density of titanium?

(b) What is the difference in melting point between copper and iron?

(c) Which metal has the highest melting point?

31. **BUSINESS AND FINANCE** The table shows sales figures for cars sold by year.

Year Manufactured	Sales (1,000s of vehicles)
2000	43.4
2001	29.1
2002	36.8
2003	19.7
2004	28.1

Create a line graph to display these data.

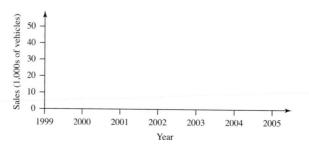

32. **MANUFACTURING TECHNOLOGY** Use the table to answer each question.

Type of wood	Tensile strength (parallel to grain), psi	Tensile strength (perpendicular to grain), psi	Compressive strength (parallel to grain), psi	Compressive strength (perpendicular to grain), psi
Elm	17,500	660	5,520	690
Maple	15,700	1,100	7,830	1,470
Oak	11,300	940	6,200	810
Cedar	6,600	320	6,020	920
Pine	10,600	310	4,800	440
Spruce	8,600	370	5,610	580

(a) What is the compressive strength of oak parallel to the grain?

(b) What is the difference in tensile strength of pine between use parallel to the grain and use perpendicular to the grain?

(c) What is the difference in tensile strength parallel to the grain between elm and pine?

(d) Make a general statement about the best way to use wood in order to obtain the greatest strength.

Answers

32. _____

Answers

1. 32,000; 10,975,000   **3.** 51.4%   **5.** 20.5%   **7.** 56.8%   **9.** 55%
11. 85%   **13.** 16,000   **15.** 9,000   **17.** 2001   **19.** December
21.

U.S. Population

23. False   **25.** always   **27.** (a) 24.6%; (b) 55.9%; (c) 8.9%
29. (a) 40°; (b) 30°

31.

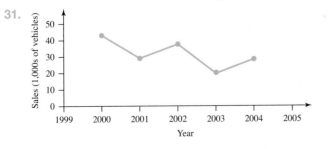

Definition/Procedure	Example	Reference
Solutions of Equations in Two Variables		Section 6.1
Solutions of Equations		
A pair of values that satisfies the equation. Solutions for equations in two variables are written as *ordered pairs*. An ordered pair has the form (x, y)	If $2x - y = 10$, $(6, 2)$ is a solution for the equation, because substituting 6 for x and 2 for y gives a true statement.	p. 411
The Rectangular Coordinate System		Section 6.2
The Rectangular Coordinate System		
A system formed by two perpendicular axes that intersect at a point called the **origin.** The horizontal line is called the **x-axis.** The vertical line is called the **y-axis.**		p. 422
Graphing Points from Ordered Pairs		
The coordinates of an ordered pair allow you to associate a point in the plane with the ordered pair. To graph a point in the plane, **1.** Start at the origin. **2.** Move right or left according to the value of the x-coordinate: to the right if x is positive or to the left if x is negative. **3.** Then move up or down according to the value of the y-coordinate: up if y is positive or down if y is negative.	To graph the point corresponding to $(2, 3)$: 	p. 424
Graphing Linear Equations		Section 6.3
Linear Equation		
An equation that can be written in the form $Ax + By = C$ in which A and B are not both 0.	$2x - 3y = 4$ is a linear equation.	p. 439
Graphing Linear Equations		
1. Find at least three solutions for the equation, and put your results in tabular form. **2.** Graph the solutions found in step 1. **3.** Draw a straight line through the points determined in step 2 to form the graph of the equation.		p. 439

Definition/Procedure	Example	Reference

The Slope of a Line

Section 6.4

Slope

The slope of a line gives a numerical measure of the steepness of the line. The slope m of a line containing the distinct points in the plane $P(x_1, y_1)$ and $Q(x_2, y_2)$ is given by

$$m = \frac{\text{vert change}}{\text{horiz change}} = \frac{y_2 - y_1}{x_2 - x_1} \quad \text{when } x_2 \neq x_1$$

To find the slope of the line through $(-2, -3)$ and $(4, 6)$:

$$m = \frac{(6) - (-3)}{(4) - (-2)}$$

$$= \frac{6 + 3}{4 + 2} = \frac{9}{6} = \frac{3}{2}$$

p. 466

The Graph of $y = mx$

A line passing through the origin with slope m.

To graph $y = -4x$:

Plot $(0, 0)$, $(-1, 4)$, and $(1, -4)$

p. 472

Direct Variation

If y is a constant positive multiple of x, we write
$y = kx$, where $k > 0$,
and say y varies directly as x.

If $y = 20$ when $x = 5$, and y varies directly as x,

$(20) = k(5)$

$k = 4 \qquad y = 4x$

p. 472

Reading Graphs

Section 6.5

Tables

A **table** is a rectangular display of information or data.

p. 485

Graphs

Graph A diagram that relates two different pieces of information.

p. 487

Pie chart A graph that shows the component parts of a whole.

Bar graph One of the most common types of graph. It relates the amounts of items to each other.

Line graph A graph in which one of the axes is usually related to time. We can use line graphs to make **predictions** about events.

This summary exercise set is provided to give you practice with each of the objectives of this chapter. Each exercise is keyed to the appropriate chapter section. When you are finished, you can check your answers to the odd-numbered exercises against those presented in the back of the text. If you have difficulty with any of these questions, go back and reread the examples from that section. Your instructor will give you guidelines on how best to use these exercises in your instructional setting.

6.1 *Tell whether the number shown in parentheses is a solution for the given equation.*

1. $7x + 2 = 16$ (2)

2. $5x - 8 = 3x + 2$ (4)

3. $7x - 2 = 2x + 8$ (2)

4. $4x + 3 = 2x - 11$ (-7)

5. $x + 5 + 3x = 2 + x + 23$ (6)

6. $\dfrac{2}{3}x - 2 = 10$ (21)

Determine which ordered pairs are solutions for the given equations.

7. $x - y = 6$ $(6, 0), (3, 3), (3, -3), (0, -6)$

8. $2x + y = 8$ $(4, 0), (2, 2), (2, 4), (4, 2)$

9. $2x + 3y = 6$ $(3, 0), (6, 2), (-3, 4), (0, 2)$

10. $2x - 5y = 10$ $(5, 0), \left(\dfrac{5}{2}, -1\right), \left(2, \dfrac{2}{5}\right), (0, -2)$

Complete the ordered pairs so that each is a solution for the given equation.

11. $x + y = 8$ $(4, \ \ ), (\ \ , 8), (8, \ \ ), (6, \ \ )$

12. $x - 2y = 10$ $(0, \ \ ), (12, \ \ ), (\ \ , -2), (8, \ \ )$

13. $2x + 3y = 6$ $(3, \ \ ), (6, \ \ ), (\ \ , -4), (-3, \ \ )$

14. $y = 3x + 4$ $(2, \ \ ), (\ \ , 7), \left(\dfrac{1}{3}, \ \ \right), \left(\dfrac{4}{3}, \ \ \right)$

Find four solutions for each equation.

15. $x + y = 10$

16. $2x + y = 8$

17. $2x - 3y = 6$

18. $y = -\dfrac{3}{2}x + 2$

6.2 *Give the coordinates of the points in the graph.*

19. *A*

20. *B*

21. *E*

22. *F*

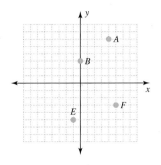

Plot points with the coordinates shown.

23. $P(6, 0)$

24. $Q(5, 4)$

25. $T(-2, 4)$

26. $U(4, -2)$

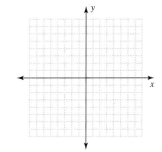

6.3 *Graph each equation.*

27. $x + y = 5$

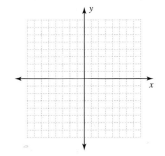

28. $x - y = 6$

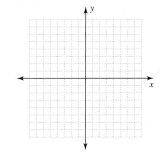

29. $y = 2x$

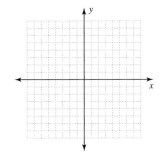

30. $y = -3x$

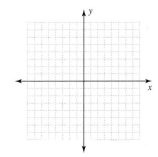

31. $y = \dfrac{3}{2}x$

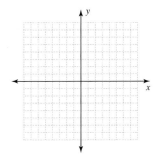

32. $y = 3x + 2$

33. $y = 2x - 3$

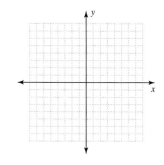

34. $y = -3x + 4$

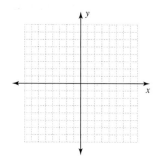

35. $y = \dfrac{2}{3}x + 2$

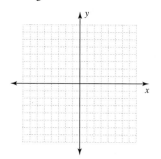

36. $3x - y = 3$

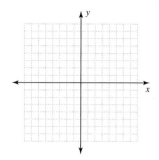

37. $2x + y = 6$

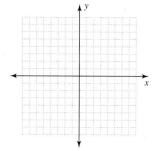

38. $3x + 2y = 12$

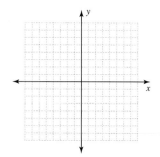

39. $3x - 4y = 12$

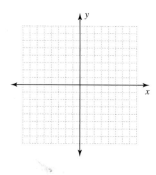

40. $x = 3$

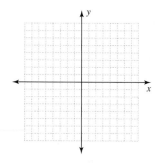

41. $y = -2$

Graph each equation.

42. $5x - 3y = 15$

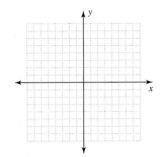

43. $4x + 3y = 12$

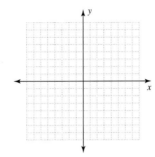

Graph each equation by first solving for y.

44. $2x + y = 6$

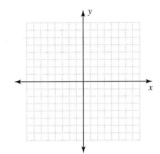

45. $3x + 2y = 6$

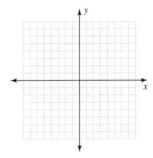

6.4 *Find the slope of the line through each pair of points.*

46. $(3, 4)$ and $(5, 8)$

47. $(-2, 3)$ and $(1, -6)$

48. $(-2, 5)$ and $(2, 3)$

49. $(-5, -2)$ and $(1, 2)$

50. $(-2, 6)$ and $(5, 6)$

51. $(-3, 2)$ and $(-1, -3)$

52. $(-3, -6)$ and $(5, -2)$

53. $(-6, -2)$ and $(-6, 3)$

In exercises 54–57, find the slope of the line graphed.

54.

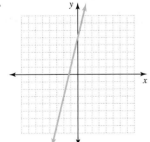

55.

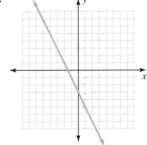

56.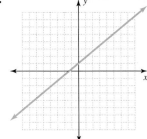

57.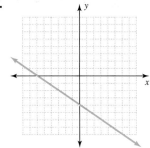

Graph each equation.

58. $y = 6x$

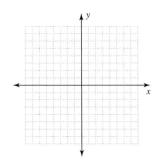

59. $y = -6x$

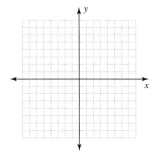

60. $y = \dfrac{2}{5}x$

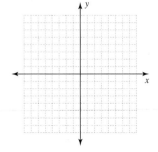

61. $y = -\dfrac{3}{4}x$

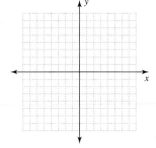

Solve for k, the constant of variation.

62. y varies directly as x; $y = 20$ when $x = 40$

63. y varies directly as x; $y = 5$ when $x = 3$

In exercises 64 and 65, y varies directly with x and the value of k is given. Graph the equation of variation.

64. $k = 4$

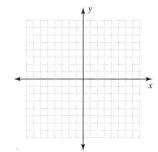

65. $k = -3.5$

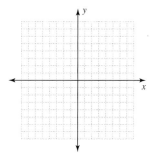

6.5 *Use the table describing technology available in public schools to answer exercises 66–69.*

Technology in U.S. Public Schools, 1995–1998

Technology	Number of Schools			
	1995	1996	1997	1998
Schools with modems[1]	30,768	37,889	40,876	61,930
Elementary	16,010	20,250	22,234	35,066
Junior high	5,652	6,929	7,417	10,996
Senior high[2]	8,790	10,277	10,781	14,540
Schools with networks[1]	24,604	29,875	32,299	49,178
Elementary	11,693	14,868	16,441	26,422
Junior high	4,599	5,590	6,035	9,003
Senior high[2]	8,159	9,166	9,565	12,853
Schools with CD-ROMs[1]	34,480	43,499	46,388	64,200
Elementary	18,343	24,353	26,377	37,908
Junior high	6,510	7,952	8,410	11,023
Senior high[2]	9,327	10,756	11,140	13,985
Schools with Internet access[1]	NA	14,211	35,762	60,224
Elementary	NA	7,608	21,026	34,195
Junior high	NA	2,707	5,752	10,888
Senior high[2]	NA	3,736	8,984	13,829

NA = Not applicable. (1) Includes schools for special and adult education, not shown separate with grade spans of K–3, K–5, K–6, K–8, and K–12. (2) Includes schools with grade spans of technical and alternative high schools and schools with grade spans of 7–12, 9–12, and 10.

Source: Quality Education Data, Inc., Denver, CO.

66. What is the increase in the number of schools with modems from 1995 to 1998?

67. How many senior high schools had either modems, networks, or CD-ROMs in 1998?

68. What is the percent increase in public schools with Internet access from 1996 to 1998?

69. What is the percent increase in elementary schools who have either modems, networks, or Internet access from 1996 to 1998?

6.5 *Use the pie charts describing U.S. car sales by size in 1993 and 2003 to answer questions 70–73.*

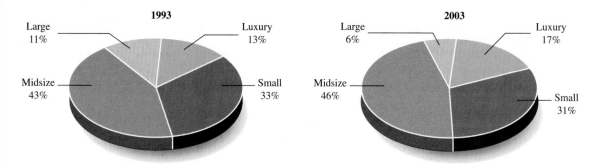

1993

Large 11% | Luxury 13%
Midsize 43% | Small 33%

2003

Large 6% | Luxury 17%
Midsize 46% | Small 31%

70. What was the percentage of midsize and small cars sold in 2003?

71. What was the percentage of large and luxury cars sold in 2003?

72. Sales of which type of car increased the most between 1993 and 2003?

73. If 21,303,000 U.S. cars were sold in 1993, how many small cars were sold?

In exercises 74 and 75, use the graph showing enrollment at Berndt Community College.

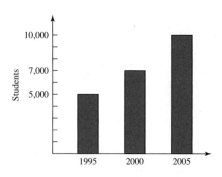

74. How many more students were enrolled in 2005 than in 1995?

75. What was the percent increase from 1995 to 2000?

6.5 *Use the line graph to complete exercises 76–78.*

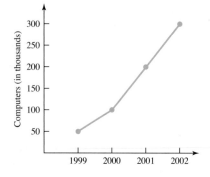

76. How many more personal computers were sold in 2002 than in 1999?

77. What was the percent increase in sales from 1999 to 2002?

78. Predict the sales of personal computers in the year 2003.

79. Use the table to create a line graph comparing annual gross income (in thousands) of 30-year-olds with their years of formal education.

Years of Education	Gross Income (in thousands)
8	16
10	21
12	23
14	28
16	31

Name _____

Section _____ Date _____

The purpose of this self-test is to help you assess your progress so that you can find concepts that you need to review before the next exam. Allow yourself about an hour to take this test. At the end of that hour, check your answers against those given in the back of this text. If you miss any, go back to the appropriate section to reread the examples until you have mastered that particular concept.

Answers

1. _____

2. _____

3. _____

4. _____

5. _____

6. _____

7. _____

8. _____

9. _____

10. _____

11. _____

12. _____

13. _____

Find four solutions for each equation.

1. $x - y = 7$

2. $5x - 6y = 30$

Plot points with the coordinates shown.

3. $S(1, -2)$

4. $T(0, 3)$

5. $U(-2, -3)$

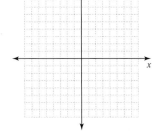

Graph each equation.

6. $2x + 5y = 10$

7. $x + 3y = 6$

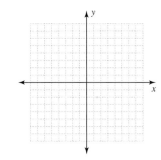
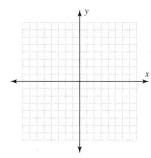

8. $x + y = 4$

9. $y = 3x$

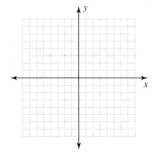
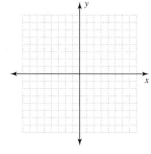

Find the slope of the line through each pair of points.

10. $(-3, 5)$ and $(2, 10)$

11. $(7, 9)$ and $(3, 9)$

12. $(-2, 6)$ and $(2, 9)$

13. $(4, 6)$ and $(4, 8)$

14. Find the slope of the line graphed.

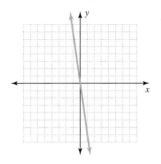

Determine which of the ordered pairs are solutions for the given equations.

15. $x + y = 9$; $(3, 6), (9, 0), (3\ 2)$ **16.** $4x - y = 16$; $(4, 0), (3, -1), (5, 4)$

Give the coordinates of the points in the graph.

17. *A*

18. *B*

19. *C*

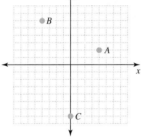

Graph each equation.

20. $y = \dfrac{3}{4}x - 4$ **21.** $y = -4$

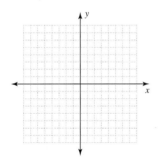   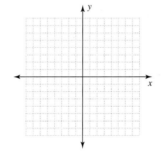

22. $y = 2x$

Answers

14. _____

15. _____

16. _____

17. _____

18. _____

19. _____

20. _____

21. _____

22. _____

Answers

23. _____

24. _____

25. _____

26. _____

27. _____

28. _____

29. _____

30. _____

23. Solve for the constant of variation if y varies directly with x and $y = 35$ when $x = 7$.

Complete the ordered pairs so that each is a solution for the given equation.

24. $4x + 3y = 12;$ (3,), (, 4), (, 3)

25. $x + 3y = 12;$ (3,), (, 2), (9,)

The pie chart below represents the portion of the $40 million tourism industry for a particular destination spent by tourists from each country. Use the chart to complete exercises 26 and 27.

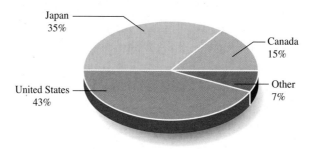

26. What percentage of the total tourism dollars is accounted for by Canada?

27. How many dollars does the United States account for?

Use the line graph to complete exercises 28–30.

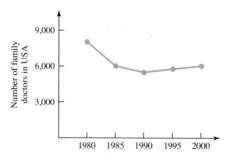

28. How many fewer family doctors were there in the United States in 1990 than in 1980?

29. What was the total change in the number of family doctors between 1980 and 2000?

30. In what 5-year period was the decrease in family doctors the greatest? What was the decrease?

Activity 6 ::
Graphing with a Calculator

The graphing calculator is a tool that can be used to help you solve many different kinds of problems. This activity will walk you through several features of the TI-84 Plus. By the time you complete this activity, you will be able to graph equations, change the viewing window to better accommodate a graph, and look at a table of values that represents some of the solutions for an equation. The first portion of this activity will demonstrate how you can create the graph of an equation. The features described here can be found on most graphing calculators. See your calculator manual to learn how to get your particular calculator model to perform this activity.

Menus and Graphing

1. To graph the equation $y = 2x + 3$ on a graphing calculator, follow these steps.

   (a) Press the $\boxed{Y=}$ key.

   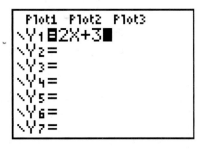

   (b) Type $2x + 3$ at the Y_1 prompt. (This represents the first equation. You can type up to 10 separate equations.) Use the $\boxed{X, T, \theta, n}$ key for the variable.

   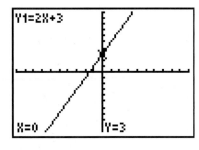

   (c) Press the $\boxed{GRAPH}$ key to see the graph.

   (d) Press the $\boxed{TRACE}$ key to display the equation. Once you have selected the $\boxed{TRACE}$ key, you can use the left and right arrows of the calculator to move the cursor along the line. Experiment with this movement. Look at the coordinates at the bottom of the display screen as you move along the line.

Frequently, we can learn more about an equation if we look at a different section of the graph than the one offered on the display screen. The portion of the graph displayed is called the **window.** The second portion of the activity explains how this window can be changed.

2. Press the WINDOW key. The **standard** graphing screen is shown.

Xmin = left edge of screen
Xmax = right edge of screen
Xscl = scale given by each tick mark on *x*-axis
Ymin = bottom edge of screen
Ymax = top edge of screen
Yscl = scale given by each tick mark on *y*-axis
Xres = resolution (do not alter this)

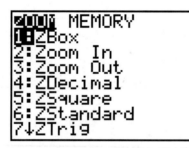

Note: To turn the scales off, enter a 0 for Xscl or Yscl. Do this when the intervals used are very large.

By changing the values for Xmin, Xmax, Ymin, and Ymax, you can adjust the viewing window. Change the viewing window so that Xmin = 0, Xmax = 40, Ymin = 0, and Ymax = 10. Again, press GRAPH . Notice that the tick marks along the *x*-axis are now much closer together. Changing Xscl from 1 to 5 will improve the display. Try it.

Sometimes we can learn something important about a graph by zooming in or zooming out. The third portion of this activity discusses this feature of the TI-84 Plus.

3.

(a) Press the ZOOM key. There are 10 options. Use the ▼ key to scroll down.

(b) Selecting the first option, ZBox, allows the user to enlarge the graph within a specified rectangle.

(i) Graph the equation $y = x^2 + x - 1$ in the standard window. *Note:* To type in the exponent, use the x^2 key or the $\wedge$ key.

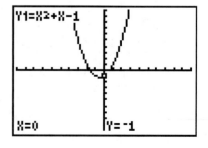

(ii) When ZBox is selected, a blinking "+" cursor will appear in the graph window. Use the arrow keys to move the cursor to where you would like a corner of the screen to be; then press the ENTER key.

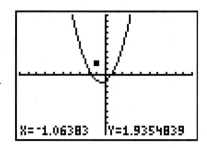

(iii) Use the arrow keys to trace out the box containing the desired portion of the graph. Do not press the ENTER key until you have reached the diagonal corner and a full box is on your screen.

After using the down arrow   *After using the right arrow*

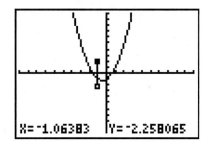   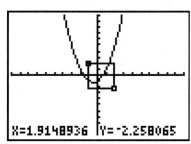

After pressing the ENTER *key a second time*

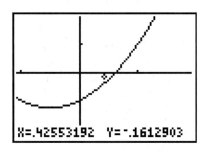

Now the desired portion of a graph can be seen more clearly.

The Zbox feature is especially useful when analyzing the roots of an equation because they correspond to the *x*-intercepts of the graph.

(c) Another feature that allows us to focus is ZoomIn. Press the ZoomIn button in the Zoom menu. Place the cursor in the center of the portion of the graph you are interested in and press the ENTER key. The window will reset with the cursor at the center of a zoomed-in view.

(d) ZoomOut works like ZoomIn, except that it sets the view larger (that is, it zooms out) to enable you to see a larger portion of the graph.

(e) ZStandard sets the window to the standard window. This is a quick and convenient way to reset the viewing window.

(f) ZSquare recalculates the view so that one horizontal unit is the same length as one vertical unit. This is sometimes necessary to get an accurate view of a graph since the width of the calculator screen is greater than its height.

Home Screen

This is where all the basic computations take place. To get to the home screen from any other screen, press $\boxed{\text{2nd}}$, $\boxed{\text{Mode}}$. This accesses the QUIT feature. To clear the home screen of calculations, press the $\boxed{\text{CLEAR}}$ key (once or twice).

Tables

The final feature that we will look at here is Table. Enter the function $y = 2x + 3$ into the $\boxed{\text{Y} =}$ menu. Then press $\boxed{\text{2nd}}$, $\boxed{\text{WINDOW}}$ to access the TBLSET menu. Set the table as shown here and press $\boxed{\text{2nd}}$, $\boxed{\text{GRAPH}}$ to access the TABLE feature. You will see the screen shown here.

The following exercises are presented to help you review concepts from earlier chapters. This is meant as a review and not as a comprehensive exam. The answers are presented in the back of the text. Section references accompany the answers. If you have difficulty with any of these exercises, be certain to at least read through the summary related to those sections.

Name _____

Section _____   Date _____

Answers

Perform the indicated operations.

1. $9 + (-6)$

2. $-4 - (-9)$

3. $25 - (-12)$

4. $-32 + (-21)$

5. $(-23)(-3)$

6. $(12)(-10)$

7. $30 \div (-6)$

8. $(-24) \div (-8)$

Evaluate the expressions if $x = -3$, $y = 4$, and $z = -5$.

9. $3x^2 y$

10. $-3z - 3y$

11. $-3(-2y + 3z)$

12. $\dfrac{3y - 2x}{5y + 6x}$

Solve each equation and check your results.

13. $5x - 2 = 2x - 6$

14. $3(x - 2) = 2(3x + 1)$

15. $\dfrac{5}{6}x - 3 = 2 + \dfrac{1}{3}x$

16. $4(2 - x) + 9 = 7 + 6x$

17. Solve the equation $F = \dfrac{9}{5}C + 32$ for C.

Solve each inequality.

18. $4x - 9 < 7$

19. $-5x + 15 \geq 2x - 6$

1. _____

2. _____

3. _____

4. _____

5. _____

6. _____

7. _____

8. _____

9. _____

10. _____

11. _____

12. _____

13. _____

14. _____

15. _____

16. _____

17. _____

18. _____

19. _____

Answers

20. _____

21. _____

22. _____

23. _____

24. _____

25. _____

26. _____

27. _____

28. _____

29. _____

30. _____

31. _____

32. _____

33. _____

34. _____

35. _____

36. _____

37. _____

38. _____

Use the properties of exponents to simplify each expression and write the results with positive exponents.

20. $(x^2 y^3)^{-2}$
21. $\dfrac{x^3 y^2}{x^4 y^{-3}}$
22. $(x^6 y^{-3})^0$

Perform the indicated operations.

23. Add $2x^2 + 4x - 6$ and $3x^2 - 4x - 4$.

24. Subtract $3a^2 - 2a + 5$ from the sum of $a^2 + 3a - 2$ and $-5a^2 + 2a + 9$.

Evaluate each polynomial for the indicated variable value.

25. $2x^2 - 5x + 7$ for $x = 4$
26. $x^3 + 3x^2 - 7x + 8$ for $x = -2$

Multiply.

27. $(3x - 5y)(2x + 4y)$
28. $3x(x - 3)(2x + 5)$

29. $(2a + 7b)(2a - 7b)$

Completely factor each polynomial.

30. $12p^2 n^2 + 20pn^2 - 16pn^3$
31. $y^3 - 3y^2 - 5y + 15$

32. $9a^2 b - 49b$
33. $6x^2 - 2x - 4$

34. $6a^2 + 7ab - 3b^2$

Solve each equation by factoring.

35. $x^2 - 8x - 33 = 0$
36. $35x^2 - 38x = -8$

Simplify each rational expression.

37. $\dfrac{-35a^4 b^5}{21ab^7}$
38. $\dfrac{2w^2 - w - 6}{2w^2 + 9w + 9}$

Add or subtract as indicated. Simplify your answer.

39. $\dfrac{2}{a - 5} - \dfrac{1}{a}$

40. $\dfrac{2w}{w^2 - 9w + 20} + \dfrac{8}{w - 4}$

Multiply or divide as indicated.

41. $\dfrac{4xy^3}{5xy^2} \cdot \dfrac{15x^3y}{16y^2}$

42. $\dfrac{m^2 - 3m}{m^2 - 9} \div \dfrac{4m^2}{m^2 - m - 12}$

Solve each equation.

43. $\dfrac{w}{w - 2} + 1 = \dfrac{w + 4}{w - 2}$

44. $\dfrac{7}{x} - \dfrac{1}{x - 3} = \dfrac{9}{x^2 - 3x}$

Graph each equation.

45. $3x + 4y = 12$

46. $y = -7$

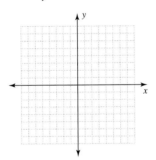

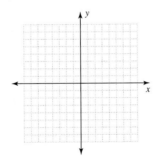

47. $x = 2y$

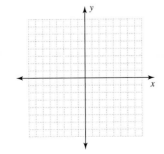

48. Find the slope of the line passing through the points $(-3, 5)$ and $(1, 13)$.

Answers

39. _____

40. _____

41. _____

42. _____

43. _____

44. _____

45. _____

46. _____

47. _____

48. _____

Answers

49. _____

50. _____

51. _____

52. _____

53. _____

49. Graph the equation of variation if $k = 4$.

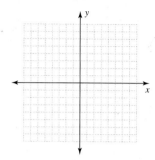

50. Find the constant of variation if y varies directly with x, and if $y = 450$ when $x = 15$.

Solve each problem.

51. The length of a rectangle is 3 in. less than twice its width. If the perimeter is 24 in., find the dimensions of the rectangle.

52. The sum of three consecutive odd integers is 129. Find the three integers.

53. The carpet outlet is selling rug remnants at 25% off. If the sale price is $150, what was the original price?

CHAPTER

7

chapter 7 > Make the Connection

INTRODUCTION

In the pharmaceutical-making process, great caution must be exercised to ensure that the medicines and drugs are pure and contain precisely what is indicated on the label. The quality control division of the pharmaceutical company is responsible for guaranteeing such purity.

A lab technician working in quality control must run a series of tests on samples of every ingredient, even simple ingredients such as salt (NaCl). One such test is a measure of how much weight is lost as a sample is dried. The technician must set up a 3-hour procedure that involves cleaning and drying bottles and stoppers and then weighing them while they are empty and again when they contain samples of the substance to be heated and dried.

The pharmaceutical company may have a standard of acceptability for this substance. For instance, the substance may not be acceptable if the loss of weight from drying is greater than 10%. The technician would then use an inequality to calculate acceptable weight loss. In the following inequality, the right side represents the percent of weight loss from drying.

$$10 \geq \frac{W_g - W_f}{W_g - T} \cdot 100$$

We will examine inequalities further in this chapter.

Graphing and Inequalities

CHAPTER 7 OUTLINE

Name _____

Section _____ Date _____

This prerequisite test provides some exercises requiring skills that you will need to be successful in the coming chapter. The answers for these exercises can be found in the back of this text. This prerequisite test can help you identify topics that you will need to review before beginning the chapter.

Answers

Solve for y.

1. $3x + 2y = 6$

2. $5x - 2y = 10$

Find the slope of the line connecting the given points.

3. $(-4, 6)$ and $(3, 20)$

4. $(5, -7)$ and $(-5, 3)$

5. $(6, 9)$ and $(3, 9)$

6. $(2, 8)$ and $(2, 5)$

Graph each inequality on a number line.

7. $\dfrac{2}{3}x \le 4$

8. $-2x \le 10$

Evaluate each expression for the given variable value.

9. $3 - 2x \quad (x = -1)$

10. $x^2 - 2 \quad (x = -2)$

1. _____

2. _____

3. _____

4. _____

5. _____

6. _____

7. _____

8. _____

9. _____

10. _____

The Streeter/Hutchison Series in Mathematics   Beginning Algebra

7.1

The Slope-Intercept Form

< 7.1 Objectives >

1 > Find the slope and *y*-intercept from the equation of a line

2 > Given the slope and *y*-intercept, write the equation of a line

3 > Use the slope and *y*-intercept to graph a line

4 > Solve an application involving slope-intercept form

In Chapter 6, we used two points to find the slope of a line. In this chapter we use the slope and *y*-intercept to sketch the graph of an equation.

First, we want to consider finding the equation of a line when its slope and *y*-intercept are known.

Suppose that the *y*-intercept of a line is $(0, b)$. Then the point at which the line crosses the *y*-axis has coordinates $(0, b)$, as shown in the sketch at left.

Now, using any other point (x, y) on the line and using our definition of slope, we can write

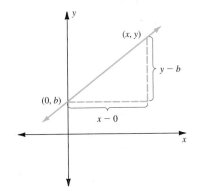

Change in *y*

$$m = \frac{y - b}{x - 0}$$

Change in *x*

or

$$m = \frac{y - b}{x}$$

Multiplying both sides of this equation by x, we have

$$mx = y - b$$

Finally, adding b to both sides of this equation gives

$$mx + b = y$$

or

$$y = mx + b$$

We summarize this discussion with a definition.

NOTE

In this form, the equation is *solved for y*. The coefficient of *x* gives the slope of the line, and the constant term gives the *y*-intercept.

Definition

The Slope-Intercept Form for a Line	An equation of the line with slope m and y-intercept $(0, b)$ is $$y = mx + b$$

525

Example 1   Finding the Slope and *y*-Intercept

< Objective 1 >

Find the slope and *y*-intercept of the equation

$$y = -\frac{2}{3}x - 5$$

$$\underset{m}{\uparrow} \qquad \underset{b}{\uparrow}$$

> **CAUTION**

Some students might say that the slope is $-\frac{2}{3}x$, but this is incorrect. The slope is the coefficient of *x*: $-\frac{2}{3}$.

The slope of the line is $-\frac{2}{3}$; the *y*-intercept is $(0, -5)$.

Check Yourself 1

Find the slope and *y*-intercept for the graph of each equation.

(a) $y = -3x - 7$ (b) $y = \frac{3}{4}x + 5$

As Example 2 illustrates, we may have to solve for *y* as the first step in determining the slope and the *y*-intercept for the graph of an equation.

 Example 2   Finding the Slope and *y*-Intercept

Find the slope and *y*-intercept of the equation

$$3x + 2y = 6$$

First, we must solve the equation for *y*.

$$3x + 2y = 6$$

NOTE

If we write the equation as
$$y = \frac{-3x + 6}{2}$$
it is more difficult to identify the slope and the intercept.

$$2y = -3x + 6 \qquad \text{Subtract } 3x \text{ from both sides.}$$

$$y = -\frac{3}{2}x + 3 \qquad \text{Divide each term by 2.}$$

The equation is now in slope-intercept form. The slope is $-\frac{3}{2}$, and the *y*-intercept is $(0, 3)$.

Check Yourself 2

Find the slope and *y*-intercept for the graph of the equation

$$2x - 5y = 10$$

As we mentioned earlier, knowing certain properties of a line (namely, its slope and *y*-intercept) will also allow us to write the equation of the line by using the slope-intercept form. Example 3 illustrates this approach.

 Example 3   Writing the Equation of a Line

< Objective 2 >

Write the equation of a line with slope $-\frac{3}{4}$ and *y*-intercept $(0, -3)$.

We know that $m = -\frac{3}{4}$ and $b = -3$. In this case,

$$m \qquad\qquad b$$

$$y = -\frac{3}{4}x + (-3)$$

or

$$y = -\frac{3}{4}x - 3$$

which is the desired equation.

 Check Yourself 3

Write the equation of each line.

(a) Slope −2 and *y*-intercept (0, 7)

(b) Slope $\frac{2}{3}$ and *y*-intercept (0, −3)

We can also use the slope and *y*-intercept of a line in drawing its graph. Consider Example 4.

Example 4   **Graphing a Line Using the Slope and *y*-intercept**

< Objective 3 >

Graph the line with slope $\frac{2}{3}$ and *y*-intercept (0, 2).

Because the *y*-intercept is (0, 2), we begin by plotting the point (0, 2). Because the horizontal change (or run) is 3, we move 3 units to the right *from that y-intercept*. Then because the vertical change (or rise) is 2, we move 2 units up to locate another point on the desired graph. Note that we will have located that second point at (3, 4). The final step is simply to draw a line through that point and the *y*-intercept.

NOTE

$m = \dfrac{2}{3} = \dfrac{\text{rise}}{\text{run}}$

The line rises from left to right because the slope is positive.

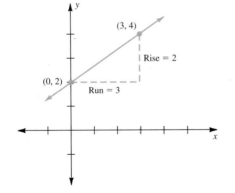

The equation of this line is $y = \frac{2}{3}x + 2$.

 Check Yourself 4

Graph the equation of a line with slope $\frac{3}{5}$ and *y*-intercept (0, −2).

When creating a graph such as the one drawn in Example 4, it is a good idea to plot several points. This is quick and easy using the idea of slope. Recall that we plotted (0, 2) and then used the slope of $\frac{2}{3}$ to plot (3, 4): We moved 3 to the right and up 2.

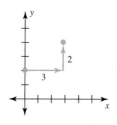

Now we simply continue this process from that point: run 3 and rise 2, repeatedly.

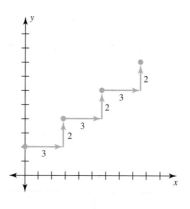

We may also move to the left and down from (0, 2): run −3 and rise −2.

NOTE

Here, $\dfrac{\text{rise}}{\text{run}} = \dfrac{-2}{-3} = \dfrac{2}{3}$, the given slope.

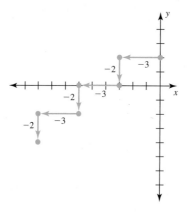

(▶)   **Example 5**   **Graphing a Line Using the Slope and y-intercept**

Sketch the graph of $y = -2x + 5$.

We begin by noting that the y-intercept is (0, 5) and the slope is −2. To view the slope as $\dfrac{\text{rise}}{\text{run}}$, we may write −2 either as $\dfrac{-2}{1}$ or $\dfrac{2}{-1}$. Now we plot (0, 5) and then use the slope to plot more points. Using $m = \dfrac{-2}{1}$, we run 1 and then rise −2; that is, as we run 1, we drop 2. We repeat this action and plot several points.

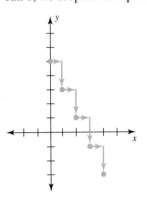

If we use $m = \dfrac{2}{-1}$, we run -1 and then rise 2; that is, as we run 1 *to the left,* we go up 2.

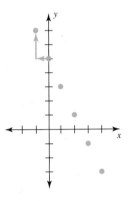

The completed graph is shown here:

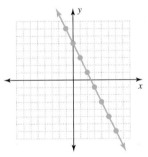

With practice, you will be able to quickly plot many points and sketch an accurate line.

 Check Yourself 5

Sketch the graph of $y = -3x + 1$.

Step by Step

Graphing by Using the Slope-Intercept Form

Step 1	Write the original equation of the line in slope-intercept form.
Step 2	Determine the slope m and the y-intercept $(0, b)$.
Step 3	Plot the y-intercept at $(0, b)$.
Step 4	Use m (the change in y over the change in x) to determine other points on the desired line.
Step 5	Draw a line through the points determined in steps 1–4 to complete the graph.

You have now seen two methods for graphing lines: the slope-intercept method (this section) and the intercept method (Section 6.3). When you graph a linear equation, you first should decide which is the appropriate method.

| Example 6 | Selecting an Appropriate Graphing Method |

Decide which of the two methods for graphing lines—the intercept method or the slope-intercept method—is more appropriate for graphing equations **(a)**, **(b)**, and **(c)**.

(a) $2x - 5y = 10$

Because both intercepts are easy to find, the better choice is the intercept method to graph this equation.

(b) $2x + y = 6$

This equation can be graphed quickly by either method. As it is written, you might choose the intercept method. It can, however, be rewritten as $y = -2x + 6$. In that case the slope-intercept method is more appropriate.

(c) $y = \dfrac{1}{4}x - 4$

Because the equation is in slope-intercept form, that is the more appropriate method to choose.

Check Yourself 6

Which would be more appropriate for graphing each equation, the intercept method or the slope-intercept method?

(a) $x + y = -2$ **(b)** $3x - 2y = 12$ **(c)** $y = -\dfrac{1}{2}x - 6$

It is important to understand what the slope of a linear equation can tell us about the relationship between x and y. For example, if $y = 2x - 5$, we note that the slope is 2. Writing this as $\dfrac{2}{1}$, we see: If x increases by 1 unit, then y increases by 2 units. Consider the next example.

| Example 7 | Interpreting the Slope |

Interpret the slope for each equation.

(a) $y = -3x + 4$

Write the slope as $\dfrac{-3}{1}$. If x increases by 1 unit, then y decreases by 3 units.

(b) $y = 0.8x + 6.2$

Writing the slope, 0.8, as $\dfrac{0.8}{1}$, we can say that as x increases by 1 unit, y increases by 0.8 (eight-tenths of a unit).

(c) $C = 25n + 120$

Write the slope as $\dfrac{25}{1}$. As n increases by 1, C increases by 25.

Check Yourself 7

Interpret the slope for each equation.

(a) $y = \dfrac{2}{3}x - 7$ (b) $y = -5x - 8$

NOTE

In the equation $y = mx + b$, b is the constant term.

The slope-intercept form lends itself easily to applications involving linear equations. Consider the cost of manufacturing a product; there are really two costs involved.

Fixed costs are those costs that are independent of the number of units produced. That is, no matter how many (or how few) items are produced, fixed costs remain the same. Fixed costs include the cost of leasing property, insurance costs, and labor costs for salaried personnel. If we write an equation relating total costs to the number of units produced, then the fixed costs are represented by the constant term.

Variable costs are those costs that change depending on the number of units produced. These include the costs of materials, labor costs for hourly employees, many utility costs, and service costs. In the slope-intercept equation, variable costs correspond to the slope m because the dollar amount increases each time another unit is produced.

Example 8	An Application of Slope-Intercept Equations

< Objective 4 >

NOTE

As x (the number of stereos produced) increases by 1, the cost y increases by 26. This says that the slope is 26.

S-Bar Electronics determines that the cost to produce each stereo is $26. In addition, the cost to keep its factory open each month is $3,500.

(a) Write an equation relating the total monthly cost of producing stereos to the number of stereos produced.

The y-intercept represents the fixed costs, so we have $b = 3,500$. The slope m is given by the variable costs, which are $26 per stereo. Putting this together gives us the equation

$$y = 26x + 3,500$$

(b) Use the cost equation to determine the total cost of producing 320 stereos in a month.

In the cost equation, we substitute 320 for x and calculate y.

$$y = 26(320) + 3,500$$
$$= 11,820$$

So it costs S-Bar Electronics $11,820 to produce 320 stereos in a month.

Check Yourself 8

A manager at the chic new restaurant Sweet Eats determines that the average dinner costs the restaurant $18 to produce. In addition, it costs the restaurant $620 to stay open each evening.

(a) Write an equation relating the total nightly cost of operation to the number of dinners served.
(b) How much does it cost the restaurant to serve 50 dinners in an evening?

The following is an application from the field of health care.

Example 9   **An Allied Health Application**

A person's body mass index (BMI) can be calculated using his or her height h, in inches, and weight w, in pounds, with the formula

$$\text{BMI} = \frac{703w}{h^2}$$

(a) Compute the BMI of a 5 ft 10 in. man who weighs 190 pounds (round your results to the nearest tenth).

At 5 ft 10 in. and 190 lb, we have $h = 70$ and $w = 190$.

$$\text{BMI} = \frac{703(190)}{(70)^2}$$

$$\approx 27.26$$

The man's BMI is about 27.3.

(b) Assume that we are looking at the BMI for a group of 5 ft 10 in. men of varying weight. For this group, the BMI formula becomes

$$\text{BMI} = \frac{703w}{h^2}$$

$$= \frac{703w}{(70)^2}$$

$$= \frac{703}{4,900}w$$

$$\approx 0.143w$$

Find the slope of the formula.
The slope is about 0.143.

(c) Interpret the slope from part (b) in the context of this application.

Generally, this means that the output increases by 0.143 when the input increases by 1. In the context of this application, the BMI of a 5 ft 10 in. man increases by 0.143 for each additional pound that he weighs.

Check Yourself 9

Using metric measurements, a person's body mass index can be calculated using his or her height h, in centimeters, and weight w, in kilograms, with the formula

$$\text{BMI} = \frac{10{,}000w}{h^2}$$

(a) Compute the BMI of a 160-cm, 70-kg woman (to the nearest tenth).

(b) Write the BMI formula for a group of women who are 160 cm tall. Round your results to the nearest tenth.

(c) Find the slope of the BMI formula for 160-cm women, to the nearest tenth.

(d) Interpret the slope of the formula from part (c) in the context of this application.

Check Yourself ANSWERS

1. **(a)** Slope is -3, y-intercept is $(0, -7)$; **(b)** Slope is $\frac{3}{4}$, y-intercept is $(0, 5)$

2. The slope is $\frac{2}{5}$; the y-intercept is $(0, -2)$

3. **(a)** $y = -2x + 7$; **(b)** $y = \frac{2}{3}x - 3$

4. 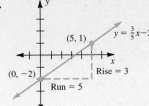   5.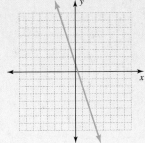

6. **(a)** Either; **(b)** intercept; **(c)** slope-intercept

7. **(a)** When x increases by 3 units, y increases by 2 units. Alternatively, when x increases by 1 unit, y increases by $\frac{2}{3}$ of a unit; **(b)** When x increases by 1 unit, y decreases by 5 units.

8. **(a)** $y = 18x + 620$; **(b)** \$1,520

9. **(a)** 27.3; **(b)** BMI $= 0.4w$; **(c)** 0.4; **(d)** The BMI of a 160-cm tall woman increases by 0.4 for each additional kilogram of weight.

Reading Your Text

The following fill-in-the-blank exercises are designed to ensure that you understand some of the key vocabulary used in this section.

SECTION 7.1

(a) In the slope-intercept form, the coefficient of x gives the _____ of the line.

(b) When graphing a line with given slope and y-intercept, begin by plotting the _____.

(c) If we write the slope as $\frac{-3}{1}$, then when x increases by 1 unit, y _____ by 3 units.

(d) _____ costs are those costs that are independent of the number of units produced.

Name _____

Section _____ Date _____

Answers

1. _____

2. _____

3. _____

4. _____

5. _____

6. _____

7. _____

8. _____

9. _____

10. _____

11. _____

12. _____

13. _____

14. _____

< Objective 1 >

Find the slope and y-intercept of the line represented by each equation.

1. $y = 3x + 5$

2. $y = -7x + 3$

3. $y = -2x - 5$

4. $y = 5x - 2$

5. $y = \dfrac{3}{4}x + 1$

6. $y = -4x$

7. $y = \dfrac{2}{3}x$

8. $y = -\dfrac{3}{5}x - 2$

9. $4x + 3y = 12$

10. $2x + 5y = 10$

11. $y = 9$

12. $2x - 3y = 6$

13. $3x - 2y = 8$ ▸ Videos

14. $x = 5$

< Objectives 2–3 >

Write the equation of the line with given slope and y-intercept. Then graph each line using the slope and y-intercept.

Answers

15. Slope: 3; *y*-intercept: (0, 5)

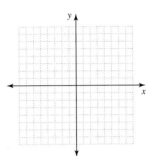

16. Slope: −2; *y*-intercept: (0, 4)

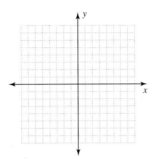

15. _____

16. _____

17. _____

18. _____

19. _____

20. _____

21. _____

22. _____

17. Slope: −3; *y*-intercept: (0, 4)

> Videos

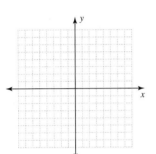

18. Slope: 5; *y*-intercept: (0, −2)

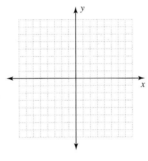

19. Slope: $\dfrac{1}{2}$; *y*-intercept: (0, −2)

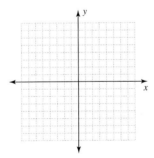

20. Slope: $-\dfrac{3}{4}$; *y*-intercept: (0, 8)

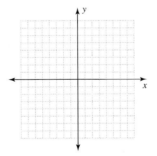

21. Slope: $-\dfrac{2}{3}$; *y*-intercept: (0, 0)

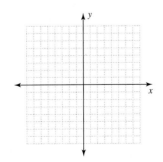

22. Slope: $\dfrac{2}{3}$; *y*-intercept: (0, −2)

> Videos

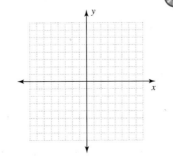

Beginning Algebra   The Streeter/Hutchison Series in Mathematics   © The McGraw-Hill Companies. All Rights Reserved.

Answers

23. _____

24. _____

25. _____

26. _____

27. _____

28. _____

29. _____

30. _____

23. Slope: $\dfrac{3}{4}$; y-intercept: $(0, 3)$

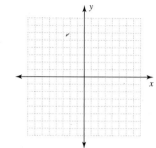

24. Slope: -3; y-intercept: $(0, 0)$

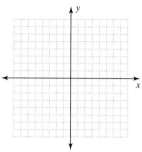

In exercises 25 to 32, match each graph with one of the equations.

(a) $y = 2x$ **(b)** $y = x + 1$ **(c)** $y = -x + 3$ **(d)** $y = 2x + 1$

(e) $y = -3x - 2$ **(f)** $y = \dfrac{2}{3}x + 1$ **(g)** $y = -\dfrac{3}{4}x + 1$

(h) $y = -4x$

25.

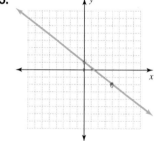

26.

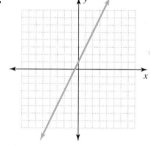

27.

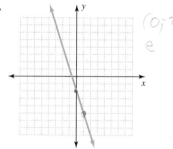

28.

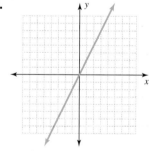

29.

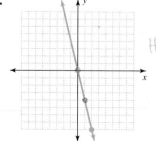

30.

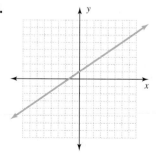

31.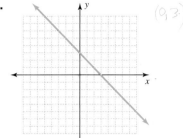

32.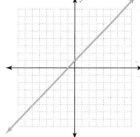

31. _____

32. _____

33. _____

34. _____

35. _____

36. _____

37. _____

38. _____

39. _____

40. _____

41. _____

42. _____

< Objective 4 >

33. SOCIAL SCIENCE The equation $y = 0.10x + 200$ describes the award money in a recycling contest. What are the slope and the y-intercept for this equation?

34. BUSINESS AND FINANCE The equation $y = 15x - 100$ describes the amount of money a high school class might earn from a paper drive. What are the slope and y-intercept for this equation?

35. SCIENCE AND MEDICINE On a certain February day in Philadelphia, the temperature at 6:00 A.M. was 10°F. By 2:00 P.M. the temperature was up to 26°F. What was the hourly rate of temperature change?   ⊙ > Videos

36. CONSTRUCTION A roof rises 8.75 ft over a horizontal distance of 15.09 ft. Find the slope of the roof to the nearest hundredth.

37. SCIENCE AND MEDICINE An airplane covered 15 mi of its route while decreasing its altitude by 24,000 ft. Find the slope of the line of descent that was followed (1 mi = 5,280 ft). Round to the nearest hundredth.

38. TECHNOLOGY Driving down a mountain, Tom finds that he has descended 1,800 ft in elevation by the time he is 3.25 mi horizontally away from the top of the mountain. Find the slope of his descent to the nearest hundredth.

| Basic Skills | **Challenge Yourself** | Calculator/Computer | Career Applications | Above and Beyond |

Determine whether each statement is **true** _or_ **false.**

39. The slope of a line through the origin can be zero.

40. A line with undefined slope is the same as a line with a slope of zero.

Complete each statement with **never, sometimes,** _or_ **always.**

41. Lines _____ have exactly one x-intercept.

42. The y-intercept of a line through the origin is _____ zero.

Answers

43. _____

44. _____

45. _____

46. _____

47. _____

48. _____

49. _____

50. _____

51. _____

In which quadrant(s) are there no solutions for each line?

43. $y = -x + 1$ › Videos

44. $y = 3x + 2$

45. $y = -2x - 5$

46. $y = -5x - 7$

47. $y = 3$

48. $x = -2$

49. Sketch both lines on the same graph.

$y = 2x - 1$ and $y = 2x + 3$

What do you observe about these graphs? Will the lines intersect?

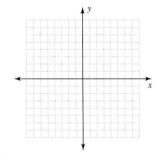

50. Repeat exercise 49 using

$y = -2x + 4$ and $y = -2x + 1$

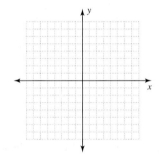

51. Sketch both lines on the same graph.

$y = \dfrac{2}{3}x$ and $y = -\dfrac{3}{2}x$

What do you observe concerning these graphs? Find the product of the slopes of these two lines.

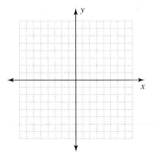

52. Repeat exercise 51 using

$$y = \frac{4}{3}x \quad \text{and} \quad y = -\frac{3}{4}x$$

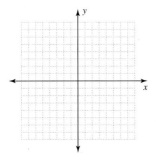

53. Based on exercises 51 and 52, write the equation of a line that is perpendicular to $y = \frac{3}{5}x$.

Basic Skills | Challenge Yourself | Calculator/Computer | **Career Applications** | Above and Beyond
▲

54. **ALLIED HEALTH** The arterial oxygen tension (P_aO_2), in millimeters of mercury (mm Hg), of a patient (lying down) can be estimated based on the patient's age (A), in years. Use the equation $P_aO_2 = 103.5 - 0.42A$.

   (a) Determine the slope of the line for this equation.

   (b) Interpret this slope verbally.

55. **ALLIED HEALTH** The arterial oxygen tension (P_aO_2), in millimeters of mercury (mm Hg), of a patient (seated) can be estimated based on the patient's age (A), in years. Use the equation $P_aO_2 = 104.2 - 0.27A$.

   (a) Determine the slope of the line for this equation.

   (b) Interpret this slope verbally.

56. **AGRICULTURAL TECHNOLOGY** The yield Y (in bushels) per acre for a cornfield is estimated from the amount of rainfall R (in inches) using the formula

$$Y = \frac{43,560}{8,000}R$$

   (a) What is the slope of the line for this equation?

   (b) Interpret the slope.

57. **AGRICULTURAL TECHNOLOGY** During a period of summer, the growth of corn plants follows a linear pattern that is approximated by the equation

$$h = 1.77d + 24.92$$

where h is the height (in inches) of the corn, and d is the number of days that have passed.

   (a) What is the slope of the line for this equation?

   (b) Interpret the slope.

Answers

52. _____

53. _____

54. _____

55. _____

56. _____

57. _____

Answers

58. _____

59. _____

60. _____

61. _____

58. Complete the statement: "The difference between undefined slope and zero slope is . . . ."

59. Complete the statement: "The slope of a line tells you . . . ."

60. STATISTICS In a study on nutrition, 18 normal adults aged 23 to 61 years old were measured for body fat, which is given as a percentage of weight. The mean (average) body fat percentage for women 40 years old was 28.6% and for women 53 years old was 38.4%. Work with a partner to decide how to show this information by plotting points on a graph. Try to find a linear equation that will tell you percentage of body fat based on a woman's age. What does your equation give for 20 years of age? For 60? Do you think a linear model works well for predicting body fat percentage in women as they age?

> chapter 7 > Make the Connection

61. BUSINESS AND FINANCE On two occasions last month, Sam Johnson rented a car on a business trip. Both times it was the same model from the same company, and both times it was in San Francisco. Sam now has to fill out an expense account form and needs to know how much he was charged per mile and the base rate. On both occasions he dropped the car at the airport booth and just got the total charge, not the details. All Sam knows is that he was charged $210 for 625 mi on the first occasion and $133.50 for 370 mi on the second trip. Sam has called accounting to ask for help. Plot these two points on a graph and draw the line that goes through them. What question does the slope of the line answer for Sam? How does the *y*-intercept help? Write a memo to Sam explaining the answers to his question and how a knowledge of algebra and graphing has helped you find the answers.

Answers

1. Slope 3, *y*-intercept $(0, 5)$ **3.** Slope -2, *y*-intercept $(0, -5)$

5. Slope $\dfrac{3}{4}$, *y*-intercept $(0, 1)$ **7.** Slope $\dfrac{2}{3}$, *y*-intercept $(0, 0)$

9. Slope $-\dfrac{4}{3}$, *y*-intercept $(0, 4)$ **11.** Slope 0, *y*-intercept $(0, 9)$

13. Slope $\dfrac{3}{2}$, *y*-intercept $(0, -4)$

15. $y = 3x + 5$ **17.** $y = -3x + 4$

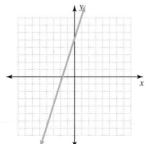

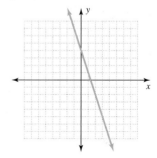

19. $y = \dfrac{1}{2}x - 2$

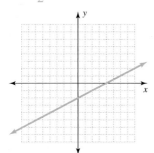

21. $y = -\dfrac{2}{3}x$

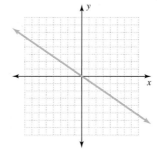

23. $y = \dfrac{3}{4}x + 3$

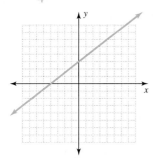

25. g   **27.** e   **29.** h   **31.** c   **33.** Slope: 0.10; y-intercept: $(0, 200)$

35. 2°F/h   **37.** -0.30 **39.** True   **41.** sometimes

43. III   **45.** 1   **47.** III and IV

49. Parallel lines; no

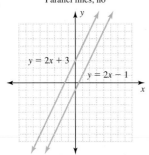

51. Perpendicular lines; -1

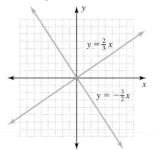

53. $y = -\dfrac{5}{3}x$ **55. (a)** $-0.27; \dfrac{\text{mm Hg}}{\text{yr}}$ **(b)** For every additional year in a patient's age, the arterial oxygen tension decreases by 0.27 mm Hg.

57. (a) 1.77; **(b)** For each additional day, the height increases by 1.77 inches.

59. Above and Beyond   **61.** Above and Beyond

7.2

Parallel and Perpendicular Lines

< 7.2 Objectives >

1 > Determine whether two lines are parallel

2 > Determine whether two lines are perpendicular

3 > Find the slope of a line parallel or perpendicular to a given line

For most inexperienced drivers, the most difficult driving maneuver to master is parallel parking. What is parallel parking? It is the act of backing into a curbside space so that the car's tires are parallel to the curb.

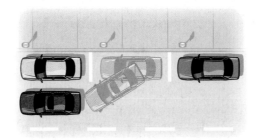

How can you tell that you've done a good job of parallel parking? Most people check to see that both the front tires and the back tires are the same distance (8 in. or so) from the curb. This is checking to be certain that the car is parallel to the curb.

How can we tell that two equations represent parallel lines? Look at the following sketch.

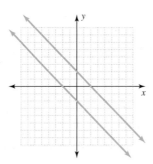

If two lines are parallel, they have the same slope. If their equations are in slope-intercept form, you simply compare the slopes.

| ▶ | **Example 1** | **Determining That Two Lines Are Parallel** |

< Objective 1 >

Which two equations represent parallel lines?

(a) $y = 2x + 3$ **(b)** $y = -\dfrac{1}{2}x - 5$

(c) $y = -2x + \dfrac{1}{2}$ **(d)** $y = -2x - 9$

Because **(c)** and **(d)** both have a slope of -2, the lines are parallel.

Check Yourself 1

Which two equations represent parallel lines?

(a) $y = -5x + 5$ (b) $y = \frac{1}{5}x - 5$

(c) $y = -5x + \frac{1}{2}$ (d) $y = 5x - 9$

More formally, we can state a property of parallel lines.

Property

Slopes of Parallel Lines

For nonvertical lines L_1 and L_2, if line L_1 has slope m_1 and line L_2 has slope m_2, then

L_1 is parallel to L_2 if, and only if, $m_1 = m_2$

Note: All vertical lines are parallel to each other.

As we discovered in Chapter 6, we can find the slope of a line from any two points on the line.

Example 2   Determining Whether Two Lines Are Parallel

NOTE

Parallel lines do not intersect unless, of course, L_1 and L_2 are actually the *same line*. In Example 2 a quick sketch shows that the lines are distinct.

Are lines L_1 through $(2, 3)$ and $(4, 6)$ and L_2 through $(-4, 2)$ and $(0, 8)$ parallel, or do they intersect?

$$m_1 = \frac{6 - 3}{4 - 2} = \frac{3}{2}$$

$$m_2 = \frac{8 - 2}{0 - (-4)} = \frac{6}{4} = \frac{3}{2}$$

Because the slopes of the lines are equal, the lines are parallel. They do *not* intersect.

Check Yourself 2

Are lines L_1 through $(-2, -1)$ and $(1, 4)$ and L_2 through $(-3, 4)$ and $(0, 8)$ parallel, or do they intersect?

Many important characteristics of lines are evident from a city map.

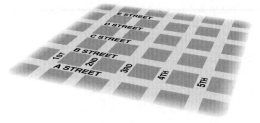

Note that 4th Street and 5th Street are parallel. Just as these streets never meet, it is true that two distinct parallel lines will never meet.

Recall that the point at which two lines meet is called their **intersection.** This is also true with two streets. We call the common area of the two streets the intersection.

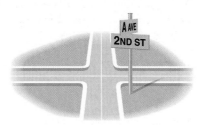

In this case, the two streets meet at right angles. When two lines meet at right angles, we say that they are **perpendicular.**

Property

Slopes of Perpendicular Lines

For nonvertical lines L_1 and L_2, if line L_1 has slope m_1 and line L_2 has slope m_2, then

L_1 is perpendicular to L_2 if, and only if, $m_1 = -\dfrac{1}{m_2}$

or equivalently

$m_1 \cdot m_2 = -1$

Note: Horizontal lines are perpendicular to vertical lines.

⊙   **Example 3**   Determining That Two Lines Are Perpendicular

< Objective 2 >

Which two equations represent perpendicular lines?

(a) $y = 2x + 3$

(b) $y = -\dfrac{1}{2}x - 5$

(c) $y = -2x + \dfrac{1}{2}$

(d) $y = -2x - 9$

Because the product of the slopes for (a) and (b) is

$$2\left(-\frac{1}{2}\right) = -1$$

these two lines are perpendicular. Note that none of the other pairs of slopes have a product of -1.

NOTE

We say that 2 and $-\dfrac{1}{2}$ are *negative reciprocals* of each other.

✓   **Check Yourself 3**

Which two equations represent perpendicular lines?

(a) $y = -5x + 5$ (b) $y = -\dfrac{1}{5}x - 5$

(c) $y = -5x + \dfrac{1}{2}$ (d) $y = 5x - 9$

Example 4

Determining Whether Two Lines Are Perpendicular

NOTE

$$\left(\frac{4}{3}\right)\left(-\frac{3}{4}\right) = -1$$

Are lines L_1 through points $(-2, 3)$ and $(1, 7)$ and L_2 through points $(2, 4)$ and $(6, 1)$ perpendicular?

$$m_1 = \frac{7 - 3}{1 - (-2)} = \frac{4}{3}$$

$$m_2 = \frac{1 - 4}{6 - 2} = -\frac{3}{4}$$

Because the slopes are negative reciprocals, the lines are perpendicular.

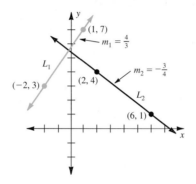

Check Yourself 4

Are lines L_1 through points $(1, 3)$ and $(4, 1)$ and L_2 through points $(-2, 4)$ and $(2, 10)$ perpendicular?

If we already have a line, we can use its slope to determine the slope of other lines that are parallel or perpendicular to the given line.

Example 5

Finding Parallel and Perpendicular Slopes

< Objective 3 >

RECALL

The slope-intercept equation of a line is given by $y = mx + b$, in which m represents the slope and $(0, b)$ is the y-intercept.

(a) Find the slope of all lines parallel to the line given by $5x + y = 1$.

We begin by writing the equation for the given line in slope-intercept form. We do this by isolating y.

$$5x + y = 1$$
$$y = -5x + 1$$

In this form, we see that the slope of this line is -5. Therefore, all lines parallel to the given line have a slope of -5.

(b) Find the slope of all lines perpendicular to the line given by $5x + y = 1$.

From part (a), we know the slope of the given line is -5. Using our property of perpendicular lines, we have

$$m_1 = -5$$

$$m_2 = -\frac{1}{m_1} = -\frac{1}{(-5)} = \frac{1}{5}$$

All lines perpendicular to the given line have a slope of $\frac{1}{5}$.

Check Yourself 5

(a) Find the slope of all lines parallel to the line given by $x - 5y = 3$.
(b) Find the slope of all lines perpendicular to the line given by $x - 5y = 3$.

We can also use the slope-intercept form to determine whether the graphs of given equations are parallel, perpendicular, or neither.

Example 6   Verifying That Two Lines Are Parallel

Show that the graphs of $3x + 2y = 4$ and $6x + 4y = 12$ are parallel lines.
First, we solve each equation for y.

$$3x + 2y = 4$$
$$2y = -3x + 4$$
$$y = -\frac{3}{2}x + 2$$
$$6x + 4y = 12$$
$$4y = -6x + 12$$
$$y = -\frac{3}{2}x + 3$$

NOTE

The slopes are the same, but the y-intercepts are different. Therefore, the lines are distinct.

Because the two lines have the same slope, here $-\frac{3}{2}$, the lines are parallel.

Check Yourself 6

Show that the graphs of the equations

$$-3x + 2y = 4 \quad \text{and} \quad 2x + 3y = 9$$

are perpendicular lines.

Many professions require people to sketch plans of one sort or another. In particular, architects are often required to sketch their designs on a rectangular coordinate system. This makes determining the relationship of adjacent objects such as walls and ceilings a fairly straightforward process, as seen in Example 7.

Example 7   An Engineering Application

Design plans for a project need to be checked by the architect Nicolas. On the sketch, one line passes through the points $(3, 6)$ and $(7, 3)$. A second line also passes through $(7, 3)$, as well as through the point $(13, 11)$. Should Nicolas approve these plans if the two lines are supposed to be perpendicular?

To determine whether the two lines are perpendicular, we compute the slope of each. The slope of the first line is given by

$$m_1 = \frac{\text{change in } y}{\text{change in } x} = \frac{y_2 - y_1}{x_2 - x_1}$$
$$= \frac{3 - 6}{7 - 3} = \frac{-3}{4}$$
$$= -\frac{3}{4}$$

The slope of the second line is found similarly.

$$m_2 = \frac{11 - 3}{13 - 7} = \frac{8}{6}$$

$$= \frac{4}{3}$$

The product of the two slopes is

$$m_1 \cdot m_2 = \left(-\frac{3}{4}\right)\left(\frac{4}{3}\right) = -1$$

so the lines are perpendicular. Nicolas should approve the plans.

Check Yourself 7

On another portion of Nicolas's plans is a line through the points (4, 9) and (6, 8). Is this line parallel to the first line [through the points (3, 6) and (7, 3)]?

Check Yourself ANSWERS

1. (a) and (c)   **2.** The lines intersect.
3. (b) and (d)   **4.** The lines are perpendicular.
5. (a) $\frac{1}{5}$; **(b)** -5

6. $y = \frac{3}{2}x + 2$ **7.** The lines are not parallel.

$$y = -\frac{2}{3}x + 3$$

$$\left(\frac{3}{2}\right)\left(-\frac{2}{3}\right) = -1$$

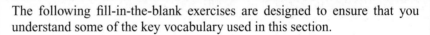

Reading Your Text

The following fill-in-the-blank exercises are designed to ensure that you understand some of the key vocabulary used in this section.

SECTION 7.2

(a) If two lines are _____, they have the same slope.

(b) The point at which two lines meet is called their _____.

(c) When two lines meet at right angles, we say that they are _____.

(d) _____ lines are perpendicular to vertical lines.

Name _____

Section _____ Date _____

Answers

1. _____

2. _____

3. _____

4. _____

5. _____

6. _____

7. _____

8. _____

Basic Skills | Challenge Yourself | Calculator/Computer | Career Applications | Above and Beyond

< Objective 1 >

In exercises 1 to 4, determine which two equations represent parallel lines.

1. (a) $y = -4x + 5$ (b) $y = 4x + 5$

(c) $y = \dfrac{1}{4}x + 5$ (d) $y = -4x + 9$

2. (a) $y = 3x - 5$ (b) $y = -3x + 5$

(c) $y = 3x + 2$ (d) $y = -\dfrac{1}{3}x - 5$

3. (a) $y = \dfrac{2}{3}x + 3$ (b) $y = -\dfrac{3}{2}x - 6$

(c) $y = 4x + 12$ (d) $y = 4x - 3$

 > Videos

4. (a) $y = \dfrac{9}{4}x - 3$ (b) $y = \dfrac{4}{9}x + 7$

(c) $y = \dfrac{4}{9}x - 7$ (d) $y = -\dfrac{9}{4}x + 7$

< Objective 2 >

In exercises 5 to 8, determine which two equations represent perpendicular lines.

5. (a) $y = 6x - 3$ (b) $y = \dfrac{1}{6}x + 3$

(c) $y = -\dfrac{1}{6}x + 3$ (d) $y = \dfrac{1}{6}x - 3$

> Videos

6. (a) $y = \dfrac{2}{3}x - 8$ (b) $y = -\dfrac{2}{3}x - 6$

(c) $y = \dfrac{2}{3}x - 6$ (d) $y = \dfrac{3}{2}x - 6$

7. (a) $y = \dfrac{1}{3}x - 9$ (b) $y = 3x - 9$

(c) $y = \dfrac{1}{3}x + 9$ (d) $y = -\dfrac{1}{3}x + 9$

8. (a) $y = \dfrac{5}{9}x - 6$ (b) $y = 6x - \dfrac{5}{9}$

(c) $y = -\dfrac{1}{6}x + \dfrac{5}{9}$ (d) $y = \dfrac{1}{6}x - \dfrac{5}{9}$

Are the pairs of lines **parallel, perpendicular,** *or* **neither?**

9. L_1 through $(-2, -3)$ and $(4, 3)$
L_2 through $(3, 5)$ and $(5, 7)$

10. L_1 through $(-2, 4)$ and $(1, 8)$
L_2 through $(-1, -1)$ and $(-5, 2)$

11. L_1 through $(8, 5)$ and $(3, -2)$
L_2 through $(-2, 4)$ and $(4, -1)$

12. L_1 through $(-2, -3)$ and $(3, -1)$
L_2 through $(-3, 1)$ and $(7, 5)$

13. L_1 with equation $x - 3y = 6$
L_2 with equation $3x + y = 3$ > Videos

14. L_1 with equation $x + 2y = 4$
L_2 with equation $2x + 4y = 5$

< Objective 3 >

15. Find the slope of any line parallel to the line through points $(-2, 3)$ and $(4, 5)$.

16. Find the slope of any line perpendicular to the line through points $(0, 5)$ and $(-3, -4)$.

17. **CONSTRUCTION** Floor plans for a building have the four corners of a room located at the points $(2, 3)$, $(11, 6)$, $(-3, 18)$, and $(8, 21)$. Determine whether the side through the points $(2, 3)$ and $(11, 6)$ is parallel to the side through the points $(-3, 18)$ and $(8, 21)$.

18. **CONSTRUCTION** For the floor plans given in exercise 17, determine whether the side through the points $(2, 3)$ and $(11, 6)$ is perpendicular to the side through the points $(2, 3)$ and $(-3, 18)$.

19. **CONSTRUCTION** Floor plans for a house have the four corners of a room located at the points $(6, 4)$, $(8, -3)$, $(-1, -6)$, and $(-3, 1)$. Determine whether the side through the points $(-1, -6)$ and $(8, -3)$ is perpendicular to the side through the points $(8, -3)$ and $(6, 4)$.

20. **CONSTRUCTION** For the floor plans given in exercise 19, determine whether the side through the points $(-1, -6)$ and $(8, -3)$ is parallel to the side through the points $(-3, 1)$ and $(6, 4)$.

Answers

9. _____

10. _____

11. _____

12. _____

13. _____

14. _____

15. _____

16. _____

17. _____

18. _____

19. _____

20. _____

Answers

21. _____

22. _____

23. _____

24. _____

25. _____

26. _____

27. _____

28. _____

29. _____

30. _____

31. _____

21. **CONSTRUCTION** Determine whether or not the room described in exercise 17 is a rectangle.

22. **CONSTRUCTION** Determine whether or not the room described in exercise 19 is a rectangle.

Basic Skills | **Challenge Yourself** | Calculator/Computer | Career Applications | Above and Beyond

Determine whether each statement is **true** *or* **false**.

23. Vertical lines are not parallel to each other.

24. If the slope of one line is $\dfrac{c}{d}$, then the slope of a perpendicular line is $-\dfrac{d}{c}$.

Complete each statement with **never, sometimes,** *or* **always.**

25. Two lines that are perpendicular _____ pass through the origin.

26. Parallel lines _____ intersect.

27. A line passing through $(-1, 2)$ and $(4, y)$ is parallel to a line with slope 2. What is the value of y?

 > Videos

28. A line passing through $(2, 3)$ and $(5, y)$ is perpendicular to a line with slope $\dfrac{3}{4}$. What is the value of y?

Use the concept of slope to determine whether the given figure is a parallelogram or a rectangle.

29.

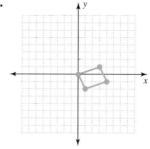

30.

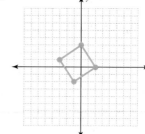

31.

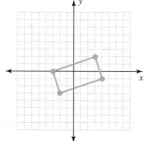

Use the concept of slope to determine whether the given figure is a right triangle (that is, does the triangle contain a right angle?). > Videos

32.

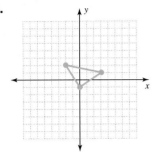

33.

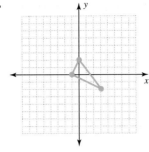

32. _____

33. _____

34. _____

35. _____

36. _____

37. _____

38. _____

34.

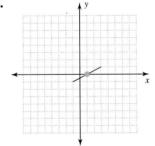

Use the concept of slope to draw a line perpendicular to the given line segment, passing through the marked point.

35.

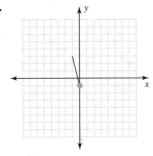

36.

Basic Skills | Challenge Yourself | Calculator/Computer | Career Applications | **Above and Beyond**
▲

Work with a partner to complete exercises 37 and 38.

37. On a piece of graph paper, draw a line segment connecting the points (0, 0) and (5, 2). Describe a procedure for drawing a square, where this line segment is one of the sides. Be sure to identify the other corners of the square, and explain why you believe the resulting figure is a square. Is there more than one way to accomplish this? Compare your solution to that of another group.

38. Follow the directions of exercise 37 using the points (−3, 1) and (2, −7).

Answers

1. a and d   **3.** c and d   **5.** a and c   **7.** b and d   **9.** Parallel

11. Neither   **13.** Perpendicular   **15.** $\dfrac{1}{3}$ **17.** Not parallel

19. Not perpendicular   **21.** Not a rectangle   **23.** False

25. sometimes   **27.** 12   **29.** Parallelogram

31. Rectangle   **33.** Yes   **35.**

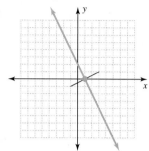

37. Above and Beyond

7.3

< 7.3 Objectives >

The Point-Slope Form

1 > Given a point and the slope, find the equation of a line

2 > Given two points, find the equation of a line

3 > Find the equation of a line from given geometric conditions

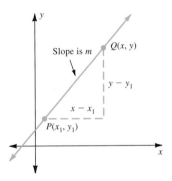

In mathematics it is often useful to be able to write the equation of a line, given its slope and *any* point on the line. In this section, we will derive a third form for the equation of a line for this purpose.

Suppose that a line has slope m and that it passes through the known point $P(x_1, y_1)$. Let $Q(x, y)$ be any other point on the line. Once again we can use the definition of slope and write

$$m = \frac{y - y_1}{x - x_1}$$

Multiplying both sides by $x - x_1$ gives

$$m(x - x_1) = y - y_1$$

or

$$y - y_1 = m(x - x_1)$$

This equation is called the *point-slope form* for the equation of a line, and all points lying on the line [including (x_1, y_1)] satisfy this equation. We can state the general result.

NOTE

We use subscripts (x_1, y_1) to indicate a fixed point on the line.

Property

Point-Slope Form for the Equation of a Line

The equation of a line with slope m that passes through point (x_1, y_1) is given by

$$y - y_1 = m(x - x_1)$$

Example 1 | **Finding the Equation of a Line**

< Objective 1 >

Write the equation for the line that passes through point $(3, -1)$ with a slope of 3.

Letting $(x_1, y_1) = (3, -1)$ and $m = 3$ in point-slope form, we have

$$y - (-1) = 3(x - 3)$$

or

$$y + 1 = 3x - 9$$

We can write the final result in slope-intercept form as

$$y = 3x - 10$$

Check Yourself 1

Write the equation of the line that passes through point $(-2, -4)$ with a slope of $\frac{3}{2}$. Write your result in slope-intercept form.

We know that two points determine a line, so it is natural that we should be able to write the equation of a line passing through two given points. Using the point-slope form together with the slope formula allows us to write such an equation.

 Example 2   **Finding the Equation of a Line**

< Objective 2 >

Write the equation of the line passing through (2, 4) and (4, 7).
    First, we find m, the slope of the line. Here

$$m = \frac{7 - 4}{4 - 2} = \frac{3}{2}$$

Now we apply the point-slope form with $m = \dfrac{3}{2}$ and $(x_1, y_1) = (2, 4)$.

$$y - 4 = \frac{3}{2}(x - 2)$$

$$y - 4 = \frac{3}{2}x - 3$$

$$y = \frac{3}{2}x + 1$$

NOTE

We could just as well have chosen to let
$(x_1, y_1) = (4, 7)$
The resulting equation is the same in either case. Take time to verify this for yourself.

 Check Yourself 2

Write the equation of the line passing through (−2, 5) and (1, 3). Write your result in slope-intercept form.

A line with slope zero is a horizontal line. A line with an undefined slope is vertical. Example 3 illustrates the equations of such lines.

Example 3   **Finding the Equations of Horizontal and Vertical Lines**

< Objective 3 >

(a) Find the equation of a line passing through (7, −2) with a slope of zero.

    We could find the equation by letting $m = 0$. Substituting the ordered pair (7, −2) into the slope-intercept form, we can solve for b.

$$y = mx + b$$
$$-2 = 0(7) + b$$
$$-2 = b$$

So,

$$y = 0 \cdot x - 2 \quad \text{or} \quad y = -2$$

It is far easier to remember that any line with a zero slope is a horizontal line and has the form

$$y = b$$

The value for b is the y-coordinate for the given point.

(b) Find the equation of a line with undefined slope passing through (4, −5).

    A line with undefined slope is vertical. It always has the form $x = a$, in which a is the x-coordinate for the given point. The equation is

$$x = 4$$

RECALL

The equation of a line with undefined slope passing through the point (x_1, y_1) is given by $x = x_1$.

Check Yourself 3

(a) Find the equation of a line with zero slope that passes through point $(-3, 5)$.

(b) Find the equation of a line passing through $(-3, -6)$ with undefined slope.

Alternate methods for finding the equation of a line through two points exist and have particular significance in other fields of mathematics, such as statistics. Example 4 shows such an alternate approach.

 Example 4 | **Finding the Equation of a Line**

Write the equation of the line through points $(-2, 3)$ and $(4, 5)$.

First, we find m, as before.

$$m = \frac{5 - 3}{4 - (-2)} = \frac{2}{6} = \frac{1}{3}$$

We now make use of the slope-intercept equation, but in a slightly different form.

Because $y = mx + b$, we can write

$$b = y - mx$$

NOTE

We substitute these values because the line must pass through $(-2, 3)$.

Now letting $x = -2$, $y = 3$, and $m = \dfrac{1}{3}$, we calculate b.

$$b = 3 - \left(\frac{1}{3}\right)(-2)$$

$$= 3 + \frac{2}{3} = \frac{11}{3}$$

With $m = \dfrac{1}{3}$ and $b = \dfrac{11}{3}$, we can apply the slope-intercept form to write the equation of the desired line. We have

$$y = \frac{1}{3}x + \frac{11}{3}$$

Check Yourself 4

Repeat the Check Yourself 2 exercise, using the technique illustrated in Example 4.

Whichever method we choose, it is an easy matter to check our equation to see if it is correct. In Example 4, we found the equation $y = \dfrac{1}{3}x + \dfrac{11}{3}$. Both points, $(-2, 3)$ and $(4, 5)$, must lie on the line, so each point must satisfy the equation. To check, we can simply substitute an x-value into the equation, and see if we obtain the expected y-value. Using $(-2, 3)$, let $x = -2$:

$$y = \frac{1}{3}(-2) + \frac{11}{3} = -\frac{2}{3} + \frac{11}{3} = \frac{9}{3} = 3 \quad \text{(the expected } y\text{-value)}$$

Using $(4, 5)$, let $x = 4$:

$$y = \frac{1}{3}(4) + \frac{11}{3} = \frac{4}{3} + \frac{11}{3} = \frac{15}{3} = 5 \quad \text{(the expected } y\text{-value)}$$

We now can write the equation of a line given appropriate geometric conditions, such as a point on the line and the slope of that line. In some applications the slope may be given not directly but through specified parallel or perpendicular lines.

 Example 5   **Finding the Equation of a Line**

Find the equation of the line passing through $(-4, -3)$ and parallel to the line determined by $3x + 4y = 12$.

First, we find the slope of the given parallel line.

$$3x + 4y = 12$$
$$4y = -3x + 12$$
$$y = -\frac{3}{4}x + 3$$

NOTE

The slope of the given line is $-\frac{3}{4}$.

The desired line is parallel, so its slope is also $-\frac{3}{4}$. We use the point-slope form to write the required equation.

$$y - (-3) = -\frac{3}{4}[x - (-4)]$$

NOTE

The line must pass through $(-4, -3)$, so let $(x_1, y_1) = (-4, -3)$.

This simplifies to

$$y = -\frac{3}{4}x - 6$$

and we have our equation in slope-intercept form.

 Check Yourself 5

Find the equation of the line passing through $(2, -1)$ and parallel to the line with equation $3x - y = 2$.

Let us look at another example using geometric information. This time we must apply what we know about perpendicular lines.

 Example 6   **Finding the Equation of a Line**

Find the equation of the line passing through $(5, 4)$ and perpendicular to the line with equation $2x - 5y = 10$.

Recall that the slopes of perpendicular lines are negative reciprocals of each other. First we find the slope of the given line.

NOTE

This is true provided that the lines are not horizontal or vertical.

$$2x - 5y = 10 \qquad \text{Solve for } y.$$
$$2x - 10 = 5y$$
$$\frac{2}{5}x - 2 = y \qquad \text{The slope is } \frac{2}{5}.$$

Since the slope of the given line is $\frac{2}{5}$, the slope of the equation we are creating must be $-\frac{5}{2}$. Because the desired line must pass through $(5, 4)$, we may use the point-slope form.

NOTE

We usually try to leave the equation in slope-intercept form.

$$y - 4 = -\frac{5}{2}(x - 5)$$
$$y - 4 = -\frac{5}{2}x + \frac{25}{2}$$
$$y = -\frac{5}{2}x + \frac{33}{2}$$

Check Yourself 6

Find the equation of the line passing through $(-2, 3)$ and perpendicular to the line with equation $4x + y = 7$.

This chart summarizes the various forms of the equation of a line.

Form	Equation for Line L	Conditions
Standard	$Ax + By = C$	Constants A and B cannot both be zero.
Slope-intercept	$y = mx + b$	Line L has y-intercept $(0, b)$ with slope m.
Point-slope	$y - y_1 = m(x - x_1)$	Line L passes through point (x_1, y_1) with slope m.
Horizontal	$y = a$	Slope is zero.
Vertical	$x = b$	Slope is undefined.

In real-world applications, we rarely begin with the equation of a line. Rather, we usually have points relating two variables based on actual data. If we believe that the two variables are linearly related, we can provide a line through the data points. This line can then be used to determine other points and make predictions.

Example 7   A Business and Finance Application

A marketing firm spent \$12,400 in advertisements in January for Alexa's Used Car Emporium. February sales at Alexa's were \$341,000. In August, the firm spent \$8,600 on advertisements, and sales in the following month were \$265,000.

(a) Write a linear equation relating the amount spent on advertisements x with sales in the following month y. Write your answer in slope-intercept form.

We identify the two points $(12,400, 341,000)$ and $(8,600, 265,000)$. The slope of the line through these two points is

$$m = \frac{265,000 - 341,000}{8,600 - 12,400}$$

$$= \frac{-76,000}{-3,800}$$

$$= 20$$

NOTE

This indicates that for each additional dollar spent on advertising, sales increase by \$20 in the following month.

We may choose either point to use with the point-slope formula. Here, we will let

$$(x_1, y_1) = (12,400, 341,000)$$

which gives

$$y - 341,000 = 20(x - 12,400)$$

We solve this equation for y and simplify to write it in slope-intercept form.

$$y - 341,000 = 20(x - 12,400)$$
$$y - 341,000 = 20x - 20 \cdot 12,400$$
$$y - 341,000 = 20x - 248,000$$
$$y = 20x - 248,000 + 341,000$$
$$y = 20x + 93,000$$

(b) Use the equation found in part (a) to predict the sales amount if $10,000 is spent on advertisements in one month.

We evaluate the slope-intercept equation found in part (a) with $x = 10,000$.

$y = 20(10,000) + 93,000$

$\quad = 200,000 + 93,000$

$\quad = 293,000$

So Alexa can assume that if she spends $10,000 on advertisements one month, she will record $293,000 in sales in the following month.

 Check Yourself 7

In 2001, the average cost of tuition and fees at public 4-year colleges was $3,351. By 2005, the average cost had risen to $4,630.

(a) Assuming that the relationship between the cost y and the year x is linear, determine an equation relating the cost to the year (write your answer in slope-intercept form). (*Hint:* If you let x be the number of years since 2000, so that $x = 1$ corresponds to 2001, then $x = 5$ corresponds to 2005.)

(b) Use the equation found in part (a) to predict the average cost of tuition and fees at public 4-year colleges in 2012.

Source: The Chronicle of Higher Education

 Check Yourself ANSWERS

1. $y = \dfrac{3}{2}x - 1$ **2.** $y = -\dfrac{2}{3}x + \dfrac{11}{3}$ **3. (a)** $y = 5$; **(b)** $x = -3$

4. $y = -\dfrac{2}{3}x + \dfrac{11}{3}$ **5.** $y = 3x - 7$ **6.** $y = \dfrac{1}{4}x + \dfrac{7}{2}$

7. (a) $y = 319.75x + 3,031.25$; **(b)** $6,868.25

Reading Your Text

The following fill-in-the-blank exercises are designed to ensure that you understand some of the key vocabulary used in this section.

SECTION 7.3

(a) The equation $y - y_1 = m(x - x_1)$ is called the _____ form for the equation of a line.

(b) A line with slope zero is a _____ line.

(c) A line with undefined slope is _____.

(d) The slopes of perpendicular lines (that are not horizontal or vertical) are negative _____ of each other.

< Objective 1 >

Write the equation of the line passing through each of the given points with the indicated slope. Give your results in slope-intercept form, where possible.

1. $(0, 2)$, $m = 3$

2. $(0, -4)$, $m = -2$

3. $(0, 2)$, $m = \dfrac{3}{2}$

4. $(0, -3)$, $m = -2$

5. $(0, 4)$, $m = 0$

6. $(0, 5)$, $m = -\dfrac{3}{5}$

7. $(0, -5)$, $m = \dfrac{5}{4}$

8. $(0, -4)$, $m = -\dfrac{3}{4}$

9. $(1, 2)$, $m = 3$

10. $(-1, 2)$, $m = 3$

11. $(-2, -3)$, $m = -3$

12. $(1, -4)$, $m = -4$

13. $(5, -3)$, $m = \dfrac{2}{5}$ > Videos

14. $(4, 3)$, $m = 0$

15. $(2, -3)$, m is undefined > Videos

16. $(2, -5)$, $m = \dfrac{1}{4}$

< Objective 2 >

Write the equation of the line passing through each of the given pairs of points. Write your result in slope-intercept form, where possible.

17. $(2, 3)$ and $(5, 6)$

18. $(3, -2)$ and $(6, 4)$

19. $(-2, -3)$ and $(2, 0)$

20. $(-1, 3)$ and $(4, -2)$

21. $(-3, 2)$ and $(4, 2)$

22. $(-5, 3)$ and $(4, 1)$

Name _____

Section _____ Date _____

Answers

1. _____
2. _____
3. _____
4. _____   5. _____
6. _____
7. _____
8. _____
9. _____   10. _____
11. _____
12. _____
13. _____
14. _____   15. _____
16. _____
17. _____   18. _____
19. _____
20. _____   21. _____
22. _____

Answers

23. _____

24. _____

25. _____

26. _____

27. _____

28. _____

29. _____

30. _____

31. _____

32. _____

33. _____

34. _____

35. _____

36. _____

37. _____

23. $(2, 0)$ and $(0, -3)$ > Videos

24. $(2, -3)$ and $(2, 4)$

25. $(0, 4)$ and $(-2, -1)$

26. $(-4, 1)$ and $(3, 1)$

< Objective 3 >

Write the equation of the line L satisfying the given geometric conditions.

27. L has slope 4 and y-intercept $(0, -2)$.

28. L has slope $-\dfrac{2}{3}$ and y-intercept $(0, 4)$.   > Videos

29. L has x-intercept $(4, 0)$ and y-intercept $(0, 2)$.

30. L has x-intercept $(-2, 0)$ and slope $\dfrac{3}{4}$.

31. L has y-intercept $(0, 4)$ and a slope of zero.

32. L has x-intercept $(-2, 0)$ and an undefined slope.

33. L passes through point $(3, 2)$ with a slope of 5.   > Videos

34. L passes through point $(-2, -4)$ with a slope of $-\dfrac{3}{2}$.

35. SCIENCE AND MEDICINE A temperature of $10°C$ corresponds to a temperature of $50°F$. Also, $40°C$ corresponds to $104°F$. Find the linear equation relating F and C.

36. BUSINESS AND FINANCE In planning for the production of a new calculator, a manufacturer assumes that the number of calculators produced x and the cost in dollars C of producing these calculators are related by a linear equation. Projections are that 100 calculators will cost $10,000 to produce and that 300 calculators will cost $22,000 to produce. Find the equation that relates C and x.

37. TECHNOLOGY A word processing station was purchased by a company for $10,000. After 4 years it is estimated that the value of the station will be $4,000. If the value in dollars V and the time the station has been in use t are related by a linear equation, find the equation that relates V and t.

38. BUSINESS AND FINANCE Two years after an expansion, a company had sales of $42,000. Four years later the sales were $102,000. Assuming that the sales in dollars S and the time in years t are related by a linear equation, find the equation relating S and t.

39. BUSINESS AND FINANCE Two years after purchase, a certain car was valued at $31,000. After 5 years, the car was valued at $20,500. Assume that the car depreciates in a linear manner.

  (a) Find the linear equation that relates the car's value V and the number of years t since it was purchased.

  (b) Give the initial purchase price of the car.

  (c) Interpret the slope.

  (d) Predict the value of the car 10 years after purchase.

40. BUSINESS AND FINANCE Three years after purchase, a certain car was valued at $16,200. After 7 years, the car was valued at $13,800. Assume that the car depreciates in a linear manner.

  (a) Find the linear equation that relates the car's value V and the number of years t since it was purchased.

  (b) Give the initial purchase price of the car.

  (c) Interpret the slope.

  (d) Predict the value of the car 10 years after purchase.

Basic Skills	**Challenge Yourself**	Calculator/Computer	Career Applications	Above and Beyond
	▲			

Determine whether each statement is **true** *or* **false.**

41. Given two points, there is exactly one line that will pass through them.

42. Given a line, there is exactly one other line that is perpendicular to it.

Complete each statement with **never, sometimes,** *or* **always.**

43. You can _____ write the equation of a line if you know two points that are on the line.

44. The equation of a line can _____ be put into slope-intercept form.

In exercises 45 to 58, write the equation of the line L that satisfies the given geometric conditions.

45. L has y-intercept $(0, 3)$ and is parallel to the line with equation $y = 3x - 5$.

46. L has y-intercept $(0, -3)$ and is parallel to the line with equation $y = \frac{2}{3}x + 1$.

Answers

38. _____

39. _____

40. _____

41. _____

42. _____

43. _____

44. _____

45. _____

46. _____

Answers

47. _____

48. _____

49. _____

50. _____

51. _____

52. _____

53. _____

54. _____

55. _____

56. _____

57. _____

58. _____

59. _____

60. _____

61. _____

47. L has y-intercept $(0, 4)$ and is perpendicular to the line with equation $y = -2x + 1$.

48. L has y-intercept $(0, 2)$ and is parallel to the line with equation $y = -1$.

49. L has y-intercept $(0, 3)$ and is parallel to the line with equation $y = 2$.

50. L has y-intercept $(0, 2)$ and is perpendicular to the line with equation $2x - 3y = 6$.

51. L passes through point $(-3, 2)$ and is parallel to the line with equation $y = 2x - 3$.

52. L passes through point $(-4, 3)$ and is parallel to the line with equation $y = -2x + 1$.

53. L passes through point $(3, 2)$ and is parallel to the line with equation $y = \dfrac{4}{3}x + 4$.

54. L passes through point $(-2, -1)$ and is perpendicular to the line with equation $y = 3x + 1$.

55. L passes through point $(5, -2)$ and is perpendicular to the line with equation $y = -3x - 2$.

56. L passes through point $(3, 4)$ and is perpendicular to the line with equation $y = -\dfrac{3}{5}x + 2$.

57. L passes through $(-2, 1)$ and is parallel to the line with equation $x + 2y = 4$.

58. L passes through $(-3, 5)$ and is parallel to the x-axis.

Basic Skills | Challenge Yourself | Calculator/Computer | **Career Applications** | Above and Beyond
▲

59. **AUTOMOTIVE TECHNOLOGY** The cost of a 3-hour car repair is $100. The cost of an 8-hour repair is $250. Write a linear equation to determine the price of a repair based on the number of hours.

60. **CONSTRUCTION** A wall that is 28 feet long requires 22 studs, while a wall that is 44 feet long requires 34 studs. Create the equation that relates the number of studs s to the length of the wall L.

61. **MECHANICAL ENGINEERING** A piece of metal at 20°C is 64 mm long. The same piece of metal expands to 67 mm at 220°C. Find the linear equation that relates the length of the piece to the temperature.

62. ALLIED HEALTH The absorbance of a solution, or the amount of light that is absorbed by the solution, is linearly related to the solution's concentration. A standard supply of glucose has a concentration of 200 milligrams per milliliter (mg/mL) and an absorbance value of 0.420, as determined by a spectrophotometer. A second supply of glucose has a concentration of 110 mg/mL and an absorbance value of 0.24. Write an equation relating the absorbance (A) to its concentration (C), in mg/mL.

chapter 7 > Make the Connection

Basic Skills | Challenge Yourself | Calculator/Computer | Career Applications | **Above and Beyond**
▲

63. Describe the process for finding the equation of a line if you are given two points on the line.

64. How would you find the equation of a line if you were given the slope and the *x-intercept*?

Answers

1. $y = 3x + 2$ **3.** $y = \frac{3}{2}x + 2$ **5.** $y = 4$ **7.** $y = \frac{5}{4}x - 5$

9. $y = 3x - 1$ **11.** $y = -3x - 9$ **13.** $y = \frac{2}{5}x - 5$ **15.** $x = 2$

17. $y = x + 1$ **19.** $y = \frac{3}{4}x - \frac{3}{2}$ **21.** $y = 2$ **23.** $y = \frac{3}{2}x - 3$

25. $y = \frac{5}{2}x + 4$ **27.** $y = 4x - 2$ **29.** $y = -\frac{1}{2}x + 2$ **31.** $y = 4$

33. $y = 5x - 13$ **35.** $F = \frac{9}{5}C + 32$ **37.** $V = -1,500t + 10,000$

39. **(a)** $V = -3,500t + 38,000$; **(b)** $38,000; **(c)** For each additional year, the value of the car drops by $3,500; **(d)** $3,000

41. True   **43.** always   **45.** $y = 3x + 3$ **47.** $y = \frac{1}{2}x + 4$

49. $y = 3$ **51.** $y = 2x + 8$ **53.** $y = \frac{4}{3}x - 2$ **55.** $y = \frac{1}{3}x - \frac{11}{3}$

57. $y = -\frac{1}{2}x$ **59.** $C = 30h + 10$ **61.** $L = 0.015T + 63.7$

63. Above and Beyond

7.4

Graphing Linear Inequalities

The Streeter/Hutchison Series in Mathematics   Beginning Algebra

< 7.4 Objectives >

1 > Graph a linear inequality in two variables

2 > Solve an application using a linear inequality in two variables

In Section 2.6 you learned to graph inequalities in one variable on a number line. We now want to extend our work with graphing to include linear inequalities in two variables. We begin with a definition.

Definition	
Linear Inequality in Two Variables	An inequality that can be written in the form $$Ax + By < C$$ in which A and B are not both 0, is called a **linear inequality in two variables**.

NOTE

The inequality symbols $\leq$, $>$, and $\geq$ can also be used.

Some examples of linear inequalities in two variables are

$$x + 3y > 6 \qquad y \leq 3x + 1 \qquad 2x - y \geq 3$$

The *graph* of a linear inequality is always a region (actually a half-plane) of the plane whose boundary is a straight line. Let's look at an example of graphing such an inequality.

 | **Example 1** | Graphing a Linear Inequality

< Objective 1 >

NOTE

The dotted line indicates that the points on the line $2x + y = 4$ are *not* part of the solution to the inequality $2x + y < 4$.

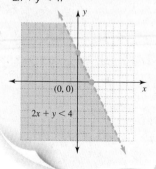

Graph $2x + y < 4$.

First, replace the inequality symbol ($<$) with an equal sign. We then have $2x + y = 4$. This equation forms the **boundary line** of the graph of the original inequality. You can graph the line by any of the methods discussed earlier.

The boundary line for our inequality is shown at the left. We see that the boundary line separates the plane into two regions, each of which is called a **half-plane.** We call $2x + y = 4$ the boundary line because:

1. Any point that is on one side of the boundary is a solution for the inequality. For example, test $(-1, -1)$:

$$2(-1) + (-1) \overset{?}{<} 4$$
$$-3 < 4 \qquad \text{A true statement}$$

2. Any point that is on the other side of the boundary line *fails* to be a solution. For example, test $(3, 3)$:

$$2(3) + (3) \overset{?}{<} 4$$
$$9 < 4 \qquad \text{A false statement}$$

3. Any point that is precisely on the boundary line *fails* in this case, because the symbol of the inequality is $<$. For example, test $(2, 0)$:

$$2(2) + (0) \overset{?}{<} 4$$
$$4 < 4 \qquad \text{A false statement}$$

So, once the boundary line is graphed, we need to choose the correct half-plane. Choose any convenient test point *not* on the boundary line. The origin (0, 0) is a good choice because it results in an easy calculation.

Substitute $x = 0$ and $y = 0$ into the inequality.

NOTE

You can always use the origin for a test point unless the boundary line passes through the origin.

$$2(0) + (0) \overset{?}{<} 4$$
$$0 + 0 \overset{?}{<} 4$$
$$0 < 4 \qquad \text{A true statement}$$

Because the inequality is *true* for the test point, we shade the half-plane containing that test point (the origin). The origin and all other points *below* the boundary line then represent solutions for our original inequality.

 Check Yourself 1

Graph the inequality *x* + 3*y* < 3.

The process is similar when the boundary line is included in the solution.

Example 2	Graphing a Linear Inequality

Graph $4x - 3y \geq 12$.

First, graph the boundary line $4x - 3y = 12$.

NOTE

Again, we replace the inequality symbol ($\geq$) with an equal sign to write the equation for our boundary line.

When equality *is included* ($\leq$ or $\geq$), use a *solid line* for the graph of the boundary line. This means the line is included in the graph of the linear inequality, because every point on the line represents a solution.

The graph of our boundary line (a solid line here) is shown on the figure.

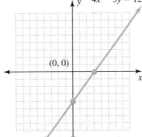

NOTE

Although any of our graphing methods can be used here, the intercept method is probably the most efficient.

Again, we use (0, 0) as a convenient test point. Substituting 0 for *x* and for *y* in the original inequality, we have

$$4(0) - 3(0) \overset{?}{\geq} 12$$
$$0 \geq 12 \qquad \text{A false statement}$$

Because the inequality is *false* for the test point, we shade the half-plane that does *not* contain that test point, here (0, 0).

NOTE

All points *on and below* the boundary line represent solutions for our original inequality.

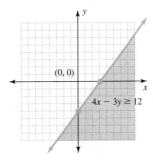

Check Yourself 2

Graph the inequality 3x + 2y ≥ 6.

▶   **Example** 3   Graphing a Linear Inequality with One Variable

Graph $x \leq 5$.

The boundary line is $x = 5$. Its graph is a solid line because equality is included. Using $(0, 0)$ as a test point, we substitute 0 for x with the result

$0 \leq 5$ A true statement

Because the inequality is *true* for the test point, we shade the half-plane containing the origin.

NOTE

If the correct half-plane is obvious, you may not need to use a test point. Did you know without testing which half-plane to shade in this example?

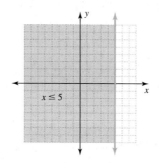

Check Yourself 3

Graph the inequality y < 2.

As we mentioned earlier, we may have to use a point other than the origin as our test point. Example 4 illustrates this approach.

▶   **Example** 4   Graphing a Linear Inequality through the Origin

Graph $2x + 3y < 0$.

The boundary line is $2x + 3y = 0$. Its graph is shown on the figure.

NOTE

We use a dotted line for our boundary line because equality is not included.

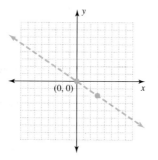

We cannot use $(0, 0)$ as our test point in this case. Do you see why?

Choose any other point *not* on the line. We chose $(1, 1)$ as a test point. Substituting 1 for x and 1 for y gives

$2(1) + 3(1) \overset{?}{<} 0$

$2 + 3 \overset{?}{<} 0$

$5 < 0$ A false statement

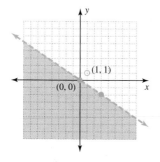

NOTE

We develop the idea of graphing "bounded" regions more fully in Chapter 8.

Because the inequality is *false* at our test point, we shade the half-plane *not* containing (1, 1). This is shown in the graph in the margin.

 Check Yourself 4

Graph the inequality $x - 2y < 0$.

The following steps summarize our work in graphing linear inequalities in two variables.

Step by Step

To Graph a Linear Inequality

Step 1	Replace the inequality symbol with an equal sign to form the equation of the boundary line of the graph.
Step 2	Graph the boundary line. Use a dotted line if equality is not included ($<$ or $>$). Use a solid line if equality is included ($\leq$ or $\geq$).
Step 3	Choose any convenient test point *not* on the line.
Step 4	If the inequality is *true* at the checkpoint, shade the half-plane including the test point. If the inequality is *false* at the checkpoint, shade the half-plane not including the test point.

Linear inequalities and their graphs may be used to represent *feasible regions* in applications. These regions include all points that satisfy some set of conditions determined by the application.

 Example 5   **An Allied Health Application**

< Objective 2 >

RECALL

"At most" means less than or equal to.

A hospital food service can serve at most 1,000 meals per day. Patients on a normal diet receive three meals per day and patients on a special diet receive four meals per day.

(a) Write a linear inequality that describes the number of patients that can be served in a day and sketch its graph.

Let x be the number of people served a normal diet. Then $3x$ represents the number of meals served to the people on a normal diet. Let y be the number of people served a special diet. Then $4y$ represents the number of meals served to the people on the special diet. We know that the total number of meals served is at most 1,000. Writing this as an inequality gives

$$3x + 4y \leq 1,000$$

or

$$y \leq -\frac{3}{4}x + 250 \qquad \text{We solved the inequality for } y.$$

We need to graph this inequality only in the first quadrant. Do you see why?

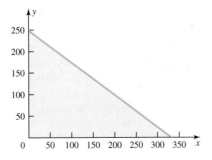

(b) Can the hospital food service serve 100 patients on a normal diet and 100 patients on the special diet in a day?

We can substitute 100 for both x and y in the inequality found in part (a) and see if a true statement results.

$$y \leq -\frac{3}{4}x + 250$$

$$(100) \overset{?}{\leq} -\frac{3}{4}(100) + 250$$

$$100 \overset{?}{\leq} -75 + 250$$

$$100 \leq 175 \qquad \text{True!}$$

Since this final inequality is true, we conclude that the hospital food service can serve 100 people on each type of diet in a single day. Graphically, we see that the point (100, 100) is in the solution region.

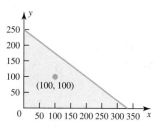

(c) If the hospital serves 200 people on a normal diet, then what is the maximum number of people who can be served the special diet?

Substitute 200 for x and calculate y.

$$y \leq -\frac{3}{4}x + 250$$

$$y \leq -\frac{3}{4}(200) + 250$$

$$y \leq -150 + 250$$

$$y \leq 100$$

We conclude that if the hospital serves 200 meals to patients on a normal diet, they can serve, at most, 100 patients on the special diet.

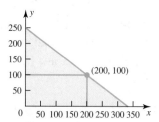

 Check Yourself 5

A manufacturer produces a standard and a deluxe model of 13-in. television sets. The standard model requires 12 h to produce, while the deluxe model requires 18 h to produce. The manufacturer has a total of 360 h of labor available in a week.

(a) Write a linear inequality to represent the number of each type of television set the manufacturer can produce in a week and graph the inequality. (*Hint:* You need to graph the inequality only in the first quadrant.)

(b) If the manufacturer needs to produce 16 standard models one week, what is the largest number of deluxe models that can be produced in that week?

Check Yourself ANSWERS

1.
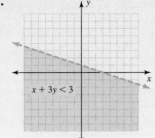

$x + 3y < 3$

2.
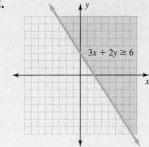

$3x + 2y \geq 6$

3.
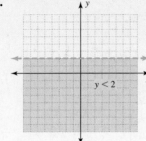

$y < 2$

4.
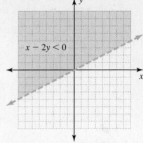

$x - 2y < 0$

5. (a) $12x + 18y \leq 360$ or $y \leq -\dfrac{2}{3}x + 20$; **(b)** nine deluxe models

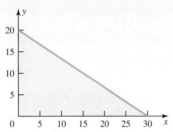

Reading Your Text

The following fill-in-the-blank exercises are designed to ensure that you understand some of the key vocabulary used in this section.

SECTION 7.4

(a) The boundary line separates the plane into two regions, each of which is called a _____.

(b) A _____ boundary line indicates that the points on the line are *not* part of the solutions to the inequality.

(c) The _____ is usually a good choice for a convenient test point.

(d) A _____ region includes all points that satisfy some set of conditions determined by an application.

7.4 exercises

Name _____

Section _____ Date _____

Answers

1. _____

2. _____

3. _____

4. _____

5. _____

6. _____

7. _____

8. _____

< Objective 1 >

In exercises 1 to 8, we have graphed the boundary line for the linear inequality. Determine the correct half-plane in each case, and complete the graph.

1. $x + y < 5$

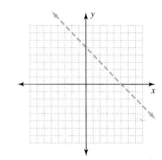

2. $x - y \geq 4$

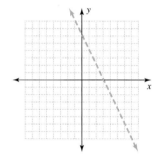

3. $x - 2y \geq 4$

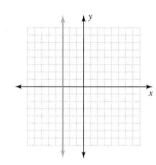

4. $2x + y < 6$

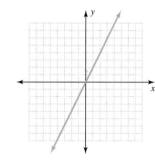

5. $x \leq -3$

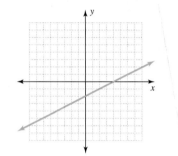

6. $y \geq 2x$

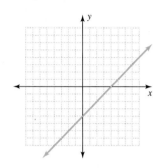

7. $y < 2x - 6$

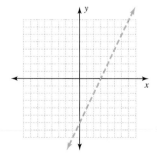

8. $y > 3$

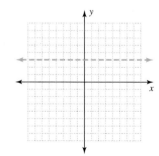

Graph each inequality.

9. $x + y < 3$

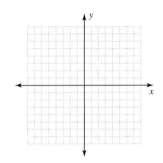

10. $x - y \geq 4$

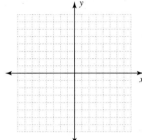

11. $x - y \leq 5$

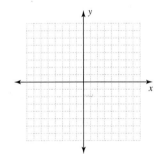

12. $x + y > 5$

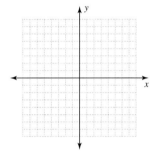

13. $2x + y < 6$

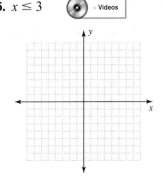

14. $3x + y \geq 6$

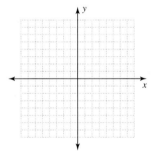

15. $x \leq 3$ 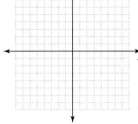 ⊙ > Videos

16. $4x + y \geq 4$

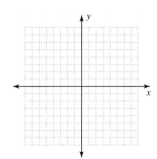

Answers

9. _____

10. _____

11. _____

12. _____

13. _____

14. _____

15. _____

16. _____

Answers

17. _____

18. _____

19. _____

20. _____

21. _____

22. _____

23. _____

24. _____

17. $x - 5y < 5$

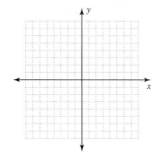

18. $x < 3$

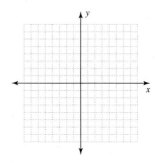

19. $y < -4$

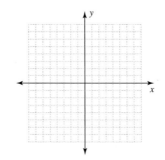

20. $4x + 3y > 12$

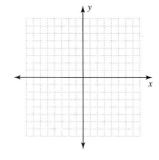

21. $2x - 3y \geq 6$ > Videos

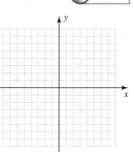

22. $x \geq -2$

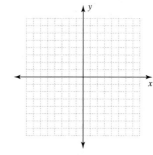

23. $3x + 2y \geq 0$

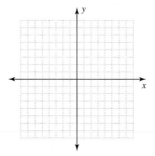

24. $3x + 5y < 15$

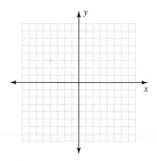

25. $5x + 2y > 10$

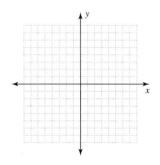

26. $x - 3y \geq 0$

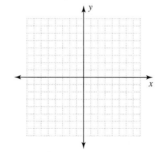

27. $y \leq 2x$ > Videos

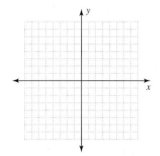

28. $3x - 4y < 12$

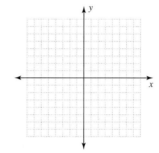

29. $y > 2x - 3$

30. $y \geq -2x$

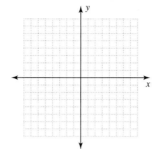

31. $y < -2x - 3$ > Videos

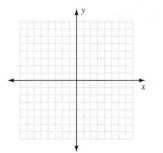

32. $y \leq 3x + 4$

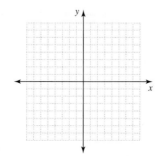

Answers

25. _____

26. _____

27. _____

28. _____

29. _____

30. _____

31. _____

32. _____

33. _____

34. _____

35. _____

36. _____

37. _____

38. _____

39. _____

40. _____

41. _____

| Basic Skills | **Challenge Yourself** | Calculator/Computer | Career Applications | Above and Beyond |

Determine whether each statement is **true** *or* **false.**

33. The boundary line for the solutions to $Ax + By < C$ is always a straight line.

34. The solution region always contains the origin $(0, 0)$.

Complete each statement with **never, sometimes,** *or* **always.**

35. The origin should _____ be used as a test point when finding the region of solutions.

36. The boundary line should _____ be drawn as a dotted line.

Graph each inequality.

37. $2(x + y) - x > 6$

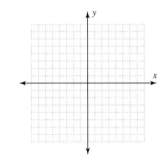

38. $3(x + y) - 2y < 3$

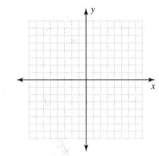

39. $4(x + y) - 3(x + y) \leq 5$

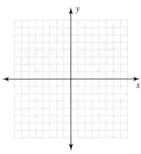

40. $5(2x + y) - 4(2x + y) \geq 4$

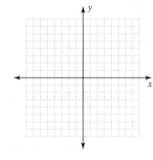

41. **BUSINESS AND FINANCE** Suppose you have two part-time jobs. One is at a video store that pays \$9 per hour and the other is at a convenience store that pays \$8 per hour. Between the two jobs, you want to earn at least \$240 per week. Write an inequality that shows the various number of hours you can work at each job.

chapter 7 > Make the Connection

42. NUMBER PROBLEM You have at least $30 in change in your drawer, consisting of dimes and quarters. Write an inequality that shows the different number of coins in your drawer.

Basic Skills | Challenge Yourself | Calculator/Computer | **Career Applications** | Above and Beyond
▲

< Objective 2 >

43. MANUFACTURING TECHNOLOGY A manufacturer produces 2-slice toasters and 4-slice toasters. The 2-slice type requires 8 hours to produce, and the 4-slice type requires 10 hours to produce. The manufacturer has at most 400 hours of labor available in a week.

chapter 7 > Make the Connection

(a) Write a linear inequality to represent the number of each type of toaster the manufacturer can produce in a week.

(b) Graph the inequality. You need to draw the graph only in the first quadrant.

(c) Is it feasible to produce 20 2-slice toasters and 30 4-slice toasters in a week?

44. MANUFACTURING TECHNOLOGY A certain company produces standard clock radios and deluxe clock radios. It costs the company $15 to produce each standard clock radio and $20 to produce a deluxe model. The company's budget limits production costs to a maximum of $3,000 per day.

chapter 7 > Make the Connection

(a) Write a linear inequality to represent the number of each type of clock radio the company can produce in a day.

(b) Graph the inequality. You need to draw the graph only in the first quadrant.

(c) Is it feasible to produce 80 of each type of clock radio in a day?

Basic Skills | Challenge Yourself | Calculator/Computer | Career Applications | **Above and Beyond**
▲

45. BUSINESS AND FINANCE Linda Williams has just begun a nursery business and seeks your advice. She has limited funds to spend and wants to stock two kinds of fruit-bearing plants. She lives in the northeastern part of

Answers

42.

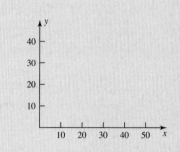

43.

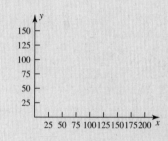

44.

45.

Answers

46. _____

Texas and thinks that blueberry bushes and peach trees would sell well there. Linda can buy blueberry bushes from a supplier for $2.50 each and young peach trees for $5.50 each. She wants to know what combinations she can buy to keep her outlay to $500 or less. Write an inequality and draw a graph to depict what combinations of blueberry bushes and peach trees she can buy for the amount of money she has. Explain the graph and her options.

46. **STATISTICS** After reading an article on the front page of *The New York Times* titled "You Have to Be Good at Algebra to Figure Out the Best Deal for Long Distance," Rafaella De La Cruz decided to apply her skills in algebra to try to decide between two competing long-distance companies. It was difficult at first to get the companies to explain their charge policies. They both kept repeating that they were 25% cheaper than their competition. Finally, Rafaella found someone who explained that the charge depended on when she called, where she called, how long she talked, and how often she called. "Too many variables!" she exclaimed. So she decided to ask one company what they charged as a base amount, just for using the service.

Company A said that they charged $5 for the privilege of using their long-distance service, whether or not she made any phone calls, and that because of this fee they were able to allow her to call anywhere in the United States after 6 P.M. for only $0.15 a minute. Complete this table of charges based on this company's plan:

Total Minutes Long Distance in 1 Month (After 6 P.M.)	Total Charge
0 minutes	
10 minutes	
30 minutes	
60 minutes	
120 minutes	

Use this table to make a graph of the monthly charges from Company A based on the number of minutes of long distance.

Rafaella wanted to compare this offer to Company B, which she was currently using. She looked at her phone bill and saw that one month she had been charged $7.50 for 30 minutes and another month she had been charged $11.25 for 45 minutes of long-distance calling. These calls were made after 6 P.M. to her relatives in Indiana and Arizona. Draw a graph on the same set of axes you made for Company A's figures. Use your graph and what you know about linear inequalities to advise Rafaella about which company is better.

Answers

1. $x + y < 5$

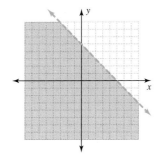

3. $x - 2y \geq 4$

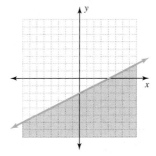

5. $x \leq -3$

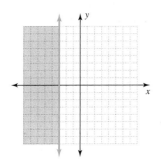

7. $y < 2x - 6$

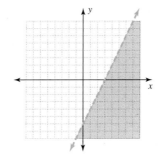

9. $x + y < 3$

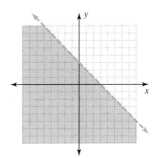

11. $x - y \leq 5$

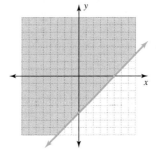

13. $2x + y < 6$

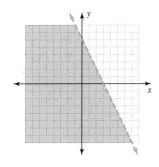

15. $x \leq 3$

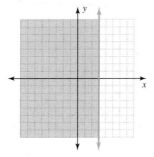

17. $x - 5y < 5$

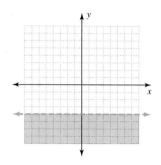

19. $y < -4$

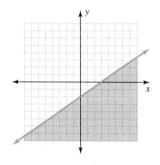

21. $2x - 3y \geq 6$

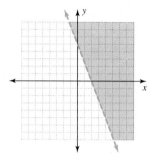

23. $3x + 2y \geq 0$

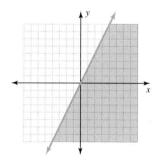

25. $5x + 2y > 10$

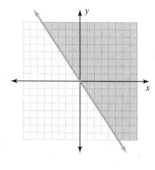

27. $y \leq 2x$

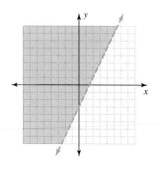

29. $y > 2x - 3$

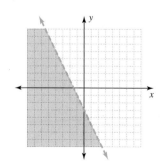

31. $y < -2x - 3$

33. True

35. sometimes

37. $x + 2y > 6$

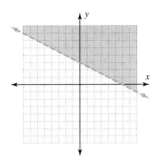

39. $x + y \leq 5$

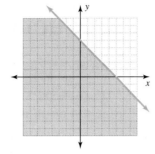

41. $9x + 8y \geq 240$

43. (a) $8x + 10y \leq 400$;

(b)

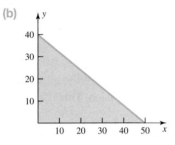

(c) no

45. Above and Beyond

7.5

An Introduction to Functions

< 7.5 Objectives >

1 > Evaluate an expression

2 > Evaluate a function

3 > Express the equation of a line as a linear function

4 > Graph a linear function

Variables can be used to represent unknown real numbers. Together with the operations of addition, subtraction, multiplication, division, and exponentiation, numbers and variables form expressions such as

$$3x + 5 \qquad 7x - 4 \qquad x^2 - 3x - 10 \qquad x^4 - 2x^2 + 3x + 4$$

Four different actions can be taken with expressions. We can

1. Substitute values for the variable(s) and **evaluate the expression.**
2. Rewrite an expression as some simpler equivalent expression. This is called **simplifying the expression.**
3. Set two expressions equal to each other and **solve for the stated variable.**
4. Set two expressions equal to each other and **graph the equation.**

Throughout this book, everything we do involves one of these four actions. We now return our focus to the first item, evaluating expressions. As we saw in Section 1.5, expressions can be evaluated for an indicated value of the variable(s). Example 1 illustrates.

Example 1 | **Evaluating Expressions**

< Objective 1 >

Evaluate the expression $x^4 - 2x^2 + 3x + 4$ for the indicated value of x.

(a) $x = 0$

Substituting 0 for x in the expression yields

$$(0)^4 - 2(0)^2 + 3(0) + 4 = 0 - 0 + 0 + 4$$
$$= 4$$

(b) $x = 2$

Substituting 2 for x in the expression yields

$$(2)^4 - 2(2)^2 + 3(2) + 4 = 16 - 8 + 6 + 4$$
$$= 18$$

(c) $x = -1$

Substituting -1 for x in the expression yields

$$(-1)^4 - 2(-1)^2 + 3(-1) + 4 = 1 - 2 - 3 + 4$$
$$= 0$$

Check Yourself 1

Evaluate the expression $2x^3 - 3x^2 + 3x + 1$ for the indicated value of x.

(a) $x = 0$ (b) $x = 1$ (c) $x = -2$

We could design a machine whose function would be to crank out the value of an expression for each given value of x. We could call this machine something simple such as f. Our *function* machine might look like this.

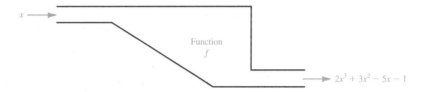

NOTE

-1 is called the *input value*, 5 is called the *output value*.

For example, when we put -1 into the machine, the machine substitutes -1 for x in the expression, and 5 comes out the other end because

$$2(-1)^3 + 3(-1)^2 - 5(-1) - 1 = -2 + 3 + 5 - 1 = 5$$

In fact, the idea of the function machine is very useful in mathematics. Your graphing calculator can be used as a function machine. You can enter the expression into the calculator as Y_1 and then evaluate Y_1 for different values of x.

Generally, in mathematics, we do not write $Y_1 = 2x^3 + 3x^2 - 5x - 1$. Instead, we write $f(x) = 2x^3 + 3x^2 - 5x - 1$, which is read as "$f$ of x is equal to. . . ." Instead of calling f a function machine, we say that f is a function of x. The greatest benefit to this notation is that it lets us easily note the input value of x along with the output of the function. Instead of "Evaluate y for $x = 4$" we say "Find $f(4)$." This means that, given the function f, $f(c)$ designates the value of the function when the variable is equal to c.

 > CAUTION

$f(x)$ does not mean f times x.

 **Example 2**   **Evaluating a Function for Different Inputs**

< Objective 2 >

Evaluate $f(x) = x^3 - 3x^2 + x + 5$ for each input.

(a) $f(0)$

Substituting 0 for x in the expression, we get

$$f(0) = (0)^3 - 3(0)^2 + (0) + 5$$
$$= 5$$

(b) $f(-3)$

Substituting -3 for x in the expression, we get

$$f(-3) = (-3)^3 - 3(-3)^2 + (-3) + 5$$
$$= -27 - 27 - 3 + 5$$
$$= -52$$

(c) $f\left(\dfrac{1}{2}\right)$

Substituting $\dfrac{1}{2}$ for x in the expression, we get

$$f\left(\frac{1}{2}\right) = \left(\frac{1}{2}\right)^3 - 3\left(\frac{1}{2}\right)^2 + \left(\frac{1}{2}\right) + 5$$

$$= \frac{1}{8} - 3\left(\frac{1}{4}\right) + \frac{1}{2} + 5$$

$$= \frac{1}{8} - \frac{3}{4} + \frac{1}{2} + 5$$

$$= \frac{1}{8} - \frac{6}{8} + \frac{4}{8} + \frac{40}{8}$$

$$= \frac{39}{8}$$

Check Yourself 2

Evaluate $f(x) = 2x^3 - x^2 + 3x - 2$ for each input.

(a) $f(0)$ (b) $f(3)$ (c) $f\left(-\dfrac{1}{2}\right)$

In Example 2 we obtained $f(0) = 5$. We can describe this by saying that when 0 is the input, 5 is the output. In part **(b),** we found $f(-3) = -52$; that is, when the input is -3, the output is -52. In general, the input value is x and the output value is $f(x)$.

Given a function f, the pair of numbers $(x, f(x))$ is very significant. We always write them in that order. The ordered pair consists of the x-value first and the function value at that x (the $f(x)$) second.

Example 3   **Finding Ordered Pairs**

Given the function $f(x) = 2x^2 - 3x + 5$, find the ordered pair $(x, f(x))$ associated with each given value for x.

NOTE

Think of the ordered pairs of a function as (input, output).

(a) $x = 0$

$f(0) = 5$

so the ordered pair is $(0, 5)$.

(b) $x = -1$

$f(-1) = 2(-1)^2 - 3(-1) + 5 = 10$

The ordered pair is $(-1, 10)$.

(c) $x = \dfrac{1}{4}$

$$f\left(\frac{1}{4}\right) = 2\left(\frac{1}{16}\right) - 3\left(\frac{1}{4}\right) + 5 = \frac{35}{8}$$

The ordered pair is $\left(\dfrac{1}{4}, \dfrac{35}{8}\right)$.

Check Yourself 3

Given $f(x) = 2x^3 - x^2 + 3x - 2$, find the ordered pair associated with each given value of x.

(a) $x = 0$ (b) $x = 3$ (c) $x = -\dfrac{1}{2}$

Any linear equation $Ax + By = C$ in which $B \neq 0$ can be written using function notation. Because the process for finding a value for y given a particular value for x is the same as finding $f(x)$ given some value for x, we know that both y and $f(x)$ always yield the same number given some value for x.

Consider the equation $y = x + 1$. If $x = 1$, then we determine y by substituting 1 for x in the equation:

$$y = (1) + 1 = 2$$

We then express this using ordered-pair notation $(x, y) = (1, 2)$ as a solution to the equation $y = x + 1$.

If we write the function $f(x) = x + 1$ and evaluate $f(1)$, we get

$$f(1) = (1) + 1 = 2$$

So, $(x, f(x)) = (1, 2)$ is an ordered pair associated with the function.

This leads us to the conclusion that $y = f(x)$ represents the relationship between function notation and equations in two variables.

To write an equation in two variables, x and y, as a function, follow these two steps.

1. Solve the equation for y, if possible.

2. Replace y with $f(x)$.

If we are unable to complete step 1, then we cannot write y as a function of x. With linear equations, $Ax + By = C$, this would happen if $B = 0$. That is, the equation for a vertical line $x = a$ does not represent a function.

On the other hand, *all* nonvertical lines may be written as functions.

| Example 4 | Writing Equations as Functions |

< Objective 3 >

Rewrite each linear equation as a function of x.

(a) $y = 3x - 4$

Because the equation is already solved for y, we simply replace y with $f(x)$.

$$f(x) = 3x - 4$$

(b) $2x - 3y = 6$

This equation must first be solved for y.

$$-3y = -2x + 6$$

$$y = \frac{2}{3}x - 2$$

We then rewrite the equation as

$$f(x) = \frac{2}{3}x - 2$$

Check Yourself 4

Rewrite each equation as a function of x.

(a) $y = -2x + 5$ (b) $3x + 5y = 15$

The process of finding the graph of a linear function is identical to the process of finding the graph of a linear equation.

  Example 5 Graphing a Linear Function

< Objective 4 >

Graph the function

$$f(x) = 3x - 5$$

We could use the slope and y-intercept to graph the line, or we can find three points (the third is a check-point) and draw the line through them. We will do the latter.

$$f(0) = -5 \qquad f(1) = -2 \qquad f(2) = 1$$

We use the three points $(0, -5)$, $(1, -2)$, and $(2, 1)$ to graph the line.

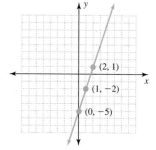

 Check Yourself 5

Graph the function

$f(x) = 5x - 3$

One benefit of having a function written in $f(x)$ form is that it makes it fairly easy to substitute values for x. In Example 5, we substituted the values 0, 1, and 2. Sometimes it is useful to substitute nonnumeric values for x.

 Example 6 Substituting Nonnumeric Values for x

Let $f(x) = 2x + 3$. Evaluate f as indicated.

(a) $f(a)$

Substituting a for x in our equation, we see that

$$f(a) = 2a + 3$$

(b) $f(2 + h)$

Substituting $2 + h$ for x in our equation, we get

$$f(2 + h) = 2(2 + h) + 3$$

Distributing the 2 and then simplifying, we have

$$f(2 + h) = 4 + 2h + 3$$
$$= 2h + 7$$

 Check Yourself 6

Let $f(x) = 4x - 2$. Evaluate f as indicated.

(a) $f(b)$ (b) $f(4 + h)$

Functions and function notation are used in many applications, as illustrated by Example 7.

Example 7 | **Applications of Functions**

NOTE

f is one name choice for a function. In applications, functions are often named in some way that relates to their usage.

The profit function for stereos sold by S-Bar Electronics is given by

$$P(x) = 34x - 3,500$$

in which x is the number of stereos sold and $P(x)$ is the profit produced by selling x stereos.

(a) Determine the profit if S-Bar Electronics sells 75 stereos.

$$P(75) = 34(75) - 3,500$$
$$= 2,550 - 3,500$$
$$= -950$$

Because the profit is negative, we determine that S-Bar Electronics suffers a *loss* of $950 if it sells only 75 stereos.

(b) Determine the profit from the sale of 150 stereos.

$$P(150) = 34(150) - 3,500$$
$$= 5,100 - 3,500$$
$$= 1,600$$

S-Bar Electronics earns a profit of $1,600 when it sells 150 stereos.

(c) Find the break-even point (that is, the number of stereos that must be sold for the profit to equal zero).

We set $P(x) = 0$ and solve for x. That is, we are trying to find x so that the profit is zero.

$$0 = 34x - 3,500$$
$$3,500 = 34x$$
$$x = \frac{3,500}{34} \approx 103$$

The break-even point for the profit equation is 103 stereos.

NOTE

If they sell 103 stereos, they earn a profit of $2; however, if they sell 102 stereos, they lose $32. Since the number of stereos sold must be a whole number, we set the break-even point to 103.

 Check Yourself 7

The manager at Sweet Eats determines that the profit produced by serving *x* dinners is given by the function

P(x) = 42x − 620

(a) Determine the profit if the restaurant serves 20 dinners.
(b) How many dinners must be served to reach the break-even point?

Here is an application from the field of electronics.

 Example 8 | **An Engineering Application**

A temperature sensor outputs a voltage for a given temperature in degrees Celsius. The output voltage is linearly related to the temperature. The function's rule for this sensor is

$$f(x) = 0.28x + 2.2$$

(a) Find $f(0)$ and interpret this result.

$$f(0) = 0.28(0) + 2.2 = 2.2$$

This says that when the temperature is 0°C, the voltage is 2.2 volts.

(b) Find $f(22)$ and interpret this result.

$$f(22) = 0.28(22) + 2.2 = 8.36$$

This says that when the temperature is 22°C, the voltage is 8.36 volts.

Check Yourself 8

The horsepower of an engine is a function of the torque (in foot-pounds) multiplied by the rpms x, multiplied by 2π, and then divided by 33,000. Given that the torque of an engine is 324 foot-pounds, the rule for the horsepower function is

$$h(x) = \frac{(2\pi)(324)x}{33,000}$$

Find and interpret $h(3,000)$.

Check Yourself ANSWERS

1. **(a)** 1; **(b)** 3; **(c)** −33   2. **(a)** −2; **(b)** 52; **(c)** −4   3. **(a)** (0, −2);

(b) (3, 52); **(c)** $\left(-\frac{1}{2}, -4\right)$ 4. **(a)** $f(x) = -2x + 5$; **(b)** $f(x) = -\frac{3}{5}x + 3$

5.

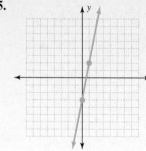

6. **(a)** $4b - 2$; **(b)** $4h + 14$ 7. **(a)** \$220; **(b)** 15   8. $h(3,000) \approx 185.07$;
At 3,000 rpm the horsepower is 185.

Reading Your Text

The following fill-in-the-blank exercises are designed to ensure that you understand some of the key vocabulary used in this section.

SECTION 7.5

(a) When we substitute values for the variable(s) in an expression, we _____ the expression.

(b) Rewriting an expression as some simpler equivalent expression is called _____ the expression.

(c) Function notation allows us to note the input value of x along with the _____ value of the function.

(d) In a "profit" application, the _____ point is the minimum number of items x that must be sold for the profit to be nonnegative.

Basic Skills | Challenge Yourself | Calculator/Computer | Career Applications | Above and Beyond

< Objectives 1–2 >

In exercises 1 to 10, evaluate each function for the value specified.

1. $f(x) = x^2 - x - 2$; find **(a)** $f(0)$, **(b)** $f(-2)$, and **(c)** $f(1)$.

2. $f(x) = x^2 - 7x + 10$; find **(a)** $f(0)$, **(b)** $f(5)$, and **(c)** $f(-2)$.

3. $f(x) = 3x^2 + x - 1$; find **(a)** $f(-2)$, **(b)** $f(0)$, and **(c)** $f(1)$.

4. $f(x) = -x^2 - x - 2$; find **(a)** $f(-1)$, **(b)** $f(0)$, and **(c)** $f(2)$.

5. $f(x) = x^3 - 2x^2 + 5x - 2$; find **(a)** $f(-3)$, **(b)** $f(0)$, and **(c)** $f(1)$.

6. $f(x) = -2x^3 + 5x^2 - x - 1$; find **(a)** $f(-1)$, **(b)** $f(0)$, and **(c)** $f(2)$.

7. $f(x) = -3x^3 + 2x^2 - 5x + 3$; find **(a)** $f(-2)$, **(b)** $f(0)$, and **(c)** $f(3)$.

 > Videos

8. $f(x) = -x^3 + 5x^2 - 7x - 8$; find **(a)** $f(-3)$, **(b)** $f(0)$, and **(c)** $f(2)$.

< Objective 3 >

In exercises 9 to 16, rewrite each equation as a function of x.

9. $y = -3x + 2$

10. $y = 5x + 7$

11. $3x + 2y = 6$

12. $4x + 3y = 12$

13. $-2x + 6y = 9$

14. $-3x + 4y = 11$

15. $-5x - 8y = -9$ > Videos

16. $4x - 7y = -10$

Boost *your* GRADE at ALEKS.com!

ALEKS®

- Practice Problems
- Self-Tests
- NetTutor
- e-Professors
- Videos

Name _____

Section _____ Date _____

Answers

1. _____
2. _____
3. _____
4. _____
5. _____
6. _____
7. _____
8. _____
9. _____
10. _____
11. _____
12. _____
13. _____
14. _____
15. _____
16. _____

7.5 exercises

Answers

17. _____

18. _____

19. _____

20. _____

21. _____

22. _____

23. _____

24. _____

< Objective 4 >

In exercises 17 to 20, graph the functions.

17. $f(x) = 3x + 7$

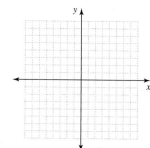

18. $f(x) = -2x - 5$

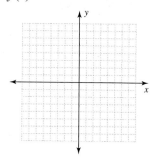

19. $f(x) = -2x + 7$ > Videos

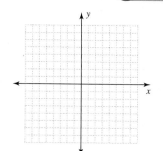

20. $f(x) = -3x + 8$

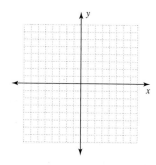

21. BUSINESS AND FINANCE The inventor of a new product believes that the cost of producing the product is given by the function

$$C(x) = 1.75x + 7{,}000$$

How much does it cost to produce 2,000 units of her invention?

22. BUSINESS AND FINANCE The inventor of a new product believes that the cost of producing the product is given by the function

$$C(x) = 3.25x + 9{,}000$$

How much does it cost to produce 5,000 units of his invention?

23. BUSINESS AND FINANCE If the inventor in exercise 21 charges $4 per unit, then her profit for producing and selling x units is given by the function

$$P(x) = 2.25x - 7{,}000$$

(a) What is her profit if she sells 2,000 units?

(b) What is her profit if she sells 5,000 units?

(c) What is the break-even point for sales?

24. BUSINESS AND FINANCE If the inventor in exercise 22 charges $8 per unit, then his profit for producing and selling x units is given by the function

$$P(x) = 4.75x - 9{,}000$$

(a) What is his profit if he sells 6,000 units?

(b) What is his profit if he sells 9,000 units?

(c) What is the break-even point for sales?

25. **SCIENCE AND MEDICINE** The height of a ball thrown in the air is given by the function

$$h(t) = -16t^2 + 80t$$

where t is the number of seconds that elapse after the ball is thrown.

(a) Find $h(2)$.

(b) Find $h(2.5)$.

(c) Find $h(5)$. What has happened here?

26. **SCIENCE AND MEDICINE** The height of a ball dropped from a roof 144 ft above the ground is given by the function

$$h(t) = -16t^2 + 144$$

where t is the number of seconds that elapse after the ball is dropped.

(a) Find $h(2)$.

(b) Find $h(2.5)$.

(c) Find $h(3)$. What has happened here?

| Basic Skills | **Challenge Yourself** | Calculator/Computer | Career Applications | Above and Beyond |

Determine whether each statement is **true** *or* **false**.

27. The equation for a vertical line could represent a function.

28. The equation for a horizontal line could represent a function.

Complete each statement with **never, sometimes,** *or* **always**.

29. It is _____ possible to use 0 as an input value for a function.

30. The break-even point for a product _____ occurs at the minimum number of units necessary for the profit function to be nonnegative.

In exercises 31 to 36, if $f(x) = 5x - 1$, *find*

31. $f(a)$

32. $f(2r)$

33. $f(x + 1)$ > Videos

34. $f(a - 2)$

35. $f(x + h)$

36. $\dfrac{f(x + h) - f(x)}{h}$

In exercises 37 to 40, if $g(x) = -3x + 2$, *find*

37. $g(m)$

38. $g(5n)$

39. $g(x + 2)$

40. $g(s - 1)$

In exercises 41 to 44, let $f(x) = 2x + 3$.

41. Find $f(1)$.

42. Find $f(3)$.

Answers

25. _____

26. _____

27. _____

28. _____

29. _____

30. _____

31. _____

32. _____

33. _____

34. _____

35. _____

36. _____

37. _____

38. _____

39. _____

40. _____

41. _____

42. _____

Answers

43. _____

44. _____

45. _____

46. _____

47. _____

48. _____

49. _____

50. _____

43. Form the ordered pairs $(1, f(1))$ and $(3, f(3))$.

44. Write the equation of the line passing through the points determined by the ordered pairs in exercise 43.

Basic Skills | Challenge Yourself | Calculator/Computer | **Career Applications** | Above and Beyond
▲

45. CONSTRUCTION TECHNOLOGY The cost of building a house is $90 per square foot plus $12,000 for the foundation.

    (a) Create a function to represent the cost of building a house and its relation to the number of square feet.

    (b) What is the cost to build a house that has 1,800 square feet?

46. AUTOMOTIVE TECHNOLOGY A car engine burns 3.4 ounces of fuel per minute plus an additional 2 ounces during start-up.

    (a) Write a function for the fuel consumption of a car as a function of the running time t, in minutes.

    (b) How much fuel does a car consume if it runs for 20 minutes?

47. MECHANICAL ENGINEERING The time it takes a welder to fill an order is a function of 2 minutes for each piece plus 23 minutes for setup.

    (a) Express this function, where p is the number of pieces.

    (b) How long would it take to complete an order for 120 parts?

48. ALLIED HEALTH Dimercaprol (BAL) is used to treat arsenic poisoning in mammals. The recommended dose is 4 milligrams (mg) per kilogram (kg) of the animal's weight.

    (a) Write a function describing the relationship between the recommended dose, in milligrams, and the animal's weight, in kilograms.

    (b) How much BAL must be administered to a 5-kg cat?

49. ALLIED HEALTH Yohimbine is used to reverse the effects of xylazine in deer. The recommended dose is 0.125 milligram (mg) per kilogram (kg) of the deer's weight.

    (a) Write a function describing the relationship between the recommended dose, in milligrams, and the deer's weight, in kilograms.

    (b) How much yohimbine must be administered to a 15-kg fawn?

50. ALLIED HEALTH A brain tumor originally weighs 41 grams (g). Every day of chemotherapy treatment reduces the weight of the tumor by 0.83 g.

    (a) Write a function describing the relationship between the weight of the tumor, in grams, and the number of days spent in chemotherapy.

    (b) How much does the tumor weigh after 2 weeks of treatment?

Basic Skills | Challenge Yourself | Calculator/Computer | Career Applications | **Above and Beyond**

Answers

51. Let $f(x) = 5x - 2$. Find **(a)** $f(4) - f(3)$; **(b)** $f(9) - f(8)$; **(c)** $f(12) - f(11)$.
(d) How do the results of parts (a) through (c) compare to the slope of the line that is the graph of f?

52. Repeat exercise 51 with $f(x) = 7x + 1$.

53. Repeat exercise 51 with $f(x) = mx + b$.

54. Based on your work in exercises 51 through 53, write a paragraph discussing how slope relates to function notation.

Answers

51. _____

52. _____

53. _____

54. _____

Answers

1. **(a)** -2; **(b)** 4; **(c)** -2 **3.** **(a)** 9; **(b)** -1; **(c)** 3
5. **(a)** -62; **(b)** -2; **(c)** 2 **7.** **(a)** 45; **(b)** 3; **(c)** -75

9. $f(x) = -3x + 2$ **11.** $f(x) = -\dfrac{3}{2}x + 3$ **13.** $f(x) = \dfrac{1}{3}x + \dfrac{3}{2}$

15. $f(x) = -\dfrac{5}{8}x + \dfrac{9}{8}$

17. $f(x) = 3x + 7$ **19.** $f(x) = -2x + 7$

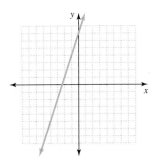   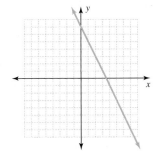

21. $\$10,500$ **23.** **(a)** Loss: $\$2,500$; **(b)** profit: $\$4,250$; **(c)** $3,112$ units
25. **(a)** 96; **(b)** 100; **(c)** 0; The ball hits the ground at 5 s.
27. False   **29.** sometimes   **31.** $5a - 1$ **33.** $5x + 4$
35. $5x + 5h - 1$ **37.** $-3m + 2$ **39.** $-3x - 4$ **41.** 5
43. $(1, 5), (3, 9)$ **45.** **(a)** $f(x) = 90x + 12,000$; **(b)** $\$174,000$
47. **(a)** $f(p) = 2p + 23$; **(b)** 263 min   **49.** **(a)** $f(x) = 0.125x$; **(b)** 1.875 mg
51. **(a)** 5; **(b)** 5; **(c)** 5; **(d)** same   **53.** **(a)** m; **(b)** m; **(c)** m; **(d)** same

Definition/Procedure	Example	Reference

The Slope-Intercept Form

The slope-intercept form for the equation of a line is

$y = mx + b$

in which the line has slope m and y-intercept $(0, b)$.

For the equation $y = \dfrac{2}{3}x - 3$,

the slope m is $\dfrac{2}{3}$ and b, which

determines the y-intercept, is -3.

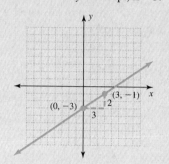

Section 7.1

p. 525

Parallel and Perpendicular Lines

Two nonvertical lines are parallel if and only if they have the same slope, that is, when

$m_1 = m_2$

$y = 3x - 5$ and
$y = 3x + 2$ are parallel.

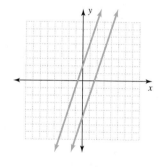

Section 7.2

p. 543

Two lines are perpendicular if and only if their slopes are negative reciprocals, that is, when

$m_1 \cdot m_2 = -1$

$y = 5x + 2$ and $y = -\dfrac{1}{5}x - 3$
are perpendicular.

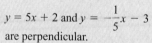
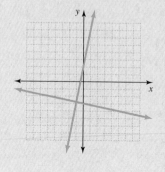

p. 544

Definition/Procedure	Example	Reference

The Point-Slope Form

The equation of a line with slope m that passes through the point (x_1, y_1) is

$$y - y_1 = m(x - x_1)$$

The line with slope $\frac{1}{3}$ passing through $(4, 3)$ has the equation

$$y - 3 = \frac{1}{3}(x - 4)$$

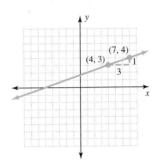

Section 7.3

p. 553

Graphing Linear Inequalities

The Graphing Steps

1. Replace the inequality symbol with an equal sign to form the equation of the boundary line of the graph.
2. Graph the boundary line. Use a dotted line if equality is not included ($<$ or $>$). Use a solid line if equality is included ($\leq$ or $\geq$).
3. Choose any convenient test point *not* on the line.
4. If the inequality is *true* at the checkpoint, shade the half-plane including the test point. If the inequality is *false* at the checkpoint, shade the half-plane that does not include the checkpoint.

To graph $x - 2y < 4$:
$x - 2y = 4$ is the boundary line.
Using $(0, 0)$ as the checkpoint, we have

$$(0) - 2(0) \overset{?}{<} 4$$
$$0 < 4 \qquad \text{(True)}$$

Shade the half-plane that includes $(0, 0)$.

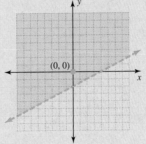

Section 7.4

p. 567

An Introduction to Functions

Given a function f, $f(c)$ designates the value of the function when the variable is equal to c.

$$f(x) = 2x^3 - x^2 + 1$$
$$f(-2) = 2(-2)^3 - (-2)^2 + 1$$
$$= 2(-8) - (4) + 1$$
$$= -19$$

Section 7.5

p. 581

This summary exercise set is provided to give you practice with each of the objectives of this chapter. Each exercise is keyed to the appropriate chapter section. When you are finished, you can check your answers to the odd-numbered exercises against those presented in the back of the text. If you have difficulty with any of these questions, go back and reread the examples from that section. Your instructor will give you guidelines on how best to use these exercises in your instructional setting.

7.1 *Find the slope and y-intercept of the line represented by each equation.*

1. $y = 2x + 5$

2. $y = -4x - 3$

3. $y = -\dfrac{3}{4}x$

4. $y = \dfrac{2}{3}x + 3$

5. $2x + 3y = 6$

6. $5x - 2y = 10$

7. $y = -3$

8. $x = 2$

Write the equation of the line with the given slope and y-intercept. Then graph each line using the slope and y-intercept.

9. Slope $= 2$; y-intercept: $(0, 3)$

10. Slope $= \dfrac{3}{4}$; y-intercept: $(0, -2)$

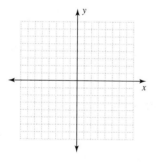

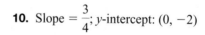
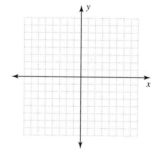

11. Slope: $-\dfrac{2}{3}$; y-intercept: $(0, 2)$

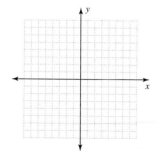

7.2 *Are the pairs of lines* **parallel, perpendicular,** *or* **neither?**

12. L_1 through $(-3, -2)$ and $(1, 3)$
L_2 through $(0, 3)$ and $(4, 8)$

13. L_1 through $(-4, 1)$ and $(2, -3)$
L_2 through $(0, -3)$ and $(2, 0)$

14. L_1 with equation $x + 2y = 6$
L_2 with equation $x + 3y = 9$

15. L_1 with equation $4x - 6y = 18$
L_2 with equation $2x - 3y = 6$

7.3 *Write the equation of the line passing through each point with the indicated slope. Give your results in slope-intercept form, where possible.*

16. $(0, -5)$, $m = \dfrac{2}{3}$

17. $(0, -3)$, $m = 0$

18. $(2, 3)$, $m = 3$

19. $(4, 3)$, m is undefined

20. $(3, -2)$, $m = \dfrac{5}{3}$

21. $(-2, -3)$, $m = 0$

22. $(-2, -4)$, $m = -\dfrac{5}{2}$

23. $(-3, 2)$, $m = -\dfrac{4}{3}$

24. $\left(\dfrac{2}{3}, -5\right)$, $m = 0$

25. $\left(-\dfrac{5}{2}, -1\right)$, m is undefined

Write the equation of the line L satisfying each set of geometric conditions.

26. L passes through $(-3, -1)$ and $(3, 3)$.

27. L passes through $(0, 4)$ and $(5, 3)$.

28. L has slope $\dfrac{3}{4}$ and y-intercept $(0, 3)$.

29. L passes through $(4, -3)$ with a slope of $-\dfrac{5}{4}$.

30. L has y-intercept $(0, -4)$ and is parallel to the line with equation $3x - y = 6$.

31. L passes through $(3, -2)$ and is perpendicular to the line with equation $3x - 5y = 15$.

32. L passes through $(2, -1)$ and is perpendicular to the line with the equation $3x - 2y = 5$.

33. L passes through the point $(-5, -2)$ and is parallel to the line with the equation $4x - 3y = 9$.

7.4 Graph each inequality.

34. $x + y \leq 4$

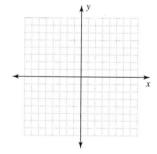

35. $x - y > 5$

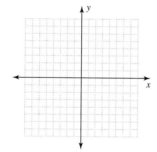

36. $2x + y < 6$

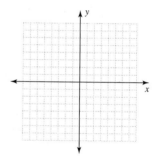

37. $2x - y \geq 6$

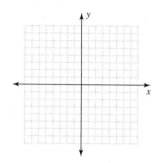

38. $x > 3$

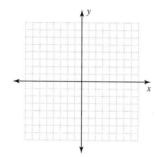

39. $y \leq 2$

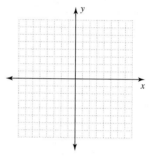

7.5 Evaluate $f(x)$ for the value specified.

40. $f(x) = x^2 - 3x + 5$; find **(a)** $f(0)$, **(b)** $f(-1)$, and **(c)** $f(1)$.

41. $f(x) = -2x^2 + x - 7$; find **(a)** $f(0)$, **(b)** $f(2)$, and **(c)** $f(-2)$.

42. $f(x) = x^3 - x^2 - 2x + 5$; find **(a)** $f(-1)$, **(b)** $f(0)$, and **(c)** $f(2)$.

43. $f(x) = -x^2 + 7x - 9$; find **(a)** $f(-3)$, **(b)** $f(0)$, and **(c)** $f(1)$.

44. $f(x) = 3x^2 - 5x + 1$; find **(a)** $f(-1)$, **(b)** $f(0)$, and **(c)** $f(2)$.

45. $f(x) = x^3 + 3x - 5$; find **(a)** $f(2)$, **(b)** $f(0)$, and **(c)** $f(1)$.

Rewrite each equation as a function of x.

46. $y = 4x + 7$

47. $y = -7x - 3$

48. $4x + 5y = 40$

49. $-3x - 2y = 12$

Graph the function.

50. $f(x) = 2x + 3$

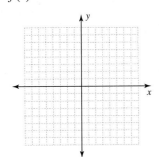

51. $f(x) = 3x - 6$

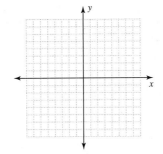

52. $f(x) = -5x + 6$

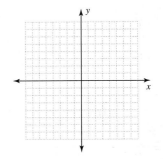

53. $f(x) = -x + 3$

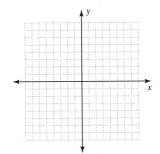

54. $f(x) = -3x - 2$

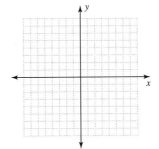

55. $f(x) = -2x + 6$

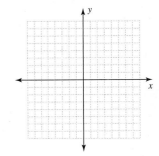

Evaluate each function as indicated.

56. $f(x) = 5x + 3$; find $f(2)$ and $f(0)$.

57. $f(x) = -3x + 5$; find $f(0)$ and $f(1)$.

58. $f(x) = 7x - 5$; find $f\left(\dfrac{5}{4}\right)$ and $f(-1)$.

59. $f(x) = -2x + 5$; find $f(0)$ and $f(-2)$.

60. $f(x) = -5x + 3$; find $f(a)$, $f(2b)$, and $f(x + 2)$.

61. $f(x) = 7x - 1$; find $f(a)$, $f(3b)$, and $f(x - 1)$.

Name _____

Section _____ Date _____

Answers

1. _____

2. _____

3. _____

4. _____

5. _____

6. _____

7. _____

8. _____

The purpose of this self-test is to help you assess your progress so that you can find concepts that you need to review before the next exam. Allow yourself about an hour to take this test. At the end of that hour, check your answers against those given in the back of this text. If you miss any, go back to the appropriate section to reread the examples until you have mastered that particular concept.

*Determine whether the pairs of lines are **parallel, perpendicular,** or **neither.***

1. L_1 through $(2, 5)$ and $(4, 9)$ L_2 through $(-7, 1)$ and $(-2, 11)$

2. L_1 with equation $y = 5x - 8$ L_2 with equation $5y + x = 3$

3. Find the equation of the line through $(-5, 8)$ and perpendicular to the line given by $4x + 2y = 8$.

Graph each inequality.

4. $x + y < 3$

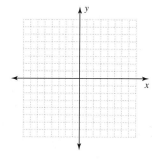

5. $3x + y \geq 9$

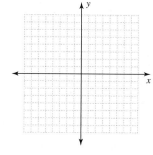

6. $x \leq 7$

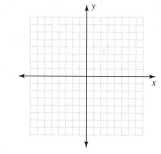

Write the equation of the line with the given slope and y-intercept. Then graph each line.

7. Slope -3 and y-intercept $(0, 6)$

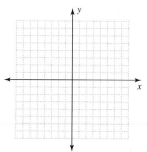

8. Slope 5 and y-intercept $(0, -3)$

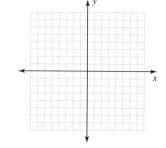

Evaluate f(x) for the value given.

9. $f(x) = x^2 - 4x - 5$; find $f(0)$ and $f(-2)$.

10. $f(x) = -x^3 + 5x - 3x^2 - 8$; find $f(-1)$ and $f(1)$.

11. $f(x) = -7x - 15$; find $f(0)$ and $f(-3)$.

12. $f(x) = 3x - 25$; find $f(a)$ and $f(x - 1)$.

Find the slope and y-intercept of the line represented by each equation.

13. $4x - 5y = 10$ **14.** $y = -\dfrac{2}{3}x - 9$

Write the equation of the line L satisfying each set of geometric conditions.

15. L passes through the points $(-4, 3)$ and $(-1, 7)$.

16. L passes through $(-3, 7)$ and is perpendicular to the line given by $2x - 3y = 7$.

17. L has y-intercept $(0, 5)$ and is parallel to the line given by $4x + 6y = 12$.

18. L has y-intercept $(0, -8)$ and is parallel to the x-axis.

Answers

9. _____
10. _____
11. _____
12. _____
13. _____
14. _____
15. _____
16. _____
17. _____
18. _____

Activity 7 ::
Graphing with the Internet

Each activity in this text is designed to either enhance your understanding of the topics of the preceding chapter, provide you with a mathematical extension of those topics, or both. The activities can be undertaken by one student, but they are better suited for a small group project. Occasionally, it is only through discussion that different facets of the activity become apparent.

I. Find a Graphing Tutorial on the Internet

Search the Internet to find a website that allows you to create a graph from an equation. You may use one of the major search engines or go through an algebra tutorial website that you are familiar with.

II. Use the Graphing Tutorial

When you find a website that allows you to enter an equation or function to be graphed, enter the following function.

$y = 2x$ (*Hint:* You may need to use $f(x) = 2x$ instead of y.)

The "viewing window" or "range" describes the limits of the coordinate system (that is, the minimum and maximum values for the variables).

1. What range is shown?

2. Does the website allow you to change the range?

To your existing graph, add the graph of the equation $y = -3x + 5$.

3. Find the point at which the lines intersect. Does the website provide you with a way to do this or do you need to estimate the point?

4. Briefly describe how the plot distinguishes between your first and second equations.

III. Evaluate the Website

5. Describe any shortcomings to the graphing capability. What improvements might you recommend?

6. Describe other algebra tutorial content available on the website.

7. Consider some topic from your algebra course that you found difficult. Does the website provide tutorial information for this topic? How useful is the tutorial provided?

The following questions are presented to help you review concepts from earlier chapters. This is meant as a review and not as a comprehensive exam. The answers are presented in the back of the text. Section references accompany the answers. If you have difficulty with any of these questions, be certain to at least read through the summary related to those sections.

Name _____

Section _____ Date _____

Answers

Perform the indicated operation.

1. $3x^2y^2 - 5xy - 2x^2y^2 + 2xy$

2. $\dfrac{36m^5n^2}{27m^2n}$

3. $(x^2 - 3x + 5) - (x^2 - 2x - 4)$

4. $(5z^2 - 3z) - (2z^2 - 5)$

Multiply.

5. $(2x - 3)(x + 7)$

6. $(2a - 2b)(a + 4b)$

Divide.

7. $(x^2 + 3x + 2) \div (x - 3)$

8. $(x^4 - 2x) \div (x + 2)$

Solve each equation and check your results.

9. $5x - 2 = 2x - 6$

10. $3(x - 2) = 2(3x + 1) - 2$

Factor each polynomial completely.

11. $x^2 - x - 56$

12. $4x^3y - 2x^2y^2 + 8x^4y$

13. $8a^3 - 18ab^2$

14. $15x^2 - 21xy + 6y^2$

Find the slope of the line through each pairs of points.

15. $(2, -4)$ and $(-3, -9)$

16. $(-1, 7)$ and $(3, -2)$

Perform the indicated operations.

17. $\dfrac{x^2 + 7x + 10}{x^2 + 5x} \cdot \dfrac{2x^2 - 7x + 6}{x^2 - 4}$

18. $\dfrac{2a^2 + 11a - 21}{a^2 - 49} \div (2a - 3)$

1. _____
2. _____
3. _____
4. _____
5. _____
6. _____
7. _____
8. _____
9. _____
10. _____
11. _____
12. _____
13. _____
14. _____
15. _____
16. _____
17. _____
18. _____

Answers

19. _____

20. _____

21. _____

22. _____

23. _____

24. _____

25. _____

26. _____

27. _____

28. _____

29. _____

30. _____

19. $\dfrac{5}{2m} + \dfrac{3}{m^2}$

20. $\dfrac{4}{x-3} - \dfrac{2}{x}$

21. $\dfrac{3y}{y^2+5y+4} + \dfrac{2y}{y^2-1}$

Solve each equation.

22. $\dfrac{13}{4x} + \dfrac{3}{x^2} = \dfrac{5}{2x}$

23. $\dfrac{6}{x+5} + 1 = \dfrac{3}{x-5}$

Solve each application.

24. If the reciprocal of 4 times a number is subtracted from the reciprocal of the number, the result is $\dfrac{1}{12}$. What is the number?

25. Kyoko drove 280 mi to attend a business conference. In returning from the conference along a different route, the trip was only 240 mi, but traffic slowed her speed by 7 mi/h. If her driving time was the same both ways, what was her speed each way?

26. A laser printer can print 400 form letters in 30 min. At that rate, how long will it take the printer to complete a job requiring 1,680 letters?

27. Write the equation of the line perpendicular to the line $7x - y = 15$ with y-intercept of $(0, 2)$.

28. Write the equation of the line with slope of -5 and y-intercept $(0, 3)$.

29. Graph the inequality $4x - 2y \geq 8$.

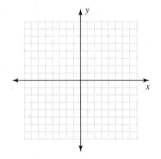

30. If $f(x) = x^2 + 3x$, find $f(-1)$.

chapter
8

> Make the
Connection

Systems of Linear Equations

INTRODUCTION

Most of the electricity in the United States is generated by burning fossil fuels (coal, oil, and gas); by nuclear fission; or by water-powered turbines in hydroelectric dams. About 65% of the electric power we use comes from burning fossil fuels. Because of this dependence on a non-renewable resource and concern over pollution caused by burning fossil fuels, there has been some urgency in developing ways to utilize other power sources. Some of the most promising projects have been in solar- and wind-generated energy.

Alternative sources of energy are expensive compared to the traditional methods of generating electricity described above. As the price per kilowatt-hour (kWh) of electric power has increased (costs to residential users have increased by about $0.0028 per kWh per year since 1970), alternative energy sources look more promising. Additionally, the cost of manufacturing and installing banks of wind turbines in windy locations has declined.

When will the cost of generating electricity using wind power be equal to or less than the cost of using our traditional energy mix? Economists use systems of equations to make projections and then advise about the feasibility of investing in wind power plants for large cities.

CHAPTER 8 OUTLINE

603

Name _____

Section _____ Date _____

This prerequisite test provides some exercises requiring skills that you will need to be successful in the coming chapter. The answers for these exercises can be found in the back of this text. This prerequisite test can help you identify topics that you will need to review before beginning the chapter.

Answers

1. _____

2. _____

3. _____

4. _____

5. _____

6. _____

7. _____

8. _____

9. _____

10. _____

Graph each equation.

1. $2x - y = 4$

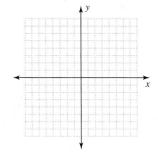

2. $y = 6$

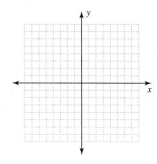

Simplify each expression.

3. $(3x + 2y) + (-3x - 3y)$

4. $3(2x - 4y) + 4(x + 3y)$

Solve each equation.

5. $2x + 3(x + 1) = 13$

6. $3(y - 1) + 4y = 18$

7. $x + 2(3x - 5) = 25$

8. $3x - 2(x - 7) = 12$

Graph the solution set for each linear inequality.

9. $2x - y \le 6$

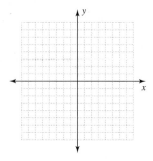

10. $y > 2x$

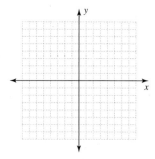

The Streeter/Hutchison Series in Mathematics   Beginning Algebra

8.1

Systems of Linear Equations: Solving by Graphing

< 8.1 Objectives >

1 > Solve a consistent system of linear equations by graphing

2 > Solve an inconsistent system of linear equations by graphing

3 > Solve a dependent system of linear equations by graphing

From our work in Section 6.1, we know that an equation of the form $x + y = 3$ is a linear equation. Remember that its graph is a straight line. When we consider two equations together they form a **system of linear equations.** An example of such a system is

$$x + y = 3$$
$$3x - y = 5$$

A solution for a linear equation in two variables is any ordered pair that satisfies the equation. Often there is just one ordered pair that satisfies both equations of a system. It is called the **solution for the system.** For instance, there is one solution for the system above, (2, 1), because replacing x with 2 and y with 1, gives

$$\begin{array}{c|c} x + y = 3 & 3x - y = 5 \\ \hline (2) + (1) \stackrel{?}{=} 3 & 3 \cdot (2) - (1) \stackrel{?}{=} 5 \\ 3 = 3 & 6 - 1 \stackrel{?}{=} 5 \\ & 5 = 5 \end{array}$$

NOTE

There is no other ordered pair that satisfies both equations.

Because both statements are true, the ordered pair (2, 1) satisfies both equations.

One approach to finding the solution for a system of linear equations is the **graphical method.** To use this, we graph the two lines on the same coordinate system. The coordinates of the point where the lines intersect is the solution for the system.

Example 1 | Solving by Graphing

< Objective 1 >

Solve the system by graphing.

$$x + y = 6$$
$$x - y = 4$$

First, we determine solutions for the equations of our system. For $x + y = 6$, two solutions are (6, 0) and (0, 6). For $x - y = 4$, two solutions are (4, 0) and (0, −4). Using these intercepts, we graph the two equations. The lines intersect at the point (5, 1).

NOTE

Check that (5, 1) is the solution by substituting 5 for x and 1 for y in both equations.

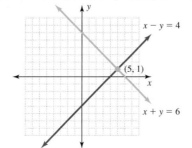

(5, 1) is the solution of the system.
It is the only point that lies on both lines.

Check Yourself 1

Solve the system by graphing.

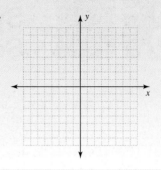

$2x - y = 4$

$x + y = 5$

Example 2 shows how to graph a system when one of the equations represents a horizontal line.

    Example 2   Solving by Graphing

Solve the system by graphing.

$3x + 2y = 6$

$y = 6$

For $3x + 2y = 6$, two solutions are $(2, 0)$ and $(0, 3)$. These represent the x- and y-intercepts of the graph of the equation. The equation $y = 6$ represents a horizontal line that crosses the y-axis at the point $(0, 6)$. Using these intercepts, we graph the two equations. The lines will intersect at the point $(-2, 6)$. So this is the solution to our system.

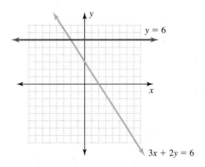

Check Yourself 2

Solve the system by graphing.

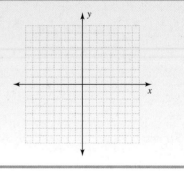

$4x + 5y = 20$

$y = 8$

The systems in Examples 1 and 2 both had exactly one solution. A system with one solution is called a **consistent system.** It is possible for a system of equations to have no solution. Such a system is called an **inconsistent system.** We present such a system here.

| Example 3 | Solving an Inconsistent System |

< Objective 2 >

NOTE

In slope-intercept form, our equations are
$y = -2x + 2$
and
$y = -2x + 4$
Both lines have slope -2.

NOTE

When solving a system, we are searching for all ordered pairs that make both of the statements true. If we determine that there are no such pairs, we have successfully solved the system.

Solve the system by graphing.

$2x + y = 2$
$2x + y = 4$

We can graph the two lines as before. For $2x + y = 2$, two solutions are $(0, 2)$ and $(1, 0)$. For $2x + y = 4$, two solutions are $(0, 4)$ and $(2, 0)$. Using these intercepts, we graph the two equations.

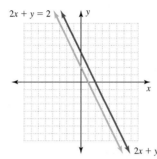

Notice that the slope for each of these lines is -2, but they have different y-intercepts. This means that the lines are parallel (they never intersect). Because the lines have no points in common, there is no ordered pair that satisfies both equations. The system has no solution. It is *inconsistent*.

Check Yourself 3

Solve the system by graphing.

$x - 3y = 3$
$x - 3y = 6$

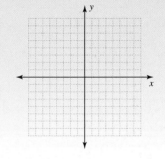

There is one more possibility for linear systems, as Example 4 illustrates.

| Example 4 | Solving a Dependent System |

< Objective 3 >

NOTE

Multiplying the first equation by 2 results in the second equation.

Solve the system by graphing.

$x - 2y = 4$
$2x - 4y = 8$

Graphing as before, we find

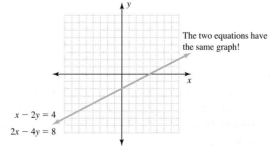

The two equations have the same graph!

$x - 2y = 4$
$2x - 4y = 8$

Because the graphs coincide, there are *infinitely many* solutions for this system. Every point on the graph of $x - 2y = 4$ is also on the graph of $2x - 4y = 8$, so any ordered pair satisfying $x - 2y = 4$ also satisfies $2x - 4y = 8$. This is called a **dependent system,** and any point on the line represents a solution.

Check Yourself 4

Solve the system by graphing.

$x +\ y = 4$
$2x + 2y = 8$

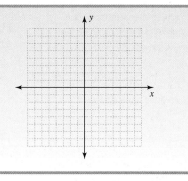

Here is a summary of our work so far.

Step by Step

To Solve a System of Equations by Graphing

Step 1   Graph both equations on the same coordinate system.
Step 2   Determine the solution to the system as follows.

a. If the lines intersect at one point, the solution is the ordered pair corresponding to that point. This is called a **consistent system.**

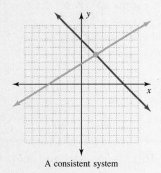

A consistent system

b. If the lines are parallel, there are no solutions. This is called an **inconsistent system.**

NOTE

There are no points that lie on both lines.

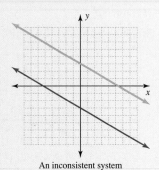

An inconsistent system

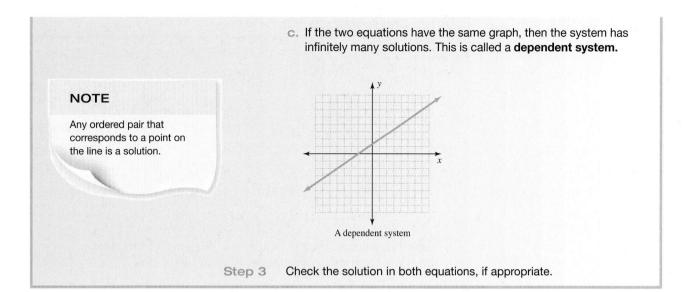

NOTE

Any ordered pair that corresponds to a point on the line is a solution.

c. If the two equations have the same graph, then the system has infinitely many solutions. This is called a **dependent system.**

A dependent system

Step 3   Check the solution in both equations, if appropriate.

In Example 5, we use the a graphical approach to solve an application from the field of medicine.

Example 5   **A Science Application**

A medical lab technician needs to determine how much 15% hydrochloric acid (HCl) solution (x) must be mixed with 5% HCl (y) to produce 50 milliliters of a 9% solution. To solve this problem, graph $x + y = 50$ and $15x + 5y = 450$ on the same set of axes, and determine the intersection point of the two lines.

The graphs of the two equations are shown here.

NOTE

If you have a graphing calculator, try solving this application on it. Use an "intersect" utility.

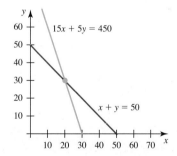

The intersection point appears to be (20, 30). Substituting into each equation verifies this. So, 20 mL of 15% HCl should be mixed with 30 mL of 5% HCl to obtain 50 mL of 9% HCl.

 Check Yourself 5

A medical lab technician needs to determine how much 6-molar (M) copper sulfate ($CuSO_4$) solution (x) must be mixed with 2 M ($CuSO_4$) (y) to produce 200 milliliters of a 3-M solution. To solve this problem, graph $x + y = 200$ and $6x + 2y = 600$ on the same set of axes, and determine the intersection point of the two lines.

Check Yourself ANSWERS

1.

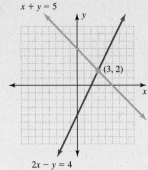

$x + y = 5$

$(3, 2)$

$2x - y = 4$

2.

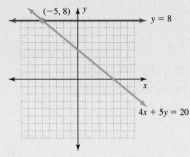

$(-5, 8)$

$y = 8$

$4x + 5y = 20$

3. There is no solution. The lines are parallel, so the system is inconsistent.

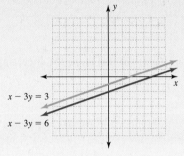

$x - 3y = 3$

$x - 3y = 6$

4. There are infinitely many solutions.

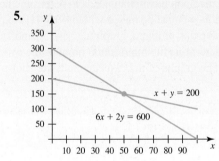

$x + y = 4$
$2x + 2y = 8$

A dependent system

5.

$x + y = 200$

$6x + 2y = 600$

Solution: (50, 150); 50 mL of 6-molar copper sulfate should be mixed with 150 mL of 2-molar copper sulfate.

Reading Your Text

The following fill-in-the-blank exercises are designed to ensure that you understand some of the key vocabulary used in this section.

SECTION 8.1

(a) When we consider two linear equations together, they form a _____ of linear equations.

(b) An ordered pair that satisfies both equations of a system is called a _____ for the system.

(c) A system with exactly one solution is called a _____ system.

(d) If there are infinitely many solutions for a system, the system is called _____.

< Objectives 1–3 >

Solve each system by graphing.

1. $2x + 2y = 12$
$\quad x - y = 4$

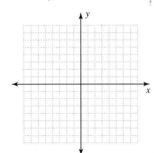

2. $x - y = 8$
$\quad x + y = 2$

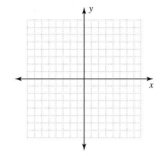

3. $-x + y = 3$
$\quad x + y = 5$

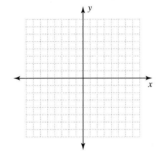

4. $x + y = 7$
$\quad -x + y = 3$

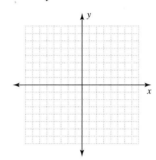

5. $x + 2y = 4$
$\quad x - y = 1$

> Videos

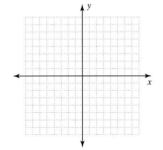

6. $3x + y = 6$
$\quad x + y = 4$

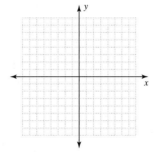

7. $2x + y = 8$
$\quad 2x - y = 0$

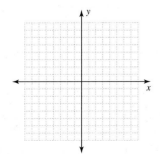

8. $x - 2y = -2$
$\quad x + 2y = 6$

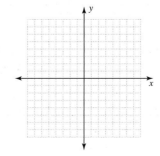

Name _____

Section _____ Date _____

Answers

1. _____

2. _____

3. _____

4. _____

5. _____

6. _____

7. _____

8. _____

Answers

9. _____

10. _____

11. _____

12. _____

13. _____

14. _____

15. _____

16. _____

9. $x + 3y = 12$
$2x - 3y = 6$

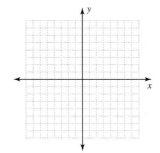

10. $2x - y = 4$
$2x - y = 6$

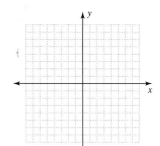

11. $3x + 2y = 12$
$ y = 3$ 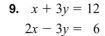 > Videos

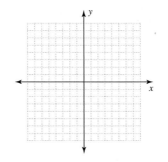

12. $x - 2y = 8$
$3x - 2y = 12$

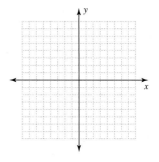

13. $x - y = 4$
$2x - 2y = 8$ > Videos

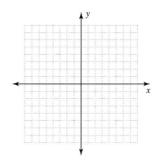

14. $2x - y = 8$
$ x = 2$

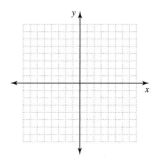

15. $x - 4y = -4$
$x + 2y = 8$

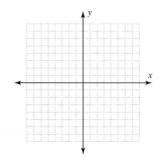

16. $x - 6y = 6$
$-x + y = 4$

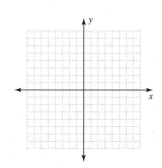

17. $3x - 2y = 6$
$2x - y = 5$

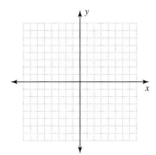

18. $4x + 3y = 12$
$x + y = 2$

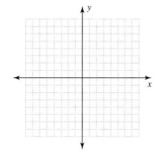

Answers

17. _____

18. _____

19. _____

20. _____

21. _____

22. _____

23. _____

24. _____

19. $3x - y = 3$
$3x - y = 6$

> Videos

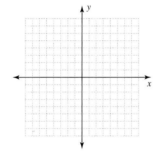

20. $3x - 6y = 9$
$x - 2y = 3$

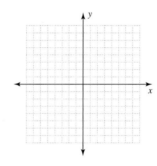

21. $2y = 3$
$x - 2y = -3$

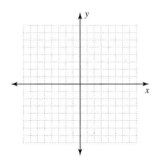

22. $x + y = -6$
$-x + 2y = 6$

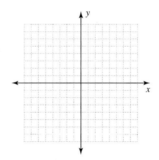

23. $x = 4$
$y = -6$

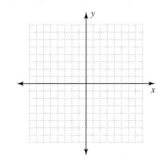

24. $x = -3$
$y = 5$

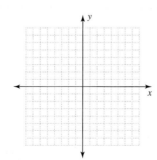

Answers

25. _____

26. _____

27. _____

28. _____

29. _____

30. _____

25. **CONSTRUCTION** The gambrel roof pictured here has several missing dimensions. These dimensions can be calculated using a system of equations.

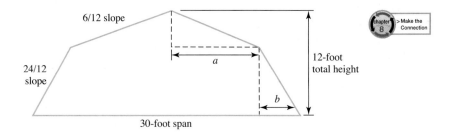

The system of equations is

$$2a + 2b = 30$$
$$\frac{24}{12}a + \frac{6}{12}b = 12$$

Solve this system of equations graphically.

26. **CONSTRUCTION** The beam shown in the figure is 15 feet long and has a load on each end.

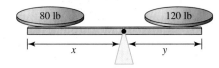

To find the point where the beam balances, we use the system of equations

$$x + y = 15$$
$$80x = 120y$$

Solve this system of equations graphically.

| Basic Skills | **Challenge Yourself** | Calculator/Computer | Career Applications | Above and Beyond |

Determine whether each statement is **true** *or* **false.**

27. A dependent system has an infinite number of solutions.

28. A linear system could have exactly two solutions.

Complete each statement with **never, sometimes,** *or* **always.**

29. A linear system _____ has at least one solution.

30. If the graphs of two linear equations in a system have different slopes, the system _____ has exactly one solution.

31. Find values for *m* and *b* in the system so that the solution to the system is (1, 2).

$$mx + 3y = 8$$
$$-3x + 4y = b$$

32. Find values for *m* and *b* in the system so that the solution to the system is (−3, 4).

$$5x + 7y = b$$
$$mx + y = 22$$

Answers

31. _____

32. _____

33. _____

34. _____

35. _____

Basic Skills | Challenge Yourself | Calculator/Computer | **Career Applications** | Above and Beyond
▲

For exercises 33 to 35, a graphing calculator with an "intersect" utility is strongly recommended.

33. **MECHANICAL ENGINEERING** At 2,100°C, a 60% aluminum oxide and 40% chromium oxide alloy separates into two different alloys. The first alloy is 78% Al_2O_3 and 22% Cr_2O_3, and the second is 50% Al_2O_3 and 50% Cr_2O_3. If the total amount of the alloy present is 7,000 grams, use a system of equations and solve by graphing to find out how many grams of each type the alloy separates into.

Use the system $0.78x + 0.50y = 4,200$ and $0.22x + 0.50y = 2,800$

where *x* is the amount of the 78%/22% alloy and *y* is the amount of the 50%/50% alloy.

34. **MECHANICAL ENGINEERING** For a plating bath, 10,000 liters of 13% electrolyte solution is required. You have 8% and 16% solutions in stock.

Use the system $\begin{aligned} x + y &= 10,000 \\ 0.08x + 0.16y &= 1,300 \end{aligned}$

where *x* is the amount of 8% solution and *y* is the amount of 16% solution.

Solve by graphing to determine how much of each type of solution to use.

35. **MANUFACTURING** A manufacturer has two machines that produce door handles. On Monday, machine A operates for 10 hours and machine B operates for 7 hours, and 290 door handles are produced. On Tuesday, machine A operates for 6 hours and machine B operates for 12 hours, and 330 door handles are produced.
  Use the system

$$10x + 7y = 290$$
$$6x + 12y = 330$$

where *x* is the number of handles produced by machine A in an hour, and *y* is the number of handles produced by machine B in an hour.

Solve the system graphically.

Answers

36. _____

37. _____

Basic Skills | Challenge Yourself | Calculator/Computer | Career Applications | **Above and Beyond**
▲

36. Complete each statement in your own words.

"To solve an equation means to . . . ."
"To solve a system of equations means to . . . ."

37. A system of equations such as the one below is sometimes called a "2-by-2 system of linear equations."

$$3x + 4y = 1$$
$$x - 2y = 6$$

Explain this phrase.

Answers

1. $\left. \begin{array}{r} 2x + 2y = 12 \\ x - y = 4 \end{array} \right\}$ (5, 1)

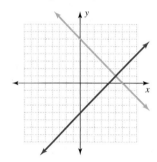

3. $\left. \begin{array}{r} -x + y = 3 \\ x + y = 5 \end{array} \right\}$ (1, 4)

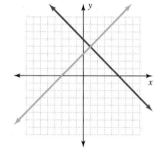

5. $\left. \begin{array}{r} x + 2y = 4 \\ x - y = 1 \end{array} \right\}$ (2, 1)

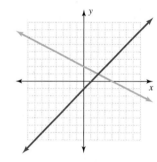

7. $\left. \begin{array}{r} 2x + y = 8 \\ 2x - y = 0 \end{array} \right\}$ (2, 4)

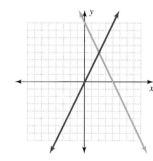

9. $\left. \begin{array}{r} x + 3y = 12 \\ 2x - 3y = 6 \end{array} \right\}$ (6, 2)

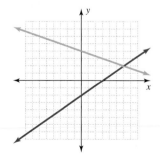

11. $\left. \begin{array}{r} 3x + 2y = 12 \\ y = 3 \end{array} \right\}$ (2, 3)

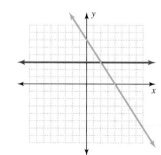

The Streeter/Hutchison Series in Mathematics   Beginning Algebra

13. $\left.\begin{array}{r} x - y = 4 \\ 2x - 2y = 8 \end{array}\right\}$ Dependent

Infinitely many solutions

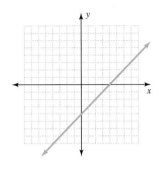

15. $\left.\begin{array}{r} x - 4y = -4 \\ x + 2y = 8 \end{array}\right\}$ $(4, 2)$

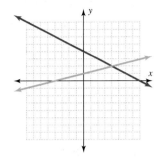

17. $\left.\begin{array}{r} 3x - 2y = 6 \\ 2x - y = 5 \end{array}\right\}$ $(4, 3)$

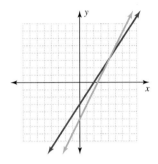

19. $\left.\begin{array}{r} 3x - y = 3 \\ 3x - y = 6 \end{array}\right\}$ Inconsistent

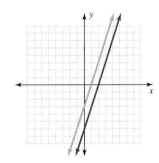

21. $\left.\begin{array}{r} 2y = 3 \\ x - 2y = -3 \end{array}\right\}$ $\left(0, \dfrac{3}{2}\right)$

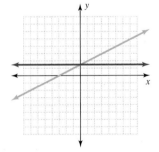

23. $\left.\begin{array}{r} x = 4 \\ y = -6 \end{array}\right\}$ $(4, -6)$

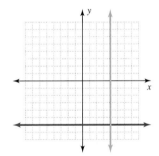

25. $a = 3, b = 12$ **27.** True   **29.** sometimes   **31.** $m = 2, b = 5$

33. $x = 2{,}500$ grams; $y = 4{,}500$ grams   **35.** A: 15 per hour; B: 20 per hour

37. Above and Beyond

Systems of Linear Equations: Solving by the Addition Method

< 8.2 Objectives >

1 > Solve systems of linear equations using the addition method

2 > Solve applications of systems of linear equations

The graphical method of solving equations, shown in Section 8.1, has two definite disadvantages. First, it is time-consuming to graph each system that you want to solve. More importantly, the graphical method is not precise. For instance, look at the graph of the system

$$x - 2y = 4$$
$$3x + 2y = 6$$

which follows:

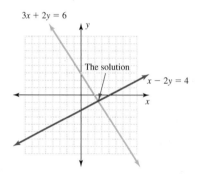

The exact solution for the system is $\left(\dfrac{5}{2}, -\dfrac{3}{4}\right)$, which is difficult to read from the graph.

Fortunately, there are algebraic methods that do not have this disadvantage and enable you to find exact solutions for a system of equations.

One algebraic method of finding a solution is called the **addition method.**

| ▷ | **Example 1** | Solving a System by the Addition Method |

< Objective 1 >

Solve the system.

$$x + y = 8$$
$$x - y = 2$$

Note that the coefficients of the y-terms are the *additive inverses* of one another (1 and -1) and that adding the two equations "eliminates" the variable y. That addition step is shown here.

$$
\begin{array}{r}
x + y = 8 \\
\underline{x - y = 2} \\
2x = 10 \\
x = 5
\end{array}
$$

By adding, we eliminate the variable y. The resulting equation contains *only* the variable x.

NOTES

This is also called **solution by elimination** for this reason.

This method uses the fact that if

$a = b$ and $c = d$

then

$a + c = b + d$

This is the **additive property** of equality. By the additive property, if equals are added to equals, the resulting sums are equal.

We now know that 5 is the x-coordinate of our solution. Substitute 5 for x into *either* of the original equations.

$$x + y = 8$$
$$(5) + y = 8$$
$$y = 3$$

So $(5, 3)$ is the solution.

To check, replace x and y with these values in *both* of the original equations.

$x + y = 8$	$x - y = 2$
$(5) + (3) \stackrel{?}{=} 8$	$(5) - (3) \stackrel{?}{=} 2$
$8 = 8$ (True)	$2 = 2$ (True)

Because $(5, 3)$ satisfies both equations, it is the solution.

Check Yourself 1

Solve the system by adding.

$$x - y = -2$$
$$x + y = 6$$

 Example 2 | **Solving a System by the Addition Method**

Solve the system.

$$-3x + 2y = 12$$
$$3x - y = -9$$

In this case, adding eliminates the x-terms.

$$-3x + 2y = 12$$
$$\underline{3x - y = -9}$$
$$y = 3$$

NOTE

It does not matter which variable is eliminated. Choose the one that requires less work.

Now substitute 3 for y in either equation. From the first equation

$$-3x + 2(3) = 12$$
$$-3x = 6$$
$$x = -2$$

and $(-2, 3)$ is the solution.

Show that you get the same x-coordinate by substituting 3 for y in the second equation rather than in the first. Remember to check your solution in both equations.

Check Yourself 2

Solve the system by adding.

$$5x - 2y = 9$$
$$-5x + 3y = -11$$

Note that in both Examples 1 and 2 we found an equation in a single variable by adding. We could do this because the coefficients of one of the variables were opposites. This gave 0 as a coefficient for one of the variables after we added the two equations.

In some systems, you will not be able to directly eliminate either variable by adding. However, an equivalent system can always be written by multiplying one or both of the equations by a nonzero constant so that the coefficients of x (or of y) are opposites. Example 3 illustrates this approach.

Example 3   **Solving a System by the Addition Method**

Solve the system.

$$2x + y = 13$$
$$3x + y = 18$$

RECALL

Multiplying both sides of an equation by some nonzero number does not change the solutions. So even though we have "altered" the equations, they are equivalent and have the same solutions.

Note that adding the equations in this form does not eliminate either variable. You would still have terms in x and in y. However, look at what happens if we multiply both sides of the second equation by -1 as the first step.

$$2x + y = 13 \xrightarrow{\text{Multiply}} 2x + y = 13$$
$$3x + y = 18 \xrightarrow{\text{by } -1} -3x - y = -18$$

Now we can add.

$$\begin{array}{r} 2x + y = 13 \\ -3x - y = -18 \\ \hline -x = -5 \\ x = 5 \end{array}$$

Substitute 5 for x into either equation. We choose the first:

$$2(5) + y = 13$$
$$y = 3$$

$(5, 3)$ is the solution. We leave it to the reader to check this solution.

Check Yourself 3

Solve the system by adding.

$$x - 2y = 9$$
$$x + 3y = -1$$

To summarize, multiplying both sides of one of the equations by a nonzero constant can yield an equivalent system in which the coefficients of the x-terms or the y-terms are opposites. This means that a variable can be eliminated by adding.

Example 4   **Solving a System by the Addition Method**

Solve the system.

$$x + 4y = 2$$
$$3x - 2y = -22$$

One approach is to multiply both sides of the second equation by 2. Do you see that the coefficients of the y-terms will then be opposites?

$$x + 4y = 2 \xrightarrow{\phantom{\text{Multiply}}} x + 4y = 2$$
$$3x - 2y = -22 \xrightarrow{\text{Multiply by 2}} 6x - 4y = -44$$

NOTE

Now the coefficients of the y-terms are opposites.

NOTE

We could substitute -6 for x in the second equation to find y.

If we add the resulting equations, the variable y is eliminated and we can solve for x.

$$
\begin{array}{rl}
x + 4y = & 2 \\
6x - 4y = & -44 \\
\hline
7x = & -42 \\
x = & -6
\end{array}
$$

Now substitute -6 for x in the first equation of this example to find y.

$$(-6) + 4y = 2$$
$$4y = 8$$
$$y = 2$$

So $(-6, 2)$ is the solution.

Again you should check this result. As is often the case, there are several ways to solve the system. For example, what if we multiply both sides of our original equation by -3? The coefficients of the x-terms will then be opposites and adding will eliminate the variable x so that we can solve for y. Try that for yourself in the Check Yourself 4 exercise.

Check Yourself 4

Solve the system by eliminating *x*.

$$x + 4y = 2$$
$$3x - 2y = -22$$

It may be necessary to multiply each equation separately so that one of the variables is eliminated when the equations are added. Example 5 illustrates this approach.

   Example 5   Solving a System by the Addition Method

Solve the system.

$$4x + 3y = 11$$
$$3x - 2y = 4$$

Do you see that, if we want to have integer coefficients, multiplying in one equation does not help in this case? We have to multiply in both equations.

To eliminate x, we can multiply both sides of the first equation by 3 and both sides of the second equation by -4. The coefficients of the x-terms will then be opposites.

NOTE

The minus sign is used with the 4 so that the coefficients of the x-term are opposites.

$$4x + 3y = 11 \xrightarrow[\text{by 3}]{\text{Multiply}} \quad 12x + 9y = 33$$

$$3x - 2y = 4 \xrightarrow[\text{by }-4]{\text{Multiply}} \quad -12x + 8y = -16$$

Adding the resulting equations gives

$$17y = 17$$
$$y = 1$$

Now substituting 1 for y in the first equation, we have

$$4x + 3(1) = 11$$
$$4x = 8$$
$$x = 2$$

NOTE

Check $(2, 1)$ in both equations of the original system.

and $(2, 1)$ is the solution.

Check Yourself 5

Solve the system by eliminating *y*.

$4x + 3y = 11$

$3x - 2y = 4$

Here is a summary of the solution steps we have illustrated.

Step by Step

To Solve a System of Linear Equations by Adding		
	Step 1	If necessary, multiply both sides of one or both equations by nonzero numbers to form an equivalent system in which the coefficients of one of the variables are opposites.
	Step 2	Add the equations of the new system.
	Step 3	Solve the resulting equation for the remaining variable.
	Step 4	Substitute the value found in step 3 into either of the original equations to find the value of the second variable.
	Step 5	Check your solution in both of the original equations.

In Section 8.1 we saw that some systems had *infinitely* many solutions. Example 6 shows how this is indicated when we are using the addition method of solving equations.

Example 6   **Solving a Dependent System**

Solve the system.

$x + 3y = -2$

$3x + 9y = -6$

We multiply both sides of the first equation by -3.

$$x + 3y = -2 \quad \xrightarrow{\text{Multiply by } -3} \quad -3x - 9y = 6$$

$$3x + 9y = -6 \quad \longrightarrow \quad \underline{3x + 9y = -6}$$

$$0 = 0$$

Adding, we see that both variables have been eliminated, and we have the true statement 0 = 0.

NOTE

The lines coincide. That is the case whenever *adding eliminates both variables* and a true statement results.

Look at the graph of the system.

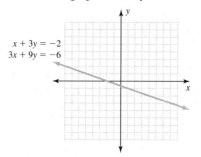

As we see, the two equations have the *same* graph. This means that the system is *dependent,* and there are *infinitely many solutions.* Any (x, y) that satisfies $x + 3y = -2$ also satisfies $3x + 9y = -6$.

Check Yourself 6

Solve the system by adding.

$x - 2y = 3$

$-2x + 4y = -6$

Earlier we encountered systems that had *no* solutions. Example 7 illustrates what happens when we try to solve such a system with the addition method.

Example 7	Solving an Inconsistent System

Solve the system.

$$3x - y = 4$$
$$-6x + 2y = -5$$

We multiply both sides of the first equation by 2.

 > CAUTION

Be sure to multiply both sides by 2.

$$3x - y = 4 \xrightarrow{\text{Multiply by 2}} 6x - 2y = 8 \quad \text{We now add the two equations.}$$
$$-6x + 2y = -5 \xrightarrow{\hspace{2cm}} \underline{-6x + 2y = -5}$$
$$0 = 3$$

Again both variables have been eliminated by addition. But this time we have the *false* statement $0 = 3$ because we tried to solve a system whose graph consists of two parallel lines. Because the two lines do not intersect, there is *no* solution for the system. It is *inconsistent*.

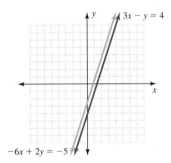

	Check Yourself 7

Solve the system by adding.

$$5x + 15y = 20$$
$$x + 3y = 3$$

RECALL

In Chapter 2, we expressed all the unknowns in each problem in terms of a single variable.

In Chapter 2 we solved word problems by using equations in a single variable. Now you have the background to use two equations in two variables to solve word problems. The five steps for solving word problems stay the same (in fact, we give them again for reference in our first application example). Many students find that using two equations and two variables makes writing the necessary equations much easier, as Example 8 illustrates.

Example 8	Solving an Application with Two Linear Equations

< Objective 2 >

Ryan bought 8 pens and 7 pencils and paid a total of $14.80. Ashleigh purchased 2 pens and 10 pencils and paid $7. Find the cost for a single pen and a single pencil.

Thompson's Stationery
3784 Main St.
Thank you for
your business

8 pens
7 pencils

Total: 14.80

RECALL

Here are the steps for using two variables:

1. Read the problem carefully. What do you want to find?
2. Assign variables to the unknown quantities.
3. Translate the problem to the language of algebra to form a system of equations.
4. Solve the system.
5. State the solution and verify your result by returning to the original problem.

Step 1   You want to find the cost of a single pen and the cost of a single pencil.

Step 2   Let x be the cost of a pen and y be the cost of a pencil.

Step 3   Write the two necessary equations.

$$8x + 7y = 14.80$$
$$2x + 10y = 7.00$$

In the first equation, $8x$ is the total cost of the pens Ryan bought and $7y$ is the total cost of the pencils Ryan bought. The second equation is formed in a similar fashion.

Step 4   Solve the system formed in step 3. We multiply the second equation by -4. Then adding will eliminate the variable x.

$$8x + 7y = 14.80$$
$$-8x - 40y = -28.00$$

Now adding the equations, we have

$$-33y = -13.20$$
$$y = 0.40$$

Substituting 0.40 for y in the first equation, we have

$$8x + 7(0.40) = 14.80$$
$$8x + 2.80 = 14.80$$
$$8x = 12.00$$
$$x = 1.50$$

Step 5   From the results of step 4 we see that the pens are $1.50 each and the pencils are 40¢ each.

To check these solutions, replace x with 1.50 and y with 0.40 in the first equation.

$$8(1.50) + 7(0.40) \stackrel{?}{=} 14.80$$
$$12.00 + 2.80 \stackrel{?}{=} 14.80$$
$$14.80 = 14.80 \quad \text{(True)}$$

We leave it to you to check these values in the second equation.

Check Yourself 8

Alana bought three digital tapes and two compact disks on sale for $66. At the same sale, Chen bought three digital tapes and four compact disks for $96. Find the individual prices for a tape and a disk.

Example 9 shows how sketches can be helpful in setting up a problem.

Example 9   **Using a Sketch to Help Solve an Application**

NOTE

You should always draw a sketch of the problem when it is appropriate.

An 18-ft board is cut into two pieces, one of which is 4 ft longer than the other. How long is each piece?

Step 1 You want to find the two lengths.

Step 2 Let x be the length of the longer piece and y the length of the shorter piece.

Step 3 Write the equations for the solution.

$x + y = 18$ ⟵ The total length is 18.
$x - y = 4$ ⟵ The difference in lengths is 4.

Step 4 To solve the system, add:

$$
\begin{array}{r}
x + y = 18 \\
\underline{x - y = 4} \\
2x = 22 \\
x = 11
\end{array}
$$

Replace x with 11 in the first equation.

$$(11) + y = 18$$
$$y = 7$$

Step 5 The longer piece has length 11 ft, and the shorter piece has length 7 ft.

We leave it to you to check this result in the original problem.

Check Yourself 9

A 20-ft board is cut into two pieces, one of which is 6 ft longer than the other. How long is each piece?

Using two equations in two variables also helps solve **mixture problems.**

Example 10   Solving a Mixture Problem

Winnifred has collected $4.50 in nickels and dimes. If she has 55 coins, how many of each kind of coin does she have?

Step 1 You want to find the number of nickels and the number of dimes.

Step 2 Let

n = number of nickels

d = number of dimes

Step 3   Write the equations for the solution.

$$n + d = \ 55 \quad \longleftarrow \text{ There are 55 coins in all.}$$
$$5n + 10d = 450$$

Value of   Value   Total value
nickels   of dimes   (in cents)

Step 4   We now have the system

$$n + \quad d = \ 55$$
$$5n + 10d = 450$$

We choose to solve this system by addition. Multiply the first equation by -5. We then add the equations to eliminate the variable n.

$$-5n - \ 5d = -275$$
$$\underline{5n + 10d = \ \ \ 450}$$
$$5d = \ \ 175$$
$$d = \ \ \ 35$$

We now substitute 35 for d in the first equation.

$$n + (35) = 55$$
$$n = 20$$

Step 5   There are 20 nickels and 35 dimes.

We leave it to you to check this result. Just verify that the value of these coins is $4.50.

Check Yourself 10

Tickets for a play cost $8 or $6. If 350 tickets were sold in all and receipts were $2,500, how many tickets of each price were sold?

We can also solve mixture problems involving percentages with two equations in two unknowns. Look at Example 11.

Example 11   **Solving a Mixture Problem**

There are two solutions in a chemistry lab: a 20% acid solution and a 60% acid solution. How many milliliters of each should be mixed to produce 200 mL of a 44% acid solution?

20%   60%   44%
x mL y mL   200 mL

Step 1   You need to know the amount of each solution to use.

Step 2   Let

$x = $ amount of 20% acid solution

$y = $ amount of 60% acid solution

Step 3 Note that a 20% acid solution is 20% acid and 80% water.

We can write equations from the total amount of the solution, here 200 mL, and from the amount of acid in that solution. Many students find a chart helpful in organizing the information at this point. Here, for example, we might have

	Amount of Solution	**% Acid**	**Amount of Acid**
	x	0.20	$0.20x$
	y	0.60	$0.60y$
Final solution	200	0.44	$(0.44)(200)$

Now we are ready to form our system.

$$x + y = 200$$
$$0.20x + 0.60y = \underline{0.44(200)}$$

Acid in 20%   Acid in 60%   Acid in
solution   solution   mixture

NOTE

The amount of acid is the amount of solution times the percentage of acid (as a decimal). That is the key to forming the third column of our table.

NOTE

The first equation is the total amount of the solution from the first column of our table.

NOTE

The second equation is the amount of acid from the third column of our table. The sum of the acid in the two solutions equals the acid in the mixture.

Step 4 If we multiply the second equation by 100 to clear it of decimals, we have

$$x + y = 200 \xrightarrow[\text{by} -20]{\text{Multiply}} -20x - 20y = -4{,}000$$

$$20x + 60y = 8{,}800 \xrightarrow{} \underline{20x + 60y = 8{,}800}$$

$$40y = 4{,}800$$

$$y = 120$$

Substituting 120 for y in the first equation, we have

$$x + (120) = 200$$
$$x = 80$$

Step 5 The amounts to be mixed are 80 mL (20% acid solution) and 120 mL (60% acid solution).

You can check this solution by verifying that the amount of acid from the 20% solution added to the amount from the 60% solution is equal to the amount of acid in the mixture.

Check Yourself 11

You have a 30% alcohol solution and a 50% alcohol solution. How much of each solution should be combined to make 400 mL of a 45% alcohol solution?

A related kind of application involves interest. The key equation involves the *principal* (the amount invested), the annual *interest rate,* the *time* (in years) that the money is invested, and the amount of *interest* you receive.

$$I = P \cdot r \cdot t$$

Interest   Principal   Rate   Time

For 1 year we have

$$I = P \cdot r \qquad \text{if } t = 1$$

Example 12   **Solving an Investment Application**

Jeremy inherits $20,000 and invests part of the money in bonds with an interest rate of 11%. The remainder of the money is in savings at a 9% rate. What amount has he invested at each rate if he receives $2,040 in interest for 1 year?

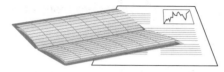

Step 1   You want to find the amounts invested at 11% and at 9%.

Step 2   Let x = the amount invested at 11% and y = the amount invested at 9%. Once again you may find a chart helpful at this point.

	Principal	**Rate**	**Interest**
	x	11%	$0.11x$
	y	9%	$0.09y$
Totals	20,000		2,040

Step 3   Form the equations for the solution, using the first and third columns of the table.

$x + y = 20{,}000$ ⟵ He has $20,000 invested in all.

$0.11x + 0.09y = 2{,}040$

The interest at 11% (rate · principal)   The interest at 9%   The total interest

Step 4   To solve the system, use addition.

$$x + \quad y = 20{,}000$$
$$0.11x + 0.09y = \quad 2{,}040$$

To do this, multiply both sides of the first equation by -9. Multiplying both sides of the second equation by 100 clears decimals. Adding the resulting equations eliminates y.

$$-9x - 9y = -180{,}000$$
$$\underline{11x + 9y = \quad 204{,}000}$$
$$2x \quad = \quad 24{,}000$$
$$x = \quad 12{,}000$$

Now, substitute 12,000 for x in the first equation and solve for y.

$$(12{,}000) + y = 20{,}000$$
$$y = \quad 8{,}000$$

Step 5   Jeremy has $12,000 invested at 11% and $8,000 invested at 9%.

To check, the interest at 11% is ($12,000)(0.11), or $1,320. The interest at 9% is ($8,000)(0.09), or $720. The total interest is $2,040, and the solution is verified.

Check Yourself 12

Jan has $2,000 more invested in a stock that pays 9% interest than in a savings account paying 8%. If her total interest for 1 year is $860, how much does she have invested at each rate?

In Chapters 2 and 5, we solved **motion problems;** they involve a distance traveled, the rate, and the time of travel. Example 13 shows the use of $d = r \cdot t$ in forming a system of equations to solve a motion problem.

     Example 13   **Solving a Motion Problem**

A boat can travel 36 mi downstream in 2 h. Coming back upstream, the trip takes 3 h. Find the rate of the boat in still water and the rate of the current.

RECALL

Distance, rate, and time of travel are related by the equation

$d = r \cdot t$

Distance   Rate   Time

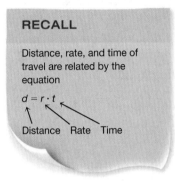

Step 1   You want to find the two rates (of the boat and the current).

Step 2   Let

x = rate of boat in still water

y = rate of current

Step 3   To write the equations, think about this: What is the effect of the current? Suppose the boat's rate in still water is 10 mi/h and the current is 2 mi/h.

The current *increases* the rate *downstream* to 12 mi/h (10 + 2). The current *decreases* the rate *upstream* to 8 mi/h (10 − 2). So here the rate downstream is $x + y$ and the rate upstream is $x - y$. At this point a chart of information is helpful.

	Distance	Rate	Time
Downstream	36	$x + y$	2
Upstream	36	$x - y$	3

From the relationship $d = r \cdot t$ we use our table to write the system

$36 = 2(x + y)$ From line 1 of our table

$36 = 3(x - y)$ From line 2 of our table

Step 4   Removing the parentheses in the equations of step 3, we have

$2x + 2y = 36$

$3x - 3y = 36$

Step 5   By either of our earlier methods, this system gives values of 15 for x and 3 for y. The rate in still water is 15 mi/h, and the rate of the current is 3 mi/h. We leave the check to you.

Check Yourself 13

A light plane flies 480 mi with the wind in 4 h. In returning against the wind, the trip takes 6 h. What is the rate of the plane in still air? What is the rate of the wind?

Check Yourself ANSWERS

1. $(2, 4)$ **2.** $(1, -2)$ **3.** $(5, -2)$ **4.** $(-6, 2)$ **5.** $(2, 1)$
6. There are infinitely many solutions. It is a dependent system.
7. There is no solution. The system is inconsistent.   **8.** Tape \$12, disk \$15
9. 7 ft, 13 ft   **10.** 150 \$6 tickets, 200 \$8 tickets
11. 100 mL (30%), 300 mL (50%)   **12.** \$4,000 at 8%, \$6,000 at 9%
13. Plane's rate in still air, 100 mi/h; wind's rate, 20 mi/h

Reading Your Text

The following fill-in-the-blank exercises are designed to ensure that you understand some of the key vocabulary used in this section.

SECTION 8.2

(a) When the coefficients of the y-terms are _____ of one another, adding the two equations eliminates the variable y.

(b) The addition method is also called solution by _____ .

(c) An equivalent system can always be obtained by _____ one or both of the equations by a nonzero constant so that the coefficients of x (or of y) are opposites.

(d) When a system is *dependent,* there are _____ solutions.

< Objective 1 >

Solve each system by addition. If a unique solution does not exist, state whether the system is inconsistent or dependent.

1. $x + y = 6$
$x - y = 4$

2. $x - y = 8$
$x + y = 2$

3. $2x - y = 1$
$-2x + 3y = 5$

4. $x - 2y = 2$
$x + 2y = -14$

5. $x + 2y = -2$
$3x + 2y = -12$

6. $4x - 3y = 22$
$4x + 5y = 6$

7. $2x + y = 8$
$2x + y = 2$

8. $5x + 4y = 7$
$5x - 2y = 19$

9. $3x - 5y = 2$
$2x - 5y = -2$

10. $2x - y = 4$
$2x - y = 6$

11. $x + y = 3$
$3x - 2y = 4$

 > Videos

12. $x - y = -2$
$2x + 3y = 21$

13. $-5x + 2y = -3$
$x - 3y = -15$

14. $x + 5y = 10$
$-2x - 10y = -20$

15. $5x + 2y = 28$
$x - 4y = -23$

16. $7x + 2y = 17$
$x - 5y = 13$

17. $3x - 4y = 2$
$-6x + 8y = -4$

18. $-x + 5y = 19$
$4x + 3y = -7$

19. $3x - 2y = 12$
$5x - 3y = 21$

20. $-4x + 5y = -6$
$5x - 2y = 16$

21. $7x + 4y = 20$
$5x + 6y = 19$

 > Videos

22. $5x + 4y = 5$
$7x - 6y = 36$

23. $2x - 7y = 6$
$-4x + 3y = -12$

24. $3x + 2y = -18$
$7x - 6y = -42$

Name _____

Section _____ Date _____

Answers

1. _____   2. _____
3. _____   4. _____
5. _____   6. _____
7. _____
8. _____   9. _____
10. _____
11. _____   12. _____
13. _____
14. _____
15. _____   16. _____
17. _____
18. _____   19. _____
20. _____   21. _____
22. _____   23. _____
24. _____

Answers

25. _____

26. _____

27. _____

28. _____

29. _____

30. _____

31. _____

32. _____

33. _____

34. _____

35. _____

36. _____

37. _____

38. _____

25. $5x - y = 20$
    $4x + 3y = 16$

26. $3x + y = -5$
    $5x - 4y = 20$

27. $3x + y = 1$
    $5x + y = 2$

28. $2x - y = 2$
    $2x + 5y = -1$

29. $5x - 2y = \dfrac{9}{5}$
    $3x + 4y = -1$

30. $2x + 3y = -\dfrac{1}{12}$
    $5x + 4y = \dfrac{2}{3}$

< Objective 2 >

Solve each problem. Be sure to show the equations used for the solution.

31. **NUMBER PROBLEM** The sum of two numbers is 40. Their difference is 8. Find the two numbers.

32. **NUMBER PROBLEM** Eight eagle stamps and two raccoon stamps cost $2.80. Three eagle stamps and four raccoon stamps cost $2.35. Find the cost of each kind of stamp.

33. **NUMBER PROBLEM** Xavier bought five red delicious apples and four Granny Smith apples at a cost of $4.81. Dean bought one of each kind at a cost of $1.08. Find the cost for each kind of apple. ⊙ > Videos

34. **NUMBER PROBLEM** Eight disks and five blank CDs cost a total of $27.50. Two disks and four blank CDs cost $16.50. Find the unit cost for each.

35. **CRAFTS** A 30-m rope is cut into two pieces so that one piece is 6 m longer than the other. How long is each piece?

36. **CRAFTS** An 18-ft board is cut into two pieces, one of which is twice as long as the other. How long is each piece?

37. **BUSINESS AND FINANCE** A coffee merchant has coffee beans that sell for $9 per pound and $12 per pound. The two types are to be mixed to create 100 lb of a mixture that will sell for $11.25 per pound. How much of each type of bean should be used in the mixture?

38. **BUSINESS AND FINANCE** Peanuts sell for $2 per pound and cashews sell for $5 per pound. How much of each type of nut is needed to create a 20-lb mixture that sells for $2.75 per pound?

39. SCIENCE AND MEDICINE A chemist has a 25% and a 50% acid solution. How much of each solution should be used to form 200 mL of a 35% acid solution?

25% acid   50% acid

40. SCIENCE AND MEDICINE A pharmacist wishes to prepare 150 mL of a 20% alcohol solution. She has a 30% solution and a 15% solution in her stock. How much of each should be used in forming the desired mixture?

41. BUSINESS AND FINANCE Otis has a total of $12,000 invested in two accounts. One account pays 8% and the other 9%. If his interest for 1 year is $1,010, how much does he have invested at each rate?

42. BUSINESS AND FINANCE Amy invests a part of $8,000 in bonds paying 12% interest. The remainder is in a savings account at 8%. If she receives $840 in interest for 1 year, how much does she have invested at each rate?

 > Videos

43. SCIENCE AND MEDICINE A light plane flies 450 mi with the wind in 3 h. Flying back against the wind, the plane takes 5 h to make the trip. What was the rate of the plane in still air? What was the rate of the wind?

44. SCIENCE AND MEDICINE An airliner made a trip of 1,800 mi in 3 h, flying east across the country with the jetstream directly behind it. The return trip, against the jetstream, took 4 h. Find the speed of the plane in still air and the speed of the jetstream.

Basic Skills	**Challenge Yourself**	Calculator/Computer	Career Applications	Above and Beyond

▲

Determine whether each statement is **true** *or* **false**.

45. Multiplying both sides of an equation by a nonzero constant will change the solutions for that equation.

46. When you have found the value of one variable, you can substitute it into either of the original equations to find the value of the other variable.

Answers

39. _____

40. _____

41. _____

42. _____

43. _____

44. _____

45. _____

46. _____

Answers

47. _____

48. _____

49. _____

50. _____

51. _____

52. _____

53. _____

54. _____

55. _____

Complete each statement with **never, sometimes,** *or* **always.**

47. Both variables are _____ eliminated when the equations of a linear system are added.

48. It is _____ possible to use the addition method to solve a linear system.

Solve the systems by adding. If a unique solution does not exist, state whether the system is inconsistent or dependent.

49. $\dfrac{x}{3} - \dfrac{y}{4} = -\dfrac{1}{2}$

$\dfrac{x}{2} - \dfrac{y}{5} = \dfrac{3}{10}$

50. $\dfrac{1}{3}x - \dfrac{1}{2}y = \dfrac{5}{6}$

$\dfrac{1}{2}x - \dfrac{2}{5}y = \dfrac{9}{10}$

51. $0.4x - 0.2y = 0.6$

$0.5x - 0.6y = 9.5$

52. $0.2x + 0.37y = 0.8$

$-0.6x + 1.4y = 2.62$

Basic Skills | Challenge Yourself | Calculator/Computer | **Career Applications** | Above and Beyond
▲

53. **CONSTRUCTION** In the vaulted split-level truss shown here, the dimensions x and y can be determined by the system of equations

$$\frac{3}{12}x + 8 = \frac{5}{12}y$$

$$x + 2 = y$$

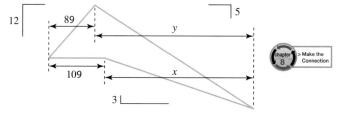

Solve this system using the addition method.

54. **MECHANICAL ENGINEERING** The forces (in pounds) on the arm of an industrial lift provide the equations

$1.5x + 2.4y = 90$

$2.7x - 1.8y = 30$

Use the addition method to solve this system. Round your results to the nearest hundredth.

55. **BUSINESS AND FINANCE** Production for this week is up by 2,600 units from last week. If total production for the two weeks is 27,200 units, the production for each week can be given by

$P_2 - 2{,}600 = P_1$

$P_1 + P_2 \quad = 27{,}200$

Solve the system of equations using the addition method.

56. **MANUFACTURING** A manufacturer produces drive assemblies and relays. The drive assemblies sell for $12 and the relays sell for $3. On a given day, 118 items are produced with a total value of $1,038. The production is described by the equations

$$12x + 3y = 1,038$$
$$x + y = 118$$

where x represents the number of drive assemblies and y represents the number of relays.

Solve this system using the addition method.

Basic Skills | Challenge Yourself | Calculator/Computer | Career Applications | **Above and Beyond**
▲

Work with a partner to solve the problem.

57. Your friend Valerie contacts you about going into business together. She wants to start a small manufacturing business making and selling sweaters to specialty boutiques. She explains that the initial investment for each of you will be $1,500 for a knitting machine. She has worked hard to come up with an estimate for expenses and thinks that they will be close to $1,600 a month for overhead. She says that each sweater manufactured will cost $28 to produce and that the sweaters will sell for at least $70. She wants to know if you are willing to invest the money you have saved for college costs. You have faith in Valerie's ability to carry out her plan. But, you have worked hard to save this money. Use graphs and equations to help you decide whether this is a good opportunity. Think about whether you need more information from Valerie. Write a letter summarizing your thoughts.

The Streeter/Hutchison Series in Mathematics   Beginning Algebra

© The McGraw-Hill Companies. All Rights Reserved.

Answers

56. _____

57. _____

Answers

1. $(5, 1)$ 3. $(2, 3)$ 5. $\left(-5, \dfrac{3}{2}\right)$ 7. Inconsistent system

9. $(4, 2)$ 11. $(2, 1)$ 13. $(3, 6)$ 15. $\left(3, \dfrac{13}{2}\right)$

17. Dependent system   19. $(6, 3)$ 21. $\left(2, \dfrac{3}{2}\right)$ 23. $(3, 0)$

25. $(4, 0)$ 27. $\left(\dfrac{1}{2}, -\dfrac{1}{2}\right)$ 29. $\left(\dfrac{1}{5}, -\dfrac{2}{5}\right)$ 31. 24, 16

33. Red delicious 49¢, Granny Smith 59¢   35. 18 m, 12 m
37. 25 lb at $9, 75 lb at $12   39. 120 mL of 25%, 80 mL of 50%
41. $7,000 at 8%, $5,000 at 9%   43. 120 mi/h, 30 mi/h
45. False   47. sometimes   49. $(3, 6)$ 51. $(-11, -25)$
53. $x = 43$ ft; $y = 45$ ft   55. $P_1 = 12,300; P_2 = 14,900$
57. Above and Beyond

Systems of Linear Equations: Solving by Substitution

< 8.3 Objectives >

1 > Solve systems using the substitution method

2 > Choose an appropriate method for solving a system

3 > Solve applications of systems of equations

In Sections 8.1 and 8.2, we looked at graphing and addition as methods of solving linear systems. A third method is called the **substitution method.**

Example 1	Solving a System by Substitution

< Objective 1 >

Solve by substitution.

$$x + y = 12$$
$$y = 3x$$

Notice that the second equation says that y and $3x$ name the same quantity. So we may substitute $3x$ for y in the first equation. We then have

Replace y with $3x$ in the first equation.

$$\downarrow$$

$$x + 3x = 12$$
$$4x = 12$$
$$x = 3$$

> **NOTE**
>
> The resulting equation contains only the variable x, so substitution is just another way of eliminating one of the variables from our system.

We can now substitute 3 for x in either original equation to find the corresponding y-coordinate of the solution. We use the first equation:

$$(3) + y = 12$$
$$y = 9$$

So $(3, 9)$ is the solution.

> **RECALL**
>
> The solution for a system is written as an ordered pair.

This last step is identical to the one you saw in Section 8.2. As before, you can substitute the known coordinate value back into either of the original equations to find the value of the remaining variable. The check is also identical.

 Check Yourself 1

Solve by substitution.

$$x - y = 9$$
$$y = 4x$$

The same technique can be readily used any time one of the equations is *already solved* for x or for y, as Example 2 illustrates.

| ▶ | **Example** 2 | Solving a System by Substitution |

Solve by substitution.

$$2x + 3y = 3$$
$$y = 2x - 7$$

Because the second equation tells us that y is $2x - 7$, we can replace y with $2x - 7$ in the first equation. This gives

NOTE

Now y is eliminated from the equation, and we can proceed to solve for x.

$$\overset{y}{2x + 3(\overline{2x - 7})} = 3 \qquad \text{Distribute the factor 3.}$$
$$2x + 6x - 21 = 3$$
$$8x = 24$$
$$x = 3$$

We now know that 3 is the x-coordinate for the solution. So substituting 3 for x in the second equation, we have

$$y = 2(3) - 7$$
$$= 6 - 7$$
$$= -1$$

The solution is $(3, -1)$. Once again you should verify this result by letting $x = 3$ and $y = -1$ in the original system.

Check Yourself 2

Solve by substitution.

$$2x - 3y = 6$$
$$x = 4y - 2$$

As we have seen, the substitution method works very well when one of the given equations is already solved for x or y. It is also useful if you can readily solve for x or for y in one of the equations.

| ▶ | **Example** 3 | Solving a System by Substitution |

Solve by substitution.

$$x - 2y = 5$$
$$3x + y = 8$$

Neither equation is solved for a variable. That is easily handled in this case. Solving for x in the first equation, we have

$$x = 2y + 5$$

Now substitute $2y + 5$ for x in the second equation.

$$\overset{x}{3(\overline{2y + 5})} + y = 8 \qquad \text{Distribute the factor 3.}$$
$$6y + 15 + y = 8$$
$$7y = -7$$
$$y = -1$$

Substituting -1 for y in the second equation yields

$$3x + (-1) = 8$$
$$3x = 9$$
$$x = 3$$

So $(3, -1)$ is the solution. You should check this result by substituting 3 for x and -1 for y in the equations of the original system.

Check Yourself 3

Solve by substitution.

$$3x - \ y = 5$$
$$x + 4y = 6$$

In Example 3, we could have solved the second equation for y instead. The second equation becomes $y = -3x + 8$. Then substituting $-3x + 8$ for y in the first equation gives

$$x - 2(-3x + 8) = 5$$
$$x + 6x - 16 = 5$$
$$7x = 21$$
$$x = 3$$

Substituting 3 for x in the second equation yields

$$3(3) + y = \ \ 8$$
$$y = -1$$

So, once again we see that the solution for the system is $(3, -1)$.

Inconsistent systems and dependent systems show up in a fashion similar to that which we saw in Section 8.2. Example 4 illustrates the approach to solving such systems.

Example 4   Solving Inconsistent or Dependent Systems

Solve each system by substitution.
(a) $4x - 2y = 6$

$$y = 2x - 3$$

From the second equation we substitute $2x - 3$ for y in the first equation.

$$4x - 2(2x - 3) = 6$$
$$4x - 4x + 6 = 6$$
$$6 = 6$$

Both variables have been eliminated, and we have the true statement $6 = 6$.

Recall from Section 8.2 that a true statement tells us that the graphs of the two equations are lines that coincide. We call this system dependent. There are an infinite number of solutions.

(b) $3x - 6y = 9$

$$x = 2y + 2$$

NOTE

Be sure to change both signs in the parentheses.

Substitute $2y + 2$ for x in the first equation.

$$3(2y + 2) - 6y = 9$$
$$6y + 6 - 6y = 9 \qquad \text{This time we have}$$
$$6 = 9 \qquad \text{a false statement.}$$

This means that the system is *inconsistent* and that the graphs of the two equations are parallel lines. There is no solution.

Check Yourself 4

Indicate whether the given system is inconsistent (no solution) or dependent (an infinite number of solutions).

(a) $5x + 15y = 10$
$\quad\; x = -3y + 1$

(b) $12x - 4y = 8$
$\qquad y = 3x - 2$

Here is a summary of our work in this section.

Step by Step

To Solve a System of Linear Equations by Substitution		
Step 1	Solve one of the given equations for x or y. If this is already done, go on to step 2.	
Step 2	Substitute this expression for x or for y into the other equation.	
Step 3	Solve the resulting equation for the remaining variable.	
Step 4	Substitute the known value into either of the original equations to find the value of the second variable.	
Step 5	State the result and check your solution in both of the original equations.	

You have seen three different ways to solve systems of linear equations: graphing, adding, and substitution. The natural question is, Which method should you use in a given situation?

Graphing is the least exact of the methods because solutions may have to be estimated.

The algebraic methods—addition and substitution—give exact solutions, and both always work for systems of linear equations. In fact, you may have noticed that several examples in this section could just as easily have been solved by adding (Example 3, for instance).

The choice of which algebraic method (substitution or addition) to use is yours and depends largely on the given system. Here are some guidelines to help you choose an appropriate method for solving a linear system.

Property

Choosing an Appropriate Method for Solving a System

1. If one of the equations is already solved for x (or for y), then substitution is the preferred method.
2. If the coefficients of x (or of y) in the two equations are the same, or opposites, then addition is the preferred method.
3. If solving for x (or for y) in either of the given equations results in fractional coefficients, then addition is the preferred method.

  Example 5   Choosing an Appropriate Method for Solving a System

< Objective 2 >

Select the most appropriate method to solve each system.

(a) $5x + 3y = 9$
$\quad\; 2x - 7y = 8$

Addition is the most appropriate method because solving for a variable results in fractional coefficients.

(b) $7x + 26 = 8$

$x = 3y - 5$

Substitution is the most appropriate method because the second equation is already solved for x.

(c) $8x - 9y = 11$

$4x + 9y = 15$

Addition is the most appropriate method because the coefficients of y are opposites.

 Check Yourself 5

Select the most appropriate method to solve each system.

(a) $2x + 5y = 3$
$8x - 5y = -13$

(b) $4x - 3y = 2$
$y = 3x - 4$

(c) $3x - 5y = 2$
$x = 3y - 2$

(d) $5x - 2y = 19$
$4x + 6y = 38$

It is quite possible that, for a given system of equations, both algebraic methods work equally well! We solved the system in Example 3 (twice!) by substitution. In Example 6, we return to the same system and use addition.

Example 6   Solving a System by Addition

Solve by the addition method.

$x - 2y = 5$
$3x + y = 8$

We multiply the second equation by 2 in order to eliminate y.

$x - 2y = 5$
$6x + 2y = 16$

Adding, we get

$7x = 21$
$x = 3$

Substituting this value for x in the second equation, we have

$3(3) + y = 8$
$y = -1$

So, again we obtain the solution $(3, -1)$. When solving a system, then, we should choose the method that seems simplest.

 Check Yourself 6

Solve the system two ways: by substitution and by addition.

$2x - 5y = 10$
$x - y = 8$

Number problems, such as those presented in Chapter 2, are sometimes more easily solved by the methods presented in this section. Example 7 illustrates this approach.

| Example 7 | Solving a Number Problem by Substitution |

< Objective 3 >

The sum of two numbers is 25. If the second number is 5 less than twice the first number, what are the two numbers?

RECALL

1. What do you want to find?
2. Assign variables. This time we use two letters, x and y.
3. Write equations for the solution. Here two equations are needed because we have introduced two variables.
4. Solve the system of equations.
5. State and check the result.

NOTE

We use the substitution method because the second equation is already solved for y.

Step 1   You want to find the two unknown numbers.

Step 2   Let $x =$ the first number and $y =$ the second number.

Step 3

$\underline{x + y} = 25$

The sum   is 25.

$y = \underline{2x - 5}$

The second   is 5 less than
number   twice the first.

Step 4

$x + y = 25$
$\quad y = 2x - 5$

Substitute $2x - 5$ for y in the first equation.

$x + (2x - 5) = 25$
$\qquad 3x - 5 = 25$
$\qquad\qquad x = 10$

From the first equation,

$(10) + y = 25$
$\qquad\quad y = 15$

Step 5   The two numbers are 10 and 15.

The sum of the numbers is 25. The second number, 15, is 5 less than twice the first number, 10. The solution checks.

Check Yourself 7

The sum of two numbers is 28. The second number is 4 more than twice the first number. What are the numbers?

Sketches are always helpful in solving applications from geometry. Let's look at such an example.

| Example 8 | Solving an Application from Geometry |

The length of a rectangle is 3 m more than twice its width. If the perimeter of the rectangle is 42 m, find the dimensions of the rectangle.

Step 1   You want to find the dimensions (length and width) of the rectangle.

Step 2   Let L be the length of the rectangle and W the width. Now draw a sketch of the problem.

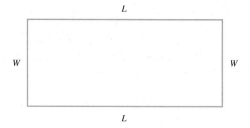

Step 3   Write the equations for the solution.

$L = 2W + 3$

3 more than twice the width

$2L + 2W = 42$

The perimeter

Step 4   Solve the system.

$$L = 2W + 3$$
$$2L + 2W = 42$$

From the first equation we can substitute $2W + 3$ for L in the second equation.

$$2(2W + 3) + 2W = 42$$
$$4W + 6 + 2W = 42$$
$$6W = 36$$
$$W = 6$$

Replace W with 6 to find L.

$$L = 2(6) + 3$$
$$= 12 + 3$$
$$= 15$$

Step 5   The length is 15 m, the width is 6 m.

Check these results. The length, 15 m, is 3 m more than twice the width, 6 m. The perimeter is $2L + 2W$, which should give us 42 m.

$$2(15) + 2(6) \stackrel{?}{=} 42$$
$$30 + 12 = 42 \qquad \text{True}$$

Check Yourself 8

The length of each of the two equal legs of an isosceles triangle is 5 in. less than the length of the base. If the perimeter of the triangle is 50 in., find the lengths of the legs and the base.

Our final example for this section comes from the field of mechanical engineering.

Example 9   An Engineering Application

The antifreeze concentration in an industrial cooling system needs to be at 45%. If the system holds 32 gallons of coolant and is currently at a concentration of 30%, how much of the solution needs to be removed and replaced with pure antifreeze to bring the concentration up to the required level?

This can be solved with the system of equations

$$0.30x + 1y = 14.4$$
$$x + y = 32$$

where x is the amount of coolant at 30%, and y is the amount removed and replaced with pure antifreeze. Solve this system of equations by substitution.

Solving the first equation for y, we have

$$y = 14.4 - 0.30x$$

We substitute the expression $14.4 - 0.30x$ for y in the second equation:

$$x + (14.4 - 0.30x) = 32$$
$$x + 14.4 - 0.30x = 32$$
$$0.7x = 17.6$$
$$x = \frac{17.6}{0.7} \approx 25.14$$

So, $y \approx 6.86$.

This means that about 6.86 gallons of antifreeze currently in the system need to be removed and replaced with pure antifreeze.

Check Yourself 9

The antifreeze concentration in an industrial cooling system needs to be at 50%. If the system holds 36 gallons of coolant and is currently at a concentration of 35%, how much of the solution needs to be removed and replaced with pure antifreeze to bring the concentration up to the required level?

This can be solved with the system of equations

$$0.35 + 1y = 18$$
$$x + y = 36$$

where x is the amount of coolant at 35%, and y is the amount removed and replaced with pure antifreeze. Solve this system of equations by substitution.

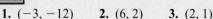

Check Yourself ANSWERS

1. $(-3, -12)$ **2.** $(6, 2)$ **3.** $(2, 1)$
4. **(a)** Inconsistent system; **(b)** dependent system
5. **(a)** Addition; **(b)** substitution; **(c)** substitution; **(d)** addition   **6.** $(10, 2)$
7. The numbers are 8 and 20.   **8.** The legs have length 15 in.; the base is 20 in.   **9.** $x = 27.7$ gal; $y = 8.3$ gal

Reading Your Text

The following fill-in-the-blank exercises are designed to ensure that you understand some of the key vocabulary used in this section.

SECTION 8.3

(a) The _____ method works very well when one of the given equations is already solved for x or y.

(b) When a system is inconsistent, the graphs of the two equations are _____ lines.

(c) _____ is the least exact of the methods for solving a system of equations.

(d) If solving for x (or for y) in either of the given equations will result in fractional coefficients, then the _____ is the preferred method.

Basic Skills | Challenge Yourself | Calculator/Computer | Career Applications | Above and Beyond
▲

< Objective 1 >

Solve each system by substitution.

1. $2x - y = 10$
    $x = -2y$

2. $x + 3y = 10$
    $3x = y$

3. $3x + 2y = 12$
    $y = 3x$

4. $4x - 3y = 24$
    $y = -4x$

5. $x - y = 4$
    $x = 2y - 2$

6. $x - y = 7$
    $y = 2x - 12$

7. $2x + y = 7$
    $y - x = -8$

8. $3x - y = -15$
    $x = y - 7$

9. $3x + 4y = 9$
    $y - 3x = 1$

10. $5x - 2y = -5$
    $y - 5x = 3$

11. $3x - 18y = 4$
    $x = 6y + 2$

12. $4x + 5y = 6$
    $y = 2x - 10$

13. $5x - 3y = 6$
    $y = 3x - 6$

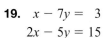 > Videos

14. $8x - 4y = 16$
    $y = 2x - 4$

15. $x + 3y = 7$
    $x - y = 3$

16. $2x - y = -4$
    $x + y = -5$

17. $6x - 3y = 9$
    $-2x + y = -3$

> Videos

18. $5x - 6y = 21$
    $x - 2y = 5$

19. $x - 7y = 3$
    $2x - 5y = 15$

20. $4x - 12y = 5$
    $-x + 3y = -1$

< Objectives 1–2 >

Solve each system by using either addition or substitution. If a unique solution does not exist, state whether the system is dependent or inconsistent.

21. $2x + 3y = -6$
    $x = 3y + 6$

22. $7x + 3y = 31$
    $y = -2x + 9$

23. $2x - y = 1$
    $-2x + 3y = 5$

24. $x + 3y = 12$
    $2x - 3y = 6$

Name _____

Section _____ Date _____

Answers

1. _____   2. _____

3. _____   4. _____

5. _____   6. _____

7. _____   8. _____

9. _____   10. _____

11. _____   12. _____

13. _____

14. _____

15. _____   16. _____

17. _____

18. _____   19. _____

20. _____   21. _____

22. _____   23. _____

24. _____

Answers

25. _____

26. _____

27. _____

28. _____

29. _____

30. _____

31. _____

32. _____

33. _____

34. _____

35. _____

36. _____

37. _____

25. $6x + 2y = 4$
$\quad\quad y = -3x + 2$

26. $3x - 2y = 15$
$\quad\quad -x + 5y = -5$

27. $x + 2y = -2$
$\quad 3x + 2y = -12$

28. $10x + 2y = 7$
$\quad\quad\quad y = -5x + 3$

29. $2x - 3y = 14$
$\quad 4x + 5y = -5$

30. $2x + 3y = 1$
$\quad 5x + 3y = 16$

< Objective 3 >

Solve each problem. Be sure to show the equation used for the solution.

31. Number Problem The sum of two numbers is 100. The second is 3 times the first. Find the two numbers.

32. Number Problem The sum of two numbers is 70. The second is 10 more than 3 times the first. Find the numbers.

33. Number Problem The sum of two numbers is 56. The second is 4 less than twice the first. What are the two numbers?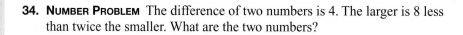

34. Number Problem The difference of two numbers is 4. The larger is 8 less than twice the smaller. What are the two numbers?

35. Number Problem The difference of two numbers is 22. The larger is 2 more than 3 times the smaller. Find the two numbers.

36. Number Problem One number is 18 more than another, and the sum of the smaller number and twice the larger number is 45. Find the two numbers.

37. Business and Finance Two packages together weigh 32 kilograms (kg). The smaller package weighs 6 kg less than the larger. How much does each package weigh?

38. **BUSINESS AND FINANCE** A washer-dryer combination costs $1,200. If the washer costs $220 more than the dryer, what does each appliance cost separately?

39. **SOCIAL SCIENCE** In a town election, the winning candidate had 220 more votes than the loser. If 810 votes were cast in all, how many votes did each candidate receive?

40. **BUSINESS AND FINANCE** An office desk and chair together cost $850. If the desk costs $50 less than twice as much as the chair, what did each cost?

41. **GEOMETRY** The length of a rectangle is 2 in. more than twice its width. If the perimeter of the rectangle is 34 in., find the dimensions of the rectangle.

42. **GEOMETRY** The perimeter of an isosceles triangle is 37 in. The length of each of the two equal legs is 6 in. less than 3 times the length of the base. Find the lengths of the three sides.

| Basic Skills | **Challenge Yourself** ▲ | Calculator/Computer | Career Applications | Above and Beyond |

Determine whether each statement is **true** *or* **false.**

43. Graphing is always the preferred method for solving a system.

44. The graphs of the equations are the same line in a dependent system.

Complete each statement with **never, always,** *or* **sometimes.**

45. It is _____ possible to use the substitution method to solve a linear system.

46. The substitution method is _____ easier to use than the addition method.

Answers

38. _____

39. _____

40. _____

41. _____

42. _____

43. _____

44. _____

45. _____

46. _____

Answers

47. _____

48. _____

49. _____

50. _____

51. _____

52. _____

53. _____

Solve each system.

47. $\dfrac{1}{3}x + \dfrac{1}{2}y = 5$

$\dfrac{x}{4} - \dfrac{y}{5} = -2$

48. $\dfrac{5x}{2} - y = \dfrac{9}{10}$

$\dfrac{3x}{4} + \dfrac{5y}{6} = \dfrac{2}{3}$

49. $0.4x - 0.2y = 0.6$

$2.5x - 0.3y = 4.7$

50. $0.4x - 0.1y = 5$

$6.4x + 0.4y = 60$

| Basic Skills | Challenge Yourself | Calculator/Computer | **Career Applications** | Above and Beyond |

51. ALLIED HEALTH A medical lab technician needs to determine how much 9% sulfuric acid (H_2SO_4) solution (x) must be mixed with 2% H_2SO_4 (y) to produce 75 milliliters of a 4% solution. From this information, the technician derives two linear equations:

$x + y = 75$
$9x + 2y = 300$

Solve this system of equations by substitution, and report how much of each type of solution is needed, to the nearest tenth of a milliliter. *chapter 8 > Make the Connection*

52. ALLIED HEALTH A medical lab technician needs to determine how much 40% alcohol solution (x) must be mixed with 25% alcohol (y) to produce 800 milliliters of a 35% solution. From this information, the technician derives two linear equations:

$x + y = 800$
$40x + 25y = 28{,}000$

Solve this system of equations by substitution, and report how much of each type of solution is needed, to the nearest tenth of a milliliter. *chapter 8 > Make the Connection*

53. ALLIED HEALTH A medical lab technician needs to determine how much 20% saline solution (x) must be mixed with 5% saline (y) to produce 100 milliliters of a 12% solution. From this information, the technician derives two linear equations:

$x + y = 100$
$20x + 5y = 1{,}200$

Solve this system of equations by substitution, and report how much of each type of solution is needed, to the nearest tenth of a milliliter. *chapter 8 > Make the Connection*

54. **ALLIED HEALTH** A medical lab technician needs to determine how much 8.2-molar (M) calcium chloride ($CaCl_2$) solution (x) must be mixed with 3.5-M $CaCl_2$ (y) to produce 400 milliliters of a 5-M solution. From this information, the technician derives two linear equations:

$$x + \quad y = 400$$
$$8.2x + 3.5y = 2,000$$

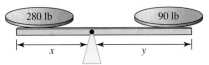

Solve this system of equations by substitution, and report how much of each type of solution is needed, to the nearest tenth of a milliliter.

55. **CONSTRUCTION TECHNOLOGY** A 24-ft beam has a weight on each end. Find the point where a pivot can be placed so that the beam balances.

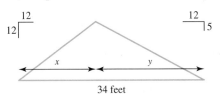

This system of equations can be used to find the balance point:

$$280x = 90y$$
$$x + y = 24$$

Solve this system of equations by substitution (round your results to the nearest hundredth).

56. **CONSTRUCTION TECHNOLOGY** A customer wants a dual-slope roof with one side having a 12/12 slope and the other side having a 5/12 slope. If the total span of the roof is to be 34 feet, find the length of the 5/12 span and the 12/12 span.

34 feet

This truss is described by this system of equations:

$$\frac{12}{12}x = \frac{5}{12}y$$
$$x + y = 34$$

Solve this system of equations.

Basic Skills | Challenge Yourself | Calculator/Computer | Career Applications | **Above and Beyond**
▲

57. You have a part-time job writing the *Consumer Concerns* column for your local newspaper. Your topic for this week is clothes dryers, and you are planning to compare the Helpmate and the Whirlgarb dryers, both readily available in stores in your area. The information you have is that the Helpmate dryer is listed at $520, and it costs 22.5¢ to dry an average-sized load at the utility rates in your city. The Whirlgarb dryer is listed at $735, and it costs 15.8¢ to run for each normal load. The maintenance costs for both dryers are about the same. Working with a partner, write a short article giving your readers helpful advice about these appliances. What should they consider when buying one of these clothes dryers?

Answers

54. _____

55. _____

56. _____

57. _____

Answers

1. $(4, -2)$ **3.** $\left(\dfrac{4}{3}, 4\right)$ **5.** $(10, 6)$ **7.** $(5, -3)$ **9.** $\left(\dfrac{1}{3}, 2\right)$

11. No solution   **13.** $(3, 3)$ **15.** $(4, 1)$

17. Infinite number of solutions   **19.** $(10, 1)$ **21.** $(0, -2)$

23. $(2, 3)$ **25.** Dependent system   **27.** $\left(-5, \dfrac{3}{2}\right)$ **29.** $\left(\dfrac{5}{2}, -3\right)$

31. 25, 75   **33.** 20, 36   **35.** 32, 10   **37.** 13 kg, 19 kg
39. Winner 515, loser 295   **41.** Width 5 in., length 12 in.
43. False   **45.** always   **47.** $(0, 10)$ **49.** $(2, 1)$
51. 21.4 mL of 9%; 53.6 mL of 2%   **53.** 46.7 mL of 20%; 53.3 mL of 5%
55. $x = 5.84$ ft; $y = 18.16$ ft   **57.** Above and Beyond

8.4

Systems of Linear Inequalities

< 8.4 Objectives >

1 > Graph a system of linear inequalities

2 > Solve an application of linear inequalities

NOTE

You might want to review graphing linear inequalities in Section 7.4 at this point.

Our previous work in this chapter dealt with finding the solution set of a system of linear equations. That solution set represented the points of intersection of the graphs of the equations in the system. In this section, we extend that idea to include systems of linear inequalities.

In this case, the solution set for each inequality is all ordered pairs that satisfy that inequality. *The graph of the solution set of a system of linear inequalities* is then the intersection of the graphs of the individual inequalities.

| Example 1 | Solving a System by Graphing |

< Objective 1 >

Solve the system of linear inequalities by graphing.

$$x + y > 4$$
$$x - y < 2$$

We start by graphing each inequality separately. The boundary line is drawn, and using $(0, 0)$ as a test point, we see that we should shade the half-plane above the line in both graphs.

RECALL

The boundary line is dashed to indicate it is *not* included in the graph.

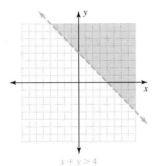

$x + y > 4$

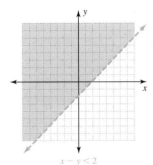

$x - y < 2$

NOTE

Points on the lines are not included in the solution set.

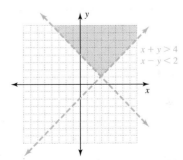

$x + y > 4$
$x - y < 2$

In practice, the graphs of the two inequalities are combined on the same set of axes, as shown in the graph at the left. This graph of the solution set of the original system is the intersection of the two original graphs.

© The McGraw-Hill Companies. All Rights Reserved.   The Streeter/Hutchison Series in Mathematics   Beginning Algebra

✓ Check Yourself 1

Solve the system of linear inequalities by graphing.

$2x - y < 4$

$x + y < 3$

In applications we often have restrictions on the possible values of x and y. We consider such restrictions in Example 2.

Example 2   **Solving a System by Graphing**

Solve the system of linear inequalities by graphing.

$1 \leq x \leq 3$

$2 \leq y \leq 5$

RECALL

The boundaries are solid because the symbol reads "less than or equal to."

The statement $1 \leq x \leq 3$ says that $1 \leq x$ *and* $x \leq 3$. We graph the vertical lines $x = 1$ and $x = 3$ as solid lines, and then shade points to the right of 1 and to the left of 3. Similarly, to graph $2 \leq y \leq 5$, we graph horizontal lines $y = 2$ and $y = 5$, and we shade between.

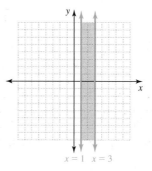
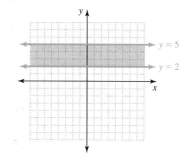

Putting this together, we have the solution.

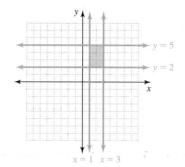

✓ Check Yourself 2

Solve the system of linear inequalities by graphing.

$-2 \leq x \leq 3$

$-4 \leq y \leq -1$

Most applications of systems of linear inequalities lead to bounded regions. This generally requires a system of three or more inequalities, as shown in Example 3.

Example 3	Solving a System by Graphing

Solve the system of linear inequalities by graphing.

$$2x + y \leq 10$$
$$x + y \leq 7$$
$$x \geq 1$$
$$y \geq 2$$

On the same set of axes, we graph the boundary line of each of the inequalities. For example, the graph shows the boundary line for $2x + y \leq 10$.

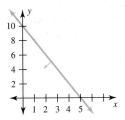

We then decide, since $(0, 0)$ is a solution, to shade below the line. Rather than shading now, we indicate the direction of solutions with an arrow.

Continuing in this fashion, we graph each boundary line on the same set of axes. The set of solutions is the intersection of the four indicated regions.

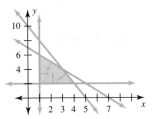

Check Yourself 3

Solve the system of linear inequalities by graphing.

$$2x - y \leq 8$$
$$x + y \leq 7$$
$$x \geq 0$$
$$y \geq 0$$

Next, we look at an application of our work with systems of linear inequalities.

Example 4	Solving a Business Application

< Objective 2 >

A manufacturer produces a standard model and a deluxe model of a 13-in. television set. The standard model requires 12 h of labor to produce, whereas the deluxe model requires 18 h. The labor available is limited to 360 h per week. Also, the plant capacity is limited to producing a total of 25 sets per week. Write a system of inequalities representing this situation. Then, draw a graph of the region representing the number of sets that can be produced, given these conditions.

We let x represent the number of standard-model sets produced and y the number of deluxe-model sets. Because the labor is limited to 360 h, we have

NOTE

The total labor is limited to (or less than or equal to) 360 h.

$$12x \quad + \quad 18y \leq 360$$

↑   ↑
12 h per   18 h per
standard set   deluxe set

The total production, here $x + y$ sets, is limited to 25, so we can write

$$x + y \leq 25$$

NOTE

We have $x \geq 0$ and $y \geq 0$ because the number of sets produced cannot be negative.

For convenience in graphing, we divide both expressions in the first inequality by 6, to write the equivalent system:

$$2x + 3y \leq 60$$
$$x + y \leq 25$$
$$x \geq 0$$
$$y \geq 0$$

NOTE

The shaded area is called the *feasible region*. All points in the region meet the given conditions of the problem and represent possible production options.

We now graph the system of inequalities as before. The shaded area represents all possibilities in terms of the number of sets that can be produced.

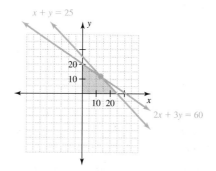

Check Yourself 4

A manufacturer produces DVD players and compact disk players. The DVD players require 10 h of labor to produce and the disk players require 20 h. The labor hours available are limited to 300 h per week. Existing orders require that at least 10 DVD players and at least 5 disk players be produced per week.

Write a system of inequalities representing this situation. Then, draw a graph of the region representing the possible production options.

Check Yourself ANSWERS

1. $2x - y < 4$

$x + y < 3$

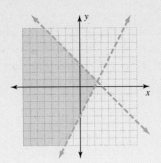

2.

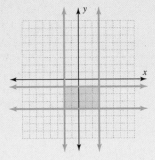

3. $2x - y \leq 8$

$x + y \leq 7$

$x \geq 0$

$y \geq 0$

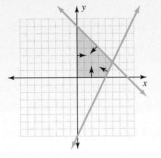

4. Let x be the number of DVD players and y be the number of CD players. The system is

$10x + 20y \leq 300$

$x \geq 10$

$y \geq 5$

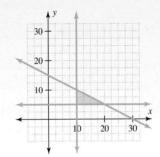

Reading Your Text

The following fill-in-the-blank exercises are designed to ensure that you understand some of the key vocabulary used in this section.

SECTION 8.4

(a) The graph of the solution set of a system of linear inequalities is the _____ of the graphs of the individual inequalities.

(b) The boundary line is _____ to indicate that it is *not* part of the solution set.

(c) Most applications of systems of linear inequalities lead to _____ regions.

(d) The shaded area in an application problem is called the _____ region.

Name _____

Section _____ Date

Answers

1. _____

2. _____

3. _____

4. _____

5. _____

6. _____

< Objective 1 >

Solve each system of linear inequalities graphically.

1. $x + 2y \le 4$
$\quad x - y \ge 1$

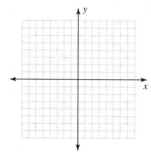

2. $3x - y > 6$
$\quad x + y < 6$

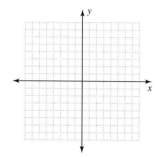

3. $3x + y < 6$
$\quad x + y > 4$

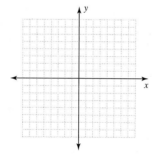

4. $2x + y \ge 8$
$\quad x + y \ge 4$

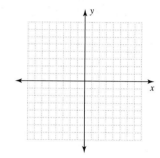

5. $x + 3y \le 12$
$\quad 2x - 3y \le 6$ 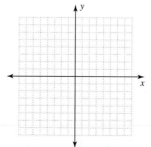 > Videos

6. $x - 2y > 8$
$\quad 3x - 2y > 12$

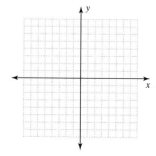

7. $3x + 2y \leq 12$
     $x \geq 2$

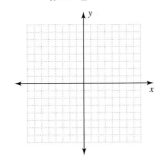

8. $2x + y \leq 6$
     $y \geq 1$

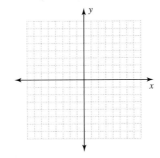

9. $2x + y \leq 8$
     $x > 1$
     $y > 2$

     > Videos

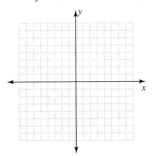

10. $3x - y \leq 6$
      $x \geq 1$
      $y \leq 3$

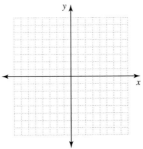

11. $x + 2y \leq 8$
    $2 \leq x \leq 6$
      $y \geq 0$

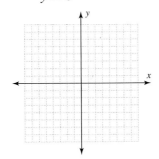

12. $x + y < 6$
    $0 \leq y \leq 3$
      $x \geq 1$

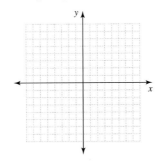

13. $3x + y \leq 6$
    $x + y \leq 4$
      $x \geq 0$
      $y \geq 0$

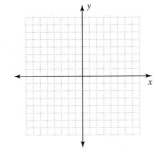

14. $x - 2y \geq -2$
    $x + 2y \leq 6$
      $x \geq 0$
      $y \geq 0$

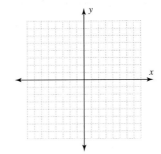

Answers

7. _____

8. _____

9. _____

10. _____

11. _____

12. _____

13. _____

14. _____

Answers

15. _____

16. _____

17. _____

18. _____

19. _____

15. $4x + 3y \leq 12$
$x + 4y \leq 8$
$x \geq 0$
$y \geq 0$

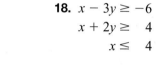

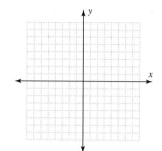

16. $2x + y \leq 8$
$x + y \geq 3$
$x \geq 0$
$y \geq 0$

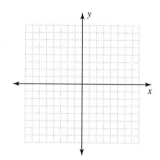

17. $x - 4y \leq -4$
$x + 2y \leq 8$
$x \geq 2$

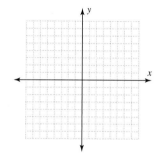

18. $x - 3y \geq -6$
$x + 2y \geq 4$
$x \leq 4$

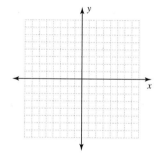

< Objective 2 >

Draw the appropriate graphs in each case.

19. **BUSINESS AND FINANCE** A manufacturer produces both two-slice and four-slice toasters. The two-slice toaster takes 6 h of labor to produce and the four-slice toaster 10 h. The labor available is limited to 300 h per week, and the total production capacity is 40 toasters per week. Write a system of inequalities representing this situation. Then, draw a graph of the feasible region given these conditions, in which x is the number of two-slice toasters and y is the number of four-slice toasters.

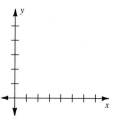

20. BUSINESS AND FINANCE A small firm produces both AM and AM/FM car radios. The AM radios take 15 h to produce, and the AM/FM radios take 20 h. The number of production hours is limited to 300 h per week. The plant's capacity is limited to a total of 18 radios per week, and existing orders require that at least 4 AM radios and at least 3 AM/FM radios be produced per week. Write a system of inequalities representing this situation. Then, draw a graph of the feasible region given these conditions, in which x is the number of AM radios and y the number of AM/FM radios.

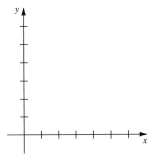

Answers

20. _____

21. _____

22. _____

23. _____

24. _____

25. _____

| Basic Skills | **Challenge Yourself** | Calculator/Computer | Career Applications | Above and Beyond |

Determine whether each statement is **true** *or* **false.**

21. The boundary lines in a system of linear inequalities are always drawn as dashed lines.

22. If the boundary lines are drawn as dashed lines, points on the lines are included in the solution set.

Complete each statement with **never, sometimes,** *or* **always.**

23. The graph of the solution set of a system of two linear inequalities _____ includes the origin.

24. The graph of the solution set of a system of two linear inequalities is _____ bounded.

25. Write the system of inequalities whose graph is the shaded region.

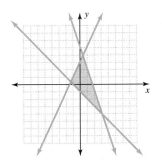

Answers

26. _____

27. _____

26. Write the system of inequalities whose graph is the shaded region.

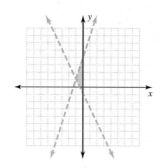

Basic Skills | Challenge Yourself | Calculator/Computer | Career Applications | **Above and Beyond**
▲

27. Describe a system of linear inequalities for which there is no solution.

Answers

1. $x + 2y \leq 4$
$x - y \geq 1$

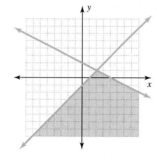

3. $3x + y < 6$
$x + y > 4$

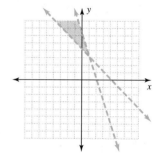

5. $x + 3y \leq 12$
$2x - 3y \leq 6$

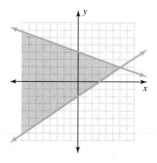

7. $3x + 2y \leq 12$
$x \geq 2$

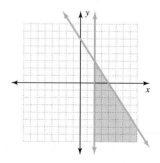

9. $2x + y \le 8$
$\quad\quad x > 1$
$\quad\quad y > 2$

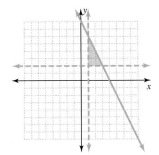

11. $x + 2y \le 8$
$\quad\quad 2 \le x \le 6$
$\quad\quad y \ge 0$

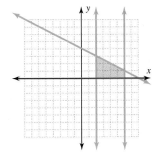

13. $3x + y \le 6$
$\quad\quad x + y \le 4$
$\quad\quad x \ge 0$
$\quad\quad y \ge 0$

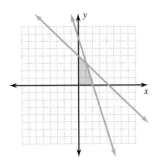

15. $4x + 3y \le 12$
$\quad\quad x + 4y \le 8$
$\quad\quad x \ge 0$
$\quad\quad y \ge 0$

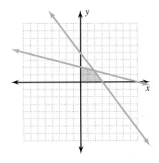

17. $x - 4y \le -4$
$\quad\quad x + 2y \le 8$
$\quad\quad x \ge 2$

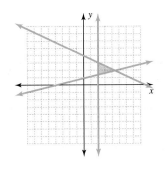

19. $6x + 10y \le 300$
$\quad\quad x + y \le 40$
$\quad\quad x \ge 0$
$\quad\quad y \ge 0$

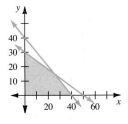

21. False   **23.** sometimes   **25.** $y \le 2x + 3$
$\quad\quad\quad\quad\quad\quad\quad\quad\quad\quad\quad\quad\quad\quad y \le -3x + 5$
$\quad\quad\quad\quad\quad\quad\quad\quad\quad\quad\quad\quad\quad\quad y \ge -x - 1$

27. Above and Beyond

Definition/Procedure	Example	Reference
Systems of Linear Equations: Solving by Graphing		Section 8.1
A System of Equations		
Two or more equations considered together.	$x + y = 4$ $2x - y = 5$	*p.* 605
Solution		
A solution for a system of two equations in two unknowns is an ordered pair that satisfies each equation of the system.	$(x, y) = (3, 1)$	*p.* 605
Solving by Graphing		

1. Graph both equations on the same coordinate system.
 2. The system may have
   a. *One solution.* The lines intersect at one point (a consistent system). The solution is the ordered pair corresponding to that point.
   b. *No solution.* The lines are parallel (an inconsistent system).
   c. *Infinitely many solutions.* The two equations have the same graph (a dependent system). Any ordered pair corresponding to a point on the line is a solution.
 3. If the system has exactly one solution, you should check that solution in both original equations.
	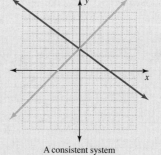
	 A consistent system
	
	 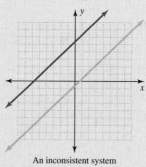
	 An inconsistent system
	
	 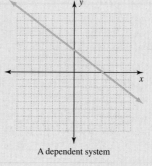
	 A dependent system
		p. 608

Definition/Procedure	Example	Reference

Systems of Linear Equations: Solving by the Addition Method

Solving by Adding

1. If necessary, multiply both sides of one or both equations by nonzero numbers to form an equivalent system in which the coefficients of one of the variables are opposites.
2. Add the equations of the new system.
3. Solve the resulting equation for the remaining variable.
4. Substitute the value found in step 3 into either of the original equations to find the value of the second variable.
5. State and check your solution in both of the original equations.

Example:

$2x - y = 4$
$3x + 2y = 13$

Multiply the first equation by 2.

$4x - 2y = 8$
$3x + 2y = 13$

Add.

$7x = 21$
$x = 3$

In the original equation,

$2(3) - y = 4$
$y = 2$

$(3, 2)$ is the solution.

Reference: Section 8.2 — p. 622

Applying Systems of Equations

Often word problems can be solved by using two variables and two equations to represent the unknowns and the given relationships in the problem.

The Solution Steps

1. Read the problem carefully. Then reread it to decide what you are asked to find.
2. Assign variables to the unknown quantities.
3. Translate the problem to the language of algebra to form a system of equations.
4. Solve the system.
5. State the solution and verify your result in the original problem.

Reference: p. 624

Systems of Linear Equations: Solving by Substitution

Solving by Substitution

1. Solve one of the given equations for x or for y. If this is already done, go on to step 2.
2. Substitute this expression for x or for y into the other equation.
3. Solve the resulting equation for the remaining variable.
4. Substitute the value found in step 3 into either of the original equations to find the value of the second variable.
5. State the solution and check your result in both of the original equations.

Example:

$x - 2y = 3$
$2x + 3y = 13$

From the first equation,

$x = 2y + 3$

Substitute in the second equation:

$2(2y + 3) + 3y = 13$
$4y + 6 + 3y = 13$
$7y + 6 = 13$
$7y = 7$
$y = 1$

$x = 2(1) + 3$
$x = 5$

$(5, 1)$ is the solution.

Reference: Section 8.3 — p. 639

Continued

Definition/Procedure	Example	Reference

Systems of Linear Inequalities

Section 8.4

A *system of linear inequalities* is two or more linear inequalities considered together. The *graph of the solution set* of a system of linear inequalities is the intersection of the graphs of the individual inequalities.

Solving Systems of Linear Inequalities Graphically

1. Graph each inequality, shading the appropriate half-plane, on the same set of coordinate axes.
2. The graph of the system is the intersection of the regions shaded in step 1.

To solve

$$x + 2y \leq 8$$
$$x + y \leq 6$$
$$x \geq 0$$
$$y \geq 0$$

graphically

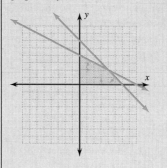

p. 651

This summary exercise set is provided to give you practice with each of the objectives of this chapter. Each exercise is keyed to the appropriate chapter section. When you are finished, you can check your answers to the odd-numbered exercises against those presented in the back of the text. If you have difficulty with any of these questions, go back and reread the examples from that section. Your instructor will give you guidelines on how best to use these exercises in your instructional setting.

8.1 *Solve each system by graphing.*

1. $x + y = 6$
$x - y = 2$

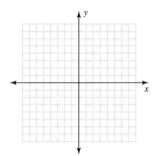

2. $x - y = 8$
$2x + y = 7$

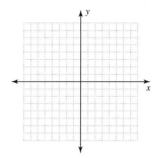

3. $x + 2y = 4$
$x + 2y = 6$

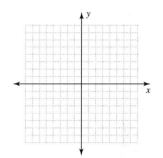

4. $2x - y = 8$
$y = 2$

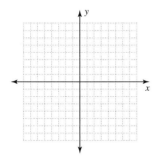

5. $2x - 4y = 8$
$x - 2y = 4$

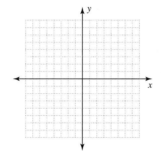

6. $3x + 2y = 6$
$4x - y = 8$

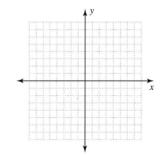

8.2 *Solve each system by addition. If a unique solution does not exist, state whether the system is inconsistent or dependent.*

7. $x + y = 8$
$x - y = 2$

8. $-x - y = 4$
$x - y = -8$

9. $2x - 3y = 16$
$5x + 3y = 19$

10. $2x + y = 7$
$3x - y = 3$

11. $3x - 5y = 14$
$3x + 2y = 7$

12. $2x - 4y = 8$
$x - 2y = 4$

13. $4x - 3y = -22$
$4x + 5y = -6$

14. $5x - 2y = 17$
$3x - 2y = 9$

15. $4x - 3y = 10$
$2x - 3y = 6$

16. $2x + 3y = -10$
  $-2x + 5y = 10$

17. $3x + 2y = 3$
  $6x + 4y = 5$

18. $3x - 2y = 23$
  $x + 5y = -15$

19. $5x - 2y = -1$
  $10x + 3y = 12$

20. $x - 3y = 9$
  $5x - 15y = 45$

21. $2x - 3y = 18$
  $5x - 6y = 42$

22. $3x + 7y = 1$
  $4x - 5y = 30$

23. $5x - 4y = 12$
  $3x + 5y = 22$

24. $6x + 5y = -6$
  $9x - 2y = 10$

25. $4x - 3y = 7$
  $-8x + 6y = -10$

26. $3x + 2y = 8$
  $-x - 5y = -20$

27. $3x - 5y = -14$
  $6x + 3y = -2$

8.3 *Solve each system by substitution. If a unique solution does not exist, state whether the system is inconsistent or dependent.*

28. $x + 2y = 10$
  $y = 2x$

29. $x - y = 10$
  $x = -4y$

30. $2x - y = 10$
  $x = 3y$

31. $2x + 3y = 2$
  $y = x - 6$

32. $4x + 2y = 4$
  $y = 2 - 2x$

33. $x + 5y = 20$
  $x = y + 2$

34. $6x + y = 2$
  $y = 3x - 4$

35. $2x + 6y = 10$
  $x = 6 - 3y$

36. $2x + y = 9$
  $x - 3y = 22$

37. $x - 3y = 17$
  $2x + y = 6$

38. $2x + 3y = 4$
  $y = 2$

39. $4x - 5y = -2$
  $x = -3$

40. $-6x + 3y = -4$
  $y = -\dfrac{2}{3}$

41. $5x - 2y = -15$
  $y = 2x + 6$

42. $3x + y = 15$
  $x = 2y + 5$

Solve each system by either addition or substitution. If a unique solution does not exist, state whether the system is inconsistent or dependent.

43. $x - 4y = 0$
$4x + y = 34$

44. $2x + y = 2$
$y = -x$

45. $3x - 3y = 30$
$x = -2y - 8$

46. $5x + 4y = 40$
$x + 2y = 11$

47. $x - 6y = -8$
$2x + 3y = 4$

48. $4x - 3y = 9$
$2x + y = 12$

49. $9x + y = 9$
$x + 3y = 14$

50. $3x - 2y = 8$
$-6x + 4y = -16$

51. $3x - 2y = 8$
$2x - 3y = 7$

Solve each problem. Be sure to show the equations used.

52. NUMBER PROBLEM The sum of two numbers is 40. If their difference is 10, find the two numbers.

53. NUMBER PROBLEM The sum of two numbers is 17. If the larger number is 1 more than 3 times the smaller, what are the two numbers?

54. NUMBER PROBLEM The difference of two numbers is 8. The larger number is 2 less than twice the smaller. Find the numbers.

55. BUSINESS AND FINANCE Five writing tablets and three pencils cost $8.25. Two tablets and two pencils cost $3.50. Find the cost for each item.

56. CONSTRUCTION A cable 200 ft long is cut into two pieces so that one piece is 12 ft longer than the other. How long is each piece?

57. BUSINESS AND FINANCE An amplifier and a pair of speakers cost $925. If the amplifier costs $75 more than the speakers, what does each cost?

58. BUSINESS AND FINANCE A sofa and chair cost $850 as a set. If the sofa costs $100 more than twice as much as the chair, what is the cost of each?

59. GEOMETRY The length of a rectangle is 4 cm more than its width. If the perimeter of the rectangle is 64 cm, find the dimensions of the rectangle.

60. **GEOMETRY** The perimeter of an isosceles triangle is 29 in. The lengths of the two equal legs are 2 in. more than twice the length of the base. Find the lengths of the three sides.

61. **NUMBER PROBLEM** Darryl has 30 coins with a value of $5.50. If they are all nickels and quarters, how many of each kind of coin does he have?

62. **BUSINESS AND FINANCE** Tickets for a concert sold for $11 and $8. If 600 tickets were sold for one evening and the receipts were $5,550, how many of each kind of ticket were sold?

63. **SCIENCE AND MEDICINE** A laboratory has a 20% acid solution and a 50% acid solution. How much of each should be used to produce 600 mL of a 40% acid solution?

64. **SCIENCE AND MEDICINE** A service station wishes to mix 40 L of a 78% antifreeze solution. How many liters of a 75% solution and a 90% solution should be used in forming the mixture?

65. **BUSINESS AND FINANCE** Martha has $18,000 invested. Part of the money is invested in a bond that yields 11% interest. The remainder is in her savings account, which pays 7%. If she earns $1,660 in interest for 1 year, how much does she have invested at each rate?

66. **SCIENCE AND MEDICINE** A boat travels 24 mi upstream in 3 h. It then takes 3 h to go 36 mi downstream. Find the speed of the boat in still water and the speed of the current.

67. **SCIENCE AND MEDICINE** A plane flying with the wind makes a trip of 2,200 mi in 4 h. Returning against the wind, it can travel only 1,800 mi in 4 h. What is the plane's rate in still air? What is the wind speed?

8.4 *Solve each system of linear inequalities.*

68. $x - y < 7$
    $x + y > 3$

69. $x - 2y \leq -2$
    $x + 2y \leq 6$

70. $x - 6y < 6$
    $-x + y < 4$

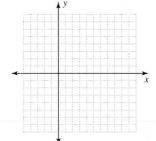
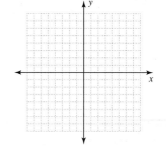

71. $2x + y \leq 8$
$ x \geq 1$
$ y \geq 0$

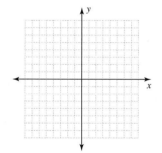

72. $2x + y \leq 6$
$ x \geq 1$
$ y \geq 0$

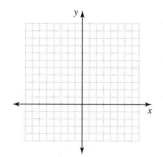

73. $4x + y \leq 8$
$ x \geq 0$
$ y \geq 2$

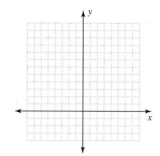

74. $4x + 2y \leq 8$
$x + y \leq 3$
$ x \geq 0$
$ y \geq 0$

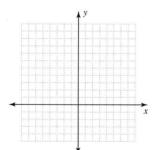

75. $3x + y \leq 6$
$x + y \leq 4$
$ x \geq 0$
$ y \geq 0$

Beginning Algebra

The Streeter/Hutchison Series in Mathematics

© The McGraw-Hill Companies. All Rights Reserved.

Name _____

Section _____ Date _____

Answers

1. _____

2. _____

3. _____

4. _____

5. _____

6. _____

7. _____

8. _____

9. _____

10. _____

11. _____

12. _____

13. _____

The purpose of this self-test is to help you assess your progress so that you can find concepts that you need to review before the next exam. Allow yourself about an hour to take this test. At the end of that hour, check your answers against those given in the back of this text. If you miss any, go back to the appropriate section to reread the examples until you have mastered that particular concept.

Solve each system by addition. If a unique solution does not exist, state whether the system is inconsistent or dependent.

1. $x + y = 5$
   $x - y = 3$

2. $x + 2y = 8$
   $x - \ y = 2$

3. $\ \ 3x + \ y = 6$
   $-3x + 2y = 3$

4. $3x + 2y = 11$
   $5x + 2y = 15$

Solve each system by substitution. If a unique solution does not exist, state whether the system is inconsistent or dependent.

5. $x + y = 8$
   $\ \ \ \ \ y = 3x$

6. $x - y = 9$
   $\ \ \ \ x = -2y$

7. $2x - \ y = 10$
   $\ \ \ \ \ \ x = y + 4$

8. $x - 3y = -7$
   $\ \ \ \ \ \ y = x - 1$

Solve each system by graphing. If a unique solution does not exist, state whether the system is inconsistent or dependent.

9. $x + y = 5$
   $x - y = 3$

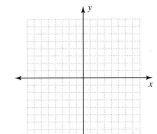

10. $x + 2y = 8$
    $x - \ y = 2$

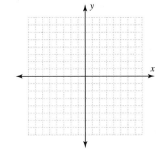

Solve each problem. Be sure to show the equations used.

11. **NUMBER PROBLEM** The sum of two numbers is 30, and their difference is 6. Find the two numbers.

12. **CONSTRUCTION** A rope 50 m long is cut into two pieces so that one piece is 8 m longer than the other. How long is each piece?

13. **GEOMETRY** The length of a rectangle is 4 in. less than twice its width. If the perimeter of the rectangle is 64 in., what are the dimensions of the rectangle?

Solve each system by addition. If a unique solution does not exist, state whether the system is inconsistent or dependent.

14. $3x - 6y = 12$
$\quad\ x - 2y = \ \ 4$

15. $4x + \ \ y = 2$
$\quad\ 8x - 3y = 9$

16. $2x - 5y = \ \ 2$
$\quad\ 3x + 4y = 26$

17. $\ \ x + 3y = 6$
$\quad 3x + 9y = 9$

Solve each system by substitution. If a unique solution does not exist, state whether the system is inconsistent or dependent.

18. $3x + y = -6$
$\quad\quad\ y = 2x + 9$

19. $4x + 2y = 8$
$\quad\quad\ \ y = 3 - 2x$

20. $5x + \ \ y = \ \ 10$
$\quad\ \ x + 2y = -7$

21. $3x - 2y = 5$
$\quad\ 2x + \ \ y = 8$

Solve each system of inequalities.

22. $x + \ \ y < 3$
$\quad\ x - 2y < 6$

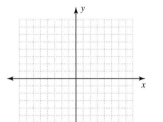

23. $4y + 3x \geq 12$
$\quad\quad\quad\ x \geq 1$

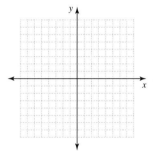

24. $2y + x \leq 8$
$\quad\ \ y + x \leq 6$
$\quad\quad\quad\ x \geq 0$
$\quad\quad\quad\ y \geq 0$

Solve each problem. Be sure to show the equations used.

25. **NUMBER PROBLEM** Murray has 30 coins with a value of $5.70. If the coins are all dimes and quarters, how many of each coin does he have?

26. **SCIENCE AND MEDICINE** Jackson was able to travel 36 miles downstream in 2 hours. In returning upstream, it took 3 hours to make the trip. How fast can his boat travel in still water? What was the rate of the river current?

Answers

14. _____

15. _____

16. _____

17. _____

18. _____

19. _____

20. _____

21. _____

22. _____

23. _____

24. _____

25. _____

26. _____

Answers

27. _____

28. _____

Solve each system by graphing. If a unique solution does not exist, state whether the system is inconsistent or dependent.

27. $x - 3y = 3$
$\quad\ x - 3y = 6$

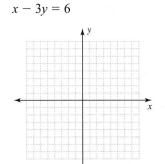

28. $4x - \ y = \ \ 4$
$\quad\ \ x - 2y = -6$

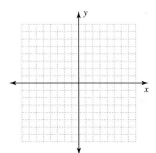

Activity 8 ::
Growth of Children—Fitting
a Linear Model to Data

When you walk into a home where children have lived, you can often find a wall with notches showing the heights of the children at various points in their lives. Nearly every time you bring a child to see the doctor, the child's height and weight are recorded, regardless of the reason for the office visit.

The National Institutes of Health (NIH) through the Centers for Disease Control and Prevention (CDC) publishes and updates data detailing heights and weights of children for the populace as a whole and for what they consider to be healthy children. These data were most recently updated in November 2000.

In this activity, we explore the graphing of trend data and its predictive capability. We use a child's height and weight at various ages for our data.

I.

1. If there is a child from 2 to 5 years old in your family, try to determine the height and weight of the child, taken every 2 months for ages 14 months to 24 months. If this is not possible, you can do this activity using the sample data set given on the next page.

2. Create a table with three columns and seven rows. Label the table "Height and Weight of [name]; Year 2."

3. Label the first column "Age (months)," the second column, "Height (in.)," and the third column, "Weight (lb)."

4. In the first column, write the numbers (one to each row): 14, 16, 18, 20, 22, 24.

5. In the second column, list the child's height at each of the months listed in the first column.

6. Do likewise in the third column with the child's weight.

II.

1. Create a scatterplot of the data in your table.

2. Describe the trend of the data. Fit a line to the data and give the equation of the line.

3. Use the equation of the line to predict the child's height and weight at 10 months of age.

4. Use the equation of the line to predict the child's height and weight at 28 months of age.

5. Use the equation of the line to predict the child's height and weight at 240 months (20 years) of age.

6. According to the model, how tall will the child be when he or she is 50 years old (600 months)?

7. Discuss the accuracy of the predictions made in steps 3 to 6. What can you discern about using scatterplot data to make predictions in general?

Sample Data Set and Solutions

I.

Median Heights and Weights for Children: Year 2		
Age (months)	Median Height for Girls (in.)	Median Weight for Boys (lb)
14	30	24
16	30.75	25
18	31.5	25.75
20	32.5	26.25
22	33	27.25
24	34	28

Source: Centers for Disease Control and Prevention.

II. 1.

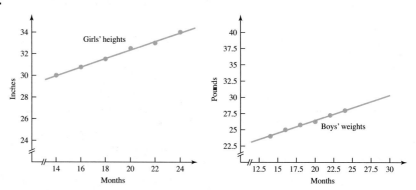

2. Both sets of data appear linear. The lines can be approximated by the equations

Girls' heights: $y = 0.4x + 24.4$

Boys' weights: $y = 0.39x + 18.6$

3–6.

Predictions				
	Age (months)			
	10	28	240	600
Girls' heights (in.)	28.4	35.6	120.4	264.4
Boys' weights (lb)	22.5	29.52	112.2	252.6

While the predictions for 10 and 28 months are reasonably accurate, 20-year-old women tend to be shorter than the 10 ft given by the prediction and 20-year-old men tend to be heavier than the predicted 112 lb. Likewise, women do not grow to be 22 ft tall by the age of 50.

7. We need to be careful that we stay close to the original data source when using data to predict. That is, we can reasonably predict a person's height at 10 months from the data of their second year of age, but we cannot extrapolate that to their height 50 years later.

The following exercises are presented to help you review concepts from earlier chapters. This is meant as a review and not as a comprehensive exam. The answers are presented in the back of the text. Section references accompany the answers. If you have difficulty with any of these exercises, be certain to at least read through the summary related to those sections.

Name _____

Section _____ Date _____

Answers

1. _____

2. _____

3. _____

4. _____

5. _____

6. _____

7. _____

8. _____

9. _____

10. _____

11. _____

12. _____

13. _____

14. _____

15. _____

16. _____

Perform each of the indicated operations.

1. $(5x^2 - 9x + 3) + (3x^2 + 2x - 7)$

2. Subtract $9w^2 + 5w$ from the sum of $8w^2 - 3w$ and $2w^2 - 4$.

3. $7xy(4x^2y - 2xy + 3xy^2)$

4. $(3s - 7)(5s + 4)$

5. $\dfrac{5x^3y - 10x^2y^2 + 15xy^2}{-5xy}$

6. $\dfrac{4x^2 + 6x - 4}{2x - 1}$

Solve the equation.

7. $5 - 3(2x - 7) = 8 - 4x$

Factor each polynomial completely.

8. $24a^3 - 16a^2$

9. $7m^2n - 21mn - 49mn^2$

10. $a^2 - 64b^2$

11. $5p^3 - 80pq^2$

12. $a^2 - 14a + 48$

13. $2w^3 - 8w^2 - 42w$

Solve each equation.

14. $x^2 - 9x + 20 = 0$

15. $2x^2 - 32 = 0$

Solve each application.

16. Twice the square of a positive integer is 35 more than 9 times that integer. What is the integer?

Answers

17. _____

18. _____

19. _____

20. _____

21. _____

22. _____

23. _____

24. _____

25. _____

26. _____

17. The length of a rectangle is 2 in. more than 3 times its width. If the area of the rectangle is 85 in.^{2}, find the dimensions of the rectangle.

Write each fraction in simplest form.

18. $\dfrac{m^2 - 4m}{3m - 12}$

19. $\dfrac{a^2 - 49}{3a^2 + 22a + 7}$

Perform the indicated operations.

20. $\dfrac{3x^2 + 9x}{x^2 - 9} \cdot \dfrac{2x^2 - 9x + 9}{2x^3 - 3x^2}$

21. $\dfrac{4w^2 - 25}{2w^2 - 5w} \div (6w + 15)$

Graph each equation.

22. $x - y = 5$

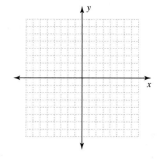

23. $y = \dfrac{2}{3}x + 3$

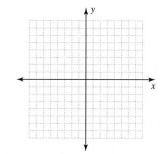

24. $2x - 5y = 10$

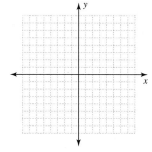

25. $y = -5$

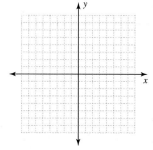

26. Find the slope of the line through the pair of points $(-2, -3)$ and $(5, 7)$.

The Streeter/Hutchison Series in Mathematics   Beginning Algebra

Answers

27. Find the slope and *y*-intercept of the line described by the equation $5x - 3y = 15$.

27. _____

28. Given the slope and *y*-intercept for the following line, write the equation of the line. Then graph the line.

28. _____

Slope $= 2$; *y*-intercept: $(0, -5)$

29. _____

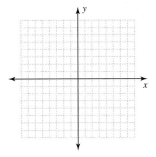

30. _____

31. _____

Graph each inequality.

29. $x + 2y < 6$

30. $3x - 4y \geq 12$

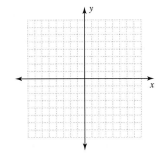 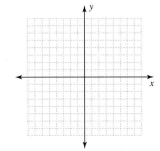

Solve the system by graphing.

31. $3x + 2y = 6$
$x + 2y = -2$

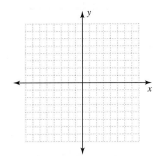

Answers

Solve each system. If a unique solution does not exist, state whether the system is inconsistent or dependent.

32. $5x + 2y = 30$
$x - 4y = 17$

33. $2x - 6y = 8$
$x = 3y + 4$

34. $4x - 5y = 20$
$2x + 3y = 10$

35. $4x + 2y = 11$
$2x + y = 5$

36. $4x - 3y = 7$
$6x + 6y = 7$

Solve each application.

37. One number is 4 less than 5 times another. If the sum of the numbers is 26, what are the two numbers?

38. Cynthia bought five blank VHS tapes and four color mini disks for $28.50. Charlie bought four VHS tapes and two color mini disks for $21.00. Find the cost of each type of media.

39. Receipts for a concert, attended by 450 people, were $27,750. If reserved-seat tickets were $70 and general-admission tickets were $40, how many of each type of ticket were sold?

40. Anthony invested part of his $12,000 inheritance in a bond paying 9% and the other part in a savings account paying 6%. If his interest from the two investments was $930 in the first year, how much did he have invested at each rate?

Beginning Algebra   The Streeter/Hutchison Series in Mathematics

CHAPTER

9

INTRODUCTION

Engineers and architects must include safety plans when designing public buildings. The Uniform Building Code states size and location requirements for exits: "If two exits are required in a building, they must be placed apart a distance not less than one-half the length of the maximum overall diagonal dimension of the building. . . ." Stated in algebraic terms, if the building is rectangular and if d is the distance between exits, l is the length of the building, and w is the width of the building, then

$$d \geq \frac{1}{2}\sqrt{l^2 + w^2}$$

For example, if a rectangular building is 50 ft by 40 ft, the diagonal dimension is $\sqrt{50^2 + 40^2}$, and the distance d between the exits must be equal to or more than half of this value. Thus,

$$d \geq \frac{1}{2}\sqrt{50^2 + 40^2}$$

In this case, the distance between the exits must be 32 ft or more.

The radical sign is often used in the measurement of distances and is based on the Pythagorean theorem, which describes the relationship between the sides of a right triangle. The algebraic expression above is an example of how algebra can make complicated statements easier to understand.

Exponents and Radicals

CHAPTER 9 OUTLINE

Name _____

Section _____ Date _____

Answers

1. _____

2. _____

3. _____

4. _____

5. _____

6. _____

7. _____

8. _____

9. _____

10. _____

This prerequisite test provides some exercises requiring skills that you will need to be successful in the coming chapter. The answers for these exercises can be found in the back of this text. This prerequisite test can help you identify topics that you will need to review before beginning the chapter.

Write the prime factorization for each number.

1. 45

2. 72

Simplify each expression.

3. $(9a^4)(5a)$

4. $(81m^4)(m^3)$

5. $9a + 7a - 12a$

6. $8x + x - 4x$

7. $\dfrac{2}{3} + \dfrac{2}{5}$

8. $2 - \dfrac{1}{5}$

9. $(m-7)(m + 7)$

10. $(3x + 5)^2$

9.1

Roots and Radicals

< 9.1 Objectives >

1 > Use radical notation to represent roots

2 > Approximate a square root

3 > Evaluate cube and fourth roots

4 > Distinguish between rational and irrational numbers

In Chapter 3, we discussed the properties of exponents. Over the next four sections, we will work with a new notation that "reverses" the process of raising to a power.

We know that a statement such as

$$x^2 = 9$$

is read as "x squared equals 9."

Here we are concerned with the relationship between the variable x and the number 9. We call that relationship the **square root** and say, equivalently, that "x is the square root of 9."

We know from experience that x must be 3 (because $3^2 = 9$) or -3 [because $(-3)^2 = 9$]. We see that 9 has two square roots, 3 and -3. In fact, every positive number has *two* square roots. In general, if $x^2 = a$, we call x a *square root* of a.

We are now ready for our new notation. The symbol $\sqrt{}$ is called a **radical sign.** We just saw that 3 is the positive square root of 9. We also call 3 the **principal square root** of 9 and write

$$\sqrt{9} = 3$$

to indicate that 3 is the principal square root of 9.

NOTE

The symbol $\sqrt{}$ first appeared in print in 1525. In Latin, "radix" means **root,** and this was contracted to a small *r*. The present symbol may have evolved from the manuscript form of that small *r*.

Definition

Square Root	$\sqrt{a}$ is the *principal* square root of a. It is the nonnegative number whose square is a.

Example 1 | **Finding Principal Square Roots**

< Objective 1 >

Find each square root.

(a) $\sqrt{49} = 7$ Because 7 is the positive number we must square to get 49.

(b) $\sqrt{\dfrac{4}{9}} = \dfrac{2}{3}$ Because $\dfrac{2}{3}$ is the positive number we must square to get $\dfrac{4}{9}$.

Check Yourself 1

Find the square roots.

(a) $\sqrt{64}$ (b) $\sqrt{144}$ (c) $\sqrt{\dfrac{16}{25}}$

NOTES

When you use the radical sign, you are referring to the *positive square root:*

$$\sqrt{25} = 5$$

$-\sqrt{x}$ is the negative square root of *x*.

Each positive number has two distinct square roots. For instance, 25 has square roots of 5 and −5 because

$$5^2 = 25 \qquad \text{and} \qquad (-5)^2 = 25$$

If you want to indicate the negative square root, you must use a minus sign in front of the radical.

$$-\sqrt{25} = -5$$

 Example 2   **Finding Square Roots**

Find the square roots.

(a) $\sqrt{100} = 10$ The principal root

(b) $-\sqrt{100} = -10$ The negative square root

(c) $-\sqrt{\dfrac{9}{16}} = -\dfrac{3}{4}$

 > CAUTION

Be Careful! Do not confuse $-\sqrt{9}$ with $\sqrt{-9}$.

The expression $-\sqrt{9}$ is −3, whereas $\sqrt{-9}$ is not a real number. Values such as $\sqrt{-9}$ will be discussed in subsequent mathematics courses.

✓ Check Yourself 2

Find the square roots.

(a) $\sqrt{16}$ **(b)** $-\sqrt{16}$ **(c)** $-\sqrt{\dfrac{16}{25}}$

Every number that we encounter in this text is a **real number.** The square roots of negative numbers are *not* real numbers. For instance, $\sqrt{-9}$ is *not* a real number because there is *no* real number *x* such that

$$x^2 = -9$$

Example 3 summarizes our discussion thus far.

 Example 3   **Finding Square Roots**

Evaluate each square root.

(a) $\sqrt{36} = 6$ **(b)** $\sqrt{121} = 11$

(c) $-\sqrt{64} = -8$ **(d)** $\sqrt{-64}$ is not a real number.

(e) $\sqrt{0} = 0$ Because 0 · 0 = 0

✓ Check Yourself 3

Evaluate, if possible.

(a) $\sqrt{81}$ **(b)** $\sqrt{49}$ **(c)** $-\sqrt{49}$ **(d)** $\sqrt{-49}$

All calculators have square-root keys, but the only integers for which the calculator gives the exact value of the square root are perfect square integers. For all other positive integers, *a calculator gives only an approximation of the correct value.* In Example 4 we use a calculator to approximate square roots.

| Example 4 | Approximating Square Roots

< Objective 2 >

 Calculator

Use your calculator to approximate each square root to the nearest hundredth.

(a) $\sqrt{45} \approx 6.708203932 \approx 6.71$ **(b)** $\sqrt{8} \approx 2.83$

(c) $\sqrt{20} \approx 4.47$ **(d)** $\sqrt{273} \approx 16.52$

Check Yourself 4

Use your calculator to approximate each square root to the nearest hundredth.

(a) $\sqrt{3}$ **(b)** $\sqrt{14}$ **(c)** $\sqrt{91}$ **(d)** $\sqrt{756}$

NOTE

The $\approx$ sign means "is approximately equal to."

NOTE

$\sqrt[3]{8}$ is read "the cube root of 8."

As we mentioned earlier, finding the square root of a number is the reverse of squaring a number. We can extend that idea to work with other roots of numbers. For instance, the **cube root** of a number is the quantity we must cube (or raise to the third power) to get the original number. For example, the cube root of 8 is 2 because $2^3 = 8$, and we write

$$\sqrt[3]{8} = 2$$

The parts of a radical expression are summarized as follows.

Definition

Parts of a Radical Expression

Every radical expression contains three parts as shown here. The principal *n*th root of *a* is written as

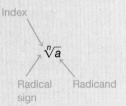

Index

$\sqrt[n]{a}$

Radical   Radicand
sign

NOTES

The index for $\sqrt[3]{a}$ is 3.

The index of 2 for square roots is generally not written. We understand that $\sqrt{a}$ is the principal square root of *a*.

To illustrate, the *cube root* of 64 is written

Index ⟶ of 3 $\sqrt[3]{64} = 4$

because $4^3 = 64$. And

Index ⟶ of 4 $\sqrt[4]{81} = 3$

is the *fourth root* of 81 because $3^4 = 81$.

We can find roots of negative numbers as long as the index is *odd* (3, 5, etc.). For example,

$$\sqrt[3]{-64} = -4$$

because $(-4)^3 = -64$.

If the index is *even* (2, 4, etc.), roots of negative numbers are *not* real numbers. For example,

$$\sqrt[4]{-16}$$

NOTE

The *even power* of a real number is always *positive* or *zero*.

is not a real number because there is no real number *x* such that $x^4 = -16$.

Here is a table showing the most common roots.

Square Roots		Cube Roots	Fourth Roots
$\sqrt{1} = 1$ $\sqrt{49} = 7$		$\sqrt[3]{1} = 1$	$\sqrt[4]{1} = 1$
$\sqrt{4} = 2$ $\sqrt{64} = 8$		$\sqrt[3]{8} = 2$	$\sqrt[4]{16} = 2$
$\sqrt{9} = 3$ $\sqrt{81} = 9$		$\sqrt[3]{27} = 3$	$\sqrt[4]{81} = 3$
$\sqrt{16} = 4$ $\sqrt{100} = 10$		$\sqrt[3]{64} = 4$	$\sqrt[4]{256} = 4$
$\sqrt{25} = 5$ $\sqrt{121} = 11$		$\sqrt[3]{125} = 5$	$\sqrt[4]{625} = 5$
$\sqrt{36} = 6$ $\sqrt{144} = 12$			

You can use this table in Example 5, which summarizes the discussion so far.

> **Example 5**   **Evaluating Roots**

< Objective 3 >

Evaluate each root.

(a) $\sqrt[5]{32} = 2$ because $2^5 = 32$.

(b) $\sqrt[3]{-125} = -5$ because $(-5)^3 = -125$.

(c) $\sqrt[4]{-81}$ is not a real number.

Check Yourself 5

Evaluate, if possible.

(a) $\sqrt[3]{64}$ **(b)** $\sqrt[4]{16}$ **(c)** $\sqrt[4]{-256}$ **(d)** $\sqrt[3]{-8}$

The radical notation helps us to distinguish between two important types of numbers: rational numbers and irrational numbers.

A **rational number** can be represented by a fraction whose numerator and denominator are integers and whose denominator is not zero. The form of a rational number is

$$\frac{a}{b} \qquad a \text{ and } b \text{ are integers, } b \neq 0$$

A whole number, such as 5, is rational because we can write it as $\frac{5}{1}$. Certain square roots are rational numbers also. For example,

$$\sqrt{4}, \qquad \sqrt{25}, \qquad \text{and} \qquad \sqrt{64}$$

represent the rational numbers 2, 5, and 8, respectively. Notice that each radicand here is a **perfect-square integer** (that is, an integer that is the square of another integer).

An **irrational number** is a number that *cannot* be written as the ratio of two integers. For example, the square root of any positive number that is not itself a perfect square is an irrational number. Because the radicands are *not* perfect squares, the expressions $\sqrt{2}$, $\sqrt{3}$, and $\sqrt{5}$ represent irrational numbers.

| **Example 6** | Identifying Rational Numbers

< Objective 4 >

Which numbers are rational and which are irrational?

$$\sqrt{\frac{2}{3}} \qquad \sqrt{\frac{4}{9}} \qquad \sqrt{7} \qquad \sqrt{16} \qquad \sqrt{25}$$

Here $\sqrt{7}$ and $\sqrt{\frac{2}{3}}$ are irrational numbers. The numbers $\sqrt{16}$ and $\sqrt{25}$ are rational because 16 and 25 are perfect squares. Also $\sqrt{\frac{4}{9}}$ is rational because $\sqrt{\frac{4}{9}} = \frac{2}{3}$.

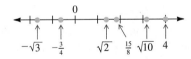

Check Yourself 6

Determine whether each root is rational or irrational.

(a) $\sqrt{26}$ (b) $\sqrt{49}$ (c) $\sqrt{\frac{6}{7}}$

(d) $\sqrt{105}$ (e) $\sqrt{\frac{16}{9}}$

NOTES

The decimal representation of a rational number always terminates or repeats. For instance,

$\frac{3}{8} = 0.375$

$\frac{5}{11} = 0.454545\ldots$

1.414 is an approximation of the number whose square is 2.

An important fact about the irrational numbers is that their decimal representations are always *nonterminating* and *nonrepeating*. We can therefore only approximate irrational numbers with a decimal that has been rounded. A calculator can be used to find roots. However, note that the values found for the irrational roots are only approximations. For instance, $\sqrt{2}$ is approximately 1.414 (to three decimal places), and we can write

$\sqrt{2} \approx 1.414$

With a calculator we find that

$(1.414)^2 = 1.999396$

The set of all rational numbers and the set of all irrational numbers together form the set of *real numbers*. The real numbers represent every point that can be pictured on the number line. Some examples are shown.

NOTE

For this reason we refer to the number line as the **real number line.**

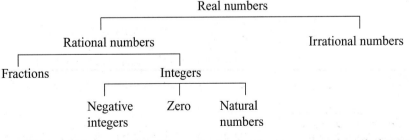

This diagram summarizes the relationships among various numeric sets.

Real numbers
— Rational numbers
— Irrational numbers
Fractions
Integers
Negative integers
Zero
Natural numbers

NOTE

This is because the principal square root of a number is always positive or zero.

We conclude our work in this section by developing a general result that we need later. Start by looking at two numerical examples.

$\sqrt{2^2} = \sqrt{4} = 2$

$\sqrt{(-2)^2} = \sqrt{4} = 2 \qquad$ because $(-2)^2 = 4$

Consider the value of $\sqrt{x^2}$ when x is positive or negative.

In the first equation, when $x = 2$:   In the second equation, when $x = -2$:

$$\sqrt{2^2} = 2$$

$$\sqrt{(-2)^2} \neq -2$$

$$\sqrt{(-2)^2} = -(-2) = 2$$

Comparing the results of these two equations, we see that $\sqrt{x^2}$ is x if x is positive (or 0) and $\sqrt{x^2}$ is $-x$ if x is negative. We can write

$$\sqrt{x^2} = \begin{cases} x & \text{when } x \geq 0 \\ -x & \text{when } x < 0 \end{cases}$$

From your earlier work with absolute values you will remember that

$$|x| = \begin{cases} x & \text{when } x \geq 0 \\ -x & \text{when } x < 0 \end{cases}$$

and we can summarize the discussion by writing

$$\sqrt{x^2} = |x| \qquad \text{for any real number } x$$

▶ **Example 7**   **Evaluating Radical Expressions**

Evaluate.

NOTE

Alternatively in part (b), we could write

$$\sqrt{(-4)^2} = \sqrt{16} = 4$$

(a) $\sqrt{5^2} = 5$

(b) $\sqrt{(-4)^2} = |-4| = 4$

 Check Yourself 7

Evaluate.

(a) $\sqrt{6^2}$ **(b)** $\sqrt{(-6)^2}$

 Check Yourself ANSWERS

1. (a) 8; (b) 12; (c) $\frac{4}{5}$ 2. (a) 4; (b) -4; (c) $-\frac{4}{5}$ 3. (a) 9; (b) 7; (c) -7;

(d) not a real number   4. (a) ≈ 1.73; (b) ≈ 3.74; (c) ≈ 9.54; (d) ≈ 27.50

5. (a) 4; (b) 2; (c) not a real number; (d) -2 6. (a) Irrational;

(b) rational (because $\sqrt{49} = 7$); (c) irrational; (d) irrational;

(e) rational $\left(\text{because } \sqrt{\frac{16}{9}} = \frac{4}{3}\right)$ 7. (a) 6; (b) 6

Reading Your Text

The following fill-in-the-blank exercises are designed to ensure that you understand some of the key vocabulary used in this section.

SECTION 9.1

(a) $\sqrt{a}$ is the positive (or _____) square root of a.

(b) To indicate a _____ square root, place a minus sign in front of the radical.

(c) A calculator gives only an _____ to the correct value for finding the square root of an integer that is not a perfect square.

(d) Every radical expression contains three parts, the radical sign, the index, and the _____.

Name _____

Section _____ Date _____

Answers

1. _____ 2. _____

3. _____ 4. _____

5. _____

6. _____

7. _____

8. _____

9. _____ 10. _____

11. _____

12. _____ 13. _____

14. _____ 15. _____

16. _____

17. _____

18. _____ 19. _____

20. _____ 21. _____

22. _____

23. _____ 24. _____

< Objectives 1–3 >

Evaluate, if possible.

1. $\sqrt{25}$

2. $\sqrt{121}$

3. $\sqrt{400}$

4. $\sqrt{64}$

5. $-\sqrt{144}$ [> Videos]

6. $\sqrt{-121}$

7. $\sqrt{-81}$

8. $-\sqrt{81}$

9. $\sqrt{\dfrac{36}{25}}$

10. $-\sqrt{\dfrac{1}{25}}$

11. $\sqrt{-\dfrac{4}{25}}$ [> Videos]

12. $\sqrt{\dfrac{4}{25}}$

13. $\sqrt{\dfrac{121}{100}}$

14. $\sqrt{\dfrac{256}{25}}$

15. $\sqrt[3]{-64}$ [> Videos]

16. $\sqrt[4]{-16}$

17. $\sqrt[4]{-81}$

18. $-\sqrt[3]{64}$

19. $-\sqrt[3]{27}$

20. $\sqrt[3]{-27}$

21. $\sqrt[4]{1{,}296}$

22. $\sqrt[3]{1{,}000}$

23. $\sqrt[3]{\dfrac{1}{27}}$

24. $\sqrt[3]{-\dfrac{8}{27}}$

< Objective 4 >

State whether each root is rational or irrational.

25. $\sqrt{21}$

26. $\sqrt{36}$

27. $\sqrt{100}$

28. $\sqrt{7}$

29. $\sqrt[3]{9}$

30. $\sqrt[3]{27}$

31. $\sqrt[4]{16}$

32. $\sqrt{\dfrac{4}{9}}$

33. $\sqrt{\dfrac{9}{15}}$

34. $\sqrt[3]{5}$

35. $\sqrt[3]{-27}$

36. $-\sqrt[4]{81}$ > Videos

37. GEOMETRY The area of a square is 32 ft^2. Find the length of a side to the nearest hundredth.

38. GEOMETRY The area of a square is 83 ft^2. Find the length of the side to the nearest hundredth.

39. GEOMETRY The area of a circle is 147 ft^2. Find the radius to the nearest hundredth.

40. GEOMETRY If the area of a circle is 72 cm^2, find the radius to the nearest hundredth.

41. SCIENCE AND MEDICINE The time in seconds that it takes for an object to fall from rest is given by $t = \dfrac{1}{4}\sqrt{s}$, in which s is the distance fallen (in feet). Find the time required for an object to fall to the ground from a building that is 800 ft high.

42. SCIENCE AND MEDICINE Find the time required for an object to fall to the ground from a building that is 1,400 ft high. (Use the formula given in exercise 41.)

Basic Skills | **Challenge Yourself** | Calculator/Computer | Career Applications | Above and Beyond
▲

Determine whether each statement is **true** *or* **false.**

43. $\sqrt{16x^{16}} = 4x^4$

44. $\sqrt{(x-4)^2} = x - 4$

45. $\sqrt{16x^{-4}y^{-4}}$ is a real number

46. $\sqrt{x^2 + y^2} = x + y$

47. $\dfrac{\sqrt{x^2 - 25}}{x - 5} = \sqrt{x + 5}$

48. $\sqrt{2} + \sqrt{6} = \sqrt{8}$

Answers

25. _____ 26. _____

27. _____ 28. _____

29. _____ 30. _____

31. _____ 32. _____

33. _____ 34. _____

35. _____ 36. _____

37. _____ 38. _____

39. _____ 40. _____

41. _____ 42. _____

43. _____

44. _____

45. _____

46. _____

47. _____

48. _____

Answers

49. _____

50. _____

51. _____

52. _____

53. _____

54. _____

55. _____

56. _____

57. _____ 58. _____

59. _____ 60. _____

61. _____ 62. _____

63. _____ 64. _____

65. _____ 66. _____

67. _____ 68. _____

69. _____

70. _____

71. _____

< Objectives 1–3 >

For exercises 49 to 54, find the two expressions that are equivalent.

49. $\sqrt{-16}, -\sqrt{16}, -4$

50. $-\sqrt{25}, -5, \sqrt{-25}$

51. $\sqrt[3]{-125}, -\sqrt[3]{125}, |-5|$

52. $\sqrt[5]{-32}, -\sqrt[5]{32}, |-2|$

53. $\sqrt[4]{10,000}, 100, \sqrt[3]{1,000}$

54. $10^2, \sqrt{10,000}, \sqrt[3]{100,000}$

| Basic Skills | Challenge Yourself | **Calculator/Computer** | Career Applications | Above and Beyond |

▲

< Objective 2 >

Use your calculator to approximate each square root to the nearest hundredth.

55. $\sqrt{11}$

56. $\sqrt{14}$

57. $\sqrt{7}$

58. $\sqrt{23}$

59. $\sqrt{65}$

60. $\sqrt{78}$

61. $\sqrt{\dfrac{2}{5}}$ > Videos

62. $\sqrt{\dfrac{4}{3}}$

63. $\sqrt{\dfrac{8}{9}}$

64. $\sqrt{\dfrac{7}{15}}$

65. $-\sqrt{18}$

66. $-\sqrt{31}$

67. $-\sqrt{27}$

68. $-\sqrt{65}$

| Basic Skills | Challenge Yourself | Calculator/Computer | Career Applications | **Above and Beyond** |

▲

In exercises 69 to 71, the area is given in square feet. Find the length of a side of the square. Round your answer to the nearest hundredth of a foot.

69.

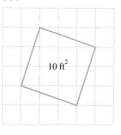

10 ft²

70.

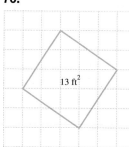

13 ft²

71.

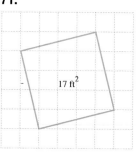

17 ft²

72. Suppose that a weight is attached to a string of length L, and the other end of the string is held fixed. If we pull the weight and then release it, allowing the weight to swing back and forth, we can observe the behavior of a simple pendulum. The period T is the time required for the weight to complete a full cycle, swinging forward and then back. The following formula may be used to describe the relationship between T and L.

$$T = 2\pi\sqrt{\frac{L}{g}}$$

If L is expressed in centimeters, then $g = 980$ cm/s^2. For each string length, calculate the corresponding period. Round to the nearest tenth of a second.

(a) 30 cm　　**(b)** 50 cm　　**(c)** 70 cm　　**(d)** 90 cm　　**(e)** 110 cm

73. Is there any prime number whose square root is an integer? Explain your answer.

74. Use your calculator to complete each exercise.

(a) Choose a number greater than 1 and find its square root. Then find the square root of the result and continue in this manner, observing the successive square roots. Do these numbers seem to be approaching a certain value? If so, what?

(b) Choose a number greater than 0 but less than 1 and find its square root. Then find the square root of the result, and continue in this manner, observing successive square roots. Do these numbers seem to be approaching a certain value? If so, what?

75. **(a)** Can a number be equal to its own square root?

(b) Other than the number(s) found in part (a), is a number always greater than its square root? Investigate.

76. Let a and b be positive numbers. If a is greater than b, is it always true that the square root of a is greater than the square root of b? Investigate.

Answers

72. _____

73. _____

74. _____

75. _____

76. _____

Answers

1. 5　　**3.** 20　　**5.** -12　　**7.** Not a real number　　**9.** $\dfrac{6}{5}$

11. Not a real number　　**13.** $\dfrac{11}{10}$　　**15.** -4　　**17.** Not a real number

19. -3　　**21.** 6　　**23.** $\dfrac{1}{3}$　　**25.** Irrational　　**27.** Rational

29. Irrational　　**31.** Rational　　**33.** Irrational　　**35.** Rational

37. 5.66 ft　　**39.** 6.84 ft　　**41.** 7.07 s　　**43.** False　　**45.** True

47. False　　**49.** $-\sqrt{16}, -4$　　**51.** $\sqrt[3]{-125}, -\sqrt[3]{125}$

53. $\sqrt[4]{10{,}000}, \sqrt[3]{1{,}000}$　　**55.** 3.32　　**57.** 2.65　　**59.** 8.06　　**61.** 0.63

63. 0.94　　**65.** -4.24　　**67.** -5.20　　**69.** 3.16 ft　　**71.** 4.12 ft

73. No; Above and Beyond　　**75.** Above and Beyond

Beginning Algebra   The Streeter/Hutchison Series in Mathematics

9.2

Simplifying Radical Expressions

< 9.2 Objectives >

1 > Simplify expressions involving numeric radicals

2 > Simplify expressions involving algebraic radicals

In Section 9.1, we introduced radical notation. For most applications, we want to make sure that all radical expressions are in *simplest form*. To accomplish this, the following three conditions must be satisfied.

Property

Square Root Expressions in Simplest Form

An expression involving square roots is in *simplest form* if

1. There are no perfect-square factors under a radical sign.

2. No fraction appears under a radical sign.

3. No radical sign appears in the denominator of a fraction.

For instance, considering condition 1,

$\sqrt{17}$ is in simplest form because 17 has *no* perfect-square factors

whereas

$\sqrt{12}$ is *not* in simplest form

because the radicand, 12, does contain a perfect-square factor.

$$\sqrt{12} = \sqrt{4 \cdot 3}$$

A perfect square

To simplify radical expressions, we need to develop two important properties. Consider the expressions

$$\sqrt{4 \cdot 9} = \sqrt{36} = 6$$
$$\sqrt{4} \cdot \sqrt{9} = 2 \cdot 3 = 6$$

Because this tells us that $\sqrt{4 \cdot 9} = \sqrt{4} \cdot \sqrt{9}$, we have a general rule for radicals.

Property

Property 1 of Radicals

RECALL

Compare **Property 1 of Radicals** to the property of exponents $(ab)^n = a^n \cdot b^n$

For any nonnegative real numbers a and b,

$$\sqrt{ab} = \sqrt{a} \cdot \sqrt{b}$$

In words, the square root of a product is the product of the square roots.

Example 1 illustrates how this property is applied in simplifying expressions when radicals are involved.

 Example 1 | **Simplifying Radical Expressions**

< Objective 1 >

Simplify each expression.

(a) $\sqrt{12} = \sqrt{4 \cdot 3}$ Now apply Property 1.

A perfect square

$= \sqrt{4} \cdot \sqrt{3}$

$= 2\sqrt{3}$

NOTE

Perfect-square factors are 1, 4, 9, 16, 25, 36, 49, 64, 81, 100, and so on.

(b) $\sqrt{45} = \sqrt{9 \cdot 5}$

A perfect square

$= \sqrt{9} \cdot \sqrt{5}$

$= 3\sqrt{5}$

NOTE

We removed the perfect-square factor from inside the radical, so the expression is in simplest form.

(c) $\sqrt{72} = \sqrt{36 \cdot 2}$ We look for the *largest* perfect-square factor, here 36. Then apply Property 1.

A perfect square

$= \sqrt{36} \cdot \sqrt{2}$

$= 6\sqrt{2}$

(d) $5\sqrt{18} = 5\sqrt{9 \cdot 2}$

A perfect square

$= 5 \cdot \sqrt{9} \cdot \sqrt{2} = 5 \cdot 3 \cdot \sqrt{2} = 15\sqrt{2}$

Be careful!   Even though

$\sqrt{a \cdot b} = \sqrt{a} \cdot \sqrt{b}$

the expression $\sqrt{a + b}$ is *not the same* as $\sqrt{a} + \sqrt{b}$

Let $a = 4$ and $b = 9$, and substitute.

$\sqrt{a + b} = \sqrt{4 + 9} = \sqrt{13}$

$\sqrt{a} + \sqrt{b} = \sqrt{4} + \sqrt{9} = 2 + 3 = 5$

Because $\sqrt{13} \neq 5$, we see that the expressions $\sqrt{a + b}$ and $\sqrt{a} + \sqrt{b}$ are not, in general, the same.

> CAUTION

The root of a product is equal to the product of the roots. The root of a sum **is not** equal to the sum of the roots.

Recall that, in our study of exponents, $(a + b)^n \neq a^n + b^n$ in general.

 Check Yourself 1

Simplify.

(a) $\sqrt{20}$ **(b)** $\sqrt{75}$ **(c)** $\sqrt{98}$ **(d)** $\sqrt{48}$

There is an alternative approach to the simplification of radical expressions. It uses the idea of **prime factorization.** Consider Example 2.

 Example 2 | **Simplifying Radical Expressions Using Prime Factorization**

Simplify each expression using prime factorization.

(a) $\sqrt{12}$

First rewrite 12 as $2 \cdot 2 \cdot 3$, or $2^2 \cdot 3$.

$$\sqrt{12} = \sqrt{2^2 \cdot 3}$$
$$= \sqrt{2^2} \cdot \sqrt{3}$$
$$= 2\sqrt{3}$$

Now apply Property 1.
Recall that $\sqrt{2^2} = 2$.

(b) $\sqrt{72}$

$$\sqrt{72} = \sqrt{2^3 \cdot 3^2}$$

Now rewrite 2^3 as $2^2 \cdot 2$, and reorder:

$$\sqrt{2^3 \cdot 3^2} = \sqrt{2^2 \cdot 2 \cdot 3^2}$$
$$= \sqrt{2^2 \cdot 3^2 \cdot 2}$$
$$= \sqrt{2^2} \cdot \sqrt{3^2} \cdot \sqrt{2}$$
$$= 2 \cdot 3 \cdot \sqrt{2}$$
$$= 6\sqrt{2}$$

Now apply Property 1.

Check Yourself 2

Simplify each expression using prime factorization.

(a) $\sqrt{20}$ (b) $\sqrt{75}$ (c) $\sqrt{98}$ (d) $\sqrt{48}$

The process is the same if variables are involved in a radical expression. In our remaining work with radicals, we will assume that all variables represent positive real numbers.

 Example 3 | **Simplifying Algebraic Radical Expressions**

< Objective 2 >

Simplify each radical.

(a) $\sqrt{x^3} = \sqrt{x^2 \cdot x}$

A perfect square

$$= \sqrt{x^2} \cdot \sqrt{x}$$
$$= x\sqrt{x}$$

NOTE

$\sqrt{x^2} = x$ (as long as x is *not* negative).

(b) $\sqrt{4b^3} = \sqrt{4 \cdot b^2 \cdot b}$

Perfect squares

$$= \sqrt{4b^2} \cdot \sqrt{b}$$
$$= 2b\sqrt{b}$$

RECALL

We want the perfect-square factor to have the largest possible even exponent, here 4. Keep in mind that

$a^2 \cdot a^2 = a^4$

(c) $\sqrt{18a^5} = \sqrt{9 \cdot a^4 \cdot 2a}$

Perfect squares

$\qquad = \sqrt{9a^4} \cdot \sqrt{2a}$

$\qquad = 3a^2\sqrt{2a}$

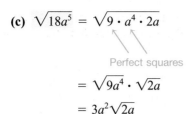

Check Yourself 3

Simplify.

(a) $\sqrt{9x^3}$ **(b)** $\sqrt{27m^3}$ **(c)** $\sqrt{50b^5}$

To develop our second property for radicals, consider the expressions

$$\sqrt{\frac{16}{4}} = \sqrt{4} = 2$$

$$\frac{\sqrt{16}}{\sqrt{4}} = \frac{4}{2} = 2$$

Because $\sqrt{\dfrac{16}{4}} = \dfrac{\sqrt{16}}{\sqrt{4}}$, we have a second general rule for radicals.

Property

Property 2 of Radicals

RECALL

Compare **Property 2 of Radicals** to the property of exponents $\left(\dfrac{a}{b}\right)^n = \dfrac{a^n}{b^n}$

For any positive real numbers a and b,

$$\sqrt{\frac{a}{b}} = \frac{\sqrt{a}}{\sqrt{b}}$$

In words, the square root of a quotient is the quotient of the square roots.

This property is used in a fashion similar to Property 1 in simplifying radical expressions. Remember that our second condition for a radical expression to be in simplest form states that no fraction should appear under a radical sign. Example 4 illustrates how expressions that violate that condition are simplified.

Example 4   **Simplifying Radical Expressions**

Write each expression in simplest form. Assume that all variables here represent positive real numbers.

NOTE

Apply Property 2 to write the numerator and denominator as separate radicals.

(a) $\sqrt{\dfrac{9}{4}} = \dfrac{\sqrt{9}}{\sqrt{4}}$ We applied Property 2.

$\qquad = \dfrac{3}{2}$

(b) $\sqrt{\dfrac{2}{25}} = \dfrac{\sqrt{2}}{\sqrt{25}}$ Apply Property 2. Then simplify.

$\qquad = \dfrac{\sqrt{2}}{5}$

NOTE

Rewrite $8x^2$ as $4x^2 \cdot 2$.

(c) $\sqrt{\dfrac{8x^2}{9}} = \dfrac{\sqrt{8x^2}}{\sqrt{9}}$　　Apply Property 2.

$= \dfrac{\sqrt{4x^2 \cdot 2}}{3}$

$= \dfrac{\sqrt{4x^2} \cdot \sqrt{2}}{3}$　　Apply Property 1 in the numerator.

$= \dfrac{2x\sqrt{2}}{3}$

Check Yourself 4

Simplify.

(a) $\sqrt{\dfrac{25}{16}}$　　(b) $\sqrt{\dfrac{7}{9}}$　　(c) $\sqrt{\dfrac{12x^2}{49}}$

In the three expressions in Example 4, the denominator of the fraction appearing in the radical was a perfect square, and we were able to write each expression in simplest radical form by removing that perfect square from the radical.

If the denominator of the fraction in the radical is *not* a perfect square, we can still apply Property 2 of radicals. As we show in Example 5, the third condition for a radical to be in simplest form is then violated, and a new technique is necessary.

Example 5　Simplifying Radical Expressions

NOTE

We begin by applying Property 2.

Write each expression in simplest form.

(a) $\sqrt{\dfrac{1}{3}} = \dfrac{\sqrt{1}}{\sqrt{3}} = \dfrac{1}{\sqrt{3}}$

Do you see that $\dfrac{1}{\sqrt{3}}$ is still not in simplest form because of the radical in the denominator? To solve this problem, we multiply the numerator and denominator by $\sqrt{3}$. Note that the denominator will become

$\sqrt{3} \cdot \sqrt{3} = \sqrt{9} = 3$

We then have

RECALL

We can do this because we are multiplying the fraction by $\dfrac{\sqrt{3}}{\sqrt{3}}$ or 1, which does not change its value.

$\dfrac{1}{\sqrt{3}} = \dfrac{1 \cdot \sqrt{3}}{\sqrt{3} \cdot \sqrt{3}} = \dfrac{\sqrt{3}}{3}$

The expression $\dfrac{\sqrt{3}}{3}$ is now in simplest form because all three of our conditions are satisfied.

NOTE

$\sqrt{2} \cdot \sqrt{5} = \sqrt{2 \cdot 5} = \sqrt{10}$
$\sqrt{5} \cdot \sqrt{5} = 5$

(b) $\sqrt{\dfrac{2}{5}} = \dfrac{\sqrt{2}}{\sqrt{5}}$

$= \dfrac{\sqrt{2} \cdot \sqrt{5}}{\sqrt{5} \cdot \sqrt{5}}$

$= \dfrac{\sqrt{10}}{5}$

and the expression is in simplest form because again our three conditions are satisfied.

NOTE

We multiply numerator and denominator by $\sqrt{7}$ to "clear" the denominator of the radical. This is also known as **rationalizing the denominator.**

(c) $\sqrt{\dfrac{3x}{7}} = \dfrac{\sqrt{3x}}{\sqrt{7}}$

$= \dfrac{\sqrt{3x} \cdot \sqrt{7}}{\sqrt{7} \cdot \sqrt{7}}$

$= \dfrac{\sqrt{21x}}{7}$

The expression is in simplest form.

Check Yourself 5

Simplify.

(a) $\sqrt{\dfrac{1}{2}}$ (b) $\sqrt{\dfrac{2}{3}}$ (c) $\sqrt{\dfrac{2y}{5}}$

In Example 6, we demonstrate that the properties of radicals introduced in this section apply to roots other than square roots.

Example 6   **Simplifying Radical Expressions**

Write each in simplest form.

NOTE

Here we are using the "prime factorization" approach seen in Example 2. We use the fact that $\sqrt[3]{3^3} = 3$.

(a) $\sqrt[3]{54} = \sqrt[3]{3^3 \cdot 2}$ Find the prime factors for the radicand. We are particularly interested in factors raised to powers that are multiples of the index, 3.

$= \sqrt[3]{3^3} \cdot \sqrt[3]{2}$

$= 3\sqrt[3]{2}$ The expression is now in simplest form.

NOTE

$\sqrt[3]{x^3} = x$, and $\sqrt[3]{2^3} = 2$.

(b) $\sqrt[3]{\dfrac{x^4}{8}} = \dfrac{\sqrt[3]{x^3 \cdot x}}{\sqrt[3]{8}}$

$= \dfrac{\sqrt[3]{x^3} \cdot \sqrt[3]{x}}{\sqrt[3]{2^3}}$ Find the prime factors for the radicand and rewrite as the quotient of two radicals.

$= \dfrac{x\sqrt[3]{x}}{2}$ The expression is now in simplest form.

Check Yourself 6

Write each in simplest form.

(a) $\sqrt[3]{128}$ (b) $\sqrt[3]{\dfrac{a^3 b^5}{27}}$

Check Yourself ANSWERS

1. (a) $2\sqrt{5}$; (b) $5\sqrt{3}$; (c) $7\sqrt{2}$; (d) $4\sqrt{3}$ 2. (a) $2\sqrt{5}$; (b) $5\sqrt{3}$;

(c) $7\sqrt{2}$; (d) $4\sqrt{3}$ 3. (a) $3x\sqrt{x}$; (b) $3m\sqrt{3m}$; (c) $5b^2\sqrt{2b}$

4. (a) $\dfrac{5}{4}$; (b) $\dfrac{\sqrt{7}}{3}$; (c) $\dfrac{2x\sqrt{3}}{7}$ 5. (a) $\dfrac{\sqrt{2}}{2}$; (b) $\dfrac{\sqrt{6}}{3}$; (c) $\dfrac{\sqrt{10y}}{5}$

6. (a) $4\sqrt[3]{2}$; (b) $\dfrac{ab\sqrt[3]{b^2}}{3}$

Reading Your Text

The following fill-in-the-blank exercises are designed to ensure that you understand some of the key vocabulary used in this section.

SECTION 9.2

(a) In simplest form radical expressions have no perfect-square _____ under a radical sign.

(b) In simplest form radical expressions contain no _____ under a radical sign.

(c) The square root of a product is the _____ of the square roots.

(d) For an expression involving square roots to be in simplest form, no radical sign appears in the _____ of a fraction.

Basic Skills | Challenge Yourself | Calculator/Computer | Career Applications | Above and Beyond

< Objectives 1–2 >

Simplify each radical expression. Assume that all variables represent positive real numbers.

1. $\sqrt{48}$ > Videos

2. $\sqrt{50}$

3. $\sqrt{28}$

4. $\sqrt{108}$

5. $\sqrt{45}$

6. $\sqrt{80}$

7. $\sqrt{54}$

8. $\sqrt{180}$

9. $2\sqrt{200}$

10. $3\sqrt{96}$

11. $4\sqrt{567}$

12. $7\sqrt{300}$

13. $3\sqrt{12}$

14. $5\sqrt{24}$

15. $\sqrt{3x^2}$

16. $\sqrt{7a^2}$

17. $\sqrt{3y^4}$

18. $\sqrt{10x^6}$

19. $\sqrt{2r^3}$

20. $\sqrt{7a^7}$

21. $\sqrt{125b^2}$ > Videos

22. $\sqrt{98m^4}$

23. $\sqrt{24x^4}$

24. $\sqrt{72x^3}$

25. $\sqrt{54a^5}$

26. $\sqrt{200y^6}$

27. $\sqrt{x^3y^2}$

28. $\sqrt{a^2b^5}$

29. $\sqrt{a^6b^4c}$ > Videos

30. $\sqrt{x^3y^4z^2}$

Answers

1.	2.
3.	4.
5.	6.
7.	8.
9.	10.
11.	12.
13.	14.
15.	16.
17.	18.
19.	20.
21.	22.
23.	24.
25.	26.
27.	28.
29.	30.

Answers

31. _____   32. _____

33. _____   34. _____

35. _____   36. _____

37. _____

38. _____

39. _____

40. _____

41. _____

42. _____

43. _____

44. _____

45. _____

46. _____

47. _____

48. _____

49. _____

50. _____

51. _____

52. _____

31. $\sqrt{\dfrac{9}{16}}$

32. $\sqrt{\dfrac{49}{25}}$

33. $\sqrt{\dfrac{3}{4}}$

34. $\sqrt{\dfrac{7}{16}}$

35. $\sqrt{\dfrac{3}{49}}$ > Videos

36. $\sqrt{\dfrac{10}{49}}$

Basic Skills	**Challenge Yourself**	Calculator/Computer	Career Applications	Above and Beyond

Determine whether each statement is **true** *or* **false.**

37. A simplified radical might have a fraction inside the radical.

38. We cannot, in general, rewrite $\sqrt{a + b}$ as $\sqrt{a} + \sqrt{b}$.

Complete each statement with **never, sometimes,** *or* **always.**

39. For positive real numbers a and b, it is _____ true that $\sqrt{ab} = \sqrt{a} \cdot \sqrt{b}$.

40. A radical in simplest form _____ has a radical in the denominator.

Use the properties for radicals to simplify each expression. Assume that all variables represent positive real numbers.

41. $\sqrt{\dfrac{8a^2}{25}}$

42. $\sqrt{\dfrac{12y^2}{49}}$

43. $\sqrt{\dfrac{1}{5}}$

44. $\sqrt{\dfrac{1}{7}}$

45. $\sqrt{\dfrac{5}{2}}$

46. $\sqrt{\dfrac{5}{3}}$

47. $\sqrt{\dfrac{3a}{5}}$

48. $\sqrt{\dfrac{2x}{5}}$

49. $\sqrt{\dfrac{3x^4}{7}}$

50. $\sqrt{\dfrac{5m^2}{2}}$

51. $\sqrt{\dfrac{8s^3}{7}}$ > Videos

52. $\sqrt{\dfrac{12x^3}{5}}$

Basic Skills | Challenge Yourself | Calculator/Computer | Career Applications | **Above and Beyond**

▲

Answers

53. _____

54. _____

55. _____

56. _____

57. _____

58. _____

Decide whether each expression is already written in simplest form. If it is not, explain what needs to be done.

53. $\sqrt{10mn}$ **54.** $\sqrt{18ab}$ **55.** $\sqrt{\dfrac{98x^2y}{7x}}$ **56.** $\dfrac{\sqrt{6xy}}{3x}$

57. Find the area and perimeter of this square:

$\sqrt{3}$

$\sqrt{3}$

One of these measures, the area, is a rational number, and the other, the perimeter, is an irrational number. Explain how this happened. Will the area always be a rational number? Explain.

58. (a) Evaluate the three expressions $\dfrac{n^2 - 1}{2}, n, \dfrac{n^2 + 1}{2}$ using odd values of n: 1, 3, 5, 7, and so forth. Complete the table shown.

n	$a = \dfrac{n^2 - 1}{2}$	$b = n$	$c = \dfrac{n^2 + 1}{2}$	a^2	b^2	c^2
1						
3						
5						
7						
9						
11						
13						
15						

(b) Check for each of these sets of three numbers to see whether this statement is true: $\sqrt{a^2 + b^2} = \sqrt{c^2}$. For how many of your sets of three did this work? Sets of three numbers for which this statement is true are called *Pythagorean triples* because $a^2 + b^2 = c^2$. Can the radical equation be written as $\sqrt{a^2 + b^2} = a + b$? Explain your answer.

Answers

1. $4\sqrt{3}$ **3.** $2\sqrt{7}$ **5.** $3\sqrt{5}$ **7.** $3\sqrt{6}$ **9.** $20\sqrt{2}$ **11.** $36\sqrt{7}$
13. $6\sqrt{3}$ **15.** $x\sqrt{3}$ **17.** $y^2\sqrt{3}$ **19.** $r\sqrt{2r}$ **21.** $5b\sqrt{5}$
23. $2x^2\sqrt{6}$ **25.** $3a^2\sqrt{6a}$ **27.** $xy\sqrt{x}$ **29.** $a^3b^2\sqrt{c}$ **31.** $\dfrac{3}{4}$
33. $\dfrac{\sqrt{3}}{2}$ **35.** $\dfrac{\sqrt{3}}{7}$ **37.** False **39.** always **41.** $\dfrac{2a\sqrt{2}}{5}$
43. $\dfrac{\sqrt{5}}{5}$ **45.** $\dfrac{\sqrt{10}}{2}$ **47.** $\dfrac{\sqrt{15a}}{5}$ **49.** $\dfrac{x^2\sqrt{21}}{7}$ **51.** $\dfrac{2s\sqrt{14s}}{7}$
53. Simplest form **55.** Remove the perfect-square factors from the radical and simplify. **57.** Above and Beyond

9.3

Adding and Subtracting Radicals

< 9.3 Objectives >

1 > Add and subtract expressions involving numeric radicals

2 > Add and subtract expressions involving algebraic radicals

Two radicals that have the same index and the same radicand (the expression inside the radical) are called **like radicals.** For example,

$2\sqrt{3}$ and $5\sqrt{3}$ are like radicals.

$\sqrt{2}$ and $\sqrt{5}$ are not like radicals—they have different radicands.

$\sqrt{5}$ and $\sqrt[3]{5}$ are not like radicals—they have different indices (2 and 3, representing a square root and a cube root).

NOTE

Indices is the plural of *index*.

Like radicals can be added (or subtracted) in the same way as like terms. We apply the distributive property and then combine the coefficients:

$$2\sqrt{5} + 3\sqrt{5} = (2 + 3)\sqrt{5} = 5\sqrt{5}$$

| Example 1 | Adding and Subtracting Like Numeric Radicals |

< Objective 1 >

Simplify each expression.

(a) $5\sqrt{2} + 3\sqrt{2} = (5 + 3)\sqrt{2} = 8\sqrt{2}$

(b) $7\sqrt{5} - 2\sqrt{5} = (7 - 2)\sqrt{5} = 5\sqrt{5}$

(c) $8\sqrt{7} - \sqrt{7} + 2\sqrt{7} = (8 - 1 + 2)\sqrt{7} = 9\sqrt{7}$

NOTE

Apply the distributive property and then combine the coefficients.

Check Yourself 1

Simplify.

(a) $2\sqrt{5} + 7\sqrt{5}$ **(b)** $9\sqrt{7} - \sqrt{7}$

(c) $5\sqrt{3} - 2\sqrt{3} + \sqrt{3}$

If a sum or difference involves terms that are *not* like radicals, we may be able to combine terms after simplifying the radicals according to our earlier rules.

| Example 2 | Adding and Subtracting Numeric Radicals |

Simplify each expression.

(a) $3\sqrt{2} + \sqrt{8}$

We do not have like radicals, but we can simplify $\sqrt{8}$. Remember that

$$\sqrt{8} = \sqrt{4 \cdot 2} = \sqrt{4} \cdot \sqrt{2} = 2\sqrt{2}$$

so

$$3\sqrt{2} + \overset{\sqrt{8}}{\sqrt{8}} = 3\sqrt{2} + 2\sqrt{2}$$
$$= (3 + 2)\sqrt{2} = 5\sqrt{2}$$

Beginning Algebra   The Streeter/Hutchison Series in Mathematics

NOTE

The radicals can now be combined. Do you see why?

(b) $5\sqrt{3} - \sqrt{12} = 5\sqrt{3} - \sqrt{4 \cdot 3}$

$\qquad\qquad\quad = 5\sqrt{3} - \sqrt{4} \cdot \sqrt{3}$

$\qquad\qquad\quad = 5\sqrt{3} - 2\sqrt{3}$

$\qquad\qquad\quad = 3\sqrt{3}$

Check Yourself 2

Simplify.

(a) $\sqrt{2} + \sqrt{18}$ 　　　　　　　　　　　 **(b)** $5\sqrt{3} - \sqrt{27}$

Example 3 illustrates the need to apply our earlier methods for adding fractions when working with radical expressions.

Example 3　　**Adding Radical Expressions**

Add $\dfrac{\sqrt{5}}{3} + \dfrac{2}{\sqrt{5}}$.

Our first step is to *rationalize the denominator* of the second fraction, to write the sum as

$$\frac{\sqrt{5}}{3} + \frac{2\sqrt{5}}{\sqrt{5} \cdot \sqrt{5}}$$

or

$$\frac{\sqrt{5}}{3} + \frac{2\sqrt{5}}{5}$$

The LCD of the fractions is 15 and rewriting each fraction with that denominator, we have

$$\frac{\sqrt{5} \cdot 5}{3 \cdot 5} + \frac{2\sqrt{5} \cdot 3}{5 \cdot 3} = \frac{5\sqrt{5} + 6\sqrt{5}}{15}$$

$$= \frac{11\sqrt{5}}{15}$$

Check Yourself 3

Subtract $\dfrac{3}{\sqrt{10}} - \dfrac{\sqrt{10}}{5}$.

If variables are involved in radical expressions, the process of combining terms proceeds in a fashion similar to that shown in Examples 1 and 2. Consider Example 4. We again assume that all variables represent positive real numbers.

Example 4　　**Simplifying Expressions Involving Variables**

< Objective 2 >

NOTES

Simplify the first term.

The radicals can now be combined.

Simplify each expression.

(a) $5\sqrt{3x} - 2\sqrt{3x} = 3\sqrt{3x}$ 　　These are like radicals.

(b) $2\sqrt{3a^3} + 5a\sqrt{3a}$

$\qquad = 2\sqrt{a^2 \cdot 3a} + 5a\sqrt{3a}$

$\qquad = 2\sqrt{a^2} \cdot \sqrt{3a} + 5a\sqrt{3a}$

$\qquad = 2a\sqrt{3a} + 5a\sqrt{3a}$

$\qquad = 7a\sqrt{3a}$

Check Yourself 4

Simplify each expression.

(a) $2\sqrt{7y} + 3\sqrt{7y}$ **(b)** $\sqrt{20a^2} - a\sqrt{45}$

▶ **Example 5**   Adding or Subtracting Algebraic Radical Expressions

Add or subtract as indicated.

(a) $2\sqrt{6} + \sqrt{\dfrac{2}{3}}$

NOTES

Multiply by $\dfrac{\sqrt{3}}{\sqrt{3}}$, or 1.

$\dfrac{\sqrt{6}}{3}$ and $\dfrac{1}{3}\sqrt{6}$ are equivalent.

We apply the quotient property to the *second term* and rationalize the denominator.

$$\sqrt{\dfrac{2}{3}} = \dfrac{\sqrt{2}}{\sqrt{3}} = \dfrac{\sqrt{2} \cdot \sqrt{3}}{\sqrt{3} \cdot \sqrt{3}} = \dfrac{\sqrt{6}}{3}$$

So

$$2\sqrt{6} + \sqrt{\dfrac{2}{3}} = 2\sqrt{6} + \dfrac{\sqrt{6}}{3}$$

$$= \left(2 + \dfrac{1}{3}\right)\sqrt{6} = \dfrac{7}{3}\sqrt{6}$$

(b) $\sqrt{20x} - \sqrt{\dfrac{x}{5}}$

NOTES

$\sqrt{20x} = \sqrt{4 \cdot 5x} = \sqrt{4} \cdot \sqrt{5x}$
$= 2\sqrt{5x}$

$\dfrac{\sqrt{5x}}{5} = \dfrac{1}{5}\sqrt{5x}$

Again we first simplify the two expressions. So

$$\sqrt{20x} - \sqrt{\dfrac{x}{5}} = 2\sqrt{5x} - \dfrac{\sqrt{x} \cdot \sqrt{5}}{\sqrt{5} \cdot \sqrt{5}}$$

$$= 2\sqrt{5x} - \dfrac{\sqrt{5x}}{5}$$

$$= \left(2 - \dfrac{1}{5}\right)\sqrt{5x} = \dfrac{9}{5}\sqrt{5x}$$

Check Yourself 5

Add or subtract as indicated.

(a) $3\sqrt{7} + \sqrt{\dfrac{1}{7}}$ **(b)** $\sqrt{40x} - \sqrt{\dfrac{2x}{5}}$

The process of adding or subtracting like radicals is identical when working with cube roots or beyond. Example 6 demonstrates this idea.

▶ **Example 6**   Adding or Subtracting Cube Roots

Perform the indicated operation.

(a) $2\sqrt[3]{7} + 3\sqrt[3]{7}$

$$2\sqrt[3]{7} + 3\sqrt[3]{7} = 5\sqrt[3]{7}$$ We can add the like radicals.

(b) $5x\sqrt[3]{2x} - \sqrt[3]{54x^4}$

$$5x\sqrt[3]{2x} - \sqrt[3]{54x^4} = 5x\sqrt[3]{2x} - \sqrt[3]{2 \cdot 3^3 \cdot x \cdot x^3}$$ Factor the radicand.

$$= 5x\sqrt[3]{2x} - \sqrt[3]{3^3 x^3} \cdot \sqrt[3]{2x}$$ Find the perfect-cube factors.

$$= 5x\sqrt[3]{2x} - 3x \cdot \sqrt[3]{2x}$$ Subtract the like terms.

$$= 2x\sqrt[3]{2x}$$

NOTE

$\sqrt[3]{3^3 x^3} = 3x$

 Check Yourself 6

Perform the indicated operation.

(a) $5\sqrt[4]{10} - 3\sqrt[4]{10}$

(b) $3a^2\sqrt[3]{5a} + \sqrt[3]{40a^7}$

Check Yourself ANSWERS

1. (a) $9\sqrt{5}$; **(b)** $8\sqrt{7}$; **(c)** $4\sqrt{3}$ **2. (a)** $4\sqrt{2}$; **(b)** $2\sqrt{3}$ **3.** $\dfrac{\sqrt{10}}{10}$

4. (a) $5\sqrt{7y}$; **(b)** $-a\sqrt{5}$ **5. (a)** $\dfrac{22}{7}\sqrt{7}$; **(b)** $\dfrac{9}{5}\sqrt{10x}$

6. (a) $2\sqrt[4]{10}$; **(b)** $5a^2\sqrt[3]{5a}$

Reading Your Text

The following fill-in-the-blank exercises are designed to ensure that you understand some of the key vocabulary used in this section.

SECTION 9.3

(a) Radicals that have the same index and the same _____ are called like radicals.

(b) Like radicals can be added by using the _____ property.

(c) In order to assure that they are real numbers, we assume that all radical variables are _____.

(d) Just as we refer to the root two as the square root, we refer to the root three as the _____ root.

Name _____

Section _____ Date _____

Answers

1. _____

2. _____

3. _____

4. _____

5. _____

6. _____

7. _____ 8. _____

9. _____ 10. _____

11. _____ 12. _____

13. _____ 14. _____

15. _____ 16. _____

17. _____ 18. _____

19. _____ 20. _____

21. _____ 22. _____

23. _____ 24. _____

25. _____ 26. _____

< Objective 1 >

Simplify by combining like terms.

1. $3\sqrt{2} + 5\sqrt{2}$

2. $\sqrt{3} + 5\sqrt{3}$

3. $11\sqrt{7} - 4\sqrt{7}$

4. $7\sqrt{3} - 5\sqrt{2}$

5. $5\sqrt{7} + 3\sqrt{6}$

6. $3\sqrt{5} - 5\sqrt{5}$

7. $3\sqrt{5} - 7\sqrt{5}$

8. $2\sqrt{11} + 5\sqrt{11}$

9. $2\sqrt{3x} + 5\sqrt{3x}$

10. $7\sqrt{2a} - 3\sqrt{2a}$

11. $2\sqrt{3} + \sqrt{3} + 3\sqrt{3}$

12. $3\sqrt{5} + 2\sqrt{5} + \sqrt{5}$

13. $6\sqrt{11} - 5\sqrt{11} + 3\sqrt{11}$

14. $3\sqrt{10} - 2\sqrt{10} + \sqrt{10}$

> Videos

15. $2\sqrt{5x} + 5\sqrt{5x} - \sqrt{5x}$

16. $8\sqrt{3b} - 2\sqrt{3b} + \sqrt{3b}$

17. $2\sqrt{3} + \sqrt{12}$

18. $7\sqrt{3} + 2\sqrt{27}$

19. $\sqrt{20} - \sqrt{5}$

20. $\sqrt{98} - 3\sqrt{2}$

21. $2\sqrt{6} - \sqrt{54}$

22. $2\sqrt{3} - \sqrt{27}$

23. $\sqrt{72} + \sqrt{50}$ > Videos

24. $\sqrt{27} - \sqrt{12}$

25. $3\sqrt{12} - \sqrt{48}$

26. $5\sqrt{8} + 2\sqrt{18}$

27. $\sqrt{12} + \sqrt{27} - \sqrt{3}$

28. $\sqrt{50} + \sqrt{32} - \sqrt{8}$

29. $3\sqrt{24} - \sqrt{54} + \sqrt{6}$

30. $\sqrt{63} - 2\sqrt{28} + 5\sqrt{7}$

Basic Skills | **Challenge Yourself** | Calculator/Computer | Career Applications | Above and Beyond
▲

31. $\sqrt{3} + \sqrt{\dfrac{1}{3}}$

32. $\sqrt{6} - \sqrt{\dfrac{1}{6}}$

33. $\dfrac{\sqrt{12}}{3} - \dfrac{1}{\sqrt{3}}$ > Videos

34. $\dfrac{\sqrt{20}}{5} + \dfrac{2}{\sqrt{5}}$

Determine whether each statement is **true** *or* **false.**

35. Unlike radicals can sometimes be added or subtracted.

36. To add or subtract like radicals, we must use the distributive property.

< Objective 2 >

Simplify by combining like terms.

37. $a\sqrt{27} - 2\sqrt{3a^2}$ > Videos

38. $5\sqrt{2y^2} - 3y\sqrt{8}$

39. $5\sqrt{3x^3} + 2\sqrt{27x}$ > Videos

40. $7\sqrt{2a^3} - \sqrt{8a}$

41. **GEOMETRY** Find the perimeter of the rectangle shown in the figure.

42. **GEOMETRY** Find the perimeter of the rectangle shown in the figure. Write your answer in radical form.

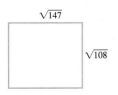

---

27. _____

28. _____

29. _____

30. _____

31. _____

32. _____

33. _____

34. _____

35. _____

36. _____

37. _____

38. _____

39. _____

40. _____

41. _____

42. _____

Answers

43.

44.

45.

46.

47.

48.

49.

50.

51.

52.

43. GEOMETRY Find the perimeter of the triangle shown in the figure. Write your answer in radical form.

$\sqrt{3} + \sqrt{2}$

3

$\sqrt{3} - \sqrt{2}$

44. GEOMETRY Find the perimeter of the triangle shown in the figure. Write your answer in radical form.

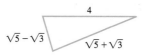

4

$\sqrt{5} - \sqrt{3}$

$\sqrt{5} + \sqrt{3}$

| Basic Skills | Challenge Yourself | **Calculator/Computer** | Career Applications | Above and Beyond |

Use a calculator to find a decimal approximation for each expression. Round your answer to the nearest hundredth.

45. $\sqrt{3} - \sqrt{2}$ **46.** $\sqrt{7} + \sqrt{11}$

47. $\sqrt{5} + \sqrt{3}$ **48.** $\sqrt{17} - \sqrt{13}$

49. $4\sqrt{3} - 7\sqrt{5}$ **50.** $8\sqrt{2} + 3\sqrt{7}$

51. $5\sqrt{7} + 8\sqrt{13}$ **52.** $7\sqrt{2} - 4\sqrt{11}$

Answers

1. $8\sqrt{2}$ **3.** $7\sqrt{7}$ **5.** Cannot be simplified   **7.** $-4\sqrt{5}$

9. $7\sqrt{3x}$ **11.** $6\sqrt{3}$ **13.** $4\sqrt{11}$ **15.** $6\sqrt{5x}$ **17.** $4\sqrt{3}$

19. $\sqrt{5}$ **21.** $-\sqrt{6}$ **23.** $11\sqrt{2}$ **25.** $2\sqrt{3}$ **27.** $4\sqrt{3}$

29. $4\sqrt{6}$ **31.** $\dfrac{4}{3}\sqrt{3}$ **33.** $\dfrac{1}{3}\sqrt{3}$ **35.** True   **37.** $a\sqrt{3}$

39. $(5x + 6)\sqrt{3x}$ **41.** 26   **43.** $2\sqrt{3} + 3$ **45.** 0.32

47. 3.97   **49.** -8.72 **51.** 42.07

The Streeter/Hutchison Series in Mathematics   *Beginning Algebra*

9.4

Multiplying and Dividing Radicals

< 9.4 Objectives >

1 > Multiply expressions involving radical expressions

2 > Divide expressions involving radical expressions

In Section 9.2 we stated the first property for radicals:

$\sqrt{ab} = \sqrt{a} \cdot \sqrt{b}$ when a and b are any nonnegative real numbers

That property has been used to simplify radical expressions. We have also used the property to multiply two radicals, as seen in Section 9.2. (You may wish to review Example 5, parts (b) and (c).) We again used this approach in Section 9.3 (see Example 5, parts (a) and (b)). Our method employed Property 1.

RECALL

The product of square roots is equal to the square root of the product of the radicands.

$\sqrt{a} \cdot \sqrt{b} = \sqrt{ab}$

For example,

$\sqrt{3} \cdot \sqrt{5} = \sqrt{3 \cdot 5} = \sqrt{15}$

We may have to simplify after multiplying, as Example 1 illustrates.

| Example 1 | Multiplying Radical Expressions |

< Objective 1 >

Multiply and then simplify each expression. Assume all variables represent positive real numbers.

(a) $\sqrt{5} \cdot \sqrt{10} = \sqrt{5 \cdot 10} = \sqrt{50}$
$$= \sqrt{25 \cdot 2} = 5\sqrt{2}$$

(b) $\sqrt{12} \cdot \sqrt{6} = \sqrt{12 \cdot 6} = \sqrt{72}$
$$= \sqrt{36 \cdot 2} = \sqrt{36} \cdot \sqrt{2} = 6\sqrt{2}$$

An alternative approach would be to simplify $\sqrt{12}$ first.
$$\sqrt{12} \cdot \sqrt{6} = 2\sqrt{3}\,\sqrt{6} = 2\sqrt{18}$$
$$= 2\sqrt{9 \cdot 2} = 2\sqrt{9}\,\sqrt{2}$$
$$= 2 \cdot 3\sqrt{2} = 6\sqrt{2}$$

(c) $\sqrt{10x} \cdot \sqrt{2x} = \sqrt{20x^2} = \sqrt{4x^2 \cdot 5}$
$$= \sqrt{4x^2} \cdot \sqrt{5} = 2x\sqrt{5}$$

Check Yourself 1

Simplify. Assume all variables represent positive real numbers.

(a) $\sqrt{3} \cdot \sqrt{6}$ **(b)** $\sqrt{3} \cdot \sqrt{18}$ **(c)** $\sqrt{8a} \cdot \sqrt{3a}$

If coefficients are involved in a product, we can use the commutative and associative properties to change the order and grouping of the factors. This is illustrated in Example 2.

 Example 2 | **Multiplying Radical Expressions**

Multiply.

NOTE

In practice, it is not necessary to show the intermediate steps.

$$(2\sqrt{5})(3\sqrt{6}) = (2 \cdot 3)(\sqrt{5} \cdot \sqrt{6})$$
$$= 6\sqrt{5 \cdot 6}$$
$$= 6\sqrt{30}$$

Check Yourself 2

Multiply $(3\sqrt{7})(5\sqrt{3})$.

The distributive property can also be applied in multiplying radical expressions, as shown in Example 3.

 Example 3 | **Multiplying Radical Expressions**

Multiply.

(a) $\sqrt{3}(\sqrt{2} + \sqrt{3})$

$$= \sqrt{3} \cdot \sqrt{2} + \sqrt{3} \cdot \sqrt{3} \qquad \text{The distributive property}$$
$$= \sqrt{6} + 3 \qquad \text{Multiply the radicals.}$$

(b) $\sqrt{5}(2\sqrt{6} + 3\sqrt{3})$

$$= \sqrt{5} \cdot 2\sqrt{6} + \sqrt{5} \cdot 3\sqrt{3} \qquad \text{The distributive property}$$
$$= 2 \cdot \sqrt{5} \cdot \sqrt{6} + 3 \cdot \sqrt{5} \cdot \sqrt{3} \qquad \text{The commutative property}$$
$$= 2\sqrt{30} + 3\sqrt{15}$$

Check Yourself 3

Multiply.

(a) $\sqrt{5}(\sqrt{6} + \sqrt{5})$ **(b)** $\sqrt{3}(2\sqrt{5} + 3\sqrt{2})$

The FOIL pattern we used for multiplying binomials in Section 3.4 can also be applied in multiplying radical expressions. This is shown in Example 4.

Example 4   **Multiplying Radical Expressions**

Multiply.

(a) $(\sqrt{3} + 2)(\sqrt{3} + 5)$

$$= \sqrt{3} \cdot \sqrt{3} + 5\sqrt{3} + 2\sqrt{3} + 2 \cdot 5$$
$$= 3 + 5\sqrt{3} + 2\sqrt{3} + 10 \qquad \text{Combine like terms.}$$
$$= 13 + 7\sqrt{3}$$

> **C A U T I O N**

Be careful! This result *cannot* be further simplified: 13 and $7\sqrt{3}$ are *not* like terms.

(b) $(\sqrt{7} + 2)(\sqrt{7} - 2) = \sqrt{7} \cdot \sqrt{7} - 2\sqrt{7} + 2\sqrt{7} - 4$
$$= 7 - 4 = 3$$

NOTE

You can use the pattern $(a + b)(a - b) = a^2 - b^2$, where $a = \sqrt{7}$ and $b = 2$, for the same result. $\sqrt{7} + 2$ and $\sqrt{7} - 2$ are called **conjugates** of each other. Note that their product is the rational number 3. The product of conjugates is *always rational*.

(c) $(\sqrt{3} + 5)^2 = (\sqrt{3} + 5)(\sqrt{3} + 5)$
$$= \sqrt{3} \cdot \sqrt{3} + 5\sqrt{3} + 5\sqrt{3} + 5 \cdot 5$$
$$= 3 + 5\sqrt{3} + 5\sqrt{3} + 25$$
$$= 28 + 10\sqrt{3}$$

 Check Yourself 4

Multiply.

(a) $(\sqrt{5} + 3)(\sqrt{5} - 2)$ **(b)** $(\sqrt{3} + 4)(\sqrt{3} - 4)$ **(c)** $(\sqrt{2} - 3)^2$

RECALL

The quotient of square roots is equal to the square root of the quotient.

We can also use our second property for radicals in a new way.

$$\frac{\sqrt{a}}{\sqrt{b}} = \sqrt{\frac{a}{b}}$$

One use of this property to divide radical expressions is illustrated in Example 5.

 Example 5   **Dividing Radical Expressions**

< Objective 2 >

Simplify.

NOTE

The clue to recognizing when to use this approach is in noting that 48 is divisible by 3.

(a) $\dfrac{\sqrt{48}}{\sqrt{3}} = \sqrt{\dfrac{48}{3}} = \sqrt{16} = 4$

(b) $\dfrac{\sqrt{200}}{\sqrt{2}} = \sqrt{\dfrac{200}{2}} = \sqrt{100} = 10$

(c) $\dfrac{\sqrt{125x^2}}{\sqrt{5}} = \sqrt{\dfrac{125x^2}{5}} = \sqrt{25x^2} = 5x$

 Check Yourself 5

Simplify.

(a) $\dfrac{\sqrt{75}}{\sqrt{3}}$ **(b)** $\dfrac{\sqrt{81s^2}}{\sqrt{9}}$

There is one final quotient form that you may encounter in simplifying expressions, and it will be extremely important in our work with quadratic equations in Chapter 10. This form is shown in Example 6.

| Example 6 | Simplifying Radical Expressions |

Simplify the expression

$$\frac{3 + \sqrt{72}}{3}$$

First, we must simplify the radical in the numerator.

> **CAUTION**

Be careful! Students are sometimes tempted to write

$$\frac{\cancel{3} + 6\sqrt{2}}{\cancel{3}} = 1 + 6\sqrt{2}$$

This is *not* correct. We must divide *both terms* of the numerator by the common factor.

$$\frac{3 + \sqrt{72}}{3} = \frac{3 + \sqrt{36 \cdot 2}}{3}$$ Use Property 1 to simplify $\sqrt{72}$.

$$= \frac{3 + \sqrt{36} \cdot \sqrt{2}}{3} = \frac{3 + 6\sqrt{2}}{3}$$

$$= \frac{3(1 + 2\sqrt{2})}{3} = 1 + 2\sqrt{2}$$ *Factor* the numerator and then divide by the *common* factor 3.

Check Yourself 6

Simplify $\dfrac{15 + \sqrt{75}}{5}$.

NOTE

We saw this briefly in Section 9.2.

In Section 9.2, we said that for an expression involving square roots to be in simplest form, no radical can appear in the denominator of a fraction. Changing a fraction that has a radical in the denominator to an equivalent fraction that has no radical in the denominator is called **rationalizing the denominator.**

The primary reason for rationalizing denominators used to be so that a better arithmetic estimation could be made. With the availability of calculators, that is now rarely necessary. However, the process continues to be useful in higher-level mathematics. So that we can review this process, we present a straightforward example below.

| Example 7 | Rationalizing a Denominator |

Rewrite each fraction so that there are no radicals in the denominator.

(a) $\dfrac{\sqrt{7}}{\sqrt{2}}$

$$\frac{\sqrt{7}}{\sqrt{2}} = \frac{\sqrt{7} \cdot \sqrt{2}}{\sqrt{2} \cdot \sqrt{2}}$$

We can always multiply the numerator and denominator of a fraction by the same nonzero value.

Here, we multiply the numerator and denominator by $\sqrt{2}$. This will make the denominator a rational number.

$$\frac{\sqrt{7} \cdot \sqrt{2}}{\sqrt{2} \cdot \sqrt{2}} = \frac{\sqrt{14}}{2}$$ We now have an equivalent fraction with a rational number for its denominator.

(b) $\dfrac{\sqrt{3x}}{\sqrt{54}}$

$$\dfrac{\sqrt{3x}}{\sqrt{54}} = \dfrac{\sqrt{3x}}{\sqrt{9 \cdot 6}} = \dfrac{\sqrt{3x}}{3\sqrt{6}}$$ First, simplify the denominator as much as possible.

$$\dfrac{\sqrt{3x}}{3\sqrt{6}} = \dfrac{\sqrt{3x} \cdot \sqrt{6}}{3\sqrt{6} \cdot \sqrt{6}} = \dfrac{\sqrt{18x}}{18}$$ Multiply numerator and denominator by $\sqrt{6}$.

$$\dfrac{\sqrt{18x}}{18} = \dfrac{\sqrt{9 \cdot 2x}}{18} = \dfrac{3\sqrt{2x}}{18}$$ Factor out the perfect square in the numerator.

$$= \dfrac{\sqrt{2x}}{6}$$ Simplify the fraction.

Check Yourself 7

Rewrite each fraction so that there are no radicals in the denominator.

(a) $\dfrac{\sqrt{17}}{\sqrt{3}}$

(b) $\dfrac{\sqrt{3a}}{\sqrt{12}}$

Check Yourself ANSWERS

1. (a) $3\sqrt{2}$; **(b)** $3\sqrt{6}$; **(c)** $2a\sqrt{6}$ **2.** $15\sqrt{21}$ **3. (a)** $\sqrt{30} + 5$;
(b) $2\sqrt{15} + 3\sqrt{6}$ **4. (a)** $-1 + \sqrt{5}$; **(b)** -13; **(c)** $11 - 6\sqrt{2}$

5. (a) 5; **(b)** $3s$ **6.** $3 + \sqrt{3}$ **7. (a)** $\dfrac{\sqrt{51}}{3}$; **(b)** $\dfrac{\sqrt{a}}{2}$

Reading Your Text

The following fill-in-the-blank exercises are designed to ensure that you understand some of the key vocabulary used in this section.

SECTION 9.4

(a) The product of two square roots is equal to the _____ of their product.

(b) The _____ pattern can be used to multiply two binomials.

(c) The product of conjugates is always _____ .

(d) The quotient of square roots is equal to the square root of the quotient of the _____ .

Name _____

Section _____ Date _____

Answers

1. _____	2. _____
3. _____	4. _____
5. _____	6. _____
7. _____	8. _____
9. _____	10. _____
11. _____	12. _____
13. _____	14. _____
15. _____	16. _____
17. _____	18. _____
19. _____	20. _____
21. _____	22. _____
23. _____	24. _____
25. _____	26. _____
27. _____	28. _____
29. _____	30. _____
31. _____	32. _____
33. _____	34. _____

< Objective 1 >

Perform the indicated multiplication. Then simplify each radical expression. Assume all variables represent positive real numbers.

1. $\sqrt{7} \cdot \sqrt{5}$

2. $\sqrt{3} \cdot \sqrt{7}$

3. $\sqrt{5} \cdot \sqrt{11}$

4. $\sqrt{13} \cdot \sqrt{5}$

5. $\sqrt{3} \cdot \sqrt{10m}$

6. $\sqrt{7a} \cdot \sqrt{13}$

7. $\sqrt{2x} \cdot \sqrt{15}$

8. $\sqrt{15} \cdot \sqrt{2b}$

9. $\sqrt{3} \cdot \sqrt{7} \cdot \sqrt{2}$

10. $\sqrt{5} \cdot \sqrt{7} \cdot \sqrt{3}$

11. $\sqrt{3} \cdot \sqrt{12}$

12. $\sqrt{8} \cdot \sqrt{8}$

13. $\sqrt{10x} \cdot \sqrt{10x}$

14. $\sqrt{5a} \cdot \sqrt{15a}$

15. $\sqrt{27} \cdot \sqrt{2}$

16. $\sqrt{8} \cdot \sqrt{10}$

17. $\sqrt{2x} \cdot \sqrt{6x}$

18. $\sqrt{3a} \cdot \sqrt{15a}$

19. $3\sqrt{2x} \cdot \sqrt{6x}$

20. $3\sqrt{2a} \cdot \sqrt{5a}$

21. $(3a\sqrt{3a})(5\sqrt{7a})$ ▷ Videos

22. $(2x\sqrt{5x})(3\sqrt{11x})$

23. $(3\sqrt{5})(2\sqrt{10})$

24. $(4\sqrt{3})(3\sqrt{6})$

25. $\sqrt{3}(\sqrt{2} + \sqrt{3})$

26. $\sqrt{3}(\sqrt{5} - \sqrt{3})$

27. $\sqrt{3}(2\sqrt{5} - 3\sqrt{3})$ ▷ Videos

28. $\sqrt{7}(2\sqrt{3} + 3\sqrt{7})$

29. $(\sqrt{3} + 5)(\sqrt{3} + 3)$

30. $(\sqrt{3} - 5)(\sqrt{3} + 2)$

31. $(\sqrt{5} - 1)(\sqrt{5} + 3)$ ▷ Videos

32. $(\sqrt{2} + 3)(\sqrt{2} - 7)$

33. $(\sqrt{5} - 2)(\sqrt{5} + 2)$

34. $(\sqrt{7} + 5)(\sqrt{7} - 5)$

35. $(\sqrt{7} + 3)(\sqrt{7} - 3)$

36. $(\sqrt{11} - 3)(\sqrt{11} + 3)$

37. $(2\sqrt{3})^2$

38. $(3\sqrt{5})^2$

39. $(\sqrt{3} + 2)^2$

40. $(\sqrt{5} - 3)^2$

41. $(\sqrt{y} - 5)^2$ > Videos

42. $(\sqrt{x} + 4)^2$

Basic Skills | **Challenge Yourself** ▲ | Calculator/Computer | Career Applications | Above and Beyond

Determine whether each statement is **true** *or* **false.**

43. The square root of a quotient is equal to the quotient of the square roots.

44. The square root of a sum is equal to the sum of the square roots.

Complete each statement with **never, sometimes,** *or* **always.**

45. The product of conjugates is _____ rational.

46. The product of two irrational numbers is _____ rational.

< Objective 2 >

Perform the indicated division. Rationalize the denominator if necessary. Then simplify each radical expression. Assume all variables represent positive real numbers.

47. $\dfrac{\sqrt{98}}{\sqrt{2}}$ > Videos

48. $\dfrac{\sqrt{108}}{\sqrt{3}}$

49. $\dfrac{\sqrt{72a^2}}{\sqrt{2}}$

50. $\dfrac{\sqrt{48m^2}}{\sqrt{3}}$

51. $\dfrac{4 + \sqrt{48}}{4}$

52. $\dfrac{12 + \sqrt{108}}{6}$

53. $\dfrac{5 + \sqrt{175}}{5}$

54. $\dfrac{18 + \sqrt{567}}{9}$

55. $\dfrac{-8 - \sqrt{512}}{4}$ > Videos

56. $\dfrac{-9 - \sqrt{108}}{3}$

57. $\dfrac{6 + \sqrt{18}}{3}$

58. $\dfrac{6 - \sqrt{20}}{2}$

59. $\dfrac{15 - \sqrt{75}}{5}$

60. $\dfrac{8 + \sqrt{48}}{4}$

Answers

35. _____

36. _____

37. _____

38. _____

39. _____

40. _____

41. _____

42. _____

43. _____

44. _____

45. _____

46. _____

47. _____

48. _____

49. _____

50. _____

51. _____

52. _____

53. _____

54. _____

55. _____

56. _____

57. _____

58. _____

59. _____

60. _____

Beginning Algebra   The Streeter/Hutchison Series in Mathematics   © The McGraw-Hill Companies. All Rights Reserved.

Answers

61. _____

62. _____

63. _____

64. _____

65. _____

61. **GEOMETRY** Find the area of the rectangle shown.

62. **GEOMETRY** Find the area of the rectangle shown.

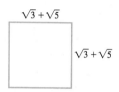

| Basic Skills | Challenge Yourself | Calculator/Computer | Career Applications | **Above and Beyond** |

63. Complete the statement "$\sqrt{2} \cdot \sqrt{5} = \sqrt{10}$ because . . . ."

64. Explain why $2\sqrt{3} + 5\sqrt{3} = 7\sqrt{3}$ but $7\sqrt{3} + 3\sqrt{5} \neq 10\sqrt{8}$.

65. When you look out over an unobstructed landscape or seascape, the distance to the visible horizon depends on your height above the ground. The equation

$$d = \sqrt{\frac{3}{2}h}$$

is a good estimate of this, in which d = distance to horizon in miles and h = height of viewer above the ground. Work with a partner to make a chart of distances to the horizon given different elevations. Use the actual heights of tall buildings or prominent landmarks in your area. The local library should have a list of these. Be sure to consider the view to the horizon you get when flying in a plane. What would your elevation have to be to see from one side of your city or town to the other? From one side of your state or county to the other?

Answers

1. $\sqrt{35}$ **3.** $\sqrt{55}$ **5.** $\sqrt{30m}$ **7.** $\sqrt{30x}$ **9.** $\sqrt{42}$ **11.** 6

13. $10x$ **15.** $3\sqrt{6}$ **17.** $2x\sqrt{3}$ **19.** $6x\sqrt{3}$ **21.** $15a^2\sqrt{21}$

23. $30\sqrt{2}$ **25.** $\sqrt{6} + 3$ **27.** $2\sqrt{15} - 9$ **29.** $18 + 8\sqrt{3}$

31. $2 + 2\sqrt{5}$ **33.** 1   **35.** -2 **37.** 12   **39.** $7 + 4\sqrt{3}$

41. $y - 10\sqrt{y} + 25$ **43.** True   **45.** always   **47.** 7   **49.** $6a$

51. $1 + \sqrt{3}$ **53.** $1 + \sqrt{7}$ **55.** $-2 - 4\sqrt{2}$ **57.** $2 + \sqrt{2}$

59. $3 - \sqrt{3}$ **61.** $\sqrt{33}$ **63.** Above and Beyond

65. Above and Beyond

9.5

Solving Radical Equations

< 9.5 Objectives >

1 > Solve an equation containing a radical expression

2 > Solve a literal equation containing a radical expression

NOTE

$$x^2 = 1$$
$$x^2 - 1 = 0$$
$$(x + 1)(x - 1) = 0$$
So the solutions are 1 and −1.

In this section, we establish some procedures for solving equations involving radical expressions. The basic technique involves raising both sides of an equation to some power. However, doing so requires some caution.

For example, begin with the equation $x = 1$. Squaring both sides gives $x^2 = 1$, which has two solutions, 1 and −1. Clearly −1 is not a solution to the original equation. We refer to −1 as an *extraneous solution*.

We must be aware of the possibility of extraneous solutions any time we raise both sides of an equation to any *even power*. Having said that, we are now prepared to introduce the power property of equality.

Property

The Power Property of Equality

Given any two expressions a and b and any positive integer n,

If $a = b$, then $a^n = b^n$.

Although we never lose a solution when applying the power property, we often find an extraneous one as a result of raising both sides of the equation to some power. Because of this, it is very important that you *check all solutions*.

Example 1

Solving a Radical Equation

< Objective 1 >

NOTE

$(\sqrt{x + 2})^2 = x + 2$
This is why squaring both sides of the equation removes the radical.

Solve $\sqrt{x + 2} = 3$.

Squaring each side, we have

$$(\sqrt{x + 2})^2 = 3^2$$
$$x + 2 = 9$$
$$x = 7$$

Substituting 7 into the original equation, we find

$$\sqrt{(7) + 2} \stackrel{?}{=} 3$$
$$\sqrt{9} \stackrel{?}{=} 3$$
$$3 = 3$$

Because 7 is the only value that makes this a true statement, the solution for the equation is 7.

Check Yourself 1

Solve the equation $\sqrt{x - 5} = 4$.

| Example 2 | Solving a Radical Equation |

NOTES

Applying the power property will remove the radical only if that radical is isolated on one side of the equation.

Notice that on the right $(-1)^2 = 1$.

Solve $\sqrt{4x + 5} + 1 = 0$.

First isolate the radical on the left side.

$$\sqrt{4x + 5} = -1$$

Then, square both sides.

$$(\sqrt{4x + 5})^2 = (-1)^2$$
$$4x + 5 = 1$$

Solve for x to get

$$x = -1$$

Now check the solution by substituting -1 for x in the original equation.

NOTE

2 is never equal to 0, so -1 is *not* a solution for the original equation.

$$\sqrt{4(-1) + 5} + 1 \stackrel{?}{=} 0$$
$$\sqrt{1} + 1 \stackrel{?}{=} 0$$
and $$2 \neq 0$$

Because -1 is an extraneous solution, there is *no solution* to the original equation.

 Check Yourself 2

Solve $\sqrt{3x - 2} + 2 = 0$.

Next, consider an example in which the procedure involves squaring a binomial.

| Example 3 | Solving a Radical Equation |

Solve $\sqrt{x + 3} = x + 1$.

We can square each side, as before.

$$(\sqrt{x + 3})^2 = (x + 1)^2$$
$$x + 3 = x^2 + 2x + 1$$

Simplifying this gives us the quadratic equation

$$x^2 + x - 2 = 0$$

Factoring, we have

$$(x - 1)(x + 2) = 0$$

which gives us the possible solutions

$$x = 1 \qquad \text{or} \qquad x = -2$$

NOTE

Verify this for yourself by substituting 1 and then -2 for x in the original equation.

Now we check for extraneous solutions and find that $x = 1$ is a valid solution, but that $x = -2$ does not yield a true statement.

Be careful!   Sometimes (as in this example), one side of the equation contains a binomial. In that case, we must remember the middle term when we square the binomial. The square of a binomial *is always a trinomial.*

 > C A U T I O N

 Check Yourself 3

Solve $\sqrt{x - 5} = x - 7$.

It is not always the case that one of the solutions is extraneous. We may have zero, one, or two valid solutions when we generate a quadratic from a radical equation.

In Example 4 we see a case in which both of the solutions derived satisfy the equation.

▶ **Example** 4	**Solving a Radical Equation**

Solve $\sqrt{7x + 1} - 1 = 2x$.

First, we must isolate the term involving the radical.

$$\sqrt{7x + 1} = 2x + 1$$

We can now square both sides of the equation.

$$7x + 1 = 4x^2 + 4x + 1$$

Now we write the quadratic equation in standard form.

$$4x^2 - 3x = 0$$

Factoring, we have

$$x(4x - 3) = 0$$

which yields two possible solutions

$$x = 0 \qquad \text{or} \qquad x = \frac{3}{4}$$

Checking the solutions by substitution, we find that both values for x give true statements, as follows.

Letting x be 0, we have

$$\sqrt{7(0) + 1} - 1 \stackrel{?}{=} 2(0)$$
$$\sqrt{1} - 1 \stackrel{?}{=} 0$$

or $\qquad\qquad 0 = 0$ *A true statement*

Letting x be $\frac{3}{4}$, we have

$$\sqrt{7\left(\frac{3}{4}\right) + 1} - 1 \stackrel{?}{=} 2\left(\frac{3}{4}\right)$$

$$\sqrt{\frac{25}{4}} - 1 \stackrel{?}{=} \frac{3}{2}$$

$$\frac{5}{2} - 1 \stackrel{?}{=} \frac{3}{2}$$

$$\frac{3}{2} = \frac{3}{2}$$ *Again, a true statement. The solutions are 0 and $\frac{3}{4}$.*

Check Yourself 4

Solve $\sqrt{5x + 1} - 1 = 3x$.

Many real-world applications of the material in this section require that a formula be solved for one of the variables. We look at an example of this next.

Example 5   Solving a Radical Equation for a Given Variable

< Objective 2 >   Solve the equation $h = \sqrt{pq}$ for the variable p.

Solving for the variable p requires that p be isolated on one side of the equation. We need to eliminate the radical in order to isolate the p.

$$h^2 = (\sqrt{pq})^2$$

$$h^2 = pq$$

$$\frac{h^2}{q} = p \qquad \text{or} \qquad p = \frac{h^2}{q}$$

 Check Yourself 5

Solve the equation $2a = \sqrt{3ab}$ for the variable b.

We summarize our work with an algorithm for solving equations involving radicals.

Step by Step

Solving Equations Involving Radicals	**Step 1**	Isolate a radical on one side of the equation.
	Step 2	Raise each side of the equation to the smallest power that eliminates the isolated radical.
	Step 3	If any radicals remain in the equation derived in step 2, return to step 1 and continue until no radical remains.
	Step 4	Solve the resulting equation to determine any possible solutions.
	Step 5	Check all solutions to determine whether extraneous solutions resulted from step 2.

 Check Yourself ANSWERS

1. 21   **2.** No solution   **3.** 9   **4.** $0, -\dfrac{1}{9}$ **5.** $\dfrac{4a}{3}$

Reading Your Text

The following fill-in-the-blank exercises are designed to ensure that you understand some of the key vocabulary used in this section.

SECTION 9.5

(a) The basic technique for solving a radical equation is to raise both _____ of an equation to some power.

(b) We must be aware of the possibility of extraneous solutions any time we raise both sides of an equation to any _____ power.

(c) We never _____ a solution when using the power property of equality.

(d) Squaring a binomial always results in _____ terms.

Basic Skills | Challenge Yourself | Calculator/Computer | Career Applications | Above and Beyond

▲

< Objective 1 >

Solve each equation. Be sure to check your solutions.

1. $\sqrt{x} = 2$

2. $\sqrt{x} - 3 = 0$

3. $2\sqrt{y} - 1 = 0$

4. $3\sqrt{2z} = 9$

5. $\sqrt{m + 5} = 3$ > Videos

6. $\sqrt{y + 7} = 5$

7. $\sqrt{x - 3} = 4$

8. $\sqrt{x + 4} = 3$

9. $\sqrt{2x - 1} = 3$

10. $\sqrt{3x + 1} = 4$

11. $\sqrt{2x + 4} - 4 = 0$

12. $\sqrt{3x + 3} - 6 = 0$

13. $\sqrt{3x - 2} + 2 = 0$

14. $\sqrt{4x + 1} + 3 = 0$

15. $x + \sqrt{x - 1} = 7$

16. $x - \sqrt{x + 2} = 10$

17. $x + \sqrt{2x - 5} = 10$ > Videos

18. $2x - \sqrt{3x - 2} = 8$

19. $\sqrt{2x - 3} + 1 = 3$

20. $\sqrt{3x + 1} - 2 = -1$

21. $2\sqrt{3z + 2} - 1 = 5$

22. $3\sqrt{4q - 1} - 2 = 7$

23. $\sqrt{6x - 8} = x$

24. $\sqrt{8y - 15} = y$

25. $\sqrt{x + 5} = x - 1$

26. $\sqrt{2x - 1} = x - 8$

27. $\sqrt{3m - 2} + m = 10$

28. $\sqrt{2x + 1} + x = 7$

29. $\sqrt{t + 9} + 3 = t$ > Videos

30. $\sqrt{2y + 7} + 4 = y$

9.5 exercises

Boost *your* GRADE at ALEKS.com!

ALEKS®

- Practice Problems
- Self-Tests
- NetTutor
- e-Professors
- Videos

Name _____

Section _____ Date _____

Answers

1. _____	2. _____
3. _____	4. _____
5. _____	6. _____
7. _____	8. _____
9. _____	10. _____
11. _____	12. _____
13. _____	
14. _____	
15. _____	16. _____
17. _____	18. _____
19. _____	20. _____
21. _____	22. _____
23. _____	24. _____
25. _____	26. _____
27. _____	28. _____
29. _____	30. _____

Answers

31. _____

32. _____

33. _____

34. _____

35. _____

36. _____

37. _____

38. _____

39. _____

40. _____

Complete each statement with **never, sometimes,** *or* **always.**

31. When applying the power property of equality, you _____ lose a solution.

32. If you square both sides of an equation, you _____ find an extraneous solution.

33. The square of a binomial is _____ a trinomial.

34. If we generate a quadratic equation from a radical, we _____ find more than one solution.

< Objective 2 >

Solve for the indicated variable.

35. $h = \sqrt{pq}$ for q **36.** $c = \sqrt{a^2 + b^2}$ for a

37. $v = \sqrt{2gR}$ for R **38.** $v = \sqrt{2gR}$ for g

39. $r = \sqrt{\dfrac{S}{2\pi}}$ for S ⊙ > Videos   **40.** $r = \sqrt{\dfrac{3V}{4\pi}}$ for V

Answers

1. 4   3. $\dfrac{1}{4}$ 5. 4   7. 19   9. 5   11. 6   13. No solution

15. 5   17. 7   19. $\dfrac{7}{2}$ 21. $\dfrac{7}{3}$ 23. 2, 4   25. 4   27. 6

29. 7   31. never   33. always   35. $q = \dfrac{h^2}{p}$ 37. $R = \dfrac{v^2}{2g}$

39. $S = 2\pi r^2$

9.6

Applications of the Pythagorean Theorem

< 9.6 Objectives >

1 > Apply the Pythagorean theorem in solving problems

2 > Use the distance formula to find the distance between two points

Perhaps the most famous theorem in all of mathematics is the **Pythagorean theorem.** The theorem was named for the Greek mathematician Pythagoras, born in 572 B.C.E. Pythagoras was the founder of the Greek society the Pythagoreans. Although the theorem bears Pythagoras's name, his own work on this theorem is uncertain because the Pythagoreans credited all new discoveries to their founder.

Property

The Pythagorean Theorem

For every right triangle, the square of the length of the hypotenuse is equal to the sum of the squares of the lengths of the legs.

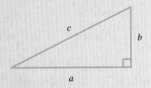

c is the hypotenuse.

a and b are the legs.

$$c^2 = a^2 + b^2$$

 Example 1   Verifying the Pythagorean Theorem

< Objective 1 >

Verify the Pythagorean theorem for the given triangles.

(a)

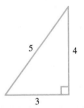

$$5^2 \stackrel{?}{=} 3^2 + 4^2$$

$$25 \stackrel{?}{=} 9 + 16$$

$$25 = 25$$

(b)

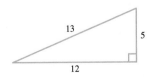

$$13^2 \stackrel{?}{=} 12^2 + 5^2$$

$$169 \stackrel{?}{=} 144 + 25$$

$$169 = 169$$

723

The Streeter/Hutchison Series in Mathematics   Beginning Algebra

Check Yourself 1

Verify the Pythagorean theorem for the right triangle shown.

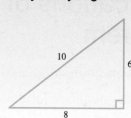

The Pythagorean theorem can be used to find the length of one side of a right triangle when the lengths of the two other sides are known.

| Example 2 | Solving for the Length of the Hypotenuse |

Find length x.

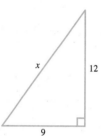

NOTE

x is longer than the given sides because it is the hypotenuse.

$$x^2 = 9^2 + 12^2$$
$$= 81 + 144$$
$$= 225$$

so

$$x = 15 \qquad \text{or} \qquad \underline{x = -15}$$

We reject this solution because lengths must be positive.

Check Yourself 2

Find length x.

Sometimes, one or more of the lengths of the sides may be represented by an irrational number.

| Example 3 | Solving for the Length of the Leg |

Calculator

Find length x. Use a calculator to approximate x to the nearest tenth.

NOTE

You can approximate $3\sqrt{3}$ (or $\sqrt{27}$) with a calculator.

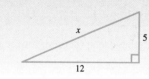

$$3^2 + x^2 = 6^2$$
$$9 + x^2 = 36$$
$$x^2 = 27$$
$$x = \pm\sqrt{27}$$

But distance cannot be negative, so

$$x = \sqrt{27}$$

So x is approximately 5.2.

Check Yourself 3

Find length x and approximate it to the nearest tenth.

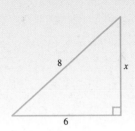

The Pythagorean theorem can be applied to solve a variety of geometric problems.

 Example 4   **Solving for the Length of the Diagonal**

 Calculator

NOTES

Always draw and label a sketch showing the information from a problem when geometric figures are involved.

Again, distance cannot be negative, so we eliminate $x = -\sqrt{89}$.

Find, to the nearest tenth, the length of the diagonal of a rectangle that is 8 cm long and 5 cm wide. Let x be the unknown length of the diagonal:

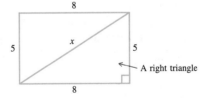
A right triangle

So

$$x^2 = 5^2 + 8^2$$
$$= 25 + 64$$
$$= 89$$
$$x = \sqrt{89}$$

Thus

$$x \approx 9.4 \text{ cm}$$

Check Yourself 4

The diagonal of a rectangle is 12 in. and its width is 6 in. Find its length to the nearest tenth.

The application in Example 5 also makes use of the Pythagorean theorem.

| ▶ | **Example** 5 | Solving an Application |

How long must a guy wire be to reach from the top of a 30-ft pole to a point on the ground 20 ft from the base of the pole? Round to the nearest foot.

Again, be sure to draw a sketch of the problem.

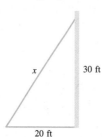

$$x^2 = 20^2 + 30^2$$
$$= 400 + 900$$
$$= 1,300$$
$$x = \sqrt{1,300}$$
$$\approx 36 \text{ ft}$$

✓ **Check Yourself** 5

A 16-ft ladder leans against a wall with its base 4 ft from the wall. How far above the floor is the top of the ladder? Round to the nearest tenth of a foot.

To find the distance between any two points in the plane, we use a formula derived from the Pythagorean theorem. First, we need an alternate form of the Pythagorean theorem.

| Property |

| **The Pythagorean Theorem** | Given a right triangle in which c is the length of the hypotenuse, we have the equation $$c^2 = a^2 + b^2$$ We can write the formula as $$c = \sqrt{a^2 + b^2}$$ |

We use this form of the Pythagorean theorem in Example 6.

| ▶ | **Example** 6 | Finding the Distance Between Two Points |

< Objective 2 >

Find the distance from (2, 3) to (5, 7).

The distance can be seen as the hypotenuse of a right triangle.

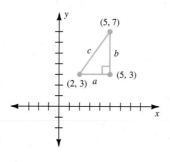

The lengths of the two legs can be found by finding the difference of the two x-coordinates and the difference of the two y-coordinates. So

$$a = 5 - 2 = 3 \qquad \text{and} \qquad b = 7 - 3 = 4$$

The distance c can then be found using the formula

$$c = \sqrt{a^2 + b^2}$$

or, in this case,

$$c = \sqrt{3^2 + 4^2}$$
$$c = \sqrt{9 + 16}$$
$$= \sqrt{25}$$
$$= 5$$

The distance is 5 units.

 Check Yourself 6

Find the distance between (0, 2) and (5, 14).

If we call our points (x_1, y_1) and (x_2, y_2), we can state the **distance formula.**

Property

Distance Formula	The distance between points (x_1, y_1) and (x_2, y_2) can be found using the formula $d = \sqrt{(x_2 - x_1)^2 + (y_2 - y_1)^2}$

Example 7   **Finding the Distance Between Two Points**

Find the distance between $(-2, 5)$ and $(2, -3)$. Simplify the radical answer.
   Using the formula,

$$d = \sqrt{[2 - (-2)]^2 + [(-3) - 5]^2}$$
$$= \sqrt{(4)^2 + (-8)^2}$$
$$= \sqrt{16 + 64}$$
$$= \sqrt{80}$$
$$= 4\sqrt{5}$$

NOTE

$\sqrt{80} = \sqrt{16 \cdot 5} = 4\sqrt{5}$

 Check Yourself 7

Find the distance between (2, 5) and (−5, 2).

In Example 7, you were asked to find the distance between $(-2, 5)$ and $(2, -3)$.

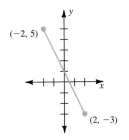

To form a right triangle, we include the point $(-2, -3)$.

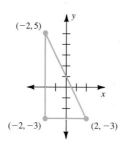

Note that the lengths of the two sides of the right triangle are 4 and 8. By the Pythagorean theorem, the hypotenuse must have length $\sqrt{4^2 + 8^2} = \sqrt{80} = 4\sqrt{5}$. The distance formula is an application of the Pythagorean theorem.

You can use the square-root key on a calculator to approximate the length of a diagonal line. This is particularly useful in checking to see if an object is square or rectangular.

| **Example 8** | **Approximating Length with a Calculator** |

Approximate the length of the diagonal of the given rectangle. The diagonal forms the hypotenuse of a triangle with legs 12.2 in. and 15.7 in. The length of the diagonal is $\sqrt{12.2^2 + 15.7^2} = \sqrt{395.33} \approx 19.88$ in. Use your calculator to confirm the approximation.

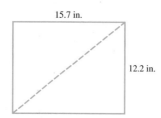

15.7 in.

12.2 in.

Check Yourself 8

Approximate the length of the diagonal of the rectangle to the nearest tenth.

13.7 in.

19.7 in.

There are many real-world applications of the Pythagorean theorem. One such application is given in Example 9.

| **Example 9** | **An Application of the Pythagorean Theorem** |

A 12-foot ladder is placed against an outside wall so that the bottom is 5 feet from the wall. How high on the wall does the ladder reach?

The ladder is the hypotenuse of a right triangle with one leg measuring 5 feet. Using the Pythagorean theorem yields

$$h^2 + 5^2 = 12^2$$
$$h^2 = 12^2 - 5^2 = 144 - 25$$
$$h = \sqrt{144 - 25}$$
$$h = \sqrt{119}$$
$$h \approx 10.9$$

The ladder rests approximately 10.9 feet up the wall.

Check Yourself 9

A 15-foot ladder is placed against an outside wall so that the bottom is 7 feet from the wall. How high on the wall will the ladder reach (to the nearest tenth of a foot)?

Check Yourself ANSWERS

1. $10^2 \overset{?}{=} 8^2 + 6^2$; $100 \overset{?}{=} 64 + 36$; $100 = 100$ **2.** 13   **3.** $2\sqrt{7}$; or approximately 5.3   **4.** The length is approximately 10.4 in.
5. The height is approximately 15.5 ft.   **6.** 13 units   **7.** $\sqrt{58}$
8. ≈ 24.0 in.   **9.** 13.3 feet

Reading Your Text

The following fill-in-the-blank exercises are designed to ensure that you understand some of the key vocabulary used in this section.

SECTION 9.6

(a) The longest side of a right triangle is called the _____.

(b) For every right triangle, the square of the length of the _____ is equal to the sum of the squares of the lengths of the legs.

(c) When solving an equation for a length, we reject solutions that are _____.

(d) To find the _____ between any two points in the plane, we use a formula derived from the Pythagorean theorem.

9.6 exercises

Name _____

Section _____ Date _____

Answers

1. _____

2. _____

3. _____

4. _____

5. _____

6. _____

7. _____

8. _____

9. _____

10. _____

11. _____

< Objective 1 >

Find the length x in each triangle. Express your answers in simplified radical form.

1.

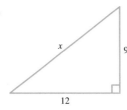

2.

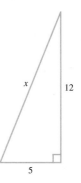

3.

4.

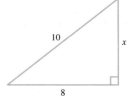

5.

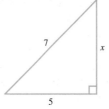

6.

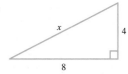

In exercises 7–12, express your answers to the nearest thousandth.

7. **GEOMETRY** Find the diagonal of a rectangle with a length of 10 cm and a width of 7 cm.

8. **GEOMETRY** Find the diagonal of a rectangle with 5 in. width and 7 in. length.

9. **GEOMETRY** Find the width of a rectangle whose diagonal is 12 ft and whose length is 10 ft.

10. **GEOMETRY** Find the length of a rectangle whose diagonal is 9 in. and whose width is 6 in.

11. **CONSTRUCTION** How long must a guy wire be to run from the top of a 20-ft pole to a point on the ground 8 ft from the base of the pole?

12. **CONSTRUCTION** The base of a 15-ft ladder is 5 ft away from a wall. How high from the floor is the top of the ladder?

13. **CONSTRUCTION** A cable is to be laid underground across a rectangular piece of land that is 240 ft by 150 ft. If the cable runs from one corner diagonally to another corner, how long is this piece of cable? Round to the nearest foot.

14. **SPORTS** Find, to the nearest tenth of a foot, the distance from home plate to second base on a baseball field. The bases are 90 feet apart.

15. **CONSTRUCTION** A homeowner wishes to insulate her attic with fiberglass insulation to conserve energy. The insulation comes in 40-cm-wide rolls that are cut to fit between the rafters in the attic. If the roof is 6 m from peak to eave and the attic space is 2 m high at the peak, how long does each of the pieces of insulation need to be? Round to the nearest tenth.

16. **CONSTRUCTION** For the home described in exercise 15, if the roof is 7 m from peak to eave and the attic space is 3 m high at the peak, how long does each of the pieces of insulation need to be? Round to the nearest tenth.

17. **MANUFACTURING** A solar collector and its stand are in the shape of a right triangle. The collector is 5.00 m long, the upright leg is 3.00 m long, and the base leg is 4.00 m long. Because of inefficiencies due to the collector's position, it needs to be raised by 0.50 m on the upright leg. How long will the new base leg be? Round to the nearest tenth.

18. **MANUFACTURING** A solar collector and its stand are in the shape of a right triangle. The collector is 5.00 m long, the upright leg is 2.00 m long, and the base leg is 4.58 m long. Because of inefficiencies due to the collector's position, it needs to be lowered by 0.50 m on the upright leg. How long will the new base leg be? Round to the nearest tenth.

< Objective 2 >

Find the distance between each pair of points. Express your answers in simplified radical form.

19. $(2, 0)$ and $(-4, 0)$ **20.** $(-3, 0)$ and $(4, 0)$

21. $(0, -2)$ and $(0, -9)$ **22.** $(0, 8)$ and $(0, -4)$

Answers

12. _____

13. _____

14. _____

15. _____

16. _____

17. _____

18. _____

19. _____

20. _____

21. _____

22. _____

Beginning Algebra   The Streeter/Hutchison Series in Mathematics   © The McGraw-Hill Companies. All Rights Reserved.

Answers

23. _____   24. _____

25. _____   26. _____

27. _____   28. _____

29. _____   30. _____

31. _____   32. _____

33. _____   34. _____

35. _____   36. _____

37. _____   38. _____

39. _____   40. _____

41. _____   42. _____

43. _____

44. _____

45. _____

46. _____

47. _____

48. _____

49. _____

50. _____

23. (2, 5) and (5, 2)

24. (3, 3) and (5, 7)

25. (5, 1) and (3, 8)

26. (2, 9) and (7, 4)

27. (−2, 8) and (1, 5)   ▶ Videos

28. (2, 6) and (−3, 4)

29. (6, −1) and (2, 2)

30. (2, −8) and (1, 0)

31. (−1, −1) and (2, 5)

32. (−2, −2) and (3, 3)

33. (−2, 9) and (−3, 3)

34. (4, −1) and (0, −5)

35. (−1, −4) and (−3, 5)

36. (−2, 3) and (−7, −1)

37. (−2, −4) and (−4, 1)

38. (−1, −1) and (4, −2)

39. (−4, −2) and (−1, −5)

40. (−2, −2) and (−4, −4)

41. (−2, 0) and (−4, −1)   ▶ Videos

42. (−5, −2) and (−7, −1)

Basic Skills | **Challenge Yourself** | Calculator/Computer | Career Applications | Above and Beyond
▲

Determine whether each statement is **true** *or* **false.**

43. It is possible to have a right triangle that has two equal sides.

44. In a right triangle, there could be an angle whose measure is more than 90°.

Complete each statement with **never, sometimes,** *or* **always.**

45. The hypotenuse is _____ the longest side of a right triangle.

46. The distance between two points is _____ positive.

Use the distance formula to show that each set of points describes an isosceles triangle (a triangle with two sides of equal length).

47. (−3, 0), (2, 3), and (1, −1)

48. (−2, 4), (2, 7), and (5, 3)

49. **Geometry** The length of one leg of a right triangle is 3 in. more than the other. If the length of the hypotenuse is 15 in., what are the lengths of the two legs?

50. **Geometry** The length of a rectangle is 1 cm longer than its width. If the diagonal of the rectangle is 5 cm, what are the dimensions (the length and width) of the rectangle?

Use the Pythagorean theorem to determine the length of each line segment. Where appropriate, round to the nearest hundredth.

51.

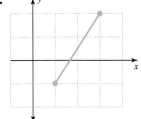

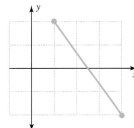

> Videos

52.

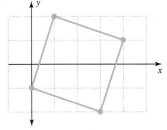

53.

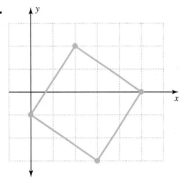

For each figure, use the slope concept and the Pythagorean theorem to show that the figure is a square. (Recall that a square must have four right angles and four equal sides.) Then give the area of the square.

54.

55.

56.

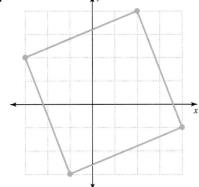

Answers

51. _____

52. _____

53. _____

54. _____

55. _____

56. _____

Answers

57. _____

58. _____

59. _____

60. _____

61. _____

62. _____

63. _____

64. _____

Find the altitude of each triangle.

57.

58.

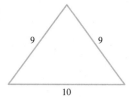

Basic Skills | Challenge Yourself | Calculator/Computer | **Career Applications** | Above and Beyond
▲

---

59. **CONSTRUCTION TECHNOLOGY** The diagram for a jetport is shown here.

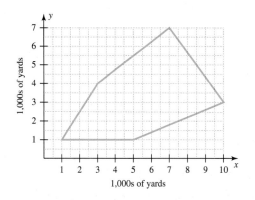

1,000s of yards

Use the distance formula to find the total length of fence around the jetport. (Note that each tick mark is 1,000 yards.)

Basic Skills | Challenge Yourself | Calculator/Computer | Career Applications | **Above and Beyond**
▲

---

60. Suppose that two legs of a right triangle each have length 5. Find the length of the hypotenuse. Express your answer as a simplified radical.

61. Repeat exercise 60 using legs of length 6.

62. Repeat exercise 60 using legs of length 7.

63. Generalize the results of exercises 60–62. That is, if the two legs of a right triangle each have length x, what do you think the length of the hypotenuse will be?

64. **BUSINESS AND FINANCE** Your architectural firm just received this memo.

To: Algebra Expert Architecture, Inc.
From: Microbeans Coffee Company, Inc.
Re: Design for On-Site Day Care Facility
Date: Aug. 10, 2005

We are requesting that you submit a design for a nursery for preschool children. We are planning to provide free on-site day care for the workers at our corporate headquarters.

The nursery should be large enough to serve the needs of 20 preschoolers. There will be three child care workers in this facility. We want the nursery to be 3,000 ft^2 in area. It needs a playroom, a small kitchen and eating space, and bathroom facilities. There should be some space to store toys and books, many of which should be accessible to children. The company plans to put this facility on the first floor on an outside wall so the children can go outside to play without disturbing workers. You are free to add to this design as you see fit.

Please send us your design drawn to a scale of 1 ft to 0.25 in., with precise measurements and descriptions. We would like to receive this design 1 week from today. Please give us some estimate of the cost of this renovation to our building.

chapter 9 > Make the Connection

Submit a design, keeping in mind that the design has to conform to strict design specifications for buildings designated as nurseries.

1. Number of exits: Two exits for the first 7 people and one exit for every additional 7 people.

2. Width of exits: The total width of exits in inches shall not be less than the total occupant load served by an exit multiplied by 0.3 for stairs and 0.2 for other exits. No exit shall be less than 3 ft wide and 6 ft 8 in. high.

3. Arrangements of exits: If two exits are required, they shall be placed a distance apart equal to but not less than one-half the length of the maximum overall diagonal dimension of the building or area to be served measured in a straight line between exits. Where three or more exits are required, two shall be placed as above and the additional exits arranged a reasonable distance apart.

4. Distance to exits: Maximum distance to travel from any point to an exterior door shall not exceed 100 ft.

Answers

1. 15   3. 15   5. $2\sqrt{6}$ 7. 12.207 cm   9. 6.633 ft
11. 21.541 ft   13. 283 ft   15. $8\sqrt{2} \approx 11.3$ m   17. 3.6 m
19. 6   21. 7   23. $3\sqrt{2}$ 25. $\sqrt{53}$ 27. $3\sqrt{2}$ 29. 5
31. $3\sqrt{5}$ 33. $\sqrt{37}$ 35. $\sqrt{85}$ 37. $\sqrt{29}$ 39. $3\sqrt{2}$ 41. $\sqrt{5}$
43. True   45. always   47. Sides have length $\sqrt{34}$, $\sqrt{17}$, and $\sqrt{17}$
49. 9 in., 12 in.   51. 4.12   53. 5   55. 13   57. 4
59. 22,991 yards   61. $6\sqrt{2}$ 63. $x\sqrt{2}$

Definition/Procedure	Example	Reference

Roots and Radicals

Section 9.1

Square Roots

$\sqrt{x}$ is the principal (or positive) square root of x. It is the nonnegative number we must square to get x.

$-\sqrt{x}$ is the negative square root of x.

The square root of a negative number is not a real number.

$\sqrt{49} = 7$

$-\sqrt{49} = -7$

$\sqrt{-49}$ is not a real number.

p. 681

Other Roots

$\sqrt[3]{x}$ is the cube root of x.

$\sqrt[4]{x}$ is the fourth root of x.

$\sqrt[3]{64} = 4$ because $4^3 = 64$.

$\sqrt[4]{81} = 3$ because $3^4 = 81$.

pp. 683–684

Simplifying Radical Expressions

Section 9.2

An expression involving square roots is in *simplest form* if

1. There are no perfect-square factors in a radical.

2. No fraction appears inside a radical.

3. No radical appears in the denominator.

p. 692

To simplify a radical expression, use the following properties.

The square root of a product is the product of the square roots.

$\sqrt{ab} = \sqrt{a} \cdot \sqrt{b}$

$\sqrt{40} = \sqrt{4 \cdot 10}$
$= \sqrt{4} \cdot \sqrt{10}$
$= 2\sqrt{10}$
$\sqrt{12x^3} = \sqrt{4x^2 \cdot 3x}$
$= \sqrt{4x^2} \cdot \sqrt{3x}$
$= 2x \cdot \sqrt{3x}$

p. 692

The square root of a quotient is the quotient of the square roots.

$\sqrt{\dfrac{a}{b}} = \dfrac{\sqrt{a}}{\sqrt{b}}$

$\sqrt{\dfrac{5}{16}} = \dfrac{\sqrt{5}}{\sqrt{16}} = \dfrac{\sqrt{5}}{4}$

$\sqrt{\dfrac{2y}{3}} = \dfrac{\sqrt{2y}}{\sqrt{3}} = \dfrac{\sqrt{2y} \cdot \sqrt{3}}{\sqrt{3} \cdot \sqrt{3}}$

$= \dfrac{\sqrt{6y}}{\sqrt{9}} = \dfrac{\sqrt{6y}}{3}$

p. 695

Adding and Subtracting Radicals

Section 9.3

Like radicals have the same index and the same radicand (the expression inside the radical).

    Like radicals can be added (or subtracted) in the same way as like terms. Apply the distributive property and combine the coefficients.

$3\sqrt{5}$ and $2\sqrt{5}$ are like radicals.

$2\sqrt{3} + 3\sqrt{3} = (2 + 3)\sqrt{3}$
$= 5\sqrt{3}$

$5\sqrt{7} - 2\sqrt{7} = (5 - 2)\sqrt{7}$
$= 3\sqrt{7}$

p. 702

Definition/Procedure	Example	Reference
Certain expressions can be combined after one or more of the terms involving radicals are simplified.	$\sqrt{12} + \sqrt{3} = 2\sqrt{3} + \sqrt{3}$ $= (2 + 1)\sqrt{3}$ $= 3\sqrt{3}$	p. 702
Multiplying and Dividing Radicals		Section 9.4
Multiplying		
To multiply radical expressions, use the first property of radicals in the following way: $\sqrt{a} \cdot \sqrt{b} = \sqrt{ab}$	$\sqrt{6} \cdot \sqrt{15} = \sqrt{6 \cdot 15} = \sqrt{90}$ $= \sqrt{9 \cdot 10}$ $= 3\sqrt{10}$	p. 709
The distributive property also applies when multiplying radical expressions.	$\sqrt{5}(\sqrt{3} + 2\sqrt{5})$ $= \sqrt{5} \cdot \sqrt{3} + \sqrt{5} \cdot 2\sqrt{5}$ $= \sqrt{15} + 10$	p. 710
The FOIL pattern allows us to find the product of binomial radical expressions.	$(\sqrt{5} + 2)(\sqrt{5} - 1)$ $= \sqrt{5} \cdot \sqrt{5} - \sqrt{5} + 2\sqrt{5} - 2$ $= 3 + \sqrt{5}$ $(\sqrt{10} + 3)(\sqrt{10} - 3)$ $= 10 - 9 = 1$	p. 710
Dividing		
To divide radical expressions, use the second property of radicals in the following way: $\dfrac{\sqrt{a}}{\sqrt{b}} = \sqrt{\dfrac{a}{b}}$	$\dfrac{\sqrt{50}}{\sqrt{2}} = \sqrt{\dfrac{50}{2}}$ $= \sqrt{25}$ $= 5$	p. 711
Solving Radical Equations		Section 9.5
Power Property of Equality		
To solve an equation involving radicals, apply the power property of equality: Given any two expressions a and b and any positive integer n, if $a = b$, then $a^n = b^n$.		p. 717

Continued

Definition/Procedure	Example	Reference

Solving Equations Involving Radicals

1. Isolate a radical on one side of the equation.
2. Raise each side of the equation to the smallest power that will eliminate the radical.
3. If any radicals remain in the equation derived in step 2, return to step 1 and continue the process.
4. Solve the resulting equation to determine any possible solutions.
5. Check all solutions to determine whether extraneous solutions may have resulted from step 2.

Solve

$$\sqrt{2x - 3} + x = 9$$
$$\sqrt{2x - 3} = -x + 9$$
$$2x - 3 = x^2 - 18x + 81$$
$$0 = x^2 - 20x + 84$$
$$0 = (x - 6)(x - 14)$$
$$x = 6 \quad \text{or} \quad x = 14$$

By substitution, 6 is the only valid solution.

p. 720

Applications of the Pythagorean Theorem

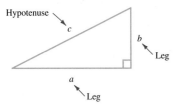

In words, for every right triangle, the square of the length of the hypotenuse is equal to the sum of the squares of the lengths of the legs.

$$c^2 = a^2 + b^2$$

Find length x.

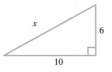

$$x^2 = 10^2 + 6^2$$
$$= 100 + 36$$
$$= 136$$
$$x = \sqrt{136} \quad \text{or} \quad 2\sqrt{34}$$

Section 9.6

p. 723

The Distance Formula

The distance between the points (x_1, y_1) and (x_2, y_2) can be found using the formula

$$d = \sqrt{(x_2 - x_1)^2 + (y_2 - y_1)^2}$$

Find the distance between $(5, 1)$ and $(-3, 4)$.

$$d = \sqrt{(-3 - 5)^2 + (4 - 1)^2}$$
$$= \sqrt{(-8)^2 + (3)^2}$$
$$= \sqrt{64 + 9}$$
$$= \sqrt{73}$$
$$\approx 8.54$$

p. 727

This summary exercise set is provided to give you practice with each of the objectives of this chapter. Each exercise is keyed to the appropriate chapter section. When you are finished, you can check your answers to the odd-numbered exercises against those presented in the back of the text. If you have difficulty with any of these questions, go back and reread the examples from that section. Your instructor will give you guidelines on how best to use these exercises in your instructional setting.

9.1 *Evaluate if possible.*

1. $\sqrt{81}$

2. $-\sqrt{49}$

3. $\sqrt{-49}$

4. $\sqrt[3]{64}$

5. $\sqrt[3]{-64}$

6. $\sqrt[4]{81}$

7. $\sqrt[4]{-81}$

9.2 *Simplify each radical expression. Assume that all variables represent positive real numbers.*

8. $\sqrt{50}$

9. $\sqrt{45}$

10. $\sqrt{7a^3}$

11. $\sqrt{20x^4}$

12. $\sqrt{49m^5}$

13. $\sqrt{200b^3}$

14. $\sqrt{147r^3s^2}$

15. $\sqrt{108a^2b^5}$

16. $\sqrt{\dfrac{10}{81}}$

17. $\sqrt{\dfrac{18x^2}{25}}$

18. $\sqrt{\dfrac{12m^5}{49}}$

19. $\sqrt{\dfrac{3}{7}}$

20. $\sqrt{\dfrac{3a}{2}}$

21. $\sqrt{\dfrac{8x^2}{7}}$

9.3 *Simplify by combining like terms.*

22. $\sqrt{3} + 4\sqrt{3}$

23. $9\sqrt{5} - 3\sqrt{5}$

24. $3\sqrt{2} + 2\sqrt{3}$

25. $3\sqrt{3a} - \sqrt{3a}$

26. $7\sqrt{6} - 2\sqrt{6} + \sqrt{6}$

27. $5\sqrt{3} + \sqrt{12}$

28. $3\sqrt{18} - 5\sqrt{2}$

29. $\sqrt{32} - \sqrt{18}$

30. $\sqrt{27} - \sqrt{3} + 2\sqrt{12}$

31. $\sqrt{8} + 2\sqrt{27} - \sqrt{75}$

32. $x\sqrt{18} - 3\sqrt{8x^2}$

9.4 *Simplify each radical expression.*

33. $\sqrt{6} \cdot \sqrt{5}$

34. $\sqrt{3} \cdot \sqrt{6}$

35. $\sqrt{3x} \cdot \sqrt{2}$

36. $\sqrt{2} \cdot \sqrt{8} \cdot \sqrt{3}$

37. $\sqrt{5a} \cdot \sqrt{10a}$

38. $\sqrt{2}(\sqrt{3} + \sqrt{5})$

39. $\sqrt{7}(2\sqrt{3} - 3\sqrt{7})$

40. $(\sqrt{3} + 5)(\sqrt{3} - 3)$

41. $(\sqrt{15} - 3)(\sqrt{15} + 3)$

42. $(\sqrt{2} + 3)^2$

43. $\dfrac{\sqrt{7x^3}}{\sqrt{3}}$

44. $\dfrac{18 - \sqrt{20}}{2}$

9.5 *Solve each equation. Be sure to check your solutions.*

45. $\sqrt{x-5} = 4$ **46.** $\sqrt{3x-2} + 2 = 5$

47. $\sqrt{y+7} = y - 5$ **48.** $\sqrt{2x-1} + x = 8$

9.6 *Find length x in each triangle. Express your answer in simplified radical form.*

49.

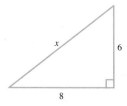

50.

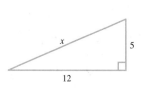

51.

52.

53.

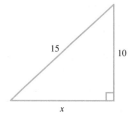

54.

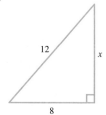

Solve each application. Approximate your answer to one decimal place where necessary.

55. Find the diagonal of a rectangle whose length is 12 in. and whose width is 9 in.

56. Find the length of a rectangle whose diagonal has a length of 10 cm and whose width is 5 cm.

57. How long must a guy wire be to run from the top of an 18-ft pole to a point on level ground 16 ft away from the base of the pole?

58. The length of one leg of a right triangle is 2 in. more than the length of the other. If the length of the hypotenuse of the triangle is 10 in., what are the lengths of the two legs?

Find the distance between each pair of points.

59. $(-3, 2)$ and $(-7, 2)$ **60.** $(2, 0)$ and $(5, 9)$

61. $(-2, 7)$ and $(-5, -1)$ **62.** $(5, -1)$ and $(-2, 3)$

63. $(-3, 4)$ and $(-2, -5)$ **64.** $(6, 4)$ and $(-3, 5)$

The purpose of this self-test is to help you assess your progress so that you can find concepts that you need to review before the next exam. Allow yourself about an hour to take this test. At the end of that hour, check your answers against those given in the back of this text. If you miss any, go back to the appropriate section to reread the examples until you have mastered that particular concept.

Simplify. Assume all variables represent positive real numbers.

1. $2\sqrt{10} - 3\sqrt{10} + 5\sqrt{10}$

2. $\sqrt{24a^3}$

3. $\sqrt{3x} \cdot \sqrt{6x}$

4. $\sqrt{\dfrac{16}{25}}$

Evaluate if possible.

5. $\sqrt[3]{27}$

6. $\sqrt{-144}$

7. $-\sqrt[3]{-64}$

8. $\sqrt{121}$

Find length x in each triangle. Write each answer in simplified radical form.

9.

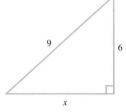

10.

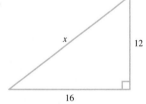

11.

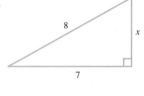

12.

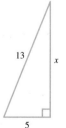

Simplify. Assume all variables represent positive real numbers.

13. $3\sqrt{8} - \sqrt{18}$

14. $\sqrt{\dfrac{5}{9}}$

15. $2\sqrt{50} - \sqrt{8} - \sqrt{50}$

16. $(\sqrt{5} + 3)(\sqrt{5} + 2)$

Answers

1. _____
2. _____
3. _____
4. _____
5. _____
6. _____
7. _____
8. _____
9. _____
10. _____
11. _____
12. _____
13. _____
14. _____
15. _____
16. _____

Answers

17. $\dfrac{\sqrt{7}}{\sqrt{2}}$

18. $\sqrt{75}$

19. $\dfrac{14 + 3\sqrt{98}}{7}$

20. $\sqrt{20} + \sqrt{45} - \sqrt{5}$

17. _____

18. _____

19. _____

20. _____

21. _____

22. _____

23. _____

24. _____

25. _____

Find the distance between each pair of points.

21. $(-2, 5)$ and $(-9, -1)$

22. $(-3, 7)$ and $(-12, 7)$

Solve each equation.

23. $\sqrt{3x + 4} + x = 8$

24. $\sqrt{x - 2} = 9$

25. If the diagonal of a rectangle is 12 cm and the width of the rectangle is 7 cm, what is the length of the rectangle? Round to the nearest thousandth.

Activity 9 ::
The Swing of the Pendulum

The action of a pendulum seems simple. Scientists have studied the characteristics of a swinging pendulum and found them to be quite useful. In 1851 in Paris, Jean Foucault (pronounced "Foo-koh") used a pendulum to clearly demonstrate the rotation of Earth about its own axis.

A pendulum can be as simple as a string or cord with a weight fastened to one end. The other end is fixed, and the weight is allowed to swing. We define the **period** of a pendulum to be the amount of time required for the pendulum to make one complete swing (back and forth). The question we pose is: How does the *period* of a pendulum relate to the *length* of the pendulum?

For this activity, you need a piece of string that is approximately 1 m long. Fasten a weight (such as a small hexagonal nut) to one end, and then place clear marks on the string every 10 cm up to 70 cm, measured from the center of the weight.

1. Working with one or two partners, hold the string at the mark that is 10 cm from the weight. Pull the weight to the side with your other hand and let it swing freely. To estimate the period, let the weight swing through 30 periods, record the time in the given table, and then divide by 30. Round your result to the nearest hundredth of a second and record it. (*Note:* If you are unable to perform the experiment and collect your own data, you can use the sample data collected in this manner and presented at the end of this activity.)

   Repeat the described procedure for each length indicated in the given table.

Length of string, cm	10	20	30	40	50	60	70
Time for 30 periods, s Time for 1 period, s							

2. Let L represent the length of the pendulum and T represent the time period that results from swinging that pendulum. Fill out the table:

L	10	20	30	40	50	60	70
T							

3. Which variable, L or T, is viewed here as the independent variable?

4. On graph paper, draw horizontal and vertical axes, but plan to graph the data points in the first quadrant only. Explain why this is so.

5. With the independent variable marked on the horizontal axis, scale the axes appropriately, keeping an eye on your data.

6. Plot your data points. Should you connect them with a smooth curve?

743

7. What period T would correspond to a string length of 0? Include this point on your graph.

8. Use your graph to predict the period for a string length of 80 cm.

9. Verify your prediction by measuring the period when the string is held at 80 cm (as described in step 1). How close did your experimental estimate come to the prediction made in step 8?

You have created a graph showing T as a function of L. The shape of the graph may not be familiar to you yet. The shape of your pendulum graph fits that of a square-root function.

Sample Data

Length of string, cm	10	20	30	40	50	60	70
Time for 30 periods, s	19	27	33	38	42	46	49

The following exercises are presented to help you review concepts from earlier chapters. This is meant as a review and not as a comprehensive exam. The answers are presented in the back of the text. Section references accompany the answers. If you have difficulty with any of these exercises, be certain to at least read through the summary related to those sections.

Simplify each expression.

1. $8x^2y^3 - 5x^3y - 5x^2y^3 + 3x^3y$

2. $(4x^2 - 2x + 7) - (-3x^2 + 4x - 5)$

Evaluate each expression when $x = 2$, $y = -1$, and $z = -4$.

3. $2xyz^2 - 4x^2y^2z$

4. $-2xyz + 2x^2y^2$

Solve each equation.

5. $-3x - 2(4 - 6x) = 10$

6. $5x - 3(4 - 2x) = 6(2x - 3)$

7. Solve the inequality $3x - 11 < 5x - 19$.

Perform the indicated operations.

8. $2x^2y(3x^2 - 5x + 19)$

9. $(5x + 3y)(4x - 7y)$

Factor each polynomial completely.

10. $36xy - 27x^3y^2$

11. $8x^2 - 26x + 15$

Perform the indicated operations.

12. $\dfrac{2}{3x + 21} - \dfrac{3}{5x + 35}$

13. $\dfrac{x^2 - x - 6}{x^2 - x - 20} \div \dfrac{x^2 + x - 2}{x^2 + 3x - 4}$

Answers

1.
2.
3.
4.
5.
6.
7.
8.
9.
10.
11.
12.
13.

Name _____
Section _____ Date _____

745

Answers

14. _____

15. _____

16. _____

17. _____

18. _____

19. _____

20. _____

21. _____

22. _____

23. _____

24. _____

25. _____

26. _____

27. _____

28. _____

29. _____

30. _____

Graph.

14. $4x + 5y = 20$

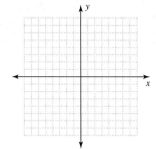

15. $5x - 4y \geq 20$

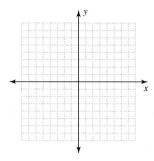

16. Find the slope of the line through the points $(2, 9)$ and $(-1, -6)$.

17. Given that the slope of a line is $-\dfrac{3}{2}$ and the *y*-intercept is $(0, 5)$, write the equation of the line.

Solve each system. If a unique solution does not exist, state whether the system is inconsistent or dependent.

18. $4x - 5y = 20$
$2x + 3y = 10$

19. $4x + 7y = 24$
$8x + 14y = 12$

Solve the application. Be sure to show the system of equations used for your solution.

20. Amir was able to travel 80 mi downstream in 5 h. Returning upstream, he took 8 h to make the trip. How fast can he travel in still water, and what was the rate of the current?

Evaluate each root, if possible.

21. $\sqrt{144}$

22. $-\sqrt{144}$

23. $\sqrt{-144}$

24. $\sqrt[3]{-27}$

Simplify each radical expression.

25. $a\sqrt{20} - 2\sqrt{45a^2}$

26. $\dfrac{\sqrt{8x^3}}{\sqrt{3}}$

27. $\dfrac{12 - \sqrt{72}}{3}$

28. $\sqrt{98x^2}$

29. $\sqrt{150m^3n^2}$

30. $\sqrt{\dfrac{12a^2}{25}}$

CHAPTER

10

Quadratic Equations

INTRODUCTION

Large cities often commission fireworks artists to choreograph elaborate displays on holidays. Such displays look like beautiful paintings in the sky, in which the fireworks seem to dance to well-known popular and classical music. The displays are feats of engineering and very accurate timing. Suppose the designer wants a second set of rockets of a certain color and shape to be released after the first set reaches a specific height and explodes. He or she must know the strength of the initial liftoff and use a quadratic equation to determine the proper time for setting off the second round.

The equation $h = -16t^2 + 100t$ gives the height in feet t seconds after the rockets are shot into the air if the initial velocity is 100 feet per second. Using this equation, the designer knows how high the rocket will ascend and when it will begin to fall. The designer can time the next round to achieve the desired effect. Displays that involve large banks of fireworks in shows that last up to an hour are programmed using computers, but quadratic equations are at the heart of the mechanism that creates the beautiful effects.

CHAPTER 10 OUTLINE

This prerequisite test provides some exercises requiring skills that you will need to be successful in the coming chapter. The answers for these exercises can be found in the back of this text. This prerequisite test can help you identify topics that you will need to review before beginning the chapter.

Solve by factoring.

1. $x^2 - 4x = 45$

Simplify each expression.

2. $\sqrt{8}$

3. $\sqrt{\dfrac{5}{3}}$

Multiply.

4. $(x - 7)^2$

5. $(2x + 5)^2$

Evaluate the expression $b^2 - 4ac$ for each set of values.

6. $a = 1, b = -2, c = -1$

7. $a = 2, b = -3, c = 4$

Evaluate each expression for the value of the variable given.

8. $x^2 - 3x - 5; x = -2$

9. $-2x^2 - 5x + 3; x = 4$

10. $-5x^2 - 5x + 6; x = -1$

10.1

More on Quadratic Equations

< 10.1 Objectives >

1 > Solve equations of the form $ax^2 = k$

2 > Solve equations of the form $(x - h)^2 = k$

We now have more tools for solving quadratic equations. In Sections 10.1–10.3 we use the ideas from Chapter 9 to extend our ability to solve equations.

In Section 4.6 we identified all equations of the form

$$ax^2 + bx + c = 0$$

as quadratic equations in standard form. In that section, we discussed solving these equations whenever the quadratic expression was factorable. In this chapter, we want to extend our equation-solving techniques so that we can find solutions for all such quadratic equations.

First, we review the factoring method that we introduced in Chapter 4.

> **Example 1** | **Solving Quadratic Equations by Factoring**

Solve each quadratic equation by factoring.

(a) $x^2 = -7x - 12$

First, we write the equation in standard form.

$$x^2 + 7x + 12 = 0$$

NOTE

Add $7x$ and 12 to both sides of the equation. The quadratic expression must be *set equal to* 0.

Once the equation is in standard form, we can factor the quadratic member.

$$(x + 3)(x + 4) = 0$$

Finally, using the zero-product principle, we solve the equations $x + 3 = 0$ and $x + 4 = 0$ to obtain

$$x = -3 \quad \text{or} \quad x = -4$$

(b) $x^2 = 16$

Again, we write the equation in standard form.

$$x^2 - 16 = 0$$

NOTE

Here we factor the quadratic expression as a difference of squares.

Factoring, we have

$$(x + 4)(x - 4) = 0$$

Finally, the solutions are

$$x = -4 \quad \text{or} \quad x = 4$$

Check Yourself 1

Solve each quadratic equation.

(a) $x^2 - 4x = 45$ **(b)** $w^2 = 25$

Certain quadratic equations can be solved by other methods, such as the square-root method. Return to the equation in part (b) of Example 1.

Beginning with

$$x^2 = 16$$

we can take the square root of each side, to write

$$\sqrt{x^2} = \sqrt{16}$$

From Section 9.1, we know that this is equivalent to

$$\sqrt{x^2} = 4$$

or

$$|x| = 4$$

4 and -4 both satisfy this last equation, and so we have the two solutions

$$x = 4 \qquad \text{or} \qquad x = -4$$

We usually write the solutions as

$$x = \pm 4$$

Two more equations solved by this method are shown in Example 2.

RECALL

By definition
$\sqrt{x^2} = |x|$

NOTE

$x = \pm 4$ is simply a convenient "shorthand" for indicating the two solutions, and we generally go directly to this form.

Example 2	Solving Equations by the Square-Root Method

< Objective 1 >

Solve each equation by the square-root method.

(a) $x^2 = 9$

By taking the square root of each side, we have

$$\sqrt{x^2} = \sqrt{9}$$
$$|x| = 3$$
$$x = \pm 3$$

(b) $x^2 = 5$

Again, we take the square root of each side to write our two solutions as

$$\sqrt{x^2} = \sqrt{5}$$
$$|x| = \sqrt{5}$$
$$x = \pm\sqrt{5}$$

✔ Check Yourself 2

Solve.

(a) $x^2 = 100$ **(b)** $t^2 = 15$

You may have to add or subtract on both sides of the equation to write an equation in the form of those in Example 2, as Example 3 illustrates.

| Example 3 | Solving Equations by the Square-Root Method |

Solve $x^2 - 8 = 0$.

First, add 8 to both sides of the equation. We have

$$x^2 = 8$$

Now take the square root of both sides.

$$x = \pm\sqrt{8}$$

Normally, the solutions are written in simplest form. In this case we have

$$x = \pm 2\sqrt{2}$$

RECALL

$\sqrt{8} = \sqrt{4 \cdot 2}$

$\quad = \sqrt{4} \cdot \sqrt{2}$

$\quad = 2\sqrt{2}$

NOTE

In the form $ax^2 = k$
a is the coefficient of x^2 and k
is some number.

Check Yourself 3

Solve.

(a) $x^2 - 18 = 0$ (b) $x^2 + 1 = 7$

To solve a quadratic equation of the form $ax^2 = k$, divide both sides of the equation by a as the first step. This is shown in Example 4.

| Example 4 | Solving Equations by the Square-Root Method |

Solve $4x^2 = 3$.

Divide both sides of the equation by 4.

$$x^2 = \frac{3}{4}$$

Now take the square root of both sides.

$$x = \pm\sqrt{\frac{3}{4}}$$

Again write your result in simplest form, so

$$x = \pm\frac{\sqrt{3}}{2}$$

RECALL

$\sqrt{\dfrac{3}{4}} = \dfrac{\sqrt{3}}{\sqrt{4}}$

$\quad = \dfrac{\sqrt{3}}{2}$

Check Yourself 4

Solve $9x^2 = 5$.

Equations of the form $(x - h)^2 = k$ can also be solved by taking the square root of both sides. Consider Example 5.

| Example 5 | Solving Equations by the Square-Root Method |

< Objective 2 >

Solve $(x - 1)^2 = 6$.

Again, take the square root of both sides of the equation.

$$x - 1 = \pm\sqrt{6}$$

Now add 1 to both sides of the equation to isolate x.

$$x = 1 \pm \sqrt{6}$$

Check Yourself 5

Solve $(x + 2)^2 = 12$.

Equations of the form $a(x - h)^2 = k$ can also be solved if each side of the equation is divided by a first, as shown in Example 6.

Example 6   **Solving Equations by the Square-Root Method**

Solve $3(x - 2)^2 = 5$.

$$(x - 2)^2 = \frac{5}{3}$$

$$x - 2 = \pm\sqrt{\frac{5}{3}} = \frac{\pm\sqrt{15}}{3}$$

$$x = 2 \pm \frac{\sqrt{15}}{3}$$

We usually write this sort of result as a single fraction.

$$x = \frac{6}{3} \pm \frac{\sqrt{15}}{3}$$

$$x = \frac{6 \pm \sqrt{15}}{3}$$

> **NOTE**
>
> $$\sqrt{\frac{5}{3}} = \frac{\sqrt{5}}{\sqrt{3}} \cdot \frac{\sqrt{3}}{\sqrt{3}} = \frac{\sqrt{15}}{3}$$

Check Yourself 6

Solve $5(x + 3)^2 = 2$.

What about an equation such as

$$x^2 + 5 = 0$$

If we apply the methods of Examples 4–6, we first subtract 5 from both sides, to write

$$x^2 = -5$$

Taking the square root of both sides gives

$$x = \pm\sqrt{-5}$$

> **RECALL**
>
> We introduced the Pythagorean theorem in Section 9.6.

But we know there are no real square roots of -5, so this equation has *no real-number solutions*. You might work with this type of equation in your next algebra course.

Many applied problem situations can be modeled with quadratic equations. Such problems may arise in geometry, construction, physics, and economics, to name but a few.

The next example relies on the Pythagorean theorem, studied previously.

Example 7   **A Construction Application**

How long must a guy wire be to reach from the top of a 30-ft pole to a point on the ground 20 ft from the base of the pole? Round to the nearest tenth of a foot.

We start by drawing a sketch of the problem.

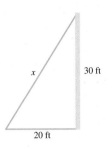

Using the Pythagorean theorem, we write

$$x^2 = 20^2 + 30^2$$
$$x^2 = 400 + 900$$
$$x^2 = 1,300$$

This is a quadratic equation that can be solved using the square-root method.

$$x = \pm\sqrt{1,300}$$

Since x must be positive, we reject $-\sqrt{1,300}$ and keep $\sqrt{1,300}$. To the nearest tenth of a foot, we have $x = 36.1$ ft.

NOTE

Always check to see if your final answer is reasonable.

Check Yourself 7

A 16-ft ladder leans against a wall with its base 4 ft from the wall. How far above the floor, to the nearest tenth of a foot, is the top of the ladder?

Check Yourself ANSWERS

1. **(a)** $-5, 9$; **(b)** $-5, 5$ 2. **(a)** ± 10; **(b)** $\pm\sqrt{15}$

3. **(a)** $\pm 3\sqrt{2}$; **(b)** $\pm\sqrt{6}$ 4. $\pm\dfrac{\sqrt{5}}{3}$ 5. $-2 \pm 2\sqrt{3}$

6. $\dfrac{-15 \pm \sqrt{10}}{5}$ 7. 15.5 ft

Reading Your Text

The following fill-in-the-blank exercises are designed to ensure that you understand some of the key vocabulary used in this section.

SECTION 10.1

(a) Equations of the form $ax^2 + bx + c = 0$ are called _____ equations in standard form.

(b) To solve an equation of the form $ax^2 = k$, _____ both sides of the equation by a as the first step.

(c) Equations of the form $(x - h)^2 = k$ can be solved by taking the _____ of both sides.

(d) The equation $x^2 + 5 = 0$ has no _____ solutions.

Name _____

Section _____ Date _____

Answers

1. _____   2. _____

3. _____   4. _____

5. _____   6. _____

7. _____   8. _____

9. _____   10. _____

11. _____   12. _____

13. _____   14. _____

15. _____   16. _____

17. _____   18. _____

19. _____   20. _____

21. _____   22. _____

23. _____   24. _____

25. _____   26. _____

27. _____   28. _____

| Basic Skills | Challenge Yourself | Calculator/Computer | Career Applications | Above and Beyond |

< Objective 1 >

Solve each equation.

1. $x^2 = 5$

2. $x^2 = 15$

3. $x^2 = 33$

4. $x^2 = 43$

5. $x^2 - 7 = 0$

6. $x^2 - 13 = 0$

7. $x^2 - 20 = 0$ > Videos

8. $x^2 = 28$

9. $x^2 = 40$

10. $x^2 - 54 = 0$

11. $x^2 + 3 = 12$

12. $x^2 - 7 = 18$

13. $x^2 + 5 = 8$

14. $x^2 - 4 = 17$

15. $x^2 - 2 = 16$

16. $x^2 + 6 = 30$

17. $9x^2 = 25$

18. $16x^2 = 9$

19. $49x^2 = 11$

20. $16x^2 = 3$

21. $4x^2 = 7$ > Videos

22. $25x^2 = 13$

< Objective 2 >

23. $(x - 1)^2 = 5$

24. $(x - 3)^2 = 10$

25. $(x + 1)^2 = 12$

26. $(x + 2)^2 = 32$

27. $(x - 3)^2 = 24$

28. $(x - 5)^2 = 27$

29. $(x + 5)^2 = 25$ > Videos

30. $(x + 2)^2 = 16$

31. $3(x - 5)^2 = 7$

32. $2(x - 5)^2 = 3$

33. $4(x + 5)^2 = 9$ > Videos

34. $16(x + 2)^2 = 25$

35. $-2(x + 2)^2 = -6$

36. $-5(x + 4)^2 = -10$

37. $-4(x - 2)^2 = -5$

38. $-9(x - 2)^2 = -11$

Solve each application. Give all answers to the nearest thousandth.

39. **CONSTRUCTION** How long must a guy wire be to run from the top of a 20-ft pole to a point on the ground 8 ft from the base of the pole?

40. **CONSTRUCTION** How long must a guy wire be to run from the top of a 16-ft pole to a point on the ground 6 ft from the base of the pole?

41. **CONSTRUCTION** The base of a 15-ft ladder is 5 ft away from a wall. How far above the floor is the top of the ladder?

42. **CONSTRUCTION** The base of an 18-ft ladder is 4 ft away from a wall. How far above the floor is the top of the ladder?

43. **GEOMETRY** One leg of a right triangle is twice the length of the other. The hypotenuse is 8 m long. Find the length of each leg.

| Basic Skills | **Challenge Yourself** | Calculator/Computer | Career Applications | Above and Beyond |

Determine whether each statement is **true** *or* **false.**

44. Not all quadratic expressions are factorable.

45. Some quadratic equations have no real-number solutions.

Complete each statement with **never, sometimes,** *or* **always.**

46. An equation of the form $(x - h)^2 = k$, where k is positive, _____ has two distinct solutions.

47. An equation of the form $x^2 = k$ _____ has real-number solutions.

Answers

29. _____

30. _____

31. _____

32. _____

33. _____

34. _____

35. _____

36. _____

37. _____

38. _____

39. _____

40. _____

41. _____

42. _____

43. _____

44. _____

45. _____

46. _____

47. _____

Answers

49.

50.

51.

52.

53.

54.

55.

56.

57.

58.

Solve each equation.

48. $(2x + 11)^2 - 9 = 0$

49. $(3x + 14)^2 - 25 = 0$

50. $x^2 - 2x + 1 = 7$
(*Hint:* Factor the left-hand side.)

51. $x^2 + 4x + 4 = 7$
(*Hint:* Factor the left-hand side.)

 > Videos

52. **NUMBER PROBLEM** The square of a number decreased by 2 is equal to the negative of the number. Find the numbers.

53. **NUMBER PROBLEM** The square of 2 more than a number is 64. Find the numbers.

54. **NUMBER PROBLEM** The square of the sum of a number and 5 is 36. Find the numbers.

Basic Skills | Challenge Yourself | Calculator/Computer | **Career Applications** | Above and Beyond
▲

55. **ALLIED HEALTH** A toxic chemical is introduced into a protozoan culture. The number of deaths per hour N is given by the equation

$N = 363 - 3t^2$

where t is the number of hours after the chemical's introduction. How long will it take before the protozoa stop dying?

56. **ALLIED HEALTH** One technique of controlling cancer is to use radiation therapy. After such a treatment, the total number of cancerous cells N, in thousands, can be estimated by the equation

$N = 121 - 4t^2$

where t is the number of days of treatment. How many days of treatment are required to kill all the cancer cells?

57. **ALLIED HEALTH** One technique of controlling cancer is to use radiation therapy. After such a treatment, the total number of cancerous cells N, in thousands, can be estimated by the equation

$N = 169 - 4t^2$

where t is the number of days of treatment. How many days of treatment are required to kill all the cancer cells?

58. **MANUFACTURING TECHNOLOGY** The volume V of structural lumber (in cubic feet) that can be harvested from a coniferous tree is given by the formula

$$V = \frac{1}{350}D^2H - \frac{1}{5}$$

where H is the height of the tree (in feet) and D is the diameter (in inches) of the tree at the base. Calculate the necessary base diameter for a 48-foot-tall tree if we desire a volume of 70 cubic feet.

Answers

59. **MECHANICAL ENGINEERING** The deflection d (in inches) of a beam loaded with a single concentrated load is described by the equation

$$d = \frac{x^2 - 64}{200}$$

Find the location x (in feet) if the deflection is equal to 0.085 inch.

59. _____

60. _____

Basic Skills | Challenge Yourself | Calculator/Computer | Career Applications | **Above and Beyond**
▲

60. In this section, you solved quadratic equations by "extracting roots," taking the square root of both sides after writing one side as the square of a binomial. But what if the algebraic expression cannot be written this way? Work with another student to decide what needs to be added to each expression to make it a "perfect-square trinomial." Label the dimensions of the squares and the area of each section.

(a)

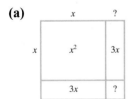

$$x^2 + 6x + \underline{\quad} = (x + ?)^2$$

(b)

$$n^2 + 10n + \underline{\quad} = (n + ?)^2$$

(c)

$$a^2 + a + \underline{\quad} = (a + ?)^2$$

(d)

$$x^2 - 12x + \underline{\quad} = (x - ?)^2$$

(e) $x^2 + 20x + \underline{\quad} = (x + ?)^2$ **(f)** $n^2 - 16n + \underline{\quad} = (n - ?)^2$

Answers

1. $\pm\sqrt{5}$ 3. $\pm\sqrt{33}$ 5. $\pm\sqrt{7}$ 7. $\pm 2\sqrt{5}$ 9. $\pm 2\sqrt{10}$

11. ± 3 13. $\pm\sqrt{3}$ 15. $\pm 3\sqrt{2}$ 17. $\pm\dfrac{5}{3}$ 19. $\pm\dfrac{\sqrt{11}}{7}$

21. $\pm\dfrac{\sqrt{7}}{2}$ 23. $1\pm\sqrt{5}$ 25. $-1\pm 2\sqrt{3}$ 27. $3\pm 2\sqrt{6}$

29. $-10, 0$ 31. $\dfrac{15\pm\sqrt{21}}{3}$ 33. $-\dfrac{13}{2}, -\dfrac{7}{2}$ 35. $-2\pm\sqrt{3}$

37. $\dfrac{4\pm\sqrt{5}}{2}$ 39. 21.541 ft   41. 14.142 ft   43. 3.578 m, 7.155 m

45. True   47. sometimes   49. $-3, -\dfrac{19}{3}$ 51. $-2\pm\sqrt{7}$

53. $6, -10$ 55. 11 hours   57. 6.5 days
59. -9 ft and $+9$ ft from the center

10.2

Completing the Square

< 10.2 Objectives >

1 > Complete the square for a trinomial expression

2 > Solve a quadratic equation by completing the square

We can solve a quadratic equation such as

$$x^2 - 2x + 1 = 5$$

very easily if we notice that the expression on the left is a perfect-square trinomial. Factoring, we have

$$(x - 1)^2 = 5$$

so

NOTE

Here, we used the square-root method, once the factoring was done.

$$x - 1 = \pm\sqrt{5} \qquad \text{or} \qquad x = 1 \pm \sqrt{5}$$

The solutions for the original equation are then $1 + \sqrt{5}$ and $1 - \sqrt{5}$.

In fact, *every* quadratic equation can be written with a perfect-square trinomial on the left. That is the basis for the **completing-the-square method** for solving quadratic equations.

First, look at two perfect-square trinomials.

$$x^2 + 6x + 9 = (x + 3)^2$$
$$x^2 - 8x + 16 = (x - 4)^2$$

There is an important relationship between the coefficient of the middle term (the x-term) and the constant.

In the first equation,

$$\left(\frac{1}{2} \cdot 6\right)^2 = 3^2 = 9$$

The x-coefficient   The constant

In the second equation,

$$\left[\frac{1}{2}(-8)\right]^2 = (-4)^2 = 16$$

The x-coefficient   The constant

It is always true that, in a perfect-square trinomial with a coefficient of 1 for x^2, the square of one-half of the x-coefficient is equal to the constant term.

| Example 1 | Completing the Square |

< Objective 1 >

(a) Find the term that should be added to $x^2 + 4x$ so that the expression is a perfect-square trinomial.

NOTE

The coefficient of x^2 must be 1 before the added term is found.

To complete the square of $x^2 + 4x$, add the square of one-half of 4 (the x-coefficient).

$$x^2 + 4x + \left(\frac{1}{2} \cdot 4\right)^2 \qquad \text{or} \qquad x^2 + 4x + 2^2 \qquad \text{or} \qquad x^2 + 4x + 4$$

The trinomial $x^2 + 4x + 4$ is a perfect square because

$$x^2 + 4x + 4 = (x + 2)^2$$

(b) Find the term that should be added to $x^2 - 10x$ so that the expression is a perfect-square trinomial.

To complete the square of $x^2 - 10x$, add the square of one-half of -10 (the x-coefficient).

$$x^2 - 10x + \left[\frac{1}{2}(-10)\right]^2 \quad \text{or} \quad x^2 - 10x + (-5)^2 \quad \text{or} \quad x^2 - 10x + 25$$

Check for yourself, by factoring, that this is a perfect-square trinomial.

Check Yourself 1

Complete the square and factor.

(a) $x^2 + 2x$ **(b)** $x^2 - 12x$

We can now use the process of Example 1 along with the solution methods of Section 10.1 to solve a quadratic equation.

▶ Example 2	Solving a Quadratic Equation by Completing the Square

< Objective 2 >

Solve $x^2 + 4x - 2 = 0$ by completing the square.
The first step is to move the constant term. Add 2 to both sides.

$$x^2 + 4x = 2$$

To visualize the next move, allow some space on the left side for completing the square.

$$x^2 + 4x + (\ ) = 2 + (\ )$$

We find the term needed to complete the square by squaring one-half of the x-coefficient.

$$\left(\frac{1}{2} \cdot 4\right)^2 = 2^2 = 4$$

NOTE

This *completes the square* on the left.

We now add 4 to both sides of the equation.

$$x^2 + 4x + 4 = 2 + 4$$

Next, factor on the left and simplify on the right.

$$(x + 2)^2 = 6$$

Solving as before, we have

$$x + 2 = \pm\sqrt{6}$$
$$x = -2 \pm \sqrt{6}$$

Check Yourself 2

Solve by completing the square.

$$x^2 + 6x - 4 = 0$$

To complete the square in this manner, the coefficient of x^2 must be 1. Example 3 illustrates the solution process when the coefficient of x^2 is not equal to 1.

Example 3   **Solving a Quadratic Equation by Completing the Square**

Solve $2x^2 - 4x - 5 = 0$ by completing the square.

$$2x^2 - 4x - 5 = 0$$ Add 5 to both sides.

$$2x^2 - 4x = 5$$ Because the coefficient of x^2 is not 1 (here it is 2), divide every term by 2. This will make the new leading coefficient equal to 1.

$$x^2 - 2x = \frac{5}{2}$$

$$x^2 - 2x + 1 = \frac{5}{2} + 1$$ Complete the square and solve as before.

$$(x - 1)^2 = \frac{7}{2}$$

$$x - 1 = \pm\sqrt{\frac{7}{2}}$$

$$x - 1 = \pm\frac{\sqrt{14}}{2}$$ Simplify the radical on the right.

$$x = 1 \pm \frac{\sqrt{14}}{2}$$

or

$$x = \frac{2 \pm \sqrt{14}}{2}$$

NOTE

$$\sqrt{\frac{7}{2}} = \frac{\sqrt{7}}{\sqrt{2}} \cdot \frac{\sqrt{2}}{\sqrt{2}}$$

$$= \frac{\sqrt{14}}{2}$$

NOTE

We have combined the terms on the right with the common denominator of 2.

Check Yourself 3

Solve by completing the square.

$$3x^2 - 6x + 2 = 0$$

The completing-the-square method is easiest to use when the coefficient of x is even. If it is odd, the method still works, but it will definitely involve fractions. Consider Example 4.

Example 4   **Solving a Quadratic Equation by Completing the Square**

Solve $2x^2 - 6x - 9 = 0$ by completing the square.

$$2x^2 - 6x - 9 = 0$$ Add 9 to both sides.

$$2x^2 - 6x = 9$$ Divide every term by 2.

$$x^2 - 3x = \frac{9}{2}$$

Be careful here. $\left[\dfrac{1}{2}(-3)\right]^2 = \left(-\dfrac{3}{2}\right)^2 = \dfrac{9}{4}$. So add $\dfrac{9}{4}$ to both sides.

$$x^2 - 3x + \dfrac{9}{4} = \dfrac{9}{2} + \dfrac{9}{4}$$

To factor the trinomial, remember how we got $\dfrac{9}{4}$: by squaring $-\dfrac{3}{2}$.

$$\left(x - \dfrac{3}{2}\right)^2 = \dfrac{27}{4}$$

$$x - \dfrac{3}{2} = \pm\sqrt{\dfrac{27}{4}}$$

$$x - \dfrac{3}{2} = \pm\dfrac{\sqrt{27}}{2}$$

$$x - \dfrac{3}{2} = \pm\dfrac{3\sqrt{3}}{2}$$

$$x = \dfrac{3}{2} \pm \dfrac{3\sqrt{3}}{2}$$

or

$$x = \dfrac{3 \pm 3\sqrt{3}}{2}$$

NOTE

We simplify $\sqrt{27}$ as follows:
$\sqrt{27} = \sqrt{9 \cdot 3} = \sqrt{9} \cdot \sqrt{3}$
$\quad\quad = 3\sqrt{3}$

Check Yourself 4

Solve by completing the square.

$2x^2 + 10x - 3 = 0$

Example 5 illustrates a geometric application using the Pythagorean theorem.

Example 5   **Solving a Geometry Application**

The length of one leg of a right triangle is 2 cm more than the other. If the length of the hypotenuse is 6 cm, what are the lengths of the two legs? Round to the nearest tenth of a centimeter.

Draw a sketch of the problem, labeling the known and unknown lengths. Here, if one leg is represented by x, the other must be represented by $x + 2$.

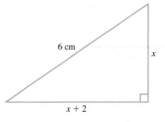

RECALL

The sum of the squares of the lengths of the legs is equal to the square of the length of the hypotenuse.

Use the Pythagorean theorem to form an equation.

$$x^2 + (x + 2)^2 = 6^2$$
$$x^2 + x^2 + 4x + 4 = 36$$
$$2x^2 + 4x - 32 = 0 \qquad \text{Divide both sides by 2.}$$
$$x^2 + 2x - 16 = 0$$

This equation can be solved by completing the square.

$$x^2 + 2x = 16$$
$$x^2 + 2x + 1 = 16 + 1$$
$$(x + 1)^2 = 17$$
$$x + 1 = \pm\sqrt{17}$$
$$x = -1 \pm \sqrt{17} \approx 3.1 \quad \text{or} \quad -5.1$$

We generally reject the negative solution in a geometric problem.

If $x \approx 3.1$, then $x + 2 \approx 5.1$. The lengths of the legs are approximately 3.1 and 5.1 cm.

Check Yourself 5

The length of one leg of a right triangle is 1 in. more than the other. If the length of the hypotenuse is 3 in., what are the lengths of the legs? Round to the nearest thousandth of an inch.

We close this section with an application that arises in the field of economics. Related to the production of a certain item, there are two important equations: a **supply** equation and a **demand** equation. Each is related to the price of the item. The **equilibrium** price is the price for which supply equals demand.

Example 6   A Business Application

The demand equation for a certain computer chip is given by

$$D = -4p + 50$$

The supply equation is predicted to be

$$S = -p^2 + 20p - 6$$

Find the equilibrium price.

We wish to know what price p results in equal supply and demand, so we write

$$-p^2 + 20p - 6 = -4p + 50$$
$$-56 = p^2 - 24p$$

or $\quad p^2 - 24p = -56$

Then, completing the square,

$$p^2 - 24p + 144 = -56 + 144$$
$$(p - 12)^2 = 88$$
$$p - 12 = \pm\sqrt{88}$$
$$p = 12 \pm \sqrt{88}$$
$$p = 21.38 \text{ or } 2.62$$

You should confirm that when $p = 2.62$, supply and demand are positive.

Here we must be careful: Supply and demand must be positive. When $p = 21.38$, they are negative, so the equilibrium price is $2.62.

Check Yourself 6

The demand equation for a certain item is given by

$$D = -2p + 32$$

The supply equation is predicted to be

$$S = -p^2 + 18p - 5$$

Find the equilibrium price.

We summarize the steps used to solve a quadratic equation by completing the square.

Step by Step

Solving a Quadratic Equation by Completing the Square

Step 1 Write the equation in the form
$$ax^2 + bx = k$$
so that the variable terms are on the left side and the constant is on the right side.

Step 2 If the coefficient of x^2 is not 1, divide both sides of the equation by that coefficient.

Step 3 Add the square of one-half the coefficient of x to both sides of the equation.

Step 4 The left side of the equation is now a perfect-square trinomial. Factor and solve as before.

Check Yourself ANSWERS

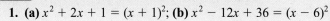

1. (a) $x^2 + 2x + 1 = (x + 1)^2$; (b) $x^2 - 12x + 36 = (x - 6)^2$

2. $-3 \pm \sqrt{13}$ **3.** $\dfrac{3 \pm \sqrt{3}}{3}$ **4.** $\dfrac{-5 \pm \sqrt{31}}{2}$ **5.** 1.562 in., 2.562 in.

6. $2.06

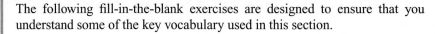

Reading Your Text

The following fill-in-the-blank exercises are designed to ensure that you understand some of the key vocabulary used in this section.

SECTION 10.2

(a) Every quadratic equation can be written with a _____ trinomial on the left.

(b) In a perfect-square trinomial with a coefficient of 1 for x^2, the square of one-half of the x-coefficient is equal to the _____ term.

(c) We find the term needed to _____ the square by squaring one-half of the x-coefficient.

(d) The completing-the-square method is easiest to use when the coefficient of x is an _____ number.

Basic Skills | Challenge Yourself | Calculator/Computer | Career Applications | Above and Beyond
▲

< Objective 1 >

Determine whether each trinomial is a perfect square.

1. $x^2 - 14x + 49$ **2.** $x^2 + 9x + 16$

3. $x^2 - 18x - 81$ **4.** $x^2 + 10x + 25$

5. $x^2 - 18x + 81$ **6.** $x^2 - 24x + 48$

Find the constant term that should be added to make each expression a perfect-square trinomial.

7. $x^2 + 6x$ **8.** $x^2 - 8x$

9. $x^2 - 10x$ **10.** $x^2 + 5x$

11. $x^2 + 9x$ **12.** $x^2 - 20x$

< Objective 2 >

Solve each quadratic equation by completing the square.

13. $x^2 + 4x - 12 = 0$ **14.** $x^2 - 6x + 8 = 0$

15. $x^2 - 2x - 5 = 0$ **16.** $x^2 + 4x - 7 = 0$

17. $x^2 + 3x - 27 = 0$ **18.** $x^2 + 5x - 3 = 0$

19. $x^2 + 6x - 1 = 0$ **20.** $x^2 + 4x - 4 = 0$

21. $x^2 - 5x + 6 = 0$ **22.** $x^2 - 6x - 3 = 0$

23. $x^2 + 6x - 5 = 0$ **24.** $x^2 - 2x = 1$

25. $x^2 = 9x + 5$ **26.** $x^2 = 4 - 7x$

Name _____

Section _____ Date _____

Answers

1. _____	2. _____
3. _____	4. _____
5. _____	6. _____
7. _____	8. _____
9. _____	10. _____
11. _____	12. _____
13. _____	14. _____
15. _____	
16. _____	
17. _____	
18. _____	
19. _____	
20. _____	
21. _____	22. _____
23. _____	24. _____
25. _____	
26. _____	

Answers

27. _____

28. _____

29. _____

30. _____

31. _____

32. _____

33. _____

34. _____

Solve each application. Give all answers to the nearest thousandth.

27. GEOMETRY The length of one leg of a right triangle is 4 in. more than the other. If the length of the hypotenuse is 8 in., what are the lengths of the two legs?

28. GEOMETRY The length of a rectangle is 2 cm longer than its width. If the diagonal of the rectangle is 4 cm, what are the dimensions (the length and the width) of the rectangle?

29. GEOMETRY The width of a rectangle is 6 ft less than its length. If the area of the rectangle is 75 ft^2, what are the dimensions of the rectangle?

30. GEOMETRY The length of a rectangle is 8 cm more than its width. If the area of the rectangle is 90 cm^2, find the dimensions.

31. SCIENCE AND MEDICINE The equation

$$h = -16t^2 - 32t + 320$$

gives the height of a ball, thrown downward from the top of a 320-ft building with an initial velocity of 32 ft/s. Find the time it takes for the ball to reach a height of 160 ft.

chapter 10 > Make the Connection

32. SCIENCE AND MEDICINE The equation

$$h = -16t^2 - 32t + 320$$

gives the height of a ball, thrown downward from the top of a 320-ft building with an initial velocity of 32 ft/s. Find the time it takes for the ball to reach a height of 64 ft.

chapter 10 > Make the Connection

33. BUSINESS AND FINANCE The demand equation for a certain type of printer is given by

$$D = -200p + 35,000$$

The supply equation is predicted to be

$$S = -p^2 + 400p - 20,000$$

Find the equilibrium price.

34. BUSINESS AND FINANCE The demand equation for a certain type of printer is given by

$$D = -80p + 7,000$$

The supply equation is predicted to be

$$S = -p^2 + 220p - 8,000$$

Find the equilibrium price.

Basic Skills | **Challenge Yourself** | Calculator/Computer | Career Applications | Above and Beyond
▲

Determine whether each statement is **true** *or* **false.**

35. To complete the square when the coefficient of x^2 is not 1, one of the first steps is to divide both sides of the equation by that coefficient.

36. If the coefficient of x is odd, the completing-the-square method will not work.

Complete each statement with **never, sometimes,** *or* **always.**

37. When completing the square for $x^2 + bx$, the number added is _____ negative.

38. A quadratic equation can _____ be solved by completing the square.

Solve each quadratic equation by completing the square.

39. $2x^2 - 6x + 1 = 0$ **40.** $2x^2 + 10x + 11 = 0$

41. $2x^2 - 4x + 1 = 0$ **42.** $2x^2 - 8x + 5 = 0$

43. $4x^2 - 2x - 1 = 0$ ⟩ Videos   **44.** $3x^2 - x - 2 = 0$

Solve each problem.

45. **NUMBER PROBLEM** If the square of 3 more than a number is 9, find the number(s).

46. **NUMBER PROBLEM** If the square of 2 less than an integer is 16, find the number(s).

47. **BUSINESS AND FINANCE** The revenue for selling x units of a product is given by

$$R = x\left(25 - \frac{1}{2}x\right)$$

Find the number of units sold if the revenue is $294.50.

48. **NUMBER PROBLEM** Find two consecutive positive integers such that the sum of their squares is 85.

Basic Skills | Challenge Yourself | Calculator/Computer | **Career Applications** | Above and Beyond
▲

49. **ALLIED HEALTH** An experimental drug is being tested on a bacteria colony. It is found that t days after the colony is treated, the number N of bacteria per cubic centimeter is given by the equation

$$N = -20t^2 - 120t + 1,000$$

In how many days will the colony be reduced to 200 bacteria?

Answers

35. _____

36. _____

37. _____

38. _____

39. _____

40. _____

41. _____

42. _____

43. _____

44. _____

45. _____

46. _____

47. _____

48. _____

49. _____

Beginning Algebra   The Streeter/Hutchison Series in Mathematics   © The McGraw-Hill Companies. All Rights Reserved.

Answers

50. _____

51. _____

52. _____

53. _____

54. _____

50. MECHANICAL ENGINEERING The rotational moment in a shaft is given by the formula

$$M = -30x + 2x^2$$

Find the value of x when the moment is equal to 152.

51. MANUFACTURING TECHNOLOGY Suppose that the cost C, in dollars, of producing x chairs is given by

$$C = 2x^2 - 40x + 2{,}400$$

How many chairs can be produced for $4,650?

52. MANUFACTURING TECHNOLOGY Suppose that the profit P, in dollars, of producing and selling x appliances is given by

$$P = -3x^2 + 240x - 1{,}800$$

How many appliances must be produced and sold to achieve a profit of $2,325?

53. MANUFACTURING TECHNOLOGY A small manufacturer's weekly profit P, in dollars, is given by

$$P = -3x^2 + 270x$$

Find the number of items x that must be produced to realize a profit of $5,775.

54. MANUFACTURING TECHNOLOGY The demand equation for a certain computer chip is given by

$$D = -5p + 62$$

The supply equation is predicted to be

$$S = -p^2 + 23p - 11$$

Find the equilibrium price.

Answers

1. Yes   **3.** No   **5.** Yes   **7.** 9   **9.** 25   **11.** $\dfrac{81}{4}$ **13.** $-6, 2$

15. $1 \pm \sqrt{6}$ **17.** $\dfrac{-3 \pm 3\sqrt{13}}{2}$ **19.** $-3 \pm \sqrt{10}$ **21.** 2, 3

23. $-3 \pm \sqrt{14}$ **25.** $\dfrac{9 \pm \sqrt{101}}{2}$ **27.** 3.292 in., 7.292 in.

29. 6.165 ft by 12.165 ft   **31.** 2.317 s   **33.** $112.92   **35.** True

37. never   **39.** $\dfrac{3 \pm \sqrt{7}}{2}$ **41.** $\dfrac{2 \pm \sqrt{2}}{2}$ **43.** $\dfrac{1 \pm \sqrt{5}}{4}$

45. $-6, 0$ **47.** 19, 31   **49.** 4 days   **51.** 45 chairs

53. 35 or 55 items

10.3

The Quadratic Formula

< 10.3 Objectives >

1 > Write a quadratic equation in standard form

2 > Use the quadratic formula to solve a quadratic equation

We are now ready to derive and use the **quadratic formula** to solve all quadratic equations. We derive the formula by using the method of completing the square.

To use the quadratic formula, the quadratic equation you want to solve must be in *standard form*. That form is

$$ax^2 + bx + c = 0 \qquad \text{in which } a \neq 0$$

Example 1	Writing Quadratic Equations in Standard Form

< Objective 1 >

Write each equation in standard form.

(a) $2x^2 - 5x + 3 = 0$

The equation is already in standard form.

$$a = 2 \qquad b = -5 \qquad \text{and} \qquad c = 3$$

(b) $5x^2 + 3x = 5$

The equation is *not* in standard form. Rewrite it by subtracting 5 from both sides.

$$5x^2 + 3x - 5 = 0 \qquad \text{Standard form}$$
$$a = 5 \qquad b = 3 \qquad \text{and} \qquad c = -5$$

Check Yourself 1

Rewrite each quadratic equation in standard form.

(a) $x^2 - 3x = 5$ (b) $3x^2 = 7 - 2x$

Once a quadratic equation is written in standard form, we can find any solutions to the equation. Remember that a solution is a value for x that makes the equation true.

What follows is the derivation of the quadratic formula, which can be used to solve quadratic equations.

Step by Step

Deriving the Quadratic Formula

Let $ax^2 + bx + c = 0$, in which $a \neq 0$.

$$ax^2 + bx = -c$$ Subtract c from both sides.

$$x^2 + \frac{b}{a}x = -\frac{c}{a}$$ Divide both sides by a.

$$x^2 + \frac{b}{a}x + \frac{b^2}{4a^2} = \frac{b^2}{4a^2} - \frac{c}{a}$$ Add $\frac{b^2}{4a^2}$ to both sides.

$$\left(x + \frac{b}{2a}\right)^2 = \frac{b^2 - 4ac}{4a^2}$$ Factor on the left and add the fractions on the right.

$$x + \frac{b}{2a} = \pm\sqrt{\frac{b^2 - 4ac}{4a^2}}$$ Take the square root of both sides.

$$x + \frac{b}{2a} = \pm\frac{\sqrt{b^2 - 4ac}}{2a}$$ Simplify the radical on the right.

$$x = -\frac{b}{2a} \pm \frac{\sqrt{b^2 - 4ac}}{2a}$$ Add $-\frac{b}{2a}$ to both sides.

$$x = \frac{-b \pm \sqrt{b^2 - 4ac}}{2a}$$ Use the common denominator, $2a$.

NOTE

This is the completing-the-square step that makes the left-hand side a perfect square.

Property

The Quadratic Formula

If $ax^2 + bx + c = 0$, then

$$x = \frac{-b \pm \sqrt{b^2 - 4ac}}{2a}$$

Let's use the quadratic formula to solve some equations.

Example 2

Using the Quadratic Formula to Solve an Equation

< Objective 2 >

Use the quadratic formula to solve $x^2 - 5x + 4 = 0$.

The equation is in standard form, so first identify a, b, and c.

$$x^2 - 5x + 4 = 0$$

$a = 1$ $b = -5$ $c = 4$

NOTE

The leading coefficient is 1, so $a = 1$.

We now substitute the values for a, b, and c into the formula.

$$x = \frac{-b \pm \sqrt{b^2 - 4ac}}{2a}$$

We find it helpful to rewrite the formula using parentheses to indicate where the values of a, b, and c will be placed.

$$x = \frac{-(\ ) \pm \sqrt{(\ )^2 - 4(\ )(\ )}}{2(\ )}$$

Then

$$x = \frac{-(-5) \pm \sqrt{(-5)^2 - 4(1)(4)}}{2(1)}$$

$$= \frac{5 \pm \sqrt{25 - 16}}{2}$$

$$= \frac{5 \pm \sqrt{9}}{2}$$

$$= \frac{5 \pm 3}{2}$$

NOTE

This result could also have been found by factoring the original equation. You should check that for yourself.

Now,

$$x = \frac{5 + 3}{2} \quad \text{or} \quad x = \frac{5 - 3}{2}$$
$$= 4 \qquad\qquad\qquad = 1$$

The solutions are 4 and 1.

 Check Yourself 2

Use the quadratic formula to solve $x^2 - 2x - 8 = 0$. Check your result by factoring.

The main use of the quadratic formula is to solve equations that *cannot* be factored.

Example 3 | **Using the Quadratic Formula to Solve an Equation**

Use the quadratic formula to solve $2x^2 = x + 4$.

First, the equation *must be written* in standard form to find a, b, and c.

$$2x^2 - x - 4 = 0$$
$$a = 2 \quad b = -1 \quad c = -4$$

NOTE

Substitute the values for a, b, and c into the formula.

$$x = \frac{-b \pm \sqrt{b^2 - 4ac}}{2a}$$

$$= \frac{-(\ ) \pm \sqrt{(\ )^2 - 4(\ )(\ )}}{2(\ )}$$

$$= \frac{-(-1) \pm \sqrt{(-1)^2 - 4(2)(-4)}}{2(2)}$$

$$= \frac{1 \pm \sqrt{1 + 32}}{4}$$

$$= \frac{1 \pm \sqrt{33}}{4}$$

 Check Yourself 3

Use the quadratic formula to solve $3x^2 = 3x + 4$.

Example 4 | **Using the Quadratic Formula to Solve an Equation**

Use the quadratic formula to solve $x^2 - 2x = 4$.

In standard form, the equation is

$$x^2 - 2x - 4 = 0$$
$$a = 1 \quad b = -2 \quad c = -4$$

NOTE

Again substitute the values into the quadratic formula.

$$x = \frac{-(-2) \pm \sqrt{(-2)^2 - 4(1)(-4)}}{2(1)}$$

$$= \frac{2 \pm \sqrt{20}}{2}$$

NOTE

20 has a perfect-square factor. So

$\sqrt{20} = \sqrt{4 \cdot 5} = \sqrt{4} \cdot \sqrt{5}$

$= 2\sqrt{5}$

Unless you are finding a decimal approximation, you should always write your solution in simplest form.

$$x = \frac{2 \pm 2\sqrt{5}}{2}$$

$$= \frac{2(1 \pm \sqrt{5})}{2}$$ Now factor the numerator and divide by the common factor 2.

$$= 1 \pm \sqrt{5}$$

 Check Yourself 4

Use the quadratic formula to solve $3x^2 = 2x + 4$.

Sometimes equations have common factors. Factoring first simplifies these equations, making them easier to solve. This is illustrated in Example 5.

▶ **Example 5**   **Using the Quadratic Formula to Solve an Equation**

Use the quadratic formula to solve $3x^2 - 6x - 3 = 0$.

Because the equation is in standard form, we could use

$a = 3 \qquad b = -6 \qquad$ and $\qquad c = -3$

in the quadratic formula. There is, however, a better approach.

Note the common factor of 3 in the quadratic member of the original equation. Factoring, we have

$$3(x^2 - 2x - 1) = 0$$

and dividing both sides of the equation by 3 gives

$$x^2 - 2x - 1 = 0$$

NOTE

The advantage to this approach is that these values require much less simplification after we substitute into the quadratic formula.

Now let $a = 1$, $b = -2$, and $c = -1$. Then

$$x = \frac{-(-2) \pm \sqrt{(-2)^2 - 4(1)(-1)}}{2(1)}$$

$$= \frac{2 \pm \sqrt{8}}{2}$$

$$= \frac{2 \pm 2\sqrt{2}}{2}$$

$$= \frac{2(1 \pm \sqrt{2})}{2}$$

$$= 1 \pm \sqrt{2}$$

 Check Yourself 5

Use the quadratic formula to solve $4x^2 - 20x = 12$.

In applications that lead to quadratic equations, you may want to find approximate values for the solutions.

▶ **Example 6**   **Using the Quadratic Formula to Solve an Equation**

Use the quadratic formula to solve $x^2 - 5x + 5 = 0$ and write your solutions in approximate decimal form. Round solutions to the nearest thousandth.

Substituting $a = 1$, $b = -5$, and $c = 5$ gives

$$x = \frac{-(-5) \pm \sqrt{(-5)^2 - 4(1)(5)}}{2(1)}$$

$$= \frac{5 \pm \sqrt{5}}{2}$$

Use your calculator to find

$$x \approx 3.618 \quad \text{or} \quad x \approx 1.382$$

Check Yourself 6

Use the quadratic formula to solve $x^2 - 3x - 5 = 0$ and approximate the solutions in decimal form to the nearest thousandth.

You may be wondering whether the quadratic formula can be used to solve all quadratic equations. It can, but not all quadratic equations will have real-number solutions, as Example 7 shows.

Example 7 | **Using the Quadratic Formula to Solve an Equation**

Use the quadratic formula to solve $x^2 - 3x = -5$.

Substituting $a = 1$, $b = -3$, and $c = 5$, we have

$$x = \frac{-(-3) \pm \sqrt{(-3)^2 - 4(1)(5)}}{2(1)}$$

$$= \frac{3 \pm \sqrt{-11}}{2}$$

NOTE

Make sure the quadratic equation is in standard form. $x^2 - 3x = -5$ is equivalent to $x^2 - 3x + 5 = 0$.

In this case, there are no real-number solutions because the radicand is negative.

Check Yourself 7

Use the quadratic formula to solve $x^2 - 3x = -3$.

The part of the quadratic formula, $b^2 - 4ac$, that is under the radical is called the **discriminant.** By computing this quantity, we can determine what type of solutions we would find if we were to solve completely. For example, for the quadratic equation in the previous example, $x^2 - 3x = -5$, we found that $a = 1$, $b = -3$, and $c = 5$. The value of the discriminant is

$$b^2 - 4ac = (-3)^2 - 4(1)(5) = 9 - 20 = -11$$

Since the discriminant is negative, we know immediately that there are no real-number solutions.

This information can be obtained simply by evaluating the discriminant.

NOTE

If $b^2 - 4ac$ is a perfect square, then the equation can be solved by factoring.

1. If the discriminant is negative, there are no real-number solutions.

2. If the discriminant is zero, there is exactly one real-number solution. In fact, the solution will be a rational number.

3. If the discriminant is positive and is a perfect square, there are two rational-number solutions.

4. And, if the discriminant is positive but is not a perfect square, there are two irrational-number solutions.

Example 8   Using the Discriminant to Determine the Type of Solutions

For each equation, evaluate the discriminant, determine how many real solutions there are, and classify them.

(a) $2x^2 - 5x + 1 = 0$

First evaluate the discriminant.

$b^2 - 4ac = (-5)^2 - 4(2)(1) = 25 - 8 = 17$

Since 17 is positive and is not a perfect square, there are two irrational solutions.

(b) $10x - x^2 = 25$

We begin by writing this in standard form.

$x^2 - 10x + 25 = 0$

Then

$b^2 - 4ac = (-10)^2 - 4(1)(25) = 100 - 100 = 0$

Since the discriminant is zero, there is exactly one rational solution.

(c) $3x^2 - x + 2 = 0$

We have

$b^2 - 4ac = (-1)^2 - 4(3)(2) = 1 - 24 = -23$

There are no real-number solutions.

 Check Yourself 8

For each equation, evaluate the discriminant, determine how many real solutions there are, and classify them.

(a) $8x - 16 = x^2$ **(b)** $2x^2 + 3x = 20$ **(c)** $x^2 + 5x + 7 = 0$

We summarize the steps used for solving equations by using the quadratic formula.

Step by Step

Solving Equations by using the Quadratic Formula

Step 1   Rewrite the equation in standard form.

$ax^2 + bx + c = 0$

Step 2   If a common factor exists, divide both sides of the equation by that common factor.

Step 3   Identify the coefficients a, b, and c.

Step 4   Substitute values for a, b, and c into the formula

$$x = \frac{-b \pm \sqrt{b^2 - 4ac}}{2a}$$

Step 5   Simplify the right side of the expression formed in step 4 to write the solutions for the original equation.

Often, applied problems lead to quadratic equations that must be solved by the methods of Section 10.2 or this section.

  Example 9   Solving a Construction Application

A rectangular garden is to be surrounded by a walkway of constant width. The garden's dimensions are 5 m by 8 m. The total area, garden plus walkway, is to be 100 m². What must be the width of the walkway? Round to the nearest hundredth of a meter.

First we sketch the situation.

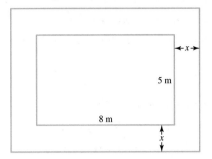

The area of the entire figure may be represented by

$$\underbrace{(8 + 2x)}_{\text{Length}}\underbrace{(5 + 2x)}_{\text{Width}} \qquad \text{(Do you see why?)}$$

Since the area is to be 100 m², we write

$$(8 + 2x)(5 + 2x) = 100$$
$$40 + 16x + 10x + 4x^2 = 100$$
$$4x^2 + 26x - 60 = 0$$
$$2x^2 + 13x - 30 = 0$$
$$x = \frac{-(13) \pm \sqrt{(13)^2 - 4(2)(-30)}}{2(2)}$$
$$= \frac{-13 \pm \sqrt{409}}{4}$$
$$x \approx 1.8059 \quad \text{or} \quad \approx -8.3059$$

Since x must be positive, we have (rounded) $x = 1.81$ m.

Check Yourself 9

A rectangular swimming pool is 16 ft by 40 ft. A tarp to go over the pool also covers a strip of equal width surrounding the pool. If the area of the tarp is 1100 ft², how wide is the covered strip around the pool? Round to the nearest tenth of a foot.

NOTE

We are neglecting air resistance in this discussion.

When a ball is thrown directly upward, or if some sort of projectile is fired upward, the height of the object is approximated by a quadratic function of time. This concept is illustrated in Example 10.

Example 10   A Science Application

Suppose that a ball is thrown upward with an initial velocity of 80 ft/s. (The initial velocity is the speed with which the ball leaves the thrower's hand. The speed will then decrease as the ball rises.) If the ball is released at a height of 5 ft, the height equation may be written as follows:

$$h = -16t^2 + 80t + 5$$

When, to the nearest hundredth of a second, will the ball be at a height of 93 ft?
    Since the desired height is 93 ft, we write

$$93 = -16t^2 + 80t + 5$$

We must solve for t:

$$93 = -16t^2 + 80t + 5$$ We need a 0 on one side.

$$16t^2 - 80t + 88 = 0$$ Divide through by 8.

$$2t^2 - 10t + 11 = 0$$ Use the quadratic formula.

$$t = \frac{-(-10) \pm \sqrt{(-10)^2 - 4(2)(11)}}{2(2)}$$

$$= \frac{10 \pm \sqrt{12}}{4} \approx 1.63 \quad \text{or} \quad 3.37$$

The ball reaches a height of 93 ft twice: first, on the way up, at 1.63 s, and second, on the way down, at 3.37 s.

Check Yourself 10

Using the same height equation provided in Example 10, determine, to the nearest hundredth of a second, when the ball will be at a height of 77 ft.

In Example 11, we look at a similar application. It highlights the importance of proper interpretation of results.

Example 11   **A Science Application**

A projectile is fired upward from a platform in such a way that the object will miss the platform (fortunately!) on the way down. If the height of the platform is 20 ft, and the initial velocity is 200 ft/s, the height is given by

$$h = -16t^2 + 200t + 20$$

To the nearest tenth of a second, when will the projectile hit the ground?

The projectile will hit the ground when the height is 0, so we write

$$0 = -16t^2 + 200t + 20$$ Divide through by −4.

$$0 = 4t^2 - 50t - 5$$

$$t = \frac{-(-50) \pm \sqrt{(-50)^2 - 4(4)(-5)}}{2(4)}$$

$$= \frac{50 \pm \sqrt{2,580}}{8} \approx 12.599 \quad \text{or} \quad -0.099$$

The projectile will hit the ground at about 12.6 s. We reject the negative solution, as it indicates a time *before* the projectile was fired.

Check Yourself 11

A projectile fired upward from a platform follows the height equation

$$h = -16t^2 + 160t + 40$$

When will it hit the ground? Round to the nearest tenth of a second.

Check Yourself ANSWERS

1. (a) $x^2 - 3x - 5 = 0$; **(b)** $3x^2 + 2x - 7 = 0$

2. $x = 4, -2$ **3.** $x = \dfrac{3 \pm \sqrt{57}}{6}$

4. $x = \dfrac{1 \pm \sqrt{13}}{3}$

5. $x = \dfrac{5 \pm \sqrt{37}}{2}$ **6.** $x \approx 4.193$ or -1.193

7. $\dfrac{3 \pm \sqrt{-3}}{2}$, no real solutions

8. (a) 1 rational solution; **(b)** 2 rational solutions; **(c)** no real solutions

9. 3.6 ft   **10.** 1.18 s, 3.82 s   **11.** 10.2 s

Reading Your Text

The following fill-in-the-blank exercises are designed to ensure that you understand some of the key vocabulary used in this section.

SECTION 10.3

(a) To use the quadratic formula, the quadratic equation you want to solve must be in _____ form.

(b) The main use of the quadratic formula is to solve equations that cannot be _____.

(c) The part of the quadratic formula that is under the radical is called the _____.

(d) If the discriminant is positive and is a perfect square, there are two _____ number solutions.

10.3 exercises

Name _____

Section _____ Date _____

Answers

Basic Skills | Challenge Yourself | Calculator/Computer | Career Applications | Above and Beyond

< Objectives 1–2 >

Use the quadratic formula to solve each quadratic equation.

1. $x^2 + 9x + 20 = 0$ > Videos

2. $x^2 - 9x + 14 = 0$

3. $x^2 - 4x + 3 = 0$

4. $x^2 - 13x + 22 = 0$

5. $3x^2 + 2x - 1 = 0$

6. $x^2 - 8x + 16 = 0$

7. $x^2 + 5x = -4$

8. $4x^2 + 5x = 6$

9. $x^2 = 6x - 9$ > Videos

10. $2x^2 - 5x = 3$

11. $2x^2 - 3x - 7 = 0$

12. $x^2 - 5x + 2 = 0$

13. $x^2 + 2x - 4 = 0$

14. $x^2 - 4x + 2 = 0$

15. $2x^2 - 3x = 3$

16. $3x^2 - 2x + 1 = 0$

17. $3x^2 - 2x = 6$ > Videos

18. $4x^2 = 4x + 5$

19. $3x^2 + 3x + 2 = 0$

20. $2x^2 - 3x = 6$

21. $5x^2 = 8x - 2$

22. $5x^2 - 2 = 2x$

23. $2x^2 - 9 = 4x$

24. $3x^2 - 6x = 2$

In exercises 25–32, evaluate the discriminant, determine how many real solutions there are, and classify them.

25. $2x^2 - 5x - 3 = 0$

26. $x^2 = 8x - 16$

27. $5x^2 - 3x + 2 = 0$ ▸ Videos

28. $4x^2 - 7x + 2 = 0$

29. $3x^2 + 7x + 4 = 0$

30. $x^2 + 3x + 4 = 0$

31. $2x^2 - 2 = 5x$

32. $6x^2 + 11x - 35 = 0$

Solve each application and round answers to the nearest thousandth where necessary.

33. GEOMETRY One leg of a right triangle is 1 in. shorter than the other leg. The hypotenuse is 4 in. longer than the shorter side. Find the length of each side.

34. GEOMETRY The hypotenuse of a given right triangle is 6 cm longer than the shorter leg. The length of the shorter leg is 2 cm less than that of the longer leg. Find the lengths of the three sides.

35. CONSTRUCTION A rectangular field is 300 ft by 500 ft. A roadway of width x ft is to be built just inside the field. What is the widest the roadway can be and still leave 100,000 ft^2 in the region?

36. CONSTRUCTION A rectangular garden is to be surrounded by a walkway of constant width. The garden's dimensions are 20 ft by 28 ft. The total area, garden plus walkway, is to be 1,100 ft^2. What must be the width of the walkway?

37. SCIENCE AND MEDICINE The equation

$$h = -16t^2 + 112t$$

gives the height of an arrow, shot upward from the ground with an initial velocity of 112 ft/s, where t is the time after the arrow leaves the ground. Find the time it takes for the arrow to reach a height of 120 ft.

chapter 10 ▸ Make the Connection

Answers

21. _____

22. _____

23. _____

24. _____

25. _____

26. _____

27. _____

28. _____

29. _____

30. _____

31. _____

32. _____

33. _____

34. _____

35. _____

36. _____

37. _____

38. SCIENCE AND MEDICINE The equation

$$h = -16t^2 + 112t$$

gives the height of an arrow, shot upward from the ground with an initial velocity of 112 ft/s, where t is the time after the arrow leaves the ground. Find the time it takes for the arrow to reach a height of 180 ft.

39. SCIENCE AND MEDICINE If a ball is thrown vertically upward from a height of 6 ft, with an initial velocity of 64 ft/s, its height h after t s is given by

$$h = -16t^2 + 64t + 6$$

How long does it take the ball to return to the ground?

40. SCIENCE AND MEDICINE If a ball is thrown vertically upward from a height of 5 ft, with an initial velocity of 96 ft/s, its height h after t s is given by

$$h = -16t^2 + 96t + 5$$

How long does it take the ball to return to the ground?

| Basic Skills | **Challenge Yourself** | Calculator/Computer | Career Applications | Above and Beyond |

Determine whether each statement is **true** *or* **false.**

41. The solutions of a quadratic equation are always rational.

42. A ball thrown vertically might reach a specified height two times.

Complete each statement with **never, sometimes,** *or* **always.**

43. The quadratic formula can _____ be used to solve a quadratic equation.

44. If the value of $b^2 - 4ac$ is negative, the equation $ax^2 + bx + c = 0$ _____ has real-number solutions.

Solve each equation. (Hint: First write each equation as a quadratic equation in standard form.)

45. $3x - 5 = \dfrac{1}{x}$ **46.** $x + 3 = \dfrac{1}{x}$

47. $(x - 2)(x + 1) = 3$ **48.** $(x - 3)(x + 2) = 5$

49. $\dfrac{3}{x} + \dfrac{5}{x^2} = 9$ **50.** $\dfrac{8}{x} - \dfrac{3}{x^2} = -6$

51. $\dfrac{x}{x + 1} + \dfrac{10x}{x^2 + 4x + 3} = \dfrac{15}{x + 3}$ **52.** $x - \dfrac{9x}{x - 2} = \dfrac{-10}{x - 2}$

Solve each quadratic equation.

53. $(x - 1)^2 = 7$

54. $(2x + 3)^2 = 5$

55. $2x^2 - 8x + 3 = 0$

56. $x^2 + 2x - 1 = 0$

57. $x^2 - 9x - 4 = 6$

58. $5x^2 + 10x + 2 = 2$

59. $4x^2 - 8x + 3 = 5$ > Videos

60. $x^2 + 4x = 21$

Basic Skills | Challenge Yourself | Calculator/Computer | **Career Applications** | Above and Beyond

▲

---

61. **ALLIED HEALTH** The concentration C, in nanograms per milliliter (ng/mL), of digoxin, a medication prescribed for congestive heart failure, is given by the equation

$$C = -0.0015t^2 + 0.0845t + 0.7170$$

where t is the number of hours since the drug was taken orally. For a drug to have a beneficial effect, its concentration in the bloodstream must exceed a certain value, the minimum therapeutic level, which for digoxin is 0.8 ng/mL.

(a) How long will it take for the drug to start having an effect?

(b) How long does it take for the drug to stop having an effect?

62. **ALLIED HEALTH** The concentration C, in micrograms per milliliter (mcg/mL), of phenobarbital, an anticonvulsant medication, is given by the equation

$$C = -1.35t^2 + 10.81t + 7.38$$

where t is the number of hours since the drug was taken orally. For a drug to have a beneficial effect, its concentration in the bloodstream must exceed a certain value, the minimum therapeutic level, which for phenobarbital is 10 mcg/mL.

(a) How long will it take for the drug to start having an effect?

(b) How long does it take for the drug to stop having an effect?

63. **ALLIED HEALTH** One technique of controlling cancer is to use radiation therapy. After such a treatment, the total number of cancerous cells N, in thousands, can be estimated by the equation

$$N = -3t^2 + 6t + 140$$

where t is the number of days of treatment. How many days of treatment are required to kill all the cancer cells?

Answers

53. _____

54. _____

55. _____

56. _____

57. _____

58. _____

59. _____

60. _____

61. _____

62. _____

63. _____

64. _____

65. _____

66. _____

67. _____

64. **MANUFACTURING TECHNOLOGY** The Scribner log rule is used to calculate the volume V (in cubic feet) of a 16-foot log given the diameter (in inches) of the smaller end (inside the bark). The formula for the log rule is

$$V = 0.79D^2 - 2D - 4$$

Find the diameter of the small end of a log that has a volume of 272 cubic feet.

Basic Skills	Challenge Yourself	Calculator/Computer	Career Applications	**Above and Beyond**

▲

65. Work with a partner to decide all values of b in each equation that will give one or more real-number solutions.

   (a) $3x^2 + bx - 3 = 0$
   (b) $5x^2 + bx + 1 = 0$
   (c) $-3x^2 + bx - 3 = 0$
   (d) Write a rule for judging whether an equation has solutions by looking at it in standard form.

66. Which method of solving a quadratic equation seems simplest to you? Which method should you try first?

67. Complete the statement: "You can tell an equation is quadratic and not linear by . . . ."

Answers

1. $-4, -5$ 3. $3, 1$ 5. $-1, \dfrac{1}{3}$ 7. $-4, -1$ 9. 3

11. $\dfrac{3 \pm \sqrt{65}}{4}$ 13. $-1 \pm \sqrt{5}$ 15. $\dfrac{3 \pm \sqrt{33}}{4}$ 17. $\dfrac{1 \pm \sqrt{19}}{3}$

19. No real-number solutions   21. $\dfrac{4 \pm \sqrt{6}}{5}$ 23. $\dfrac{2 \pm \sqrt{22}}{2}$

25. 49; 2 rational   27. -31; no real solutions   29. 1; 2 rational

31. 41; 2 irrational   33. 7.899 in., 8.899 in., 11.899 in.

35. 34.168 ft   37. 1.321 or 5.679 s   39. 4.092 s   41. False

43. always   45. $\dfrac{5 \pm \sqrt{37}}{6}$ 47. $\dfrac{1 \pm \sqrt{21}}{2}$ 49. $\dfrac{1 \pm \sqrt{21}}{6}$

51. 5 53. $1 \pm \sqrt{7}$ 55. $\dfrac{4 \pm \sqrt{10}}{2}$ 57. $-1, 10$

59. $\dfrac{2 \pm \sqrt{6}}{2}$ 61. (a) 1 hour; (b) 55.3 hours   63. 8 days

65. Above and Beyond   67. Above and Beyond

10.4

Graphing Quadratic Equations

< 10.4 Objectives >

1 > Graph a quadratic equation by plotting points

2 > Find the axis of symmetry of a parabola

3 > Find the *x*-intercepts of a parabola

In Section 6.3 you learned to graph first-degree (or linear) equations. We use similar methods to graph quadratic equations of the form

$$y = ax^2 + bx + c \qquad a \neq 0$$

The first thing to notice is that the graph of an equation in this form is not a straight line. The graph is always a curve called a **parabola.**

Here are some examples:

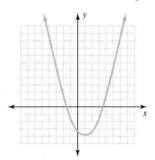

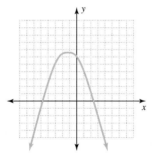

To graph quadratic equations, start by finding solutions for the equation. We begin by completing a table of values. Choose any convenient values for *x*. Then use the given equation to compute the corresponding values for *y*. Example 1 illustrates this process.

| | Example 1 | Completing a Table of Values |

< Objective 1 >

RECALL

A solution is a pair of values that makes the equation a true statement.

Complete the ordered pairs to form solutions for the equation $y = x^2$. Then show these results in a table of values.

$(-2, \ ), (-1, \ ), (0, \ ), (1, \ ), (2, \ )$

For example, to complete the pair $(-2, \ )$, substitute -2 for *x* in the given equation.

$$y = (-2)^2 = 4$$

So $(-2, 4)$ is a solution.

Substituting the other values for x in this manner gives a table of values for $y = x^2$.

x	y
-2	4
-1	1
0	0
1	1
2	4

Check Yourself 1

Complete the ordered pairs to form solutions for $y = x^2 + 2$ and form a table of values.

$(-2,\ ), (-1,\ ), (0,\ ), (1,\ ), (2,\ )$

We can now plot points in the Cartesian coordinate system that correspond to solutions to the equation.

 Example 2   **Plotting Some Solution Points**

Plot the points from the table of values corresponding to $y = x^2$ from Example 1.

x	y
-2	4
-1	1
0	0
1	1
2	4

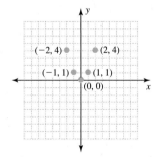

Notice that the y-axis acts as a mirror. Do you see that any point graphed in quadrant I is "reflected" in quadrant II?

Check Yourself 2

Plot the points from the table of values formed in Check Yourself 1.

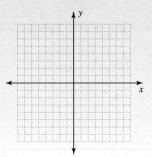

The graph of the equation can be drawn by joining the points with a smooth curve.

| Example 3 | Completing the Graph of the Solution Set |

Draw the graph of $y = x^2$.

We can now draw a smooth curve between the points found in Example 2 to form the graph of $y = x^2$.

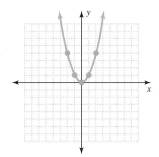

Check Yourself 3

Draw a smooth curve between the points plotted in Check Yourself 2.

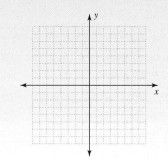

You can use any convenient values for x in forming your table of values. You should use as many pairs as are necessary to get the correct shape of the graph (a parabola).

| Example 4 | Graphing the Solution Set |

Graph $y = x^2 - 2x$. Use integer values of x from -1 to 3.

First, determine solutions for the equation. For instance, if $x = -1$,

$$y = (-1)^2 - 2(-1)$$
$$= 1 + 2$$
$$= 3$$

and $(-1, 3)$ is a solution for the given equation.

Substituting the other values for x, we can form a table of values. We then plot the corresponding points and draw a smooth curve to form our graph.

The graph of $y = x^2 - 2x$.

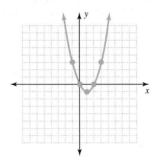

x	y
-1	3
0	0
1	-1
2	0
3	3

Check Yourself 4

Graph $y = x^2 + 4x$. Use integer values of x from -4 to 0.

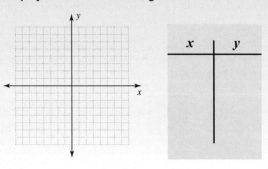

x	y

Choosing values for x is also a valid method of graphing a quadratic equation that contains three terms.

Example 5   Graphing the Solution Set

Graph $y = x^2 - x - 2$. Use integer values of x from -2 to 3. We show the computation for two of the solutions.

If $x = -2$:
$$y = (-2)^2 - (-2) - 2$$
$$= 4 + 2 - 2$$
$$= 4$$

If $x = 3$:
$$y = (3)^2 - (3) - 2$$
$$= 9 - 3 - 2$$
$$= 4$$

You should substitute the remaining values for x into the given equation to verify the other solutions shown in our table of values.

The graph of $y = x^2 - x - 2$.

x	y
-2	4
-1	0
0	-2
1	-2
2	0
3	4

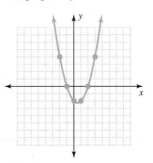

Check Yourself 5

Graph $y = x^2 - 4x + 3$. Use integer values of x from -1 to 4.

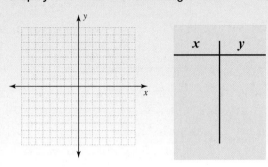

In Example 6, the graph looks significantly different from previous graphs.

| Example 6 | Graphing the Solution Set |

Graph $y = -x^2 + 3$. Use integer values from -2 to 2.

Again, we show two computations.

If $x = -2$:

$$y = -(-2)^2 + 3$$
$$= -4 + 3$$
$$= -1$$

If $x = 1$:

$$y = -(1)^2 + 3$$
$$= -1 + 3$$
$$= 2$$

Verify the remaining solutions shown in the table of values for yourself.

The graph of $y = -x^2 + 3$.

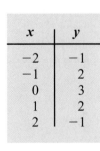

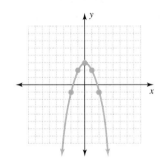

x	y
-2	-1
-1	2
0	3
1	2
2	-1

There is an important difference between this graph and the others we have seen. This time the parabola opens downward! Can you guess why? The answer is in the coefficient of the x^2-term.

If the coefficient of x^2 is *positive,* the parabola opens *upward.*

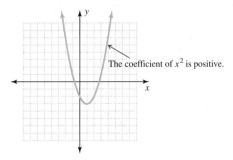

The coefficient of x^2 is positive.

If the coefficient of x^2 is *negative,* the parabola opens *downward.*

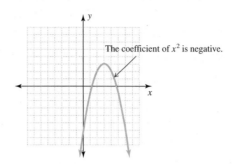

The coefficient of x^2 is negative.

Check Yourself 6

Graph $y = -x^2 - 2x$. Use integer values from -3 to 1.

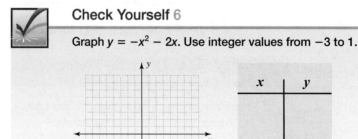

x	y

There are two other terms we would like to introduce before closing this section on graphing quadratic equations. As you may have noticed, all the parabolas that we graphed are symmetric about a vertical line. This is called the **axis of symmetry** for the parabola.

The point at which the parabola intersects that vertical line (this will be the lowest—or the highest—point on the parabola) is called the **vertex.**

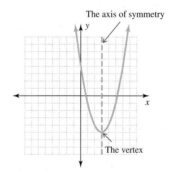

The axis of symmetry

The vertex

As we construct a table of values, we want to include x-values that will span the parabola's turning point: the vertex. For a quadratic equation $y = ax^2 + bx + c$, the equation for the axis of symmetry is

$$x = \frac{-b}{2a}$$

Once we have this, we can choose x-values on either side of the axis of symmetry to help us build a table. Look at Example 7.

Example 7	**Using the Axis of Symmetry to Create a Table and Graph**

< Objective 2 >

Given $y = -x^2 - 8x - 12$, complete each exercise.

(a) Write the equation for the axis of symmetry.

Since $a = -1$ and $b = -8$, we have

$$x = \frac{-b}{2a} = \frac{-(-8)}{2(-1)} = \frac{8}{-2} = -4$$

Thus the vertical line $x = -4$ is the axis of symmetry.

(b) Use the equation to create a table of values for the quadratic equation.

Finding values on both sides of -4 gives

x	y
-6	0
-5	3
-4	4
-3	3
-2	0

(c) Sketch the graph. Show the axis of symmetry as a dotted vertical line.

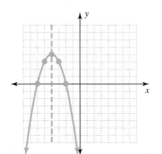

NOTE

The parabola opens downward.

 Check Yourself 7

For $y = x^2 - 7x + 4$, find the axis of symmetry, create a table of values, and sketch the graph.

The x-intercepts of a parabola are the points where the parabola touches the x-axis. We can locate these directly without actually graphing the parabola. The key is to set y equal to 0 and then solve for x.

▶ **Example 8**	**Finding the x-Intercepts of a Parabola**

< Objective 3 >

Find the x-intercepts for the graph of each equation.

(a) $y = x^2 - x - 6$

We set $y = 0$, and solve.

$0 = x^2 - x - 6$ We solve this by factoring.

$0 = (x - 3)(x + 2)$

$x = -2$ or 3

There are two x-intercepts: $(-2, 0)$ and $(3, 0)$.

(b) $y = x^2 - 5x + 2$

We set $y = 0$ and solve.

$0 = x^2 - 5x + 2$ Use the quadratic formula.

$$x = \frac{5 \pm \sqrt{(-5)^2 - 4(2)}}{2} = \frac{5 \pm \sqrt{17}}{2}$$

There are two x-intercepts: $\left(\dfrac{5 - \sqrt{17}}{2}, 0\right)$ and $\left(\dfrac{5 + \sqrt{17}}{2}, 0\right)$.

(c) $y = x^2 + 6x + 9$

$0 = x^2 + 6x + 9$

$0 = (x + 3)(x + 3)$

$x = -3$

There is only one x-intercept: $(-3, 0)$. The parabola touches (but does not cross) the x-axis at this point.

(d) $y = x^2 - 4x + 5$

$0 = x^2 - 4x + 5$

$$x = \frac{4 \pm \sqrt{(-4)^2 - 4(5)}}{2}$$

$$= \frac{4 \pm \sqrt{-4}}{2}$$

Because these values are not real numbers, there are no x-intercepts. The parabola does not touch the x-axis.

Check Yourself 8

Find the x-intercepts for the graph of each equation.

(a) $y = x^2 + 3x - 5$ **(b)** $y = x^2 - 10x + 25$

In addition to searching for the x-intercepts, we should also look for the y-intercept. This is the point where the parabola passes through the y-axis. For $y = ax^2 + bx + c$, there will always be a y-intercept, and, since it occurs when $x = 0$, we have

$y = a(0)^2 + b(0) + c$

$y = c$

The y-intercept is $(0, c)$.

Example 9   **Finding the y-Intercept of a Parabola**

Find the y-intercept for the graph of each equation.

(a) $y = x^2 - x - 6$

Since $c = -6$, the y-intercept is $(0, -6)$.

(b) $y = x^2 + 6x$

Since $c = 0$, the y-intercept is $(0, 0)$. (Note that $(0, 0)$ is also one of the x-intercepts for this equation.)

Check Yourself 9

Find the *y*-intercept for the graph of each equation.

(a) $y = x^2 - 10x + 25$ (b) $y = x^2 - 2x$ (c) $y = x^2 + 3$

RECALL

For the equation
$y = ax^2 + bx + c$
the axis of symmetry has
equation
$x = \dfrac{-b}{2a}$.

In many applications involving quadratic graphs, the point on the graph of a parabola that is often desired is the vertex. As we saw earlier, the vertex must lie on the axis of symmetry. This means that, if we determine the equation of the axis of symmetry, we know the *x*-coordinate of the vertex. If we then substitute this *x*-value into the equation to determine the corresponding *y*-value, we will have the vertex.

Example 10   **A Biology Application**

An experimental drug is being tested on a bacteria colony. It is found that *t* days after the colony is treated, the number *N* of bacteria per cubic centimeter is given by the equation

$$N = 15t^2 - 100t + 700$$

(a) When will the bacteria colony be at a minimum?

We know that the graph of this equation is a parabola, and that the graph opens up. (Why?) So the vertex represents a minimum point for the graph.

To find the vertex, we first find the equation of the axis of symmetry.

$$t = \frac{-b}{2a} = \frac{-(-100)}{2(15)} = \frac{100}{30} = \frac{10}{3}$$

This means the bacteria colony will be at a minimum after $\dfrac{10}{3}$, or $3\dfrac{1}{3}$, days.

(b) What is that minimum number?

We now find *N* that corresponds to this *t*-value.

$$N = 15\left(\frac{10}{3}\right)^2 - 100\left(\frac{10}{3}\right) + 700$$

Using a calculator, we obtain $N = 533\dfrac{1}{3}$.

So the minimum number of bacteria is approximately 533.

NOTE

On a parabola, the vertex represents either a minimum point (if $a > 0$) or a maximum point (if $a < 0$).

NOTE

The vertex of the parabola is
$\left(3\dfrac{1}{3}, 533\dfrac{1}{3}\right)$.

Check Yourself 10

The number of people who are sick *t* days after the outbreak of a flu epidemic is given by the equation

$$P = -t^2 + 120t + 20$$

(a) On what day will the maximum number of people be sick?
(b) How many people will be sick on that day?

We summarize below the key features of parabolas. These features will aid in drawing the graphs of equations of the form $y = ax^2 + bx + c, a \neq 0$.

Property

The Graph of
$y = ax^2 + bx + c$,
$a \neq 0$

1. If $a > 0$, the parabola opens upward; if $a < 0$, the parabola opens downward.

2. The equation of the **axis of symmetry** is $x = \dfrac{-b}{2a}$.

3. When the x-value $\dfrac{-b}{2a}$ is determined, the **vertex** can be located by computing the corresponding y-value. If $a > 0$, the vertex represents a **minimum** point on the parabola. If $a < 0$, the vertex represents a **maximum** point on the parabola.

4. The x-values may be chosen on either side of the axis of symmetry to build a table of ordered-pair solutions.

5. The **x-intercepts** can be found by setting y equal to 0 and then solving for x. There can be 0, 1, or 2 x-intercepts.

6. The **y-intercept** is the point $(0, c)$. There will always be a y-intercept.

Check Yourself ANSWERS

1.

x	y
-2	6
-1	3
0	2
1	3
2	6

2.

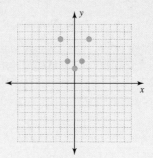

3. $y = x^2 + 2$

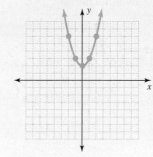

4. $y = x^2 + 4x$

x	y
-4	0
-3	-3
-2	-4
-1	-3
0	0

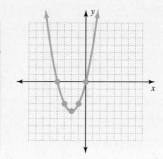

5. $y = x^2 - 4x + 3$

x	y
-1	8
0	3
1	0
2	-1
3	0
4	3

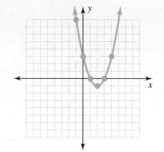

6. $y = -x^2 - 2x$

x	y
-3	-3
-2	0
-1	1
0	0
1	-3

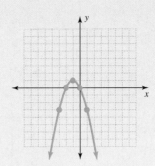

7. $y = x^2 - 7x + 4$; axis: $x = \dfrac{7}{2}$

x	y
1	-2
2	-6
3	-8
4	-8
5	-6
6	-2

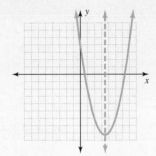

8. **(a)** $\left(\dfrac{-3 - \sqrt{29}}{2}, 0 \right)$ and $\left(\dfrac{-3 + \sqrt{29}}{2}, 0 \right)$; **(b)** $(5, 0)$

9. **(a)** $(0, 25)$; **(b)** $(0, 0)$; **(c)** $(0, 3)$

10. **(a)** Day 60; **(b)** 3,620 people

Reading Your Text

The following fill-in-the-blank exercises are designed to ensure that you understand some of the key vocabulary used in this section.

SECTION 10.4

(a) The graph of $y = ax^2 + bx + c$, in which $a \neq 0$, is always a curve called a _____.

(b) If the coefficient of x^2 is negative, the parabola opens _____.

(c) All the parabolas that we graphed are symmetric about a vertical line, which is called the _____.

(d) The point at which a parabola intersects its axis of symmetry is called the _____.

Name _____

Section _____ Date _____

Answers

1. _____

2. _____

3. _____

4. _____

< Objective 1 >

Graph each quadratic equation after completing the given table of values.

1. $y = x^2 + 1$

x	y
-2	
-1	
0	
1	
2	

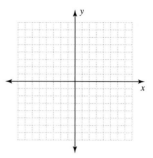

2. $y = x^2 - 2$

x	y
-2	
-1	
0	
1	
2	

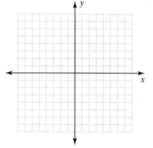

3. $y = x^2 - 4$

x	y
-2	
-1	
0	
1	
2	

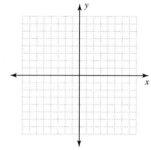

4. $y = x^2 + 3$

x	y
-2	
-1	
0	
1	
2	

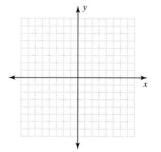

5. $y = x^2 - 4x$

x	y
0	
1	
2	
3	
4	

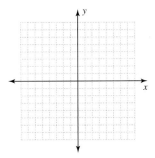

Answers

5. _____

6. _____

7. _____

8. _____

6. $y = x^2 + 2x$

x	y
−3	
−2	
−1	
0	
1	

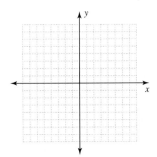

7. $y = x^2 + x$

x	y
−2	
−1	
0	
1	
2	

8. $y = x^2 - 3x$

x	y
−1	
0	
1	
2	
3	

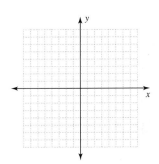

Answers

9. _____

10. _____

11. _____

12. _____

9. $y = x^2 - 2x - 3$

 > Videos

x	y
-1	
0	
1	
2	
3	

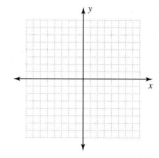

10. $y = x^2 - 5x + 6$

x	y
0	
1	
2	
3	
4	

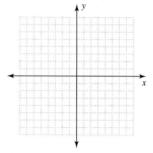

11. $y = x^2 - x - 6$

x	y
-1	
0	
1	
2	
3	

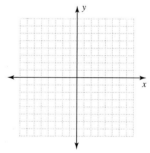

12. $y = x^2 + 3x - 4$

x	y
-4	
-3	
-2	
-1	
0	

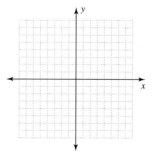

The Streeter/Hutchison Series in Mathematics   Beginning Algebra

13. $y = -x^2 + 2$

x	y
−2	
−1	
0	
1	
2	

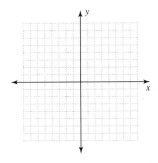

Answers

14. $y = -x^2 - 2$

x	y
−2	
−1	
0	
1	
2	

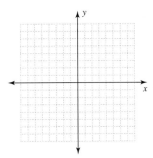

15. $y = -x^2 - 4x$ > Videos

x	y
−4	
−3	
−2	
−1	
0	

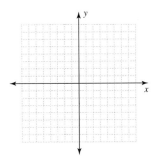

16. $y = -x^2 + 2x$

x	y
−1	
0	
1	
2	
3	

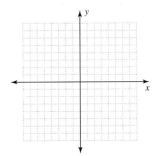

Answers

17. _____

18. _____

19. _____

20. _____

21. _____

22. _____

23. _____

24. _____

25. _____

26. _____

27. _____

28. _____

| Basic Skills | **Challenge Yourself** | Calculator/Computer | Career Applications | Above and Beyond |

Complete each statement with **never, sometimes,** *or* **always.**

17. The vertex of a parabola is _____ located on the axis of symmetry.

18. The vertex of a parabola is _____ the highest point on the graph.

19. The graph of $y = ax^2 + bx + c$ _____ intersects the x-axis.

20. The graph of $y = ax^2 + bx + c$ _____ has more than two x-intercepts.

21. The graph of $y = ax^2 + bx + c$ _____ intersects the y-axis.

22. The graph of $y = ax^2 + bx + c$ _____ intersects the y-axis more than once.

Match each graph with the correct equation.

(a) $y = -x^2 + 1$ **(b)** $y = 2x$ **(c)** $y = x^2 - 4x$ **(d)** $y = -x + 1$

(e) $y = -x^2 + 3x$ **(f)** $y = x^2 + 1$ **(g)** $y = x + 1$ **(h)** $y = 2x^2$

23.

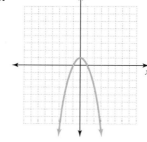

24.

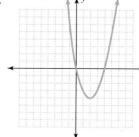

25.

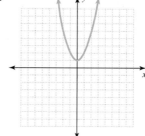

26.

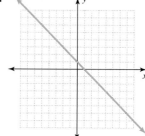

27.

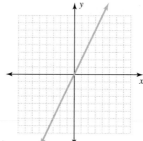

28.

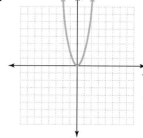

29.

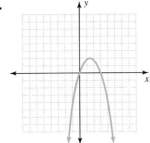

30.

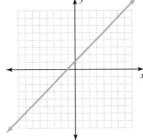

Answers

29. _____

30. _____

31. _____

32. _____

33. _____

< Objective 2 >

For each equation in exercises 31 to 34, identify the axis of symmetry, create a suitable table of values, and sketch the graph (including the axis of symmetry).

31. $y = x^2 + 4x$

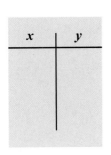

32. $y = x^2 - 5x + 3$

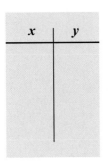

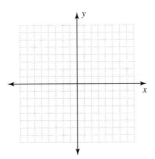

33. $y = -x^2 - 3x + 3$ > Videos

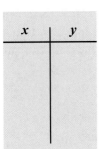

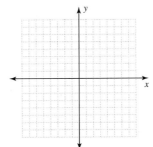

Answers

34. _____

35. _____

36. _____

37. _____

38. _____

39. _____

40. _____

41. _____

42. _____

43. _____

44. _____

45. _____

46. _____

47. _____

48. _____

49. _____

50. _____

51. _____

52. _____

53. _____

34. $y = -x^2 + 6x - 2$

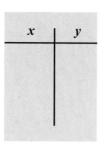

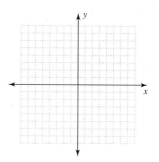

< Objective 3 >

Find the x-intercepts.

35. $y = x^2 - 2x - 8$ **36.** $y = x^2 + x - 6$

37. $y = x^2 + 3x - 6$ **38.** $y = x^2 - 5x - 4$

39. $y = x^2 - 2x + 1$ 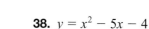   **40.** $y = x^2 - x + 2$

Find the y-intercept.

41. $y = x^2 - 5x - 4$ **42.** $y = -x^2 + 2x$

43. $y = 2x^2 - 12x + 5$ **44.** $y = x^2 - x + 2$

Find the vertex.

45. $y = x^2 - 4x$ **46.** $y = x^2 + 5x$

47. $y = x^2 + 3x - 4$ > Videos   **48.** $y = x^2 - 2x - 3$

49. $y = -x^2 + 2x$ **50.** $y = -x^2 - 3x + 3$

51. $y = 2x^2 - 12x + 5$ **52.** $y = -2x^2 - 10x + 7$

Basic Skills | Challenge Yourself | Calculator/Computer | **Career Applications** | Above and Beyond

53. ALLIED HEALTH The concentration C, in nanograms per milliliter (ng/mL), of digoxin, a medication prescribed for congestive heart failure, is given by the equation

$$C = -0.0015t^2 + 0.0845t + 0.7170$$

where t is the number of hours since the drug was taken orally.

(a) When will the drug's concentration be at a maximum?

(b) Determine the drug's maximum concentration in the bloodstream.

54. ALLIED HEALTH A patient's body temperature T, in degrees Fahrenheit, t hours after taking acetaminophen, a commonly prescribed analgesic, can be approximated by the equation

$$T = 0.4t^2 - 2.6t + 103$$

(a) When will the patient's temperature be at a minimum?

(b) Determine the patient's minimum temperature.

Answers

1. $y = x^2 + 1$

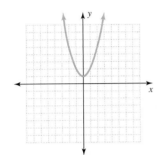

3. $y = x^2 - 4$

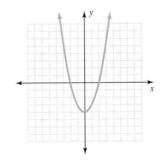

5. $y = x^2 - 4x$

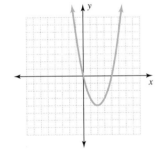

7. $y = x^2 + x$

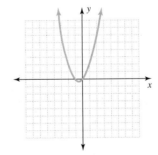

9. $y = x^2 - 2x - 3$

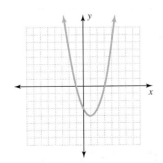

11. $y = x^2 - x - 6$

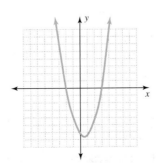

13. $y = -x^2 + 2$

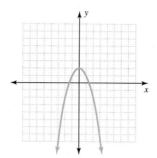

15. $y = -x^2 - 4x$

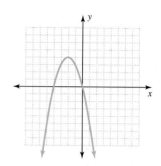

17. always   **19.** sometimes   **21.** always
23. f   **25.** a   **27.** b   **29.** e
31. Axis: $x = -2$

x	y
−4	0
−3	−3
−2	−4
−1	−3
0	0

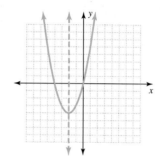

33. Axis: $x = -\dfrac{3}{2}$

x	y
−4	−1
−3	3
−2	5
−1	5
0	3
1	−1

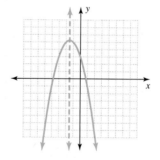

35. $(-2, 0), (4, 0)$ **37.** $\left(\dfrac{-3 - \sqrt{33}}{2}, 0\right), \left(\dfrac{-3 + \sqrt{33}}{2}, 0\right)$

39. $(1, 0)$ **41.** $(0, -4)$ **43.** $(0, 5)$ **45.** $(2, -4)$

47. $\left(-1\dfrac{1}{2}, -6\dfrac{1}{4}\right)$ **49.** $(1, 1)$ **51.** $(3, -13)$

53. (a) 28.2 hours; **(b)** 1.91 ng/mL

Definition/Procedure	Example	Reference

More on Quadratic Equations

Solving Quadratic Equations by Factoring

Section 10.1

1. Add or subtract the necessary terms on both sides of the equation so that the equation is in standard form (set equal to 0).
2. Factor the quadratic expression.
3. Set each factor equal to 0.
4. Solve the resulting equations to find the solutions.
5. Check each solution by substituting in the original equation.

Solve

$$x^2 + 7x = 30$$
$$x^2 + 7x - 30 = 0$$
$$(x + 10)(x - 3) = 0$$
$$x + 10 = 0 \quad \text{or} \quad x - 3 = 0$$

$x = -10$ and $x = 3$ are solutions.

p. 749

Square-Root Property

If $x^2 = k$, then $x = \sqrt{k}$ or $x = -\sqrt{k}$.
If $(x - h)^2 = k$, then $x = h \pm \sqrt{k}$.

Solve

$$(x - 3)^2 = 5$$
$$x - 3 = \pm\sqrt{5}$$
$$x = 3 \pm \sqrt{5}$$

p. 750

Completing the Square

Completing the Square

Section 10.2

To solve a quadratic equation by completing the square:

1. Write the equation in the form

   $$ax^2 + bx = k$$

   so that the variable terms are on the left side and the constant is on the right side.
2. If the leading coefficient of x^2 is not 1, divide both sides by that coefficient.
3. Add the square of one-half the coefficient of x to both sides of the equation.
4. The left side of the equation is now a perfect-square trinomial. Factor and solve as before.

Solve

$$2x^2 + 2x - 1 = 0$$
$$2x^2 + 2x = 1$$
$$x^2 + x = \frac{1}{2}$$
$$x^2 + x + \left(\frac{1}{2}\right)^2 = \frac{1}{2} + \left(\frac{1}{2}\right)^2$$
$$\left(x + \frac{1}{2}\right)^2 = \frac{3}{4}$$
$$x + \frac{1}{2} = \pm\sqrt{\frac{3}{4}} = \pm\frac{\sqrt{3}}{2}$$
$$x = \frac{-1 \pm \sqrt{3}}{2}$$

p. 764

Continued

Definition/Procedure	Example	Reference

The Quadratic Formula

The Quadratic Formula

To solve an equation by formula:

1. Rewrite the equation in standard form.

   $ax^2 + bx + c = 0$

2. If a common factor exists, divide both sides of the equation by that common factor.

3. Identify the coefficients a, b, and c.

4. Substitute the values for a, b, and c into the quadratic formula.

   $$x = \frac{-b \pm \sqrt{b^2 - 4ac}}{2a}$$

5. Simplify the right side of the expression formed in step 4 to write the solutions for the original equation.

Example:

Solve

$x^2 - 2x = 4$

Write the equation as

$x^2 - 2x - 4 = 0$

$a = 1 \qquad b = -2 \qquad c = -4$

$$x = \frac{-(-2) \pm \sqrt{(-2)^2 - 4(1)(-4)}}{2(1)}$$

$$= \frac{2 \pm \sqrt{20}}{2}$$

$$= \frac{2 \pm 2\sqrt{5}}{2} = \frac{2(1 \pm \sqrt{5})}{2}$$

$$= 1 \pm \sqrt{5}$$

Reference: Section 10.3 — p. 774

Graphing Quadratic Equations

To graph equations of the form

$y = ax^2 + bx + c$

1. Find the axis of symmetry using

   $$x = \frac{-b}{2a}$$

2. Form a table of values by choosing x-values that span the axis of symmetry.

3. Plot the points from the table of values.

4. Draw a smooth curve between the points.

The graph of a quadratic equation will always be a parabola. The parabola opens upward if a, the coefficient of the x^2-term, is positive. The parabola opens downward if a is negative.

Example:

$y = x^2 - 4x$

$$x = \frac{-(-4)}{2(1)} = \frac{4}{2} = 2$$

Choose x-values to the right and left of 2.

x	y
-1	5
0	0
1	-3
2	-4
3	-3
4	0
5	5

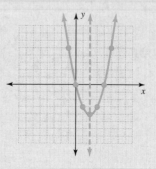

Reference: Section 10.4 — p. 788

Definition/Procedure	Example	Reference
The vertex can be found by substituting the value $x = \dfrac{-b}{2a}$ into the original equation and computing y.	To find the vertex for $$y = x^2 - 4x - 6$$ $$x = \frac{-b}{2a} = \frac{-(-4)}{2(1)} = \frac{4}{2} = 2$$ Then $$y = (2)^2 - 4(2) - 6 = 4 - 8 - 6 = -10$$ The vertex is $(2, -10)$.	*p.* 788
The x-intercepts of a parabola are the points where the parabola intersects the x-axis. They may be found by setting y equal to 0, and then solving for x.	To find the x-intercepts of $$y = x^2 - 4x - 6$$ set $y = 0$ and solve. $$0 = x^2 - 4x - 6$$ $$x = \frac{4 \pm \sqrt{(-4)^2 - 4(-6)}}{2}$$ $$= \frac{4 \pm \sqrt{40}}{2} = \frac{4 \pm 2\sqrt{10}}{2}$$ $$= 2 \pm \sqrt{10}$$ The x-intercepts are $(2 - \sqrt{10}, 0)$ and $(2 + \sqrt{10}, 0)$.	*p.* 789
The y-intercept is $(0, c)$.	The y-intercept of $$y = x^2 - 4x - 6$$ is $(0, -6)$.	*p.* 790

This summary exercise set is provided to give you practice with each of the objectives of this chapter. Each exercise is keyed to the appropriate chapter section. When you are finished, you can check your answers to the odd-numbered exercises against those presented in the back of the text. If you have difficulty with any of these questions, go back and reread the examples from that section. Your instructor will give you guidelines on how best to use these exercises in your instructional setting.

10.1 *Solve each equation by the square-root method.*

1. $x^2 = 10$

2. $x^2 = 48$

3. $x^2 - 20 = 0$

4. $x^2 + 2 = 8$

5. $(x - 1)^2 = 5$

6. $(x + 2)^2 = 8$

7. $(x + 3)^2 = 5$

8. $64x^2 = 25$

9. $4x^2 = 27$

10. $9x^2 = 20$

11. $25x^2 = 7$

12. $7x^2 = 3$

10.2 *Solve each equation by completing the square.*

13. $x^2 - 3x - 10 = 0$

14. $x^2 - 8x + 15 = 0$

15. $x^2 - 5x + 2 = 0$

16. $x^2 - 2x - 2 = 0$

17. $x^2 - 4x - 4 = 0$

18. $x^2 + 3x = 7$

19. $x^2 - 4x = -2$

20. $x^2 + 3x = 5$

21. $x^2 - x = 7$

22. $2x^2 + 6x = 12$

23. $2x^2 - 4x - 7 = 0$

24. $3x^2 + 5x + 1 = 0$

10.3 *Solve each equation using the quadratic formula.*

25. $x^2 - 5x - 14 = 0$

26. $x^2 - 8x + 16 = 0$

27. $x^2 + 5x - 3 = 0$

28. $x^2 - 7x - 1 = 0$

29. $x^2 - 6x + 1 = 0$

30. $x^2 - 3x + 5 = 0$

31. $3x^2 - 4x = 2$

32. $2x - 3 = \dfrac{3}{x}$

33. $(x - 1)(x + 4) = 3$

34. $x^2 - 5x + 7 = 5$

35. $2x^2 - 8x = 12$

36. $5x^2 = 15 - 15x$

Solve by any method in this chapter.

37. $5x^2 = 3x$

38. $(2x - 3)(x + 5) = -11$

39. $(x - 1)^2 = 10$

40. $2x^2 = 7$

41. $2x^2 = 5x + 4$

42. $2x^2 - 4x = 30$

43. $2x^2 = 5x + 7$

44. $3x^2 - 4x = 2$

45. $3x^2 + 6x - 15 = 0$

46. $x^2 - 3x = 2(x + 5)$

47. $x - 2 = \dfrac{2}{x}$

48. $3x + 1 = \dfrac{5}{x}$

10.4 *Graph each quadratic equation after completing the table of values.*

49. $y = x^2 + 3$

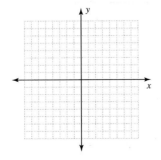

50. $y = x^2 - 2$

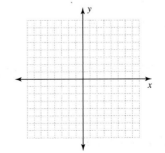

51. $y = x^2 - 3x$

x	y
−1	
0	
1	
2	
3	

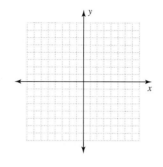

52. $y = x^2 + 4x$

x	y
−4	
−3	
−2	
−1	
0	

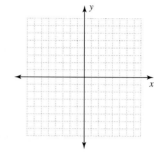

53. $y = x^2 - x - 2$

x	y
−1	
0	
1	
2	
3	

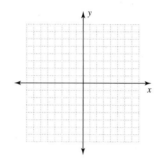

54. $y = x^2 - 4x + 3$

x	y
0	
1	
2	
3	
4	

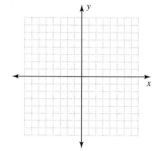

55. $y = x^2 + 2x - 3$

x	y
−3	
−2	
−1	
0	
1	

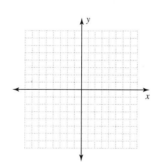

56. $y = 2x^2$

x	y
−2	
−1	
0	
1	
2	

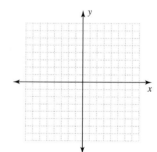

57. $y = 2x^2 - 3$

x	y
−2	
−1	
0	
1	
2	

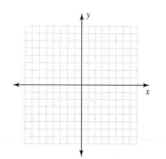

58. $y = -x^2 + 3$

x	y
−2	
−1	
0	
1	
2	

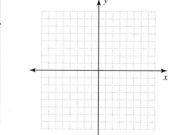

59. $y = -x^2 - 2$

x	y
−2	
−1	
0	
1	
2	

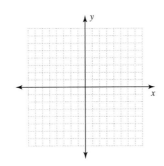

60. $y = -x^2 + 4x$

x	y
0	
1	
2	
3	
4	

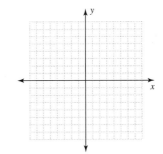

For each equation, identify the axis of symmetry, create a suitable table of values, and sketch the graph.

61. $y = x^2 + 6x + 3$

x	y
−5	
−4	
−3	
−2	
−1	

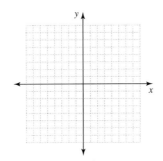

62. $y = -x^2 - 4x + 3$

x	y
−4	
−3	
−2	
−1	
0	

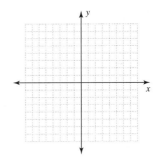

63. $y = -x^2 + 5x + 2$

x	y
0	
1	
2	
3	
4	
5	

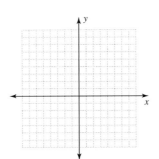

64. $y = x^2 - 3x + 4$

x	y
−1	
0	
1	
2	
3	
4	

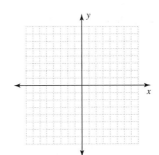

For each equation, find the x-intercepts and the y-intercept of the graph.

65. $y = x^2 + 2x - 15$

66. $y = x^2 - 3x - 4$

67. $y = x^2 + 6x + 3$

68. $y = x^2 - 3x + 4$

69. $y = -x^2 + 2x - 3$

70. $y = -x^2 - 4x + 1$

For each equation, find the vertex of the graph.

71. $y = x^2 + 2x - 15$

72. $y = x^2 - 3x - 4$

73. $y = x^2 + 6x + 3$

74. $y = x^2 - 3x + 4$

75. $y = -x^2 + 2x - 3$

76. $y = -x^2 - 4x + 1$

Solve each application. Round results to the nearest hundredth.

77. GEOMETRY The longer leg of a right triangle is 3 cm less than twice the shorter leg. The hypotenuse is 18 cm long. Find the length of each leg.

78. CONSTRUCTION A rectangular garden is to be surrounded by a walkway of constant width. The garden's dimensions are 15 ft by 24 ft. The total area, garden plus walkway, is to be 750 ft². What must be the width of the walkway?

79. SCIENCE AND MEDICINE If a ball is thrown vertically upward from a height of 5 ft, with an initial velocity of 80 ft/s, its height h after t seconds is given by $h = -16t^2 + 80t + 5$. Find the time it takes for the ball to reach a height of 75 ft.

80. SCIENCE AND MEDICINE If a ball is thrown vertically upward from a height of 5 ft, with an initial velocity of 80 ft/s, its height h after t seconds is given by $h = -16t^2 + 80t + 5$. How long does it take the ball to return to the ground?

81. BUSINESS AND FINANCE Suppose that the cost C, in dollars, of producing x items is given by $C = 3x^2 - 50x + 1,800$. How many items can be produced for $13,000?

82. BUSINESS AND FINANCE The demand equation for a certain computer chip is given by

$D = -2p + 25$

The supply equation is predicted to be

$S = -p^2 + 16p + 5$

Find the equilibrium price.

The purpose of this self-test is to help you assess your progress so that you can find concepts that you need to review before the next exam. Allow yourself about an hour to take this test. At the end of that hour, check your answers against those given in the back of this text. If you miss any, go back to the appropriate section to reread the examples until you have mastered that particular concept.

Solve each equation by completing the square.

1. $x^2 + 2x - 5 = 0$ **2.** $2x^2 - 5x + 1 = 0$

Solve each equation by using the quadratic formula.

3. $x^2 - 2x - 3 = 0$ **4.** $2x^2 = 2x + 5$

Graph each quadratic equation after completing the given table of values.

5. $y = x^2 - 2x$

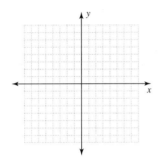

x	y
-1	
0	
1	
2	
3	

6. $y = x^2 + x - 2$

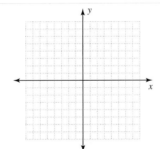

x	y
-2	
-1	
0	
1	
2	

7. $y = x^2 + 4$

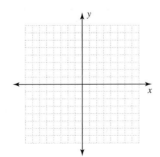

x	y
-2	
-1	
0	
1	
2	

Name _____

Section _____ Date _____

Answers

1. _____

2. _____

3. _____

4. _____

5. _____

6. _____

7. _____

Answers

8. _____

9. _____

10. _____

11. _____

12. _____

13. _____

14. _____

15. _____

16. _____

17. _____

18. _____

19. _____

20. _____

Solve each equation.

8. $x^2 = 15$ **9.** $(x - 1)^2 = 7$

Solve each equation by completing the square.

10. $x^2 - 2x - 8 = 0$ **11.** $x^2 + 3x - 1 = 0$

Give the axis of symmetry and the x-intercepts.

12. $y = x^2 - 4x - 21$

Solve each equation by using the quadratic formula.

13. $x^2 - 6x + 9 = 0$ **14.** $2x - 1 = \dfrac{4}{x}$

Solve each application. Round results to the nearest hundredth.

15. **GEOMETRY** If the length of a rectangle is 4 ft less than 3 times its width and the area of the rectangle is 96 ft², what are the dimensions of the rectangle?

16. **SCIENCE AND MEDICINE** If a pebble is dropped toward a pond from the top of a 180-ft building, its height h after t seconds is given by $h = -16t^2 + 180$. How long does it take the pebble to pass through a height of 30 ft?

Solve each equation.

17. $x^2 - 8 = 0$ **18.** $9x^2 = 10$

Solve each equation by using the quadratic formula.

19. $x^2 - 5x = 2$ **20.** $(x - 1)(x + 3) = 2$

chapter 10 > Make the Connection

Activity 10 ::
The Gravity Model

Each activity in this text is designed to either enhance your understanding of the topics of the preceding chapter, provide you with a mathematical extension of those topics, or both. The activities can be undertaken by one student, but they are better suited for a small-group project. Occasionally it is only through discussion that different facets of the activity become apparent.

This activity requires two to three people and an open area outside. You will also need a ball (preferably a baseball or some other small, heavy ball), a stopwatch, and a tape measure.

1. Designate one person as the "ball thrower."

2. The ball thrower should throw the ball straight up into the air, as high as possible. While this is happening, another group member should take note of exactly where the ball thrower releases the ball. Be careful that no one gets hit by the ball as it comes back down.

3. Use the tape measure to determine the height above the ground that the thrower released the ball. Convert your measurement to feet, using decimals to indicate parts of a foot. For example, 3 inches is 0.25 ft. Record this as the "initial height."

4. Repeat steps 2 and 3 to ensure that you have the correct initial height.

5. The thrower should now throw the ball straight up, as high as possible. The person with the stopwatch should time the ball until it lands.

6. Repeat step 5 twice more, recording the time.

7. Take the average (mean) of your three recorded times. We will use this number as the "hang time" of the ball for the remainder of this activity.

According to Sir Isaac Newton (1642–1727), the height of an object with initial velocity v_0 and initial height s_0 is given by

$$s = -16t^2 + v_0 t + s_0$$

in which the height s is measured in feet and t represents the time, in seconds.

The initial velocity v_0 is positive when the object is thrown upward, and it is negative when the object is thrown downward.

For example, the height of a ball thrown upward from a height of 4 ft with an initial velocity of 50 ft/s is given by

$$s = -16t^2 + 50t + 4$$

To determine the height of the ball 2 seconds after release, we evaluate the polynomial for $t = 2$.

$$= -16(2)^2 + 50(2) + 4$$
$$= 40$$

The ball is 40 feet high after 2 seconds.

Take note that the height of the ball when it lands is zero.

8. Substitute the time found in step 7 for t, your initial height for s_0, and 0 for s in the height equation. Solve the resulting equation for the initial velocity, v_0.

9. Now write your height equation using the initial velocity found in step 8 along with the initial height. Your equation should have two variables, s and t.

10. What is the height of the ball after 1 second?

11. The maximum height of the ball will be attained when $t = \dfrac{v_0}{32}$. Determine this time and find the maximum height of the thrown ball.

12. If a ball is dropped from a height of 256 feet, how long will it take to hit the ground? (*Hint:* If a ball is dropped, its initial velocity is 0 ft/s.)

The following exercises are presented to help you review concepts from earlier chapters. This is meant as a review and not as a comprehensive exam. The answers are presented in the back of the text. Section references accompany the answers. If you have difficulty with any of these exercises, be certain to at least read through the summary related to those sections.

Name _____

Section _____   Date _____

Simplify.

1. $6x^2y - 4xy^2 + 5x^2y - 2xy^2$

2. $(3x^2 + 2x - 5) - (2x^2 - 3x + 2)$

Evaluate each expression when $x = 2$, $y = -3$, and $z = 4$.

3. $4x^2y - 3z^2y^2$

4. $-3x^2y^2z^2 - 2xyz$

5. Solve: $4x - 2(3x - 5) = 8$.

6. Solve the inequality $4x + 15 > 2x + 19$.

Perform the indicated operations.

7. $3xy(2x^2 - x + 5)$

8. $(2x + 5)(3x - 2)$

9. $(3x + 4y)(3x - 4y)$

Factor completely.

10. $16x^2y^2 - 8xy^3$

11. $8x^2 - 2x - 15$

12. $25x^2 - 16y^2$

Perform the indicated operations.

13. $\dfrac{7}{4x + 8} - \dfrac{5}{7x + 14}$

14. $\dfrac{5x + 5}{x - 2} \cdot \dfrac{x^2 - 4x + 4}{x^2 - 1}$

15. $\dfrac{3x^2 + 8x - 3}{15x^2} + \dfrac{3x - 1}{5x^2}$

Answers

1. _____

2. _____

3. _____

4. _____

5. _____

6. _____

7. _____

8. _____

9. _____

10. _____

11. _____

12. _____

13. _____

14. _____

15. _____

Answers

16. _____

17. _____

18. _____

19. _____

20. _____

21. _____

22. _____

23. _____

24. _____

25. _____

Graph each equation.

16. $3x - 2y = 6$

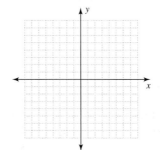

17. $y = 4x - 5$

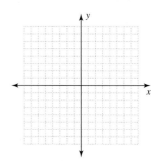

18. Find the slope of the line through the points $(2, 9)$ and $(-1, -6)$.

19. Given that the slope of a line is 2 and the *y*-intercept is $(0, -5)$, write the equation of the line.

20. Graph the inequality $x + 2y < 6$.

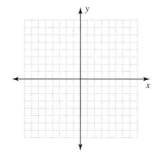

Solve each system. If a unique solution does not exist, state whether the system is inconsistent or dependent.

21. $2x - 3y = 6$
$\quad\ x - 3y = 2$

22. $2x + y = 4$
$\quad\quad\ y = 2x - 8$

23. $5x + 2y = 8$
$\quad\ x - 4y = 17$

24. $2x - 6y = 8$
$\quad\quad x = 3y + 4$

Solve each application. Be sure to show the system of equations used for your solution.

25. One number is 4 less than 5 times another. If the sum of the numbers is 26, what are the two numbers?

26. Receipts for a concert attended by 450 people were $2,775. If reserved-seat tickets were $7 and general admission tickets were $4, how many of each type of ticket were sold?

27. A chemist has a 30% acid solution and a 60% solution already prepared. How much of each of the two solutions should be mixed to form 300 mL of a 50% solution?

Evaluate each root, if possible.

28. $\sqrt{169}$

29. $-\sqrt{169}$

30. $\sqrt{-169}$

31. $\sqrt[3]{-64}$

Simplify each radical expression.

32. $\sqrt{12} + 3\sqrt{27} - \sqrt{75}$

33. $3\sqrt{2a} \cdot 5\sqrt{6a}$

34. $(\sqrt{2} - 5)(\sqrt{2} + 3)$

35. $\dfrac{8 - \sqrt{32}}{4}$

Solve each equation.

36. $x^2 - 72 = 0$

37. $x^2 + 6x - 3 = 0$

38. $2x^2 - 3x = 2(x + 1)$

Graph each quadratic equation.

39. $y = x^2 - 2$

40. $y = x^2 - 4x$

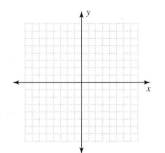

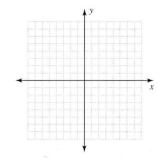

Answers

26. _____

27. _____

28. _____

29. _____

30. _____

31. _____

32. _____

33. _____

34. _____

35. _____

36. _____

37. _____

38. _____

39. _____

40. _____

Answers

41. _____

42. _____

Solve each application.

41. The equation $h = -16t^2 - 64t + 250$ gives the height of a ball, thrown downward from the top of a 250-ft building with an initial velocity of 64 ft/s. Find the time it takes for the ball to reach a height of 100 ft. Round your answer to the nearest thousandth.

42. The demand equation for a certain type of printer is given by

$$D = -120p + 16,000$$

The supply equation is predicted to be

$$S = -p^2 + 260p - 9,000$$

Find the equilibrium price.

The Streeter/Hutchison Series in Mathematics   Beginning Algebra

Evaluate each expression.

1. $|-25| - |-11|$

2. $16 + (-22)$

3. $(-41) - (-15)$

4. $(-5)(-3)(-7)$

5. $\dfrac{3(-2) - 8}{-7 - (-4)(3)}$

6. $6 - 2^3 \cdot 5$

Evaluate the expressions for the given values of the variables.

7. $b^2 - 4ac$ for $a = -3$, $b = -4$, and $c = 2$ **8.** $-x^2 - 7x - 3$ for $x = -2$

9. Write the expression $9 \cdot p \cdot p \cdot p \cdot q \cdot q \cdot q \cdot q \cdot q$ in exponential form.

Simplify the expressions using the properties of exponents. Write all answers using positive exponents only.

10. $z^{-11}z^5$

11. $(5c^8 d^7)^2$

12. $\dfrac{4x^8 y^5 z^3}{2x^6 y^9 z^7}$

Perform the indicated operations. Write each answer in simplified form.

13. $2x(x + 3) + 5$

14. $(6x^2 - 3x - 20) - 2(4x^2 - 16x + 11)$

15. $(7x - 9)(4x + 5)$

16. $(3x + 4y)(3x - 4y)$

17. $(5x - 2y)^2$

18. $x(x - 3y) - 2y(y + 6x)$

Solve each equation.

19. $5x - 9 = -7x - 3$

20. $2x - 3(x - 2) = 8$

21. $\dfrac{5 - x}{-2} = 3x$

22. Solve the inequality $4x - 3 > 6x - 2$.

Name _____

Section _____ Date _____

Answers

1. _____ 2. _____

3. _____ 4. _____

5. _____ 6. _____

7. _____ 8. _____

9. _____

10. _____

11. _____

12. _____

13. _____

14. _____

15. _____

16. _____

17. _____

18. _____

19. _____

20. _____

21. _____

22. _____

819

Answers

23. _____

24. _____

25. _____

26. _____

27. _____

28. _____

29. _____

30. _____

31. _____

32. _____

33. _____

34. _____

35. _____

36. _____

37. _____

38. _____

Factor each expression completely.

23. $10x^2 - 490$

24. $x^2 - 12x + 36$

25. $4p^2 - p - 18$

26. $3xy + 3xz - 5y - 5z$

Simplify each expression.

27. $\dfrac{2y - 36}{3y^2 - 54y}$

28. $\dfrac{4z}{z + 8} + \dfrac{32}{z + 8}$

29. $\dfrac{6a}{a^2 - 9} - \dfrac{5a}{a^2 + a - 6}$

30. $\dfrac{y^2 + y - 2}{y + 5} \cdot \dfrac{3y + 3}{9y - 9}$

31. $\dfrac{x^2 - 3x - 10}{3x} \div \dfrac{5x - 25}{15x^2}$

Solve each equation by the indicated method.

32. $3x^2 + 5x - 2 = 0$ by factoring

33. $x^2 + x - 3 = 0$ by using the quadratic formula

34. $x^2 + 6x = 5$ by completing the square

35. Solve the equation $\dfrac{1}{x - 1} + \dfrac{x + 1}{x^2 + 2x - 3} = \dfrac{1}{x + 3}$.

36. Find the slope of the line through the points $(2, -3)$ and $(5, 9)$.

37. Find the slope of the line whose equation is $3x - 4y = 12$.

38. Find the equation of the line that passes through the point $(4, -2)$ and is parallel to the line $2x + y = 6$.

39. Graph the line whose equation is $4x + 5y = 20$.

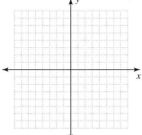

39. _____

40. _____

41. _____

42. _____

43. _____

44. _____

45. _____

46. _____

47. _____

In exercises 40–43, perform the indicated operations and simplify the result.

40. $-\sqrt{\dfrac{64}{25}}$

41. $\sqrt{18x^5 y^6}$

42. $3\sqrt{20} - 2\sqrt{125}$

43. $(\sqrt{5} + 2)(\sqrt{5} - 8)$

44. Solve the system of equations.

$2x + 3y = 4$

$4x - 2y = 8$

45. Determine the equation of the graphed line.

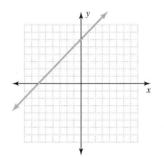

46. The length of a rectangle is 2 cm more than 3 times the width. The perimeter is 44 cm. Find the dimensions of the rectangle.

47. Find the equation of the line that passes through the points $(-1, -3)$ and $(2, 6)$.

Answers

48. _____

49. _____

50. _____

48. One number is 3 more than 6 times another. If the sum of the numbers is 38, find the two numbers.

49. A store marks up items to make a 30% profit. If an item sells for $3.25, what does it cost before the markup?

50. Graph the equation $y = x^2 - 3x + 2$.

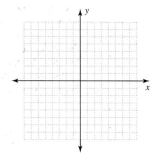

Answers to Prerequisite Tests, Reading Your Text, Summary Exercises, Self-Tests, and Cumulative Reviews

Prerequisite Test for Chapter 1

1. $10 - 8 = 2$ **2.** $3 + 5 \times 6 = 33$ **3.** $\dfrac{1}{12}$ **4.** $\dfrac{8}{37}$
5. 1 **6.** 1 **7.** 1 **8.** 23 **9.** 64 **10.** 26
11. $14,117.65 **12.** $1.55

Reading Your Text

Section 1.1 (a) commutative; (b) parentheses; (c) associative; (d) area
Section 1.2 (a) negative; (b) absolute; (c) zero; (d) opposite
Section 1.3 (a) negative; (b) positive; (c) one; (d) zero
Section 1.4 (a) variables; (b) difference; (c) multiplication; (d) expression
Section 1.5 (a) variable; (b) evaluating; (c) operations; (d) principal
Section 1.6 (a) term; (b) coefficient; (c) first; (d) one
Section 1.7 (a) add; (b) base; (c) multiply; (d) subtract

Summary Exercises for Chapter 1

1. Associative property of addition
3. Associative property of multiplication **5.** $72 = 72$
7. $20 = 20$ **9.** $80 = 80$ **11.** $3 \cdot 7 + 3 \cdot 4$
13. $\dfrac{1}{2} \cdot 5 + \dfrac{1}{2} \cdot 8$ **15.** -11 **17.** 0 **19.** -18
21. -4 **23.** -5 **25.** 17 **27.** 0 **29.** 3
31. 157 **33.** -96 **35.** 4.24 **37.** 2.048
39. 29.75 **41.** -70 **43.** 45 **45.** 0 **47.** $-\dfrac{3}{2}$
49. 5 **51.** 9 **53.** -4 **55.** -1 **57.** -7
59. $y + 5$ **61.** $8a$ **63.** $5mn$ **65.** $17x + 3$
67. Yes **69.** No **71.** 3 **73.** 80 **75.** 41 **77.** 20
79. 324 **81.** 11 **83.** -7 **85.** 15 **87.** 19
89. 25 **91.** 1 **93.** -3 **95.** $4a^3, -3a^2$
97. $5m^2, -4m^2, m^2$ **99.** $12c$ **101.** $2a$ **103.** $3xy$
105. $19a + b$ **107.** $3x^3 + 9x^2$ **109.** $10a^3$ **111.** x^7
113. x **115.** $2p^2$ **117.** $5m^5n^2$ **119.** $8p^2q^2$ **121.** $20x^7$
123. $24x^5y^5$ **125.** $12x^6y$ **127.** $25 - x$ **129.** $2x + 5$
131. $6n - 7$ **133.** $0.1x + 0.25q$

Self-Test for Chapter 1

1. -13 **2.** -3 **3.** -21 **4.** 1 **5.** -6
6. -21 **7.** 9 **8.** 0 **9.** -40 **10.** 63 **11.** -27
12. -24 **13.** -25 **14.** 3 **15.** -5 **16.** Undefined
17. 17 **18.** 65 **19.** 144 **20.** -9 **21.** $13a$
22. $19x + 5y$ **23.** a^{14} **24.** $8x^7y^3$ **25.** $3x^6$
26. $4ab^3$ **27.** x^9 **28.** $8a^2$ **29.** $a - 5$ **30.** $6m$
31. $4(m + n)$ **32.** $\dfrac{a + b}{3}$ **33.** -4
34. Commutative property of multiplication **35.** Distributive property **36.** Associative property of addition **37.** 21
38. $20x + 12$ **39.** Not an expression **40.** Expression
41. $2x - 8$ **42.** $2w + 4$

Prerequisite Test for Chapter 2

1. $8x + 12$ **2.** $-6x + 16$ **3.** $-\dfrac{1}{10}$ **4.** $-\dfrac{4}{3}$ **5.** 1
6. 1 **7.** -49 **8.** 49 **9.** $4x^2 - 3x$ **10.** $x + 2y$
11. $239.40 **12.** $15.96

Reading Your Text

Section 2.1 (a) equation; (b) solution; (c) one; (d) solutions
Section 2.2 (a) equivalent; (b) multiplication; (c) multiplying; (d) reciprocal
Section 2.3 (a) isolate; (b) addition; (c) original; (d) denominators
Section 2.4 (a) formula; (b) coefficient; (c) equal; (d) original
Section 2.5 (a) geometry; (b) Mixture; (c) amount; (d) identifying
Section 2.6 (a) inequality; (b) smaller; (c) solution; (d) negative

Summary Exercises for Chapter 2

1. Yes **3.** Yes **5.** No **7.** 2 **9.** -7 **11.** 5
13. 1 **15.** -7 **17.** 7 **19.** -4 **21.** 32 **23.** 27
25. 3 **27.** -2 **29.** $\dfrac{7}{2}$ **31.** 18 **33.** 6 **35.** $\dfrac{2}{5}$
37. 6 **39.** 6 **41.** 4 **43.** 5 **45.** $\dfrac{1}{2}$ **47.** $\dfrac{P - 2W}{2}$ or $\dfrac{P}{2} - W$ **49.** $\dfrac{2A}{b}$ **51.** $mq + p$ **53.** 8 **55.** 17, 19, 21
57. Susan: 7 yr; Larry: 9 yr; Nathan: 14 yr **59.** 22%
61. $18,200 **63.** $2,800 **65.** 6.5%
67. Before: $3,150; after: $3,276 **69.** 500 s **71.** $114.50
73. $x > -5$
75. $x \geq -3$
77. $x \leq -15$
79. $x \geq 3$
81. $x \leq 7$
83. $x > -6$

Self-Test for Chapter 2

1. No **2.** Yes **3.** 11 **4.** 12 **5.** 7
6. 7 **7.** -12 **8.** 25 **9.** 3 **10.** 4 **11.** $-\dfrac{2}{3}$
12. $-\dfrac{9}{4}$ **13.** $\dfrac{C}{2\pi}$ **14.** $\dfrac{3V}{B}$ **15.** $\dfrac{6 - 3x}{2}$
16. $x \leq 14$
17. $x < -4$

18. $x \geq \dfrac{4}{3}$

19. $x > -1$

20. 7   **21.** 21, 22, 23   **22.** Juwan: 6; Jan: 12; Rick: 17
23. 10 in., 21 in.   **24.** 5%   **25.** \$35,000

Cumulative Review Chapters 1–2

[1.2–1.3] 1. 4   **2.** -12 **3.** 8   **4.** 3   **5.** -18
6. 44   **7.** -5 **8.** 10   **9.** 0   **10.** Undefined
[1.5] 11. 20   **12.** -11 **13.** 27   **14.** -28 **15.** -4
16. 2   **[1.6–1.7] 17.** $3x^2y$ **18.** $6x^4 - 10x^3y$
19. $x - 2y + 3$ **20.** $12x^2 + 3x$ **[2.1–2.3] 21.** 5
22. -24 **23.** $\dfrac{5}{4}$ **24.** $-\dfrac{2}{5}$ **25.** 5

[2.4] 26. $\dfrac{I}{Pt}$ **27.** $\dfrac{2A}{b}$ **28.** $\dfrac{c - ax}{b}$

[2.6] 29. $x < 3$

30. $x \leq -\dfrac{3}{2}$

31. $x > 4$

32. $x \geq \dfrac{4}{3}$

[2.4–2.5] 33. 13   **34.** 42, 43   **35.** 7   **36.** \$420
37. 5 cm, 17 cm   **38.** 8 in., 13 in., 16 in.   **39.** 2.5%
40. 7.5%

Prerequisite Test for Chapter 3

1. 625   **2.** 432   **3.** -81 **4.** 81   **5.** 230,000
6. 0.000023   **7.** $-x + 8$ **8.** $2x + 4y$ **9.** 0   **10.** 58
11. 9, 11   **12.** 0.792 watts

Reading Your Text

Section 3.1 **(a)** multiplication; **(b)** term; **(c)** coefficient; **(d)** degree
Section 3.2 **(a)** one; **(b)** positive; **(c)** exponents; **(d)** ten
Section 3.3 **(a)** plus; **(b)** sign; **(c)** first; **(d)** like
Section 3.4 **(a)** exponents; **(b)** first; **(c)** outer; **(d)** three
Section 3.5 **(a)** exponents; **(b)** term; **(c)** degree; **(d)** zero

Summary Exercises for Chapter 3

1. x^7 **3.** x **5.** $2p^2$ **7.** $5m^5n^2$ **9.** $8p^2q^2$ **11.** $4a^2b^2$
13. $72x^{12}y^8$ **15.** x **17.** $27y^{12}$ **19.** -4 **21.** -19
23. Binomial   **25.** Trinomial   **27.** Binomial   **29.** $9x, 1$
31. $x + 5, 1$ **33.** $7x^6 + 9x^4 - 3x, 6$ **35.** 1   **37.** 1
39. $\dfrac{1}{3^3}$ **41.** $\dfrac{4}{x^4}$ **43.** $\dfrac{1}{m^2}$ **45.** $\dfrac{x^5}{y^5}$ **47.** $\dfrac{1}{a^{18}}$
49. 5.1×10^4 cps   **51.** 3.22×10^9 **53.** 2×10^{17}
55. $21a^2 - 2a$ **57.** $5y^3 + 4y$ **59.** $5x^2 + 3x + 10$
61. $x - 2$ **63.** $-9w^2 - 10w$ **65.** $9b^2 + 8b - 2$
67. $2x^2 - 2x - 9$ **69.** $5a^5$ **71.** $54p^5$ **73.** $15x - 40$
75. $-10r^3s^2 + 25r^2s^2$ **77.** $x^2 + 9x + 20$ **79.** $a^2 - 49b^2$
81. $a^2 + 7ab + 12b^2$ **83.** $6x^2 - 19xy + 15y^2$
85. $y^3 - y + 6$ **87.** $x^3 - 8$ **89.** $2x^3 - 2x^2 - 60x$
91. $x^2 + 14x + 49$ **93.** $4w^2 - 20w + 25$
95. $a^2 + 14ab + 49b^2$ **97.** $x^2 - 25$ **99.** $4m^2 - 9$
101. $25r^2 - 4s^2$ **103.** $2x^3 - 20x^2 + 50x$
105. $3a^3$ **107.** $3a - 2$ **109.** $-3rs + 6r^2$ **111.** $x - 5$

113. $x - 3 + \dfrac{2}{x - 5}$ **115.** $x^2 + 2x - 1 + \dfrac{-4}{6x + 2}$
117. $x^2 + x + 2 + \dfrac{1}{x + 2}$

Self-Test for Chapter 3

1. a^{14} **2.** $15x^3y^7$ **3.** $2x^3$ **4.** $4ab^3$ **5.** $27x^6y^3$
6. $\dfrac{4w^4}{9t^6}$ **7.** $16x^{18}y^{17}$ **8.** $\dfrac{25m^2}{2n}$ **9.** $10x^2 - 12x - 7$
10. $7a^3 + 11a^2 - 3a$ **11.** $3x^2 + 11x - 12$ **12.** $b^2 - 7b - 5$
13. $7a^2 - 10a$ **14.** $4x^2 + 5x - 6$ **15.** $2x^2 - 7x + 5$
16. $15a^3b^2 - 10a^2b^2 + 20a^2b^3$ **17.** $3x^2 + x - 14$
18. $2x^3 + 7x^2y - xy^2 - 2y^3$ **19.** $8x^2 - 14xy - 15y^2$
20. $12x^3 + 11x^2y - 5xy^2$ **21.** $9m^2 + 12mn + 4n^2$
22. $a^2 - 49b^2$ **23.** $2x^2 - 3y$ **24.** $4c^2 - 6 + 9cd$
25. $x - 6$ **26.** $x + 2 + \dfrac{10}{2x - 3}$
27. $2x^2 - 3x + 2 + \dfrac{7}{3x + 1}$ **28.** $x^2 - 4x + 5 + \dfrac{-4}{x - 1}$
29. Binomial   **30.** Trinomial   **31.** 6   **32.** $8x^4 - 3x^2 - 7$;
8, $-3, -7$; 4   **33.** $\dfrac{1}{y^5}$ **34.** $\dfrac{3}{b^7}$ **35.** $\dfrac{1}{y^4}$ **36.** $\dfrac{1}{p^{10}}$
37. 1   **38.** 6   **39.** 1.68×10^{20} **40.** 3.12×10^{-10}
41. 2.5×10^{-12} **42.** 5.2×10^{19}

Cumulative Review Chapters 1–3

[1.2–1.3] 1. 17   **2.** 6   **3.** 150   **4.** 4   **[1.5] 5.** 55
6. $-\dfrac{26}{21}$ **[3.1] 7.** $9x^{16}$ **8.** $\dfrac{x^{10}}{y^6}$ **9.** $8x^9y^3$ **[3.2] 10.** 7
11. 1   **12.** $\dfrac{1}{x^4}$ **13.** $\dfrac{3}{x^2}$ **14.** $\dfrac{1}{x^4}$ **15.** $\dfrac{1}{x^3y^3}$
[1.6] 16. $4x^5y$ **[3.3] 17.** $x^2 + 7x$ **18.** $2x - 2y$
[3.4] 19. $x^2 - 2x - 15$ **20.** $x^2 + 2xy + y^2$
21. $9x^2 - 24xy + 16y^2$ **[3.5] 22.** $x + 4$
[3.4] 23. $x^3 - xy^2$ **[2.3] 24.** -2 **25.** -2 **26.** 84
27. 1   **[2.4] 28.** $B = 2A - b$ **[2.6] 29.** $x \geq -2$
30. $x < -22$ **[2.5] 31.** Sam: \$510; Larry: \$250   **32.** 37, 39
33. \$2,120   **34.** \$645

Prerequisite Test for Chapter 4

1. $2^2 \cdot 3 \cdot 11$ **2.** $2^3 \cdot 5 \cdot 31$ **3.** $4x - 32$ **4.** $-6x^2 + 6x - 2$
5. $6x^2 - 12x$ **6.** $21x^4 + 28x^3 - 63x^2$ **7.** $2x^2 + 5x - 3$
8. $15x^2 - 13x - 20$ **9.** $3x^2 + 4x - 1$ **10.** $2x + 1$

Reading Your Text

Section 4.1 **(a)** distributive; **(b)** GCF; **(c)** multiplying; **(d)** grouping
Section 4.2 **(a)** multiplication; **(b)** middle; **(c)** positive; **(d)** factors
Section 4.3 **(a)** plus; **(b)** opposite; **(c)** GCF; **(d)** ac test
Section 4.4 **(a)** multiples; **(b)** factored; **(c)** GCF; **(d)** sum
Section 4.5 **(a)** GCF; **(b)** never; **(c)** cubes; **(d)** polynomial
Section 4.6 **(a)** quadratic; **(b)** zero-product; **(c)** zero; **(d)** repeated

Summary Exercises for Chapter 4

1. $6(3a + 4)$ **3.** $8s^2(3t - 2)$ **5.** $7s^2(5s - 4)$
7. $9m^2n(2n - 3 + 2n^2)$ **9.** $8ab(a + 3 - 2b)$
11. $(x + y)(2x - y)$ **13.** $(x + 4)(x + 5)$ **15.** $(a - 4)(a + 3)$
17. $(x + 6)(x + 6)$ **19.** $(b - 7c)(b + 3c)$
21. $m(m + 7)(m - 5)$ **23.** $3y(y - 7)(y - 9)$
25. $(3x + 5)(x + 1)$ **27.** $(2b - 3)(b - 3)$

29. $(5x - 3)(2x - 1)$ **31.** $(3y - 5z)(3y + 4z)$
33. $4x(2x + 1)(x - 5)$ **35.** $3x(2x - 3)(x + 1)$
37. $(p + 7)(p - 7)$ **39.** $(m + 3n)(m - 3n)$
41. $(5 + z)(5 - z)$ **43.** $(5a + 6b)(5a - 6b)$
45. $3w(w + 2z)(w - 2z)$ **47.** $2(m + 6n^2)(m - 6n^2)$
49. $(x + 4)^2$ **51.** $(2x + 3)^2$ **53.** $x(4x + 5)^2$
55. $(x - 4)(x + 5)$ **57.** $(3x + 2)(2x - 5)$
59. $x(2x + 3)(3x - 2)$ **61.** $1, -\dfrac{3}{2}$ **63.** $0, 10$ **65.** $-3, 5$
67. $2, \dfrac{5}{4}$ **69.** $0, 3$ **71.** $-4, 4$

Self-Test for Chapter 4

1. $6(2b + 3)$ **2.** $3p^2(3p - 4)$ **3.** $5(x^2 - 2x + 4)$
4. $6ab(a - 3 + 2b)$ **5.** $(a - 5)(a - 5)$
6. $(8m + n)(8m - n)$ **7.** $(7x + 4y)(7x - 4y)$
8. $2b(4a + 5b)(4a - 5b)$ **9.** $(a - 7)(a + 2)$
10. $(b + 3)(b + 5)$ **11.** $(x - 4)(x - 7)$
12. $(y + 10z)(y + 2z)$ **13.** $(x + 2)(x - 5)$
14. $(2x - 3)(3x + 1)$ **15.** $(2x - 1)(x + 8)$
16. $(3w + 7)(w + 1)$ **17.** $(4x - 3y)(2x + y)$
18. $3x(2x + 5)(x - 2)$ **19.** $3, 5$ **20.** $-1, 4$
21. $-1, \dfrac{2}{3}$ **22.** $0, 3$ **23.** $0, 4$ **24.** $-3, 8$ **25.** 7
26. $\dfrac{5}{2}$ **27.** 20 cm   **28.** 2 s

Cumulative Review Chapters 1–4

[1.2–1.3] 1. 17 **2.** -2 **[3.3] 3.** $9x^2 - x - 5$
4. $-4a^2 - 2a - 5$ **5.** $6b^2 + 8b - 3$
[3.4] 6. $15r^3s^2 - 12r^2s^2 + 18r^2s^3$ **7.** $6a^3 - 5a^2b + 3ab^2 - b^3$
[3.5] 8. $-y^2 + 3xy - 2x^2$ **9.** $3a + 2$ **10.** $x^2 - 2x + \dfrac{5}{2x + 4}$
[2.3] 11. $x = -2$ **[2.6] 12.** $x \le \dfrac{33}{5}$ **[2.4] 13.** $t = \dfrac{2S - na}{n}$
[3.1] 14. x^{17} **15.** $6x^5y^7$ **16.** $9x^4y^6$ **17.** $4xy^2$
18. $108x^8$ **[4.5] 19.** $12w^4(3w - 4)$ **20.** $5xy(x - 3 + 2y)$
21. $(5x + 3y)^2$ **22.** $4p(p + 6q)(p - 6q)$ **23.** $(a + 3)(a + 1)$
24. $2w(w^2 - 2w - 12)$ **25.** $(3x + 2y)(x + 3y)$ **[4.6] 26.** $3, 4$
27. $-4, 4$ **28.** $\dfrac{2}{3}, -1$ **29.** 6 **30.** 5 in. by 21 in.

Prerequisite Test for Chapter 5

1. $\dfrac{2}{3}$ **2.** $\dfrac{13}{6}$ **3.** $\dfrac{2}{3}$ **4.** $-\dfrac{3}{7}$ **5.** $\dfrac{35}{8}$ **6.** $\dfrac{49}{32}$ **7.** $\dfrac{21}{40}$
8. $\dfrac{4}{7}$ **9.** $\dfrac{15}{14}$ **10.** $\dfrac{25}{63}$ **11.** $\dfrac{25}{24}$ **12.** $10\dfrac{5}{6}$ **13.** $-\dfrac{2}{15}$
14. $\dfrac{8}{15}$ **15.** $24x + 32$ **16.** $-4x - 1$ **17.** $-3x^2 + 21x$
18. $-x + 1$ **19.** $\dfrac{5}{8}$ in.   **20.** 48 lots

Reading Your Text

Section 5.1 (a) equivalent; (b) factors; (c) simplest; (d) negative
Section 5.2 (a) multiplying; (b) simplify; (c) invert; (d) zero
Section 5.3 (a) like; (b) simplest; (c) parentheses; (d) factor
Section 5.4 (a) rational; (b) common; (c) simplest; (d) opposite
Section 5.5 (a) invert; (b) nonzero; (c) fraction; (d) LCD
Section 5.6 (a) expressions; (b) LCD; (c) excluded; (d) original
Section 5.7 (a) original; (b) time; (c) rate; (d) variable

Summary Exercises for Chapter 5

1. $\dfrac{2}{3a}$ **3.** $\dfrac{w^2 - 25}{2w - 8}$ **5.** $\dfrac{-m - 1}{m + 3}$ **7.** $\dfrac{2}{3x}$ **9.** $\dfrac{x}{2}$
11. $\dfrac{2}{3p}$ **13.** $\dfrac{2x - 3}{x}$ **15.** $\dfrac{a + b}{4a}$ **17.** $\dfrac{x}{3}$ **19.** $\dfrac{11}{x + 2}$
21. $\dfrac{2r - s}{r}$ **23.** 2 **25.** $\dfrac{7x}{6}$ **27.** $\dfrac{5m - 6}{2m^2}$
29. $\dfrac{3x + 3}{x(x - 3)}$ **31.** $\dfrac{3w - 5}{(w - 5)(w - 3)}$ **33.** $-\dfrac{11}{6(x - 1)}$
35. $\dfrac{5a}{(a + 4)(a - 1)}$ **37.** $\dfrac{2}{3x}$ **39.** $\dfrac{y + x}{y - x}$ **41.** $\dfrac{n - m}{n + m}$
43. $\dfrac{a + 3}{a - 3}$ **45.** None   **47.** $-1, 2$ **49.** $-1, -2$ **51.** 7
53. 40 **55.** 6 **57.** 7 **59.** 8 **61.** $4, 8$ **63.** $4, 12$
65. 48 mi/h, 40 mi/h   **67.** 120 mi/h   **69.** 150 mL

Self-Test for Chapter 5

1. $-\dfrac{3x^4}{4y^2}$ **2.** $\dfrac{4}{a}$ **3.** $\dfrac{x + 1}{x - 2}$ **4.** a **5.** 2 **6.** 5
7. $\dfrac{17x}{15}$ **8.** $\dfrac{3s - 2}{s^2}$ **9.** $\dfrac{4x + 17}{(x - 2)(x + 3)}$ **10.** $\dfrac{15}{w - 5}$
11. $\dfrac{4p^2}{7q}$ **12.** $\dfrac{2}{x - 1}$ **13.** $\dfrac{3}{4y}$ **14.** $\dfrac{3}{m}$ **15.** $\dfrac{2}{3x}$
16. $\dfrac{n}{2n + m}$ **17.** 4 **18.** $-3, 3$ **19.** 36 **20.** $2, 6$
21. 6 **22.** 4 **23.** $4, 12$ **24.** 50 mi/h, 45 mi/h
25. 20 ft, 35 ft

Cumulative Review Chapters 1–5

[1.6–1.7] 1. $-2xy$ **2.** $\dfrac{4a^2}{3}$ **3.** $2x^2 - 5x + 6$
4. $3a^2 + 6a + 1$ **[0.4] 5.** 31 **[0.5] 6.** 1
[3.4] 7. $2x^2 - xy - 6y^2$ **8.** $x^2 + 11x + 28$
[3.5] 9. $2x - 1 + \dfrac{1}{x + 2}$ **10.** $x + 1 - \dfrac{4}{x - 1}$ **[2.3] 11.** 4
12. -2 **[4.5] 13.** $(x - 7)(x + 2)$ **14.** $3mn(m - 2n + 3)$
15. $(a + 3b)(a - 3b)$ **16.** $2x(x - 6)(x - 8)$ **[2.4] 17.** 7
[5.7] 18. 4 **19.** 264 ft/s   **[4.6] 20.** 5 in. by 7 in.
[5.1] 21. $\dfrac{m}{3}$ **22.** $\dfrac{a - 7}{3a + 1}$ **[5.4] 23.** $\dfrac{8r + 3}{6r^2}$
24. $\dfrac{x + 33}{3(x - 3)(x + 3)}$ **[5.2] 25.** $\dfrac{3}{x}$ **26.** $\dfrac{1}{3w}$
[5.5] 27. $\dfrac{x - 1}{2x + 1}$ **28.** $\dfrac{n}{3n + m}$ **[5.6] 29.** $\dfrac{6}{5}$ **30.** $-\dfrac{9}{2}, 7$

Prerequisite Test for Chapter 6

1. $x = -2$ **2.** $x = \dfrac{4}{5}$ **3.** $x = 4$ **4.** $y = -3$
5. $x = -\dfrac{2}{3}$ **6.** $y = \dfrac{3}{4}$ **7.** 2 **8.** $\dfrac{3}{2}$ **9.** 1 **10.** 0

Reading Your Text

Section 6.1 (a) solution; (b) linear; (c) ordered-pair; (d) first
Section 6.2 (a) origin; (b) quadrants; (c) negative; (d) zero
Section 6.3 (a) linear; (b) vertical; (c) horizontal; (d) intercept
Section 6.4 (a) slope; (b) run; (c) undefined; (d) variation
Section 6.5 (a) table; (b) circle; (c) legend; (d) line

Summary Exercises for Chapter 6

1. Yes   **3.** Yes   **5.** No   **7.** $(6, 0), (3, -3), (0, -6)$
9. $(3, 0), (-3, 4), (0, 2)$ **11.** $(4, 4), (0, 8), (8, 0), (6, 2)$
13. $(3, 0), (6, -2), (9, -4), (-3, 4)$ **15.** $(0, 10), (2, 8),$
$(4, 6), (6, 4)$ **17.** $(0, -2), (3, 0), (6, 2), (9, 4)$ **19.** $(4, 6)$
21. $(-1, -5)$

23.–25.

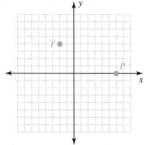

27.

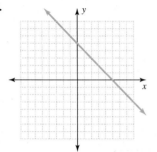

29.

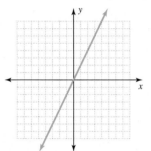

31.

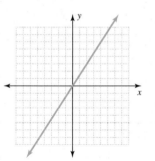

33.

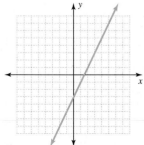

35.

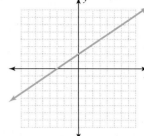

37.

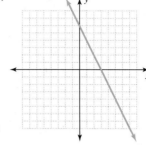

39.

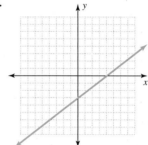

41.

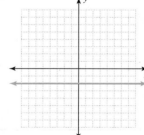

43.

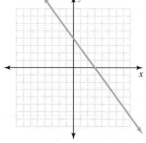

45. $y = -\dfrac{3}{2}x + 3$

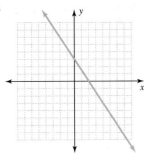

47. -3 **49.** $\dfrac{2}{3}$ **51.** $-\dfrac{5}{2}$ **53.** Undefined

55. -2 **57.** $-\dfrac{2}{3}$

59.

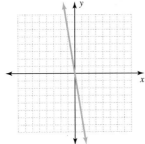

61.

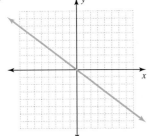

63. $\dfrac{5}{3}$

65.

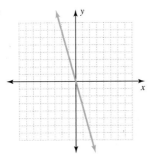

67. 41,378   **69.** 52%   **71.** 23%   **73.** 7,029,990
75. 40%   **77.** 500%

79.

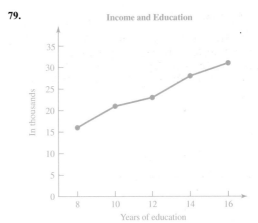

Self-Test for Chapter 6

1. Answers will vary.   **2.** Answers will vary.

3–5.

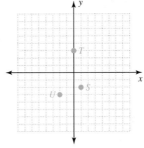

6.

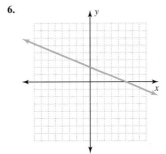

7.

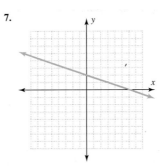

8.

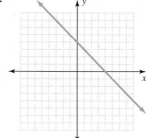

9.

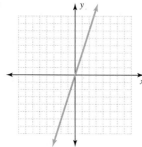

10. 1   **11.** 0   **12.** $\frac{3}{4}$ **13.** Undefined   **14.** -7

15. $(3, 6), (9, 0)$ **16.** $(4, 0), (5, 4)$ **17.** $(4, 2)$

18. $(-4, 6)$ **19.** $(0, -7)$

20.

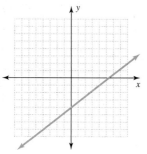

21.

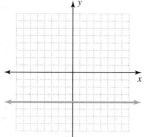

22.

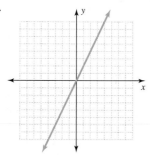

23. 5   **24.** $(3, 0), (0, 4), \left(\frac{3}{4}, 3\right)$ **25.** $(3, 3), (6, 2), (9, 1)$

26. 15%   **27.** $17,200,000   **28.** 2,500

29. Decreased by 2,000   **30.** 1980 to 1985; 2,000

Cumulative Review Chapters 1–6

[1.2–1.3] 1. 3   **2.** 5   **3.** 37   **4.** -53 **5.** 69

6. -120 **7.** -5 **8.** 3   **[1.5] 9.** 108   **10.** 3

11. 69   **12.** 9   **[2.3] 13.** $-\frac{4}{3}$ **14.** $-\frac{8}{3}$

15. 10   **16.** 1   **17.** $C = \frac{5}{9}(F - 32)$ **[2.6] 18.** $x < 4$

19. $x \le 3$ **[3.1] 20.** $\frac{1}{x^4 y^6}$ **21.** $\frac{y^5}{x}$ **22.** 1

[3.3] 23. $5x^2 - 10$ **24.** $-7a^2 + 7a + 2$ **25.** 19   **26.** 26

[3.4] 27. $6x^2 + 2xy - 20y^2$ **28.** $6x^3 - 3x^2 - 45x$

29. $4a^2 - 49b^2$ **[4.5] 30.** $4pn^2(3p + 5 - 4n)$

31. $(y - 3)(y^2 - 5)$ **32.** $b(3a + 7)(3a - 7)$

33. $2(3x + 2)(x - 1)$ **34.** $(2a + 3b)(3a - b)$

[4.6] 35. $-3, 11$ **36.** $\frac{4}{5}, \frac{2}{7}$ **[5.1] 37.** $\frac{-5a^3}{3b^2}$

38. $\frac{w - 2}{w + 3}$ **[5.3] 39.** $\frac{a + 5}{a(a - 5)}$ **40.** $\frac{10}{w - 5}$

[5.2] 41. $\frac{3x^3}{4}$ **42.** $\frac{m - 4}{4m}$ **[5.6] 43.** 6   **44.** 5

[6.3] 45.

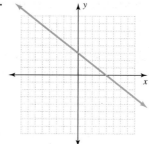

46.

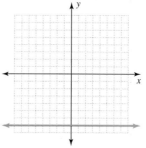

47.

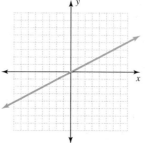

[6.4] 48. 2

49.

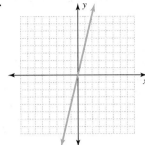

50. 30   **[2.5] 51.** Width: 5 in.; length: 7 in.   **52.** 41, 43, 45

[2.5] 53. $200

Prerequisite Test for Chapter 7

1. $y = -\dfrac{3}{2}x + 3$ **2.** $y = \dfrac{5}{2}x - 5$ **3.** 2   **4.** -1

5. 0   **6.** Undefined

7.   **8.**

9. 5   **10.** 2

Reading Your Text

Section 7.1 (a) slope; **(b)** y-intercept; **(c)** decreases; **(d)** Fixed

Section 7.2 (a) parallel; **(b)** intersection; **(c)** perpendicular;
  (d) Horizontal

Section 7.3 (a) point-slope; **(b)** horizontal; **(c)** vertical;
  (d) reciprocals

Section 7.4 (a) half-plane; **(b)** dotted; **(c)** origin; **(d)** feasible

Section 7.5 (a) evaluate; **(b)** simplifying; **(c)** output; **(d)** break-even

Summary Exercises for Chapter 7

1. Slope: 2; y-intercept: $(0, 5)$ **3.** Slope: $-\dfrac{3}{4}$; y-intercept: $(0, 0)$

5. Slope: $-\dfrac{2}{3}$; y-intercept: $(0, 2)$ **7.** Slope: 0; y-intercept: $(0, -3)$

9. $y = 2x + 3$

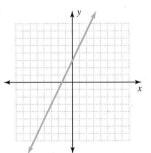

11. $y = -\dfrac{2}{3}x + 2$

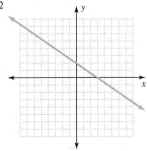

13. Perpendicular   **15.** Parallel   **17.** $y = -3$

19. $x = 4$ **21.** $y = -3$ **23.** $y = -\dfrac{4}{3}x - 2$

25. $x = -\dfrac{5}{2}$ **27.** $y = -\dfrac{1}{5}x + 4$ **29.** $y = -\dfrac{5}{4}x + 2$

31. $y = -\dfrac{5}{3}x + 3$ **33.** $y = \dfrac{4}{3}x + \dfrac{14}{3}$

35.

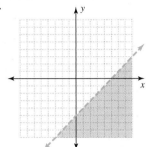

37.

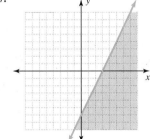

39.

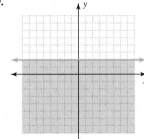

41. (a) -7; **(b)** -13; **(c)** -17 **43. (a)** -39; **(b)** -9; **(c)** -3

45. (a) 9; **(b)** -5; **(c)** -1 **47.** $f(x) = -7x - 3$

49. $f(x) = -\dfrac{3}{2}x - 6$

51.

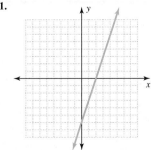

53.

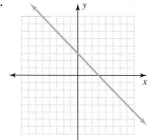

55.

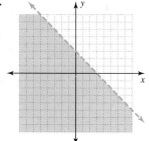

57. 5, 2   **59.** 5, 9   **61.** $7a - 1, 21b - 1, 7x - 8$

Self-Test for Chapter 7

1. Parallel   **2.** Perpendicular   **3.** $y = \frac{1}{2}x + \frac{21}{2}$

4.

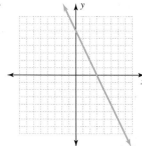

5.

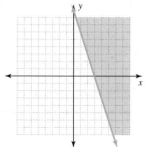

6.

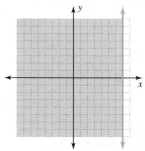

7. $y = -3x + 6$

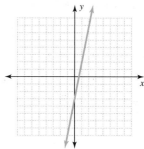

8. $y = 5x - 3$

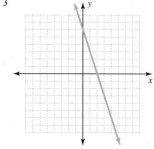

9. $-5; 7$ **10.** $-15; -7$ **11.** $-15; 6$
12. $3a - 25; 3x - 28$ **13.** Slope: $\frac{4}{5}$; y-intercept: $(0, -2)$
14. Slope: $-\frac{2}{3}$; y-intercept: $(0, -9)$ **15.** $y = \frac{4}{3}x + \frac{25}{3}$
16. $y = -\frac{3}{2}x + \frac{5}{2}$ **17.** $y = -\frac{2}{3}x + 5$ **18.** $y = -8$

Cumulative Review Chapters 1–7

1. [3.3] $x^2y^2 - 3xy$ **2.** $\frac{4m^3n}{3}$ **3.** $-x + 9$

4. $3z^2 - 3z + 5$ [3.4] **5.** $2x^2 + 11x - 21$

6. $2a^2 + 6ab - 8b^2$ [3.5] **7.** $x + 6 + \frac{20}{x - 3}$

8. $x^3 - 2x^2 + 4x - 10 + \frac{20}{x - 2}$ [2.3] **9.** $-\frac{4}{3}$

10. -2 [4.5] **11.** $(x - 8)(x + 7)$ **12.** $2x^2y(2x - y + 4x^2)$

13. $2a(2a + 3b)(2a - 3b)$ **14.** $3(5x - 2y)(x - y)$

[6.4] **15.** 1   **16.** $-\frac{9}{4}$ [5.2–5.4] **17.** $\frac{2x - 3}{x}$ **18.** $\frac{1}{a - 7}$
19. $\frac{5m + 6}{2m^2}$ **20.** $\frac{2x + 6}{x(x - 3)}$ **21.** $\frac{5y}{(y + 4)(y - 1)}$
[5.6] **22.** -4 **23.** 7, -10 [5.7] **24.** 9
25. Going: 49 mi/h; returning: 42 mi/h   **26.** 126 min

[7.2] **27.** $y = -\frac{1}{7}x + 2$ [7.1] **28.** $y = -5x + 3$

[7.4] **29.**

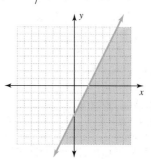

[7.5] **30.** -2

Prerequisite Test for Chapter 8

1.

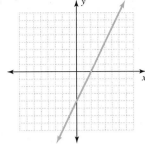

2.

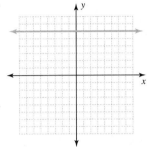

3. $-y$ **4.** $10x$ **5.** $x = 2$ **6.** $y = 3$ **7.** $x = 5$
8. $x = -2$
9.

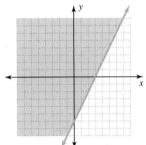

10.

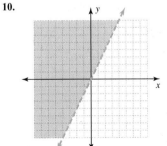

Reading Your Text

Section 8.1 **(a)** system; **(b)** solution; **(c)** consistent; **(d)** dependent
Section 8.2 **(a)** opposites; **(b)** elimination; **(c)** multiplying;
              (d) infinitely many
Section 8.3 **(a)** substitution; **(b)** parallel; **(c)** Graphing;
              (d) addition
Section 8.4 **(a)** intersection; **(b)** dashed; **(c)** bounded; **(d)** feasible

Summary Exercises for Chapter 8

1. $(4, 2)$

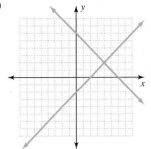

3. No solution

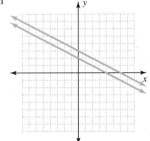

5. Infinite number of solutions   **7.** $(5, 3)$ **9.** $(5, -2)$

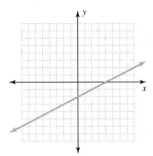

11. $(3, -1)$ **13.** $(-4, 2)$ **15.** $\left(2, -\dfrac{2}{3}\right)$

17. Inconsistent system   **19.** $\left(\dfrac{3}{5}, 2\right)$ **21.** $(6, -2)$

23. $(4, 2)$ **25.** Inconsistent system   **27.** $\left(-\dfrac{4}{3}, 2\right)$

29. $(8, -2)$ **31.** $(4, -2)$ **33.** $(5, 3)$
35. Inconsistent system   **37.** $(5, -4)$ **39.** $(-3, -2)$
41. $(-3, 0)$ **43.** $(8, 2)$ **45.** $(4, -6)$ **47.** $\left(0, \dfrac{4}{3}\right)$
49. $\left(\dfrac{1}{2}, \dfrac{9}{2}\right)$ **51.** $(2, -1)$ **53.** 4, 13
55. Tablet $1.50, pencil $0.25   **57.** Speakers $425, amplifier $500
59. Width 14 cm, length 18 cm   **61.** 10 nickels, 20 quarters
63. 200 mL of 20%, 400 mL of 50%   **65.** $10,000 at 11%,
$8,000 at 7%   **67.** Plane 500 mi/h, wind 50 mi/h

69.
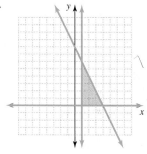

71.
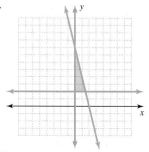

73.
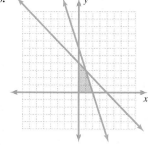

75.
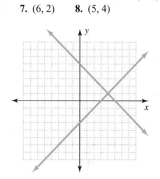

Self-Test for Chapter 8

1. $(4, 1)$ **2.** $(4, 2)$ **3.** $(1, 3)$ **4.** $\left(2, \dfrac{5}{2}\right)$ **5.** $(2, 6)$

6. $(6, -3)$ **7.** $(6, 2)$ **8.** $(5, 4)$

9. $(4, 1)$

10. $(4, 2)$
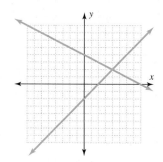

11. 12, 18   **12.** 21 m, 29 m   **13.** Width: 12 in.; length: 20 in.

14. Dependent system   **15.** $\left(\dfrac{3}{4}, -1\right)$ **16.** $(6, 2)$

17. Inconsistent system   **18.** $(-3, 3)$

19. Inconsistent system   **20.** $(3, -5)$ **21.** $(3, 2)$

22.
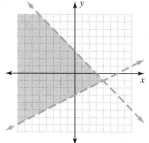

23.
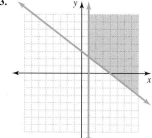

24.
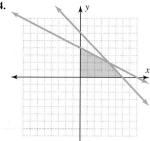

25. 12 dimes, 18 quarters   **26.** Boat: 15 mi/h; current: 3 mi/h
27. Inconsistent system

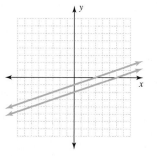

The Streeter/Hutchison Series in Mathematics   Beginning Algebra

28. $(2, 4)$

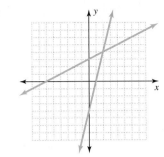

25.

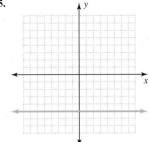

[6.4] 26. $\dfrac{10}{7}$ **[7.1] 27.** Slope: $\dfrac{5}{3}$; y-intercept: $(0, -5)$

28. $y = 2x - 5$

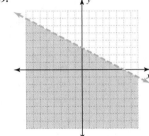

Cumulative Review Chapters 1–8

[3.3] 1. $8x^2 - 7x - 4$ **2.** $w^2 - 8w - 4$

[3.4] 3. $28x^3y^2 - 14x^2y^2 + 21x^2y^3$ **4.** $15s^2 - 23s - 28$

[3.5] 5. $-x^2 + 2xy - 3y$ **6.** $2x + 4$

[2.3] 7. 9   **[4.1–4.4] 8.** $8a^2(3a - 2)$ **9.** $7mn\,(m - 3 - 7n)$

10. $(a + 8b)(a - 8b)$ **11.** $5p(p + 4q)(p - 4q)$

12. $(a - 6)(a - 8)$ **13.** $2w(w - 7)(w + 3)$

[4.6] 14. $4, 5$ **15.** $-4, 4$ **16.** 7   **17.** 5 in. by 17 in.

[5.1] 18. $\dfrac{m}{3}$ **19.** $\dfrac{a - 7}{3a + 1}$ **[5.2] 20.** $\dfrac{3}{x}$ **21.** $\dfrac{1}{3w}$

[7.1] 22.

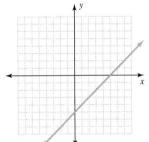

[7.4] 29.

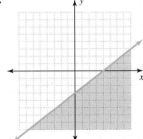

23.

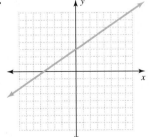

30.

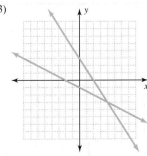

24.

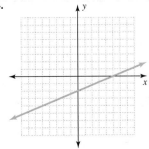

[8.1] 31. $(4, -3)$

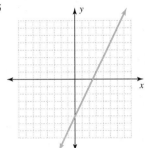

[8.2–8.3] 32. $\left(7, -\dfrac{5}{2}\right)$ **33.** Dependent system **34.** $(5, 0)$

35. Inconsistent system **36.** $\left(\dfrac{3}{2}, -\dfrac{1}{3}\right)$ **37.** $5, 21$

38. VHS \$4.50, cassette \$1.50 **39.** 325 at \$70 and 125 at \$40
40. \$5,000 at 6%, \$7,000 at 9%

Prerequisite Test for Chapter 9

1. $3^2 \cdot 5$ **2.** $2^3 \cdot 3^2$ **3.** $45a^5$ **4.** $81m^7$ **5.** $4a$

6. $5x$ **7.** $\dfrac{16}{15}$ **8.** $\dfrac{9}{5}$ **9.** $m^2 - 49$ **10.** $9x^2 + 30x + 25$

Reading Your Text

Section 9.1 **(a)** principal; **(b)** negative; **(c)** approximation;
  (d) radicand
Section 9.2 **(a)** factors; **(b)** fractions; **(c)** product; **(d)** denominator
Section 9.3 **(a)** radicand; **(b)** distributive; **(c)** positive; **(d)** cube
Section 9.4 **(a)** square root; **(b)** FOIL; **(c)** rational; **(d)** radicand
Section 9.5 **(a)** sides; **(b)** even; **(c)** lose; **(d)** three
Section 9.6 **(a)** hypotenuse; **(b)** hypotenuse; **(c)** negative;
  (d) distance

Summary Exercises for Chapter 9

1. 9 **3.** Not a real number **5.** -4 **7.** Not a real number
9. $3\sqrt{5}$ **11.** $2x^2\sqrt{5}$ **13.** $10b\sqrt{2b}$ **15.** $6ab^2\sqrt{3b}$
17. $\dfrac{3x\sqrt{2}}{5}$ **19.** $\dfrac{\sqrt{21}}{7}$ **21.** $\dfrac{2x\sqrt{14}}{7}$ **23.** $6\sqrt{5}$
25. $2\sqrt{3a}$ **27.** $7\sqrt{3}$ **29.** $\sqrt{2}$ **31.** $2\sqrt{2} + \sqrt{3}$
33. $\sqrt{30}$ **35.** $\sqrt{6x}$ **37.** $5a\sqrt{2}$ **39.** $2\sqrt{21} - 21$
41. 6 **43.** $\dfrac{x\sqrt{21x}}{3}$ **45.** 21 **47.** 9 **49.** 10 **51.** 15
53. $5\sqrt{5}$ **55.** 15 in. **57.** 24.1 ft **59.** 4 **61.** $\sqrt{73}$
63. $\sqrt{82}$

Self-Test for Chapter 9

1. $4\sqrt{10}$ **2.** $2a\sqrt{6a}$ **3.** $3x\sqrt{2}$ **4.** $\dfrac{4}{5}$ **5.** 3

6. Not real **7.** 4 **8.** 11 **9.** $3\sqrt{5}$ **10.** 20 **11.** $\sqrt{15}$

12. 12 **13.** $3\sqrt{2}$ **14.** $\dfrac{\sqrt{5}}{3}$ **15.** $3\sqrt{2}$ **16.** $11 + 5\sqrt{5}$

17. $\dfrac{\sqrt{14}}{2}$ **18.** $5\sqrt{3}$ **19.** $2 + 3\sqrt{2}$ **20.** $4\sqrt{5}$ **21.** $\sqrt{85}$

22. 9 **23.** 4 **24.** 83 **25.** 9.747 cm

Cumulative Review Chapters 1–9

[1.6] 1. $3x^2y^3 - 2x^3y$ **[3.3] 2.** $7x^2 - 6x + 12$ **[3.1] 3.** 0
4. -8 **[2.3] 5.** 2 **6.** 6 **[2.6] 7.** $x > 4$
[3.4] 8. $6x^4y - 10x^3y + 38x^2y$ **9.** $20x^2 - 23xy - 21y^2$
[4.5] 10. $9xy(4 - 3x^2y)$ **11.** $(4x - 3)(2x - 5)$
[5.4] 12. $\dfrac{1}{15(x + 7)}$ **[5.2] 13.** $\dfrac{x - 3}{x - 5}$

[6.3] 14.

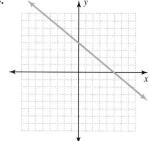

[7.4] 15.

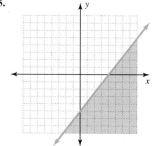

[6.4] 16. 5 **[7.3] 17.** $y = -\dfrac{3}{2}x + 5$ **[8.2] 18.** $(5, 0)$
19. Inconsistent system **20.** Boat: 13 mi/h; current: 3 mi/h
[9.1] 21. 12 **22.** -12 **23.** Not a real number **24.** -3
[9.2–9.4] 25. $-4a\sqrt{5}$ **26.** $\dfrac{2x\sqrt{6x}}{3}$ **27.** $4 - 2\sqrt{2}$

28. $7x\sqrt{2}$ **29.** $5mn\sqrt{6m}$ **30.** $\dfrac{2a\sqrt{3}}{5}$

Prerequisite Test for Chapter 10

1. $-5, 9$ **2.** $2\sqrt{2}$ **3.** $\dfrac{\sqrt{15}}{3}$ **4.** $x^2 - 14x + 49$

5. $4x^2 + 20x + 25$ **6.** 8 **7.** -23 **8.** 5 **9.** -49 **10.** 6

Reading Your Text

Section 10.1 **(a)** quadratic; **(b)** divide; **(c)** square root;
  (d) real-number
Section 10.2 **(a)** perfect-square; **(b)** constant; **(c)** complete;
  (d) even
Section 10.3 **(a)** standard; **(b)** factored; **(c)** discriminant;
  (d) rational
Section 10.4 **(a)** parabola; **(b)** downward; **(c)** axis of symmetry;
  (d) vertex

Summary Exercises for Chapter 10

1. $\pm\sqrt{10}$ **3.** $\pm 2\sqrt{5}$ **5.** $1 \pm \sqrt{5}$ **7.** $-3 \pm \sqrt{5}$
9. $\pm\dfrac{3\sqrt{3}}{2}$ **11.** $\pm\dfrac{\sqrt{7}}{5}$ **13.** $-2, 5$ **15.** $\dfrac{5 \pm \sqrt{17}}{2}$
17. $2 \pm 2\sqrt{2}$ **19.** $2 \pm \sqrt{2}$ **21.** $\dfrac{1 \pm \sqrt{29}}{2}$ **23.** $\dfrac{2 \pm 3\sqrt{2}}{2}$
25. $-2, 7$ **27.** $\dfrac{-5 \pm \sqrt{37}}{2}$ **29.** $3 \pm 2\sqrt{2}$ **31.** $\dfrac{2 \pm \sqrt{10}}{3}$
33. $\dfrac{-3 \pm \sqrt{37}}{2}$ **35.** $2 \pm \sqrt{10}$ **37.** $0, \dfrac{3}{5}$ **39.** $1 \pm \sqrt{10}$
41. $\dfrac{5 \pm \sqrt{57}}{4}$ **43.** $-1, \dfrac{7}{2}$ **45.** $-1 \pm \sqrt{6}$ **47.** $1 \pm \sqrt{3}$

49.

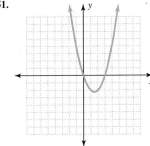

51.

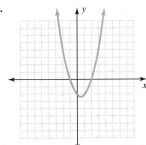

53.

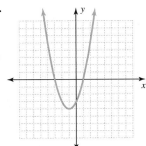

55.

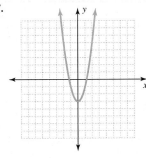

57.

59.

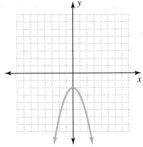

61. Axis: $x = -3$

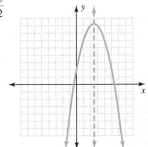

63. Axis: $x = \dfrac{5}{2}$

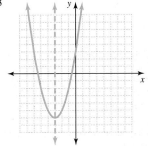

65. x-intercepts: $(-5, 0), (3, 0)$; y-intercept: $(0, -15)$

67. x-intercepts: $(-3 - \sqrt{6}, 0), (-3 + \sqrt{6}, 0)$; y-intercept: $(0, 3)$

69. x-intercepts: None; y-intercept: $(0, -3)$ **71.** $(-1, -16)$

73. $(-3, -6)$ **75.** $(1, -2)$ **77.** 9.23 cm, 15.46 cm

79. 1.13 s, 3.87 s   **81.** 70 items

Self-Test for Chapter 10

1. $-1 \pm \sqrt{6}$ **2.** $\dfrac{5 \pm \sqrt{17}}{4}$ **3.** $-1, 3$ **4.** $\dfrac{1 \pm \sqrt{11}}{2}$

5.

x	y
-1	3
0	0
1	-1
2	0
3	3

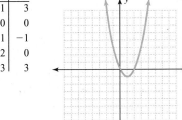

A-14   ANSWERS

6.

x	y
−2	0
−1	−2
0	−2
1	0
2	4

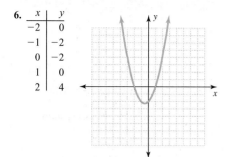

7.

x	y
−2	8
−1	5
0	4
1	5
2	8

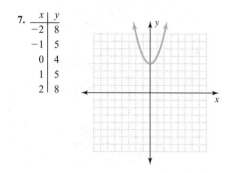

8. $\pm\sqrt{15}$ **9.** $1 \pm \sqrt{7}$ **10.** $4, -2$ **11.** $\dfrac{-3 \pm \sqrt{13}}{2}$

12. $x = 2; (-3, 0), (7, 0)$ **13.** 3 **14.** $\dfrac{1 \pm \sqrt{33}}{4}$

15. 6.36 ft by 15.08 ft   **16.** 3.06 s   **17.** $\pm 2\sqrt{2}$

18. $\dfrac{\pm\sqrt{10}}{3}$ **19.** $\dfrac{5 \pm \sqrt{33}}{2}$ **20.** $-1 \pm \sqrt{6}$

Cumulative Review Chapters 1–10

[1.6] 1. $11x^2y - 6xy^2$ **2.** $x^2 + 5x - 7$ **[1.5] 3.** -480
4. $-1,680$ **[2.3] 5.** 1 **[2.6] 6.** $x > 2$
[3.4] 7. $6x^3y - 3x^2y + 15xy$ **8.** $6x^2 + 11x - 10$
9. $9x^2 - 16y^2$ **[4.1–4.4] 10.** $8xy^2(2x - y)$
11. $(2x - 3)(4x + 5)$ **12.** $(5x - 4y)(5x + 4y)$

[5.2, 5.4] 13. $\dfrac{29}{28(x + 2)}$ **14.** $\dfrac{5(x - 2)}{x - 1}$

15. $\dfrac{(3x - 1)(x + 6)}{15x^2}$

[6.4] 16.

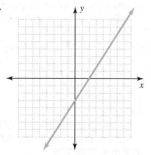

17.

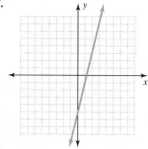

[6.5] 18. 5   **[7.1] 19.** $y = 2x - 5$
[7.4] 20.

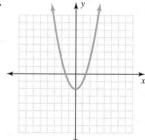

[8.2–8.3] 21. $\left(4, \dfrac{2}{3}\right)$ **22.** $(3, -2)$ **23.** $\left(3, -\dfrac{7}{2}\right)$
24. Dependent system   **25.** 5, 21
26. 325 reserved seats, 125 general admissions
27. 100 mL of 30%, 200 mL of 60%   **[9.1] 28.** 13   **29.** -13
30. Not a real number   **31.** -4 **[9.2–9.4] 32.** $6\sqrt{3}$
33. $30a\sqrt{3}$ **34.** $-13 - 2\sqrt{2}$ **35.** $2 - \sqrt{2}$

[10.1–10.3] 36. $\pm 6\sqrt{2}$ **37.** $-3 \pm 2\sqrt{3}$ **38.** $\dfrac{5 \pm \sqrt{41}}{4}$

[10.4] 39.

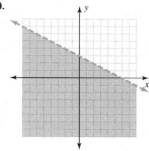

40.

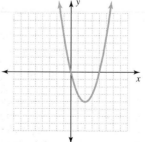

41. 1.657 s   **42.** $84.64

HUTCHISON'S BEGINNING ALGEBRA, EIGHTH EDITION

Published by McGraw-Hill, a business unit of The McGraw-Hill Companies, Inc., 1221 Avenue of the Americas, New York, NY 10020. Copyright © 2010 by The McGraw-Hill Companies, Inc. All rights reserved. Previous editions © 2008, 2005, and 2001. No part of this publication may be reproduced or distributed in any form or by any means, or stored in a database or retrieval system, without the prior written consent of The McGraw-Hill Companies, Inc., including, but not limited to, in any network or other electronic storage or transmission, or broadcast for distance learning.

Some ancillaries, including electronic and print components, may not be available to customers outside the United States.

This book is printed on acid-free paper.

3 4 5 6 7 8 9 0 DOW/DOW 1 0 9 8 7 6 5 4 3 2

ISBN 978–0–07–338418–4
MHID 0–07–338418–6

ISBN 978–0–07–729210–2 (Annotated Instructor's Edition)
MHID 0–07–729210–3

Editorial Director: *Stewart K. Mattson*
Executive Editor: *David Millage*
Director of Development: *Kristine Tibbetts*
Developmental Editor: *Adam Fischer*
Marketing Manager: *Victoria Anderson*
Senior Project Manager: *April R. Southwood*
Senior Production Supervisor: *Kara Kudronowicz*
Senior Media Project Manager: *Sandra M. Schnee*
Senior Designer: *David W. Hash*
Cover Designer: *John Joran*
(USE) Cover Image: *Dahlia, ©iStockphoto/Ramona Heim*
Senior Photo Research Coordinator: *Lori Hancock*
Supplement Producer: *Mary Jane Lampe*
Compositor: *Macmillan Publishing Solutions*
Typeface: *10/12 New Times Roman*
Printer: *RR Donnelley*

All credits appearing on page or at the end of the book are considered to be an extension of the copyright page.

Chapter 1 Opener: © Copyright IMS Communications Ltd./Capstone Design. All Rights Reserved; p. 85: © Vol. 80/PhotoDisc/Getty RF; Chapter 2 Opener: © Comstock/Alamy RF; p. 130, 140, 142: © PhotoDisc/Getty RF; p. 177: © Comstock/Alamy RF; Chapter 3 Opener: © Corbis RF; p. 203, 207: © PhotoDisc/Getty RF; p. 254: © Corbis RF; Chapter 4 Opener: © Getty RF; p. 317: © PhotoDisc/Getty RF; p. 325: © Getty RF; Chapter 5 Opener: © Getty RF; p. 389, p. 395: © Corbis RF; p. 406: © Getty RF; Chapter 6 Opener: © PhotoDisc/Getty RF; p. 426: © Image Source RF; p. 429: © PhotoDisc/Getty RF; p. 480: © American Vignette/Corbis RF; p. 515: © 2006 Texas Instruments; Chapter 7 Opener: © Corbis RF; p. 537, 575, 600: © PhotoDisc/Getty RF; Chapter 8 Opener: © Corbis RF; p. 673: © PhotoDisc/Getty RF; Chapter 9 Opener: © Corbis RF; p. 716, p. 731: © PhotoDisc/Getty RF; p. 743: © Stockbyte/Punchstock RF; Chapter 10 Opener: © Corbis RF; p. 813: © ImageSource/Punchstock RF.

Library of Congress Cataloging-in-Publication Data

Baratto, Stefan.
  Hutchison's beginning algebra.—8th ed. / Stefan Baratto, Barry Bergman.
    p. cm.
  Includes index.
  ISBN 978–0–07–338418–4—ISBN 0–07–338418–6 (hard copy : alk. paper)
  ISBN 978–0–07–729210–2—ISBN 0–07–729210–3 (annotated instructor's ed.) 1. Algebra—Textbooks.
I. Bergman, Barry. II. Hutchison, Donald, 1948- Elementary algebra. III. Title. IV. Title: Beginning algebra.
  QA152.3.B367   2010
  512.9—dc22
                                                                    2009015543

www.mhhe.com

Beginning Algebra

eighth edition

Stefan Baratto
Clackamas Community College

Barry Bergman
Clackamas Community College

Don Hutchison
Clackamas Community College

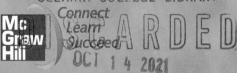

Connect
Learn
Succeed